GENES

The gene [is] a huge molecule, capable only of discontinuous change, which consists in a rearrangement of the atoms and leads to an isomeric molecule. The rearrangement may affect only a small region of the gene, and a vast number of different rearrangements may be possible.

Erwin Schrödinger, 1945

GENES

BENJAMIN LEWIN
Editor, *Cell*

John Wiley & Sons
New York   Chichester   Brisbane
Toronto   Singapore

Production supervised by Linda Indig
Manuscript editor was Mondy Dana,
supervised by Vivian Kahane Cooper
Cover and text designed by Laura Ierardi
Illustrations by John Balbalis with
the assistance of the Wiley
illustration department.

Library of Congress Cataloging in Publication Data:

Lewin, Benjamin
  Genes.

  Includes bibliographies and indexes.
  1. Genetics.   I. Title.

QH430.L487   1983   575.1   82-20022
ISBN 0-471-09316-5

Printed in the United States of America

10 9 8 7 6 5 4 3 2 1

PREFACE

The title of this book almost makes a preface super-fluous. *Genes* is simply about genes, recognizing what amounts to a new field whose extraordinary progress has all but overwhelmed the traditional discipline of genetics. My aim is to cut through the enormous mass of information that has accumulated recently, to discern general principles and describe the state of the art in this exciting area. This text asks: what is a gene, how is it reproduced, how is it expressed, what controls its expression?

The underlying theme of this book is that the gene has at its disposal a vast repertoire of strategies for survival, different examples of which are displayed in various systems. Reflecting the perspective of current research, procaryotic and eucaryotic molecular biology are given equal weight. Both are now part of the same story. The starting point is the issue of how a gene is represented in protein; and from the protein, we work backward at the molecular level, as it were, to the DNA itself.

As a comprehensive introduction to the molecular biology of the gene, this book assumes no prior knowledge and is up to date with current research. I hope that this overview will make this rapidly advancing subject more accessible and readily allow readers to proceed to more advanced works. In view of the size of the relevant literature, it would only be confusing to rely on the citation of individual research articles. Each chapter therefore concludes with a bibliography to suggest useful reviews and some research articles that lead more deeply into the subject.

I have tried to illustrate all important points dia-gramatically, and where appropriate the illustrations attempt to give some feeling for the scale and relationships of the elements involved. This preface would certainly be incomplete without acknowledgment of the considerable artistic endeavors of John Balbalis to realise this aim.

One of the pleasures of writing this book has been the ensuing discussions with my friends and colleagues who have commented on it. Many improvements have resulted from the generous efforts of Sankar Adhya, Sidney Altman, French Anderson, David Clayton, Nicholas Cozzarelli, Bernard Davis, Igor Dawid, Arg Efstratiadis, Nina Federoff, Alice Fulton, Joe Gall, Nicholas Gillham, Philip Hanawalt, Ira Herskowitz, Lee Hood, Joel Huberman, George Khoury, Nancy Kleckner, Marilyn Kozak, Charles Kurland, Art Landy, Jeffrey Miller, Masayasu Nomura, Charles Radding, Jeff Roberts, Rich Roberts, Gerry Rubin, Robert Schimke, David Schlessinger, David Shafritz, Phil Sharp, Allen Smith, Phang-C. Tai, Susumu Tonegawa, Harold Varmus, Alex Varshavsky, and Hal Weintraub. Finally, it hardly needs saying that the book would have been much less fun to write without the enthusiastic participation of my family.

Benjamin Lewin

CONTENTS

GENES

PART 1
THE NATURE OF
GENETIC INFORMATION

In calling the structure of the chromosome fibers a code-script we mean that the all-penetrating mind could tell from their structure whether the egg would develop, under suitable conditions, into a black cock or into a speckled hen, into a fly or a maize plant, a beetle, a mouse or a woman. . . . But the term code-script is, of course, too narrow. The chromosome structures are at the same time instrumental in bringing about the development they foreshadow. They are law-code and executive power—or, to use another simile, they are architect's plan and builder's craft—in one.

Erwin Schrödinger, 1945

CHAPTER 1
WHAT IS A GENE?
A GENETIC VIEW

The concept of the gene has been the focus of some hundred years of work to establish the basis of heredity. It has been scrutinized from so many perspectives that there is no simple one-line answer to satisfy the question: what is a gene?

A gene is a sequence of DNA that carries the information representing a protein. Until very recently this would have been an adequate (if incomplete) biochemical description. The sequence of DNA could be identified as a continuous stretch of nucleotides, related to the protein sequence by a colinear readout. But now it is clear that the sequence representing protein is not always continuous; it may be interrupted by sequences not concerned with specifying the protein. So genes may be in pieces that are put together during the process of gene expression.

A gene can be identified as a cluster of mutations all of which prevent the production of the protein that it represents. Although this remains true, the properties of these clusters become more complex for interrupted genes. This issue has been moot, because most interrupted genes are found in situations in which detailed genetic analysis is not possible; but in principle, the genetic view of the gene now needs modification.

The large number of genes that make up the genome of any species are organized into a comparatively small number of chromosomes. The genetic material of each chromosome consists of an extremely long stretch of DNA, containing many genes in a linear order. How many genes are present *in toto* has been a puzzle for a long time. Recently the view that each gene may reside by itself as a unique entity has been superseded by the realization that, in many cases, there may be clusters of related genes that constitute small families.

The genome generally has been viewed as rather stable, subject to changes in overall constitution and organization only on an extraordinarily slow evolutionary time scale. This contrasts with recent evidence that in some instances there may be rearrangements that occur regularly; and there may be components of the genome that are relatively mobile.

The concept of the gene has therefore undergone an evolution in which, although many of its traditional properties remain, exceptions have been found to show that none constitutes an absolute rule. Starting from the discovery of the gene as a fixed unit of inheritance, its properties were defined in terms of its residence at a definite position on the chromosome, this in turn leading to the view that the genetic material of the chromosome is a continuous length of DNA representing many genes. By tracing the development of these ideas, we can arrive at an operational description of the gene, although it is impractical to provide a capsule definition.

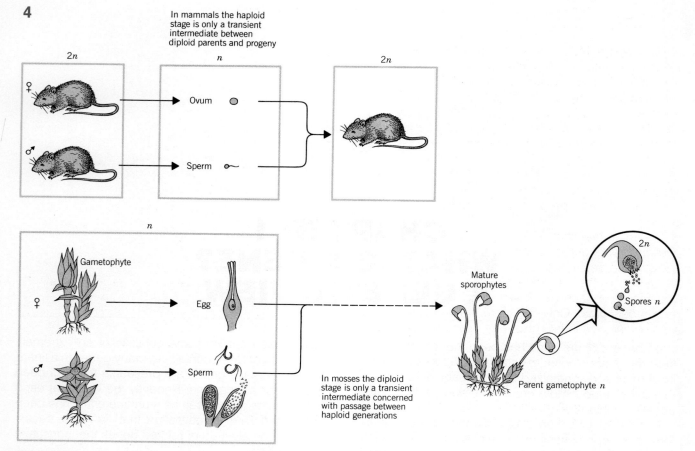

In mammals the haploid stage is only a transient intermediate between diploid parents and progeny

In mosses the diploid stage is only a transient intermediate concerned with passage between haploid generations

Figure 1.1

When eucaryotes perpetuate their genes through an alternation of diploid and haploid states, either type of state may provide the predominant "adult" type while the other is concerned solely with gamete and zygote formation. The mammals and the mosses are extreme examples in which the adult generations are diploid and haploid, respectively.

Red indicates diploid (2n) tissue; grey indicates haploid (n) tissue.

A PARTICULATE FACTOR OF INHERITANCE

"The constant characters that appear in the group of plants may be obtained in all the associations that are possible according to the mathematical laws of combination. Those characters which are transmitted visibly, and therefore constitute the characters of the hybrid, are termed the dominant, and those that become latent in the process are termed recessive. The expression 'recessive' has been chosen because such characters withdraw or disappear entirely in the hybrids, but nevertheless reappear unchanged in their progeny in predictable proportions, without any essential alterations. Transitional forms were not observed in any experiment."

This (abbreviated) quotation from Mendel's remarkable paper of 1865 introduces the fundamental concept of genetics: there is a unit of heredity consisting of some factor that is passed from parent to progeny. This particulate factor is, of course, what we now know as the gene. Mendel's work elucidated the general behavior of the gene in inheritance. Implicit in this is the concept that the organism is the gene's way of expressing and perpetuating itself, a point illustrated by the alternation of generations shown in **Figure 1.1.**

In striking contrast to the confusion of earlier thought

about heredity, Mendel's work established genetics as an experimental science through the introduction of several critical features. First, the initial set of plants was **true breeding** for the character under consideration: each gave rise to plants showing only that characteristic and none of the alternatives. Then the inheritance of only the one character was initially followed. With this established, the simultaneous behavior of two or more traits could be investigated. The experiments were placed on a quantitative basis by counting the numbers of each progeny type resulting from each mating.

The characteristics of the gene discernible from Mendel's work are summarized in what are known as his first and second laws. The first law deals with the properties of the individual gene. The organism has two copies of each gene: in modern terminology it is said to be **diploid.** Only one of the two copies is passed from parent to offspring through the **gametes** (sex cells). Thus when the gametes unite, the **zygote** (fertilized egg) gains one copy from each parent, restoring the situation in which every organism has one copy of paternal origin and one of maternal origin.

A gene may exist in alternative forms that result in the expression of a different characteristic (such as red versus white flower color). These forms are called **alleles.** The *law of independent segregation* states that these alleles do not affect each other when present in the same plant, but segregate unchanged by passing into different gametes when the next generation forms.

In a true-breeding organism, a **homozygote,** both alleles are the *same*. But a mating between two parents each of which is homozygous for a *different* allele generates a hybrid or **heterozygote.** If one allele is **dominant** and the other is **recessive,** the organism will have the appearance or **phenotype** only of the dominant type (so the heterozygote is indistinguishable from the true-breeding dominant parent).

But Mendel's first law recognizes that the genetic constitution or **genotype** of the hybrid comprises the presence of both alleles. This is revealed, as shown in **Figure 1.2,** when the hybrid is crossed with another hybrid to form the second generation. The critical point is that the alleles do not mix ("no transitional forms were observed"), but are physical entities whose interaction is at the level of expression.

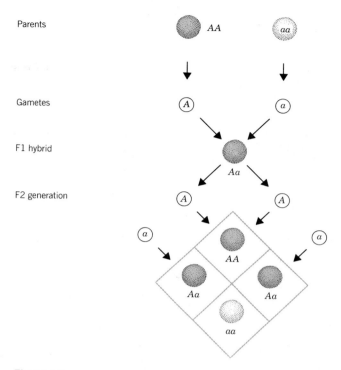

Figure 1.2
Alleles show independent segregation.

The two parents are homozygous: *AA* has two copies of the dominant allele; *aa* has two copies of the recessive allele. Each forms only one type of gamete, so that the **F1** (first hybrid generation) is uniformly hybrid as *Aa*. Because *A* is dominant over *a*, the *phenotype* of *Aa* is the same as that of *AA* (indicated by the color). The phenotype of the recessive homozygote *aa* is indicated by the lack of color.

Each F1 hybrid forms both *A* and *a* gametes in equal amounts. Upon mating these unite randomly to generate an **F2** (second hybrid generation) consisting of: 1 *AA* : 2 *Aa* : 1 *aa*. Since the *AA* and *Aa* have the same phenotype, this gives the classic 3 : 1 ratio of dominant : recessive types.

In cases in which the heterozygote *Aa* has a phenotype intermediate between the parental *AA* and *aa*, the F1 would be distinct, and in the F2 the ratio of phenotypes would be 1 dominant : 2 intermediate : 1 recessive.

Although the characteristics studied by Mendel (fortunately) showed complete dominance, this is not necessarily always the case. Alleles may exhibit **incomplete (partial) dominance** or no dominance (sometimes known as **codominance**). In the latter case, the heterozygote is distinguished from the homozy-

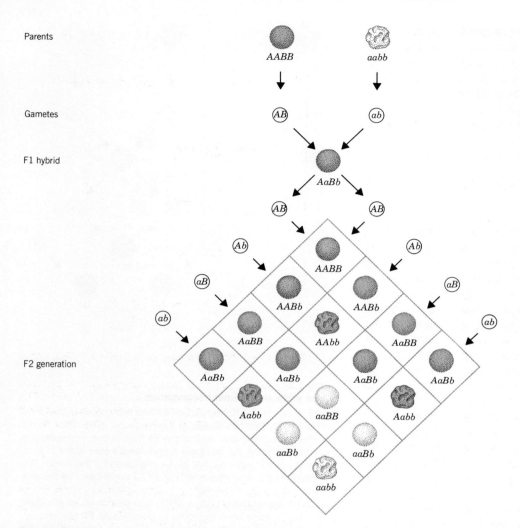

Figure 1.3
Different genes assort independently.

One parent is homozygous for two dominant genes, *A* determining color, and *B* determining shape (shown by round structure). The other parent is homozygous for the recessive alleles *a* and *b* (characteristics shown by no color and wrinkled shape). The F1 is uniform with the dominant characteristics.

The F1 parents produce gametes in which there is independent segregation of alleles and independent assortment of genes, so that equal amounts are produced of each of the four possible types of gamete. These unite randomly to form 9 genotypic classes, which because of the dominance relationships appear as the four phenotypic classes: 9 colored-smooth : 3 colored-wrinkled : 3 noncolored-smooth : 1 noncolored-wrinkled.

Note that the same numbers are present of each reciprocal genotype; for example, the two parents (one each of *AABB* and *aabb*) or the recombinant classes (one each of *AAbb* and *aaBB*). The 3 : 1 ratios are maintained for each individual segregating character.

The number of phenotypic classes will be greater if one or both of the characters is not dominant (so that heterozygotes appear different from either homozygote). It will be less if two genes affect a single characteristic. So if both *A* and *B* were needed for the production of color, a ratio of 9 colored : 7 noncolored would be seen in the F2.

gotes by properties that are intermediate between them. In the snapdragon, for example, a cross between red and white generates hybrids with pink flowers. Although there is no dominance, the same rule is observed that the first hybrid (F1) generation is uniform; and the same ratios are generated in a further hybrid cross, although three phenotypes can be distinguished instead of two.

THE INDEPENDENCE OF DIFFERENT GENES

Mendel's second law summarizes the *independent assortment of different genes.* When a plant that is dominant for two different characters is crossed with a parent that is recessive for both, as before the F1 consists of plants uniformly of the dominant type. But in the next hybrid cross, two classes of plants are seen. One consists of the two **parental types.** The other consists of new phenotypes, representing plants with the dominant feature of one parent and the recessive feature of the other. These are called **recombinant types;** and they occur in both possible (**reciprocal**) combinations.

Figure 1.3 shows that the ratios in which these occur can be explained by supposing that gamete formation involves an entirely random association of one of the alleles for the first character with one of the alleles for the other. All four possible types of gamete are formed in equal proportion, and associate at random in forming the zygotes of the next generation.

Once again, the characteristic ratio of phenotypes conceals a greater variety of genotypes, which can be confirmed by the **backcross** to the recessive parent. Essentially the backcross provides a method for looking directly at the genotype of the organism that is being examined, as illustrated in **Figure 1.4.**

The law of independent assortment therefore says that the behavior of any pair (or greater number) of genes can be predicted overall by the rules of mathematical combination. Thus the assortment of one gene does not influence the assortment of the other. Implicit in this concept is the view that assortment is a matter of *statistical probability* and not an exact result. The ratios of progeny types will approximate increasingly closely to the predicted numbers as a greater number of crosses are performed.

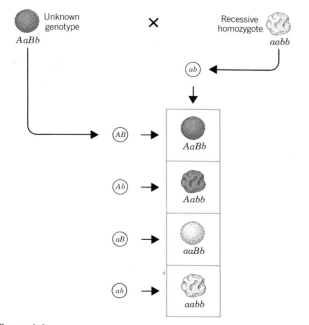

Figure 1.4
Backcross to recessive homozygote is used to assign genotypes.

The gametes formed by an unknown genotype represent the usual independent segregation of alleles and assortment of genes. The genotype of each gamete is revealed directly by the result of a mating with a gamete that brings in only the recessive alleles.

THE ROLE OF CHROMOSOMES IN HEREDITY

Appreciation of Mendel's discoveries was much inhibited by the lack of any physical basis for the postulated factors. When the chromosomal theory of inheritance was simultaneously proposed by Sutton and by Boveri, it resolved the discussion that had been continuing for some time on the role of chromosomes. They had already been implicated in heredity, although in a somewhat hazy manner; in 1903 it was realized that their properties corresponded exactly with those ascribed to Mendel's particulate units of inheritance.

The cell theory established in the middle of the nineteenth century proposed that all organisms are composed of cells, and that these can arise only from preexisting cells. Early cytology showed that a "typical" cell consists of a dense nucleus separated by a membrane from the less-dense surrounding cytoplasm. Within the nucleus, the granular region of **chro-**

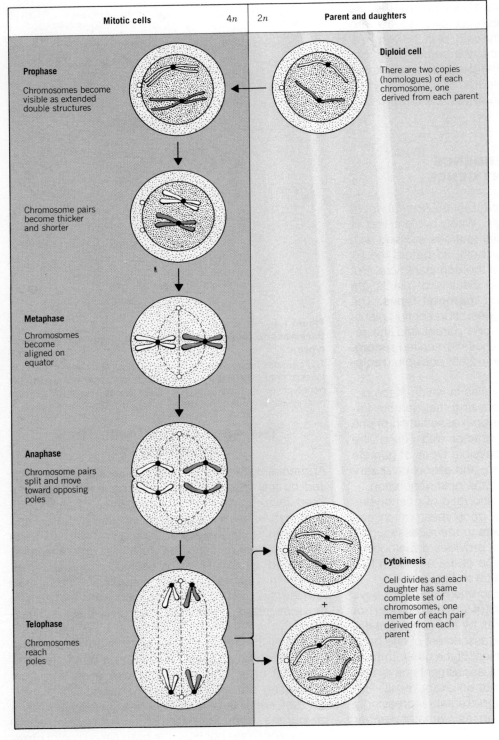

Mitotic cells   4n   2n   Parent and daughters

Prophase

Chromosomes become visible as extended double structures

Chromosome pairs become thicker and shorter

Metaphase

Chromosomes become aligned on equator

Anaphase

Chromosome pairs split and move toward opposing poles

Telophase

Chromosomes reach poles

Diploid cell

There are two copies (homologues) of each chromosome, one derived from each parent

Cytokinesis

Cell divides and each daughter has same complete set of chromosomes, one member of each pair derived from each parent

Figure 1.5

Mitosis perpetuates the chromosome constitution of the cell.

A pair of homologous chromosomes is indicated as white (derived from one parent) and colored (derived from the other parent). The diploid cell has two copies of each chromosome (total number = 2n). These are duplicated before mitosis, so that prophase starts with the constitution 4n. The two members of each pair divide into their individual chromosomes, which are then segregated into different progeny cells. This restores the diploid number 2n, so that the daughter cells have the same chromosome constitution as the parent cell.

matin could be recognized by its reaction with certain stains. Not long after Mendel's work, it was found that the chromatin consists of a discrete number of thread-like particles, the **chromosomes.**

Chromosomes can be visualized in most cells only during the process of cell division. The two types of division in sexually reproducing organisms explain both the perpetuation of the genetic material and the process of inheritance as predicted by Mendel's laws.

During growth of the organism, the **cell cycle** falls into two parts. The long period of **interphase** represents the time during which the cell engages in its synthetic activities and reproduces its components. Then the short period of **mitosis** is an interlude during which the actual process of division into two daughter cells is accomplished. The products of the series of mitotic divisions that generate the entire organism are called the **somatic cells.**

At the end of mitosis, each daughter cell can be seen to start its life with two copies of each chromosome. These are called **homologues.** The total number of chromosomes in what is known as the **diploid set** can therefore be described as $2n$. The typical somatic cell exists in the diploid state (except when it is preparing for or actually engaged in division).

During interphase, a growing cell duplicates its chromosomal material. This is not evident at the time and becomes apparent only during the subsequent mitosis. During mitosis, each chromosome appears to split longitudinally to generate two copies. These are called **sister chromatids.** At this point, the cell contains $4n$ chromosomes, organized as $2n$ pairs of sister chromatids. In other words, there are two (homologous) copies of each sister chromatid pair.

Figure 1.5 illustrates the series of events by which mitotic division is accomplished. Essentially the sister chromatids are pulled toward opposite poles of the cell, so that each daughter cell receives one member of each sister chromatid pair. Now these are chromosomes in their own right. The $4n$ chromosomes present at the start of division have been divided into two sets of $2n$ chromosomes. This process is repeated in the next cell cycle. Thus mitotic division ensures the constancy of the chromosomal complement in the somatic cells.

A different purpose is served by the production of **germ cells** (gametes; eggs or spermatozoa), which is accomplished by another type of division. This is illustrated in **Figure 1.6. Meiosis** generates cells that contain the **haploid** chromosome number, **n.** This involves two, successive divisions. Once again, the chromosomes have been duplicated previously, so that the cell enters division with $4n$.

When the first division begins, the homologous pairs of sister chromatids **synapse** or **pair** to form **bivalents.** Each bivalent contains all four of the cell's copies of one homologue. The first division causes each bivalent to segregate into its two component sister chromatid pairs. This generates two sets of $2n$ chromosomes, each set consisting of n sister chromatid pairs.

Now the second meiotic division follows, in which both of the sets of $2n$ divide again. This division is formally like a mitotic division, since one member of each sister chromatid pair segregates to a different daughter cell.

The overall result of meiosis is to divide the starting number of $4n$ chromosomes into four haploid cells. These may then give rise to mature eggs or sperm.

In forming these gametes, homologues of paternal and maternal origin are separated, so that each gamete gains *only one of the two homologues* of its parent. A critical feature in relating this process to the predictions of Mendel's laws was the realization that *nonhomologous* chromosomes undergo segregation *independently,* so that either member of one homologous pair enters the gamete at random with either member of a different homologous pair.

To summarize the parallels between chromosomes and Mendel's units of inheritance. Genes occur in allelic pairs, one member of each pair having been contributed by each parent; the diploid set of chromosomes results from the contribution of a haploid set by each parent. The assortment of nonallelic genes into gametes should be independent of origin; nonhomologous chromosomes undergo independent segregation. The critical proviso is that each gamete obtains a complete haploid set, and this is fulfilled whether viewed in terms of factors or chromosomes.

GENES LIE ON CHROMOSOMES

The proof that genes reside on chromosomes required a demonstration that a particular gene is always present

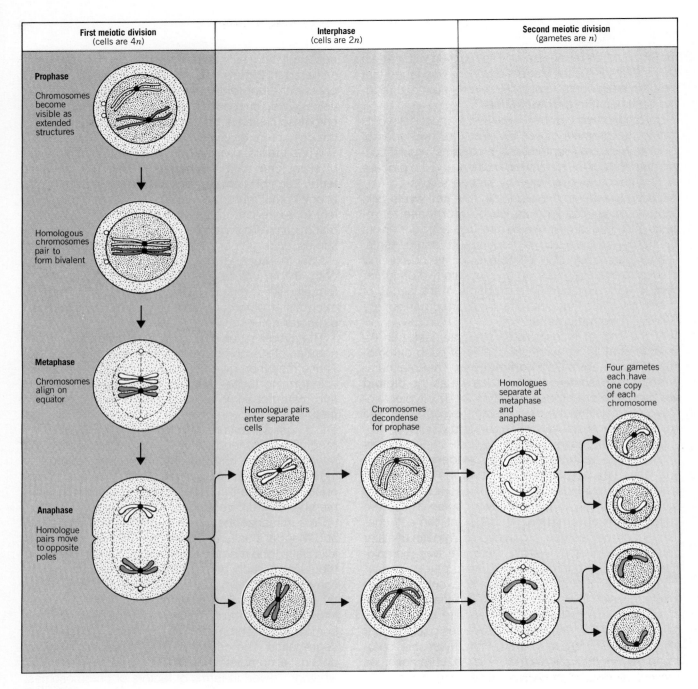

Figure 1.6
Meiosis halves the chromosome number.

The figure shows the behavior of a single homologous pair. Both members of the pair have been duplicated before the start of prophase. During the first division, the homologues pair together, and then each pair moves into a different cell. During the second division, the duplicate members of each pair are segregated into dif-ferent cells, so that each gamete obtains one copy. When several homologous pairs are present, the members of each pair assort independently, so that a random process governs which homologue of any chromosome (maternal or paternal) is represented in each gamete.

on a particular chromosome. This was provided by the properties displayed by a variant of the fruit fly *Drosophila melanogaster* obtained by Morgan in 1910. This white-eyed male appeared spontaneously in a line of flies of the usual red eye color.

Since the red eye color is usual, it is referred to as the **wild type.** The white eye is the **mutant** phenotype. The event responsible for generating the mutant phenotype is a genetic change called a **mutation.** Most (but not all) mutations prevent the gene in which they occur from functioning properly or at all; and so the study of mutations is for the most part the study of inactive alleles. Sometimes it is possible for a mutation to be reversed by a genetic change that restores the original state of the genetic material. This is called a **reversion** to the wild type.

The *white* mutation could be located on a particular chromosome because of its association with sexual type. In many sexually reproducing organisms there is an exception to the rule that chromosomes occur in homologous pairs whose separation (disjunction) at meiosis produces identical haploid sets. Male and female sets may differ visibly in chromosome constitution, the most common form of difference being the replacement of one member of a homologue pair with a different chromosome in *one* of the sexes.

This pair is referred to as the **sex chromosomes,** and the remaining homologous pairs are called the **autosomes.** The chromosome complements of the two sexes can be described as *2A + XX* and *2A + XY,* where the haploid set of autosomes is denoted *A* and the two sex chromosomes are *X* and *Y*. The sex with the complement of *2A + XX* is called **homogametic;** it forms gametes only of the type *A+ X*. The sex with the complement of *2A + XY* is called **heterogametic;** it forms equal proportions of gametes of the types *A + X* and *A + Y*. The random union of gametes from one sex with gametes from the other sex perpetuates the equal sex ratio at zygote formation. In *Drosophila,* the homogametic sex happens to be the female.

A critical prediction of Mendel's laws is that the results of a genetic cross should be the same regardless of orientation—that is, irrespective of which parent introduces which allele. But the reciprocal crosses with white eye in Drosophila give different results, as shown in **Figure 1.7.**

The cross of white male × red female gives the

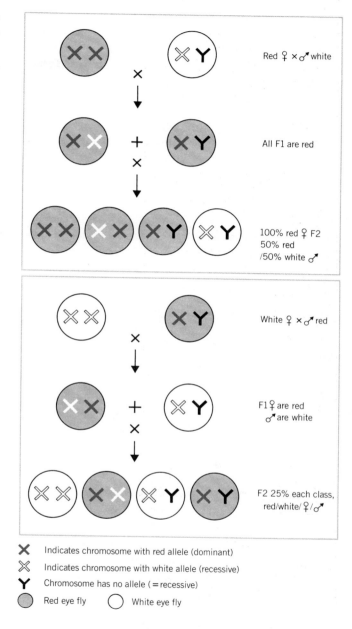

Red ♀ × ♂ white

All F1 are red

100% red ♀ F2
50% red
/50% white ♂

White ♀ × ♂ red

F1 ♀ are red
♂ are white

F2 25% each class,
red/white/♀/♂

X Indicates chromosome with red allele (dominant)
X Indicates chromosome with white allele (recessive)
Y Chromosome has no allele (= recessive)
⬤ Red eye fly   ◯ White eye fly

Figure 1.7
Genes on the X chromosome show sex-linked inheritance.
Red/white eye color of a male fly depends only on the X chromosome received from its mother. The phenotype of the female is determined by whether it receives a dominant allele from *either* parent. This generates the characteristic "crisscross" pattern of sex-linked inheritance.

entirely red F1 expected if red is dominant and white is recessive. But in the F2 all the white-eye flies that reappear are *males*. In the reciprocal cross of red male × white female, all the F1 males are white eyed and all the females are red eyed. Crossing these gives an F2 with equal proportions of white and red eyes in each sex.

This pattern of inheritance exactly follows that of the sex chromosomes. If the alleles for red and white eyes are carried on the X chromosome, with no locus for eye color present on the Y chromosome, the pheno-type of a male will be determined by the single allele present on its X chromosome. This allele will be transmitted to all of its daughters and none of its sons. This is the typical pattern of **sex linkage.**

CHROMOSOMES CONTAIN A LINEAR ARRAY OF GENES

The independent segregation of chromosomes seen at meiosis explains the independent assortment of

Figure 1.8
Linkage can be measured by a backcross with a double heterozygote.

The chromosomes are indicated by shaded bars. Genes from the heterozygote are dark letters; genes from the double recessive parent are light letters. This is the same procedure outlined in Figure 1.4.

In the upper cross, one chromosome of the heterozygote carries both dominant alleles; the other carries both recessive alleles (this is a "coupling" cross). Thus the parental types are *AB* and *ab*. The recombinant types are *Ab* and *aB*.

In the lower cross, one chromosome of the heterozygote carries one dominant and one recessive allele; the other chromosome carries the reverse combination (this is a "repulsion" cross). This makes *Ab* and *aB* the parental types; the recombinants are *AB* and *ab*.

In each case the progeny contain an increase in the proportion of parental types (70%) and a decrease in the proportion of recombinant types (30%) compared with the 50% for each that is expected from independent assortment. Note that both parental types are present in the same amount, and both recombinant types are present in equal amount. The linkage between *A* and *B* is measured as 30% or 30 map units.

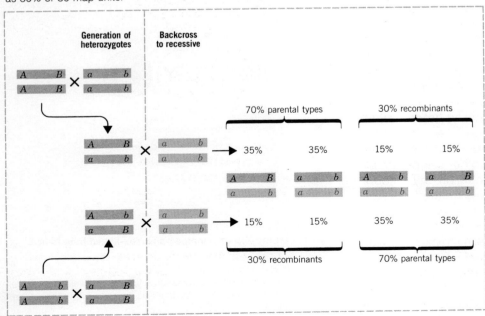

genes that are carried on different chromosomes. But the number of genetic factors is much greater than the number of chromosomes. If all genes lie on chromosomes, many genes must be carried by each chromosome. What is the relationship between these genes?

The quickening pace of genetics after the turn of the century was shown by the passage of only three years between the report of the first eye-color mutant and Sturtevant's study in 1913 of the pattern of inheritance of six sex-linked mutations. Morgan had shown in 1911 that each of these factors shows the same sex-linked pattern as white/red eye color. By the same logic, each must therefore be carried by the X chromosome.

Some of these factors are associated. Instead of obeying Mendel's second law, the proportion of parental genotypes in the F2 is greater than expected, due to a reduction in the generation of recombinant genotypes. The propensity of some characters to remain associated instead of assorting independently is given the name **linkage. Figure 1.8** shows an example of the measurement of linkage.

Morgan proposed that the cause of genetic linkage is the "simple mechanical result of the location of the (factors) in the chromosomes." He suggested that the production of genetic recombinant classes can be equated with the process of **crossing-over** that is visible during meiosis. Early in meiosis, at a stage when there are four copies of each chromosome organized in a bivalent, pairwise exchanges of material occur between the closely associated (synapsed) homologue pairs. This exchange is called a **chiasma;** it is illustrated diagrammatically in **Figure 1.9.**

Jumping some years ahead, it was not until 1931 that formal proof was obtained of the responsibility of crossing-over for recombination. In maize and in *Drosophila*, suitable **translocations** had occurred in which a part of one chromosome had broken off and become attached to another. This allows the translocation chromosome to be distinguished by its appearance from the normal chromosome. In suitable crosses, as indicated in **Figure 1.10**, it is then possible to show that the formation of genetic recombinants occurs only when there has been a physical crossing-over between the appropriate chromosome regions.

If the likelihood that a chiasma will form between two points on a chromosome depends on their dis-

tance apart, genes located near each other will tend to remain together. As the distance increases, the probability of crossing-over between them will increase. Thus if crossing-over is responsible for recombination, genes lying near each other will be tightly

Figure 1.9
Chiasma formation is responsible for recombinant formation.
Each bar represents a chromosome, and there are two homologues derived from each parent. These associate to form a **bivalent** at the prophase of meiosis. The top diagram is an enlargement of the second depicted in Figure 1.6. A **crossing-over** occurs between two of the chromosomes, as shown in the middle diagram. This generates the two recombinant chromosomes (carrying genes *Ab* and *aB*), as shown in the lower diagram. The other two chromosomes remain of the parental types (*AB* and *ab*).

Some features of chiasma formation parallel aspects of genetic recombination. The occurrence of one chiasma reduces the likelihood that another will occur nearby. This is called **positive interference.** This is seen in a reduction in the number of double recombinants in a triple cross. (The cross *ABC* × *abc* gives the double recombinants *AbC* and *aBc* when crossovers occur between *A* and *B* and again between *B* and *C* on the same chromosome.) The reduction is revealed because the actual frequency of double crossovers is less than the product of the frequencies of the individual crossovers (which would prevail if the two crossover events were independent).

This phenomenon extends the distance over which genetic maps can be constructed by direct linkage, because it counters the approach to a plateau of 50% recombination. When adjacent crossovers do occur, which two of the four copies are involved in one crossover is independent of which two are involved in the other event.

Parental chromosomes	Chromosomes of progeny

Noncrossover   Recombinant

Figure 1.10
Genetic recombination is caused by physical crossing-over.
In a mutant of maize, the chromosome carrying the alleles C and wx has gained a "knob" at one end, while the other end has a long translocation of material from another chromosome. In a heterozygote with a normal chromosome carrying the alleles c and Wx, recombination between the genetic markers is always associated with the formation of a new type of chromosome. Thus it must have occurred by a physical exchange of parts.

linked, and genetic linkage will decrease with physical distance apart. Reversing the argument, genetic linkage can be taken to be a measure of physical distance.

The concept that genes on the same chromosome are linked was extended by Sturtevant with the proposal that the extent of recombination between them can be used as a **map distance** to measure their relative locations. This is expressed as the percent recombination; that is,

$$\text{Map distance} = \frac{\text{Number of recombinants} \times 100}{\text{Total number of progeny}}$$

Distance is given in terms of **map units,** defined by 1 map unit (or centiMorgan) equals 1% recombination.

A critical feature was observed when the six sexlinked characters were followed together. The individual map distances are *additive*. Thus if two genes A and B are 10 units apart, and it is 5 units from B to a further gene C, the direct measure of distance between A and C will be close to 15. The genes can therefore be placed in a *linear order*.

A crucial point in the construction of maps is that the distance between genes does not depend on the

alleles that are used, but only on the gene **loci.** The locus defines the *position* on the chromosome at which the gene for a particular trait resides; the various alternative forms of the gene—that is, the alleles used in mapping—all reside at the *same* location. So genetic mapping is concerned with identifying the locations of gene loci, which are fixed and in a linear order. In a mapping experiment the same result is obtained irrespective of the particular combination of alleles (see Figure 1.8). Sturtevant concluded that his results "form a new argument in favor of the chromosome view of inheritance, since they strongly indicate that the factors investigated are arranged in a linear series, at least mathematically."

This last qualification is interesting. Although a genetic map can be constructed that represents a chromosome as a linear array of genes, this does not prove that the genetic content of the chromosome is *physically* a continuous array of genes. Many other models were considered before it became clear that a chromosome contains a single thread of genetic material.

Linkage is not displayed between all pairs of genes located on a single chromosome. The maximum recombination that occurs is the 50% predicted by Mendel's second law. (Although there is a high probability that recombination will occur between two genes lying far apart on a chromosome, each individual recombination event involves only two of the four associated chromosomes, thus generating 50% crossover between the genetic factors.)

So factors that are well separated may show no direct linkage, in spite of their presence on the same chromosome. However, each can be linked via intermediate factors, and so a genetic map can be extended beyond the limit of 50 map units directly measurable between any pair of genes. In fact, because of the distortion of linkage measured directly between markers that are not closely linked, map distances usually are based on measurements between fairly close factors, and may be subject to corrections from the simple percent recombination.

The **linkage group** includes all those genes that can be connected either directly or indirectly by linkage relationships. Those close together show direct linkage; those more than 50 map units apart in practice assort independently. As linkage relationships are extended, each gene identified in an organism can be placed into linkage with a group of previously mapped genes. Thus the genes fall into a discrete number of linkage groups. Genes in one linkage group always show independent assortment with regard to genes located in other linkage groups.

The number of linkage groups is the same as the number of chromosomes. The relative lengths of the linkage groups are similar to the actual relative sizes of the chromosomes. The example of *Drosophila* (where it happens to be particularly easy to measure the chromosome lengths) is shown in **Figure 1.11.**

Mendel's concept of the gene as a discrete particulate factor can therefore be extended into the concept that the chromosome constitutes a linear unit, divided into many genes, whose physical arrangement may underlie their genetic behavior.

THE GENETIC MAP IS CONTINUOUS

On the genetic maps of higher organisms established during the first half of this century, the genes are arranged like beads on a string. They occur in a fixed order, and genetic recombination involves the transfer of corresponding portions of the string between homologous chromosomes. The gene is, to all intents and purposes, a mysterious object (the bead), whose relationship to its surroundings (the string) is not clear.

Because of the small numbers of progeny that can be obtained from matings, the resolution of the recombination map is poor. Recombination occurs so infrequently between points that are close together that it rarely is observed between two mutations that affect the same character and may lie in the same gene. This poses the general question: is the structure of the gene the same as the regions between genes? In practical terms, we need to ask: can recombination occur within a gene; and can this be used to generate a linear order for mutations in a single gene?

To answer these questions it was necessary to turn to microbial systems in which very large numbers of progeny can be obtained from each genetic cross. The bacteriophage T4 is a virus that infects and then kills the bacterium *Escherichia coli*. Infection of a single bacterium leads to the production of about 100 progeny phages in less than 30 minutes. The phages are released when the bacterium is lysed (bursts open).

The constitution of an individual locus was investigated in a series of intensive studies of the *rII* genes of the phage, which are responsible for a change in

Figure 1.11
The linkage map can be related to the physical size of the chromosome.

In *D. melanogster,* each chromosome can be visualized as a number of "bands" (in an unusual state that is discussed in Chapter 28). The relative number of bands reflects the physical length of each chromosome. The relative numbers of map units per chromosome are similar.

Chromosome IV — No recombination — 45 bands

X chromosome — 66 map units / 1012 bands — 15.3 map units/band

Chromosome II — 108 map units / 1940 bands — 18.0 map units/band

Chromosome III — 106 map units / 2062 bands — 19.4 map units/band the pattern of bacterial killing, known as **rapid lysis.** When two different *rII* mutant phages are used to infect a bacterium simultaneously, it is possible to arrange the conditions so that progeny phages will be produced *only* if recombination has occurred between the two mutations to generate a wild-type recombinant. The frequency of recombination depends on distance just as in the eucaryotic chromosome.

The selective power of this technique allows even the rarest recombination events to be quantitated, so that the map distance between *any* pair of mutations can be measured. (A limit is placed on this by the frequency at which mutants "revert" spontaneously to wild type. This is about 0.0001%, so in order to achieve a minimum level of measurement ten-fold above background, only recombination frequencies greater than 0.001% can be assayed.)

About 2400 mutations were analyzed, and these identified 304 different mutant sites. (When two mutations failed to recombine, they were assumed to represent different spontaneous occurrences at the *same* genetic site.) The mutations could be arranged into a linear order, showing that the gene itself has the same linear construction as the array of genes on a chromosome. This suggests that the genetic map is linear within as well as between loci; that is, it consists of an unbroken sequence within which the genes reside.

ONE GENE—ONE PROTEIN

Until 1945 the gene was considered to be the fundamental unit of inheritance, but how it achieved this function was unknown. Genes could be identified only by mutations that produced some aberration in the phenotype of the organism. The apparent complexity of these aberrations varied from alterations of single features (such as eye color) to quite drastic morphological changes in more than one tissue of the organism. In almost all cases the effects could be analyzed only descriptively. However, it was assumed that there must be some common basis for these effects.

The idea that hereditary factors act through enzymes dates almost from the rebirth of genetics. In studies between 1902 and 1908, Garrod suggested that the human disease alkaptonurea results from the absence of an enzymatically catalyzed metabolic

reaction. His phrase "inborn errors of metabolism" carries the concept that a genetic defect may cause a metabolic failure that is responsible for the observed phenotype. Over the next three decades, further examples accumulated in which particular biochemical reactions could be associated with the effects of specific mutations. The main difficulty in these studies was the need to rely on those mutants that happened to be available, and which were not necessarily suitable for biochemical studies.

A systematic attempt to associate genes with enzymes was started by Beadle and Ephrussi in the 1930s, when they were able to conclude that the development of the normal red eye color in *Drosophila* passes through a discrete series of stages. Blockage at different stages results in the production of different mutant colors. But it was not until much later that the complete pathway could be worked out. Beadle and Tatum thus attempted to approach the problem from the other direction. As Beadle recollected, "it suddenly occurred to me that it ought to be possible to reverse the procedure we had been following and instead of attempting to work out the chemistry of known genetic differences we should be able to select mutants in which known chemical reactions were blocked. *Neurospora* was an obvious organism on which to try this approach, for it probably could be grown in a culture medium of known composition."

So mutants of *Neurospora* were induced by X-irradiation and selected for their inability to grow on a minimal medium. The biochemical nature of the defect could be identified by finding which growth factor allowed the mutant strain to grow. Each mutant turned out to be blocked in a particular metabolic step, undertaken in the wild-type strain by a single enzyme. When a given substrate is utilized through a whole series of metabolic reactions, mutation can block the series at any point, as shown by the accumulation of the metabolic intermediate immediately prior to this step. An example of the eye pigment pathway in *Drosophila,* which was subsequently shown to conform with this principle, is illustrated in **Figure 1.12.**

By 1945 the results of this analysis had become known in common parlance as the **one-gene : one-enzyme hypothesis.** This proposed that each metabolic step is catalyzed by a particular enzyme, whose production is the responsibility of a single gene. A

Tryptophan

Vermilion mutant

Formylkynurenin

Cinnabar mutant

Hydroxykynurenin

Brown pigment

Figure 1.12
Genes control metabolic steps.

Two mutations in the eye color of *D. melanogaster* act by blocking different stages in the pathway for conversion of tryptophan to brown pigment. Each mutation results in the absence of activity of the particular enzyme responsible for the corresponding conversion, causing accumulation of an intermediate whose contribution to eye color (or lack thereof) determines the phenotype.

mutation in the gene may abolish the activity of the protein for which it is responsible. Since a mutation is a random event with regard to the structure of the gene, the greatest probability is that it will damage gene function. Thus the majority of mutations create a nonfunctional gene; but this is by no means the only consequence of mutation, which sometimes alters rather than abolishes gene function.

These ideas explain the nature of recessive mutations: they represent an absence of function, because the mutant gene has been prevented from producing the usual enzyme. As illustrated in **Figure 1.13,** however, in a heterozygote containing one wild type and one mutant gene, the wild-type gene is able to direct production of the enzyme. The wild-type gene is therefore dominant. (This assumes that an adequate *amount* of protein is made by the one wild-type allele. When this is not true, the smaller amount made by one allele as compared to two may be seen in the intermediate phenotype of a partially dominant or codominant wild-type gene.)

The direct proof that a gene actually is responsible for controlling the structure of a protein had to wait until 1957, when Ingram showed that the single-gene trait of sickle cell anemia can be accounted for by a change in the amino acid composition of the protein hemoglobin.

Some proteins consist of more than one subunit. These are said to be **multimeric.** If the subunits are all the same, the protein is a **homomultimer,** coded by a single gene. If the subunits are different, the protein is a **heteromultimer.** Hemoglobin is an example of a protein consisting of more than one type of polypeptide chain; a heme group is associated with two α subunits and two β subunits. Each type of subunit is a different polypeptide chain, and is the product of a different gene. Thus the function of hemoglobin may be inhibited by mutation in either of the genes coding for the α and β subunits, respectively. Stated as a more general rule applicable to any heteromultimeric protein, the one-gene : one-enzyme hypothesis becomes more precisely expressed as **one-gene : one polypeptide chain.**

A NEW DEFINITION: THE CISTRON

If a recessive mutation is produced by every change in a gene that prevents the production of an active protein, there should be a large number of such mutations in any one gene; for many amino acid replacements may be able to change the structure of the protein sufficiently to impede its function. When a series

Figure 1.13
Genes code for proteins.

In the wild type, both genes are active (solid bars) and produce protein. In the heterozygote the dominant allele is active in producing protein, but the recessive allele (open bar) does not produce any active protein. In a homozygote with two recessive alleles, no protein is produced; the function is therefore absent.

Wild type   Heterozygote   Mutant homozygote

of mutations is available, all affecting the same characteristic and mapping together, they may represent **multiple alleles** of a gene. In this case, it is possible to have a heterozygote for two mutant alleles.

But how are multiple alleles to be distinguished from mutations in genes that are close together and affect the same characteristic of the phenotype? The basic test is that mutations in different genes show **complementation** in the heterozygote. In practical terms, the test is to make a genetic cross between two parents that are homozygous for the different mutations.

If the mutations lie in the same gene, each parent provides a different mutant allele, so that the heterozygote has the constitution

$$\frac{m_1}{m_2}$$

in which no wild-type gene is present.

But if the mutations lie in different genes, the parental genotypes can be represented as $(m_1 +)$ and $(+ m_2)$, where each has a wild-type allele of one gene and a mutant allele of the other. (A plus is used to signify wild type.) Then the heterozygote has the constitution

$$\frac{m_1 +}{+ m_2}$$

in which the two parents between them have provided a wild-type copy of each gene. The heterozygote thus has wild phenotype: the two genes are said to **complement.**

We can elaborate this description of the complementation test as shown in **Figure 1.14,** which represents the genotypes in either of two possible configurations. In the **cis** configuration, both mutations are present on the *same* chromosome. In the **trans** configuration, they are present on *opposite* chromosomes.

First consider the situation in which the mutations lie in the same gene. The *trans* configuration corresponds to the test that we have just described. Both copies of the gene are mutant. In the *cis* configuration, however, one genome elaborates a double-mutant protein, while the other produces a wild-type protein. Thus when two mutations lie in the same gene, the phenotype of a heterozygote is determined by the

configuration. It is mutant when the mutations lie in *trans,* and wild type when they lie in *cis.*

By contrast, when the mutations lie in different genes, the configuration is irrelevant. In each case there is one copy of each mutant gene and one copy of each wild-type gene.

This means that when mutations are tested in the *cis* configuration, the same result occurs whether they lie in the same or in different genes. Thus the *cis* configuration is used essentially as a formal control to demonstrate the presence of wild-type function (and against which the failure of complementation in *trans* could be judged).

The procedure for testing the complementation of two mutations is therefore to determine whether the *trans* heterozygote shows wild-type phenotype (the mutations lie in different genes) or mutant phenotype (the mutations lie in the same gene). For eucaryotes this is done by constructing the appropriate double heterozygote; for viruses it is done by simultaneously infecting a host cell with the two mutant types. If they complement in *trans,* each must specify a different product that is able to cooperate with the product of the other gene so that together they create the wild phenotype. The products are said to be **trans-acting,** and it is inferred that they represent diffusible products that can act together irrespective of their origins in different genomes.

When two mutations *fail* to complement in *trans,* the inference is that both affect the same function. They are therefore assigned to the same **complementation group.** In the analysis of the *rII* region of phage T4, all of the mutations fell into one of two complementation groups, each constituting a contiguous group of mutant sites. No pair of mutations in group *rIIA* could complement, no pair of mutations in group *rIIB* could complement; but any mutation in *rIIA* could complement any mutation in *rIIB*. This showed that the *rII* mutations identify two adjacent complementation groups, each concerned with the same function of the phenotype.

The full description of this assay is the **cis/trans complementation test.** It defines the complementation group as a unit all of whose parts must be wild type on the same chromosome. A single mutation in any part of this unit may abolish its function. The genetic unit defined by this assay is called the **cistron.**

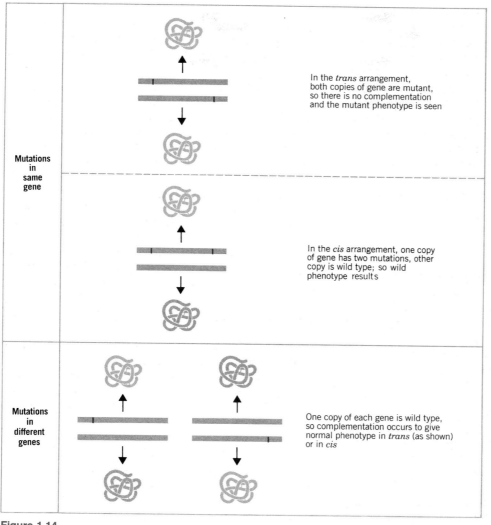

In the *trans* arrangement,
both copies of gene are mutant,
so there is no complementation
and the mutant phenotype is seen

In the *cis* arrangement, one copy
of gene has two mutations, other
copy is wild type; so wild
phenotype results

One copy of each gene is wild type,
so complementation occurs to give
normal phenotype in *trans* (as shown)
or in *cis*

**Mutations
in
same
gene**

**Mutations
in
different
genes**

Figure 1.14
The cistron is defined by the *cis/trans* complementation test.
Wild-type genes are continuous; mutations are indicated by colored vertical lines. Wild-type proteins
are grey; mutant proteins are colored.

In this terminology, two mutations in the *same* cistron *cannot* complement; the occurrence of complementation implies that two mutations lie in *different* cistrons.

The reason for the introduction of a new term was that there had been much emphasis on defining genetic units: the unit of mutation, the unit of recombination, the unit of function. The work with the *rII* system established a firm basis for the view that the units of mutation and recombination are smaller than the unit of function, defined by the cistron. This term now is generally reckoned to be equivalent with the older term "gene" and, in fact, is passing into disuse.

An exception to the rule that only different genes can complement is sometimes found when a gene codes for a polypeptide that is the subunit of a homomultimeric protein. In the wild-type cell, the active protein consists of several identical subunits. In a cell containing two different mutant alleles, however, their products can mix to form multimeric proteins that contain both types of subunit. Sometimes the two mutations compensate, so that the mixed-subunit protein is active, even though the proteins consisting solely of either type of mutant subunit are inactive. This is called **interallelic complementation.**

To summarize, this genetic analysis defined the gene as a contiguous series of sites in the genome. This series of sites is involved in the production of a single *trans*-acting product, that is, a polypeptide chain. But we shall see shortly that this definition is no longer sufficiently precise.

A NOTE ABOUT TERMINOLOGY

Before proceeding to the molecular analysis of the gene, we should mention the terminology that is used to describe genetic loci and their effects.

Genetic marker is sometimes used to describe a gene that is of interest, for example, one being used in a mapping experiment or identifying a particular region. Thus a cell may be said to carry a particular set of markers, that is, alleles.

Features of the genotype are *always italicized;* the phenotype is described by the same terms, but not italicized. "Wild type" can be used in the context of referring either to the usual active form of a gene or to a phenotypic feature. It may be indicated by a plus superscript to the name of the locus. Sometimes + is used by itself to describe the wild-type allele, and only the mutant alleles are indicated by the name of the locus. An entirely defective form of the gene (or absence of phenotype) may be indicated by a minus superscript (in some systems). When there is a variety of mutant alleles, other superscripts may be introduced to make finer distinctions. For example, the *w* (white eye) locus of *D. melanogaster* may be occupied by the wild-type allele, w^+ (confering red eye color), by the w^i allele (ivory eye color), by the w^a allele (apricot eye color), or by any one of many other alleles.

Genes are given abbreviations, whose nature depends on the type of system. In most eucaryotic systems, there is no restriction on the form of the abbreviation. A common convention is that the first letter of the abbreviation is capitalized for a dominant allele (in yeast the entire locus name is capitalized), while a recessive allele is entirely in lower case.

In bacteria, the standard nomenclature is to use a three-letter, lowercase abbreviation for all genes that affect a particular characteristic. Each individual gene is given an additional capital letter. Thus there is a series of *lac* genes, known as *lacZ, lacY,* and *lacA*. The wild-type allele of *lacZ* would be *lacZ*$^+$; a defective mutant would be *lacZ*$^-$. The same abbreviations are used to describe the phenotype, but the first letter is capitalized. Thus wild-type or mutant *lac* alleles would give rise to the respective bacterial phenotypes of Lac$^+$ or Lac$^-$. The product of a gene is a feature of the phenotype, and is referred to accordingly; for example, the LacZ protein is the enzyme β-galactosidase.

FURTHER READING

The modern view of the gene/cistron starts with **Benzer's** paper (*Proc. Nat. Acad. Sci. USA* **41,** 344, 1955). This work has been reviewed frequently, not least effectively by **Hayes** in *Genetics of Bacteria and Their Viruses* (Wiley, New York, 1968).

CHAPTER 2
WHAT IS A GENE?
A BIOCHEMICAL VIEW

Formal genetics demonstrate that the gene is a particulate factor, part of a chromosome, that is passed from parent to progeny. A critical step toward defining its nature was the realization that the gene is distinct from the characteristic that it represents. The diverse effects of mutations on the phenotype are explained by supposing that each mutation prevents the synthesis of a protein. But the physical nature of the genetic material responsible for the production of the protein was a matter of controversy and, when finally discovered, disbelief.

A great impetus to studies of the nature of the gene was given by Schrödinger's review in 1945 of its properties from the perspective of the physicist. "Incredibly small groups of atoms, much too small to display exact statistical laws, do play a dominating role in the very orderly and lawful events within a living organism. . . . The gene is much too small . . . to entail an orderly and lawful behavior according to physics." From this perspective, Schrödinger went on to develop from the previous views of Delbrück the concept that the laws of physics might be inadequate to account for the properties of the genetic material, in particular its stability during untold generations of inheritance. It was clear that the gene should be expected to obey the laws of physics so far established, but it was thought that studies of its properties might lead to the discovery of new laws of physics, a prospect that brought many physicists into biology.

Schrödinger summarized the properties of the gene in terms that, with modest redefinition, correspond remarkably well with our present view. "We shall assume the structure of the gene to be that of a huge molecule, capable only of discontinuous change, which consists in a rearrangement of the atoms and leads to an isomeric molecule. The rearrangement [mutation] may affect only a small region of the gene, and a vast number of different rearrangements may be possible." Although the gene was then visualized as an aperiodic crystal, because this seemed the only known physical form consistent with the demands placed on the genetic material, the concept that it might be a continuous part of the chromosome also was clear in this early view.

Now, of course, we know that the gene consists of DNA and that its structure is not maintained and perpetuated in isolation, but depends on enzymes that influence the accuracy of the process as well as undertake the necessary catalytic functions. All of these activities obey the known laws of physics and chemistry; and it has not, in the end, been necessary to invoke new regulations.

THE GENETIC MATERIAL IS DNA

In contrast with the concentration of early genetics on higher organisms, the first direct transfer of genetic information to be discovered was in bacteria. But the same principle since has been extended to higher organisms. The genetic material takes the same form in both procaryotes and eucaryotes, the two major divisions of living organisms.

Procaryotes (bacteria) are recognized by the lack of a nucleus (or of any other compartments, such as mitochondria, chloroplasts, etc.). The bacterial cell is a single compartment in which chromosomes are not evident, although a **nucleoid body** containing the genetic material can be seen. **Eucaryotic** cells have a nuclear membrane, which contains their genetic material, and whose dissolution at cell division is followed by the visible appearance of chromosomes. There are also other discrete compartments in addition to the nucleus.

Eucaryotes, of course, include the so-called microbial eucaryotes, which exist as single cells, as well as multicellular plants and animals. It was assumed that the basis of heredity would be similar in all eucaryotes, though there was doubt whether bacteria would have the same features.

Transformation was first observed in the study of pneumococcal infection of mice reported by Griffith in 1928. The virulence of this bacterium is determined by the capsular polysaccharide, a component of the surface. Virulent bacteria have a smooth (S) appearance. Several "types" of pneumococcus have different capsular polysaccharides. Any of these types can give rise to mutants that fail to produce the capsular polysaccharide; and these bacteria have a rough (R) surface. They are avirulent: they do not kill the mice (because in the absence of the polysaccharide they are subject to phagocytosis in the animal).

But when live R bacteria are injected into mice together with heat-killed S bacteria (which by themselves are unable to harm the animal), the mice die as the result of pneumococcal infection. Virulent S bacteria can be recovered from the animals postmortem, as illustrated in **Figure 2.1.**

The dead S bacteria had been of type I. The live R bacteria had been derived from type II. The virulent bacteria recovered from the mixed infection were type I S. Thus some property of the dead type I S bacteria can **transform** the live R bacteria so that they make the type I capsular polysaccharide.

As soon as these results could be repeated with a cell-free system, in which extracts of the S bacteria

Figure 2.1
The transforming principle is DNA.
Neither heat-killed smooth bacteria nor live rough (mutant) bacteria can kill mice. But the mixture kills mice from which live smooth bacteria can be recovered. The transformation of inactive rough bacteria into virulent smooth bacteria can be accomplished *in vitro* by the addition of DNA that was extracted from smooth bacteria.

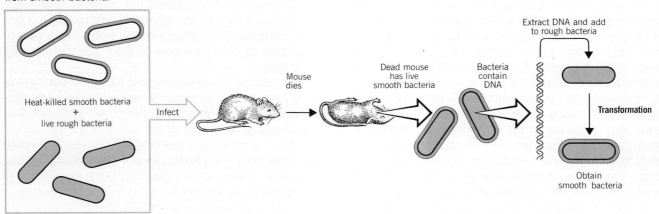

were used instead of living cells, it became practical to purify the **transforming principle,** as the active component of the system was called. The classic studies of Avery and his colleagues showed in 1944 that this was **deoxyribonucleic acid (DNA).** This was confirmed shortly after with the purification of an enzyme that degrades DNA, deoxyribonuclease, followed by the demonstration that its addition irreversibly destroyed the transforming principle. The possibility that transformation might be due to the presence of some protein contaminant in the DNA preparation was excluded by showing that treatment with the enzyme trypsin (which degrades all proteins) had no effect on the activity.

The surprise of this result is indicated by the fact that at this time DNA was not even known to be a component of pneumococcus, although of course it had been recognized as a major component of eucaryotic chromosomes for many decades. As Avery remarked in describing the isolation: "When alcohol reaches a concentration of about 9/10 volume there separates out a fibrous substance which on stirring the mixture wraps itself about the glass rod like thread on a spool and the other impurities stay behind as a granular precipitate. The fibrous material is redissolved and the process repeated several times. In short, this substance is highly reactive and on elementary analysis conforms *very* closely to the theoretical values of pure DNA (who could have guessed it). This type of nucleic acid has not to my knowledge been recognized in pneumococcus before, although it has been found in other bacteria."

The implications of the result are precisely captured by the discussion in the original paper reporting the research. "The inducing substance, on the basis of its chemical and physical properties, appears to be a highly polymerized and viscous form of DNA. On the other hand, the type III capsular substance, the synthesis of which is evoked by this transforming agent, consists chiefly of a nonnitrogenous polysaccharide. . . . Thus it is evident that the inducing substance and the substance produced in turn are chemically distinct and biologically specific in their action and that both are requisite in determining the type specificity of the cells of which they form a part." The distinction between the genetic material and the products of its expression was implicit in subsequent studies.

A reluctance to accept DNA as the genetic material led to some increasingly strained arguments against the obvious conclusion, which was then buttressed by further lines of research. With transformation itself, the generality of the conclusion was confirmed by showing that other capsular phenotypes in addition to the original types can be transferred by DNA in pneumococcus; and also by showing that transformation occurs with other bacteria and for a wide variety of phenotypic properties. This disposed of arguments that transformation might reflect some particular role of DNA in capsular polysaccharide formation rather than a general genetic role. In fact, we now know that the gene transferred in the original experiments was that responsible for production of UDPG dehydrogenase, an enzyme that catalyzes one of the steps of capsular polysaccharide synthesis. The next, quite different, property to be transformed was that of resistance to penicillin.

MORE EVIDENCE FOR THE ROLE OF DNA

From bacterial transformation, the next step was to demonstrate that DNA provides the genetic material in a quite different system. Phage T2 is a virus that infects the bacterium *E. coli.* When phages are added to bacteria, they adsorb to the outside surface, some material enters the bacterium, and then about 20 minutes later the bacterium bursts open (lyses) to release a large number of progeny phages.

In 1952, Hershey and Chase infected *E. coli* bacteria with T2 phages that had been radioactively labeled *either* in their DNA component (with ^{32}P) *or* in their protein component (with ^{35}S). The phages were mixed with bacteria, and any unadsorbed phage particles were removed by centrifugation. The infected bacteria were agitated in a blender, and two fractions were separated by centrifugation. One contained the empty phage coats that were released from the surface of the bacteria. The other consisted of the bacteria themselves.

Figure 2.2 shows that the ^{35}S label was associated with the phage coats. But most of the ^{32}P label was present in the infected bacteria. When progeny phage particles produced by the infection were examined, they contained some 30% of the original ^{32}P label. But

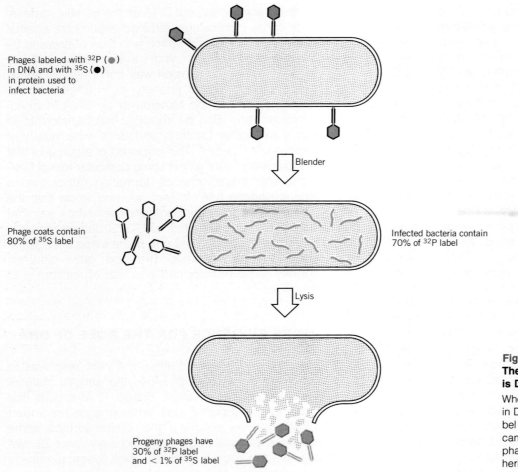

Phages labeled with ^{32}P (●)
in DNA and with ^{35}S (●)
in protein used to
infect bacteria

Blender

Phage coats contain
80% of ^{35}S label

Infected bacteria contain
70% of ^{32}P label

Lysis

Progeny phages have
30% of ^{32}P label
and < 1% of ^{35}S label

Figure 2.2
**The genetic material of phage T2
is DNA.**

When phages are labeled with ^{32}P
in DNA or ^{35}S in protein, the ^{32}P label enters the infected bacteria and
can be recovered from the progeny
phages. But the ^{35}S label is not inherited by the progeny.

very little—actually less then 1%—of the protein of the
original phages could have entered their progeny. This
experiment therefore shows directly that the DNA of
parent phages enters the bacteria and then becomes
part of the progeny phages, exactly the pattern of inheritance expected of genetic material.

Whether viruses are to be classified as "living" has
been a debating point for a long time. The classic
example of phage T2 (remember that phages are simply viruses that infect bacteria) shows that, on infection, the virus is able to commandeer the machinery
of the host cell to manufacture more copies of itself.
The phage possesses genetic material whose behavior is analogous to that of cellular genomes: its traits

are faithfully reproduced, and they are subject to the
same rules that allow mutations to be mapped in a
linear order. Thus the case of T2 reinforces the general
conclusion that the genetic material is DNA.

It is interesting that, although clearly demonstrating
the role of DNA as genetic material, these experiments
technically are not quite as precise as transformation
in excluding protein as a contaminant of the DNA.
Increased purification of transforming DNA reduced
to ridiculously small proportions the amount of contaminant that could be present. In the phage infection,
however, a small amount of protein actually *is* transferred into the bacteria with the DNA; but, of course,
it is only the DNA that is recovered from the progeny.

The climate of opinion had much to do with the immediate acceptance of the Hershey-Chase experiment compared with the disbelief in Avery's work.

Bacteria and bacteriophages clearly have DNA as their genetic material. But what about higher organisms? For a long time the evidence was only inferential. DNA is present in the chromosomes; its amount is constant in all somatic cells; and only half the amount is found in the gametes. This is exactly the behavior expected of the genetic material, but falls short of direct proof.

In discussing his results, Avery made a comment with wider-ranging implications than he could have known. "If we are right," he wrote, "it means that nucleic acids are not merely structurally important but functionally active substances in determining the biochemical activities and specific characteristics of cells and that by means of a known chemical substance it is possible to induce predictable and hereditary changes in cells." He had in mind the system of bacterial transformation, but similar results now have been obtained with the cells of higher organisms in culture.

The first discovery that opened these possibilities was that chromosomes could be prepared from one cell and added to another just by mixing; at very low frequencies, recipient cells could be found that express new functions provided by the donor chromosomes. Then it became possible to perform these experiments with purified DNA; and now, by increasing the purity of the transforming preparation, it is possible in effect to add pure genes—that is, DNA containing only a single gene—to cells that lack the gene. Transformants then can be obtained that contain and express the gene. This can now be done for any gene whose corresponding enzyme activity can be assayed. An example of one of the standard systems is depicted in **Figure 2.3.**

Although for historical reasons these experiments are described as **transfection** when performed with eucaryotic cells, they are a direct counterpart to bacterial transformation. Such experiments show directly not only that DNA is the genetic material in eucaryotes, but also that genes can be transferred between different species and yet remain functional. At first, these experiments could be performed only with individual cells adapted to growth in a culture medium. Now, however, genes also have been introduced into mouse eggs by microinjection; and then they are found as stable components of the adult mouse genome. (Transfection is discussed in Chapter 38.)

Figure 2.3
Eucaryotic cells can be transfected with DNA.

In a procedure very similar to transformation of bacteria, eucaryotic cells growing in culture can take up added DNA and as a result display new phenotypes.

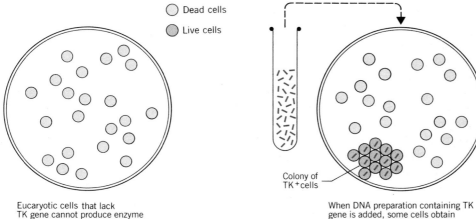

Dead cells
Live cells

Colony of
TK+cells

Eucaryotic cells that lack
TK gene cannot produce enzyme
thymidine kinase and therefore die
in absence of thymidine

When DNA preparation containing TK
gene is added, some cells obtain
the gene and thus can grow—
they pile up into a colony

There are some viruses in which the alternative **ribonucleic acid (RNA)** is found as the genetic material. Although its chemical formula is slightly different from that of DNA, RNA exercises the same role in these circumstances. The general principle of the nature of the genetic material, then, is that cells always use DNA; but while some viruses use DNA, others rely on RNA. All genetic material is nucleic acid.

THE COMPONENTS OF DNA

One of the reasons for scepticism about the genetic function of DNA was the misunderstanding of its structure. Although known to consist of four types of nucleotide, DNA was thought to comprise a regularly repeating tetranucleotide unit. By the 1930s Caspersson had shown that DNA consists of extremely large molecules—much larger than proteins—but the monotony of its structure appeared to preclude any possible genetic role, for which, after all, variability of form would be expected.

But then the continuing analysis carried out by Chargaff demonstrated that the four bases found in DNA are present in variable amounts. These are different and characteristic for each species. This observation led to the concept that the sequence of bases in DNA might be the form in which **genetic information** was carried, this sequence in some way specifying, or **coding** for, the sequence of amino acids in a protein. By the 1950s the concept of genetic information was common; the twin problems that it posed were working out the structure of the nucleic acid, and explaining how a sequence in DNA could represent the sequence of a protein.

Nucleic acids consist of a chemically linked sequence of **nucleotides.** Each nucleotide contains a heterocyclic ring of carbon and nitrogen atoms (the **nitrogenous base**), a five-carbon sugar in ring form (a **pentose**), and a **phosphate** group.

Two types of pentose are found in nucleic acids. These distinguish DNA and RNA, and give rise to the general names for the two types of nucleic acid. In DNA the pentose is 2-deoxyribose; whereas in RNA it is ribose. The difference lies in the absence/presence of the hydroxyl group at position 2 of the sugar ring, as illustrated in **Figure 2.4.**

The nitrogenous bases fall into the two types shown in the figure: **pyrimidines** and **purines.** Pyrimidines have a six-member ring; purines have fused five- and six-member rings. Each nucleic acid is synthesized from only four types of base. The same two purines, adenine and guanine, are present in both DNA and RNA. The two pyrimidines in DNA are cytosine and thymine; in RNA uracil is found instead of thymine. The only difference between uracil and thymine is the presence of a methyl substituent at C_5. The bases are usually referred to by their initial letters; so DNA contains A, G, C, T, while RNA contains A, G, C, U. (Certain modifications are made to these bases in some cases, after their incorporation into nucleic acid, as described for RNA in Chapter 7 and for DNA in Chapter 30.)

The nitrogenous base is linked to position 1 on the pentose ring by a glycosidic bond from N_3 of pyrimidines or N_9 of purines. To avoid ambiguity between the numbering systems of the heterocyclic rings and the sugar, positions on the pentose are given a prime (′).

Nucleotides are linked together into a **polynucleotide chain** by a backbone consisting of an alternating series of sugar and phosphate residues. The 5′ position of one pentose ring is connected to the 3′ position of the next pentose ring via a phosphate group, as shown in **Figure 2.5.** Thus the phosphodiester-sugar backbone is said to consist of 5′ - 3′ linkages. The terminal nucleotide at one end of the chain has a free 5′ group; the terminal nucleotide at the other end has a free 3′ group. It is conventional to write nucleic acid sequences in this direction—that is, from the 5′ terminus toward the 3′ terminus.

As can be seen from the figure, the nitrogenous bases "stick out" from the sugar-phosphate backbone. Note that the phosphate groups carry negative charges that require neutralization. This may be accomplished by metal ions and/or positively charged proteins.

When DNA or RNA is broken into its constituent nucleotides, the cleavage may take place on either side of the phosphodiester bonds. Depending on the circumstances, nucleotides may therefore have their phosphate group attached to either the 5′ or the 3′ position of the pentose, as shown in **Figure 2.6.** The

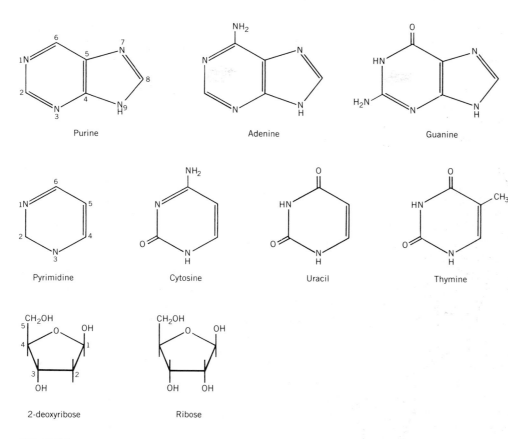

Figure 2.4
The components of nucleic acids are nitrogenous bases (purines or pyrimidines), sugar (deoxyribose or ribose), and phosphate (not shown).

The base and sugar positions are numbered as shown. A base-sugar moiety is a **nucleoside;** a base-sugar-phosphate moiety is a **nucleotide.** With "d" representing the deoxy form and its absence implying the presence of 2'-OH, their nomenclature is

Base	Nucleoside	Nucleotide and Abbreviation
Adenine	Adenosine	Adenylic acid = AMP (or dAMP)
Guanine	Guanosine	Guanylic acid = GMP (or dGMP)
Cytosine	Cytidine	Cytidylic acid = CMP (or dCMP)
Thymine	Thymidine	Thymidylic acid = dTMP
Uracil	Uridine	Uridylic acid = UMP

sugar-base moiety by itself is described as a **nucleoside;** so the two types of nucleotide released from nucleic acids are nucleoside-3'- monophosphates and nucleoside-5'-monophosphates.

All the nucleotides can exist in a form in which there is more than one phosphate group linked to the 5' position. An example is shown in Figure 2.6. The bonds between the first (α) and second (β) and between the second (β) and third (γ) phosphate groups are energy-rich and are used to provide an energy source for various cellular activities. The triphosphates are the forms from which nucleic acids are synthesized.

Bases

Sugar-phosphate backbone

Figure 2.5
A polynucleotide chain consists of a series of 5'-3' sugar-phosphate links that form a backbone from which the bases protrude.

DNA IS A DOUBLE HELIX

The discovery of the double helix was one of the more dramatic events in molecular biology. Two notions converged in the construction of this model by Watson and Crick in 1953.

First, X-ray diffraction data showed that DNA has the form of a regular helix, making a complete turn every 34Å, and with a diameter of about 20Å. Since the distance between adjacent nucleotides is 3.4Å, this means that there must be 10 nucleotides per turn.

The density of DNA suggested that the helix must contain two polynucleotide chains. The constant diameter of the helix can be explained if the bases in each chain face inward and are restricted so that a purine is always opposite a pyrimidine, avoiding purine-purine (too thick) or pyrimidine-pyrimidine (too thin) partnerships.

Figure 2.6
Nucleotides may carry phosphate in the 5′ or 3′ position.

On the left is a 3′ monophosphate of cytosine. On the right is a 5′ triphosphate of guanosine. Diphosphate forms are abbreviated as NDP, and triphosphates are written as NTP.

The second critical feature that explained how this form of organization is accomplished lay in the previous observation of Chargaff that, irrespective of the actual amounts of each base, the proportion of G and C is always the same in DNA, and the proportion of A and T is always the same. (Thus any DNA can be characterized by its **(G + C)/(A + T) ratio,** which ranges from 26% to 74%, but is a characteristic of each species, a clear refutation of the repeating-tetranucleotide theory.)

Watson and Crick proposed that the two polynucleotide chains in the double helix are not connected by covalent bonds, but associate through **hydrogen bonding** between the bases. **Figure 2.7** shows that in their usual forms, G can hydrogen bond specifically only with C, while A can pair specifically only with T. These reactions are described as **base pairing,** and the paired bases (G with C or A with T) are said to be

Figure 2.7
Specific base pairing involves the formation of two hydrogen bonds between A and T, while three hydrogen bonds are formed between G and C. No other pairs form in DNA.

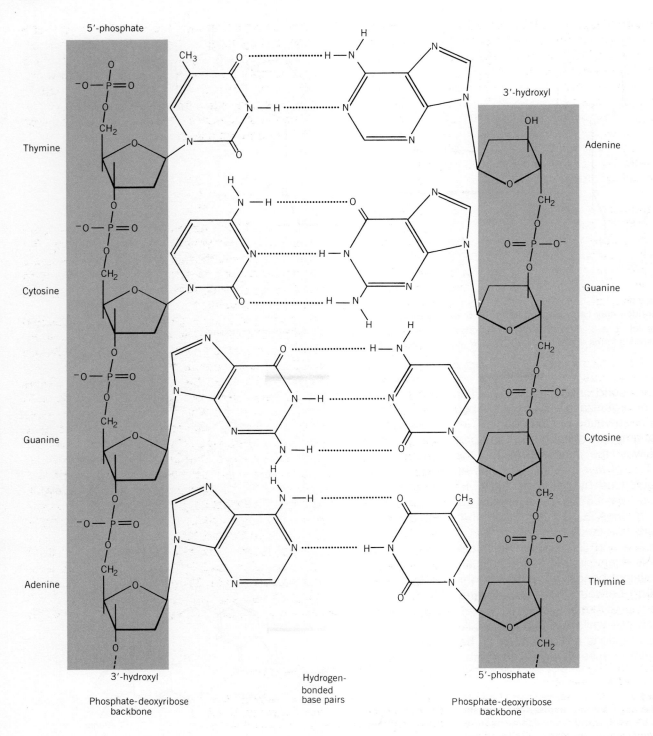

5'-phosphate

3'-hydroxyl

Thymine

Adenine

Cytosine

Guanine

Guanine

Cytosine

Adenine

Thymine

3'-hydroxyl

5'-phosphate

Phosphate-deoxyribose backbone

Hydrogen-bonded base pairs

Phosphate-deoxyribose backbone

complementary. To permit specificity in base pairing, the use of the appropriate form of the base is critical. The movement of a hydrogen atom allows each base to exist in **tautomeric** forms. Those present in the double helix have amino groups (NH_2) and keto groups ($C=O$), as opposed to the tautomeric alternative of imino groups (NH) and enol groups (COH).

The model requires the two polynucleotide chains to run in opposite directions (**antiparallel**), as shown in **Figure 2.8.** Looking along the helix, therefore, one strand runs in the 5'-3' direction, while its partner runs 3'-5'.

The bases are flat structures lying in pairs perpendicular to the axis of the helix. The energy released by hydrogen bonding between the paired bases contributes to the thermodynamic stability of the duplex. Proceeding up the helix, bases are **stacked** above each other, and part of the energy for maintaining the helix comes from hydrophobic **base-stacking** interactions.

Each base pair is rotated 36° around the axis of the helix, relative to the next base pair. Thus 10 base pairs make a complete turn of 360°. The twisting of the two strands around each other forms a double helix which has a **narrow groove** (about 12Å across) and a wide groove (about 22Å across), as can be seen from the scale model shown in **Figure 2.9.** The double helix is **right-handed:** the turns run clockwise looking along the helical axis. These features represent the accepted model for what is known as the **B-form** of DNA.

One of the crucial requirements for the genetic material is that it must be reproduced accurately with every cell division. In what has become a classic understatement, Watson and Crick concluded their paper by saying that "it has not escaped our notice that the specific pairing we have postulated immediately suggests a possible copying mechanism for the genetic material." This idea was taken up in a subsequent paper, where they pointed out that, because the two polynucleotide strands are joined only by hydrogen bonds, they might be able to separate. The specificity of base pairing suggests that each strand could act as a **template** for the synthesis of a complementary strand, as shown in **Figure 2.10.** This mechanism is called **semiconservative synthesis** (see Chapter 31). Thus the structure of DNA carries the information needed to perpetuate its sequence.

ABOUT ALTERNATIVE DOUBLE-HELICAL STRUCTURES

The structure of DNA generally has been considered to be one of the certainties of molecular biology. But recently it has become clear some of the parameters of the classic B-form helix need to be adjusted and that DNA even may be able to exist in other types of double-helical structures.

The problem in assigning specific values for parameters of the double helix is that all the original structural evidence comes from X-ray diffraction studies on fibers. These data indicate basic features, such as the number of residues per turn or the spacing of residues along the helical axis, but do not assign positions to specific atoms in the way that is possible with X-ray crystallography. This leaves it to model building to provide structures consistent with the experimental data. One consequence is that (in theory) it is always possible that another model may be constructed to explain the same data. Another is that the model simply provides a fit for what may be *average* data, with variations occurring in particular regions. For example, so far we have treated the DNA as though it were a long, rigid double helix; but we know that in reality it must be coiled up and tightly compacted in order to fit in the cell. This could change the details of its structure.

At least it now seems that double-stranded DNA is always a double helix. Another model, apparently con-

Figure 2.8
The double helix maintains a constant width because purines always face pyrimidines in specific G-C and A-T base pairs.

The sugar–phosphate backbone is on the outside and carries negative charges; the bases face inward and are stacked above each other. In the direction from top to bottom, the strand on the left runs 5'–3', and the strand on the right runs 3'5' This figure is diagrammatic and does not show the winding of the two helices around each other.

Figure 2.9
The two strands of DNA form a double helix.
(Photograph kindly provided by Maurice Wilkins.)

sistent with the X-ray diffraction data, was constructed in which the two antiparallel DNA strands remain associated via complementary base pairing, but lie "side by side" instead of being wound into a continuous double helix. But the existence of DNA as a double helix has been confirmed by experiments to measure directly the number of base pairs per turn. This proves to be 10.4 instead of the 10.0 predicted by the classic B-model. The change requires a slight adjustment in the angle of rotation between adjacent base pairs along

the helix, to 34.6°, so that it takes slightly longer to accomplish the full 360° turn.

It is especially significant that the value of 10.4 is an average for DNA as a whole under certain conditions. Changes in the conditions or even in the particular base sequence could lead to tighter or looser helical structures in particular regions. Indeed, the structure of a particular 12 base-pair molecule has been shown by X-ray crystallography to have 10.1 base pairs per turn, achieved by a slight twist of each base

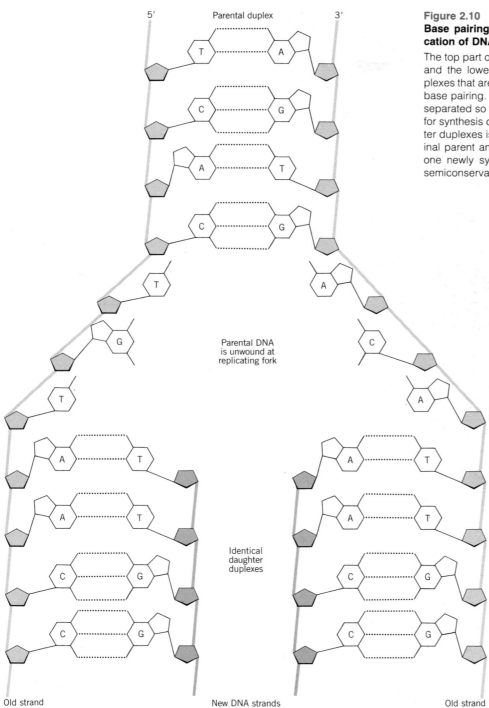

5′   Parental duplex   3′

Parental DNA
is unwound at
replicating fork

Identical
daughter
duplexes

Old strand

New DNA strands

Old strand

Figure 2.10

Base pairing provides a mechanism for replication of DNA

The top part of the figure shows a parental duplex, and the lower part shows the two daughter duplexes that are being produced by complementary base pairing. The two parental strands have been separated so that each can be used as a template for synthesis of a complement. Each of the daughter duplexes is identical in sequence with the original parent and contains one parental strand and one newly synthesized strand (this is known as semiconservative replication).

pair that improves base stacking relative to the original model.

Because of this variation, the idea that there is a single structure for the DNA double helix has been replaced with the view that there are families of structures, each of a characteristic type, but showing differences in the governing parameters n (the number of nucleotides per turn) and h (the distance between adjacent repeating units). The variation is achieved by changes in the rotation of groups about bonds with rotational freedom. Within each family, the parameters can vary slightly; for example, for B-DNA n could be 10.0–10.6.

Three structural forms of DNA have been known for a long time, and transitions between them occur when appropriate changes are made in the conditions. Their general characteristics are summarized in **Table 2.1.**

The **B-form** for which Watson and Crick constructed their model is found in fibers of very high (92%) relative humidity and in solutions of low ionic strength. This corresponds to the form that is thought to prevail in the living cell.

The **A-form** is found in fibers at 75% relative humidity and requires the presence of sodium, potassium, or cesium as the counterion. Instead of lying flat, the bases are tilted with regard to the helical axis; and there are more base pairs per turn. The A-form is biologically interesting because it is probably very close to the conformation adopted by DNA-RNA hybrids or by RNA-RNA double-stranded regions (see later). The reason is that the presence of the 2′ hydroxyl group prevents RNA from lying in the B-form.

The **C-form** occurs when DNA fibers are maintained in 66% relative humidity in the presence of lithium ions. It has fewer base pairs per turn than B-DNA.

These three forms are available to all DNAs, irrespective of sequence. Some further forms have been found that appear to represent options open only to DNA molecules with particular quirks in their base composition.

The **D-form** and **E-form** (actually possibly extreme variants of the same form) have the fewest base pairs per turn (only 8 and 7½) and are taken up only by certain DNA molecules that lack guanine.

The **Z-form** provides the most striking contrast with the classic structure families. It is a *left-handed* helix, whereas all the others are *right-handed*. It has the most base pairs per turn, and so has the least twisted structure; it is very skinny, and its name is taken from the zigzag path that the sugar-phosphate backbone follows along the helix, quite different from the smoothly curving path of the backbone of B-form DNA. The structures of B-DNA and Z-DNA are compared in **Figure 2.11.**

The Z-form double helix has been found in polymers that have an alternating purine-pyrimidine sequence. The two that have been examined have a dinucleotide pair that is repeated many times: poly $d\left(\dfrac{GC}{CG}\right)_n$ and poly $d\left(\dfrac{AC}{TG}\right)_n$. (The "d" indicates that these are the deoxy forms, that is, this is DNA and not RNA. The upper and lower rows of letters indicate the two strands of DNA.) The Z-form DNA exists only at very high salt

Table 2.1
DNA Can Exist in Several Types of Structure Families

Helix Type	Base Pairs per Turn	Rotation per Base Pair, deg	Vertical Rise per Base Pair, Å	Helical Diameter, Å
A	11	+32.7	2.56	23
B	10	+36.0	3.38	19
C	9⅓	+38.6	3.32	19
Z	12	−30.0	3.71	18

The rotation per base pair is indicated as (+) for a right-handed helix and (−) for a left-handed helix.

Groove

Major groove

Minor groove

Z DNA   B DNA

Figure 2.11
B-DNA and Z-DNA present different double-helical structures for DNA.

concentration (the need for which is explained by the requirement to counter the increased electrostatic repulsion between the nucleotides compressed into the slimmer double helix of Z-DNA).

The possible biological importance of Z-DNA is indicated by the observation that replacing the C residue in the polymer poly $d\left(\dfrac{GC}{CG}\right)_n$ with 5-methylcytosine makes the Z-DNA much more stable at lower salt concentrations. The only difference between 5-Me-C and C is the presence of the methyl group at position 5; and this modification of cytosine is a reaction that occurs at $\dfrac{CG}{GC}$ dinucleotide sequences at specific lo-

cations *in vivo*. This makes it seem plausible that Z-DNA could exist *in vivo* in the right circumstances, that is, if a sequence of appropriate base composition finds itself in conditions promoting Z-DNA formation.

So far we have treated the DNA as a structure of nucleic acid in splendid isolation. In reality, it is associated with proteins. These may substantially affect its ability to undergo the transition from B-form to Z-form. For example, DNA associated with histones (the basic chromosomal proteins of eucaryotic nuclei) does not display the transition under conditions when free DNA can do so. Thus one of the features necessary for the formation of Z-DNA *in vivo* is likely to be the presence of particular proteins that stabilize its structure.

Is Z-DNA the only left-handed form of the double helix? Models have been constructed for left-handed versions of the traditional A, B and C forms; so it is possible that these may exist. We should remember that all of these forms retain the critical biological feature of the double helix: complementary base pairing is responsible for the specificity of strand association.

The model building considers each form of DNA independently, as constituting the structure for the entire molecule under appropriate conditions. But this is an unlikely scenario *in vivo*. The majority of cellular DNA is likely to take the B-form, as modified by recent changes in the description of the helical parameters, and possibly with specific regional variation. Only rather short stretches are likely to be involved in transitions to other forms.

Is it possible for a transition to occur from one form to another in the *same molecule* of DNA? Under certain conditions, a stretch of 26–32 $\dfrac{GC}{CG}$ doublets *can* convert to Z-DNA, while the regions on either side remain in the classic B-form. This strengthens the idea that conformational transitions could occur *in vivo* at specific sites.

Antibodies can distinguish the Z-form of DNA from the B-form. These antibodies bind to certain chromosomes of *D. melanogaster*. The reaction can be visualized because the chromosomes have an unusual structure, in which compact regions (bands) alternate with less compact regions (interbands) (see Chapter 28). The Z-DNA resides in the interbands. This demonstrates that the existence of Z-DNA is a natural phenomenon, although we do not yet know the length of the individual stretch of Z-DNA.

SUPERCOILING MAY BE SUPERIMPOSED ON THE DOUBLE HELIX

So far we have dealt with DNA in terms of a linear, double-stranded molecule. However, DNA also exists as circular molecules or as simulacra of them. For small viruses, the genome may actually consist of **circular DNA,** in which both strands of the double helix run continuously around the circle. For bacterial and eucaryotic genomes, the DNA may exist as somewhat larger loops; a critical feature is that each loop is held together at the base in such a way that again there are effectively no free ends (see Chapter 28). The importance of this form of organization is that it permits an additional structural constraint to be superimposed on the double helix.

By taking a duplex of DNA and twisting it around its axis, **supercoils** can be introduced. The usual analogy is to consider a rubber band twisted about itself, to generate a tightly coiled structure in which the rubber band (the double helix) crosses over itself in space. This places it under torsion, giving the type of structure depicted in **Figure 2.12.** Of course, the torsion can

be maintained only when the lack of free ends prevents unwinding. We shall consider how supercoils are to be measured and how they are actually introduced into DNA in Chapter 32. All we need to remember at present is that, in principle, one supercoiled turn is introduced for every time the thread is twisted about its axis, and the more supercoils there are, the greater the torsion.

For a rubber band, it does not matter in which *direction* we apply the twist that generates the supercoils. (The two edges of the rubber band are equivalent.) But because the double helix is itself a twisted structure (as seen in the intertwining of the two strands), its response to torsion depends on the direction of the supercoiling.

Negative supercoils twist the DNA about its axis in the *opposite* direction from the clockwise turns of the right-handed double helix. This means that the DNA can in principle partially relieve the torsional pressure by adjusting the structure of the double helix itself. This could take the form of reducing the rotation per base pair, or it could even lead to a limited disruption of base pairing. Thus DNA with negative supercoils is said to be **underwound.** (The extreme case would be the conversion of the right-handed helix into a region of left-handed helix.)

It is important to remember that these events do not

Figure 2.12
Supercoiling causes a DNA duplex to be twisted about itself in space; negative supercoiling can be relieved by disrupting base pairs.

Circular DNA with zero supercoiling   Negatively supercoiled DNA   Negative supercoiling may be converted into strand separation occur spontaneously, but happen in an environment in which DNA is under a variety of constraints. A practical way to look at the issue is to consider that a supercoiled molecule has acquired excess energy through the introduction of the supercoils. In this sense, they are a store of energy. So if some change is to be made in the structure of DNA—for example, because a protein binds to it to keep some region in the single-stranded state, or to convert it into another form such as Z-DNA—*less energy* will be required to make this change if the DNA is negatively supercoiled. Thus the degree of supercoiling may influence the equilibrium of such reactions. Indeed, a propensity toward these types of conformational change has been found in negatively supercoiled DNA.

Can we quantitate the magnitude of this effect? All genomes that have been examined exhibit some negative supercoiling. A typical level of supercoiling *in vivo* is about one negative turn for every 200 base pairs. This confers an energy on the molecule of about 9 kcal/mol. The energy needed to unwind DNA depends on the sequence of base pairs (see next section), but is in the range of 12–50 kcal/mol per 10 base pairs separated. So the actual level of supercoiling probably corresponds to enough energy to assist the unwinding of just a few base pairs.

The opposite types of effect are caused if DNA is supercoiled in the *same* direction as the intrinsic winding of the double helix. The introduction of **positive supercoils** tightens the structure, applying torsional pressure to wind the double helix more tightly. Positively supercoiled DNA is said to be **overwound.** This can be accomplished by treatment *in vitro,* but does not occur as a natural state.

RNA ALSO HAS SECONDARY STRUCTURE

What structural options are available to RNA? Its primary structure is the same as that of DNA: a polynucleotide chain with 5' - 3' sugar-phosphate links. But generally it exists as a *single* polynucleotide chain rather than a double helix of antiparallel strands. Thus the base composition is not restricted to the overall G = C, A = U equalities; and its biophysical properties show much less secondary structure. However, base pairing can take place within and between RNA mol-

ecules. We have mentioned that such double-stranded regions of RNA probably exist in the A-form.

When a sequence of bases is followed by a complementary sequence in the same chain, the polynucleotide may fold back on itself to generate an antiparallel duplex structure. This is called a **hairpin.** It consists of a base-paired, double-helical region, the **stem,** often with a **loop** of unpaired bases at one end. An example is shown later in Figure 2.13.

When two more distant sequences of an RNA chain are complementary, they may come into juxtaposition to base pair to form a double-helical region. Essentially this creates a stem with a very long single-stranded loop. Also, two *different* RNA molecules may have regions that are complementary and that base pair in the appropriate environment; usually these regions comprise rather short sequences.

These interactions are of crucial biological significance, and examples of each type will be found throughout this book. They make the point that the concept of base pairing is central to all processes that involve nucleic acids.

In view of the irregular occurrence of secondary structure in RNA, how is it to be predicted from the sequence or to be measured in practice? Sometimes an RNA molecule may have several regions that are able potentially to base pair with one another in alternative arrangements, so it is necessary to resolve which actually occurs.

The *overall* extent of base pairing can be determined from measurements of the biophysical properties of the molecule, as described in the next section. However, this does not reveal the details of the individual regions that are involved. Single-stranded and double-stranded regions have different susceptibilities to some nucleases, and this provides a test for analyzing the particular regions involved in base pairing. However, such data have limited applicability, and for the most part the analysis of RNA secondary structure depends more on theory than on experimental information.

The *plausibility* of a particular base-paired structure can be predicted by **rules** that describe the interactions of the base pairs. When alternative structures exist, their relative stabilities can be assessed. Of course, this treats the RNA as an isolated and stable structure, ignoring any other factors that may intervene to influence the conformation *in vivo* (for example, binding by proteins). However, the application of these rules provides a first step to seeing whether a particular structure is able in principle to exist.

The basis of these rules is to calculate the **free energy** for the formation of each structure. The free energy is (briefly) a thermodynamic constant that gives the available energy that can be released by the reaction. Formally, this is the parameter ΔG, which has a *negative* value corresponding to the energy that is available. Thus if there are alternative structures whose formations have free energy values of $\Delta G = -21$ and $\Delta G = -35$, the latter is more likely to be formed, other things being equal.

The overall free energy of a structure is calculated by summing the free energies of the individual base-pairing reactions that are involved. At first, this was done on the basis of assigning a value to each base pair. Since G-C pairs have three hydrogen bonds (and the formation of each hydrogen bond releases a certain amount of energy), they are the most stable, with a ΔG of -2.4 kcal/mole. The ΔG for A-U pairs is -1.2 kcal/mole (since they are maintained by only two hydrogen bonds). Another base pair that can form in RNA is the irregular partnership G·U (remember that G·T is excluded from DNA). This has a ΔG of O; in other words, a G may face a U in a double-helical RNA structure without either disrupting the helix or contributing to its formation.

Actually, it turns out that a major force for providing free energy in double-helical pairing comes from the hydrophobic interactions of bases with one another as they are stacked along the helix. The result of this base stacking is that the free energy differs for each combination of adjacent base pairs. It is influenced not just by their composition but also by the sequence. For example, the doublet sequences $\frac{GG}{CC}$ and $\frac{GC}{CG}$ release more free energy ($\Delta G = -5.0$) than the sequence $\frac{CG}{GC}$ ($\Delta G = -3.2$), although they all have the same composition of two G-C pairs. Similarly, the sequence $\frac{AA}{UU}$ releases less energy ($\Delta G = -1.2$) than

the alternatives of the same composition, $\frac{AU}{UA}$ and $\frac{UA}{AU}$ ($\Delta G = -1.6$). (The sequences of the doublets are written following the convention that the top row represents $5' - 3'$ linkage; the bottom row is the complementary $3' - 5'$ order.)

For this reason, the free energy usually is calculated for each doublet of adjacent base pairs in a potential double-helical structure. Each base pair is involved in the calculation for two doublets, one for the base pair on each side. (It is involved in only one calculation if it lies at the end of the helix, so that it participates in only one doublet pair.) The sum of all the doublet free energies gives the overall free energy of formation. It remains true that combinations involving G-C pairs release more energy than those involving A·U, so that the stability of a double-helical region increases with the proportion of G-C pairs, but the exact value depends on the order of the base pairs.

An example of a free-energy calculation is shown in **Figure 2.13.** Unlike DNA, which is maintained as a perfect duplex structure, double-helical regions of RNA usually form between single strands that exhibit less than perfect complementarity. This means that there may be interruptions in the integrity of double-helical regions. These take the form of **hairpin loops** (occurring when the double helix is formed by looping back to pair with an immediately complementary sequence), **interior loops** (when there are corresponding regions in each potential complement that do not correspond), and **bulge loops** (when one of the potential complements contains a base not corresponding to the other). All of these unpaired regions *hinder* the formation of the double helix; and this is taken into account in the calculation by assigning each type of interruption a *positive* free-energy value that must be included in the sum total. In other words, to hold these regions in a particular structure requires the *input* of energy, and this must be set against the free energy that is released from the base pairing and stacking. Details of these values are given in the figure. Overall, the free-energy calculation must have a sufficiently negative value or the secondary structure will not form.

Figure 2.13

The free energy of formation for a potential base-paired region can be calculated by summing the ΔG values released (by base-pair formation) or demanded (to maintain single-stranded regions in a restricted conformation).

DNA CAN BE DENATURED AND RENATURED

The hydrogen-bonded structure of the double helix of DNA (or RNA) can be disrupted by heating. Because there are no covalent bonds connecting the partner strands, the two polynucleotide chains of a duplex DNA separate entirely when all the hydrogen bonds are broken. The process of strand separation is called **denaturation** or (more colloquially) **melting.**

Denaturation occurs over a narrow temperature range and is reflected in striking changes in many of the physical properties of DNA. A particularly useful change occurs in the optical density. The heterocyclic rings of nucleotides adsorb light strongly in the ultraviolet range (with a maximum close to 260mμ that is characteristic for each base). But the adsorption of DNA itself is some 40% less than would be displayed by a mixture of free nucleotides of the same composition. This is described as the **hypochromic effect.** It re-

Figure 2.14
The denaturation of DNA can be followed by the increase in optical density and is described by the T_m.

sults from interactions between the electron systems of the bases, made possible by their stacking in the parallel array of the double helix. Any departure from the duplex state is immediately reflected by a decline in this effect—that is, by an increase in optical density toward the value characteristic of free bases. Thus the denaturation of DNA can be followed by this **hyperchromicity.**

The midpoint of the temperature range over which the strands of DNA separate is called the **melting temperature,** denoted **T_m.** An example of a melting curve followed by optical adsorbance is shown in **Figure 2.14.** The curve always takes the same form, but its position on the temperature scale (that is, its T_m) is influenced by both the base composition of DNA and the conditions employed to obtain denaturation.

When DNA is in solution at approximately physiological conditions, the T_m usually lies in a range of 85–95°C. The value depends on the base composition. The greater free energy of formation of the 3-H-bonded G-C base pair makes it more resistant to disruption than the 2-H-bonded A-T base pair. The dependence of T_m on base composition is linear; it increases about 0.4°C for every percent increase in G-C content. So a DNA that is 40% G-C (for example, as typical of mammalian genomes) will denature with a T_m of about 87°C under the usual conditions, whereas a DNA that is 60% G-C will have a T_m of about 95°C under the same conditions.

A major effect on T_m is exerted by the ionic strength of the solution. The T_m increases 16.6°C for every tenfold increase in monovalent cation concentration. The most commonly used condition is to perform the reaction in 0.12 M phosphate buffer, which provides a monovalent Na^+ concentration of 0.18 M.

The T_m can be greatly varied by performing the reaction in the presence of reagents, such as formamide, that destabilize hydrogen bonds. This allows the T_m to be reduced to as low as 40°C, which has the advantage that the DNA does not suffer other damage (such as strand breakage) from the increase in temperature.

An extremely useful property of denatured DNA is that, under appropriate conditions, the reaction can be reversed, so that the two separated complementary strands can reform into a double helix. This is called **renaturation** and is illustrated in **Figure 2.15.** It depends on the specificity of base pairing between

Figure 2.15
Denatured single strands of DNA can renature to the duplex form.

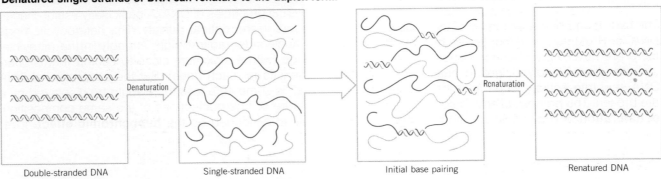

Double-stranded DNA   Single-stranded DNA   Initial base pairing   Renatured DNA

the complementary strands. The reaction takes place in two stages. First, two short complementary sequences in the reacting strands encounter each other (by chance) and pair to form a double-helical region. Then the region of base pairing extends along the molecule by a zipper-like effect to form a lengthy duplex molecule. The reconstruction of the double helix is accompanied by a restitution of the original properties that were lost when the DNA was denatured.

NUCLEIC ACIDS HYBRIDIZE BY BASE PAIRING

Renaturation involves the reaction together of two complementary sequences that were separated by denaturation. However, this technique can be extended to allow any two complementary sequences of nucleic acid to **anneal** with each other to form a duplex structure. This is generally referred to as **hybridization** when nucleic acids from different sources are involved, as in the case when one preparation consists of DNA and one of RNA. It depends on the same feature of complementary base pairing that provides for replication of DNA base by base or for renaturation en masse. Indeed, the ability of two nucleic acid preparations to hybridize constitutes a precise test for their sequence complementarity.

The principle of the reaction is to expose two single-stranded nucleic acid preparations to each other and then to measure the amount of double-stranded material that is formed. There are two principal ways of performing the reaction: **solution** (or **liquid**) **hybridization** and **filter hybridization.**

Solution hybridization is described by its name: the single-stranded preparations are mixed together in solution. For larger amounts of material, the reaction can be followed by the change in optical density. When smaller amounts of material are involved, one of the components may carry a radioactive label, whose entry into duplex form is followed by determining directly the amount of double-stranded DNA containing the label. This can be done either by degrading all the single strands that have not reacted or by separating the double strands from the single strands by chromatography.

A problem with solution hybridization occurs when the relationship of two preparations is to be investigated and one or both of them consists of duplex DNA. For example, if two duplex DNA molecules are denatured to single strands and the preparations are mixed together, two types of reaction can occur. The *original* complementary single strands can renature. Or each single strand can hybridize with a complementary sequence in the *other* DNA. Because the two reactions compete, it is difficult to assess the extent of hybridization.

This difficulty can be overcome by immobilizing one of the DNA preparations so that it cannot renature.

Figure 2.16

Filter hybridization establishes whether a solution of denatured DNA (or RNA) contains sequences complementary to the strands immobilized on the filter.

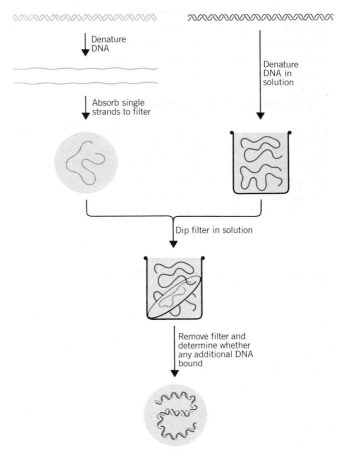

Denature DNA

Absorb single strands to filter

Denature DNA in solution

Dip filter in solution

Remove filter and determine whether any additional DNA bound

Nitrocellulose filters have the useful property that single strands of DNA adsorb, while RNA does not; and once a filter has been used to adsorb DNA, any further adsorption of single strands can be prevented by certain treatments. **Figure 2.16** illustrates the resulting procedure in which a DNA preparation is denatured and the single strands are adsorbed to the filter. Then a second denatured DNA preparation (or an RNA preparation) is added. This will adsorb to the filter only if it is able to base pair with the DNA that was originally adsorbed.

This technique has been used in many of the experiments that we shall discuss in the next few chapters. Usually the form it takes is to add to the filter an RNA preparation that carries a radioactive label. Then the extent of hybridization can be measured by quantitating the radioactivity retained on the filter. This is important because, with the exception of the RNA that constitutes some viral genomes, all RNA is made by copying DNA. The hybridization technique provides a vital tool for determining which segment of DNA has been used to produce a particular RNA.

THE MOLECULAR BASIS OF MUTATION

The isolation of mutants has been the pivot of the analysis of the gene. Until very recently, genes have been identifiable only by characterizing mutations that either are lethal or block the development of some visible or otherwise measurable feature of the phenotype. The mapping of mutations leads to the central concept that a group of mutations all affecting the same characteristic may lie close together; and the use of the complementation test shows that each such group constitutes a functional genetic unit. What is the nature of these mutations?

From the discovery that DNA is the genetic material, the concept that a mutation is a change in the sequence of base pairs follows naturally. We shall see shortly that a given sequence of nucleotides in a gene is responsible for producing a particular sequence of amino acids in the corresponding protein. A change in the nucleotide sequence may lead to a change in the amino acid sequence, thus altering or abolishing the activity of the protein.

Some mutations occur as the result of normal cellular operations or interactions with the environment. These are called **spontaneous mutations,** and they occur at a characteristic rate in any particular organism. The occurrence of mutations can be increased by treatment with certain compounds. These are termed **mutagens,** and the changes that they cause are referred to as **induced mutations.** Most mutagens act directly by virtue of an ability either to act on a particular base of DNA or to become incorporated into the nucleic acid. The effectiveness of a mutagen is judged by the degree to which it increases the spontaneous "background" rate of mutation.

Any base pair of DNA may be mutated. An alteration that changes only a single base pair is termed a **point mutation.** This can take two forms. A **transition** describes the substitution of one pyrimidine by the other or of one purine by the other; thus a G-C pair is exchanged with an A-T pair or vice versa. This is the most common class of point mutation. The other is a **transversion,** in which a purine is replaced by a pyrimidine or vice versa, so that an A-T pair becomes a T-A or a C-G pair.

One source of transitions is the direct *chemical change* of one base into another. A common class of mutagens consists of compounds that can interact directly with one or more bases to change their pairing properties. For example, a major effect of nitrous acid is to cause the oxidative deamination of cytosine, which converts it into uracil. As shown in **Figure 2.17,** in the next replication cycle the U pairs with an A, instead of with the G with which the original C would have paired. So the C-G pair is replaced by a T-A pair (the T is inserted subsequently, by pairing with the A in the next replication cycle). Nitrous acid also deaminates adenine, causing the reverse transition from A-T to G-C.

Another way that transitions may occur is by **base mispairing,** when unusual partners pair in defiance of the usual restriction to the Watson-Crick pairs. In the normal replication cycle, an error may occasionally occur when the wrong base is inserted. The spontaneous error frequency is primarily a function of the enzyme (DNA polymerase) responsible for replication (see Chapter 32). More limited synthesis of DNA also occurs as the result of "repair" mechanisms that become active as a consequence of genetic recombination or of damage to DNA (see Chapter 34). Different repair systems have different characteristic error rates; for example, one of the repair systems in *E. coli* is especially prone to permit errors, and so its activation may result in mutations. We do not have very much information on the regularity with which spontaneous mutations of this sort occur.

Some mutagens consist of analogs of the usual nucleotide bases that have ambiguous pairing properties. These may be incorporated into DNA during replication. Because of mistaken pairing either during their insertion or in subsequent cycles of replication, they may perpetuate a series of transitions, as shown in **Figure 2.18** for the example of bromouracil (BrdU).

This is incorporated into DNA in the place of thymine. But because BrdU also can mispair reasonably well with guanine (instead of with adenine), the presence of the bromine leads to substitution of the original A-T pair by G-C pairs. This occurs with a certain probability in each replication cycle, so once BrdU has been incorporated into DNA in place of thymine, it continues to cause transitions in subsequent cycles.

Another class of mutagen is provided by the acridines. These bind to DNA to distort the structure of the double helix in such a way that additional bases are incorporated or bases are omitted during replication. Each mutagenic event sponsored by an acridine results in the addition or the removal of a single base pair. Because of the effect such changes have on the function of a gene, they are often called **frameshift mutations,** in contrast with the point mutations that substitute one base for another (see Chapter 4).

For a long time, point mutations were thought to be the principal means of change in individual genes. However, we now know that **insertions** of stretches of additional material may be quite frequent, at least in bacteria. The source of the inserted material lies with **transposable elements,** sequences of DNA that have the ability to move from one site to another (see Chapter 36). Analogous events also occur in eucaryotes, and, in fact, the very first mutation to have been found (*white* eye in *Drosophila*) arose by the insertion of an extraneous piece of DNA into a locus governing production of eye pigment. The insertion abolishes the activity of the gene. Where such insertions have occurred, **deletions** of part or all of the inserted material, and sometimes of the adjacent regions, are common.

A significant difference between point mutations on the one hand and these insertions/deletions on the other is that the frequency of point mutation can be increased by mutagens, whereas the occurrence of changes caused by transposable elements is indifferent to these reagents. However, insertions and deletions also can occur by other mechanisms—for example, involving mistakes made during replication or recombination—though probably these are less common. The smallest insertions and deletions are those induced by the acridines (the same terms are used to describe these single base-pair changes as to describe the insertion/deletion of long segments of DNA).

Figure 2.17
Mutations can be induced by chemical modification of a base.
Nitrous acid oxidatively deaminates cytosine to uracil. Following DNA replication, one daughter duplex has the original C-G pair, but the other has a U-A pair, which is replicated to give A-T pairs in subsequent generations.

Figure 2.18

Mutations can be induced by the incorporation of base analogs into DNA.

The bromine atom in bromouracil replaces the methyl group of thymine. This allows the keto → enol shift to occur more frequently, with the result that BrdU may pair with guanine as well as with adenine. Thus if BrdU is incorporated in DNA in place of thymine, it may cause the replacement of an A by a G at a certain frequency in each generation. This causes A-T pairs to be replaced by G-C pairs.

MUTATIONS ARE CONCENTRATED AT HOTSPOTS

So far we have dealt with mutations in terms of individual changes in the sequence of DNA that influence the activity of the genetic unit within which they occur. When we consider mutations in terms of the inactivation of genes, most genes show more or less similar rates of mutation, whether spontaneous or induced. Actually, to be more precise, the rate is a function of the size of the gene, that is, of the target for mutation. But when we come to consider the sites of mutation within the sequence of DNA, we have to ask whether all base pairs in a gene are equally susceptible or some are more likely to become mutated than others.

This question is approached by isolating a large number of mutations independently in a gene. This means that many mutants are obtained, each of which has suffered an individual mutational event. Then the site of each mutation is determined (conventionally by genetic mapping, now often by direct analysis of the DNA sequence). Most mutations will lie at different sites. But some will lie at the same site. Two independently isolated mutations at the same site may constitute exactly the same change (in which case the same mutational event has happened on more than one occasion), or they may constitute different changes (three different point mutations are possible at each base pair).

The statistical probability that more than one mutation occurs at a particular site (in this context, "site" means base pair) is given by random-hit kinetics (as seen in the Poisson distribution). So some sites will gain one, two, or three mutations, while others will not gain any. This depends on the total number of mutations analyzed. But some sites gain far more than the number of mutations expected from a random distribution; they may have 10 or even 100 times more mutations than predicted by random hits. These sites are called **hotspots.**

For a long time, the origin of hotspots was mysterious. It was assumed that a hotspot must represent some sequence particularly susceptible to spontaneous or induced mutation, but the reason was not clear. Indeed, one set of sites provides hotspots for spontaneous point mutation and for mutations induced by some mutagens, while different sets of hotspots are displayed by other mutagens.

A major cause of spontaneous point mutation in *E. coli* has been pinned down to an unusual base in DNA. In addition to the four bases that are inserted into DNA by the enzymes of replication and repair, **modified bases** are sometimes found. The name reflects their origin: they are produced by chemically modifying one of the four bases *present in DNA.* They are not inserted into DNA in the modified form. The most common modified base is 5-methylcytosine, which is present in the DNA of both procaryotes and eucaryotes (and which we have already mentioned in the context of the structure of Z-DNA). It is generated by a methylase enzyme that adds a methyl group to cytosine only at certain sites in the DNA to modify a small proportion of the total number of C residues.

In one gene that has been examined in detail (the *lacI* gene of *E. coli),* the hotspots for spontaneous point mutation all occur at sites at which the wild-type sequence contains a 5-methylcytosine. In each case the mutation takes the form of a G-C to A-T transition. In strains of *E. coli* that are unable to perform the methylation reaction, these hotspots do not exist.

The reason is that 5-methylcytosine suffers a spontaneous deamination at an appreciable frequency; as can be seen from Figure 2.4, the replacement of the amino group by a keto group converts it to thymine. This creates a mispaired G·T partnership, whose separation at the subsequent replication produces one wild-type G-C pair and one mutant A-T pair.

The same deamination reaction occurs with cytosine itself, to generate uracil. However, *E. coli* contains an enzyme, uracil-DNA-glycosidase, that removes uracil residues from DNA (see Chapter 32). This leaves an unpaired G residue, and a repair system then uses the usual base pairing to insert a cytosine. Presumably this system is adequate to protect DNA against the consequences of the spontaneous deamination (although it is not active enough to prevent nitrous acid from acting as a mutagen; see Figure 2.17.) However, because the deamination of 5-methycytosine leaves thymine, which is a perfectly respectable constituent of DNA in its own right, the system does not operate in these circumstances, and a mutation results.

Figure 2.19 compares the results of deamination of

Figure 2.19
The deamination of 5-methylcytosine produces thymine (leading to C-G to T-A transitions), while the deamination of cytosine produces uracil (which is removed and replaced by cytosine).

cytosine and 5-methylcytosine. This casts an interesting light on the use of T in DNA compared with U in RNA. Perhaps it relates to the need of DNA for stability of sequence; the use of T means that any deaminations of C are immediately recognized, because they generate a base (U) not usually present in the DNA.

Another major hotspot in the *lacI* gene occurs at a site where the sequence CTGG is repeated three times in succession. Mutations result from the addition or deletion of one unit of four bases from this sequence, thus

Mutant	C T G G C T G G
	↑ deletion
Wild type	C T G G C T G G C T G G
	↓ insertion
Mutant	C T G G C T G G C T G G C T G G

The origin of these mutations may lie in a "slippage" in which one strand of DNA becomes misaligned with the other during replication; for example, the first 4-base unit of one strand could become paired with the second 4-base unit of the other strand. This could lead to additions or losses of 4-base units.

Some transposable elements have preferred sites for insertion that act as hotspots for mutation. However, this effect varies with the particular element, and most elements appear to insert at random (see Chapter 36).

THE RATE OF MUTATION

Spontaneous mutations that inactivate gene function occur in bacteria at a rate of about 10^{-6} to 10^{-7} events per locus per generation. We have no really accurate measurement of the rate of mutation in eucaryotes, although usually it is thought to be somewhat similar to that of bacteria on a per-locus per-generation basis.

We do not know what proportion of the spontaneous events in bacteria result from point mutations as opposed to insertions by transposable elements. It is clear that both are important sources of mutation. The same is true of eucaryotes, where at present very few genes have been investigated in sufficient detail to distinguish the origins of mutations.

If a bacterial gene consists of 1000 base pairs, the mutation rate corresponds to changes at individual nucleotides of 10^{-9} to 10^{-10} per generation. Considering only point mutations, this calculation is an oversimplification, because not all mutations in DNA actually lead to a detectable change in the phenotype. Mutations without apparent effect are called **silent mutations.** They are of two types. Some involve base changes in DNA that do not cause any change in the amino acid present in the corresponding protein. Others change the amino acid, but the replacement in the protein does not affect its activity. These are called **neutral** substitutions.

Those mutations that inactivate a gene represent **forward mutations.** Their effects may be reversed by **back mutations,** which are of two types. An exact reversal of the original mutation is called **true reversion.** Thus if an A-T pair has been replaced by a G-C pair, another mutation to restore the A-T pair will exactly regenerate the wild-type sequence. Alternatively, another mutation may occur elsewhere in the gene, and its effects may compensate for the first mutation. This is called **second-site reversion.** For example, one amino acid change in a protein may abolish gene function, but a second alteration may compensate for the first and restore protein activity.

A forward mutation results from any change that inactivates a gene, whereas a back mutation must restore function to a protein damaged by a particular forward mutation. Thus the demands for back mutation are much more specific than those for forward mutation. The rate of back mutation is correspondingly lower than that of forward mutation, typically by a factor of about 10.

Mutations may also occur in other genes to circumvent the effects of mutation in the original gene. This is called **suppression** (or, more formally, extragenic suppression).

The isolation of revertants is an important characteristic that distinguishes point mutations and insertions from deletions. A point mutation can revert by reversal of the base-pair change or by a second-site change; an insertion can revert by deletion of the additional material. However, a deletion of part of a gene cannot be reverted. Such deletions have a critical use in genetic mapping. The genetic extent of a deletion is defined by its inability to complement with a series of adjacent mutations (because it lacks the entire wild-type stretch of corresponding DNA). Thus once a series of deletions in a gene has been established, any new mutation can be mapped simply by testing its ability to complement with them.

We should mention one final point about the nature of mutations. Ideally, we should like to be able to mutate every single gene in an organism. But we are restricted in two ways.

First, we need a criterion for distinguishing the mutant from the wild type. This applies whether a mutant is isolated by serendipity or whether we have set out to **select** it. The simplest criterion is visible change: for example, change in eye color. Other criteria can be established by devising suitable selective procedures: for example, adjusting the conditions of growth so that the absence or presence of some metabolic enzyme is required. The problem in all of this is that there may be functions whose existence we do not suspect, and which we therefore neither see nor think to devise selective tests for. Taken to an extreme, there is the possibility that genes exist whose products are not even necessary, and whose absence has no effect. How are these to be detected?

Our second difficulty concerns functions that are *essential* for viability. A mutation in such a gene is likely simply to kill the organism. To allow mutants to be selected in essential functions, they must be obtained in the form of **conditional lethal** mutations. These are lethal under one set of conditions, but show no deficiency, or a much reduced deficiency, under alternative conditions that allow the cell to be perpetuated. In other words, the same (mutant) cell can be grown under two conditions: one (**permissive**) in which it does not display the mutant phenotype; the other (**nonpermissive**) in which it dies or becomes severely ill, but nonetheless can be studied if only in the brief period before the lethality takes effect. (Incidentally, conditional mutations are not restricted to essential functions, but in principle can be found in any gene.)

FURTHER READING

The original spark for molecular biology was provided by **Schrödinger** in his book *What is Life?* (Cambridge University Press, Cambridge, England, 1945). Historical recollections of subsequent events including the discovery of DNA are contained in a volume edited by **Cairns, Stent, & Watson,** *Phage and the Origins of Molecular Biology* (Cold Spring Harbor Laboratory, New York, 1966). Historical accounts of this period have been written by **Olby,** *The Path to the Double Helix* (MacMillan, London, 1974) and **Judson,** *The Eighth*

Day of Creation (Knopf, New York, 1978). Recent work on the structure of DNA has been elegantly discussed by **Cantor** (*Cell* **26**, 293–295, 1981). The perspective for the original work on mutations can be obtained from **Drake** & **Balz's** review (*Ann. Rev. Biochem* **45**, 11–37, 1976). The discovery of hotspots comes from **Benzer's** work (*Proc. Nat. Acad. Sci. USA* **47**, 403–416, 1961), and their equation with modified bases was reported by **Coulondre** et al. (*Nature* **274**, 775–780, 1978). Transposable elements are discussed in Chapters 36—38 of this book.

CHAPTER 3
WHAT IS A GENE?
MOLECULAR STRUCTURE

A chromosome contains a long, uninterrupted thread of DNA along which lie many genes. Identifiable sites in the genetic material are provided by mutations that create changes in the sequence of base pairs. On the linkage map of the chromosome, distances between mutations are represented by their frequencies of recombination.

How accurately does the genetic map represent the actual length of the genetic material? At the gross level of the chromosome, the genetic map is usually a reasonable representation of the underlying molecular structure. However, the map unit depends on the relative frequency of recombination, which may be different in each species, so that genetic maps of different organisms cannot be compared.

By comparing the total map length of all the linkage groups in a genome with its content of DNA, we can obtain some idea of how frequently genetic exchange occurs relative to length of DNA. We may take 50 map units (50% recombination) to correspond to the average distance between independent recombination events. Values for a range of genomes are summarized in **Table 3.1.**

In bacteriophages there are *exceedingly* frequent genetic exchanges (though any quantitative estimate of the frequency is only *very* approximate). Also, it is possible to detect recombinants at very low frequency (that is, from among enormous numbers of progeny). These features allow detailed genetic mapping within a gene. We have in fact seen in Chapter 1 that it is possible to obtain recombinants between mutations in adjacent base pairs. In bacteria also there is a high rate of genetic exchange, and it is possible to identify recombinants from very large numbers of progeny, allowing detailed genetic mapping within a gene.

In lower eucaryotes, at least as typified by yeast, the frequency of recombination remains high, but in higher eucaryotes it is roughly a hundredfold lower. We do not yet really have adequate information to es-

Table 3.1.

The Frequency of Genetic Recombination Is Much Less in Eucaryotes than in Procaryotes

Species	Average Distance Between Independent Crossovers (in base pairs of DNA)
Bacteriophage (T4)	1.0×10^4
Bacterium (*E. coli*)	1.2×10^5
Yeast (*S. cerevisiae*)	2.5×10^5
Fruit fly (*D. melanogaster*)	2.5×10^7
Mammal (*M. musculus*)	9.0×10^7

tablish a range for the higher eucaryotes, but the two examples of fruit fly and mouse illustrate the difficulties of detailed genetic mapping. Suppose a gene in the fruit fly is 5000 base pairs long. This corresponds to 0.01% recombination (taking 2.5×10^7 base pairs to equal 50% recombination). This means that to detect recombinants between mutations located at opposite ends of the gene, it would be necessary to isolate 1 fly in 10,000 progeny. The result is that genetic mapping within genes of the fruit fly is barely possible. It is not at all practical with mice, in which the frequency of recombination is lower and the number of progeny from each mating is less.

The conclusion to which this forces us is that fine-structure mapping is possible only with procaryotes and (perhaps) lower eucaryotes. With this approach, we cannot obtain information about the organization of the gene in higher eucaryotes.

A DIRECT APPROACH
TO GENE STRUCTURE

Even in procaryotes, the information to be gained from fine-structure mapping is limited by the nature of the recombination event. At the level of mapping within a gene, recombination frequencies are not entirely independent of the particular mutations that are used in each cross, but may be influenced quite strongly by the local sequence of DNA. In other words, they do not have the ideal property of allele-independence that we discussed in Chapter 1, but show allele-specific effects. Our view of the gene in terms of the genetic map is therefore distorted by the characteristics of the recombination systems.

The map distances between mutations therefore do not necessarily correspond with the distance that actually separates them on DNA. Also, of course, this approach is limited by the availability of mutations; a gap in the genetic map can represent a region in which no mutations have been found, or it could be caused ─── ─ ─ease in recombination frequency. We ─── to suppose that the situation would be ─── eucaryotes if we were able to carry out ─── napping. So for both theoretical and ─── ns, we need another source of infor-

mation. How are we to establish the molecular structure of the gene and relate it to the structure of the protein product? How far apart are adjacent genes and how are we to recognize the regions between them?

The ultimate aim of genetic mapping is to establish the nucleotide sequence of the gene and to extend it to the neighboring genes on either side. First, this requires the ability to isolate the DNA corresponding to the particular gene. This was achieved originally for viruses, in which the DNA comprising the entire genome usually is readily available. Since then, however, it has become almost routine to isolate the cellular DNA for which any protein product is available (although this can be a lengthy process if the product is scarce).

Then we must obtain a map of the DNA. This is accomplished by breaking the DNA at defined points, whose distance apart can be accurately determined. This breakage is made possible by the ability of certain **restriction enzymes** to recognize rather short, specific sequences of DNA as targets for cleavage. For this reason, a map of DNA obtained by identifying these points of breakage is known as a **restriction map.** It represents a linear sequence of the sites at which particular restriction enzymes find their targets, the distance between the sites being measured directly in base pairs of DNA (abbreviated **bp** for short distances, or given in **kb,** corresponding to kilobases, for longer regions).

From this physical map it is possible to proceed to determine the sequence of DNA between points that are sufficiently close together (within about 300 bp for practical purposes). By suitable choice of points, these regions can be connected into a sequence of the entire gene. The sequence of the DNA then can be compared with the sequence of the protein that the gene represents. This allows the regions corresponding to the protein to be delineated; and, by extending the sequence in either direction, the distance to the next gene can be determined.

By determining the sequence of a wild-type DNA and of its mutant allele, the nature of the mutation and its exact site of occurrence can be determined. This provides equivalence between the genetic map (based entirely on sites of mutation) and the physical map (based on or even comprising the sequence of DNA). Ultimately, the map of a region of the genome can be expressed in base pairs of DNA rather than in the relative map units of formal genetics. Indeed, genes can now be identified by virtue of their protein product or even sometimes by their sequence alone, so we are no longer entirely dependent on mutations to provide the raw material for constructing a map of the genome. However, mutants remain essential for identifying the functions of gene products.

RESTRICTION ENZYMES CLEAVE DNA INTO SPECIFIC FRAGMENTS

The key to restriction mapping lies in the properties of a class of enzymes found in bacteria (and discussed in the context of their natural habitat in Chapter 34). We have mentioned that these restriction enzymes cut duplex DNA at specific sites. Each restriction enzyme has a particular target, usually a specific sequence of a length between 4 and 6 base pairs. The enzyme cuts the DNA at every point at which this short sequence occurs. Different restriction enzymes have different target sequences, and a large range of these activities now is available. (Their use is described in detail in Chapter 19.)

If we take a particular DNA molecule and cut it with a suitable restriction enzyme, a limited number of breaks are introduced, producing a number of distinct fragments. These fragments can be separated on the basis of their size by **gel electrophoresis.** The cleaved DNA is placed on top of an agarose or polyacrylamide gel. When an electric current is applied, fragments move down the gel at a rate that depends on their length. The smaller a fragment, the more rapidly it moves. (The distance moved depends on the log of the fragment length.)

This produces a series of **bands,** those at the top of the gel representing large fragments, those at the bottom representing smaller fragments. The length of any particular DNA fragment can be determined by **calibrating** the gel. This is done by running a control in parallel in another slot of the same gel. The control has a mixture of standard fragments all of known size (often called **markers**). The migration of the markers

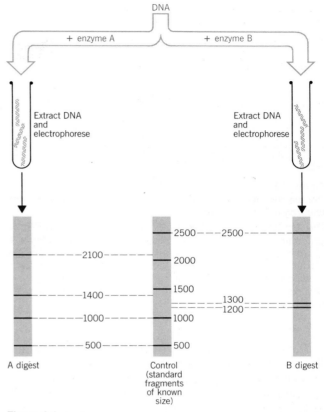

Figure 3.1
DNA can be cleaved by restriction enzymes into fragments that can be separated by agarose gel electrophoresis.

The sizes of the individual fragments generated by enzyme A (left) or enzyme B (right) are determined by comparison with the positions of fragments of known size, such as the control shown in the center.

defines the relationship between fragment length and distance moved.

An example of this technique is shown in **Figure 3.1.** A particular DNA molecule of length 5000 bp is incubated separately with two restriction enzymes, A and B. When the cleaved DNA is electrophoresed, we see that enzyme A has cut the substrate DNA into four fragments, of lengths 2100, 1400, 1000, and 500 bp; while enzyme B has generated three fragments, of 2500, 1300, and 1200 bp. Can we generate a map from these data to place the sites of breakage in a defined order?

Figure 3.2
Double digests define the cleavage positions of one enzyme with regard to the other.

The four fragments obtained by digestion with enzyme A can be eluted from the gel shown in Figure 3.1 and then degraded with enzyme B (four gels on left). Similarly, the three bands originally produced by enzyme B can be obtained and then digested with enzyme A (three bands in center). A complete double digest is obtained by subjecting the intact DNA to both enzymes simultaneously (band at right).

There are several ways to relate the patterns of cutting by the two enzymes. The principle of one method is illustrated in **Figure 3.2,** and involves making **double digests.** This means that the DNA is cleaved with two enzymes instead of with only one. The most complete way to approach this analysis is to extract each fragment produced in the individual digests with either enzyme A or enzyme B and then to cleave it with the other enzyme. The result of this cleavage is then analyzed in the same way by electrophoresis.

Each gel in Figure 3.2 is labeled according to the fragment that was isolated from the gel in Figure 3.1. Thus A-2100 identifies the fragment of 2100 bp produced by degrading the original DNA molecule with enzyme A. When this fragment is retrieved and subjected to enzyme B, it is cut into fragments of 1900 and 200 bp. In other words, one of the cuts made by enzyme B lies 200 bp from the nearest cut made by enzyme A, and is 1900 bp away from the A cut on the other side. A related pattern of cuts is seen when we examine the susceptibility of fragment B-2500 to enzyme A. It is cut into fragments of 1900 and 600 bp. So the 1900 bp fragment is generated by double cuts, with an A site at one end and a B site at the other end. It can be released from either of the single-cut frag-

ments (A-2100 or B-2500) that contain it. These single-cut fragments therefore must **overlap** in the region of the 1900 bp of the common fragment that can be generated from them.

When all of the fragments are analyzed in this manner, we can see that every fragment produced by cutting an original A fragment with the B enzyme *also* is found in one of the double digests in which an original B fragment was cut with the A enzyme. The entire pattern can be seen in the complete double digest (the gel at the right), in which every double-cut fragment occurs once. These data allow the sites of cutting to be placed into an unequivocal map.

CONSTRUCTING A RESTRICTION MAP

The key to restriction mapping is the use of overlapping fragments. Let us start with the fragment of 1900 bp that we have already discussed. This is produced either by cutting fragment A-2100 with enzyme B, or by cutting fragment B-2500 with enzyme A. From the cleavage of A-2100 we can see that it must be 200 bp from one end of the 1900 double-cut fragment to the next A target. From the cleavage of B-2500 we can see that it must be 600 bp from the other end of the 1900 bp sequence to the next B target. In other words, we can produce the map

where the vertical arrows indicate sites of cleavage by enzyme A or enzyme B. Note that when considering the reconstruction of larger fragments from smaller fragments, we can rely on the *complete additivity* of lengths (within experimental limits) that is characteristic of restriction mapping.

Now we can extend the map by seeing which other fragments generate the 200 and 600 bp double-cut fragments at the left and right ends, respectively. The same 200 bp fragment is produced by cutting B-1200 with enzyme A, so the next B site must lie 1000 bp to the left. The same 600 bp fragment is produced by cutting A-1400 with enzyme B, so the next A site must lie 800 bp to the right. This gives the map

We can now complete the map by identifying the source of the two fragments at each end. At the left end, the 1000 bp fragment arises from B-1200 or from A-1000. In the latter, enzyme B does not cut the A fragment. This fragment must therefore lie at the end of the map; in other words, proceeding from the left end, it is 1000 bp to the first A site and 1200 bp to the first B site. This is why we have omitted any B cut from the left end of the preceding map (although formally we treated this as a B cutting site in the previous paragraph).

At the right end of the map, the 800 bp double-cut fragment is generated by cutting B-1300 with enzyme A, which places a fragment of 500 bp to the right. This is the terminal fragment, as seen by the failure of enzyme B to cut the A-500 fragment. So one last segment must be added to the map.

Together these data allow us to relate the two patterns of cutting by enzymes A and B into the restriction map shown in **Figure 3.3.** This divides the 5000 bp DNA into a series of regions of defined length that lie between the sites recognized by the two enzymes.

SOME NICETIES
OF RESTRICTION MAPPING

The actual construction of a restriction map usually requires recourse to several enzymes, so that it is necessary to resolve quite a complex pattern of overlapping fragments generated by the various enzymes. Some further techniques can be used to assist in the mapping.

One is to make **partial digests,** by using conditions in which an enzyme does not recognize every target in every DNA molecule, but instead sometimes fails to cleave some targets. For example, in Figure 3.3, enzyme A might give a partial digest of fragments of 3100, 1400, and 500 bp. By comparing this with the

Figure 3.3
A restriction map is a linear sequence of sites separated by defined distances on DNA.

The map identifies the sites cleaved by enzymes A and B as defined by the individual fragments produced by the single and double digests.

complete digest pattern, the 1000 and 2100 bp fragments can be assigned as adjacent components of the 3100 partial fragment. This allows the sites of a single enzyme to be ordered with respect to one another. Another partial fragment might be generated of 3500 bp, allowing the 2100 and 1400 fragments to be placed as adjacent.

Another useful technique is to label the *ends* of the DNA molecule with a radioactive phosphate (this is called **end labeling** and can be performed by certain enzymes, for either 5′ or 3′ termini). Then the fragments containing these ends can be identified by their radioactive label. Thus in the fragment A preparation, A-1000 and A-500 would be placed immediately at opposite ends of the map; and similarly, fragments B-1200 and B-1300 would be picked out as ends.

Often fragments are compared by the technique of nucleic acid hybridization. For example, the A-2100 and B-2500 fragments contain overlapping sequences, and therefore should be able to hybridize with each other.

All of these techniques require that we have at our disposal a complete set of unique restriction fragments that together account for the entire DNA region being mapped. There are some circumstances in which this expectation may fail to be fulfilled.

If several sites for one enzyme lie *very close* together (say, within 50 bp), very small fragments may be generated that are lost from the agarose gel. Of course, this will result in a discrepancy when the molecular weights of the other fragments are totaled, but this would not necessarily be considered sinister by itself, since there are always modest discrepancies when fragment sizes are compared.

By combining the technique of end labeling with a partial digest, all of the fragments generated by a given enzyme can be displayed directly. By using this technique, and by constructing a map in which several restriction enzymes are used to provide sufficient overlaps between the fragments, it is possible to ensure that all of the DNA is contained within the map, and that no small regions have been lost.

A complication is provided by the assumption that every restriction fragment is seen as an individual band on the gels. This will not be true when more than one fragment has the same molecular weight (or, more practically, whenever fragments exist with sizes so similar that they cannot be resolved). To check whether each band is unique, its intensity can be quantitated. Bands that contain more than one fragment should be correspondingly more intense. Multiple fragments of the same size can be resolved by using further restriction enzymes.

Once a restriction map has been obtained, it can be equated with the genetic map. This can be done on the basis of restriction mapping alone when large changes are known on the genetic map, for example, deletions or insertions that affect substantial regions of the map. Such changes should be seen as reductions or increases in the sizes of the particular restriction fragments in which they lie. Sometimes they may even remove or introduce restriction sites.

Locating point mutations on the restriction map is more difficult. Occasionally they may change target sites for restriction enzymes, but otherwise they remain undetectable, since the sizes of the restriction fragments remain the same in wild-type and mutant DNA. To locate these base substitutions, it may therefore be necessary to determine the sequence of the DNA.

One final point about restriction enzymes concerns nomenclature. The enzymes are named by a three letter (sometimes four letter) abbreviation that identifies their origin. Roman numerals are added to distinguish multiple enzymes that have the same origin. Examples are HpaI (derived from *Haemophilus parainfluenzae*), and EcoRI (derived from *E. coli*).

SEQUENCING DNA

We have already seen that, by relying on the specificity of base pairing (G = C, A = T), a double helix of DNA maintains a constant structure irrespective of its particular sequence of base pairs. This means that the difference between individual molecules of DNA lies in their particular sequences of base pairs, not in any gross changes of structure. (This contrasts with proteins, where the sequence of amino acids is needed to construct the overall form of the protein, but it is the gross structure of the molecule that is responsible for its biological functions.) Very small differences in DNA sequence can be highly significant, as is evident from the demonstration that mutations consist of changes in DNA sequence as small as one base pair.

Can the exact nucleotide pair sequence of DNA be determined? The first approach to nucleic acid sequencing followed that of protein sequencing: break the molecule into small fragments, determine their base composition, and by obtaining overlapping fragments deduce the exact sequence. With proteins this is practical because of the variety offered by the 20 amino acids; with nucleic acids, this poses a problem because there are only four bases. This effectively restricts such techniques to determining the sequences of very short nucleic acids.

But now it is possible to determine DNA sequences *directly.* In fact, DNA sequences can be obtained much more rapidly than protein sequences.

The various sequencing protocols all generate a series of single-stranded DNA molecules, each one base longer than the last. DNA molecules of the *same sequence,* but differing in length by as little as one base at one end, can be separated by electrophoresis on acrylamide gels. Bands corresponding to molecules of increasing length form a "ladder" that can be followed for up to about 300 nucleotides.

Two types of approach have been used to obtain these sets of bands. One is to use chemical reactions that cleave DNA at individual bases. The other is to synthesize the DNA *in vitro* in such a way that the reaction terminates specifically at the position corresponding to a given base.

To determine the sequence of a DNA molecule, it is treated in four separate reactions, each reaction specific for one of the bases. The products are electrophoresed in four parallel slots of the same gel. The band corresponding to a particular length appears in only one of the four slots. This identifies the nucleotide that is present at the corresponding position in DNA. By proceeding from each band to the band of the next size, the series of nucleotides can be "read" according to which slot they appear in. This gives the DNA sequence.

In the chemical method, a DNA fragment previously labeled at one (unique) end with radioactive ^{32}P is broken specifically at one of the four types of base. The key to using this as a sequencing technique is to ensure that breakage at any particular susceptible position is only partial, occurring with a probability of, say, I–2%.

Thus we start with a preparation that contains a large number of identical molecules, all labeled at the same end. When subjected to the chemical treatment, each molecule is likely to be cleaved at only a small number of the sites at which the appropriate base occurs. But in the entire preparation of DNA molecules, every position will be attacked on some occasion, as illustrated in **Figure 3.4.**

The treated molecules are electrophoresed, and those carrying the radioactive label are identified (by autoradiography). Each will form a band whose length depends on the distance from the labeled end to the site of cleavage. There will therefore be a whole series of bands for any chemical treatment, identifying every position at which the particular base occurs relative to the end of the molecule.

In the original method, specificity for the purines was provided by using dimethylsulfate. This methylates the N7 position of guanine about 5 times more effectively than the N3 position of adenine. Upon heating, the methylated base is lost from the DNA, generating a break in the polynucleotide chain. Because the reaction is more effective with guanine, G residues generate dark bands and A residues generate light bands (that is, the cleavage has been less frequent). The pattern can be reversed by using acid instead of heat to release the methylated bases. This allows G and A to be identified individually in the appropriate reactions.

Pyrimidines are analyzed by treatment with hydra-

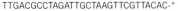

TTGACGCCTAGATTGCTAAGTTCGTTACAC-*

Treat to break
chain at G residues

TT ACGCCTAGATTGCTAAGTTCGTTACAC-*
TTGAC CCTAGATTGCTAAGTTCGTTACAC-*
TTGACGCCTA ATTGCTAAGTTCGTTACAC-*
TTGACGCCTAGATT CTAAGTTCGTTACAC-*
TTGAC CCTAGATTGCTAA TTCGTTACAC-*
TTGACGCCTA ATTGCTAAGTTC TTACAC-*

Electrophorese and
identify labeled
fragments

27
24
19
15
10
6

Bands Length

Figure 3.4
The locations of a particular base in single-stranded DNA can be determined by electrophoresing end-labeled fragments after treatment with a suitable reagent.

In the example, the chain is broken at G residues. Some chains are broken only once, some more than once, but in each case the fragment that is analyzed is the one possessing the labeled end (indicated by the asterisk). Under the conditions of reaction, the labeled molecules extend from the common end to every position at which a G was present in the original DNA. They can be separated by electrophoresis; each gives a band whose position on the gel depends on its length.

zine, which acts equally effectively on cytosine and thymine; but these can be distinguished in high salt, when only cytosine reacts. So two series of bands are generated, one representing C only and one representing C + T.

A set of four reaction mixtures is set up as indicated in **Figure 3.5.** The products of the reactions are analyzed in parallel by electrophoresis. Each of the four contains a series of bands identifying molecules in which breakage has occurred at the various positions at which the target base was located. The sequence of DNA is obtained by proceeding from each band to

DNA labeled at one end

| G breakage (DMS + heat) | A breakage (DMS + acid) | C breakage (hydrazine + salt) | C + T breakage (hydrazine) |

Electrophorese to identify end-labeled fragments

Figure 3.5
DNA can be sequenced by a set of reactions that together are specific for all four bases.

Four reactions are performed in parallel and the products are electrophoresed. The sequence is read up from the bottom of the gel following the bands successively across as indicated by the zigzag. G, A, and C are identified by the left three gels, both C and T appear on the right, so the difference between the C and the C + T bands gives the T positions. The DNA sequence is

end-N_{20}GAGCATGACGGTAGCTAGAGTA . . .

At the bottom of the gel, the shortest fragments are usually about 20 bases long, so the sequence starts this distance from the labeled end. The bands become progressively closer together proceeding up the gel, until at the top it usually becomes impossible to read the sequence.

the band that is one base longer. The shortest bands are present at the bottom of the gel, so the DNA sequence is read upward.

How accurate is DNA sequencing? As we have implied, the set of reactions gives the sequence of one strand of the DNA. There is always a risk that there will be occasional error. However, the accuracy can be improved by independently sequencing *both* strands of DNA. Any sites at which they are not complementary are identified as possible errors that can be cor-

rected. With this precaution, errors in the sequence should be exceedingly rare.

What we have discussed so far has been the generation of a sequence from a defined but short restriction fragment. How do we proceed from this to obtain the sequence of much longer regions? Just as with restriction mapping itself, the key concept is the use of overlapping fragments. Suppose we have a series of adjacent fragments, ordered on a restriction map as

| A | | B | | A | | B | | A | | B |
| | 200 | | 200 | | 150 | | 200 | | 150 | |

We can directly determine the sequence of each of these fragments. But to be sure that there are no small missing fragments generated by closely adjacent cleavages of the same enzyme (the problem we discussed in the last section), it is essential to sequence *across each junction*. This means we require another set of fragments, one in which all the A-B junctions are intact. Then just as we construct the restriction map from overlapping fragments, we can construct the overall DNA sequence by overlapping the ends of each individually sequenced fragment.

ARE GENES AND PROTEINS COLINEAR?

The concept that each strand of DNA serves as a template for synthesis of its complement (during replication) also is critical in defining the nature of the gene and its relationship with protein. Each gene is expressed via the production of a **messenger RNA,** which is the actual template used for synthesis of protein, as we shall see in Chapter 5. For our present purposes, it is sufficient to remember that the sequence of nucleotides along the messenger RNA corresponds to the sequence of amino acids in the protein. In bacteria, the sequence of the messenger RNA corresponds exactly with the sequence of one strand of DNA, and is complementary to the other strand, the one that directs its synthesis by the same mechanism of complementary base pairing used in replication.

Two experiments showed formally that the bacterial gene and its protein product are **colinear;** that the sequence of nucleotides in the gene corresponds exactly with the sequence of amino acids in the protein. This is equivalent to saying that the linear structure of the chromosome extends within as well as between genes; it is a counterpart to the demonstration that recombination frequencies apply in the same way to sites within as well as between genes. This suggests the view of the genome as a long molecule of DNA, successive segments of which represent different proteins.

The first experiment compared the distance between sites of mutation in the gene with the distance between sites of alteration in the protein. Genetic distance is measured by the percent recombination, protein distance by the number of amino acids separating the sites of replacement. **Figure 3.6** shows that the order of seven sites of mutation is the same as the order of the corresponding sites of amino acid replacement. The recombination distances between the

mutations were (surprisingly) relatively the same as the actual distances in the protein.

The second approach used mutations that prematurely terminate protein synthesis. If gene and protein are colinear, the proportion of the protein remaining in a mutant should correspond to the distance of the mutation from the start of the gene. In each case analyzed, the first part of the protein was present, but the last part was missing. The parts present in each mutant can be ordered into a hierarchy; as the site of mutation moves along the gene, the proportion of the protein that is present increases.

Both these approaches are superseded by our ability to determine directly the nucleotide sequence of a stretch of DNA. When the sequence of the DNA is compared with the sequence of the corresponding protein, bacterial genes display an exact equivalence. Each gene contains a continuous stretch of DNA whose

Figure 3.6
The recombination map of the tryptophan synthetase gene corresponds with the amino acid sequence of the protein.

Black bars indicate the sites of mutation, as identified by amino acid substitutions in the protein sequence or by mapping of mutations in the gene. The distance between the bars indicates their percent separation on the map (as detailed by the numbers).

The total map distance of the gene is 3.9 units, and there are 268 amino acids in the protein, so on average there are 0.015 map units/amino acid. (There are deviations from this in considering individual pairs of mutations.)

Note that in two cases there are mutations that can be separated on the lower map, but that affect the same amino acid in the upper map (the connecting lines converge). In one case, glycine is changed to arginine or valine; in the other, glycine is changed to cysteine or aspartic acid. Formally this indicates that the unit of genetic recombination (actually one base pair) is smaller than the unit coding for the amino acid (actually three base pairs).

DNA

RNA 5′ 3′

Protein

Region representing protein

Additional regions at each
end of gene, represented
in messenger RNA, but
not used to make
protein

Figure 3.7
**The gene may be longer than the
sequence coding for protein.**

length is directly related to the size of the protein that
it represents.

Actually the unit of DNA may be somewhat longer
than is needed just to produce the protein, because
the messenger RNA may contain some additional sequences at either end that are not used to produce
the protein. Usually the "gene" is considered to include the entire sequence that is represented in mRNA.
The situation for a gene that is expressed as a single
unit is illustrated in **Figure 3.7.** (The regions become
more complex to describe when several genes are
expressed as part of one unit; see Chapter 9.) None
of this materially affects our definition of the gene as
a sequence of DNA that is concerned with producing
a protein, and in which the region that actually represents the protein is a continuous part.

EUCARYOTIC GENES
CAN BE INTERRUPTED

The simple view that a complementation group always
corresponds to a set of mutations lying on a continuous stretch of DNA that corresponds directly with its

protein product was upset in 1977 with the discovery
of **interrupted genes.** The primary evidence for the
existence of these interruptions was a comparison between the structure of DNA and the corresponding
messenger RNA. A discrepancy may be found in the
form of sequences that lie within the gene (as defined
earlier), but which are missing from the messenger
RNA. This situation is found only in eucaryotes.

The sequences of the DNA comprising the gene are
thus divided into the two categories depicted in **Figure 3.8.** The **exons** comprise the regions that are represented in the messenger RNA used to produce the
protein product. The **introns** are missing from the
messenger RNA. The process of gene expression involves a new step, one that does not occur in bacteria.
The DNA gives rise to an RNA copy that exactly represents the genome sequence. But this RNA is only a
precursor; it cannot be used for producing protein.
First the introns must be removed from the RNA to
give a messenger RNA that consists only of the series
of exons. This process is called **RNA splicing.**

How does this change our view of the gene? The
exons are always joined together in the same order in
which they are represented in DNA. This means that

Figure 3.8
**Interrupted genes are expressed
via a precursor RNA, from which
the introns are removed when the
exons are spliced together to give
the messenger RNA.**

the colinearity of gene and protein continues between the individual exons and the corresponding parts of the protein chain. The *order* of mutations in the gene remains the same as the order of amino acid replacements in the protein. But the *distances* in the gene may not correspond at all with the distances in the protein.

The consequences for the construction of the complementation group may be more serious. But this depends on the functions of the introns. The mutations in the gene that prevent synthesis of active protein will lie in clusters, each cluster corresponding to an exon. (The reason why this has not been observed in practice is the difficulty of mapping sufficient numbers of mutations in higher organisms.) None of these mutations can complement any other, because they are all represented on the same molecule of RNA, and the splicing together of the exons occurs only as an *intra-molecular* reaction. Thus there is no joining of exons carried by *different* RNA molecules. This excludes any splicing together of sequences representing different alleles. So the existence of splicing does not change the definition of the complementation group.

What is the nature of mutations in the introns? Intron mutations cannot directly affect protein structure, since the introns are not part of the messenger RNA, but they could prevent the production of the messenger

RNA—for example, by inhibiting the splicing together of exons. Such mutations could prevent production of protein by the particular allele in which they reside. They therefore would fail to complement any other mutation in that allele and constitute part of the same complementation group as the exons.

But if an intron should happen to code for some other protein, then a mutation in it that prevents the function of this protein will be part of an independent complementation group. In this case, the mutations in the exons will form a series of clusters defining one complementation group. And between two of these clusters will be another group of mutations defining a different complementation group.

In the interrupted genes of eucaryotic genomes, most introns do not appear to be capable of exercising any independent function in the production of protein. So intermingling of complementation groups is not common. However, this situation does arise in mitochondrial genes, and thus is not merely hypothetical.

Eucaryotic genes are not necessarily interrupted. Some correspond directly with their protein product in the same manner as bacterial genes. We do not yet have a good idea of the relative proportions of interrupted and uninterrupted genes in eucaryotes, though as more data are obtained, it begins to seem that the former are in the majority.

OVERLAPPING AND ALTERNATIVE GENES

There are some cases in which a single sequence of DNA codes for more than one protein. These situations will be discussed in detail later, but we should keep in mind the types of relationship that can be shown between DNA and proteins.

Overlapping genes may occur in the relatively simple situation in which one gene is part of the other. Thus the first half (or the second half) of a gene may be used independently to specify a protein that represents the first (or second) half of the protein specified by the full gene. This does not present any particular *genetic* problem (although it does require adjustments to be made in the act of synthesizing the protein; see Chapter 7). The end result is much the same as though a partial cleavage took place in the protein product to generate part-length as well as full-length forms. This relationship has been found in viral and in bacterial genomes.

In some cases two genes overlap in a more subtle manner. In both procaryotic and eucaryotic viruses, genes have been found in which the same sequence of DNA is shared between two *nonhomologous* proteins. This is accomplished by reading the sequence of DNA in different ways, as we shall see in Chapter 4. A mutation in the shared sequence may, depending on its type, affect either gene or both. It may therefore be part of one or of two complementation groups.

In the usual form of the interrupted gene, each exon codes for a single sequence of amino acids representing an appropriate part of the protein, and each intron plays no part in the final production of protein. Their roles are distinct. But in some cases a sequence of DNA may be used in more than one way, so it cannot simply be characterized as an exon or intron.

In the viruses of higher organisms, and in the occasional case in the eucaryotic cell itself, *alternative* patterns of gene expression are found in which there may be switches in the pathway for connecting exons.

Figure 3.9
Genes may be difficult to define when there are alternative pathways for expression.

In the first pathway, *exon 1* is joined to *exon 2* by removing from RNA the regions marked *"segment"* and *"intron."* In the second pathway, the *exon-1–segment* region is joined directly to *exon 2,* removing only the *intron.* Thus in the first pathway the *segment* region is an intron, but in the second pathway it is an exon. The result of these pathways is to produce two proteins that are the same at their ends, but one of which has an additional sequence in the middle. Thus this region of DNA may code for more than one protein.

Thus a particular exon may be connected to any one of several alternative exons to form a messenger RNA. The alternative forms produce proteins in which one part is common while the other part is different. An example of such a pathway is shown in **Figure 3.9,** and this demonstrates that there may be regions that behave as exons when expressed via one pathway, but are introns when expressed via another. The complementation properties of such mutations can in principle be complex, and it is not necessarily possible to assign all mutations to one or another independent complementation group.

WHAT'S IN A GENE?

The question of what's in a name is especially appropriate for the gene. Clearly we can no longer say that a gene is a sequence of DNA that continuously and uniquely codes for a particular protein. For situations in which a stretch of DNA is responsible for production of one particular protein, current usage is to regard the entire sequence of DNA, from the first point represented in the messenger to the last point corresponding to its end, as comprising the "gene," exons, introns, and all.

In cases where the sequences representing proteins are overlapping or have alternative forms of expression, we may reverse the note on which we concluded Chapter 1. Instead of saying "one gene–one polypeptide," we may describe the relationship as "one polypeptide–one gene." Thus we may regard the sequence actually responsible for production of the polypeptide (including introns as well as exons) as the gene, while recognizing that from the perspective of another protein, part of this same sequence may also belong to *its* gene. This allows the use of descriptions such as "overlapping genes."

Given this use of "gene" as a term to mean a sequence of DNA responsible for the synthesis of a given protein, the circumstances mean that we can no longer attribute to it those precise connotations of complementation and uniqueness of representation that would have been implicit only a short time ago.

FURTHER READING

The principle of restriction mapping was first adumbrated by **Danna, Sack,** & **Nathans** (*J. Mol. Biol.* **78,** 363-376, 1973) and was reviewed by **Nathans** & **Smith** (*Ann. Rev. Biochem.* **46,** 273–293, 1975). Methods for sequence analysis of DNA have been reviewed by **Wu** (*Ann. Rev. Biochem.* **47,** 607–734, 1978). The two methods in most popular use, the first of which is described in this book, were developed by **Maxam** & **Gilbert** (*Proc. Nat. Acad. Sci. USA* **74,** 560–564, 1977) and **Sanger, Nicklen,** & **Coulson** (*Proc. Nat. Acad. Sci. USA* **74,** 5463–5467, 1977). The colinearity of gene and protein pursues directly the perspective of traditional molecular genetics and was brought to fruition by **Yanofsky** et al. (*Proc. Nat. Acad. Sci. USA* **57,** 296–298, 1967) and **Sarabhai** et al. (*Nature* **201,** 13–17, 1964). The genetic implications of interrupted gene structure are discussed in Chapter 7 of **Lewin's** *Gene Expression,* **2,** *Eucaryotic Chromosomes* (Wiley, New York, 1980).

CHAPTER 4
BREAKING THE GENETIC CODE

The concept that the gametes provide the *information* for development of the embryo rather than comprising miniature versions of the body parts actually was introduced by Aristotle 2500 years ago (although he appreciated the significance only of the egg, not of the sperm). It became possible to utilize this idea as a principle of modern work only when Mendel's analysis showed that genes are particulate factors. But how do these factors give rise to the various features of the phenotype?

The discovery that DNA is the genetic material provided a basis for the concept that what is inherited is **genetic information.** A critical feature of the structure of DNA is that it is independent of the particular sequence of base pairs. Thus the sequence of bases in the polynucleotide chain is important not in the sense of structure per se, but because it **codes** for a corresponding sequence of amino acids that constitutes a protein. The concept that each protein consists of a fixed series of particular amino acids dates from Sanger's characterization of insulin in the 1950s.

Together the various protein products of a cell undertake the catalytic and structural activities that are responsible for establishing its phenotype. Of course, in addition to gene sequences that code for proteins, DNA also contains sites that consist of a specific sequence of base pairs whose function is to be recognized by a regulator molecule, usually a protein. Here the function of the DNA is determined by its sequence directly, not via any intermediary code. Both types of sequence, genes expressed as proteins and regions recognized as such, constitute the genetic information of the organism.

One of the dictates of molecular biology is that every somatic cell in an organism contains the identical set of chromosomes and therefore carries the same genetic information. Not all of this information is expressed in every cell, so the genetic material can be viewed as consisting of a *store* of genetic information, the appropriate parts of which are expressed in a given cell, while the remainder is silent. However, recent developments have shown that in certain instances changes may occur in the content of genetic information as well as in its expression (see Part 10).

DNA NEEDS ONLY TO CODE FOR A SEQUENCE OF AMINO ACIDS

The **genetic code** describes the relationship between the sequence of the gene and the sequence of the protein that it specifies. How can a sequence of nucleotides represent a sequence of amino acids?

Twenty amino acids are commonly found in proteins, as summarized in **Figure 4.1.** They are joined together by **peptide bonds,** formed by the condensation of the amino (NH_2) group of one amino acid with the carboxyl $\left(C \diagdown_{OH}^{\diagup O} \right)$ group of the next, as shown in **Figure 4.2.** Protein sequences are conventionally written in order from the amino acid with a free NH_2 group (the N-terminus) to the amino acid with a free COOH group (the C-terminus).

Any additional amino acids—for example, the (relatively) rare hydroxyproline—are introduced by special modification of one of the fundamental twenty moieties. Other modifications include phosphorylation, methylation, and acetylation. Each is carried out by specific modifying enzymes after the protein has been synthesized.

The series of amino acids that constitutes a polypeptide chain comprises its **primary structure.** But the critical feature of a protein is its folding into a **conformation** that creates active sites or structural features that allow it to play its role in the cell. The **secondary structure** is generated by the folding of the primary sequence (using movement about chemical bonds of free rotation); this term refers to the spatial organization of the polypeptide backbone. The **tertiary structure** is the complete organization in three dimensions of all the atoms in the polypeptide chain, including the side groups as well as the polypeptide backbone. Yet a higher level of organization is recognized in **multimeric proteins,** which consist of aggregates of more than one polypeptide chain. The **quaternary structure** is the conformation assumed by the multimeric protein.

A fundamental principle is that higher-order structures are determined directly by lower-order structures. This means that the primary sequence of amino acids carries the information that is necessary for folding into the correct conformation. This may involve the formation of both noncovalent bonds and covalent bonds. For example, one of the important features in establishing the secondary structure may be the formation of S-S bridges between two cysteine residues (forming cystine). Each cysteine is separately placed into the polypeptide chain at the appropriate location, with the condensation of the two -SH groups occurring later, when they are brought into apposition by the chain folding into the appropriate conformation.

What conditions are necessary for the primary sequence to fold into the correct higher-order structures? Is this an intrinsic feature of the entire primary sequence? This would imply that the final structure is always the most stable in thermodynamic terms. Or can the correct structure be generated only during the original synthesis of the protein? This allows the possibility that an intrinsically less stable structure could prevail because the protein becomes established in it during synthesis.

In some cases, the higher-order structure follows ineluctably from the primary sequence. This was first demonstrated by experiments in which the protein ribonuclease was **denatured.** This term describes the loss of conformation that results from heating or from chemical treatments, such as disruption of S-S bridges. Reversal of the denaturing procedure allows the active conformation to be regained, showing that *all* the necessary information resides in the primary sequence. Thus in order to produce active ribonuclease, it is necessary only to synthesize the correct sequence of amino acids. (Denaturation is the same term applied to DNA, where, as described in Chapter 2, it has the specific meaning of conversion from double-stranded to single-stranded form.)

In other cases, denaturation is irreversible. Thus the protein has alternative stable conformations. Which is achieved depends on the conditions. It is possible that the correct conformation can be attained *only* during biosynthesis, as specific regions of the protein interact in the absence of the regions that have not yet been synthesized. Proteins are synthesized only in one direction, and this could be important in determining higher-order structures. (The same principle, that the direction of synthesis itself conveys information that is necessary to acquire the correct conformation, applies also to some RNAs; for example, see Chapter 15.)

In some cases, a cofactor that is part of the protein (such as the iron-binding heme group in the cytochromes) must be present in order for the polypeptide chain to take up its proper conformation.

Proteins show enormous diversity of form, the structure and/or enzymatic activity of each protein following from its primary sequence of amino acids. Thus by determining the sequence of amino acids in each protein, the gene is able to carry all the information that

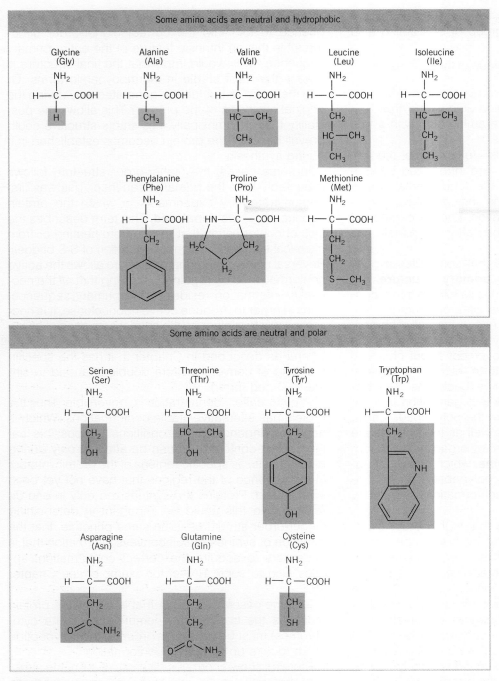

Figure 4.1
The 20 common amino acids in proteins fall into several classes, depending on the nature of their side groups (R).

Figure 4.1 (continued)

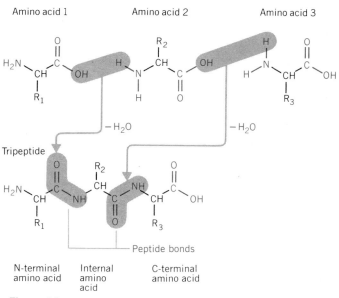

Figure 4.2
A peptide bond is formed by a condensation reaction.

is necessary for production of an active polypeptide chain. In this way, a single constant type of structure—the gene—is able to represent itself in innumerable polypeptide forms.

THE GENETIC CODE IS READ IN TRIPLETS

So the sequence of nucleotides must contain sufficient coding units to represent 20 amino acid residues. But there are only 4 bases in DNA. This means that the **coding ratio** must be greater than 1: it takes more than 1 base to specify each amino acid. If 2 bases coded for each amino acid, the DNA could specify 16 (i.e., 4^2) types of amino acid. This is inadequate: so the coding ratio must be at least 3.

If the genetic code is triplet, three adjacent bases must represent each amino acid. Since there are 64 possible triplet combinations (i.e., 4^3), a triplet code implies either that only some of the triplets actually specify amino acids, or that some amino acids must be specified by more than one type of triplet.

The first ideas about the nature of the genetic code envisaged some form of direct relationship between the nucleic acid and protein sequences. This could take the form of a stereochemical fit. Actually, there are still proposals that the present code may have evolved through some such mechanism, but in its current operation, at least, the genetic code is deciphered by a complex apparatus that stands between nucleic acid and protein (and which is therefore necessary if the information carried in DNA is to have meaning).

Genetic experiments provided the first demonstration that the genetic code is read in groups of three nucleotides, each representing one amino acid. Each trinucleotide sequence is called a **codon.** The sequence of codons is read continuously from a fixed starting point at one end of the gene to a termination point at the other end. Writing the nucleotide sequence in the conventional 5'–3' direction, we shall see that this corresponds to the amino acid sequence of the protein written in the N-terminal to C-terminal direction.

The genetic analysis involved the use of deletions induced by acridine in the *rII* region of phage T4. In 1961, Crick and his colleagues showed that acridine-induced mutations have two critical differences from the mutations induced by base substitution (such as those illustrated in Figures 2.17 and 2.18). Base substitution mutants often are **leaky:** they show some residual function. Acridine mutants, on the other hand, usually lack gene function completely. Base substitution mutants can be induced to revert by mutagens causing base substitutions. But acridine-induced mutants can be induced to revert only by further treatment with acridine.

Acridines exert their mutagenic effect by causing nucleotides to be inserted or deleted in DNA. Their effect in abolishing gene function is explained if the genetic code is read in *nonoverlapping triplets from a fixed starting point.*

"Nonoverlapping" implies that each codon consists of three nucleotides and that successive codons are represented by successive trinucleotides.

The use of a fixed starting point means that assembly of a protein must start at one end and work to the other, so that different parts of the coding sequence cannot be read independently.

If the genetic code is read in nonoverlapping triplets, there are three possible ways of translating a nucleotide sequence into protein, depending on the starting point. These are called **reading frames.** Thus for the sequence

ACGACGACGACGACGACG

the three possible frames of reading are

ACG ACG ACG ACG ACG ACG
CGA CGA CGA CGA CGA CGA
GAC GAC GAC GAC GAC GAC

If only one reading frame is used, a point mutation that alters only a single base will change only the one codon in which that base is located. Thus only one amino acid is altered in the protein. While this may reduce the activity of the protein, it does not necessarily abolish it altogether; this explains why base-substitution mutants often are leaky. By contrast, the insertion or deletion of a single base will change the reading frame for the entire subsequent sequence. This is called a **frameshift.** Since the new reading frame is completely different in sequence from the old one, this means that the entire amino acid sequence of the protein is altered beyond the site of mutation. Thus the function of the protein is likely to be lost completely, as usually seen with acridines.

If an acridine mutant is produced by, say, addition of a nucleotide, it should revert to wild type by deletion of the nucleotide. But reversion can also be caused by the deletion of a different base, at a site close to the first, either just before or after it. The second mutation is described as a **suppressor. Figure 4.3** shows that in such a situation the code is read in the incorrect frame only between the two sites of mutation; but on either side, the correct reading frame is read. (We note that in the context of this work, "suppressor" is used in a different sense from the usual general meaning defined in Chapter 2.)

When genetic recombination is used to separate acridine mutants and their suppressors, the suppressor (as a mutation by itself) turns out to have properties that are related to those of the mutation on which it acts. In fact, stated as a more general rule, all acridine mutations can be classified into one of two sets, described as (+) and (−). Either type of mutation by itself causes a frameshift, the (+) type by virtue of a base addition, the (−) type by virtue of a base deletion. Double mutant combinations of the types (+ +) or (− −) continue to show mutant behavior. But combinations of the types (+ −) or (− +) suppress one another.

This shows that the genetic code is read as a sequence in a reading frame that is fixed by the starting point, so that additions or deletions compensate each other, but double additions or double deletions remain mutant. But this does not reveal how many nucleotides make up each codon. However, when triple mutants are constructed, (+ + +) and (− − −) combinations

THE APPARATUS FOR SEQUENTIAL PROTEIN SYNTHESIS

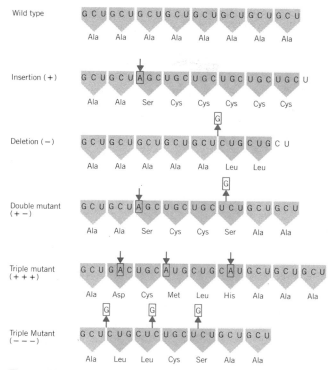

Wild type

Insertion (+)

Deletion (−)

Double mutant (+ −)

Triple mutant (+ + +)

Triple Mutant (− − −)

Figure 4.3

Frameshift mutations show that the genetic code is read in triplets from a fixed starting point.

Bases inserted into or deleted from the wild-type sequence are indicated by boxes and arrows. The triplet reading frame is indicated by pentagons, with grey indicating wild-type codons, and color indicating mutant codons. The corresponding amino acids are shown (see later in text).

If the genetic code is read as a series of triplets, a protein should be assembled sequentially from one end to the other. This was first demonstrated in 1961, when Dintzis investigated the synthesis of the red blood cell protein, hemoglobin. In cells synthesizing hemoglobin, one of the amino acids was replaced with a radioactively labeled form. Then completed protein chains were extracted, broken with the enzyme trypsin into peptides representing different parts of the molecule, and the radioactivity of each peptide was determined.

Figure 4.4 shows that the radioactivity enters protein chains at all possible positions, corresponding to

Figure 4.4

Proteins are synthesized from the N-terminus toward the C-terminus.

A radioactive label enters all parts of the growing chains simultaneously. When completed chains are extracted, they contain a label in the *last* region to be completed. So the gradient of label from C-toward N-terminus indicates that synthesis proceeds in the opposite direction.

Protein chains are at various stages of completion

Radioactive label is added (0000) for short period

Completed chains are extracted; last end to be made (C-terminus) is labeled

After longer period of labeling, more chains are completed, so this fraction includes label in gradient decreasing from C to N-terminus

show the wild phenotype; other combinations are mutant. This suggests that the code is read in triplets, since three insertions or three deletions correspond respectively to the addition or omission overall of only a single amino acid. The wrong sequence of amino acids represents only the distance between the first and third sites of mutation, as indicated in Figure 4.3.

The prediction that active double or triple mutants should have the same amino acid sequence as wild type except between the sites of mutation could not be confirmed with the *rII* system. Unfortunately, at this time it was not possible to obtain the protein products for analysis. However, comparable analysis with other systems was able to confirm the prediction.

whatever point the synthesis of each individual molecule has reached. But in *completed* chains, the radioactivity should first appear at the end that is synthesized last. With longer periods, it is found in the parts of the chain that were made earlier. Thus the radioactivity works its way back, as it were, from the end to the start. By this criterion, proteins are assembled only in the direction from the N-terminus to the C-terminus.

As we have already mentioned in Chapter 3, the first step in gene expression is the formation of a messenger RNA sequence that represents one strand of the DNA. This accords with the concept that, since each strand of the DNA double helix is complementary to the other, the sequence of either (rather than both) should suffice for coding purposes. The strand that bears the *same* sequence as the messenger RNA (except for possessing T instead of U) is sometimes described as the **coding strand.** The other, which provides the template for RNA synthesis via complementary base pairing, may correspondingly be known as the **anticoding strand.** Since the genetic code is actually *read* on the messenger RNA, usually it is described in terms of the four bases present in RNA: U, C, A, and G.

The synthesis of the single strand of messenger RNA on the DNA is known as **transcription.** The role of the messenger RNA is to provide the template on which the amino acids actually are assembled into a polypeptide chain; this process is known as **translation,** since it is at this step that the triplet nucleotides are deciphered as corresponding to individual amino acids.

The process of translation is undertaken by a cellular organelle known as the **ribosome.** Ribosomes are particles consisting of RNA and protein; they have roughly a squashed spherical shape. As we shall see in the next chapter, they attach to messenger RNA at its 5′ end, and move along the polynucleotide toward the 3′ end, translating each triplet codon into an amino acid en route.

The incongruity of structure between trinucleotide and amino acid immediately raises the question of how each codon is matched to its particular amino acid. Even before the exact form of the code had been discovered, Crick suggested that translation might be mediated by an "adaptor" molecule. This since was discovered to be **transfer RNA,** abbreviated **tRNA,** consisting of a chain of 75–85 nucleotides, providing the smallest RNA molecule in the cell.

Each transfer RNA has two crucial properties. It is able to recognize only one amino acid, to which it is *covalently linked* to form an **aminoacyl-tRNA.** Aminoacyl-tRNAs have the property of associating with the ribosome, where they exercise the second function of tRNA. Each tRNA contains a trinucleotide sequence, the **anticodon,** which is complementary to the codon representing its amino acid. The transfer RNA can therefore recognize the codon via complementary base pairing. A ribosome allows the transfer RNA species corresponding to two successive codons to be present at one time, so that peptide bonds can form between the two appropriate amino acids.

CODONS REPRESENTING AMINO ACIDS

When the triplet nature of the code was discovered, there seemed to be no approach available that would allow codons to be assigned systematically to amino acids. But it was only a very short time before two methods were able to assign meanings to most of the codons.

Protein-synthesizing systems can be prepared in the form of cell-free extracts (see Chapter 6). The original procedure consisted essentially of breaking cells of the bacterium *E. coli* and centrifuging the mixture to remove cell wall and membrane fragments; DNA was destroyed by adding DNAase (and bacterial messenger RNA is too unstable to provide an endogenous template). The supernatant is active in translating any messenger RNA that is added to it, so long as it is provided with the necessary precursors and an energy source. Since the introduction of this system, others consisting of more purified components have been developed; and comparable systems have been developed from many other cell types, eucaryotic as well as procaryotic.

The use of *synthetic* polynucleotides to direct protein synthesis followed from the report by Nirenberg and Matthaei in 1961 that poly(U)—polyuridylic acid—can act as a messenger RNA. It directs the assembly of phenylalanine into polyphenylalanine *in vitro*. This means that UUU must be a codon for phenylalanine. Similarly, the homopolymers poly(C) and poly(A) direct synthesis respectively of polyproline and poly-lysine.

The next step was to use *heteropolymers,* which contain more than one base, and may direct the incorporation of several amino acids into protein. Such heteropolymers can be prepared by using the enzyme polynucleotide phosphorylase (normally a nucleic acid *metabolic* enzyme) to link together a polynucleotide sequence from a preparation of nucleoside diphosphates. (This artificial situation is the sole exception to the rule that triphosphates provide the precursors for nucleic acid synthesis.)

If the nucleotide substrates are incorporated randomly into a polynucleotide, then the frequencies with which the various triplet combinations occur can be matched with the relative amounts of each amino acid incorporated into protein. A more precise technique was developed by Khorana, with the ability to synthesize polynucleotides of *known* sequence, so that the codons may be correlated directly with the amino acids in the polypeptide. Such techniques allowed the meaning of about half of the 64 codons to be assigned.

Another technique for making codon assignments was developed by Nirenberg and Leder in 1964. A trinucleotide of known sequence causes specific molecules of aminoacyl-tRNA to bind to ribosomes. Essentially this mimics the process of codon-anticodon recognition that usually occurs between transfer RNA and messenger RNA within the ribosome. The triple complexes of trinucleotide•aminoacyl-tRNA•ribosome can be isolated by virtue of the ability of ribosomes to bind to nitrocellulose filters. The aminoacyl-tRNA itself does not bind to such filters, but it is retained as part of the triple complex. Its retention is followed by means of a radioactive label in the amino acid component. Then the meaning of each trinucleotide can be determined by testing which one of 20 labeled aminoacyl-tRNA preparations is retained on the filter. This is called the **ribosome binding assay.**

The two techniques together assigned meaning to 61 of the 64 codons. Of course, both make use of *in vitro* reactions, so it remained important to confirm the meanings *in vivo.* The first line of evidence to confirm that the codons actually have the same meaning in the living cell was provided by analysis of mutations. The types of mutation caused by base substitutions

could be correlated with the amino acid substitutions that resulted. More productively, the sequences of wild-type and mutant proteins could be compared between two sites of frameshift.

Such experiments generally tended to confirm the codon assignments, but fell short of complete proof of their use. But since then, the direct evaluation of codon meanings has become possible with the ability to determine the actual sequence of DNA. In fact, as we have mentioned, a gene now can be sequenced far more rapidly than its protein product. Where protein sequences are known, gene sequences can be related directly to them by reading the code in triplets. The sequence of the coding strand of DNA, read in the direction from 5′ to 3′, consists of triplets corresponding to the amino acid sequence of the protein, read from N-terminus to C-terminus. The entire genetic code has been confirmed in overwhelming detail from such analyses.

The genetic code is shown in **Figure 4.5.** It is at once obvious that it is **degenerate:** 61 codons rep-

Figure 4.5
All the triplet codons have meaning.

Sixty-one of the codons represent amino acids, and all of the amino acids except tryptophan and methionine have more than one codon. The synonym codons usually form groups in which the base in the third position has the least meaning. Three codons cause termination (TERM). The order of bases in a codon is written (as are other sequences) in the direction from 5′ to 3′.

		SECOND BASE			
		U	C	A	G
FIRST BASE	U	UUU UUC Phe UUA UUG Leu	UCU UCC UCA UCG Ser	UAU UAC Tyr UAA UAG TERM	UGU UGC Cys UGA TERM UGG Trp
	C	CUU CUC CUA CUG Leu	CCU CCC CCA CCG Pro	CAU CAC His CAA CAG Gln	CGU CGC CGA CGG Arg
	A	AUU AUC Ile AUA AUG Met	ACU ACC ACA ACG Thr	AAU AAC Asn AAA AAG Lys	AGU AGC Ser AGA AGG Arg
	G	GUU GUC GUA GUG Val	GCU GCC GCA GCG Ala	GAU GAC Asp GAA GAG Glu	GGU GGC GGA GGG Gly

resent the 20 amino acids. Almost every amino acid is represented by several **synonym codons.** In fact, the frequency with which each amino acid is found in proteins seems to accord reasonably well with the total number of codons representing it. **Figure 4.6** shows the correlation, which is good except for arginine.

Another feature of the code is that codons tend to be clustered in groups representing a single amino acid. Often the base in the third position of a codon is not significant, since the four codons differing only in the third base all represent the same amino acid. Sometimes a distinction is made only between a purine versus pyrimidine in this position. The reduced specificity at this position is known as **third-base degeneracy.**

This feature, together with a tendency for like amino acids (that is, polar, hydrophobic, etc.) to be represented by related codons, minimizes the effects of mutations; it ensures that a single random base change

has an increased probability (compared with random assignment of codons) of resulting in no amino acid substitution or in one involving amino acids of similar character.

THE NATURE OF SIGNALS FOR TERMINATION

Only 61 of 64 triplets can be assigned to amino acids. What is the function of the other three codons?

The triplets UAG, UAA, and UGA all are **termination codons** that end protein synthesis. When present in synthetic polynucleotides, they cause the polypeptide synthesized so far to be terminated. (This is not strong evidence, however, because it could represent some deficiency in the *in vitro* systems.)

The effects of mutations show that these codons have a different type of role *in vivo* from that of the other codons.

A point mutation that causes a codon to be replaced by a triplet representing a different amino acid is called a **missense mutation.** The effect on the protein depends on the nature of the amino acid replacement.

When the mutation creates one of the three termination codons, however, it causes **premature termination** of protein synthesis at the mutant codon. This is likely to abolish protein function, since it means that only the first part of the protein will be made in the mutant cell. Such changes are described as **nonsense mutations.** (Sometimes the term "nonsense codon" is used to describe these triplets. "Nonsense" is really a misnomer, since the codons actually do have meaning, albeit an unpalatable one for a mutant gene.)

Nonsense and missense mutations were first distinguished by a genetic test constructed by Benzer and Champe in 1961. There is a variant of phage T4 in which a deletion has joined the *rIIA* and *rIIB* genes together, so that instead of individual proteins they synthesize a single joint protein. This has B activity in spite of the attachment of the A region. **Figure 4.7** shows that nonsense and missense mutations in *rIIA* can be distinguished by constructing a double mutant that carries the deletion as well as the nonsense or missense mutation. If the mutation is missense, it will not prevent B activity, because the sequence of the A

Figure 4.6
The number of codons for each amino acid correlates with its frequency of use in proteins.

Arginine is an exception, because there is a reduction in the occurrence of the doublet sequence CG in eucaryotic DNA. The four arginine codons that start with this doublet therefore are present in DNA at lower frequency than would be predicted by their base composition per se.

Figure 4.7
The r1589 deletion of phage T4 provides a test for distinguishing nonsense mutations and missense mutations.

segment does not affect the activity of the B segment in the joint protein. But if the mutation is nonsense, it will terminate protein synthesis, so that no B segment is produced.

The nonsense mutants defined by this test showed an unexpected property: their behavior depended on which strain of *E. coli* was used as host for the phage. This constitutes a **conditional lethal** system of the sort we defined at the end of Chapter 2. In **nonpermissive** strains, the phage mutants cannot grow, because the nonsense mutation terminates protein synthesis. But the *same phage* is able to grow on **permissive** strains of the bacterium. This occurs because the permissive strains possess **suppressor** mutations that overcome the effects of the nonsense mutation in the phage. The suppressor takes the form of a mutant aminoacyl-tRNA which is able to recognize the termination codon and to insert an amino acid in response, thus allowing protein synthesis to continue beyond the site of mutation (see Chapter 7).

The suppressor mutations define three classes of nonsense mutation. The original class isolated in the phage experiments all respond to the same suppressor and are termed **amber** mutations. By testing phages for their ability to grown on amber-suppressor strains of *E. coli,* the class of **ochre** nonsense mutations was found. Although ochre nonsense mutations are not suppressed by amber suppressors, they can be suppressed by a further class, the ochre suppressors. Actually, the ochre suppressors *are* able to suppress both ochre and amber nonsense codons, which ar-

gues that, although different, the two types of nonsense mutant must be related. Later a third class of nonsense mutation was discovered (occasionally referred to as opal). It is not suppressed by ochre or amber suppressors; and its suppressors do not act on either ochre or amber termination codons.

The nucleotide sequences of the termination codons were originally deduced by the characteristics of the mutations that induce them. From the relationships between the wild-type amino acids that can be mutated to nonsense types, it is possible to show that amber codons are UAG, while ochre codons are UAA; the third class is UGA. This has since been confirmed by direct sequencing of mutant genes to show that a termination codon has been introduced at the site of mutation.

All of this work shows that, when created by a nonsense mutation, any one of the three codons UAG, UAA, or UGA causes protein synthesis to terminate. Are the same signals used to terminate protein synthesis naturally at the end of the gene? In all sequenced genes, one (or sometimes more than one) of these codons lies immediately after the codon representing the C-terminal amino acid of the wild-type sequence. So these are natural termination signals that retain their usual meaning when generated within a gene by mutation.

None of the termination codons is represented by a transfer RNA. This excludes any mechanism for termination in which there exist special types of transfer RNA able to respond to nonsense signals to cause the cessation of protein synthesis. Instead there are protein factors that act when a ribosome reaches a termination codon. So the termination codons are **punctuation signals** that act in a different manner from the codons that represent amino acids.

IS THE CODE UNIVERSAL?

The genetic code was originally elucidated in the bacterium *E. coli.* We must therefore ask whether it is the same in all other living organisms.

The messenger RNA of one species can be correctly translated *in vitro* or *in vivo* by the protein synthetic apparatus of a different species. Many of these **heterologous translation systems** now have been developed. Their successful operation indicates that the codons used in the messenger RNA of one species must have the same meaning for the ribosomes and transfer RNAs of another species. Thus the code is unchanged.

The direct evidence made available by comparing DNA sequences with the corresponding protein sequences since has confirmed that the *identical set of codon assignments* is used in all bacteria and eucaryotic nuclei. Of course, since there are drastic variations in the base compositions of genomes, contrasted with some stability in the amino acid contents of proteins, different species must use the various synonym codons representing a given amino acid to different and characteristic extents. Indeed, it is only the degeneracy of the code that permits it to be maintained unchanged in such circumstances.

Recently the first exceptions to the general genetic code have been found in mitochondria from several species. The only common change is that in each case UGA has the same meaning as UGG and therefore represents tryptophan instead of termination. Other changes are found that are characteristic for each species. In yeast, CUA, and possibly the entire CUN family, codes for threonine instead of for the usual leucine. In mammals, AUA appears to have the same meaning as AUG and represents methionine instead of isoleucine. The codons AGA and AGG cause termination instead of representing arginine. Some of these changes make the code simpler, by replacing two co-

dons that had different meanings with a pair that has a single meaning.

The universality of the code (ignoring these rare exceptions) argues that it must have been established very early in evolution. Some models suppose that there must originally have been some stereochemical relationship between amino acids and the codons representing them. Then the system now used for protein synthesis may have evolved by selection for features such as greater efficiency or minimizing errors.

Possibly the code started in a primitive form in which a small number of codons was used to represent comparatively few amino acids, perhaps even with one codon corresponding to any member of a group of amino acids. More precise codon meanings and more amino acids could have been introduced later. One possibility is that at first only the first two out of the three bases in each codon were used, with discrimination at the third position evolving subsequently. The pattern of recognition between tRNA and codons is consistent with this in mitochondria, where the changes in the code mean that in most cases there is complete, instead of the usual partial, third-base degeneracy; so the third base is irrelevant to meaning.

At some point, the present code might have been "frozen," because the system was becoming sufficiently sophisticated for any changes in codon meaning to cause drastic disruption of existing proteins by the substitution of new amino acids. Its universality implies that this must have happened at such an early stage that all living organisms are descended from a single pool of primitive cells in which this occurred.

Why has the mitochondrion been able to evolve changes in the code? Since the mitochondrial codon assignments are simplified (relative to the general code), one argument is that this represents a more primitive pattern. This could be true, for example, for the use of UGA as a pair with UGG to code for tryptophan. Alternatively, the simplification may have evolved as a response to special needs of the system for protein synthesis in the mitochondrion (see Chapter 7). But the presence of changes that are species-specific implies that there must be more flexibility here than in the whole organism. The mitochondrial protein synthesis system is concerned with producing only a very small number of proteins (about 10), so that the problem of disruption by changes in meaning is much less severe. Probably the codons that are altered were not used extensively in locations where amino acid substitutions would have been deleterious.

TRANSLATION IN OVERLAPPING READING FRAMES

The definition that a gene represents a single polypeptide chain came to fruition with the breaking of the genetic code to provide the concept that the sequence of DNA can be related to the sequence of protein as a series of triplet codons. The idea that only one of the three possible reading frames is utilized in any gene was implicit in this work.

A reading frame that contains a sustained series of codons that represent amino acids is said to be **open.** We have dealt so far with the question of codon assignments within the context of an open frame that is being read from a fixed point. But what provides the starting signal? Just as nonsense codons are used to terminate protein synthesis, so is another set of codons used to initiate it. The common initiation signal is an AUG codon (representing methionine); in conjunction with a preceding sequence that causes ribosomes to recognize the messenger RNA, this serves to start protein synthesis. In some cases, GUG also is used, in which case it is translated as methionine instead of the usual valine. So a coding region is defined by an AUG (or GUG) **initiation codon,** followed by a series of triplets constituting an open reading frame, terminated by UAA, UAG, or UGA nonsense codons.

The two reading frames that represent the other phases of an open reading frame usually cannot in principle be translated into protein, because termination codons occur frequently. Such reading frames are said to be **blocked.** A typical situation in which an open reading frame overlaps with two blocked reading frames is shown in **Figure 4.8.** What presumably happens is that selective pressures fall on the open reading frame as reflected in the amino acid sequence, so there is no pressure against the accumulation of termination codons in the out-of-phase reading frames; indeed, they might even be favored for preventing synthesis of unwanted proteins. In a random sequence of DNA, termination codons would comprise 3/64 = 1 in 20 of the triplets (depending on the exact base composition of the DNA). (Thus frameshift mutations often are followed shortly after by termination at a nonsense codon in the out-of-phase reading frame.)

The existence of an open reading frame is therefore taken to establish a prima facie case in favor of its probable translation. Only selective pressure against the presence of termination codons is likely to prevent them from occurring frequently enough to prevent translation of proteins longer than a few residues. Because of our present capacity to sequence long stretches of DNA, open reading frames are periodically discovered where the protein product is not known; their actual utilization then has to be established by identifying the protein.

An extraordinary situation was first discovered in the phage φX174, where two open reading frames in different phases overlap for a substantial part of their length. A (hypothetical) example of such an overlap is shown in **Figure 4.9.** Within the first open reading frame, there is an AUG codon in *another reading frame* that is succeeded by a series of codons representing amino acids. The two open reading frames thus coexist in the same sequence of DNA. In the case of the

Opening reading frame

RNA sequence

Blocked reading frames

Met Lys Ala Cys Tyr Ser Arg Leu Lys

A U G A A G G C U U G C U A U A G U A G G C U G A A G

TERM TERM

TERM

Figure 4.8

An open reading frame usually is accompanied by two frames in the alternative phases that are blocked by the frequent presence of termination codons.

Protein 1

RNA sequence

Protein 2

Blocked reading frame

Figure 4.9
Two open reading frames can coexist in the same sequence of DNA (the third frame is blocked).

phage—and in other cases since, in viruses and in mitochondria—proteins corresponding to alternative reading frames have been identified. Since the proteins are coded by different reading frames, they can start and finish at different locations, so the distance of overlap varies widely in such situations.

So far we have dealt with the genetic code entirely in terms of one strand of DNA. If we consider both strands, there are actually six alternative reading frames, three on each strand, read (of course) in opposite directions since the strands are antiparallel. But in *almost* every case, only one strand of DNA has coding function at any point, and the other is blocked in all potential reading frames. In the case of the overlapping reading frames in phage φX174 (and indeed in most such cases), the overlap is between two of the reading frames on the same strand of DNA. However, once the admissibility of such situations is allowed, we can see that it is possible in principle for an open reading frame on one strand to overlap with an open reading frame on the other strand. Some overlaps of this sort have been found.

The use of overlapping reading frames means that a single sequence of DNA may code for two proteins. This immediately eliminates the flexibility usually dis-

played in a coding region, since changes in third-base positions may not affect one protein but are likely to affect the other. So most mutations in such a region should be detectable as causing amino acid changes in one, and sometimes in both, proteins. Mutations affecting only one protein will define a single complementation group, mutations affecting the other protein will define a different though interspersed group, and some mutations may be members of both groups.

This situation is rare and probably results from the need of a virus to code for the proteins without increasing the length of its genome. Its occurrence is harder to explain in (mammalian) mitochondria. How overlapping reading frames have been able to evolve, and under what restraints, is not at all clear.

FURTHER READING

The breaking of the genetic code has been reviewed innumerable times. An extensive record is given by **Ycas** in *The Biological Code* (North Holland, Amsterdam, 1969). A briefer but complete account is in Chapter 1 of **Lewin's** *Gene Expression, 1, Bacterial Genomes* (Wiley, New York, 1974). The basis of frameshift mutations is analyzed by **Roth** (*Ann. Rev. Genet.* **8,** 319–346, 1974).

CHAPTER 5
FROM GENE TO PROTEIN

"The central dogma states that once 'information' has passed into protein it cannot get out again. The transfer of information from nucleic acid to nucleic acid, or from nucleic acid to protein, may be possible, but transfer from protein to protein, or from protein to nucleic acid, is impossible. Information means here the precise determination of sequence, either of bases in the nucleic acid or of amino acid residues in the protein."

With this formal declaration of the current view in 1958, Crick enunciated a rule that has continued to be known as the **central dogma** of molecular biology. Genetic information is carried by the sequence of DNA (or in the case of some viruses by RNA). Through the stages of transcription and translation, it may be converted into the amino acid sequences of proteins.

The situation described by the central dogma is illustrated in **Figure 5.1.** For some time it was thought that the entire process was irreversible: once information was transcribed into RNA, retrieval for use as genetic material did not seem possible, and its only further informational use could be for translation into protein. Since then the process of **reverse transcription** has been discovered as an action undertaken by RNA tumor viruses. The single-stranded RNA genome of the virus may be converted into a duplex DNA that is inserted into the genome, and then is inherited like any other gene. Also, RNA viruses have been discovered that replicate purely through the RNA form, using complementary base pairing, sometimes with and sometimes without the formation of a stable RNA duplex. Although neither reverse transcription nor the replication of RNA appears to be a mechanism that is usually employed by the cell itself, their existence establishes firmly the principle that the appropriate en-

Figure 5.1
The central dogma states that information in nucleic acid can be transferred or perpetuated, but the transfer of information into protein is irreversible.

The processes used by the cell are indicated as continuous lines, while those apparently used only in viral infection are broken.

75

zymes may undertake the conversion and perpetuation of sequence information between and within either form of nucleic acid.

RIBOSOMES ARE THE SITE OF PROTEIN SYNTHESIS

The ribosomes were first identified as the site of protein synthesis by experiments in which radioactively labeled amino acids were injected into rats. Liver tissue was removed at various times after the injection and homogenized to give various cellular fractions. Within a couple of minutes, the label can be found in the fraction containing the ribosomes. Then it is released into the cell sap (a term indicating the general cytoplasm). This suggests that amino acids are assembled into polypeptides at the ribosomes, which they leave upon completion.

Ribosomes are compact **ribonucleoprotein particles** that consist of two subunits. Each subunit comprises several proteins associated with an RNA molecule; the RNAs are known as **ribosomal RNA,** abbreviated **rRNA.**

The ribosomal subunits separate when the concentration of magnesium ions is reduced, typically at a level of about 1.5 mM for bacterial ribosomes. The ribosomes (and their subunits) of bacteria and eucaryotic cytoplasm are distinguished by their sizes, as first revealed by their rates of sedimentation. (The unit of sedimentation is the Svedberg, denoted as **S.** The greater the mass, the faster the rate of sedimentation, and the higher the S value. The rate is also influenced by shape, because more compact bodies sediment faster.)

Bacterial ribosomes generally sediment at about 70S; the two subunits sediment at about 50S and 30S when separated. The larger subunit is roughly spherical, and is about twice the size of the smaller, which is more asymmetrical.

The ribosomes of higher eucaryotic cytoplasm are larger, and usually sediment at about 80S. The individual subunits have sedimentation rates of 60S and 40S.

Further types of ribosomes are found in mitochondria and in chloroplasts. These display quite a wide range of sizes, and may actually be smaller than bacterial ribosomes (as in the human or frog mitochondrion, where the size is 60S), or may lie between bacterial and eucaryotic cytoplasmic ribosomes (as in fungal mitochondria, where the size is about 74S).

The sizes of these ribosomes are "nominal." For convenience, bacterial and eucaryotic cytoplasmic ribosomes are referred to as "70S" and "80S" types, although the actual sedimentation rates may vary from this depending on the circumstances.

SEARCHING FOR THE MESSENGER

The breaking of the genetic code showed that genetic information is stored in the form of nucleotide triplets, but this does not reveal how each codon is translated to give its corresponding amino acid. Indeed, the concept that there must be a code evolved pari passu with the idea that the process of translation must involve a **template.** From the physical separation between the genetic material in the nucleus and the occurrence of protein synthesis in the cytoplasm of eucaryotic cells, it was clear that DNA could not *itself* provide the template.

A general relationship between RNA levels and protein synthesis had been apparent for some time, and this led to the idea that RNA might provide an intermediate between the DNA and protein. First thoughts about the location of this intermediate were that it might constitute part of the ribosome, an idea that was almost correct, but which missed the point in a subtle way. Since ribosomes contain RNA as well as protein, the idea was that different ribosomes might be specialized by virtue of their possession of different RNAs, used for synthesizing the corresponding proteins. Now we know that all ribosomes in the cell are the same (in both protein and rRNA content), and that it is another RNA species, the **messenger RNA** or **mRNA,** that binds transiently to the ribosomes, providing them with the instructions for synthesizing particular proteins.

The search for the messenger was conducted in bacteria, where it proved somewhat elusive. Bacterial mRNA is unstable and has only a brief existence. Evidence for its occurrence was found in bacteria in-

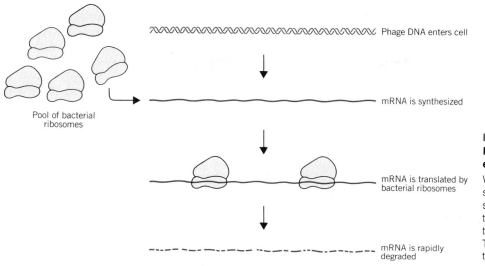

Pool of bacterial
ribosomes

Phage DNA enters cell

mRNA is synthesized

mRNA is translated by
bacterial ribosomes

mRNA is rapidly
degraded

Figure 5.2
Messenger RNA has only a transient existence in bacteria.
When a phage DNA enters the cell, new species of messenger RNA are transcribed from it. They are translated by the preexisting bacterial ribosomes, but then the mRNA is quickly degraded. The ribosomes remain available to translate further messengers.

fected with phage T2, and then confirmed in uninfected bacteria.

Because there is no change in the ribosomes, which contain most of the mass of the RNA in the cell (typically about 80%), no overall change can be detected in the *total* content of RNA during phage infection. But following the entry of phage DNA, a minor proportion of RNA is rapidly synthesized and then just as rapidly degraded. The base composition of this *new* RNA reflects that of the phage DNA rather than that of the host bacterium.

In other experiments performed at the same time, the use of very brief radioactive labels made it possible to detect the incorporation of nucleotide precursors into RNA in the uninfected bacterium. (The reason why very brief labels are required is that otherwise the radioactivity enters the RNA, but because of the instability, most of the label is released before it can be measured; however, with a *very* short labeling period, there is no time for the mRNA to be degraded, so that the label can be detected in it.)

In both infected and uninfected bacteria, the newly synthesized mRNA associates with the ribosomes, but soon is released from them and is degraded, as illustrated in **Figure 5.2.** The critical demonstration here, then, was that in phage infection the *preexisting*

ribosomes of the bacterial cell are responsible for protein synthesis; and in both phage-infected and uninfected bacteria, they are associated with an unstable RNA.

The phage mRNA was characterized by showing that it has the property of specifically representing the sequences of the DNA of the phage. The experiments were performed by testing the ability of the radioactively labeled RNA to hybridize with the phage DNA. Because the phage T2 DNA can be denatured into single strands that can be separated, it was possible to show further that the mRNA represents only one of the strands of the duplex DNA. The RNA did not hybridize with DNA from the bacterial host.

Without ever seeing an intact mRNA, it was therefore possible to demonstrate the existence of an RNA that specifically represents the sequence of one strand of DNA, is very unstable, and associates with ribosomes. We may infer that its transcription follows a pathway generally like that illustrated in **Figure 5.3.** The DNA must be unwound to permit base pairing; one strand is used as template for transcription of RNA by a process that involves movement of the unwound region along the DNA, and then the original DNA duplex reforms as the RNA is released. All nucleic acids are synthesized in the direction from 5′ to 3′. This

RNA synthesis starts with unwinding of DNA	
RNA is synthesized by base pairing with one strand of DNA	
Unwound region moves along DNA	DNA has rewound 5′ mRNA
Unwound region reaches end of gene	
RNA is freed entirely and DNA is now reformed duplex	

Figure 5.3

Messenger RNA is synthesized by base pairing with one strand of DNA in a region that is temporarily unwound.

The region that is unwound moves from left to right, and the DNA duplex reforms behind it, displacing the mRNA in the form of a single polynucleotide chain.

means that mRNA is transcribed in the same direction in which it will be translated. In bacteria, in fact, this makes it possible for ribosomes to begin translating an mRNA even before its transcription has been completed.

In eucaryotes, the search for the messenger also at first encountered difficulties. As with bacteria, mRNA constitutes only a small proportion of the total cellular RNA (roughly 3% of the mass). The first specific mRNA to be isolated was a globin messenger from red blood cells. Since then, a large number of individual messengers have been obtained; and now it is almost routine to isolate a particular mRNA. These messengers are generally stable, at least for a period of hours. They can therefore be isolated intact and translated *in vitro* when ribosomes and other necessary components are added. In fact, the proof that a given mRNA represents a particular protein now is provided by

showing that this protein is synthesized under its direction *in vitro*. This can be extended by determining the nucleotide sequence of the RNA for comparison with the amino acid sequence of the protein.

TRANSFER RNA IS THE ADAPTOR

How is a trinucleotide sequence in mRNA translated into its corresponding amino acid in protein? This question lies at the heart of the process of information transfer. We have already mentioned that, once the idea that there might be some stereochemical fit had been disposed of, attention turned to possible intermediates in the process. Transfer RNA provides the "adaptor," with the twin functions of being used to recognize both the codon and the amino acid.

Transfer RNA was first identified as a fraction of

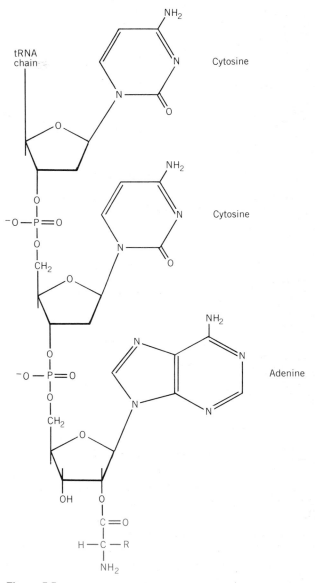

Figure 5.4
Transfer RNA has the dual properties of an adaptor.

The sugar-base moieties are represented by the boxes; the lines connecting them are phosphodiester bonds. The stems of each arm are maintained by complementary base pairing (indicated by small open circles); the bases at the end of each arm form a single-stranded loop. At the loop shown at the bottom, a triplet sequence is complementary to the codon for the amino acid that the tRNA represents. This anticodon sequence recognizes the codon by base pairing. At the 3′ end of the molecule, at the arm shown at the top, the amino acid is linked to the last nucleotide in the tRNA.

Figure 5.5
Aminoacyl-tRNA consists of an amino acid linked through an ester bond (usually) to the 2′-OH terminal position of the tRNA, which always ends in the three-base sequence -CCA.

RNA sedimenting at 4S, and was known as "soluble RNA," a property stemming from its small size. Typical tRNAs are 75–85 nucleotides in length.

The nucleotide sequence of every tRNA can be written out in the form of a **cloverleaf,** as shown in **Figure 5.4,** in which complementary base pairing forms **stems** for single-stranded **loops.** The stem-loop structures are known as the **arms** of tRNA.

Each tRNA can be **charged** only with the amino acid for which its anticodon is appropriate. The amino acid is linked by an ester bond involving its carboxyl group to one of the hydroxyl groups of the ribose of the last base of the tRNA (which is always adenine). This forms the structure shown in **Figure 5.5.**

Usually the amino acid is linked to the 2′-OH, but in some cases, it may be placed on the 3′-OH group.

Figure 5.6
The aminoacyl-tRNA synthetase charges tRNA with an amino acid.

The enzyme has sites that can be filled by tRNA, the amino acid, and ATP. The first reaction requires the amino acid and ATP; this generates aminoacyl-AMP. Then tRNA is bound and the amino acid is transferred to it. When the aminoacyl-tRNA is released, the enzyme is ready for another cycle. There is an aminoacyl-tRNA synthetase for each amino acid, so that the enzyme sites are specific for the R group of each amino acid and for the corresponding tRNA(s).

The question of which hydroxyl group is used in any particular case is moot, because the 2′ and 3′ forms of aminoacyl-tRNA are in rapid equilibrium with each other in solution—and therefore presumably in the cell.

The process of charging the tRNA to form aminoacyl-tRNA is catalyzed by a specific enzyme, **aminoacyl-tRNA synthetase.** One enzyme recognizes each amino acid: it has the property of recognizing only that amino acid and also *all the tRNAs* onto which it can legitimately be placed. There may be many such tRNAs: in addition to the several tRNAs that may be needed to respond to various synonym codons, sometimes there may also be multiple species of tRNA reacting with the same codon.

The charging process involves a reaction between the amino acid and ATP to form aminoacyl~adenylate; then the activated amino acid is transferred to tRNA. This releases AMP and pyrophosphate, as illustrated in **Figure 5.6.** The energy for the reaction is provided by the cleavage of phosphate. The two-step mechanism has been confirmed by isolating the complex produced by the first reaction, which is catalyzed when the enzyme is provided only with ATP and the amino acid; then upon addition of tRNA, the second reaction takes place.

The function of recognizing the codon is accomplished by the presence in the tRNA of the trinucleotide **anticodon** sequence that is complementary to the codon. Is this sufficient for codon recognition? An early experiment showed that, once aminoacyl-tRNA has been formed, it is the anticodon that is solely responsible for recognizing the correct codon. **Figure 5.7** shows that when cysteinyl-tRNA is subjected to reductive desulfuration by treatment with Raney Nickel, the product is alanyl-tRNA$_1^{Cys}$. [The convention for naming tRNAs is to use the three-letter (superscript) abbreviation for their amino acid, accompanied by (subscript) numerals if there is more than one tRNA for the same amino acid.] The alanine residue is incorporated into protein in place of cysteine. Thus once

Figure 5.7

The meaning of tRNA is determined by its anticodon and not by its amino acid.

Cysteinyl-tRNA has an anticodon that responds to the codon UGU. When the cysteinyl residue is chemically converted to an alanyl moiety, the tRNA continues to respond to the UGU codon, but now places alanine instead of cysteine in the protein.

a tRNA has been charged, it functions in protein synthesis strictly according to its anticodon.

RIBOSOMES TRAVEL IN CONVOY

The interaction between tRNA and mRNA is sponsored by the ribosome, which provides an environment that controls the recognition between codons and anticodons. To accomplish the sequential synthesis of a protein, the ribosome **moves** along the mRNA, "reading" successive triplets by admitting the aminoacyl-tRNAs that possess the corresponding anticodons. At each step, the growing polypeptide chain becomes longer by one amino acid.

When active ribosomes are isolated in the form of the fraction associated with radioactive amino acids, they sediment more rapidly than 70S (or 80S). The rapidly sedimenting unit is a complex of an mRNA associated with several ribosomes. This is the **polyribosome** or **polysome.**

A classic characterization is shown in the electron micrograph of **Figure 5.8.** This shows pentasomes engaged in synthesizing globin protein in red blood cells. Each pentasome consists of five ribosomes connected by a thread of mRNA. The ribosomes are located at various positions along the messenger. Those at one end will just have started protein synthesis; while those at the other end will almost have completed production of the polypeptide chain.

So as a ribosome moves along the messenger, it carries a polypeptide chain that lengthens progressively with each amino acid that is added. Each ribosome synthesizes a single polypeptide during its traverse of the message sequence. Thus the mRNA has a series of ribosomes that carry increasing lengths of the protein product, moving from the 5′ to the 3′ end. This is illustrated in **Figure 5.9.** Roughly the last 30–35 amino acids added to a growing polypeptide chain are protected by the structure of the ribosome from the environment. Probably all of the preceding part of the polypeptide protrudes and is free to start folding into its proper conformation. (Thus proteins can be recognized by specific antibodies even before their synthesis has been completed on the polysomes.)

The size of the polysome depends on a variety of

Figure 5.8
Protein synthesis occurs on polyribosomes.

The electron micrograph shows pentasomes synthesizing globin. The ribosomes are (roughly) spherical objects of about 7 nm (70 Å) in diameter, connected by a thread of messenger RNA. Depending on the conditions, the polysomes may appear compact or elongated.
Photograph kindly provided by Alex Rich.

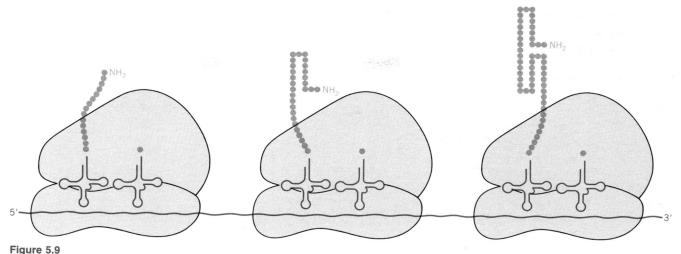

Figure 5.9
Polyribosomes consist of several ribosomes translating a messenger by moving in the direction from 5′ to 3′.

Each ribosome has two tRNA molecules, the first carrying the polypeptide chain synthesized so far from the N-terminus, the second carrying the next amino acid to be added to the chain.

factors. In bacteria it may be very large, with tens of ribosomes simultaneously engaged in translation. Partly this is due to the large size of the mRNA (which may code for more than one protein), partly it is due to the efficiency with which the ribosomes translate the mRNA. Since the ribosomes may have attached to bacterial mRNA even before its transcription has been completed, the polysome is likely still to be attached to DNA.

In eucaryotic cytoplasm, the mRNA must be transported from the nucleus to reach the ribosomes in the cytoplasm. The polysomes are likely to be smaller than those in bacteria; again, their size is a function both of the length of the mRNA and of the characteristic frequency with which ribosomes attach to it. One thinks of eucaryotic mRNAs generally as having fewer than 10 ribosomes attached at any one time.

The number of ribosomes on each mRNA molecule synthesizing a particular protein is not precisely determined, in either bacteria or eucaryotes, but is a matter of statistical fluctuation determined by the variables of mRNA size and efficiency.

An overall view of the attention devoted to protein synthesis in the intact bacterium is given in **Table 5.1.** The 20,000 or so ribosomes account for an appreciable proportion of the cell mass. The tRNA molecules

Table 5.1.
Considering *E. coli* in Terms of the Components of the Bacterium

Component	Proportion of Cell Dry Weight, %	Number/Cell
Wall	10	
Membrane	10	
DNA	2	1
mRNA	2	
tRNA	3	160,000
rRNA	21	
Ribosomal protein	9	20,000
Soluble protein	42	10^6
Small Metabolites	1	6.5×10^6

(rRNA, Ribosomal protein bracketed as 30)

outnumber the ribosomes by almost tenfold; most of them are present as aminoacyl-tRNA, that is, ready to be used at once in protein synthesis. Because of their instability, it is difficult to calculate the number of mRNA molecules, but a reasonable guess would be 2000–3000, in varying states of synthesis and decomposition.

FURTHER READING

Current reviews of the general processes involved in protein synthesis are rare (most concentrate on particular aspects, and these are cited in Chapters 6–9). A general overview can be found in Chapters 2 and 3 of **Lewin's** *Gene Expression, 1, Bacterial Chromosomes* (Wiley, New York, 1974).

PART 2
HOW PROTEINS
ARE SYNTHESIZED

The results of [our] labeling experiments were even more surprising in view of the ideas then current [in 1955] on the biosynthesis of proteins and their state within the cell. The work of Schoenheimer [in 1942] had actually persuaded most biochemists that in an organism proteins are inherently in a "dynamic" state, each molecule being perpetually destroyed and reconstructed by exchange of amino acid residues. Our experiments, however, showed that β-galactosidase is entirely stable *in vivo,* as are other bacterial proteins, under conditions of normal growth. . . . These investigations were given new meaning by the perspectives opened to biology around 1955. It was in 1953 that Watson and Crick . . . proposed their model of the structure of DNA. . . . Meanwhile, one year earlier, Sanger had described the peptide sequence of insulin, and it was already known . . . that a genetic mutation can cause a limited modification in the structure of a protein. In 1954, the genetic code theory [proposed that] the primary structure of proteins is determined and defined by the linear sequence of nucleotides in DNA.

Jacques Monod, 1965

CHAPTER 6
THE ASSEMBLY LINE FOR POLYPEPTIDE SYNTHESIS

The synthesis of proteins is somewhat like an assembly line in which the ribosomes proceed inexorably along the messenger, bringing in the aminoacyl-tRNAs that provide the actual building blocks of the protein product. The ribosome itself constitutes a small factory, in which a compact package of proteins and rRNAs forms several active centers able to undertake various catalytic activities. Different sets of accessory factors assist the ribosome in each of the three stages of protein synthesis: initiation, elongation, and termination. The energy for protein synthesis is provided by hydrolysis of GTP.

Initiation involves the reactions prior to forming the peptide bond between the first two amino acids of the protein. It requires the ribosome to bind to the messenger RNA, forming an initiation complex together with the first aminoacyl-tRNA. This is a relatively slow step in protein synthesis.

Elongation includes all the reactions from synthesis of the first peptide bond to addition of the last amino acid to the chain. It is the most rapid stage of protein synthesis and involves movement of the ribosome from the first to the last codon on the messenger RNA.

Termination encompasses the subsequent steps that are needed to release the completed polypeptide chain; at the same time, the ribosome dissociates from the mRNA. This also is slow compared with the time required to add an amino acid to the chain.

It is only relative to the rate of elongation that the punctuation stages of initiation and termination seem slow. Protein synthesis is a rapid process (although the rate depends strongly on temperature). In bacteria at 37°C, elongation varies from 12 to 17 amino acids added to the growing polypeptide chain every second, the exact value depending on the growth conditions. So it takes only about 20 seconds to synthesize an average protein of 300 amino acids. About 80% of the bacterial ribosomes are engaged in protein synthesis, so the pool of free units is only a minor proportion. In eucaryotes, the rate of protein synthesis is slower; in red blood cells at 37°C, elongation typically sees about 2 amino acids added to the chain per second.

Most of the experiments to define the stages of protein synthesis have been performed with *in vitro* systems, consisting of ribosomes, aminoacyl-tRNAs, accessory factors, and an energy source. In these systems, the rate of protein synthesis may be slower by an order of magnitude or so than the rate *in vivo*.

THE RIBOSOMAL SITES OF ACTION

Messenger RNA is associated with the small subunit of the ribosome. A stretch of about 30 bases of the mRNA is bound. But only two molecules of tRNA can

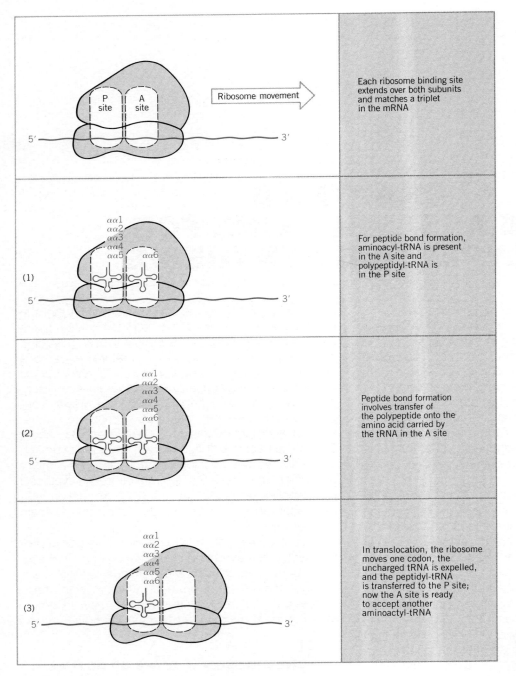

Each ribosome binding site extends over both subunits and matches a triplet in the mRNA

For peptide bond formation, aminoacyl-tRNA is present in the A site and polypeptidyl-tRNA is in the P site

Peptide bond formation involves transfer of the polypeptide onto the amino acid carried by the tRNA in the A site

In translocation, the ribosome moves one codon, the uncharged tRNA is expelled, and the peptidyl-tRNA is transferred to the P site; now the A site is ready to accept another aminoactyl-tRNA

Figure 6.1
The ribosome has two sites at which tRNA may be bound.

be in place on the ribosome at any moment. Thus protein synthesis involves reactions at just two of the ten codons covered by the ribosome.

Figure 6.1 shows that each tRNA lies in a distinct site. These two sites have different features.

The only site that can be entered by an incoming

aminoacyl-tRNA is the **A site** (or **entry site**). Prior to the entry of aminoacyl-tRNA, the site exposes the codon representing the next amino acid that is due to be added to the chain.

The codon representing the *last* amino acid to have been added to the chain lies in the **P site** (or **donor site**). This is occupied by **peptidyl-tRNA,** a tRNA carrying the entire polypeptide chain that has been synthesized up to this point (see step 1).

When both sites are occupied, peptide bond formation occurs by a reaction in which the polypeptide carried by the peptidyl-tRNA is transferred to the amino acid carried by the aminoacyl-tRNA. This transfer takes place on the large subunit of the ribosome.

Thus the end of the tRNA that carries an amino acid is located on the large subunit, while the other end is the anticodon that interacts with the mRNA bound by the small subunit. So the P and A sites must extend across both ribosomal subunits, as drawn in Figure 6.1

Transfer of the polypeptide generates a ribosome in which a tRNA now devoid of any amino acid lies in the P site, while a (new) peptidyl-tRNA has been created in the A site (see step 2). This peptidyl-tRNA is one amino acid residue longer than the peptidyl-tRNA that had been in the P site in step 1.

Then the ribosome moves one triplet along the messenger. This expels the uncharged tRNA and moves the peptidyl-tRNA into the P site (see step 3). The next codon to be translated now lies in the A site, so that a new aminoacyl-tRNA can enter; and the cycle can be repeated.

INITIATION: A SPECIAL INITIATOR tRNA

What initiates protein synthesis: how is the first codon of the gene recognized as providing the starting point for translation?

Two features on messenger RNA are necessary for the initiation reaction. First, the mRNA possesses a sequence of bases that precedes the actual coding region and functions as a signal for the ribosome to bind. We shall discuss the nature of such signals in Chapter 9. Second, the signal that marks the start of the reading frame is a special codon, the triplet AUG

(in bacteria, occasionally GUG is used instead). It is the conjunction of these two signals that establishes the starting point for translation.

The first clue that initiation cannot take place at *any* codon, but requires a special mechanism, was provided by the observation that the amino acid composition of E. coli proteins at their N-terminal ends is highly biased. Methionine occurs as the first amino acid in almost 50% of the proteins. Then the species **N-formyl-methionyl-tRNA** was discovered; and this seemed to be a good candidate for a specific initiator, because the blocked amino group would prevent it from participating in chain elongation, but would permit it to initiate a protein.

There are two types of tRNA able to carry methionine in E. coli. One is used for initiation, the other for recognizing AUG codons during elongation.

The initiator tRNA is known as $tRNA_f^{Met}$. After it has been charged with its amino acid to generate Met-$tRNA_f$, the formylation reaction shown in **Figure 6.2** occurs to block the free NH_2 group. The product is the active initiator; its name is usually abbreviated to fMet-$tRNA_f$. It is used *only* for initiation, and recognizes either of the codons AUG or GUG.

The species responsible for recognizing AUG codons in internal locations is $tRNA_m^{Met}$. This responds

Figure 6.2

The initiator N-formyl-methionyl-tRNA (fMet-tRNA$_f$) is generated by formylation of methionyl-tRNA$_f$ using formyl-tetrahydrofolate as cofactor.

only to internal AUG codons; and its methionine cannot be formylated.

The meaning of the AUG codon remains the same, in representing methionine, whether the triplet is present at the beginning or within a messenger, although it is recognized by different tRNAs in these two circumstances. But the meaning of the GUG codon depends on its location. When present as the *first* codon, it is read via the initiation reaction as formyl-methionine. Yet when present *within* a gene, it is read by Val-tRNA, one of the regular members of the tRNA set, to mean valine as required by the genetic code (see Figure 4.5).

The precise meaning of both AUG and GUG codons is therefore established by their **context.** Initiation pre-

vails when the codon is juxtaposed with a signal to indicate that ribosomes should bind de novo to the mRNA; the usual internal reading occurs when the codons are encountered by a ribosome that is *continuing* to translate an mRNA (see next· section).

The initiation reaction is slightly different in eucaryotic cytoplasm. First, only the codon AUG is used as an initiator; GUG is not found naturally. In fact, if mutation causes a GUG codon to replace an AUG codon, the initiation reaction is prevented. Second, the initiator tRNA continues to be a distinct species, but it is known as tRNA$_i^{Met}$ because its methionine does *not* become formylated. Thus the difference between initiating and elongating Met-tRNAs lies solely in the tRNA moiety, with Met-tRNA$_i$ used for initiation and Met-tRNA$_m$

Figure 6.3
Newly synthesized bacterial proteins start with formyl-methionine.

The formyl group is removed by a deformylase to generate a normal NH$_2$ terminus. In about half the proteins, this is followed by removal of the methionine by an aminopeptidase, leaving as the N-terminal amino acid R$_2$ (which was originally the second amino acid incorporated into the chain).

used for elongation. (However, the Met-tRNA$_i$ can be recognized as a substrate *in vitro* by the *E. coli* formylase.)

The same existence of two tRNAs responding to AUG is found in mitochondria, where the initiator species may be used with formylated methionine as in bacteria. Changes in the pattern of codon recognition seem to allow AUA, and sometimes AUG and AUU also, to be used as initiation codons in mammalian mitochondria.

There is always a system to remove unwanted residues from the N-terminus. In bacteria and in mitochondria, the formyl residue is removed by a specific deformylase enzyme. If methionine is to be the N-terminal amino acid of the protein, this is the only step that is necessary. For either bacterial or eucaryotic proteins that have some other amino acid at the N-terminus, the methionine must be removed by an aminopeptidase. When both steps are necessary, they occur sequentially, as depicted in **Figure 6.3.** The removal reaction(s) occur rather rapidly, probably when the nascent polypeptide chain has reached a length of 15–30 amino acids.

INITIATION NEEDS 30S SUBUNITS AND ACCESSORY FACTORS

Moving along mRNA to elongate the polypeptide chain, the intact (70S or 80S) ribosome catalyzes protein synthesis. But the intact ribosome cannot bind as such to an mRNA to initiate protein synthesis. The initial contact with the mRNA occurs by binding of a small (30S or 40S) subunit to form an **initiation complex** at the site of the first (AUG) codon.

The small subunit sits on the mRNA in such a way that the initiation codon lies within the part of the P site carried by the subunit. The *only* aminoacyl-tRNA that can become part of the initiation complex is the initiator, which has the unique property of being able to enter directly into the partial P site to recognize its codon. This is shown in **Figure 6.4.**

Then the large subunit joins the complex to generate a complete ribosome, in which the initiator aminoacyl-tRNA lies in the now-intact P site; and the A site is available for entry of the aminoacyl-tRNA complementary to the next codon (the second triplet in the gene).

Through this procedure, the initiation codons are encountered only by initiator aminoacyl-tRNA (via the small subunit), while triplets *within* the coding region are encountered only by the regular aminoacyl-tRNAs (via the complete ribosome).

Although it is the 30S subunit that is involved in initiation, this is not by itself competent to undertake the reactions of binding mRNA and initiator tRNA. The inadequacy of ribosome subunits in the initiation reaction was first discovered by the effects of "washing" 30S subunits with ammonium chloride. This treatment leaves them unable to sponsor initiation of new chains. The reason is that active 30S subunits contain **initiation factors** (**IF**). These are proteins that are bound

Only fmet-tRNA$_f$ enters partial P site on 30S subunit bound to mRNA

50S subunit joins to complete P and A sites and next $\alpha\alpha$-tRNA enters A site

fmet

fmet $\alpha\alpha$

AUG   XXX

AUG   XXX

Figure 6.4
Only fMet-tRNA$_f$ can be used for initiation by 30S subunits; only other aminoacyl-tRNAs can be used for elongation by 70S ribosomes.

loosely enough to be released by the ammonium chloride wash (which does not have any effect on the "permanent" ribosomal proteins).

There are three initiation factors in bacteria, numbered **IF1, IF2,** and **IF3.** They are needed for each of the interactions of the small ribosome subunit with the other components: mRNA and initiator tRNA. The first stage involves the binding of 30S subunits specifically at initiation sites in mRNA, and not elsewhere; IF3 participates in this reaction. The second stage is involved in ensuring that only the initiator fMet-tRNA$_f$ is placed in the partial P site of the subunit-mRNA complex; IF2 is required for this reaction.

The role of IF1 in initiation has not yet been pinned down. The smallest initiation factor, a protein of about 9500 daltons, it binds to 30S subunits only as an integral part of the initiation complex. IF1 may be involved in stabilizing the complex, perhaps by causing some change in the conformation of the 30S subunit, rather than playing a direct role in recognition of any of the components. There is enough IF1 in the cell to bind about 3% of the total number of 30S subunits.

The initiation factors all can be found only on "free" 30S subunits. During the conversion of 30S●mRNA complexes to 70S ribosomes, the factors are released. This distinguishes them from the ribosomal structural proteins. Thus the initiation factors are concerned solely with the formation of the initiation complex, they are absent from 70S ribosomes, and they play no part in the stages of elongation.

THE BRIEF FREEDOM OF 30S SUBUNITS

Ribosomes engaged in protein synthesis exist as 70S (or 80S) particles, and they are released from the mRNA in this form. Then they enter a pool of free ribosomes. How are the necessary "free" 30S subunits generated for the initiation reaction?

The dissociation of subunits from ribosomes in the free 70S pool in bacteria is assisted by a protein factor, originally known as the **dissociation factor,** but later recognized as IF3. This factor had previously been discovered by virtue of its involvement in the binding of 30S subunits to mRNA. Thus it has dual functions: it is needed first to generate 30S subunits; and then must remain present for them to bind to mRNA. In this

sense, it is a factor that controls the freedom of 30S subunits, which lasts from their dissociation from the free 70S pool to their reassociation with a 50S subunit at initiation.

Figure 6.5 shows that the 70S ribosomes in the free pool are in a dynamic equilibrium with 30S and 50S subunits. When IF3 binds to a 30S subunit, it upsets this equilibrium by generating a stable free subunit that does not reassociate with a 50S subunit. In this capacity, its role is to act as an antiassociation factor. The reaction between IF3 and the 30S subunit is stoichiometric: one molecule of IF3 (a single polypeptide chain of about 23,000 daltons) binds per subunit. The factor is present in relatively minor amounts in the cell, so that about 5% of the 30S subunits carry it (this is about 25% of the subunits not engaged in protein synthesis).

The presence of IF3 is absolutely necessary if the 30S subunit is to be able to bind to mRNA. Small subunits that lack the factor can be obtained either by washing free 30S subunits (with ammonium chloride) or by artificially dissociating 70S subunits (with low Mg^{2+} concentration). These subunits cannot form initiation complexes with mRNA. However, the *specificity* with which initiation sites are selected on mRNA appears to be entirely a function of the ribosome subunit.

Figure 6.5

IF3 is involved in upsetting the equilibrium with 70S ribosomes to generate stable 30S subunits with the ability to bind at initiation sites on mRNA.

Free subunits

IF-3

50S
+
30S

Dynamic equilibrium

Pool of 70S ribosomes

Stable 30S subunit

Initiation complex at AUG codon

mRNA

Thus IF3 is necessary for the reaction, but does not seem to be involved in selecting the site.

The factor is released before the small subunit is joined by the 50S subunit. A 30S subunit cannot simultaneously bind IF3 and a 50S subunit. The options open to the 30S subunit are therefore either to be "free" when associated with IF3, or to be part of a 70S ribosome. The 30S subunit is driven around its life cycle by these alternatives. When a ribosome is a member of the free pool, the dynamic equilibrium with subunits allows IF3 to replace the 50S subunit. When an initiation complex has been formed, the IF3 is released, and the 30S subunit is joined by a 50S subunit. Upon its release, IF3 immediately recycles by finding another 30S subunit; this rapid reutilization is displayed by all the accessory factors.

PICKING OUT THE INITIATOR tRNA

A role for IF2 in binding the fMet-tRNA$_f$ initiator to the 30S•mRNA complex is illustrated in **Figure 6.6.** The factor, a protein of about 73,000 daltons, is able to form a complex specifically with fmet-tRNA$_f$. It is not clear which features of the initiator tRNA are involved in distinguishing it from other aminoacyl-tRNAs, except that the presence of a blocked NH$_2$ group is essential. This discrimination occurs *in vitro* only when there is a low concentration of magnesium ions.

The binary complex of fmet-tRNA$_f$ and IF2 could represent the first stage in the entry of the initiator tRNA into the initiation complex. Then the 30S•mRNA complex binds the IF2•fmet-tRNA$_f$ complex so that the tRNA lies in the partial P site. This order of events is not proven; IF2 can bind directly to 30S subunits, so that the fmet-tRNA$_f$ could be recognized by IF2 already on the ribosome subunit. Whatever the order of events, the IF2 protein remains part of the 30S subunit at this stage; it has a further role to play.

It has been known for a long time that IF2 has a ribosome-dependent GTPase activity: the factor sponsors the hydrolysis of GTP in the presence of ribosomes, releasing the energy stored in the high-energy bond. Although GTP is known to be a component of the complete initiation complex, it is not yet clear exactly at which stage the nucleotide is incorporated. The most likely model is that this happens at or im-

Figure 6.6
Bacterial initiation factors are needed for 30S subunits to bind mRNA and for fMet-tRNA$_f$ to bind to the 30S-mRNA complex; they are released with cleavage of GTP when the 50S subunit arrives.

mediately after the point at which the binary complex associates with the 30S subunit. Then when the 50S subunit joins to generate a complete ribosome, the GTP is cleaved in an action that depends on the presence of IF2. Probably IF2 is not itself the GTPase, but activates a ribosomal protein with this function.

Although the exact function of the GTP cleavage is not clear, probably it is connected with changing the conformation of the assembly so that the independent subunits are converted into a jointly acting 70S particle. Following the GTP cleavage, the ribosome has its P and A sites in the right attitude. The fmet-tRNA$_f$ is in the P site, and the A site is ready to accept aminoacyl-tRNA in the usual way. This model assigns a role to the GTP hydrolysis that is somewhat similar to that of the high-energy bond release that occurs during ribosome movement (see later).

EUCARYOTIC INITIATION INVOLVES MANY FACTORS

The process of initiation in eucaryotes appears to be generally analogous to that in *E. coli*. There are more initiation factors—at least eight already have been found in reticulocytes (immature red blood cells), in which the most work has been done—and their exact roles are not yet clear. The factors are named similarly to those in bacteria, but with a prefix "e" to indicate their eucaryotic origin. The present set of factors comprises eIF1, eIF2, eIF3, eIF4A, eIF4B, eIF4C, eIF4D, and eIF5. The eIF2 and eIF3 fractions each contain multiple polypeptide chains, and their names reflect the fact that their roles seem to be analogous to IF2 and IF3, respectively. The other factors are mostly single polypeptides with roles that are not well characterized.

Eucaryotic initiation proceeds through the formation of a **ternary complex** containing Met-tRNA$_i$, eIF2 and GTP. This is formed in two stages: first, GTP binds to eIF2; this increases its affinity for Met-tRNA$_i$, which then is bound. **Figure 6.7** shows that the ternary complex associates directly with free 40S subunits. The association is stimulated by the presence of eIF3 and eIF4C, which probably stabilize it.

Mammalian eIF2 is a protein of three subunits, α, β, and γ (of molecular sizes 35,000, 38,000, and 55,000 daltons, respectively). The structure is similar in wheat germ, which suggests that the function of this factor may be similar in many eucaryotes. The exact roles of all the subunits are not yet defined, but it seems that eIF2α binds GTP, while eIF2γ binds Met-tRNA$_i$; β could be a recycling factor.

Binding of the 40S ternary complex to mRNA depends on eIF3, which may have a role analogous to IF3 in maintaining 40S subunits in the dissociated state. The binding to mRNA also requires some further factors (eIF1, eIF4A, eIF4B), and a high-energy bond (in ATP) is hydrolyzed. Junction of 60S subunits with this initiation complex requires the factor eIF5; and the GTP

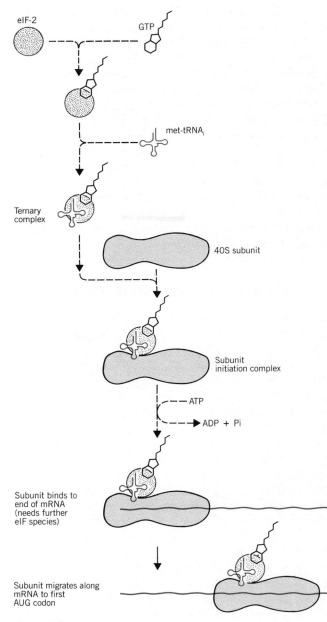

Figure 6.7

In eucaryotic initiation, eIF2 forms a ternary complex with Met-tRNA$_i$ that binds to free 40S subunits; then the initiation complex binds to mRNA at its 5′ end, from where it must migrate to the AUG initiation codon (see Chapter 9).

(brought in by the ternary complex) is hydrolyzed when the 60S subunit joins.

With radiolabeled factors it is possible to demonstrate directly the presence of eIF2 and eIF3 as components of the 40S initiation complex. Other factors have not been identified in this way, but the protein synthesis inhibitor edeine freezes the reaction at the initiation stage and causes eIF4A, eIF4B, and eIF4C to be retained in 40S-mRNA complexes. Probably all the eucaryotic initiation factors are released from the small subunit when it is joined by the large subunit to form a complete ribosome.

It is important to realize that, in the context of characterizing initiation factors, "eucaryotic" refers to only a small number of systems. Others could follow a pathway that is different in detail. (Similarly, there are some variations in the initiation pathway in bacteria other than *E. coli*.)

THE IMPORTANCE OF BEING eIF2

The factor eIF2 has been implicated as a possible control point for protein synthesis in two circumstances. In both cases, the small subunit, eIF2α, is phosphorylated; and this modification in some (unknown) way is connected with a failure in the initiation of protein synthesis.

One situation is provided by depriving reticulocytes of hemin, the cofactor needed to assemble the red blood cell protein, hemoglobin. In the absence of hemin, the initiation of protein synthesis ceases. This happens because, when the level of hemin falls below a critical amount, an inhibitor of protein synthesis becomes active. This inhibitor is a protein kinase (an enzyme that catalyzes the addition of phosphate groups to proteins); eIF2α is the substrate for the enzyme. The phosphorylation of eIF2α seems to be involved in the failure of hemin-depleted cells to initiate protein synthesis, probably through an indirect mechanism. Phosphorylation does not directly prevent eIF2 from functioning once in protein synthesis, but does interfere with its interaction with another factor. The second factor seems to be involved in regenerating eIF2-GTP from the eIF2-GDP moiety that is released after initiation. This is necessary for repeated rounds of initiation to occur.

The other situation occurs in a variety of cells (including reticulocytes). The addition of double-stranded RNA (dsRNA) activates another protein kinase, which makes exactly the same phosphorylation of eIF2α. Synthesis of the kinase may be induced by interferon (which is itself induced very potently by dsRNA). Interferon specifically inhibits the protein synthetic activity of infecting viruses, while leaving host-cell protein synthesis unaffected. However, it is not clear how the phosphorylation of eIF2 is related to this discriminatory inhibition, because eIF2 itself appears to be equally involved in initiation of protein synthesis for both viral and host mRNAs.

THE ORDER OF EVENTS IN PROCARYOTES AND EUCARYOTES

It is not clear whether there is an obligatory order in which the bacterial 30S subunits bind the mRNA and initiator tRNA; a current view is that either can bind first. By contrast, as illustrated in Figure 6.7, eucaryotic 40S subunits first bind the initiator tRNA (in an action that must therefore be independent of codon-anticodon interaction). The later binding of mRNA requires the presence of the initiator tRNA.

There is a difference in the way that bacterial 30S and eucaryotic 40S subunits find their binding sites for initiating protein synthesis on mRNA. In bacteria, the initiation complex forms directly at a sequence surrounding the AUG initiation codon. In eucaryotes, small subunits first recognize the 5′ end of the mRNA, and then move to the initiation codon, where they are joined by large subunits.

A bacterial mRNA often consists of several coding regions, each of which has its own initiation sequence, able to bind 30S subunits directly. The position of the initiation sequence per se within the mRNA is irrelevant to its function (although sometimes an initiation site may be masked by the secondary structure of the RNA molecule). Thus proximity to the 5′ end of the mRNA does not influence the recognition of initiation sites.

Each eucaryotic mRNA usually codes only for a single polypeptide. The 5′ end of the mRNA is an important feature in binding the 40S subunit, which *cannot bind directly* to internal sites. Often the AUG initiation codon does not lie within immediate reach of the 5′ end. This suggests that 40S subunits may "migrate" along the mRNA until they encounter the initiation site. This migration depends on ATP hydrolysis, and may involve some sort of "scanning" along the sequence for AUG codons. We shall return to this issue in Chapter 9.

ELONGATION: BRINGING AMINOACYL-tRNA INTO THE A SITE

Once the complete ribosome is formed at the initiation codon, the stage is set for a cycle in which aminoacyl-tRNA enters the A site of a ribosome whose P site is occupied by peptidyl-tRNA. For the formation of the first peptide bond, the initiator tRNA that has been introduced directly into the P site serves as an analog for peptidyl-tRNA. We know already that any aminoacyl-tRNA except the initiator can enter the A site. Its entry does not occur spontaneously, but is mediated by an **elongation factor** (**EF**). Just like its counterpart in initiation, this factor associates with the ribosome when it places aminoacyl-tRNA in the A site; and then its dissociation is necessary before protein synthesis can proceed past the next stage. It is indeed this cyclic association with, and dissociation from, the ribosome that is the hallmark of the accessory factors.

The factors concerned with the entry of aminoacyl-tRNA were originally obtained from *E. coli* as a single fraction (called **T factor** as an abbreviation for transfer factor). Later this was separated into the two components **EF-Tu** and **EF-Ts.** Their names reflect the characteristic that EF-Tu is **u**nstable when heated, whereas EF-Ts is **s**table.

Aminoacyl-tRNA can bind to ribosomes in either of two ways. Some association occurs even in the absence of any of the proper factors. This is called **non-enzymatic binding;** it is not controlled in the manner characteristic of protein synthesis in the cell. This reflects an intrinsic affinity of aminoacyl-tRNA for the ribosome. Proper binding occurs in the presence of EF-Tu, which is responsible for placing the aminoacyl-tRNA in the A site.

The series of events by which aminoacyl-tRNA is handled by bacterial ribosomes is illustrated in **Figure**

6.8. The EF-Tu protein carries a guanine nucleotide. The active form of EF-Tu has GTP. It is therefore the EF-Tu●GTP **binary complex** that binds aminoacyl-tRNA to form the **ternary complex** of aminoacyl-tRNA ● EF-Tu●GTP. The ternary complex binds *only* to the A site of ribosomes whose P site is already occupied by peptidyl-tRNA. This is the critical reaction in ensuring that the aminoacyl-tRNA and peptidyl-tRNA are correctly positioned for peptide bond formation.

After binding to the ribosome, the GTP is cleaved; and then the binary complex EF-Tu●GDP is released. This form of EF-Tu is relatively ineffective in binding aminoacyl-tRNA. Probably the function of the guanine nucleotide in the binding reaction is to ensure that EF-Tu is in the proper conformation to bind aminoacyl-tRNA only when the triphosphate is present. Thus GTP and not GDP is brought onto the ribosome in prepa-

Figure 6.8

EF-Tu●GTP places aminoacyl-tRNA on the ribosome, is released as EF-Tu●GDP, and then is regenerated as EF-Tu●GTP by the action of EF-Ts.

ration for the subsequent cleavage of the high-energy bond.

The active form, EF-Tu•GTP, is regenerated from the used form, EF-Tu•GDP, by the mediation of the factor EF-Ts. This is able to displace the GDP from EF-Tu, to form the combined factor EF-Tu•EF-Ts. (This is the original complete T factor.) Then the EF-Ts is in turn displaced by GTP, reforming EF-Tu•GTP. This binary complex can now bind another aminoacyl-tRNA; and the released EF-Ts can recycle.

The interactions between EF-Ts, EF-Tu, and GTP are reversible *in vitro;* but the reaction with aminoacyl-tRNA is irreversible, and it is this step that drives the reaction sequence only in the forward direction. From the perspective of the elongation factors, EF-Ts cycles between freedom and binding EF-Tu, while EF-Tu oscillates between being bound to GDP or GTP in the binary and ternary complexes or being associated with EF-Ts in the T factor.

The protein EF-Tu is a single polypeptide chain of 393 amino acids, molecular weight 43,225. In *E. coli* there are two genes, *tufA* and *tufB,* that code for EF-Tu. It is one of the major proteins of *E. coli.* There are about 70,000 molecules per cell, which corresponds to 5% of the total protein. This is sufficient to bind the entire aminoacyl-tRNA pool of the cell, roughly a tenfold excess over the number of ribosomes. The amount of EF-Ts is close to the number of ribosomes. This means that most of the EF-Tu in the cell exists as binary or ternary complexes.

In forming the ternary complex with EF-Tu•GTP, the most important structural feature of the tRNA seems to be its acceptor stem (the duplex sequence ending in the -ACCA-OH to which the amino acid is attached). In particular, changes in the structure of the terminal adenosine may prevent recognition of the aminoacyl-tRNA. The EF-Tu•GTP complex can recognize aminoacyl-tRNA in either the 2′ or 3′ configuration, so the position at which the amino acid is attached to the sugar ring is not relevant. However, at some stage before binding to the ribosome, the ternary complex stabilizes the ester bond of aminoacyl-tRNA in the 2′ configuration.

The only aminoacyl-tRNA that cannot be recognized by the EF-Tu•GTP binary complex is fMet-tRNA$_f$. Its failure to bind excludes it from responding to internal AUG or GUG codons. This ensures that these codons are read only by the appropriate aminoacyl-tRNAs (tRNA$_m^{Met}$ and tRNA^{Val}). One reason for the lack of reaction may be the structure of the acceptor stem of the tRNA$_f^{Met}$. Preceding the terminal -ACCA-amino acid sequence, the stem ends in an unpaired C·A instead of the base pair present in all other tRNAs. Other structural features in the initiator tRNA also may be involved; and the blocking of the NH$_2$ group is important, since aminoacyl-tRNAs with blocked termini are prevented from binding to EF-Tu•GTP. (The features responsible for discrimination may be peculiar to bacteria, and are different in eucaryotes.)

In eucaryotes, as typified by reticulocytes, there are two elongation factors, eEF1 and eEF2, which are equivalent respectively to the bacterial factors EF-T and EF-G (see later).

The role of eEF1 is to bring aminoacyl-tRNA to the ribosome, again in a reaction that involves cleavage of a high-energy bond by conversion of GTP to GDP. The active factor consists of aggregates of polypeptide chains of various sizes. The details of how GTP is regenerated after cleavage are not known, but there may be components of the factor that are analogous to EF-Tu and EF-Ts (which is why eEF1 is described as a counterpart to the combined factor EF-T). However, these may not be the only components of the eEF1 preparation, so it would be premature to draw exact parallels in the relationships of the components. Quantitatively, the situation may be similar in eucaryotes and procaryotes, with eEF1 (like EF-Tu) constituting a major protein of the cell.

Probably there are independent elongation factors involved in aminoacyl-tRNA•ribosome binding in mitochondria and in chloroplasts, but they have not yet been characterized.

GTP CLEAVAGE OCCURS AFTER RIBOSOME BINDING

The exact point at which the GTP is cleaved occurs some time after the aminoacyl-tRNA•EF-Tu•GTP complex binds to the ribosome. The role of GTP has been studied by substituting an analog that cannot be hydrolyzed. The compound **GMP-PCP** has a structure in which a methylene bridge replaces the oxygen that links the β and γ phosphates in GTP. In the presence of GMP-PCP, a ternary complex can be formed that binds aminoacyl-tRNA to the ribosome. But the peptide bond cannot be formed. This means that the *presence* of GTP is needed for aminoacyl-tRNA to be bound at the A site; but the GTP *hydrolysis* is not required until later. Quantitative experiments show that one GTP is cleaved for every aminoacyl-tRNA that enters the ribosome.

Much information about the individual steps of bacterial protein synthesis has been gained from the use of antibiotics that inhibit the process at specific stages. *Kirromycin* binds to EF-Tu and interferes with its function. One of the effects of the binding is to activate a GTPase activity that is independent of the ribosomes. This means that the GTP hydrolysis that follows the binding of the ternary complex to the ribosome could be catalyzed by EF-Tu rather than by a ribosomal protein (and see later).

The effect of kirromycin on protein synthesis is to block the process prior to peptide bond formation. When EF-Tu is bound by kirromycin, it remains able to bind aminoacyl-tRNA to the A site. A single round of GTP hydrolysis then can occur. But the EF-Tu•GDP complex cannot be released from the ribosome. Its continued presence prevents the peptide bond reaction from occurring between peptidyl-tRNA and aminoacyl-tRNA, with the result that the ribosome becomes "stalled" on mRNA. The inability of one ribosome to move prevents translation by others on the mRNA, and thus brings protein synthesis to a halt. The effects of this antibiotic demonstrate that blocking one step in protein synthesis blocks the subsequent steps; in this case, the right ribosomal conformation for peptide bond formation cannot be achieved when EF-Tu remains bound.

Resistance to kirromycin can be conferred by mutations that prevent the activity of EF-Tu from being impeded by the antibiotic. Because there are two genes for the protein, both must be mutated, since the presence of any sensitive EF-Tu in the cell allows kirromycin to block translation.

PEPTIDE BOND FORMATION IS A RIBOSOMAL FUNCTION

The growing polypeptide chain is elongated by transferring the polypeptide attached to the tRNA in the

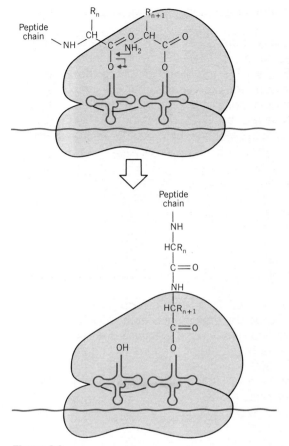

Figure 6.9
Peptide bond formation takes place by transferring the polypeptide from the tRNA in the P site onto the aminoacyl-tRNA in the A site.

ribosomal P site to the aminoacyl-tRNA present in the A site. The reaction is shown in **Figure 6.9.** Synthesis of the peptide bond is catalyzed by a **peptidyl transferase** activity, which prefers aminoacyl-3′-tRNA as substrate. Possibly a transfer of the amino acid from 2′ to 3′ position may precede peptide bond formation.

Peptidyl transferase is a function of the 50S (or 60S) subunit. Although the active site is located entirely on the large subunit, usually the activity can be detected only when the subunit is part of a complete 70S (or 80S) ribosome. Perhaps this restriction is necessary to prevent free large subunits from making random linkages between aminoacyl-tRNAs. However, peptidyl transferase of either procaryotic or eucaryotic large subunits can be activated *in vitro* by various abuses, such as addition of appreciable amounts of ethanol.

The peptidyl transferase must be part of a ribosomal site at which the ends of the peptidyl-tRNA and amino-acyl-tRNA are brought into close juxtaposition. Actually, one of the problems still unsolved in protein synthesis is how two rather bulky tRNA molecules are able to be base paired at adjacent triplets on mRNA, with their acceptor stems close enough to allow peptide bond formation. It is not clear how the expected steric hindrance is avoided.

THE TRANSLOCATION STEP

The cycle of addition of amino acids to the growing polypeptide chain is completed by the **translocation** illustrated in **Figure 6.10,** in which the ribosome advances three nucleotides along the mRNA.

Following peptide bond formation, the ribosome carries uncharged tRNA in the P site and peptidyl-tRNA in the A site. As part of the same mechanism, the translocation simultaneously expels the uncharged tRNA and moves the peptidyl-tRNA into the P site. The ribosome then has an empty A site that is accessible to the next aminoacyl-tRNA for another cycle of amino acid addition.

Figure 6.10
After peptide bond formation, EF-G and GTP are needed for translocation, but then dissociate from the ribosome (unless fusidic acid is present to stabilize the posttranslocation state).

Aminoacyl-tRNA   Puromycin

Figure 6.11
Puromycin mimics aminoacyl-tRNA because it resembles an aromatic amino acid (shaded part) joined to a sugar-base moiety.

The need for additional factors in translocation was revealed by some of the original experiments that distinguished the A and P sites. These involved the use of the antibiotic **puromycin. Figure 6.11** shows that it closely resembles an amino acid attached to the terminal adenosine of tRNA. The N in puromycin that corresponds to the O in aminoacyl-tRNA prevents cleavage. The antibiotic is treated by the ribosome as thought it were an incoming aminoacyl-tRNA. This mimicry is followed by peptide bond formation, which releases the nascent polypeptide in the form of polypeptidyl-puromycin. The premature termination of protein synthesis is responsible for the lethal action of the antibiotic against bacteria.

Because puromycin enters the ribosome by the same route as aminoacyl-tRNA, it can react with a peptidyl-tRNA only when the P site is occupied and the A site is available (this is the bottom condition of Figure 6.10). When peptide bond formation has been completed,

but translocation has not yet occurred, the puromycin is unable to enter, because the A site remains blocked by the peptidyl-tRNA (this is the situation in the top state of Figure 6.10). Ribosomes in this condition can be rendered puromycin-reactive (which is accomplished by translocation) only when GTP and another elongation factor, **EF-G,** are provided (The name of the protein reflects its original isolation as a GTP-requiring factor.)

Translocation involves a concerted set of events triggered by the association of EF-G with the ribosome, and requiring hydrolysis of GTP. Factor EF-G is a major constituent of the cell; a protein of 72,000 daltons, coded by the single gene *fus,* it represents 2% of the soluble protein. It is present in amounts that are roughly equimolar with the ribosomal proteins.

The factor binds to the ribosome in order to sponsor translocation; and then it is released when the GTP hydrolysis occurs. Binding requires the presence of

GTP, but also can occur when GMP-PCP is used, which suggests that hydrolysis is not needed for binding per se, although then the translocation step is unable to occur. Any complex that EF-G may form with GTP (or GDP) is unstable and cannot be characterized directly. Thus we do not know whether the GTP molecule is brought in by EF-G or associates directly with the ribosome.

Ribosomes cannot bind EF-Tu and EF-G simultaneously, so protein synthesis follows a cycle in which these factors alternately are bound to, and released from, the ribosome. Thus EF-Tu●GDP must be released before EF-G can bind; and then EF-G must be released before aminoacyl-tRNA●EF-Tu ● GTP can bind.

The need for EF-G release was discovered by the effects of the steroid antibiotic fusidic acid, which "jams" the ribosome in its posttranslocation state. In the presence of fusidic acid, one round of translocation occurs: EF-G binds to the ribosome, GTP is hydrolyzed, and the ribosome moves three nucleotides. But then EF-G and GDP remain on the ribosome instead of being released (see Figure 6.10). The reason is that fusidic acid stabilizes the ribosome●EF-G●GDP complex. The failure to release EF-G prevents further progress, because the ribosome cannot bind aminoacyl-tRNA●EF-Tu●GTP to the A site. This prevents the addition of any further amino acids to the chain.

We do not know whether the ability of each elongation factor to exclude the other is mediated via an effect on the overall conformation of the ribosome or by direct competition for the same or overlapping binding sites. (The drug thiostrepton prevents both EF-Tu and EF-G from binding.) The need for each factor to be released before the other can bind ensures that the events of protein synthesis proceed in an orderly fashion.

The hydrolysis of GTP that is associated with the action of EF-G is not catalyzed by the factor itself, but is a ribosomal function. The property resides in 50S subunits (see Chapter 8). We have already mentioned that it is not certain whether the GTP of the aminoacyl-tRNA●EF-Tu●GTP ternary complex is cleaved by EF-Tu or by a ribosomal protein.

The distance of three bases that the ribosome moves is probably determined by the interaction between the codon on mRNA and the anticodon on tRNA. Certain mutant tRNAs that recognize anticodons of four bases allow the ribosomes to move four bases in the translocation step. So the stereochemistry of the codon-anticodon reaction in the A site may provide the means for measuring out the distance for movement (see Chapter 8).

The eucaryotic counterpart to EF-G is the protein eEF2, which seems to function in a similar manner, as a translocase dependent on GTP hydrolysis. Its action also is inhibited by fusidic acid, although less effectively than is the case with bacterial ribosomes. This factor is a large protein, roughly 100,000 daltons, and represents a major component of the cell, probably about 1% of the soluble protein. Unlike the bacterial factor, a stable complex of eEF2 with GTP can be isolated. The complex can bind to ribosomes with consequent hydrolysis of its GTP.

A unique reaction of eEF2 is its susceptibility to diphtheria toxin. The action of the toxin, which uses NAD (nicotinamide adenine dinucleotide) as a cofactor, is to transfer the adenosine diphosphate ribosyl moiety (ADPR) onto the eEF2. The resulting ADPR-eEF2 is inactive in protein synthesis. The substrate for modification is an unusual amino acid, which is produced in eEF2 by alteration of a histidine residue; it is present in the eEF2 of many species. The ADP-ribosylation is responsible for the lethal effects of diphtheria toxin. The reaction is extremely effective: a single molecule of toxin is adequate to modify sufficient eEF2 molecules to kill a cell. Mutants of cultured cells (although not of intact organisms) have been obtained in which the eEF2 is resistant to the ADP-ribosylation.

GETTING THE ENERGY FOR RIBOSOME ACTION

Altogether a total of three high-energy bonds are cleaved for every amino acid added to the protein chain. Cleavage of an ATP is involved in charging tRNA with its amino acid. A cleavage of GTP occurs after aminoacyl-tRNA is bound into the A site via the ternary complex. Another cleavage of GTP occurs when the ribosome translocates one triplet along the mRNA. (See also Chapter 8.)

The GTP moieties involved in ribosome action probably fulfill two roles. They maintain the elongation factors with which they are associated in the appropriate conformation; when the factor is associated with GDP, it is unable to exercise its function. And they provide the energy for the corresponding step in protein synthesis, in each case probably involving a conformational change in the ribosome.

In the aminoacyl-tRNA binding reaction, the energy is required to place the ribosome in the proper conformation, in which the aminoacyl-tRNA lies in the A site and is available for reaction. This is not accomplished just by the binding per se. In the translocation step, the hydrolysis of GTP provides the energy for ribosome movement. In each case, a feature of the reaction is the release of the elongation factor from the ribosome.

Although detailed bookkeeping is more difficult with eucaryotic systems, probably the same energy requirements apply as with bacterial ribosomes.

TERMINATION: COMPLETING PROTEIN SYNTHESIS

The completion of protein synthesis is an unusual reaction because it involves the direct recognition of a termination codon by a protein factor. Since the reaction is quite distinct from the codon-anticodon interactions that are involved in initiation and elongation, there seems to be no particular reason why a triplet sequence should be involved in termination. Presumably this reflects some aspect of the evolution of the genetic code.

In *E. coli* two proteins have been found that catalyze the termination. They are called **release factors (RF)**. The factors cause ribosomes to bind to nucleotide triplets; **RF1** recognizes UAA and UAG; **RF2** recognizes UGA and UAA. The release reaction requires polypeptidyl-tRNA to be present in the P site. The release factors seem to act at the A site, since any mutant tRNAs able to recognize termination codons can compete with them for entry to the ribosome (see Chapter 7). A third factor, **RF3,** appears to stimulate the action of both the others.

In eucaryotic systems only a single release factor, **eRF,** has been found. With this factor, GTP seems to be necessary for ribosome binding (it is not involved in bacteria). Probably the GTP is cleaved after the termination step has occurred; and this may be necessary to allow the eRF to dissociate from the ribosome.

The termination reaction involves the release of the polypeptide from the last tRNA, the expulsion of the tRNA from the ribosome, and the dissociation of the ribosome from the mRNA. The separation of polypeptide and tRNA, with release of both into the cell, could occur by reactions analogous to those involved in chain elongation, although diverted here to accomplish termination. We do not yet understand the mechanics of the reaction by which the ribosome dissociates from mRNA. The dissociation may result from a conformational change triggered by the RF protein, but it is possible that another protein factor(s) could be involved.

FURTHER READING

Most current reviews concentrate on the roles of the factors that essentially control the stages of protein synthesis. Among these are the papers from **Weissbach** & **Ochoa** (*Ann. Rev. Biochem.* **45,** 191–216, 1976), **Lodish** (*Ann. Rev. Biochem.* **45,** 39–72, 1976), and **Ochoa** & **de Haro** (*Ann. Rev. Biochem.* **48,** 549–580, 1979). A more recent discussion of eucaryotic initiation and its regulation is provided by **Jagus, Anderson,** & **Safer** (*Prog. Nucleic Acids Res.* **25,** 127–185, 1981).

CHAPTER 7
TRANSFER RNA:
THE TRANSLATIONAL ADAPTOR

Transfer RNA occupies a pivotal position in protein synthesis, providing the adaptor molecule that accomplishes the translation of each nucleotide triplet into an amino acid. In Crick's phrase, tRNA represents nature's attempt to make a nucleic acid fulfill the sort of functions usually performed by protein. For tRNAs are involved in a multiplicity of reactions in which it is necessary for them to share certain characteristics in common, yet be distinguished by others. The crucial feature that allows tRNA to accomplish this is its ability to fold into a specific tertiary structure.

Working backward from the final action of tRNA in protein synthesis, all tRNAs are able to sit in the P and A sites of the ribosome, where at one end they are associated with mRNA via codon-anticodon pairing, while at the other end the polypeptide is being transferred. For the P and A sites to welcome all tRNAs, the entire set must conform to the same general dictates for size and shape.

Similarly, all tRNAs (except the initiator) share the ability to be recognized by the protein synthesis factors (EF-Tu or eEF1) for binding to the ribosome. The initiator instead is recognized by IF2 or eIF2. So the tRNA set must possess common features for interaction with elongation factors, but the initiator must lack this feature or possess some contradictory feature.

There must be critical differences between small groups of tRNAs. In most cases, multiple tRNAs represent each amino acid. These are called **isoaccepting tRNAs.** They must be charged only by the single aminoacyl-tRNA synthetase specific for their amino acid. Thus isoaccepting tRNAs must share some common features enabling their enzyme to pick them out from the other tRNAs, which must lack the distinguishing characteristic. The entire complement of tRNAs must be divided into 20 isoaccepting groups, each group carrying some feature recognized specifically by its synthetase.

What are the features of tRNAs that allow common recognition by the ribosome and by elongation factors? How are the groups representing each amino acid distinguished? Early attempts to deduce common features focused on seeking out short nucleotide stretches that were common to all tRNAs or to a particular group of tRNAs. However, although some features in the primary sequence of tRNA are highly conserved, and are presumably essential for one or more of its functions, attempts to correlate individual regions or sequences with particular protein recognition reactions have not been successful. The common (or distinctive) features of tRNAs may transcend the primary and secondary structures and be conveyed at least in part by the tertiary structure. To analyze this in any detail requires comparison of several (perhaps all) of the complement of tRNAs; this will be a long job, since at present only a few individual tRNA molecules have been analyzed at the level of tertiary structure.

THE UNIVERSAL CLOVERLEAF

The sequences of more than 150 tRNAs from a wide variety of sources now have been determined, and they all conform to the general secondary structure originally deduced from the sequence of the very first tRNA to be determined. Each sequence can be written in the form of a **cloverleaf,** which is maintained by base pairing between short regions that are complementary (see Figure 5.4). The reality of this conformation, rather than possible alternatives, was first authenticated for several tRNAs by showing that the bases expected to be in the more exposed (unpaired) positions indeed are those more susceptible to attack by nucleases.

A general form of the cloverleaf is illustrated in **Figure 7.1.** There are four major **arms,** named for their structure or function. The **acceptor arm** consists of a base-paired stem that ends in an unpaired sequence whose free 2′/3′-OH group is aminoacylated. The other arms consist of base-paired stems and unpaired loops. The **TψC arm** is named for the presence of this triplet sequence, the **anticodon arm** always contains the anticodon triplet in the center of the loop, and the **D arm** is named for its content of the base dihydrouridine. (ψ stands for pseudouridine and D stands for dihydrouridine, two of the "unusual" bases in tRNA that are discussed later.)

A numbering system that illustrates the constancy of the structure is shown in Figure 7.1. Positions are numbered from 5′ to 3′ according to the most common tRNA structure, which has 76 residues. The overall range of tRNAs is from 74 to 95 bases. The variation in length is due to differences in only two of the arms.

In the D loop there is variation of up to four residues. The extra nucleotides relative to the most common structure are denoted 17:1 (lying between 17 and 18) and 20:1 and 20:2 (lying between 20 and 21). However, in the smallest D loops, residue 17 as well as these three may be absent.

The most variable feature of tRNA is the so-called **extra arm,** which lies between the TψC and anticodon arms. Depending on the nature of the extra arm, tRNAs can be divided into two classes. **Class 1 tRNAs** have a small extra arm, possessing only 3–5 bases. These represent about 75% of all tRNAs. **Class 2 tRNAs** have a large extra arm (it may even be the longest in the tRNA), with 13–21 bases, and about 5 base pairs in the stem. The additional bases are numbered from 47:1 through 47:16.

The base pairing that maintains the structure is virtually invariant. Going clockwise around the cloverleaf, there are always 7 base pairs in the acceptor stem, 5 in the TψC arm, 5 in the anticodon arm, and usually 3 (sometimes 4) in the D arm. Within a given tRNA, most of the base pairings will be conventional partnerships of A-U and G-C, but occasional G-U, G-ψ, or A-ψ pairs will be found. The additional types of base pairs are less stable than the regular pairs, but allow a double-helical structure to form in RNA.

When the sequences of tRNAs are compared, the bases found at some positions are **conserved** (or **invariant**); there is almost always a particular base found at the position. Actually, as more tRNAs are sequenced, positions that seemed entirely invariant do display occasional exceptions. So for practical purposes, the description of any position as conserved means that the specified base is present in more than 90–95% of the tRNAs that have been sequenced. Sometimes the exceptions are individual; sometimes they fall into groups representing some peculiarity of a particular cell.

Some positions are described as **semiconserved** (or **semiinvariant**) because they seem to be restricted to one type of base (purine versus pyrimidine), but either base of that type may be present.

tRNA CONTAINS MANY MODIFIED BASES

Transfer RNA is unique among nucleic acids in its content of "unusual" bases. An unusual base is any purine or pyrimidine ring except for the usual A, G, C, and U from which all RNAs are synthesized. All other bases are produced by **modification** of one of the four bases after it has been incorporated into the polyribonucleotide chain. All classes of RNA display some degree of modification, but in all cases except tRNA this is confined to rather simple events, such as the addition of methyl groups. Only in tRNA is there a vast

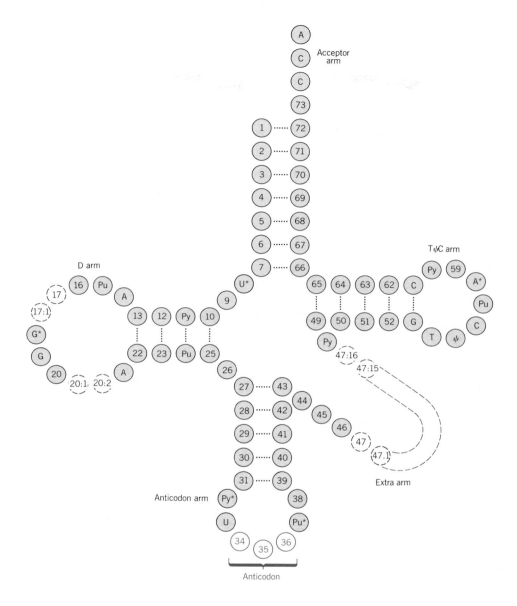

Figure 7.1

Transfer RNA forms a cloverleaf.
Circles indicate base positions. Numbers in shaded circles indicate positions that are always present but may be filled by any base. When a position is always occupied by a particular base, this is indicated in place of its number. For conserved bases, the particular species is indicated; for semiconserved bases, Py indicates either pyrimidine, and Pu indicates either purine. An asterisk indicates that the base is modified but the form of modification may vary. Open circles identify positions present in some but not all tRNAs. Dots indicate base pairing.

range of modifications, ranging from simple methylations to large-scale restructuring of the purine rings. These modifications confer on tRNA a much greater range of structural versatility, which presumably is important for its various functions.

Figure 7.2 shows some of the more common modified bases. There is a tendency for the modifications of the pyrimidines (C and U) to be less complex than

those of the purines (A and G). Some of the modifications occur in the positions involved in base pairing and therefore alter the stability of pairing, or even the partner with which the base pairs. Others are located elsewhere and have no effect on pairing.

The most common modifications of uridine are straightforward. Methylation at position 5 creates ribothymidine (T). The base is the same commonly found

Ribothymidine (T)   Dihydrouridine (D)   Pseudouridine (ψ)   4-thiouridine (S^4U)

3-methylcytidine (m^3C)   5-methylcytidine (m^5C)

Inosine (I) N^6 methyladenosine (m^6A) N^6 isopentenyladenosine (i^6A)

7-methyl guanosine (m^7G)   Queuosine (Q)   Wyosine (Y)

Figure 7.2
Some of the modified bases found in tRNA (modifications are indicated by shading).

in DNA; but here it is attached to ribose, not deoxyribose. In RNA, thymine constitutes an unusual base, originating by modification of U. (In DNA, uracil would be an unusual base, and sometimes is generated by mutagenic treatments, as described in Chapter 2.)

Dihydrouridine (D) involves the saturation of a double bond, changing the ring structure. Pseudouridine (ψ) has N and C atoms interchanged in position. And 4-thiouridine has sulfur substituted for oxygen. Other changes that occur to uridine may involve combinations of these changes, or the addition of more complex groups; for example, oxyacetic acid may be added at position 5.

The most common alterations of cytidine are methylations; also, acetylation at position 4, and the generation of 2-thiocytidine.

Inosine is a base that is found normally in the cell as a constituent of the purine biosynthetic pathway. However, it is not incorporated directly into RNA, where instead its existence depends on deamination of A. Other modifications of A include the addition of complex groups.

Two complex series of nucleotides depend on modification of G. The Q bases, of which queuosine is the simplest example, have an additional pentenyl ring added via an NH linkage to the methyl group of 7-methylguanosine. The pentenyl ring may carry various further groups. The Y bases, of which wyosine is the simplest example, have an additional ring fused with the purine ring itself, and this carries a long carbon chain, again to which further groups are added in different cases.

In addition to these modifications of the bases themselves, methylations at the 2′-0 position of the ribose ring also may be found.

In all cases except one, the modification reaction involves the alteration of, or addition to, existing bases in the tRNA. The exception is the synthesis of Q bases, where an enzyme has been found that is able to exchange free queuosine with a guanine residue in the tRNA. The reaction involves breaking and remaking bonds on either side of the nucleoside; and this is unprecedented in nucleic acid synthesis.

The list of modified nucleosides found in tRNA extends to more than 40 residues at present. They are synthesized by specific tRNA-modifying enzymes. The original nucleoside present at each position can be determined either by comparing the sequence of tRNA with that of its gene or (less efficiently) by isolating precursor RNA molecules that lack some or all of the modifications. This latter approach shows that different modifications are introduced at different stages during the maturation of tRNA from a precursor RNA.

The features that are recognized by the tRNA-modification enzymes in their substrates are not known. In some cases, a particular modification is found at more than one position in a tRNA; for example, pseudouridine may be located at several positions. A different modification enzyme may be needed to synthesize the pseudouridine at each location. We do not know what controls the specificity of the modifying enzymes, but it is clear that there are a large number of enzymes with varying specificities. Some enzymes may undertake single reactions with individual tRNAs; others may have a greater range of substrate molecules. Some reactions may involve more than one enzyme in making successive modifications.

Some modifications are constant features of all tRNA molecules—for example, the D residues that give rise to the name of the D arm, and the ψ that is found in the TψC sequence. Next to the anticodon (on its 3′ side), there is always a modified purine, although the modification varies widely.

So the modifications are found in all parts of the tRNA molecule. There are some species-specific patterns and also some tRNA-specific patterns; for example, wyosine bases are characteristic of tRNA[Phe] in bacteria, yeast, and mammals.

At present, we have only an inadequate appreciation of the functions of modification. The most direct effect is seen in the anticodon itself, in which modification may influence ability to pair with the codon, thus determining the meaning of the tRNA. Other modifications in the vicinity of the anticodon also may influence its pairing.

AN L-SHAPED TERTIARY STRUCTURE

The cloverleaf form in which the secondary structure of tRNA is written out should not be allowed to convey a misleading view of the tertiary structure, which actually is compact. To determine the tertiary structure, it is necessary to grow crystals of the tRNA for X-ray crystallography. This is difficult and actually has been done for only a few tRNAs, most from yeast. However, the similarities of the structures, and the nature of the interactions that maintain each tertiary structure, suggest that there may be a common tertiary theme honored by all tRNAs, although each will present its own variation.

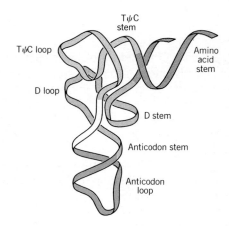

Figure 7.3
Transfer RNA folds into a compact L-shaped tertiary structure.

The left drawing shows the code used to depict the different arms of the cloverleaf, the center shows a two-dimensional projection illustrating how two double-helical regions form at right angles, the right shows a schematic drawing in which the backbone can be traced through the base-paired and unpaired regions. This is the structure of yeast tRNA^{Phe}, but others are similar.

Figure 7.4
A space-filling model shows that tRNA tertiary structure is compact.

This model of yeast tRNA^{Phe} is shown from two views separated by 90°. The view on the right corresponds with the two-dimensional projection in the preceding figure.
Photograph kindly provided by S. H. Kim.

The base-paired double-helical stems of the secondary structure are maintained in the tertiary structure, but their arrangement in three dimensions essentially creates two double helices at right angles to each other, as illustrated in **Figure 7.3.** The acceptor stem and the TψC stem form one continuous double helix with a single gap; the D stem and anticodon stem form another continuous double helix, also with a gap. The region between the double helices, where the turn is made, contains the TψC loop and the D loop. Thus the amino acid resides at the extremity of one arm of the L, and the anticodon loop forms the other end.

This model was worked out in detail for tRNA^Phe of yeast, the first tRNA to be crystallized, a molecular model of which is shown in **Figure 7.4.** In another case, tRNA^Asp, the angle between the two axes is slightly greater, so the molecule has a similar, but slightly more open, conformation. Detailed differences affecting one region or another are found in other tRNAs. This accommodates the dilemma that all tRNAs must have a similar shape, yet it must be possible to recognize differences between them.

In addition to the base pairing that is represented in the secondary structure, further H bonds form in the tertiary structure. The bonds of the cloverleaf are described as **secondary H bonds;** the additional bonds of the tertiary structure are called **tertiary H bonds.** Many of the conserved and semiconserved bases are involved in the formation of tertiary H bonds, which both explains their conservation and suggests that the general form of the tertiary structure may be common to all tRNAs. The pattern of additional associations for tRNA^Phe is summarized in **Figure 7.5,** which shows that some of them involve unusual bases or arrange-

Figure 7.5
Most tertiary interactions in tRNA structure involve conserved or semiconserved bases.

The molecule is drawn as a cloverleaf (left) or L-shape (right), with tertiary H bonds indicated by the lighter lines. The heavy lines indicate the covalent bonds joining the two arms of the L-shape. Colored circles indicate conserved or semiconserved bases.

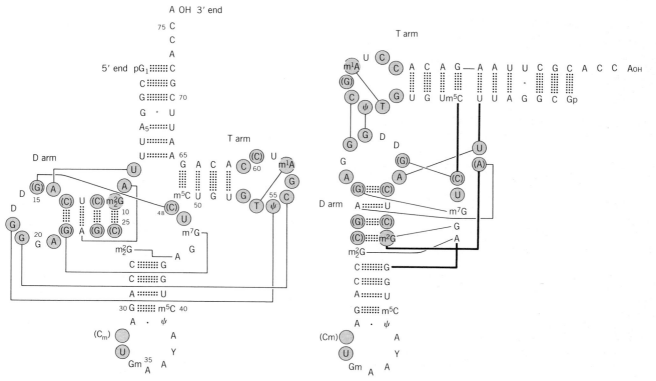

ments for triple pairing. The greatest concentration of these tertiary interactions occurs between the D loop and the TψC loop. Not every one of these interactions is necessarily universal—for example, some are absent from tRNA^{Asp}—but probably this is always the *general* pattern for establishing tRNA structure.

The structure that is visualized by X-ray crystallography represents a stable organization of the molecule. In solution, or in association with proteins, some flexibility may be shown. The conformation of tRNA can change in both circumstances, but we do not know how these changes are related to the crystal structure.

The structure suggests some general conclusions about the function of tRNA. First, its sites for exercising particular functions are maximally separated. The amino acid is as far distant from the anticodon as possible, which is consistent with the need for the aminoacyl group to be located near the peptidyl transferase site on the large subunit of the ribosome, while the anticodon pairs with mRNA on the small subunit (see Chapter 8). The TψC sequence lies at the junction of the arms of the L, possibly a critical location for controlling changes in tertiary structure. (There have been proposals that such changes might occur during protein synthesis; see Chapter 8.) The structure may therefore accommodate the various demands on tRNA by providing several different sites, analogous to the active sites of a protein.

SYNTHETASES SORT THROUGH tRNAs AND AMINO ACIDS

Amino acids enter the protein synthesis pathway through the aminoacyl-tRNA synthetases, which provide the interface for connection with nucleic acid. The synthetases are intriguing enzymes that are able to sort the tRNAs and amino acids into corresponding sets. In bacteria, there is one aminoacyl-tRNA synthetase for each amino acid; and this single enzyme recognizes all the (isoaccepting) tRNAs corresponding to the amino acid. Probably the same situation prevails generally in eucaryotic cytoplasm, although there may be some exceptions where there is more than one enzyme able to charge tRNAs representing a particular amino acid. Organelles possess their own sets of enzymes.

The function of the aminoacyl-tRNA synthetase has been conserved. It is to bind three substrates: tRNA, amino acid, and ATP. The energy from ATP cleavage is used to place the amino acid on the tRNA via an AMP-amino acid intermediate. To recapitulate the information of Figure 5.6, the same two-step reaction appears to be the basis for action of all the synthetases.

But the structures of the enzymes vary extensively. At present, four classes are recognized, described as α, α_2, α_4, and $\alpha_2\beta_2$, where α and β each represent a unique polypeptide chain. So the synthetase enzymes may be proteins that are monomers (α) of from 50,000 to 120,000 daltons, or multimers of either one or two polypeptide chains with a total molecular weight up to 300,000 daltons. From the point of view of the enzyme, the same function can therefore be exercised in several different ways.

How each synthetase recognizes the set of tRNAs representing its amino acid is a critical question in the area of protein-nucleic acid interactions. The multiple tRNAs that correspond to a particular amino acid may be related in two ways. First, there may be tRNAs that possess the *same anticodon,* and therefore respond to exactly the same codon(s), but that differ elsewhere in the molecule. Second, there may be tRNAs with *different anticodons,* needed to respond to the various synonym codons for the amino acid; usually these will also have other differences in structure. The tRNAs properly recognized by a synthetase are described as its **cognate** tRNAs.

A large number of attempts to deduce similarities in sequence between the isoaccepting tRNAs, or to induce chemical alterations that affect their charging, all have led to the realization that the basis for recognition does not lie in some constant feature of primary or secondary structure alone. However, three general areas of the tRNA molecule have been consistently implicated in contacting the synthetase. The acceptor stem, of course, is involved by virtue of its role in being charged on the 3' terminus with amino acid. The D stem is implicated by cross-linking reactions, in which enzymes are cross-linked photochemically to their tRNAs; this area is always one of the regions of contact. Often the anticodon stem also is linked to the enzyme. This suggests the general model

Figure 7.6
Aminoacyl-tRNA synthetase enzymes may contact tRNA along the inside of the L-structure, especially at the acceptor stem, D stem, and anticodon arm.

for tRNA•synthetase binding shown in **Figure 7.6,** in which the protein binds the tRNA along the "inside" of the L-shaped structure.

The recognition reaction is extraordinarily precise. Mutations that alter only single bases may change the recognition of a tRNA, preventing it from being recognized by its synthetase, and/or allowing recognition by another synthetase. These mutations are clustered in two regions of the molecule: the acceptor stem and the anticodon arm. So the differences in features that provide the basis for discrimination may be excruciatingly fine, with single points of contact distinguishing one tRNA from another. (Of course, not all mutations have this effect; for example, the "suppressor" mutations discussed later in this chapter change a base in the anticodon, and therefore the codons to which a tRNA responds, without altering its charging with amino acid.) The reverse mutational experiment can also be performed: substitutions in the synthetase enzyme may alter its discrimination of tRNAs. These results suggest the general view that recognition depends on the interaction between a few amino acids constituting an active site in the protein and a few

points of contact in the tRNA, concentrated at the extremities and in the D stem.

THE CHARGING STEP

The nature of discriminatory events is a general issue that is raised by several steps in gene expression. How do synthetases recognize just the corresponding tRNAs and amino acids? How does a ribosome recognize only the tRNA corresponding to the codon in the A site? How do DNA or RNA polymerases recognize only the base complementary to the template? In each case, there is a similar problem: to distinguish one particular member from the entire set, all of which share the same general features.

Probably there is a random-hit process in which any member initially can contact the active center, but then the wrong members are rejected and only the appropriate one is accepted. The appropriate member is always in a minority (1 of 20 amino acids, 1 of ~40 tRNAs, 1 of 4 bases), so the criteria for discrimination must be strict. We can distinguish two general ways in which the decision whether to reject or accept might be taken.

First, the cycle of admittance, scrutiny, rejection/acceptance could represent a single binding step that precedes all other stages of whatever reaction is involved. This is tantamount to saying that the affinity of the binding site is controlled in such a way that only the appropriate species is comfortable there. In the case of synthetases, this would mean that only the cognate tRNAs could form a stable attachment at the site.

Alternatively, the reaction may proceed through some of its stages, after which a decision is reached on whether the correct species is present. If it is present, the reaction continues to completion. If some other species is present, the reaction is reversed, or a bypass route is taken, and the wrong member is expelled. This sort of postbinding scrutiny is generally described as **proofreading.** In the example of synthetases, it would constitute allowing the charging reaction to proceed through certain stages even if the wrong tRNA or amino acid were present.

The intrinsic abilities of synthetases to sort through

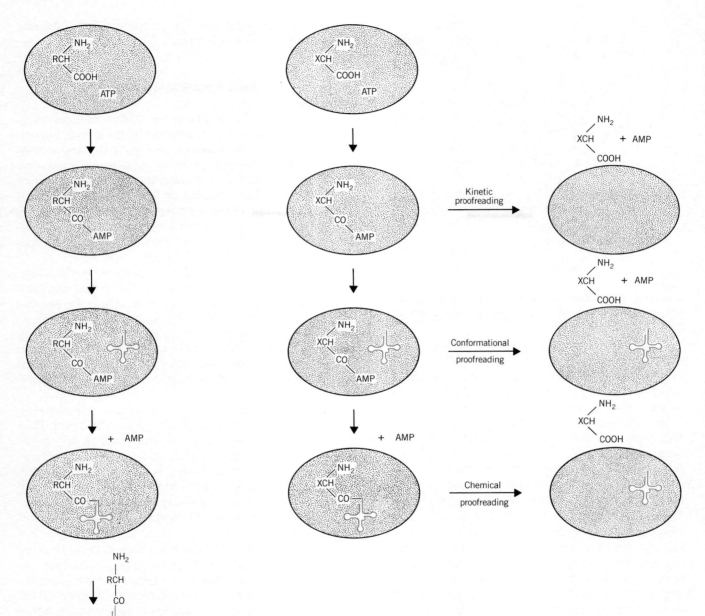

Figure 7.7

Discrimination of amino acids could occur at several stages in synthetase action.

The correct amino acid (indicated by R) may be bound in preference to most (perhaps all) other amino acids (indicated by X). If an incorrect amino acid is bound, it may be proofread kinetically, by a conformational change induced by tRNA, or chemically via an abortive transfer to tRNA.

tRNAs and amino acids show some variation, and different enzymes rely more or less heavily on discrimination at particular stages. Thus synthetases differ in the extent to which they distinguish initially between related amino acids. Some show highly specific binding for a single amino acid; but others can activate any one of a number of amino acids closely related to the proper substrate. But although the analog amino acid can be converted to the adenylate form, in none of these cases is a misactivated amino acid actually used to form a stable aminoacyl-tRNA.

At what stage is the false aminoacyl-adenylate rejected? **Figure 7.7** summarizes the options for a hypothetical enzyme in rejecting an amino acid. One possibility is that the structure of the aminoacyl-adenylate is subjected to closer scrutiny than was the structure of the amino acid itself. Thus the wrong aminoacyl-adenylate would be hydrolyzed. This could happen as an intrinsic process of closer examination (**kinetic proofreading**); or because tRNA binding induces a conformational change in the enzyme, establishing more exacting requirements for the proper aminoacyl-adenylate (**conformational proofreading**). However, present opinion favors the idea that selection occurs at a subsequent step by **chemical proofreading.** In this process, the wrong amino acid is actually transferred to tRNA, is then recognized as incorrect by its structure in the tRNA binding site, and so is hydrolyzed and released. This requires a continual cycle of linkage and hydrolysis until the correct amino acid is transferred to the tRNA.

The classic example in which discrimination between amino acids depends on the presence of tRNA is provided by the Ile-tRNA synthetase of *E. coli.* The enzyme can charge valine with AMP, but hydrolyzes the valyl-adenylate when tRNA^{Ile} is added. The overall error rate in aminoacyl-tRNA charging depends on the specificities of the individual steps. In this example, the erroneous activation of valine occurs with a frequency of about 1/225. Chemical proofreading allows only about 1/270 of the Val-tRNA^{Ile} molecules to be released. The overall error rate is the product, 1/60,000 = 1.5×10^{-5}. This compares with a measured rate of substitution of valine for isoleucine (in rabbit globin) of $2-5 \times 10^{-4}$, which suggests that mischarging provides only a small fraction of the errors that actually occur.

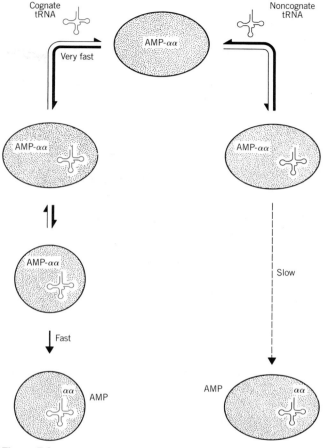

Figure 7.8

Recognition of the correct tRNA by synthetase is controlled at two steps.

First, the enzyme has a greater affinity for its cognate tRNA, whose binding can be stabilized by a conformational transition in the enzyme. Second, the aminoacylation of the incorrect tRNA is very slow (V_{max} reduced by 10^4 compared with the cognate tRNA), which further increases the probability that the tRNA will dissociate.

The binding of tRNA to the synthetase involves the steps depicted in **Figure 7.8.** First, tRNA is bound rapidly; then, only if the correct tRNA is present, binding is stabilized by a conformational change in the enzyme. This is followed by aminoacylation, but when the wrong tRNA is present, the reaction proceeds much more slowly; this increases the chance that the tRNA will dissociate from the enzyme before it is charged. Possibly the critical element in this slowdown is the location of the 3′ terminus of the tRNA, which may lie in the correct site for aminoacylation only when all the proper contacts are made elsewhere.

So there may be a dual system for discrimination, in which first the accomplishment of stable binding depends on making proper contacts with the tRNA; and then the catalytic reaction itself is influenced by the tRNA present.

CODON-ANTICODON RECOGNITION INVOLVES WOBBLING

The function of tRNA in protein synthesis is fulfilled when it recognizes the codon in the ribosome A site. The interaction between anticodon and codon takes place by base pairing, but under rules that vary from the usual restriction to only G-C and A-T partnerships.

Experimental data on the forces involved in codon-anticodon recognition are indecisive; there is even some doubt about the extent to which G-C pairing stabilizes the recognition reaction compared with A-U pairing. Also, interactions measured in solution may not be entirely appropriate, since the structure of tRNA and the environment of the A site play their role in establishing the conditions for the association of tRNA with mRNA.

We must therefore deduce the general rules governing the interaction from information about the sequences of anticodons corresponding to particular codons. The ability of any tRNA to respond to a given codon can be measured by the trinucleotide binding assay or by its use in an *in vitro* protein synthetic system (the procedures used originally to define the genetic code, as discussed in Chapter 4).

The genetic code itself yields some important clues about the process of codon recognition. The pattern of third-base degeneracy is drawn in **Figure 7.9,** which shows that in almost all cases either the third base is irrelevant or a distinction is made only between purines and pyrimidines. There are eight **codon families** in which all four codons sharing the same first two bases have the same meaning, so that the third base has no role at all in specifying the amino acid. There are seven **codon pairs** in which the meaning is the

Figure 7.9
Third bases have the least influence on codon meanings.
The code is drawn with boxes to indicate groups of codons within which third-base degeneracy ensures that the meaning is the same. These can be divided into four classes.

Third base irrelevant

- any of U, C, A, G codes same amino acids: 32 such codons in 8 families.

Only purines and pyrimidines distinguished

- Either U or C code same amino acid: 14 such codons form 7 pairs.
- Either A or G code same amino acid: 12 such codons form 6 pairs (including UAA/UAG terminators).

Three out of four

- any of U, C, A have same meaning: only 3 Ile codons like this
  U
  (AUC).
  A

Unique definitions

- only G can be present: only 2 codons like this (AUG = Met, UGG = Trp).
- only A can be present: this is the UGA codon (not recognized by tRNA).

same whichever pyrimidine is present, and there are five codon pairs in which either purine may be present without changing the amino acid that is coded (another pair of this type has two termination codons).

It may be more significant to look at the code from

the reverse perspective. There are only three cases in which a unique meaning is conferred by the presence of a particular base at the third position: AUG (for methionine), UGG (for tryptophan), and UGA (nonsense). This means that C and U never have a unique meaning in the third position, and A never signifies a unique amino acid.

Because the anticodon is complementary to the codon, it is the *first* base in the anticodon sequence written conventionally in the direction from 5' to 3' that pairs with the *third* base in the codon sequence written by the same convention. Thus the combination

Codon	5' A C G	3'
Anticodon	3' U G C	5'

is usually written as codon ACG/anticodon CGU, where the anticodon sequence must be read backward for complementarity with the codon. To avoid confusion, we shall use the usual convention in which all sequences are written 5' → 3', but we shall indicate anticodon sequences with a backward arrow as a reminder of the relationship with the codon. Thus the codon/anticodon pair shown will be written as ACG and C̆GU, respectively.

Does each triplet codon demand its own tRNA with a complementary anticodon? Or can single tRNAs respond to both members of a codon pair and to all (or at least some) of the four members of a codon family? The pattern of codon recognition by isolated tRNAs from many sources shows that often one tRNA can recognize more than one codon. This means that the base in the first position of the anticodon must be able to partner alternative bases in the corresponding third position of the codon. In other words, base pairing at this position cannot be limited to the usual G-C and A-U partnerships.

The rules governing the recognition patterns are summarized in the **wobble hypothesis,** which states that the pairing between codon and anticodon at the first two codon positions always follows the usual rules, but that exceptional "wobbles" may occur at the third position. This happens because the conformation of the tRNA anticodon loop permits unusual flexibility at the first base of the anticodon. Some permitted wobble pairings are summarized in **Figure 7.10.**

The rules for recognition of the third codon base admit pairing between G and U in addition to the usual pairs. This single change creates a pattern of base pairing in which A can no longer have a unique meaning in the codon (because the U that recognizes it also must recognize G). Similarly, C also no longer has a unique meaning (because the G that recognizes it also must recognize U).

To summarize the recognition pattern,

U in the anticodon recognizes either A or G in the codon.

C in the anticodon recognizes only G in the codon.

A in the anticodon recognizes only U in the codon.

G in the anticodon recognizes either C or U in the codon.

It therefore is possible to recognize unique codons only when the third bases are G or U; UGG and AUG are examples of the first type, although there is none of the second type.

BASE MODIFICATION MAY CONTROL CODON RECOGNITION

Because the bases in tRNA are not restricted to the four standard purines and pyrimidines, further pairing patterns are possible in addition to those predicted by the regular and wobble pairing involving A, C, U, and G. Actually, some of the regular combinations do not occur, because some bases are always modified; in particular, U and A are not employed at the first position of the anticodon.

Usually, U at this position is converted to a modified form that may have altered pairing properties. In regular tRNA sets in bacteria and eucaryotic cytoplasm, only a single exception has been found to this rule. This is a tRNA with unusual pairing properties, able to use the anticodon ŪAG to recognize all *six* leucine codons!

There seems to be an absolute ban on the employment of A, which has never been found; usually it is converted to I. Inosine (I) often is present at the first position of the anticodon, where it is able to pair with

Figure 7.10

Wobble in base pairing allows some bases at the first position of the anticodon to recognize more than one base in the third position of the codon.

Some wobble pairings of both the usual and modified bases are compared with the usual G-C, A-U pairing. G can pair with U, but modification to 2-thiouridine restricts pairing of the modified U to only A, since only one H bond could form with G. Inosine can pair with U, C, and A.

three bases, U, C, and A. Note that this ability is especially important in the isoleucine codons, where AUA codes for isoleucine while AUG codes for methionine. Remembering that with the usual bases it is not possible to recognize A alone in the third position, any tRNA with U in its anticodon would have to recognize AUG as well as AUA. So AUA must be read together with AUU and AUC, a problem that is solved by the existence of tRNA with I in the anticodon.

Two other modified nucleosides that can pair with three bases in the third codon position are uridine-5-oxyacetic acid and 5-methoxyuridine. These recognize A and G efficiently, and recognize U less efficiently.

Another case in which multiple pairings can occur, but with some preferred to others, is provided by the series of queuosine and its derivatives. This modified G base continues to recognize both C and U, but pairs with U more readily.

A restriction not allowed by the usual rules can be achieved by the employment of 2-thiouridine in the anticodon. This modification allows the base to continue to pair with A, but prevents it from indulging in wobble pairing with G.

All of these pairing relationships make the general point that there is more than one way to construct a set of tRNAs able to recognize all the 61 codons representing amino acids. No particular pattern predominates in any given organism, although the absence of a certain pathway for modification can prevent the use of some recognition patterns. Thus a particular codon family may be read by tRNAs with different anticodons in different cells. Often the tRNAs will have overlapping responses, so that a particular codon may be read by more than one tRNA. In such cases there may be differences in the efficiencies of the alternative recognitions. And in addition to the construction of a set of tRNAs able to recognize all the codons, there will often be multiple tRNAs that respond to the same codons.

The predictions of wobble base pairing accord very well with the observed abilities of almost all tRNAs. But there are exceptions in which a tRNA pairs with an additional codon not predicted by the wobble rules, or fails to pair with one that is predicted. Probably such effects result from the influence of neighboring bases and/or the conformation of the anticodon loop in the overall tertiary structure of the tRNA. Indeed, the importance of the structure of the anticodon loop is inherent in the idea of the wobble hypothesis itself. Further support for the influence of the surrounding structure is provided by the isolation of occasional mutants in which a change in a base in some other region of the molecule alters the ability of the anticodon to recognize codons (see later).

Another situation in which there is an unexpected pairing reaction is the ability of the bacterial initiator, fmet-tRNA$_f$, to recognize both AUG and GUG. This involves the third base of the anticodon in misbehavior.

MITOCHONDRIA HAVE MINIMAL tRNA SETS

According to the wobble hypothesis, a minimum of 31 tRNAs (not including the initiator) are required to recognize all 61 codons (because at least 2 tRNAs are required for each codon family and 1 tRNA is needed per codon pair or single codon). But an unusual situation exists in at least some mitochondria (of both fungi and mammals) in which there may be as few as 23–24 different tRNAs. How does this limited set of tRNAs accommodate all the codons?

The critical feature lies in a simplification of codon-anticodon pairing, in which one tRNA recognizes all four members of a codon family. This reduces to 23 the minimum number of tRNAs required to respond to all internal codons. In all eight codon families, the sequence of the tRNA contains an unmodified U at the first position of the anticodon.

The remaining codons are grouped into pairs in which all the codons ending in pyrimidines are read by G in the anticodon, and all the codons ending in purines are read by U in the anticodon, as predicted by the wobble hypothesis. The complication of the single codons UGG and AUG is avoided completely in mammalian mitochondria, where the reading pattern changes the genetic code, so that UGA is read with UGG as tryptophan, while AUA ceases to represent isoleucine and instead is read with AUG as methionine (see Chapter 4). This allows all the nonfamily codons to be read as 14 pairs.

The 23 identified tRNA genes therefore code for an initiator, 14 tRNAs representing pairs, and 8 tRNAs representing families. This leaves the two usual nonsense codons UAG and UAA unrecognized by tRNA, together with an arginine codon pair (AG^A_G) that also may be used for termination (another change in the code). Similar rules are followed in the mitochondria of fungi, except that AUA remains an isoleucine codon (necessitating a G in the anticodon, probably modified to read U, C, or A), while a C in the anticodon provides specific reading of AUG for methionine.

Now an immediate problem is presented by the use of U in the anticodons responding both to the entire codon families and to the purine-ending codon pairs. It seems that the difficulty is overcome by modifying the U in the tRNAs that respond to codon pairs only. An unmodified U in the anticodon may be able to wobble with U and C as well as with the A and G shown in Figure 7.10. If this is an inherent ability, it may explain why U is always modified in the first position of anticodons in bacteria and eucaryotic cytoplasm (although we do not know the reason why U is never used to recognize entire codon families there).

MUTANT tRNAs MAY READ DIFFERENT CODONS

The isolation of mutant tRNAs has been one of the most potent tools for analyzing the ability of a tRNA to respond to its codon(s) in mRNA, and for determining the effects that different parts of the tRNA molecule may have on codon-anticodon recognition.

Mutant tRNAs are isolated by virtue of their ability to overcome the effects of mutations in genes coding for proteins. We have already mentioned the terminology in which a mutation that is able to overcome the effects of another is called a **suppressor** (see Chapter 4). In the tRNA suppressor systems, the primary mutation changes a codon in the mRNA so that the protein product is no longer functional; the secondary, suppressor mutation changes the anticodon of a tRNA, so that it recognizes the mutant codon instead of (or as well as) its original target codon. This restores protein function. The suppressors are named

as **nonsense** or **missense**, depending on the nature of the original mutation.

The premature termination of protein synthesis caused by a nonsense mutation can be suppressed by a tRNA that has been mutated so that it recognizes the termination codon. In the normal cell, the termination codon is recognized only by a release factor. The mutation of a tRNA to recognize the termination codon therefore adds a new capacity to the translational system. **Figure 7.11** shows that this restores the ability to insert an amino acid at the mutant site, and thus allows synthesis of a full-length protein. If the amino acid that is inserted by suppression is different from the amino acid that was originally present at this site in the wild-type protein, the activity of the protein may be reduced.

Nonsense suppressors fall into three classes, one for each termination codon. **Table 7.1** describes the properties of some of the best characterized suppressors.

The easiest to characterize have been amber suppressors. In *E. coli*, at least 6 tRNAs have been mu-

Table 7.1

Nonsense Suppressor tRNAs Are Generated by Mutations in the Anticodon

Locus	tRNA	Wild Type		Suppressor	
		Codons Recognized	Anticodon	Anticodon[a]	Codons Recognized
supD (su1)	Ser	UCG	CGA	**C**UA	UAG
supE (su2)	Gln	CAG	CUG	CU**A**	UAG
supF (su3)	Tyr	UA^C_U	GUA	**C**UA	UAG
supC (su4)	Tyr	UA^C_U	GUA	**U**UA	UA^A_G
supG (su5)	Lys	AA^A_G	UUU	UU**A**	UA^A_G
supU (su7)	Trp	UGG	CCA	**U**CA	UG^A_G

[a]Boldface indicates the base in the mutant anticodon that has been changed.

Nomenclature. The loci were originally known as *su1, su2,* etc., as indicated in parentheses, with su^+ indicating the suppressor mutant form of the gene, and su^- indicating the wild-type form. Now the loci have been renamed *sup* and lettered from A to V. All the loci described here are structural genes coding for tRNAs.

tated to recognize UAG codons. All of the amber suppressor tRNAs have the anticodon $\overleftarrow{CUA}$, in each case derived from wild type by a single base change. The site of mutation can be any one of the three bases of

Figure 7.11
Nonsense mutations can be suppressed by a tRNA with a mutant anticodon.

The wild-type gene contains a UUG codon specifying leucine, which is recognized by a tRNA with the anticodon $\overleftarrow{CAA}$. A mutation converts the codon to UAG, which is now recognized by release factor. Mutation of a tyrosine tRNA to change its anticodon from $\overleftarrow{GUA}$ to $\overleftarrow{CUA}$ allows it to recognize the amber codon. This causes synthesis of a full-length protein in which the original Leu residue has been replaced by Tyr.

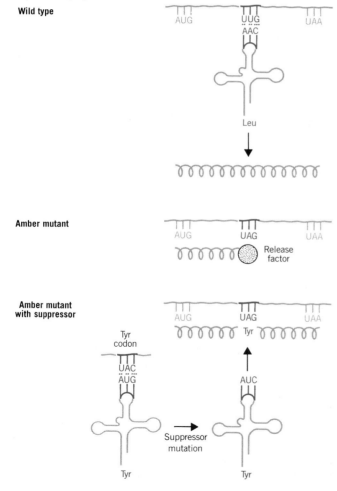

the anticodon, as seen from *supD*, *supE*, and *supF*. Each suppressor tRNA now recognizes only the UAG codon, instead of its former codon(s). The amino acids inserted are serine, glutamine, or tyrosine, the same as those carried by the wild-type tRNA.

Ochre suppressors also arise by mutations in the anticodon. The best known are *supC* and *supG*, which insert tyrosine or lysine in response to both ochre (UAA) and amber (UAG) codons. This conforms exactly with the prediction of the wobble hypothesis that UAA cannot be recognized alone. Indeed, all ochre suppressors act on both ochre and amber codons; and this is nicely demonstrated by the introduction of further mutations in the *supE* or *supF* amber suppressors. If their anticodons are mutated from $\overleftarrow{CUA}$ to $\overleftarrow{UUA}$, the tRNAs become ochre suppressors, acting on both ochre and amber codons, instead of merely on amber codons as before. (We write U as the first anticodon base, but it could be modified.)

The suppressors of UGA codons are interesting. One type *(supU)* is a mutant of tryptophan tRNA. The anticodon of the wild-type tRNA^{Trp} is $\overleftarrow{CCA}$, which recognizes only the UGG codon. The suppressor tRNA has the mutant anticodon $\overleftarrow{UCA}$, which recognizes both its original codon and the termination codon UGA.

Another UGA suppressor has an entirely unexpected property. It is derived from the same tRNA^{Trp}, but its only mutation is the substitution of A in place of G at position 24. This replaces a G-U base pair in the D stem with an A-U pair, increasing the stability of the helix. The sequence of the anticodon remains the same as the wild type, $\overleftarrow{CCA}$. So the D-stem mutation must in some way alter the conformation of the anticodon loop, allowing $\overleftarrow{CCA}$ to pair with UGA in an unusual wobble pairing of C with A. The suppressor tRNA continues to recognize its usual codon, UGG.

Actually, this ability of the tRNA^{Trp} to recognize the UGA termination codon may not be entirely unique to the suppressor form, because UGA mutants are always somewhat "leaky"—there is some level of residual expression even in the absence of a mutant tRNA suppressor. This suggests that perhaps the wild-type tRNA^{Trp} reacts (rather poorly) with UGA as well as UGG; the mutation at position 24 may enhance this effect rather than create it de novo. This implies that, if a coding region ends in a UGA codon, it is likely that termination will not be complete at this site, so that some ribosomes will continue to the next termination codon. The result is to produce two proteins with the same N-terminal region, one of which extends past the C-terminus of the other. Thus partial nonsense suppression can be used as a natural mechanism to produce related proteins.

A related anticodon-codon response is seen with a eucaryotic tRNA. Bovine liver contains a tRNA that is aminoacylated with serine. Its anticodon is ^{m}CCA, which would be predicted by the wobble rules to respond to the tryptophan codon UGG. But in fact it responds only to the termination codon UGA. Thus the ^{m}C pairs specifically with A. This result raises the possibility that the use of UGA as a termination codon is suppressed in this situation.

The general importance of these observations lies in the demonstration that codon-anticodon recognition of either wild-type or mutant tRNA cannot be predicted entirely from the relevant triplet sequences, but may well be influenced by other features of the molecule.

All of these nonsense suppressors are created by changes in the anticodon (or elsewhere) that result from point mutation in the structural gene coding for the tRNA. However, in principle, such suppressor tRNAs also could be caused by changes in the modification of bases that affect the coding response. This could result from mutation of the gene coding for the modifying enzyme.

Missense mutations alter a codon representing one amino acid into a codon representing another amino acid, one that cannot function in the protein in place of the original residue. (Formally any substitution of amino acids constitutes a missense mutation, but in practice these are detected only if they change the activity of the protein.) The mutation can be suppressed by the insertion either of the original amino acid or of some other that is acceptable to the protein. **Figure 7.12** shows that this can be accomplished in the same way as nonsense suppression, by mutating the anticodon of a tRNA carrying an acceptable amino acid so that it responds to the mutant codon. Thus missense suppression involves a substitution in the meaning of the codon from one amino acid to another.

Figure 7.12

Missense suppression occurs by mutation in the anticodon of a tRNA so that it responds to the "wrong" codon.

A mutation of a codon from GGA to AGA causes the wild-type Gly to be replaced by Arg in the mutant protein. If the anticodon of a tRNA^{Gly} is mutated fron UCC to UCU, it inserts Gly in response to the codon AGA that should represent Arg. Note that both the tRNA^{Arg} and the suppressor tRNA^{Gly} will respond to AGA, so that suppression only happens some of the time.

Missense suppressors have been isolated from several tRNAs that insert glycine in response to one of the codons GGN. In one case, a mutation in the tryptophan synthetase gene of *E. coli* changes a glycine codon to an arginine codon. Suppressor mutations in either tRNA$_1^{Gly}$ or tRNA$_2^{Gly}$ (coded by the loci *supT* and *supA,* respectively) change the original anticodon $\overset{\frown}{CCC}$ or $\overset{\frown}{UCC}$ (recognizing GGG and GG$_G^A$) to the anticodon $\overset{\frown}{UCU}$, which now recognizes the arginine codons AG$_G^A$.

An alternative way to provide missense suppression is for a mutation in the tRNA (or synthetase) to change the amino acid with which a tRNA is charged. Then it would respond to the same codon as previously, but the insertion of a different amino acid might cause suppression. This is rare, but some mutants of this type have been isolated from the *supF* tyrosine suppressors. Several further mutations in the acceptor stem of the amber suppressor result in the insertion of glutamine in response to the UAG codon; some ability to insert tyrosine usually remains. The mutations must therefore allow the tRNA to be recognized by the Gln-tRNA synthetase as well as by its cognate Tyr-tRNA synthetase.

All of these suppressors have been characterized in *E. coli,* although some comparable mutants have been obtained with other bacteria (notably *S. typhimurium*), which suggests that the situation is essentially similar there. Much less is known about the occurrence and possible suppression of nonsense and missense mutants in eucaryotes. Ochre and amber suppressors inserting tyrosine, serine, or leucine have been isolated in yeast, where each suppresses only its own target codon. Probably this results from the use of modified bases in the anticodons of the ochre suppressors.

SUPPRESSOR tRNAs COMPETE FOR THEIR CODONS

There is an interesting difference between the usual recognition of a codon by its proper aminoacyl-tRNA and the situation in which mutation allows a suppressor tRNA to recognize a new codon. In the wild-type cell, only one meaning can be attributed to a given codon, which represents either a particular amino acid or a signal for termination. But in a cell carrying a suppressor mutation, the mutant codon may have the alternatives of being recognized by the suppressor tRNA or of being read with its usual meaning.

A nonsense suppressor tRNA must compete with the release factors to recognize its target termination codon(s). A missense suppressor tRNA must compete with the tRNA that usually recognizes its new codon. The extent of competition will influence the efficiency of suppression; so the effectiveness of a particular suppressor may depend not only on the affinity between its anticodon and the target codon, but also on its concentration in the cell, and the comparable parameters governing the competing termination or insertion reactions. This may partially explain why the efficiency of suppression is never much greater than 50%; also, there are wide differences in the efficiencies that can be found among different types of suppressors, so the efficiency may be an intrinsic feature of each class.

A nonsense suppressor is isolated by its ability to respond to a mutant nonsense codon, one that has deleterious effects by virtue of its (mutant) location. But the same triplet sequence constitutes one of the normal termination signals of the cell! The mutant tRNA that suppresses the nonsense mutation must in principle be able to suppress natural termination at the end of any gene that uses this codon. This **read-through** will result in the synthesis of a longer protein, with additional C-terminal material. Thus any extensive suppression of termination is likely to be deleterious to the cell by producing extended proteins whose functions are thereby altered.

Similarly, one gene's missense suppressor is likely to be another's mutator: if a suppressor corrects a mutation by substituting one amino acid for another at the mutant site, it will have the same effect elsewhere. But in other locations it will be introducing a new amino acid instead of the normal wild-type residue. This may inhibit normal protein function. So there is a dilemma for the cell, which must suppress what is a mutant codon at one location, while failing to change too extensively its normal meaning at other locations.

Amber suppressors tend to be quite efficient, working at rates measured from 10 to 50%, depending on the system. If we suppose that the cell could not tolerate such a high level of suppression of termination of natural proteins, this may indicate that amber co-dons are used less frequently than the other nonsense triplets at the ends of genes.

Ochre suppressors are difficult to isolate, and are always much less efficient, with activities usually be-low 10%. All ochre suppressors grow rather poorly, which indicates that suppression of both UAA and UAG is damaging to the cell, probably because the ochre codon is used most frequently as a natural termination signal. The absence of any strong missense suppres-sors similarly is explained by the damaging effects that efficient substitution of amino acids would have on other genes.

The efficiency with which any particular codon is read may be influenced by its location. Thus the effi-ciency of nonsense suppression by a given tRNA can vary quite widely, depending on the neighboring bases. We do not really understand the influence of **context,** but it can change the frequency with which a codon is recognized by a particular tRNA by about an order of magnitude. Thus although amber codons are read more efficiently than ochre codons, the ratio of effi-ciencies may vary between different locations. The base on the 3′ side of a codon seems to have a particularly strong effect.

The mutation that creates a suppressor tRNA can have two consequences. First, of course, it allows the tRNA to recognize a new codon. Second, it may sometimes prevent the tRNA from recognizing the co-dons to which it previously responded. It is significant that all the high-efficiency amber suppressors are de-rived by mutation of one copy of a redundant tRNA set. In other words, the cell has several tRNAs able to respond to the original codon recognized by the wild-type tRNA; this means that the mutation does not abol-ish recognition of the old codons, which continue to be served adequately by the other tRNAs of the set.

In contrast with this situation, another amber sup-pressor tRNA is a **recessive lethal.** This means that it can act as a suppressor when the cell has two copies of the gene, one wild type and one mutant. But if there is only a single copy that is mutant, the cell dies. The reason is that this gene *(supU)* is the only structural gene for tryptophan tRNA in *E. coli.* A mutation changes its anticodon from $\overleftarrow{\text{CCA}}$ to $\overleftarrow{\text{CUA}}$. But this leaves the cell with no tRNA able to respond to UGG, a lethal change in its capacity. (By contrast, this tRNA *can* be mutated

to recognize UGA *and* UGG, as described earlier.) Limitations of this sort may explain the failure to isolate all possible nonsense suppressors—that is, deriva-tives of every tRNA recognizing a codon differing by just one base from a termination codon.

TRANSFER RNA MAY INFLUENCE THE READING FRAME

The reading frame of a messenger usually is invariant. Translation starts at an AUG codon and continues in triplets to a termination codon. Reading takes no cog-nizance of sense: insertion or deletion of a base causes a **frameshift mutation,** in which the reading frame is changed beyond the point of mutation. Ribosomes and tRNAs continue ineluctably in triplets, synthesizing an entirely different series of amino acids.

Frameshift suppression involves restoration of the original reading frame. We have already discussed the way this can be done by compensating base dele-tions and insertions in the gene (see Chapter 4). How-ever, some *extragenic* suppressors also can be iso-lated; these do not act on termination codons, nor do they substitute one amino acid for another.

Several external frameshift suppressors have been isolated in bacteria. One type corrects the reading frame when the mutation is caused by the insertion of an additional base within a stretch of identical resi-dues; for example, the insertion of a G in a run of several contiguous G bases. The frameshift suppres-sor is a tRNA^{Gly} that has an extra base inserted in its anticodon loop, converting the anticodon from the usual triplet sequence $\overleftarrow{\text{CCC}}$ to the quadruplet sequence $\overleftarrow{\text{CCCC}}$. So the tRNA recognizes a 4-base "codon," in some cases like this with the usual wobble at the last (fourth) position.

A related type of frameshift suppression has been found in the yeast mitochondrion, in which mutations that insert or delete a single T residue in a run of five T residues all are leaky. The reason seems to be that the tRNA^{Phe} responding to the UUU codon sometimes allows the ribosome to slip a base in either direction, thus suppressing the mutation. This slippage seems to be a normal, if infrequent, occurrence (less than 5%).

Another situation in which an occasional frameshift

is a normal event is presented by the RNA phage MS2. The mechanism of the frameshift is not known, but its occurrence is revealed because it causes the ribosomes to terminate at an early position in the new reading frame. This permits them to initiate translation of another gene, the lysis gene, whose initiation codon lies a few bases farther along. Indeed, the frameshift-dependent termination event is a prerequisite for the lysis gene initiation event. This makes the important point that the rare occurrence of "misreading" events can be relied on as a necessary step in natural translation.

Some frameshift suppressors can recognize more than one 4-base "codon." For example, a bacterial tRNA^{Lys} suppressor is able to respond to either AAAA or AAAU, instead of to the usual codon AAA. Another suppressor can read any 4-base "codon" with ACC in the first three positions; the next base is irrelevant. In these cases, the alternative bases that are acceptable in the fourth position of the longer "codon" are not related by the usual wobble rules. What probably happens here is that the suppressor tRNA actually recognizes a 3-base codon, but for some other reason—most likely steric hindrance—the adjacent base is blocked. This forces one base to be skipped before the next tRNA can find a codon.

A related phenomenon that may cast some light on the basis of this effect is the ability to cause a shift in phase by manipulating *in vitro* systems that contain only normal components. A vast excess or a deficiency of a particular aminoacyl-tRNA can produce frameshifts. Probably the long delay in responding to the 3-base codon in the A site allows a tRNA that recognizes an overlapping (out-of-phase) codon to gain access.

This suggests a uniform model for control of ribosome translocation. When the ribosome moves, it shifts the tRNA that was in the A site so that now it properly occupies the P site. When a tRNA is bound to a 4-base "codon," or if it extends sterically beyond the usual 3-base codon, the next aminoacyl-tRNA binds to a triplet that is out of phase in the A site. But upon translocation, the ribosome places this species properly into the P site, so that subsequent aminoacyl-tRNAs encounter a triplet codon in the A site in the usual way. Thus the geometry of the tRNA/mRNA interaction is used to count distance for the ribosome.

FURTHER READING

The literature on tRNA has been well served by some review books. In **Altman's** (Ed.) *Transfer RNA* (MIT Press, Cambridge, 1978) the primary and secondary structure are discussed by **Clark** (pp. 14–47) and the tertiary structure by **Kim** (pp. 248–293). Modified bases are discussed by **Nishimura** (pp. 168–195). One viewpoint of aminoacyl-tRNA synthetases is given by **Igloi** and **Cramer** (pp. 294–340); another is to be found in **Schimmel & Soll** (*Ann. Rev. Biochem.* **48**, 601–648, 1979). A two-volume set of reviews and research papers also gives a broad view of tRNA. The first is edited by **Schimmel, Soll, & Abelson,** *Transfer RNA: Structure, Properties and Recognition;* the second is edited by **Soll, Abelson, & Schimmel,** *Transfer RNA: Biological Aspects* (Cold Spring Harbor Laboratory, New York, 1979). The coding properties of tRNA and their change via mutation are dealt with in **Lewin's** *Gene Expression, 1, Bacterial Genomes* (Chapter 5) (Wiley, New York, 1974). Suppression is reviewed by **Korner, Feinstein, & Altman** (pp. 105–135 of Altman's book), and recent developments are briefly covered by **Roth** (*Cell* **24**, 601–602, 1981).

CHAPTER 8
THE RIBOSOME
TRANSLATION FACTORY

Ribosomes were originally obtained in several guises, and they have been variously described as *cellular organelles, microsomal particles, ribonucleoproteins,* or just the *protein-synthesizing apparatus.* The fact is that they constitute a macromolecular assembly with a defined structure that, together with some accessory factors, possesses the range of enzymatic activities required for the various steps in protein synthesis. In short, the ribosome is a small migrating factory, one that travels along the template engaging in rapid cycles of peptide bond synthesis; aminoacyl-tRNAs shoot in and out of the particle at a fearsome rate, depositing amino acids, and elongation factors cyclically associate and dissociate.

Ribosomes constitute a major cellular component. In an actively growing bacterium, there may be roughly 20,000 ribosomes per genome (see Table 5.1). They contain about 10% of the total bacterial protein and account for 80% or so of the total mass of cellular RNA. In eucaryotic cells, their proportion of total protein is less, but their absolute number is greater, and still they contain most of the mass of RNA of the cell. The number of ribosomes (in procaryotes or in eucaryotes) is directly related to the protein-synthesizing activity of the cell.

Bacterial ribosomes are attached to mRNAs that are themselves still connected with the DNA. In eucaryotic cytoplasm, the ribosomes commonly are associated with the cytoskeleton, a fibrous matrix. In some eucaryotic cells, some ribosomes are associated with the membranes of the endoplasmic reticulum, and may be obtained in the form of a "microsomal" fraction. The common feature is that ribosomes engaged in the act of translation are not free in the cell, but are associated, directly or indirectly, with cellular structures.

All ribosomes can be dissociated into two subunits, one roughly twice the size of the other. Each subunit contains a major rRNA component and many different protein molecules, almost all present in only one copy per subunit. The larger subunit may contain a smaller RNA molecule(s) in addition to the major rRNA.

A pointer to the importance of rRNA is that its sequence remains almost invariant within a cell, although it is coded by many genes. This argues that there is selection against changes in sequence. And mutations that affect rRNA can influence ribosome activities.

The role of the major rRNAs has been a subject of debate. A possible direct function in ribosome activity could be an involvement in binding mRNA and tRNA (via complementary base pairing). A major function of the rRNA clearly is structural. Proteins bind to it at particular sites and in a specific order that is required for assembly of the subunit. Indeed, ribosomes are interesting not only for their ability to translate mRNA, but also for the process by which they self-assemble from the constituent RNA and protein molecules.

The ribosomal proteins and the accessory factors together provide the enzymatic activities that are necessary for protein synthesis. Some functions can be identified with individual proteins; other activities require several proteins. The ribosome possesses several active centers, each of which is constructed from a particular group of proteins. In this sense, the ribosome represents a collection of several enzymes, active only in the context of the ribosome structure, whose coordinated activities together accomplish the act of translation.

Only a small proportion of the ribosomal proteins have been identified with particular enzymatic or structural functions involved in protein synthesis. Bacterial mutations exist in most of the ribosomal proteins; this shows that their functions are necessary, and allows their genes to be characterized. Perhaps many of the proteins (and also the rRNA) are concerned principally with establishing the overall structure that brings the various active sites into the right relationship and supports ribosomal movement; they need not necessarily participate directly in the synthetic reactions.

RIBOSOMES ARE COMPACT RIBONUCLEOPROTEIN PARTICLES

There are appreciable variations in the overall size and proportions of RNA and protein in the ribosomes of bacteria, eucaryotic cytoplasm, and organelles. Thus the ribosome does not take a fixed form, but can be constructed in several ways. **Table 8.1** summarizes the components of the bacterial ribosome, as exemplified for *E. coli*.

The structure of every component is known for *E. coli* ribosomes. The ribosomal RNAs have been sequenced, and the amino acid sequences of most of the proteins have been determined. The two subunits have masses of 0.93×10^6 and 1.59×10^6 daltons, respectively. The proportion of RNA by mass is 60% in the small subunit and 70% in the large subunit. The 52 proteins are divided into 21 in the small subunit (named S1-S21), with 31 in the large subunit (named L1-L34). (The numbering goes to 34 because of some early misassignments.)

The ribosomal proteins show no relationship to one another: each is distinct as judged by amino acid sequence or by immunological comparison. There are two exceptions to this rule. Protein S20 is identical with protein L26; it is mostly (80%) associated with the small subunit, but a small amount fractionates with the large subunit, possibly reflecting a location at the junction

Table 8.1
The Components of the *E. coli* Ribosome

	Ribosome	Small Subunit	Large Subunit
Sedimentation Mass, daltons	70S 2,520,000	30S 930,000	50S 1,590,000
RNA 　Major 　Smaller Mass of RNA Proportion of mass	 66%	16S = 1541 bases 560,000 60%	23S = 2904 bases 5S = 120 bases 1,104,000 70%
Proteins Mass of protein Proportion of mass	 34%	21 polypeptides 370,000 40%	31 polypeptides 487,000 30%

of the subunits. Protein L7 differs from L12 only by its possession of an acetyl group on the N-terminus (some other proteins also are acetylated). This protein (known as L7/L12) forms dimers in solution; there are two dimers per ribosome. Protein S6 is present in two copies. All other proteins (including S20/L26) are present in only one copy per ribosome. This means that all ribosomes in a bacterium have exactly the same components: each consists of a complex between the ribosomal RNAs and a specific set of unique proteins.

The ribosomes of higher eucaryotic cytoplasm are larger than those of bacteria, with subunits of 1.4×10^6 and 2.8×10^6 daltons. **Table 8.2** summarizes the parameters of a mammalian example. The total content of both RNA and protein is greater: the RNA molecules are longer, and the number of proteins is greater, about 82 in all. The proportion of RNA is less than in bacteria, 50% and 65% of the mass in the small and large subunits, respectively. (Also, the complexity of the accessory factors involved in initiation and elongation is much greater than in bacteria, as described in Chapter 6). Probably most or all of the proteins are present in stoichiometric amounts, although this has not been established in detail. There is no relationship between the ribosomal proteins of bacteria and mammalian cytoplasm, with the exception of L7/L12, which can exchange with its eucaryotic counterpart.

Ribosomes take varied forms in organelles. In the mitochondria of lower eucaryotes (such as the fungi), their overall size is slightly larger than in E. coli. In plant mitochondria, their size is only a little less than that of the surrounding cytoplasm. In mammalian or amphibian mitochondria, however, they are smaller, with a total size of 60S, and a low proportion by mass of RNA (25–31%). In chloroplasts, the ribosomes are roughly the same size as in bacteria, although with a greater proportion of nucleic acid.

The shapes of bacterial and eucaryotic cytoplasmic ribosomes are generally similar (organelle ribosomes have not been characterized). **Figure 8.1** shows a diagrammatic representation of the structure of the small bacterial subunit, which is best described as an oblate ellipsoid with a somewhat flat shape of roughly $55 \times 220 \times 220$ Å. **Figure 8.2** shows that the large subunit has a more spherical structure, with dimensions of about $150 \times 200 \times 200$ Å.

The complete 70S ribosome has not been characterized directly in as great detail as the subunits, but one model for its construction is illustrated in **Figure 8.3**. This supposes that the two subunits are attached principally by an association between the cleft of the 30S subunit and the central protuberance of the 50S subunit. There could be a space or sort of tunnel between them.

Some electron micrographs of subunits and complete bacterial ribosomes are shown in **Figure 8.4,**

Table 8.2
The Components of the Rat Liver Cytoplasmic Ribosome

	Ribosome	Small Subunit	Large Subunit
Sedimentation	80S	40S	60S
Mass, daltons	4,420,000	1,400,000	2,820,000

RNA
  Major
  Smaller
		18S = 1900 bases	28S = 4700 bases
			5.8S = 160 bases
			5S = 120 bases

Mass of RNA Proportion of mass	 60%	700,000 50%	1,820,000 65%
Proteins Mass of protein Proportion of mass	 40%	33 polypeptides 700,000 50%	49 polypeptides 1,000,000 35%

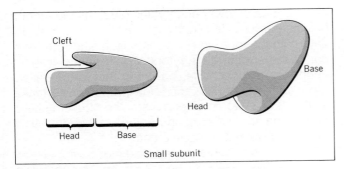

Figure 8.1

The 30S subunit has an elongated and asymmetrical shape, with a constricted region and a "cleft" that divide the "head" from the "base."

The two views display the subunit rotated by 90°.

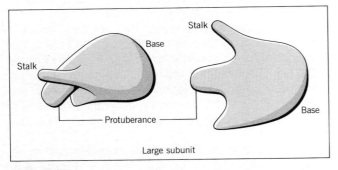

Figure 8.2

The 50S subunit has a fairly compact body, from which both a "central protuberance" and a "stalk" stick out.

The two views are rotated by 90°.

Figure 8.3

70S ribosomes may involve associations between several discrete areas of the two subunits.

The two views are rotated by 90°.

Figure 8.4

Electron microscope images of bacterial ribosomes and subunits reveal their shapes.

The four examples of each structure are shown in the same orientation. S denotes small subunit; L indicates large subunit; B is the body of the large subunit; C is the "crown" of the large subunit; l, m, and r give the orientation from left, middle, and right; h indicates the head of the small subunit; p is the side protrusion.

Photographs kindly provided by Miloslav Boublik.

Figure 8.6
Size comparisons show that the ribosome is large enough to bind two tRNAs (as well as 40 bases of mRNA).

Figure 8.5
Electron micrographs of 80S ribosomes and their subunits show conservation of structure among phylogenetically distant species, and a relationship with bacterial ribosomes.

The left column shows complete ribosomes, the center column has large (60S) subunits, and the right column shows small (40S) subunits. The top row is from brine shrimp (*A. salina*), the second row from slime mold (*D. discoideum*), the third row from rabbit reticulocytes, and the lowest row from HeLa (human) cells.
Photographs kindly provided by Miloslav Boublik.

Much attention has been paid to the location of the mRNA. One popular idea is that it fits between or close to the junction of the subunits, possibly in the "tunnel" between them. Some models depict the mRNA as being "threaded" through the ribosome, but its location is just as likely to be superficial. And, as is evident from **Figure 8.6,** the two tRNAs are quite large relative to the ribosome; they seem as likely to be inserted into large "clefts" opening from the surface as to fit completely into holes in the interior.

THE STRUCTURE OF RIBOSOMAL RNA

The major ribosomal RNAs are synthesized in the form of a single precursor molecule that contains both their sequences. In eucaryotes, the joint precursor is sufficiently stable to be isolated. In bacteria, this is more difficult and requires conditions in which normal maturation is interrupted; usually the two bacterial rRNA sequences are separated by cleavage before transcription has been completed (see Chapter 25). However, the cleavage generates molecules that are longer than the mature rRNAs. For example, the first detectable rRNA for the small bacterial subunit is a species known as p16, which is some 94 nucleotides longer than 16S rRNA. The p16 molecule is incorporated into a complex with ribosomal proteins, and its additional sequences are removed later.

Ribosomal RNA is methylated. In mammalian cells, the 18S and 28S rRNAs carry 43 and 74 methyl groups, respectively, so that roughly 2% of the nucleotides are methylated. About 80% of the methyl groups are car-

together with models in the corresponding orientation.

The apparent structure of mammalian cytoplasmic ribosomes is similar, but with greater dimensions. Some electron micrographs of subunits and complete 80S ribosomes are presented in **Figure 8.5.** Although a detailed model has not yet been constructed, the relationship between the subunits looks similar to that of procaryotic ribosomes. There is a suggestive similarity between the structures of bacterial and eucaryotic small subunits; the widest divergences appear to be presented by the structure of the large subunit.

ried on the 2′ position of ribose; the remainder are modifications of the bases (mostly adenine). In bacterial rRNA, there are 10 methylated nucleotides in 16S rRNA, and probably about 20 in 23S rRNA—a total proportion of about 0.7% of the nucleotides. The methyl groups in 16S rRNA are located mostly toward the 3′ end of the molecule; the methylated nucleotides include dimethyladenine, methyladenine, and methylguanine. Another modification is the conversion of some uridines into pseudouridine.

Some conservation of sequence is seen when the rRNAs of different bacteria are compared; and there

Figure 8.7

The secondary structure of 16S *E. coli* rRNA has many short double-stranded regions, falling into four general domains.

is some homology with organelle rRNAs. The significance of these relationships is not yet clear. There is no extensive conservation of sequence with eucaryotic rRNAs, although among eucaryotes conserved regions can be found, mostly located in the vicinities of the methylated groups. This is true of species as distant as yeasts and mammals.

The sequence of rRNA can be used to predict the formation of base-paired regions. However, in a molecule this large, there are many alternative conformations. It is not possible a priori to predict which would be chosen even if the molecule were effectively free in solution in the bacterial cell. Different models can be distinguished by their predictions on the relative stabilities of particular base-paired regions; but data of this type are somewhat limited in scope.

The most penetrating approach is to compare the sequences of the rRNA in related organisms (as determined directly by the analysis of the DNA of the genes). Those regions that are important in the secondary structure are conserved; so if a base pair is important, one can form at the same relative position in each rRNA. From such comparisons, models have been constructed for both 16S and 23S rRNA.

The current model for *E. coli* 16S rRNA is illustrated in diagrammatic form in **Figure 8.7.** Just under half of the sequence is base paired. The individual double-helical regions tend to be short (less than eight base pairs long). Often the duplex regions are not perfect, but contain bulges of unpaired bases. The molecule as a whole forms four general domains.

Does the structure of free rRNA correspond with its organization in the ribosomal subunit, in which its extensive binding to proteins could influence the availability of regions for base pairing? Methods that have been used to investigate structure in the subunit include determining which bases can react with kethoxal (a reagent that attacks unpaired guanines), which regions become cross-linked to each other on treatment with the reagent psoralen, where preferential sites of cleavage by nucleases are located, and also which regions of the rRNA can be linked to which proteins. So far as they go, the results are reasonably consistent with the model for free rRNA structure, but there seems to be less base-paired structure in the subunit than in the free form. Also, some differences in the reactivity of the 16S rRNA are found when 30S subunits are

compared with 70S ribosomes, which suggests that a conformational change may occur when the subunits join together.

In addition to the two major rRNAs, which form what is sometimes thought of as the backbone of each subunit, the large subunit also contains a molecule of **5S RNA** (in all ribosomes except those of mitochondria). The 5S RNAs of the procaryotes show some conservation of sequence, especially in the regions that bind to ribosomal proteins. Similarly, there is conservation of eucaryotic 5S RNAs, with one sequence (for example) predominating in the mammals. However, there is no relationship between the two groups.

All 5S RNA molecules display a highly base-paired structure, although there has been some difficulty in settling on an exact model (more than 20 have been proposed). This emphasizes the difficulty in distinguishing between alternative pairing possibilities even in quite small molecules, unless there is some further source of information (for example, such as the general conformity with a cloverleaf displayed by tRNA). **Figure 8.8** shows a current model that fits all known procaryotic sequences to a structure with four base-paired regions; its counterpart for eucaryotic 5S RNA is a model with five helical regions. As with tRNA, the secondary structure is unlikely to convey an accurate impression of the form the molecule takes when viewed in three dimensions.

In eucaryotic cytoplasmic ribosomes, another small RNA is present in the large subunit. This is the 5.8S RNA. Its sequence may correspond to the 5′ end of the procaryotic 23S rRNA.

RIBOSOMAL PROTEINS OCCUPY SPECIFIC LOCATIONS

Two approaches have gone hand in hand in analyzing the construction of the ribosomal subunits: the study of assembly (with which we shall deal later); and the direct analysis of structure.

Biochemical studies have concentrated on identifying the locations of individual ribosomal proteins by immune electron microscopy, determining the relationships of adjacent proteins by cross-linking them, and analyzing the interactions between particular ribosomal proteins and rRNA.

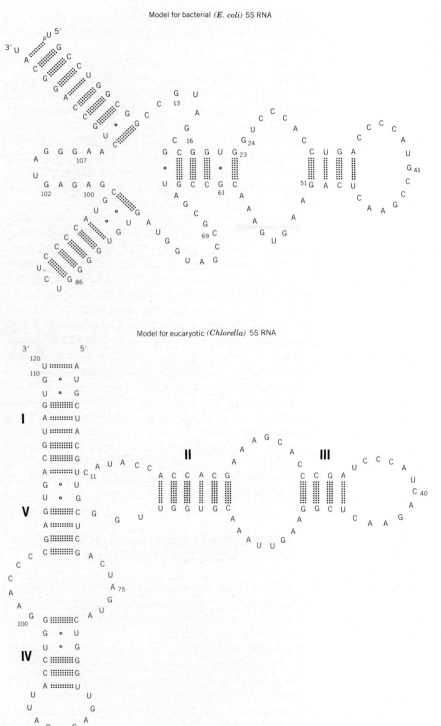

Model for bacterial *(E. coli)* 5S RNA

Model for eucaryotic *(Chlorella)* 5S RNA

Figure 8.8
Procaryotic 5S RNAs fit a 4-helix secondary structure, while eucaryotic 5S RNAs have a 5-helix structure.

Immune electron microscopy is performed by preparing antibodies against particular ribosomal proteins, binding the antibodies to intact subunits, and then visualizing the sites of binding by electron microscopy. The locations of virtually all the proteins of the 30S subunit have been determined in this way, and many of the 50S.

When pairwise relationships are analyzed by cross-linking, the closeness of two proteins can be ascertained by using reagents with varying cross-linking spans. The protocol is to treat intact subunits with the cross-linking reagent, and then to analyze the proteins present in the cross-linked structures (which generally restricts the approach to pairs). Cross-linking of proteins to rRNA allows the binding sites on the nucleic acid to be determined. Other techniques that have yielded comparable results on protein distribution include neutron scattering of subunits in which specific ribosomal proteins are deuterated, and the measurement of energy transfer between fluorescently labeled pairs of ribosomal proteins.

A somewhat similar picture emerges from all of these techniques. Each ribosomal protein can be located at a particular position in the structure. Some particular proteins now can even be equated with particular features of ribosomal appearance. Probably the organization of each subunit is highly defined. A detailed picture is becoming available for the 30S subunit; the results are similar in nature for the 50S subunit, although this is taking longer to analyze because of its greater size.

INTERACTIONS BETWEEN RIBOSOMAL PROTEINS AND rRNA

The ribosomal RNAs constitute the major part of the mass of the 30S and 50S bacterial subunits. Their presence is pervasive, and probably most or all of the ribosomal proteins associate with rRNA. Thus the organization of the rRNA is the central thread in the structure of each subunit. The problem in defining the conformation of the rRNAs and the way in which they bind ribosomal proteins has been that the ability of the proteins to bind to the rRNAs depends on the circumstances of the reaction.

Figure 8.9
Binding sites for ribosomal proteins are clustered in the 5′ half of 16S rRNA.

The first studies of the ability of individual ribosomal proteins to bind to isolated rRNA showed that some proteins bind strongly. These include the proteins that are assembled first into the particle *in vitro,* as we shall see later. The positions on the rRNA at which these proteins bind can be defined by characterizing the parts of the nucleic acid that are protected against degradation by nucleases when the protein is present. This gives a linear map of sites on the rRNA; for 16S rRNA, the attachment sites are grouped in the 5′ half of the molecule, as shown in **Figure 8.9.**

Other proteins did not appear to bind directly to the rRNA, but this result depends on the way that the rRNA itself is isolated. Using different methods of preparation, it was later shown that several other proteins also can bind quite strongly. This suggests that the conformation of rRNA (which is influenced by the method of isolation) is an important feature in determining whether binding sites exist for some proteins. As each set of proteins binds, it may induce changes in the conformation of rRNA that influence the binding of subsequent proteins (that previously were unable to bind). A common feature of the binding sites identified so far on rRNA is the presence of secondary structure, often hairpins whose duplex stems contain unpaired bulges. With the 30S subunit of *E. coli,* it now seems that almost or perhaps even all of the ribosomal proteins can bind (albeit with varying affinities) to the 16S rRNA. In the 50S subunit, similar, although at present less detailed, results have been obtained.

RIBOSOMAL SUBUNITS CAN BE DISSOCIATED AND RECONSTITUTED

When bacterial ribosomal subunits are centrifuged in CsCl, a discrete group of proteins (the **split** proteins) is lost from each subunit. This generates 23S and 42S particle **cores** from the 30S and 50S subunits, respectively. In fact, when ribosomes are exposed to increasing concentrations of either CsCl or LiCl, groups of proteins are lost in successive disruptions. This suggests that the ribosome has groups of cooperatively organized proteins, so that disruption of a group effectively causes loss of all its members together.

The stepwise dissociation can be reversed by removing the CsCl and providing Mg^{2+}. The reconstituted subunits are active in protein synthesis. Detailed reconstitution procedures were developed first for the 30S subunit of *E. coli* and now have been extended to the 50S subunit. They involve a two-stage reaction *in vitro,* which includes a temperature-dependent transition. This is illustrated in **Figure 8.10.**

The first stage in 30S subunit assembly is the association of the 16S rRNA with a group of about 15 proteins. This occurs in the cold to form **RI particles.**

Before the remaining 6 proteins can join, these particles must be heated; this allows a unimolecular rearrangement to take place, generating what are known as **RI* particles.**

The nature of this rearrangement is indicated by the effects of EDTA on 30S subunits: this chelating agent removes magnesium ions and causes the subunit to unfold and lose some proteins. The resulting particles are similar to the RI particles in being unable to regain the missing proteins to form active subunits until after they have been heated. So the limiting step in assembly appears to be a temperature-dependent conversion in which the assembling particle folds into a more compact structure.

MUTANTS IN RIBOSOME ASSEMBLY

Because the reconstitution of ribosomal subunits *in vitro* requires incubation at high temperature, it seemed that mutations blocking the assembly process might be detected as **cold-sensitive** mutants. These should be recovered as bacteria in which protein synthesis is normal at the usual growth temperature, but is prevented at low temperature because ribosomes cannot be assembled. Using this rationale, Nomura and his colleagues isolated several types of *sad* (subunit assembly defective) mutants. Also, some mutants isolated originally for other properties (such as resistance to antibiotics that act on the ribosome) proved to have similar properties.

When the temperature of these mutant cells is reduced, production of the affected class of subunit ceases. Some of the mutations constitute alterations in individual ribosomal proteins; others may lie in genes whose products directly or indirectly affect assembly.

The *sad* mutants accumulate smaller particles that are precursors in the assembly of the affected ribosome subunit. The precursor particles contain the rRNA associated with some, but not all, of the subunit proteins. In each case, the deficiency of a particular protein blocks assembly at a specific stage. The major precursors to each subunit have been identified *in vivo.*

For the 50S subunit there are 43S and 32S particles, each containing 23S rRNA and some of the 50S proteins. The 32S particle is the first identified stage of

Figure 8.10
30S subunits can be dissociated into subparticles that are related to the intermediate in assembly.

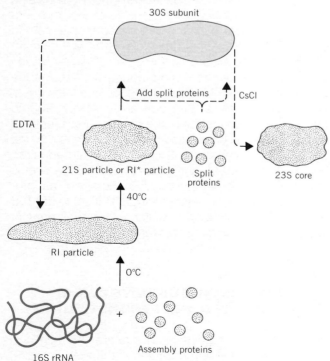

assembly. The 43S particle represents a later stage; it has all the proteins present in the 32S particle plus some additional species.

For the 30S subunit, there is a precursor sedimenting at 26S or 21S, depending on the conditions. This has also been identified as a minor (and transient) component of wild-type bacteria.

SUBUNIT ASSEMBLY IS LINKED TO TOPOLOGY

The small subunit precursor found *in vivo* and the RI particles reconstituted *in vitro* have very similar (or identical) protein components, as implied in Figure 8.10. The RNAs are slightly different, since the *in vivo* precursor has the longer p16, while the *in vitro* particle has mature 16S rRNA; it is interesting (and fortunate) that mature rRNA can be used for assembly.

Figure 8.11 shows that the proteins present in the precursors and RI particles also are very similar to the proteins that are the last to be removed by treatment with LiCl or CsCl. This suggests that the topology of the 30S subunit reflects its assembly process. Those proteins that are part of the precursor (which include all the species binding most strongly to the rRNA) tend to be the more secure components of the 30S subunit, and are harder to remove by abusive treatment.

The roles of particular components in ribosome subunit assembly can be analyzed by attempting to reconstitute subunits in the absence of individual proteins. When some proteins are left out of the assembly mixture, no reconstitution occurs. These are therefore essential for assembly. Other proteins *can* be omitted. The subunits that assemble may appear biophysically normal, but when tested in protein synthesis they prove to be deficient in some function.

Partial reconstitutions are used to examine whether the omission of one protein affects the ability of other proteins to assemble into the particle. Essentially the protocol is to leave out one protein and then to see which other proteins are unable to bind to the assembling particle as the result of the omission. This can be done irrespective of whether the partial assembly that occurs generates a functional particle or not.

The *in vitro* assembly map of **Figure 8.12** was constructed by determining which proteins must already be bound to 16S rRNA for each particular protein to bind. The results show that ribosomal proteins must bind to the assembling particle in a certain order. Some of the proteins that show dependent relationships in the assembly map in fact lie physically close to one another. This is consistent with a scheme in which dependence mostly involves physical interactions; clearly this is not the whole story, however, since (for example) the binding of one protein may open a site

Figure 8.11

Related sets of ribosomal proteins are present in precursor particles assembled *in vivo* or *in vitro* and in core particles.

Proteins that are usually present are indicated by shaded circles, those that are sometimes present by open circles.

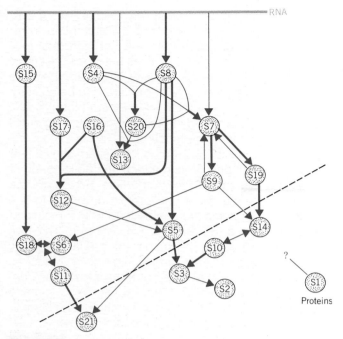

Figure 8.12
Ribosome assembly involves a series of sequential reactions.

Arrows between proteins indicate that the first assists the binding to the particle of the second (direction of effect is indicated by the arrowhead; double arrows indicate mutual effects). Major effects are indicated by heavy lines and lesser effects by thinner lines. When two proteins act independently on another, separate arrows are shown; when the effect is cooperative, there is a joint arrow. Arrows from the rRNA indicate proteins able to bind most readily to rRNA, but other proteins also bind.

The proteins closest to the rRNA are those found most persistently in precursor particles; those farther away are assembled into the particle later and are removed more easily by salt. The dashed line separates those proteins often found in precursor particles from those never found in the particles. This last set is equivalent to the split proteins.

elsewhere on rRNA for another to bind. All the proteins that are defined as essential for assembly are located in the early part of the assembly map.

The functional roles of some proteins can be studied by leaving them out of the reconstitution mixture. Obviously, this can be done only for those proteins located later in the assembly pathway, and not for those whose inclusion is necessary for other proteins to be added. Some 30S subunit proteins have been shown

in this way to be involved in particular stages of protein synthesis, because the reconstituted particle is specifically deficient in some activity. In the 50S subunit, one protein, L24, which binds strongly to several sites at the 5′ end of the rRNA, and is one of the early assembly proteins needed to form the first precursor particle, can be removed from the subunit by treatment with LiCl. But the subunits lacking L24 remain perfectly active in protein synthesis. This implies that L24 may be needed only for assembly, and does not play any subsequent role in ribosome function.

Assembly of the 30S subunit probably starts while the rRNA is being transcribed. In this context, it may be significant that most of the strongest rRNA binding sites are clustered in the 5′ region of the molecule, which is transcribed first. Probably by the time the rRNA is released by cleavage from its partner in the joint precursor, it already has several ribosomal proteins attached, as illustrated in **Figure 8.13.** Proteins continue to attach in the order indicated by the assembly pathway.

The only discrete intermediate that can be found is the single particle containing most of the proteins. We have mentioned that this particle contains an immature form of the 16S RNA, which has some extra nucleotides. Also, it is only slightly methylated at this stage. Probably the conformational change is associated with cleavage of the additional sequences and with methylation of the rRNA; then the remaining proteins are added. Methylation cannot be achieved *in vitro* with free 16S rRNA, but core particles are substrates for methylase activity, which is consistent with this scheme.

ALL RIBOSOMAL COMPONENTS CAN BE MUTATED

Ribosomes play a central role in cell life, and obtaining bacterial mutants for all their components has taken a long time, although now almost every ribosomal protein has been subject to mutation, and its gene mapped on the chromosome. Some mutations have been isolated by virtue of their (relatively) minor effects on protein synthesis; but others have been isolated only by resorting to the use of conditional lethals (as defined in Chapter 2).

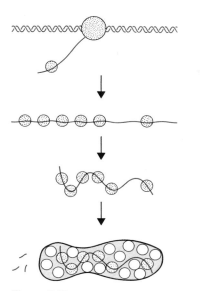

Ribosomal proteins start to associate with 16S rRNA while it is being transcribed.

The rRNA is released as a longer precursor, p16 and about half the 30S subunit protein bind to the rRNA, causing it to change conformation; in the 21S particle, the rRNA has about 10-20% of the methyl groups.

The remaining proteins bind, surplus sequences of the rRNA are cleaved off, and the remaining methyl groups are added

Figure 8.13
Ribosome subunit assembly is an ordered process in which changes in conformation occur as each protein is added.

Some antibiotics act by preventing bacterial protein synthesis. Mutants can be obtained that are resistant to their action. Often the mutations alter ribosomal proteins (or enzymes that act on ribosomal components). Five 30S proteins and four 50S proteins have been mutated to confer antibiotic resistance. These variants may be useful in investigating the role of the protein in the ribosome.

Direct attempts to isolate mutants in protein synthesis arise from changes in the protein-synthetic apparatus that alter the accuracy of translation. When a mutation has occurred in some protein, it may be suppressed by a tRNA mutation in the way that we have described in Chapter 7; another type of suppression is conferred by mutation in the ribosome. Thus several ribosomal mutants have been obtained in the form of second-site revertants that overcome defects caused by original mutations elsewhere. Sometimes a known step in translation is affected, so the role of the particular ribosomal protein can be determined. About six proteins have been mutated in this way.

We have already seen that mutants can be obtained in ribosome assembly by virtue of their cold-sensitivity. This phenotype also is shown by some antibiotic-resistant mutants. So far, four of the 30S proteins and one 50S protein have been identified in assembly mutants. This type of mutation is a conditional lethal, since lack of ribosomes kills the cell.

Some ribosomal proteins have been identified simply by virtue of their alteration in mutants that are temperature-sensitive. This means that the bacteria show reduced growth at higher temperature, and the effect is due to some alteration in a ribosomal protein that (presumably) malfunctions when the temperature is raised. The generality of this type does not leave any clue as to the function of the protein.

There are still quite a few ribosomal proteins whose genes have not yet provided sites in which mutations are damaging. It remains possible that these proteins are dispensable for ribosome assembly and activity.

Usually mutations involving the bacterial ribosome take the form of alterations in proteins. In a few cases (six so far), the relevant gene may have been deleted. The ribosomes of such mutants could be particularly useful for investigating whether the functions of these proteins are essential for ribosome action.

By using a procedure to identify altered proteins directly via changes in their electrophoretic mobility, mutants in all of the ribosomal proteins have been obtained; in these cases it is not known whether the mutation has any effect on the function of the protein.

The 16S rRNA has been shown to play a direct role in protein synthesis by the properties of kasugamycin-resistant mutants. Kasugamycin blocks the initiation reaction. Resistant mutants of the type *ksgA* lack a methylase enzyme that introduces four methyl groups into two adjacent adenines at a point near the 3′-terminus of the 16S rRNA. This generates the sequence $G-m_2^6 A-m_2^6 A$, which is highly conserved and is found in both procaryotic and eucaryotic small rRNA.

This means that in the wild-type ribosome, the presence of the four methyl groups renders the ribosome sensitive to the action of kasugamycin. Why does the absence of the methyl groups confer resistance to the drug? Both wild-type and mutant 30S subunits are equally sensitive to inhibition by kasugamycin of the binding of fMet-tRNA$_f$. But in the presence of 50S subunits, the fMet-tRNA$_f$ is released from the sensitive (wild-type) ribosomes; whereas it is retained by the mutants, thus allowing protein synthesis to continue. This suggests that the methyl groups are involved in the association of the 30S subunit with the 50S subunit, this being connected also with the retention of the initiator tRNA. The influence of the large subunit on the reaction is shown by the isolation of some 50S protein mutations that make ribosome dependent on kasugamycin.

Another indication of the importance of the 3′ end of 16S rRNA is given by its susceptibility to the lethal agent colicin E3. Produced by some bacteria, this cleaves about 50 nucleotides from the 3′ end of the 16S rRNA of *E. coli*. The cleavage entirely abolishes initiation of protein synthesis.

The large rRNA shows mutations conferring resistance to chloramphenicol in mitochondria. Chloramphenicol blocks protein synthesis in bacteria and in mitochondria by interfering with the peptidyl transferase reaction. In bacteria, resistant mutants have altered 50S proteins. In mitochondria, they possess single-base changes at sites about 243 nucleotides from the 3′ end of the large rRNA.

RIBOSOMES HAVE SEVERAL ACTIVE CENTERS

We can distinguish several ribosomal activities on the basis of function or location. The 30S subunits bind mRNA and the initiator-tRNA●initiation factor complex; then they bind 50S subunits. The 70S ribosomes have the functionally distinct P and A sites at which tRNA is bound; the peptidyl transferase center is carried on the 50S subunit. A simplified view of these sites is shown in **Figure 8.14.** How are these sites related to the actual topology of the ribosome? Which ribosomal components are involved in its various functions? Several approaches have been used to analyze the relationship between structure and function.

One way to identify sites involved in particular functions is to use affinity labeling. This relies on analogs of components that bind to the ribosome; these are either themselves chemically reactive or can be activated *in situ*. For example, tRNA with appropriate modifications can be used; then it is possible to identify the ribosomal components to which a label is transferred. A related technique is to use compounds

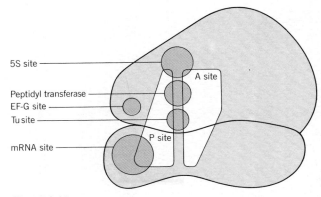

Figure 8.14
Active centers on the ribosome lie in several places.

The locations of the various sites are strictly diagrammatic and cannot yet be related to the precise three-dimensional structure of the ribosome.

mRNA binding site is located on the 30S subunit and includes proteins S1, S18 and S21, as well as S3, S4, S5, S12, and the 3′ terminal region of the 16S rRNA; also, this is the site where IF3 binds. It is close to the P site.

P site is probably located mostly on the 30S subunit and is able to bind initiator tRNA. The 16S rRNA is a component, probably providing a region fairly close to its 3′ end, but not the last 50 bases that include the mRNA binding site. On the 50S subunit, the P site includes L2 and L27, as well as L14, L18, L24, and L33.

A site must lie close to the P site, but the exact geometry is unknown. The 16S rRNA influences the A site and could be a component. Most of the functions identified with the A site are on the 50S subunit. They include L1, L5, L7/L12, L20, L30, and L33.

Peptidyl transferase site should lie somewhere in the region connecting the A and P sites, close to the terminus of tRNA. Proteins needed for this activity are L2, L3, L4, L15, and L16.

5S RNA site may lie near to peptidyl transferase site, including L5, L8, and L25.

Tu site includes several proteins lying close together that can be cross-linked to EF-Tu; these are L5, L1, L20, and L7/L12.

L7/L12 site is the stalk of Figure 8.2; this is not shown here, but the stalk consists of the two dimers of L7/L12.

EF-G site may be located on the large subunit close to the interface with the small subunit, near S12.

that are susceptible to photochemical cross-linking and to identify the components that become cross-linked.

Peptidyl-tRNA derivatives (all labeled in the amino acid acceptor end) react with the proteins L2 and L27, which thus appear to be part of the P site. By using derivatives with greater reach, it is possible also to label L14, L18, L24, and L33, which thus are identified as surrounding proteins that also may belong to this site. These results show that the end of tRNA carrying the peptidyl chain lies on the large subunit in an area bounded by a few particular proteins.

Modification of complete subunits has been extensively used. One approach is to treat subunits with reagents that damage protein or RNA and then to try to correlate particular damage with specific loss of function. For example, kethoxal has been used to damage rRNA, and tetranitromethane to nitrate proteins. The problem is to limit the damage so that a particular event can be closely correlated with a given loss of function.

Treatment with kethoxal prevents ribosomes from binding aminoacyl-tRNA (although it leaves them able to bind mRNA, which provides a control to show that the treatment has not simply irreversibly damaged the entire function of the particle). About six guanine residues of the 16S rRNA are modified. This suggests that the 16S rRNA constitutes a functional part of the A site or at least influences it. More direct evidence is provided by photochemical cross-linking studies, which show that peptidyl-tRNA can be linked to a point not far from the 3' end of 16S rRNA; this suggests the rRNA is also part of the P site. Cross-linking does not occur between rRNA and aminoacyl-tRNA lying in the A site.

The incorporation of 5S RNA into 50S subunits that are assembled *in vitro* depends on the presence of three proteins, L5, L8, and L25, which form a stoichiometric complex with it. The complex can bind to 23S rRNA, although none of the isolated components can do this. It may lie in the vicinity of the A and P sites.

The conserved sequence in the TψC loop of tRNA is complementary to a sequence in 5S RNA. There have been suggestions that base pairing between 5S RNA and tRNA could be involved in stabilizing aminoacyl-tRNA in the A site. However, experiments *in vitro* show that the removal of the relevant segment from the 5S RNA does not impede the ability of ribosomes to engage in translation. So this putative interaction does not seem to be necessary for protein synthesis.

Cross-linking studies using EF-Tu show that the A site lies in the context of proteins L1, L5, L7/L12, L20,

L30, and L33. We do not know much about how these (and other) proteins interact to provide the tRNA-binding site.

The group of proteins needed for peptidyl transferase activity comprises L2, L3, L4, L15, and L16. Attempts to identify a particular one of these proteins with the enzymatic activity have not been successful. The two best candidates are L2 and L16.

When L7/L12 is removed from the 50S subunit, the remaining particles cannot perform GTP hydrolysis with EF-Tu, EF-G, or IF2. This does not necessarily mean that L7/L12 is the GTPase; it may instead be necessary for the action of some other protein. The situation with EF-Tu is not entirely clear, because there is evidence that this factor itself possesses GTPase activity, and therefore may not necessarily rely on the ribosomal GTPase (see Chapter 6).

The translocation factor EF-G binds to the 50S subunit; but translocation involves movement of the mRNA through the 30S subunit. The binding site for EF-G is very close to S12, one of the proteins that is in the 30S mRNA-binding site. This places EF-G at the interface between subunits, in the vicinity of the L7/L12 dimers.

The P and A sites must lie close together, since their tRNAs are bound to adjacent triplets on mRNA. Also, there is a functional connection, since the P site must be occupied before aminoacyl-tRNA can enter the A site. It has always been a problem to see how two tRNA molecules might fit into the ribosome next to each other. The distance between the anticodons cannot be greater than about 10 Å, yet the thickness of the tRNA is about 20 Å.

The conformation of tRNA seems to be the same in both the P site and A site, although it is not certain that this conformation is the same as the tertiary structure displayed in a tRNA-crystal. One solution to the stereochemical problem would be to have a twist or kink in the mRNA between the codons, so that the two tRNAs fit onto the mRNA from different directions. Then ribosome movement might be a matter of flicking the tRNA out of the A site, as it were, around the corner into the P site. This would allow mRNA advancement to be a function of tRNA movement, which would let anticodon-codon pairing determine the distance for translocation, as discussed in Chapter 7.

The functional relationships between the various ribosomal sites have yet to be defined. The ribosome may well prove to be a highly interactive structure, in which a change at one point could greatly affect the activity of another site elsewhere.

BINDING 30S SUBUNITS TO mRNA

The initial binding of 30S subunits to mRNA requires the protein S1, whose true structural assignment has been debated. S1 appears almost to be a final assembly protein; it is readily removed from the ribosome, and under some conditions can be found free in excess amounts. Thus there was some question about whether it should be regarded as a factor or as a constituent of the ribosome. Under proper conditions, however, it is present in one copy per subunit; and it is not released from the ribosome during protein synthesis, so now it is generally considered to be a subunit component.

S1 has a strong affinity for single-stranded nucleic acid; this is exercised in binding to mRNA, which consequently is maintained in the single-stranded state. This interaction may be responsible for preventing the mRNA from taking up a base-paired conformation that would be unsuitable for ribosome binding.

Analyzing the location of S1 in the ribosome is complicated by its structure, which is extremely elongated. Cross-linking studies show that it is closely related to S18 and S21. These proteins also are among those that react in affinity labeling experiments when initiator tRNA is bound to an AUG codon. They are located in the small head of the subunit. These data suggest that the three proteins constitute a relatively small domain that is involved both in the initial binding of mRNA and in binding the initiator tRNA to it.

The 3' end of the 16S rRNA also is located near this domain and is involved in the binding reaction. It possesses a short sequence that is complementary to a sequence found just prior to each AUG initiation codon in all bacterial mRNAs. It is likely that complementary base pairing between these sequences is involved in the initial binding of ribosomes to initiation sites on mRNA (see Chapter 9).

The initiation factor IF3 binds in the same region of the ribosome. IF3 can be cross-linked to the 3' end of the rRNA, as well as to several ribosomal proteins, including those thought to be involved in binding mRNA. If this region of the small subunit is the same region involved in binding the large subunit, the role of IF3 could be to bind here to stabilize mRNA•30S subunit binding, then being displaced when the 50S subunit joins.

Not very much is known about the mechanism of subunit joining. There is some complementarity between a part of 23S rRNA and (again) the 3' region of 16S rRNA. One possibility is that base pairing is involved in bringing the subunits together. In this case, the ability of IF3 to bind to 16S rRNA might be incompatible with subunit association.

THE ACCURACY OF TRANSLATION

The lack of detectable variation when the sequence of a protein is analyzed demonstrates that protein synthesis must be extremely accurate: very few mistakes are apparent in the form of substitutions of one amino acid for another. There are two stages in protein synthesis at which errors could be made.

Charging a tRNA only with its correct amino acid clearly is critical. We have seen in Chapter 7 that this is a function of the aminoacyl-tRNA synthetase enzyme. Probably the error rate will vary with the particular enzyme, but current estimates are that mistakes are made in less than 1 in 10^5 aminoacylations.

The correct recognition of a codon by its anticodon is crucial. There may be variations in the binding constants for different codon-anticodon reactions, since different base pairs are involved. The frequency of incorrect binding may depend therefore on the particular codon-anticodon pair. An estimate in one instance, for the substitution of arginine by cysteine in bacterial flagellin, gives a value of roughly 1 in 10^4 misreadings per codon (all of the misreadings are likely to occur at codon recognition, since tRNA mischarging is unlikely in this particular case).

One of the puzzling features of protein synthesis is its accuracy. When free in solution, tRNAs can bind to their trinucleotide codon sequences only fairly weakly; and related, but erroneous triplets (with two correct bases out of three) are recognized only some 10–100 times less effectively than the correct triplets. Thus codon-anticodon base pairing per se is not nearly specific enough to account for the 10^{-4} error rate.

This suggests that the ribosome has some function that directly or indirectly acts as a "proofreader," distinguishing correct and incorrect codon-anticodon pairs, and thus amplifying the rather modest intrinsic difference. Now suppose that there is no specificity in the initial collision between the aminoacyl-tRNA•EF-Tu•GTP complex and the ribosome. If any complex, irrespective of its tRNA, is able to enter the A site, the number of incorrect entries must far exceed the number of correct entries. So there must be some mechanism for stabilizing the correct aminoacyl-tRNA, allowing its amino acid to be accepted as a substrate to which the polypeptide chain is transferred; incorrect contacts must be rapidly broken, so that the complex leaves without reaction. How does the ribosome assess the codon-anticodon reaction in the A site to determine whether a proper fit has been achieved?

The ability of the ribosome to influence the accuracy of translation was first shown by the effects of mutations that confer resistance to streptomycin. One of the effects of streptomycin on protein synthesis is to increase the level of misreading of the pyrimidines U and C (usually one is mistaken for the other, occasionally for A). The site at which streptomycin acts may be the S12 protein; in resistant mutants, it is this protein whose sequence is altered. Ribosomes with an S12 protein derived from resistant bacteria show a reduction in the level of misreading compared with wild-type ribosomes. This compensates for the effect of streptomycin on misreading.

Mutations at two other loci, coding for proteins S4 and S5, also influence misreading, since revertants showing the usual level of misreading can be isolated at either locus. Thus the accuracy of translation is controlled by the interactions of these three proteins. The level of misreading and the nature of the response to streptomycin both depend on the versions of these proteins that are present. (Some combinations even make the ribosome **dependent** on the presence of streptomycin for correct translation.)

A direct mechanism for controlling accuracy would be via the stereochemistry of the A site; this could

change so as to determine the latitude with which codon-anticodon recognition takes place. The original results obtained with streptomycin were interpreted in terms of ribosome geometry, supposing that codon-anticodon recognition is scrutinized in such a way that the criteria for accepting aminoacyl-tRNA could be made more or less precise.

An alternative is that the effect could be indirect; the speed with which the ribosome functions could determine the time available for tRNA recognition, and thus the efficiency of the process. So the effects of streptomycin also can be explained by making adjustments in the kinetics of chain elongation. The relevant parameter is the speed of ribosome action relative to the time required to make and break contacts. If the velocity of peptide bond formation is increased, incorrect aminoacyl-tRNAs are more likely to be trapped by bond formation before the aminoacyl-tRNA escapes. Slowing the rate of protein synthesis gives more time to correct errors. There is some evidence that the rate of polypeptide chain elongation is related to the level of misreading.

One idea has been that the making of a correct contact between codon and anticodon could be signaled to the other end of the tRNA by a change in conformation. This could be needed to place the amino acid in the appropriate location to accept the polypeptide from the peptidyl-tRNA.

A question that is important in calculating the cost of protein synthesis is the stage at which the decision is taken whether to accept a tRNA. If this occurs immediately upon association of the aminoacyl-tRNA•EF-Tu•GTP complex, which is then just released if improperly recognized, there is little extra cost for rejecting the large number of incorrect tRNAs that are likely (statistically) to enter the A site before the correct tRNA is recognized. But if the GTP is hydrolyzed on binding of the complex, an additional high-energy bond will be cleaved for every incorrectly associating tRNA. This would increase the cost of protein synthesis well above the three high-energy bonds known to be used in adding every (correct) amino acid to the polypeptide chain. There is some evidence that use of GTP *in vitro* is greater than had been expected, possibly involving roughly an extra 3–4 GTP cleavages per amino acid.

FURTHER READING

A major reference work on ribosomes is a book edited by
Nomura, Tissieres, & **Lengyel:** *Ribosomes* (Cold Spring
Harbor Laboratory, New York, 1974). Although now somewhat outdated, it still gives a useful general view of structural
and functional aspects. Ribosome structure has been the
subject of several reviews: bacterial ribosomes by **Kurland**
(*Ann. Rev. Biochem.* **46,** 173–200, 1977) and by **Brima-
combe, Stoffler,** & **Wittmann** (*Ann. Rev. Biochem.* **47,**
217–250, 1978), and eucaryotic ribosomes by **Wool** (*Ann.
Rev. Biochem.* **48,** 719–754, 1979). Ribosomal RNA has been
analyzed by **Noller** & **Woese** (*Science* **212,** 403–411, 1981).
The structure and function of ribosome sites have been described by **Pongs** (pp 78–104 in Altman (Ed), *Transfer RNA,*
MIT Press, Cambridge, 1978.) The assembly of 50S subunits
has caught up to earlier work with the report of **Rohl** &
Nierhaus (*Proc. Nat. Acad. Sci. USA* **79,** 729–737, 1982).
The last word on the lack of a role for 5S RNA in interacting
with tRNA comes from **Pace et al.** (*Proc. Nat. Acad. Sci.
USA* **79,** 36–40, 1982). The parameters involved in the accuracy of translation have been summarized briefly by **Kurland** (*Cell* **28,** 201–202, 1982).

CHAPTER 9
THE MESSENGER RNA TEMPLATE

The existence of mRNA was first suspected because of the evident need for some intermediary in eucaryotic cells to carry genetic information from the nucleus where DNA resides into the cytoplasm where proteins are synthesized. Conceived as providing a template on which amino acids would be assembled into polypeptides, the messenger was first sought in bacteria. The ribosomes, established as the site of protein synthesis, were excluded from the role of providing the template, and then it was found that mRNA is a transient species that associates with ribosomes to be translated into proteins (see Chapter 5).

Although the transient existence of bacterial mRNA prevented isolation of the molecule, the properties of the messenger were deduced in detail from the features of translation into protein. It was some time until mRNA could be isolated from eucaryotes, but then it proved to be a relatively more stable component of the cytoplasm, found again as part of the polyribosomes. Because of its stability, eucaryotic mRNA can be isolated with some facility; these days, it is possible in principle to isolate the mRNA for any particular protein. By an irony of circumstance, therefore, the wheel has come full circle, and mRNA has been characterized directly in the greatest detail in eucaryotes, where its existence was first imagined.

Messenger RNA is fixed of purpose in all living cells: to be translated via the genetic code into protein. Yet there are differences in the details of the synthesis and structure of procaryotic and eucaryotic mRNA. The most evident is that in eucaryotes the mRNA is synthesized as a large precursor molecule in the nucleus. After an involved process of maturation, often involving a considerable reduction in size as well as other modifications, the mRNA is exported to the cytoplasm to be translated. Its synthesis and expression thus occur in different cellular compartments. In bacteria, on the other hand, mRNA is transcribed and translated in the single cellular compartment; and the two processes are so closely linked that they may occur simultaneously on a single template. The principal difference in the process of translation is that a bacterial mRNA may code for several proteins, whereas a eucaryotic mRNA invariably is translated into only one polypeptide chain.

THE TRANSIENCE OF BACTERIAL MESSENGERS

The instability of bacterial mRNAs is very pronounced. It may be followed in two ways. Both rely on halting the transcription of mRNA and then following the fate of the existing mRNA molecules in the cell.

The **functional half-life** measures the ability of the mRNA to serve as template for the synthesis of its protein product. It is usually about 2 minutes. In other words, the amount of new protein synthesized by an individual mRNA is halved about every 2 minutes.

The **chemical half-life** is determined by measuring the decline in the amount of the mRNA able to hybridize with the DNA representing its gene. This is comparable to the functional half-life, but generally the chemical decline lags slightly behind the functional decline.

This suggests that the degradation of mRNA involves an initial step that prevents its use as a template, but that is not detected as a structural alteration sufficient to impede its hybridization. For example, a single cleavage in the mRNA would not affect its ability to hybridize with DNA, but could prevent its translation.

The initial step is followed by breakdown of the mRNA into its constituent nucleotides. The details of the molecular events responsible for the degradation have been hard to pin down, but breakdown seems to proceed more or less sequentially along the mRNA from the 5' end to the 3' end—that is, in the same direction in which the mRNA is transcribed and translated.

Transcription and translation are intimately related in bacteria. The exact rates vary with the temperature, but usually are similar to each other. For example, at 37°C, transcription of mRNA occurs at about 2500 nucleotides/minute, which corresponds to a synthesis of about 14 codons/second. This is very close to the rate of protein synthesis, which is roughly 15 amino acids/second. And when the expression of a new gene is initiated, its mRNA typically will appear in the cell within about 2.5 minutes. The corresponding protein will appear within perhaps another 0.5 minute.

The coincidence of these rates, and the close timing of the initial appearance of mRNA and protein, implies that transcription and translation occur simultaneously. Gene expression is initiated when an RNA polymerase enzyme binds to DNA and starts transcribing the mRNA. Ribosomes will attach to the 5' end of the mRNA and start translation, even before the rest of the message has been synthesized. In effect, a bunch of ribosomes will follow closely behind the RNA polymerase.

The degradation of mRNA in turn very closely follows its translation. Probably it begins within 1 minute of the start of transcription. Again, the 5' end of the mRNA may have started to decay before the 3' end has been synthesized or translated. Degradation seems to follow the last ribosome of the convoy along the mRNA. But the degradation proceeds more slowly, probably at about half the speed of transcription or translation. This has the result that the more distant (3') parts of the mRNA survive for longer than the initial (5') regions.

This series of events is only possible, of course, because transcription, translation, and degradation all occur in the same direction. **Figure 9.1** gives some idea of the timing of events for the expression of a typical bacterial gene.

We should not allow our need to visualize the "average" case obscure the harsh reality that every mRNA is in statistical jeopardy at all times, with a constant probability that its decay will begin. Thus "young" mRNAs are as likely to be attacked as "old" mRNAs. Some copies of an mRNA therefore will be translated many times; others may function hardly at all. This random life expectancy is a feature of individual mRNA molecules of both procaryotes and eucaryotes. But the overall translational yield of any messenger sequence is predictable.

THE CONSTITUTION OF BACTERIAL MESSENGERS

Because of their instability, bacterial mRNAs can rarely be isolated intact. Information about their constitution can be deduced in some detail, however, from the structures of the genes that they represent. The region of DNA that is found in the mRNA can be defined by hybridization assays, which measure the ability of radioactively labeled mRNA to hybridize with DNA fragments representing specific parts of the gene. Because it is relatively easy to sequence DNA, the organization of the mRNA can be deduced from the nucleotide sequence of the DNA transcription unit. In this way, a detailed picture of the structure and function of bacterial mRNA has been built up without characterizing the mRNA molecules directly.

Useful exceptions where the mRNA can be obtained are provided by phage infections. In some cases,

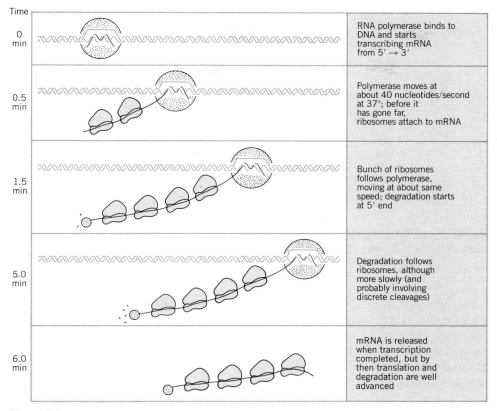

Time		
0 min		RNA polymerase binds to DNA and starts transcribing mRNA from 5′ → 3′
0.5 min		Polymerase moves at about 40 nucleotides/second at 37°; before it has gone far, ribosomes attach to mRNA
1.5 min		Bunch of ribosomes follows polymerase, moving at about same speed; degradation starts at 5′ end
5.0 min		Degradation follows ribosomes, although more slowly (and probably involving discrete cleavages)
6.0 min		mRNA is released when transcription completed, but by then translation and degradation are well advanced

Figure 9.1

Transcription, translation, and degradation of mRNA all proceed simultaneously in bacteria.

The figure shows a hypothetical transcription unit of length about 12,000 bp, whose expression starts at time 0. For simplicity, the unit is drawn as though it consisted of a single gene.

mRNA is more stable in the phage-infected cell than in the bacterial host, and this allows physical characterization of the molecule. The example par excellence is provided by the RNA phages, where the mRNA also serves as the genome, and therefore is both stable and can be obtained in large amounts.

An important approach for working directly with bacterial mRNA is the use of cell-free systems for both transcription and translation. The mRNA can be transcribed *in vitro* from the appropriate template, usually a "cloned" copy of the gene. It can be translated with *E. coli* ribosomes, aminoacyl-tRNAs, etc., into the protein. Thus both the structure and function of the mRNA are accessible by this means.

Bacterial mRNAs vary greatly in the number of proteins for which they code. Some mRNAs represent only a single gene: they are **monocistronic.** Others (the majority) carry sequences coding for several proteins: these are **polycistronic.** In these cases, a single mRNA is transcribed from a group of adjacent genes. (As we shall see in Chapter 14, such a cluster of genes constitutes an *operon* that is controlled as a single genetic unit.)

All mRNAs contain two types of region. The **coding region** consists of a series of codons representing the amino acid sequence of the protein, starting (usually) with AUG and ending with a termination codon. But the mRNA is always longer than the coding region. In a monocistronic mRNA, extra regions may be present at both ends. Additional sequences at the 5′ end that precede the start of the coding region are described as **leaders.** Additional sequences that follow the ter-

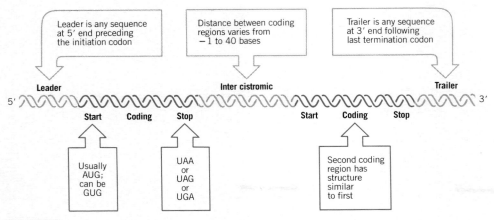

Figure 9.2
Bacterial mRNA includes nontranslated as well as translated regions. Each coding region possesses its own initiation and termination signals. A typical mRNA may have several coding regions.

mination signal and form the 3′ end are described as **trailers.** Although part of the transcription unit, these sequences are not used to code for protein.

The structure of a polycistronic mRNA is illustrated in **Figure 9.2.** The **intercistronic regions** that lie between the various coding regions vary greatly in size. In the example of the RNA phages, they may be quite large, up to 100 or so bases. In some bacterial mRNAs they may constitute a few (up to 30) nucleotides, but they can be even shorter, with as little as 1 or 2 nucleotides separating the termination codon for one protein from the initiation codon for the next. In an extreme case, two genes actually overlap, so that the last base of the UGA termination codon at the end of one coding region is also the first base of the AUG initiation codon at the start of the next gene.

TRANSLATION OF POLYCISTRONIC MESSENGERS

Are the several coding regions in a polycistronic mRNA translated independently, or is there a connection between them? Is the mechanism of initiation the same for all cistrons, or does it vary between the first cistron and the internal cistrons?

For polycistronic bacterial mRNAs, the timing of events indicates that translation must proceed se-

quentially through the cistrons. At the time when ribosomes attach to the first coding region, the subsequent coding regions may not yet even have been transcribed. By the time the second ribosome binding site is available, translation will be well on the way through the first cistron.

But what happens between coding regions? This is not entirely clear and may depend on the individual mRNA. Probably in most cases the ribosomes bind independently at the beginning of each cistron. This is consistent with evidence that initiation of internal cistrons shows the same sensitivity to kasugamycin (a specific inhibitor of initiation) as initiation at the first cistron.

The most likely series of events is illustrated in the upper part of **Figure 9.3.** Translation of the first cistron terminates in the usual way; the ribosomes dissociate into subunits that leave the mRNA. Then a new 30S subunit must attach at the next initiation codon, be joined by 50S subunit, and set out to translate the next cistron.

This is not to say that the translation of one cistron is necessarily without effect on the translation of the next cistron. There are cases in which a nonsense mutation in one gene prevents expression of a gene located farther along a polycistronic mRNA. This is known as **polarity.** The principal cause of polarity is an (indirect) effect on transcription, which stops soon

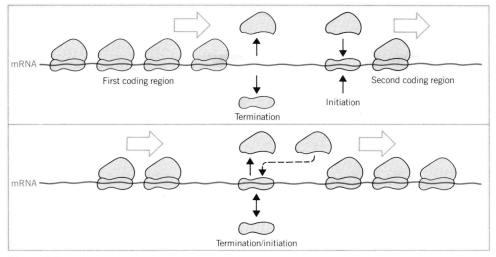

Figure 9.3
The form of reinitiation within polycistronic mRNAs may depend on the intercistronic region.

Upper. With intercistronic regions longer than the span of the ribosome, dissociation at the termination site is followed by independent initiation at the next cistron.

Lower. With very short intercistronic regions, the 30S subunit might dissociate transiently or even remain on the mRNA during termination and reinitiation.

after the site of mutation (see Chapter 13). However, polarity may also result from a relationship between the translation of the two cistrons in mRNA.

One form of relationship between adjacent cistrons may occur when the space between coding regions is small. A ribosome physically spans about 35 bases of mRNA, so that it may simultaneously contact a termination codon and the next initiation site even if they are separated by up to a few base pairs—certainly when they are adjacent or overlap. This juxtaposition could allow some of the usual intercistronic events to be bypassed. For example, the 30S subunit of a terminating ribosome may fail to dissociate from mRNA. It could be retained because it is instantly attracted to the initiation site of which it is already virtually in possession. As illustrated in the lower part of Figure 9.3, this would mean that the 50S subunit and the completed polypeptide chain would be released, but the 30S subunit would remain *in situ* to reinitiate translation for the next cistron. It is even possible that there are circumstances in which 70S ribosomes remain attached, although this seems less likely.

Another possible relationship between cistrons in a polycistronic mRNA is mediated by secondary struc-

ture. Usually this will not apply in bacterial mRNAs, because the ribosomes follow the RNA polymerase closely, and the regions ahead of them do not base pair to form stable double-stranded regions. However, when the ribosome dissociates at a nonsense mutation in an early cistron, the subsequent regions of mRNA could form a secondary structure; and base pairing of some other region with an initiation site might prevent it from being recognized by the 30S subunits. This may be responsible for polarity at the level of translation.

An effect of this nature is also seen in the normal translation of the RNA phages, whose cistrons always are expressed in a set order. The reason is that the phage RNA takes up a secondary structure in which only one initiation sequence is accessible; the others cannot be recognized by ribosomes because they are base paired with other regions of the RNA. However, translation of the first cistron disrupts the secondary structure, allowing ribosomes to bind to the initiation site of the next cistron. So in this mRNA, the secondary structure controls translatability.

The number of ribosomes engaged in translating a particular mRNA at any moment depends on the ef-

Figure 9.4
Simultaneous transcription and translation can be visualized in bacteria.

The thin central line is DNA, and the lines extending from it are RNA molecules being transcribed.
There are about 15 RNAs being synthesized from the transcription unit. They increase in length from
left to right, presumably the direction of synthesis. Each RNA is covered in ribosomes.
Photograph kindly provided by Oscar Miller.

ficiency with which the binding site(s) is recognized. In the case of the tryptophan genes of *E. coli,* whose kinetics of expression have been analyzed in detail, on average at any one time about 15 RNA polymerase molecules are likely to be present on the entire 7000 bp of the transcription unit. Each mRNA probably is translated by about 30 ribosomes in the interval between its transcription and degradation. So if about 5 initiation events occur for transcription every minute, roughly 150 molecules of protein are translated per minute in the steady state.

Electron microscopy of *E. coli* cells has identified the (unknown) transcription units shown in **Figure 9.4,** in which several mRNAs can be seen to be under synthesis simultaneously, and each has many ribosomes engaged in translation.

A FUNCTIONAL DEFINITION
FOR EUCARYOTIC mRNA

Polyribosomes constitute a stable fraction of the eucaryotic cytoplasm and can be isolated by centrifugation of the cellular components. But because the mRNA is only a minor component of the total mass of the RNA in this fraction (the overwhelming proportion

is rRNA), it cannot be isolated directly by the usual fractionation techniques. Attempts to label the mRNA preferentially by using radioactive nucleotide precursors (the approach used in bacteria) proved difficult, because some contaminating cytoplasmic RNA fractions also are labeled just as rapidly and efficiently. This problem was responsible for the delay in isolating eucaryotic mRNAs after the discovery of mRNA in bacteria.

The first technique to distinguish mRNA from other RNAs took advantage of the results of treating polysomes with the chelating agent EDTA (whose main effect in these circumstances is to remove Mg^{2+} ions from ribosomes). This causes the polysomes to dissociate into individual ribosomal subunits. The mRNA is released in the process in the form of a fraction sedimenting at ~18S. This consists of mRNA associated with proteins in the form of a ribonucleoprotein particle (**mRNP**). The contaminating ribonucleoprotein material that is not involved in translation is not disrupted by EDTA; it continues to sediment rapidly at ~200S.

In spite of considerable advances since this technique was developed, the release of mRNA by EDTA remains important for two reasons. First, it is a *functional* assay: it identifies mRNA actually in the process of translation by the ribosomes. The presence of an mRNA in this fraction therefore is taken as proof that it is indeed used to direct protein synthesis in the cell from which it was obtained. Second, the assay allows the mRNA to be isolated in what is presumably its natural form, the ribonucleoprotein particle.

Usually only a small number of proteins are found in the mRNP. In almost all the cases that have been examined in mammalian cells, two of the major proteins have constant molecular weights, roughly of 52,000 and 78,000 daltons. Usually not more than two or three of the other proteins are present in similar amounts; the remainder are minor components. In the best-characterized case, that of the globin mRNA of red blood cells, the amount of the 52,000- and 78,000- dalton proteins is sufficient for there to be one or two molecules of each associated with each mRNA molecule.

The functional significance of these proteins is not known. They could be concerned with transporting the mRNA from nucleus to cytoplasm or possibly with influencing its translation.

The contaminating ribonucleoprotein fraction is "contaminating" only in the sense that it is an unwanted component of the polysomal fraction extracted from the cells. The RNP is presumably a legitimate component of the cell in its own right. As part of this fraction, there are RNA molecules whose properties are similar to mRNA. But according to the EDTA-release assay, these RNAs are not active in translation. Their relationship with the mRNP is not clear; but it is possible that this fraction may include legitimate mRNAs that (although currently inactive) will be used for translation later or under different circumstances.

Some embryonic cells contain "stored" mRNAs. These are ribonucleoprotein particles not under translation, but whose mRNA is used to direct protein synthesis at later stages of embryogenesis. Although we do not know whether comparable storage mechanisms are used in adult cells, it is now well established (especially in marine organisms) that an appreciable part of newly synthesized mRNA is not immediately translated in the egg or early embryo, but is stored for use later.

Messenger RNA in the eucaryotic cytoplasm is reasonably stable. Measurements of its stability often identify more than one distinct component. Typically about half of the mRNA of mammalian cells in tissue culture has a half-life of a few hours, 6 hours or so, while the other half has a stability roughly equivalent to the length of the cell cycle, about 24 hours. In differentiated cells devoted to the synthesis of specific products, some mRNAs may be even more stable.

EUCARYOTIC mRNAs MAY BE POLYADENYLATED AT THE 3′ END

Most eucaryotic mRNAs have a sequence of polyadenylic acid at the 3′ end. This terminal stretch of A residues is often described as the **poly(A) tail;** and mRNA with this feature is denoted **poly(A)$^+$.**

The poly(A) sequence is not coded in the DNA, but is added to the RNA in the nucleus after transcription. The addition of poly(A) is catalyzed by the enzyme poly(A) polymerase, which recognizes the free 3′-OH end of the mRNA and adds about 200 A residues (as characterized in detail for mammalian cells). The added stretch consists exclusively of A residues. We do not know how its length is controlled.

When mRNA first enters the cytoplasm, it has approximately the same length of poly(A) tail that was added in the nucleus. The tail gradually is shortened, perhaps in discrete steps involving endonucleolytic cleavage. Thus the cellular population of mRNA includes both "new" and "old" molecules, with relatively longer and shorter stretches of poly(A). However, the length of the poly(A) tail present on any particular molecule does not seem to influence its ability to be translated or its stability in the cytoplasm.

When certain mRNAs are injected into *Xenopus laevis* oocytes, their stability increases with the length of the poly(A) tail; but the correlation does not hold for other mRNAs. Preventing the addition of poly(A) to adenovirus mRNA also reduces its stability in the host cell.

In locations other than mammalian cytoplasm, the precise length of poly(A) may be somewhat shorter, but the same general property is seen of reduction in length with age. Poly(A) is present on mRNA of all eucaryotic cytoplasms that have been investigated, and also in the mitochondrion.

The poly(A) of mammalian mRNA is associated with the common 78,000-dalton protein that is a predominant component of the mRNP. This suggests that the structure of the 3' end of the mRNA consists of a stretch of poly(A) bound to roughly an equal mass of protein. (And this location for the protein explains why at least this component of the mRNP does not interfere with its translation by the ribosomes.)

When poly(A) was first discovered, we thought that all cellular mRNAs might possess it. Although the fraction of mRNA that contains poly(A) always appears to be less than 100%, the discrepancy was attributed to breakage during preparation. A single break in a poly(A)$^+$ mRNA would generate a 5' end lacking poly(A); this fragment could be mistakenly taken for an authentic mRNA. However, using extremely careful techniques to prepare mRNA in which breakage is minimized, there still remains a fraction of the mRNA that lacks poly(A). Typically this constitutes up to one third of the total mRNA.

A predominant component of this **poly(A)$^-$** fraction is provided by histone mRNAs, which code for the histone proteins present in the chromosomes. These mRNAs are short; and they are synthesized only during a specific part of the cell cycle. The remaining (roughly two thirds) part of the poly(A)$^-$ fraction is identical with the poly(A)$^+$ fraction in all structural and functional parameters except for the presence of poly(A). Thus the length, stability, translational efficiency, and nucleocytoplasmic transport all appear the same for poly(A)$^-$ and poly(A)$^+$ mRNA.

Do the poly(A)$^+$ and poly(A)$^-$ mRNAs code for the same or for different proteins? The histone mRNAs are unique to the poly(A)$^-$ fraction, since none of the histone mRNAs usually is found in the poly(A)$^+$ fraction. But the other components of the poly(A)$^-$ fraction show considerable overlap with those of the poly(A)$^+$ fraction. It is possible that all of these mRNAs also exist in polyadenylated form. Thus a particular gene may be represented in transcripts some of which possess poly(A), and some of which do not. There may be differences in the extent to which any particular mRNA is likely to be polyadenylated; but at present it is difficult to form a general conclusion on the significance of the polyadenylation (or its absence). In some cases the presence of poly(A) does seem to affect the stability of the mRNA; no other effects have yet been found.

ISOLATION OF mRNA VIA POLY(A)

The presence of poly(A) has an extremely important practical consequence. The poly(A) region of mRNA can bind by base pairing to oligo(U) or oligo(dT), and this reaction can be used to isolate poly(A)$^+$ mRNA. The most convenient technique is to immobilize the oligo(U or dT) on a solid support material, for example, by chemical linkage to Sepharose. Then when an RNA population is applied to the column, only the poly(A)$^+$ mRNA is retained. It can be retrieved by treating the column with a salt solution, which breaks the hydrogen bonds to release the RNA, as illustrated in **Figure 9.5.**

The only complication in this procedure is that it does isolate *all* the RNA that contains poly(A). If RNA of the whole cell is used, for example, both nuclear and cytoplasmic poly(A)$^+$ RNA will be retained. If

Figure 9.5

Poly(A)$^+$ RNA can be separated from the other RNAs by fractionation on Sepharose-oligo(dT).

Most of RNA population is rRNA that lacks poly (A)

mRNA with poly (A) is small proportion of total RNA

Apply to column in low salt

Oligo (dT) Sepharose

Poly (A)$^+$ RNA sticks to column

rRNA flows through

Elute with high salt

Poly (A)$^+$ RNA

preparations of polysomes are used (a common procedure), most of the isolated poly(A)$^+$ RNA will be active mRNA; but some of the RNA in the contaminating RNP fraction also carries poly(A) and therefore will be included. This makes it necessary to retain the EDTA-release assay as a functional test when it is important to isolate precisely the active mRNA population.

Another approach for purifying mRNA also makes use of the poly(A) tail. The procedure is to anneal a small oligo(dT) fragment to the poly(A) stretch; the fragment functions as a starting point or "primer" for the enzyme **reverse transcriptase,** which copies the mRNA to make a complementary DNA strand (known as **cDNA**). Then this cDNA can be used as template to synthesize a DNA strand that is identical with the original mRNA sequence. The product of these reactions is a double-stranded DNA that corresponds to the sequence of the mRNA. This DNA can readily be "cloned" by the techniques described in more detail in Chapter 19.

The availability of large amounts of any cloned DNA makes it easy to isolate the corresponding mRNA by hybridization techniques. The mRNAs that are present in only a very few copies per cell can be isolated by this technique. Indeed, only mRNAs that are present in relatively large amounts can be isolated directly, without involving a cloning step.

EUCARYOTIC mRNA HAS
A METHYLATED CAP AT THE 5′ END

As with other nucleic acids, only the usual four ribonucleotides are incorporated into mRNA during its transcription. Modifying enzymes then are responsible for introducing additional groups at specific locations. Two types of modification event occur at the 5′ end of mRNA found in the eucaryotic cytoplasm (but not in the mitochondrion or chloroplast). Probably these reactions are common to all eucaryotes.

Transcription starts with a nucleoside triphosphate (usually a purine, A or G). The first nucleotide retains its 5′ triphosphate group, and makes the usual phosphodiester bond from its 3′ position to the 5′ position of the next nucleotide. Thus the initial sequence of the transcript can be represented as

$$5'\quad ppp^A_GpNpNpNpNp\ldots$$

But when the mature mRNA is treated *in vitro* with enzymes that should degrade it into individual nucleotides, the 5′ end does not give rise to the expected nucleoside triphosphate. Instead it contains *two* nucleotides, connected by a *5′-5′ triphosphate linkage* and also bearing methyl groups. The terminal base is always a guanine; and this residue is added to the original RNA molecule *after transcription.*

The addition of the 5′ terminal G is catalyzed by a nuclear enzyme, guanylyl transferase. The reaction occurs so soon after transcription has started that it is not possible to detect more than trace amounts of the original 5′ triphosphate end in the nuclear RNA. The overall reaction can be represented as a condensation between GTP and the original 5′ triphosphate terminus of the RNA. Thus

$$\begin{array}{c} 5'\ 5' \\ \mathbf{Gppp} + pppApNpNp \ldots \\ \downarrow \\ 5'\ 5' \\ \mathbf{Gp}ppApNpNp \ldots + \mathbf{pp} + p \end{array}$$

The reaction may take place in two stages. First, the enzyme becomes covalently linked to the donor GMP moiety (the pyrophosphate is lost). Then the GMP is added to the RNA, which loses its terminal γ phosphate group. The result is that the new **G** residue now at the end of the RNA is in the reverse orientation from all the other nucleotides.

This structure is called the **cap** of mRNA. It is a substrate for methylations at certain positions that occur in a specific order. The full structure of a cap after all the possible methylations have been accomplished is shown in **Figure 9.6.**

The first methylation occurs in all eucaryotes, and consists of the addition of a methyl group to the 7 position of the terminal guanine. The enzyme responsible for this, guanine-7-methyl-transferase, is present in the cytoplasm. As with all the other methylases, it uses S-adenosyl-methionine as cofactor to provide the methyl group. A cap that possesses this single methyl

Figure 9.6
The cap blocks the 5′ end of mRNA and may be methylated at several positions.

group is known as a **cap 0.** This is as far as the reaction proceeds in the unicellular eucaryotes.

The next step is addition of another methyl group, to the 2′-0 position of the penultimate base (this is actually the original first base of the transcript before any modifications were made). This reaction is catalyzed by another enzyme (2′-0-methyl-transferase). A cap with the two methyl groups is called **cap 1.** This is the predominant form of the cap in all eucaryotes except the unicellular organisms.

In a small minority of cases in higher eucaryotes, another methyl group is added to the second base. This event happens only when the position is occupied by adenine, and the reaction involves the addition of a second methyl group at the N^6 position. The enzyme responsible, 2′-0-methyladenosine-N^6-methyl-transferase, acts only on an adenosine substrate that already has the methyl group in the 2′-0 position.

In some species, a methyl group may be added to the third base of the capped mRNA. The substrate for this reaction is the cap 1 mRNA that already possesses two methyl groups. The third-base modification is always a 2′-0 ribose methylation. It creates the **cap 2** type. In cases where this occurs, this cap is usually 10–15% of the total capped population.

In a population of eucaryotic mRNAs, every molecule is capped. The proportions of the different types of cap are characteristic for a particular organism. Is the structure of a particular mRNA invariant or may it have more than one type of cap? Insufficient data are available for a final answer.

In addition to the methylation involved in capping,

a low frequency of internal methylation occurs in the mRNA only of higher eucaryotes. This is accomplished by the generation of N^6 methyladenine residues at a frequency of about one modification per thousand bases. There may be 1–2 methyladenines in a typical higher eucaryotic mRNA, although their presence is not obligatory, since some mRNAs (for example, globin) do not have any at all.

THE POWER OF *IN VITRO* TRANSLATION SYSTEMS

The ability to translate mRNA *in vitro* provides an assay that is crucial in defining the process of translation. First, it provides unequivocal evidence for the coding function of any particular mRNA, by showing that the product of translation *in vitro* has the authentic properties of a protein found *in vivo* (generally as judged by electrophoretic mobilities). [This often can be confirmed in more detail by determining the nucleotide sequence of the mRNA (usually via its conversion into a DNA copy), and showing that it has an appropriate coding region. Sometimes, of course, the nucleotide sequence of the mRNA can be used to predict the amino acid sequence of a protein that is not known.]

In vitro systems (for transcription as well as for translation) provide the only approach for defining the *mechanism* of the process. In both cases, the critical step is initiation. The translation systems allow the binding of ribosomes to mRNA to be characterized in terms of the relevant features on each component.

Although one generally expects *in vitro* translation systems to mimic the *in vivo* process, there are two important situations in which *differences* between the products of translation *in vitro* and *in vivo* provide interesting information.

In some cases, it is possible to isolate species of mRNA that direct protein synthesis *in vitro,* although the corresponding proteins are not synthesized in the cells from which the mRNA was taken. The ability of the mRNA to be translated *in vitro* demonstrates that it is capable in principle of functioning as a template. So its failure to be translated *in vivo* can be taken as evidence for **translational control.** Some mechanism must act *in vivo* to prevent translation. The "stored"

RNP particles of marine embryos that we mentioned previously are the best-known example of this phenomenon. (Of course, a condition of this conclusion is that the *in vitro* products must be authentic proteins, not results of aberrant translation. Thus they must be shown to represent proteins that actually are synthesized under other conditions *in vivo.*)

Sometimes the product of translation *in vitro* is related to the authentic *in vivo* protein, but possesses additional amino acids, usually a short length at the N-terminal end. This discrepancy usually means that the protein found *in vivo* is not the initial product of translation, but is derived by cleavage of a precursor that contains the extra residues. Processing of such precursors *in vivo* generally is rapid, and this makes it difficult to detect them, unless inhibitors are used to block the cleavage reaction.

Isolated mRNAs have been characterized by translation in two types of system. These are depicted in **Figure 9.7.**

Reconstituted **cell-free systems** include ribosomes, protein synthesis factors, and tRNA. This is the traditional *in vitro* approach that we have already discussed. Several such systems have been developed: the best known are derived from wheat germ, rabbit reticulocyte, and mouse ascites cells. The systems are somewhat inefficient, since each mRNA is translated a relatively low number of times at a rate much below that found *in vivo*. At best the systems function for 90–120 minutes before stopping. In all cases, there is some background due to translation of endogenous mRNAs that were not removed; its level varies with the system.

An alternative translation system is presented by the intact **Xenopus oocyte.** Injected mRNAs are translated by the natural protein synthetic apparatus. The system is efficient and treats the injected mRNAs as though they were endogenous mRNAs, so that they are used for repeated rounds of translation. The only limit is that too much mRNA may saturate the translation system (which is clearly present in great excess relative to the demands placed on it by endogenous mRNAs). Generally the system continues to be active for 24–48 hours.

Neither type of system displays any tissue or species specificity. This indicates that mRNAs and the

		Advantages	Disadvantages
Cell-free systems	Contain: Ribosomes Protein synthesis factors Aminoacyl-tRNA ATP — Energy	Can be used to test roles of individual components Easy to purify protein product Can isolate precursor proteins	Work slowly and inefficiently (relative to rate of protein synthesis *in vivo*)
Xenopus oocyte	Inject mRNA prep with micropipette Oocyte	Very efficient, works just like *in vivo*	More difficult to purify protein products

Figure 9.7
Exogenous mRNAs can be translated by cell-free systems or by injection into *X. laevis* oocytes.

protein-synthetic apparatus of (perhaps) any cytoplasm are interchangeable. The implication is that there is no translational control in the sense that there are factors or ribosomes that function with one set of mRNAs but not with another. The translational machinery is not preprogrammed, but will accept any mRNA as template. Thus any translational control must take the form of preventing mRNAs from reaching the protein-synthesizing apparatus, most probably by sequestering them in a form in which they are physically unavailable.

The oocyte system is able to process at least some proteins that usually suffer cleavage reactions during or shortly after synthesis. In some cases, it can even direct protein products to enter facsimiles of the appropriate cell compartment. The processing ability implies that signals for processing may be common in different cell types and species; and as a practical consequence, it means that the isolation of precursor proteins must be undertaken in cell-free translation systems.

INITIATION MAY INVOLVE PAIRING BETWEEN mRNA AND rRNA

The sites on mRNA at which ribosomes initiate protein synthesis can be identified by binding ribosomes to the mRNA to form a stable complex under conditions that block elongation of the polypeptide chain. If ribonuclease is added to the initiation complex, all the regions of mRNA outside the ribosome are degraded, but those actually bound to it are protected, as illustrated in **Figure 9.8.** The protected fragments can be recovered and characterized. This type of experiment has been performed with many individual mRNAs, both bacterial and eucaryotic.

The initiation sequences protected by bacterial ribosomes are about 35–40 bases long. They always include the AUG (or GUG) initiation codon, which is located about two thirds along the protected sequence. Very little sequence homology exists between the ribosome binding sites of different bacterial mRNAs. In fact, the only common feature is the presence of a

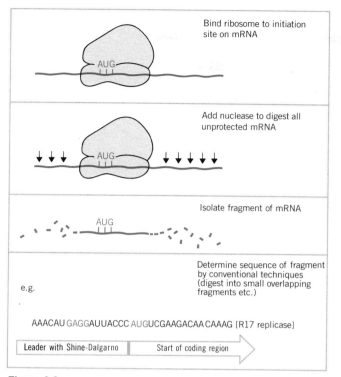

Figure 9.8
Ribosome binding sites on mRNA can be recovered from initiation complexes.

short sequence that is complementary to part of the region close to the 3′ end of the 16S rRNA.

The 3′ terminal region of bacterial 16S rRNA is highly conserved; it is even almost identical in the maize chloroplast. It has two features that may be significant in protein synthesis. First, it is self-complementary, and could form the base-paired hairpin shown in **Figure 9.9.** Second, it contains the sequence highlighted in the figure, which written in reverse direction is

$$3'\ldots\text{U C C U C C}\ldots\quad 5'$$

In all but one of the known *E. coli* mRNA initiation sites, there is a sequence complementary to at least a trinucleotide part of this, and more usually to a length of 4–5 bases. Thus bacterial mRNA contains part or all of the oligonucleotide

$$5'\ldots\text{A G G A G G}\ldots\quad 3'$$

This polypurine stretch is often known as the **Shine-Dalgarno** sequence. It lies 4–7 bases before the AUG codon. Does the Shine-Dalgarno sequence pair with its complement in rRNA during mRNA–ribosome binding?

The complementary sequence in the rRNA cannot simultaneously be part of the intramolecular hairpin and bind to mRNA. These could be alternatives, in which case the initiation reaction may involve disruption of the terminal hairpin to allow base pairing be-

Figure 9.9
Hairpins can be formed by base pairing near the 3′ ends of rRNA, and include the regions that might be involved in recognizing mRNA (indicated in red).

Figure 9.10
The 3′ ends of bacterial and mammalian small rRNAs are well conserved. The differences are indicated in the boxes, and the regions that may be involved in recognizing in mRNA are shown in red.

tween rRNA and mRNA. Following this, the rRNA–mRNA duplex may be broken by reconstitution of the rRNA base-paired hairpin. This mechanism could reconcile the need to form a stable initiation complex at a particular site with the need later to move along the mRNA.

The 3′ terminal region of rRNA is implicated in protein synthesis by its location as the target site for the antibiotic kasugamycin. Also, this region can be cross-linked to initiation factors. A principal reason for believing that the Shine-Dalgarno sequence is involved in initiation is its ubiquity in procaryotic initiation sites. There is direct evidence in one case (the mRNA of gene 0.3 of phage T7), in which ribosome binding is destabilized by a mutation that changes the sequence from GAGG to GAAG. The only mRNA from which the Shine-Dalgarno sequence is absent is the lambda P_m transcript, which has the 5′ terminal sequence pppAUG. . . . This mRNA is translated rather poorly. Probably the unprecedented location of the AUG codon directly at the 5′ terminus allows it to be recognized by the ribosomes. We do not know whether the AUG initiation codon itself is recognized by the small subunit in the binding reaction.

When eucaryotic 40S subunits bind to mRNA, they protect a region of up to 60 bases; when the 60S subunits join the complex, the protected region contracts somewhat. The stable binding site of 80S ribosomes extends for about the same length of 30–40 bases seen in procaryotes, with the AUG codon usually close to the center. The reduction in the length of the region protected by 80S ribosomes could be caused by a conformational change in the small subunit when it is joined by the large subunit, or it could represent

the loss of initiation factors able directly to protect some additional region of the mRNA.

The sequence at the 3′ end of rRNA is well conserved between procaryotes and eucaryotes, as shown in **Figure 9.10.** In the 20 nucleotides between the adjacent double-methylated adenines and the 3′ end, there are only two significant changes. In bacteria, there are two U residues, whereas in higher eucaryotes there are two A residues (in lower eucaryotes the sequence is an intermediate AU). And in all eucaryotes there is a deletion of the five-base sequence CCUCC that is the principal complement to the Shine-Dalgarno sequence. (This means that if binding to the Shine-Dalgarno sequence is essential for initiation, eucaryotic ribosomes should be unable to initiate translation of bacterial mRNAs.)

Within the highly conserved 3′ terminal region (and present in bacterial as well as in eucaryotic rRNA), there is a highly purine-rich sequence,

$$3′ . . . U A G G A A G G C G U . . . 5′$$

that in eucaryotes is about the same distance from the 3′ end as the CCUCC sequence is in bacteria. Again, it could participate in hairpin formation, as shown in **Figure 9.9.** Some eucaryotic mRNAs have a tetra- or pentanucleotide sequence able in theory to pair with part of this (if the rRNA hairpin were disrupted). However, the distance of this potential site from the AUG codon is variable; often it lies outside the expected span of the ribosome binding site. We do not know whether this sequence is involved in initiation, but if it is, its use cannot be obligatory, since some eucaryotic mRNAs entirely lack it.

There is one case in which the general mechanism for initiation cannot involve pairing between mRNA and the 3′ terminal region of rRNA. In mammalian mitochondrial 12–13S rRNA, the 3′ terminal region does resemble that found in eucaryotic cytoplasm and in bacteria, but is less well conserved. It remains able to form a terminal hairpin, but no sequence complementarity with any region prior to the AUG codons on mRNA has been noticed.

SMALL SUBUNITS MAY MIGRATE TO INITIATION SITES ON EUCARYOTIC mRNA

The average mRNA in the eucaryotic cytoplasm is between 1000 and 2000 bases long, has a methylated cap at the 5′ terminus, and has a length of poly(A) of 100–200 bases at the 3′ terminus. When the lengths of individual mRNAs are compared with the proteins for which they code, the mRNA is invariably longer than is necessary to code for the protein. But the extra length is not used to code for a second protein. Each mRNA is therefore monocistronic, consisting of a short (usually but not always less than 100-base) leader, a single coding region, and a trailer (often long and sometimes up to 1000 bases or so). By virtue of its location, the leader sequence is not ignored during initiation, but we do not know of any function for the trailer in translation.

The ribosomes of eucaryotic cytoplasm do not bind directly to the initiation site at the start of the coding region. Instead, the first feature to be recognized is the methylated cap that marks the 5′ end. Messengers that lack caps, either because their synthesis has been prevented or because they have been enzymatically removed from mRNA, are not translated efficiently in the *in vitro* systems. With the sole exception of a few viral mRNAs that are not capped, the modification is made to the 5′ end of *all* mRNAs in eucaryotic cytoplasm (although not in organelles). Only the exceptional viral mRNAs can be translated *in vitro* without caps. This suggests that in most mRNAs the cap is obligatory for translation; the viral mRNAs must have some alternative feature that renders it unnecessary.

The cap is recognized by the 40S ribosome subunit. Which of its features are necessary? The introduction

of the first methyl group onto the 7 position of the terminal G is essential. The subsequent methylations (to progress from cap 0 to cap 2) probably improve the efficiency of ribosome binding but are not of primary importance. The 40S subunits can bind to capped poly(U) or other synthetic polynucleotides, which suggests that the cap per se (not the following sequence) is the main recognition feature; but 60S subunits cannot associate with the initiation complex without some sequence information from the mRNA itself.

In some cases, the AUG initiation codon lies within 40 bases of the 5′ terminus of the mRNA, so that both the cap and AUG sites lie within the span of ribosome binding. But several mRNAs are known in which the cap and AUG are farther apart, even 200–300 bases distant. Yet the presence of the cap still is necessary for a stable complex to be formed at the initiation codon. How can the ribosome rely on two sites so far apart?

One possibility is that the mRNA takes up a secondary structure in which base pairing brings the cap into juxtaposition with the AUG codon, so that they can be contacted simultaneously. There is no evidence for this.

Another model supposes that the 40S subunit initially recognizes the 5′ cap, and then "migrates" along the mRNA until it encounters the AUG initiation codon (see Figure 6.7). Usually, although not in every case, this will be the first AUG triplet sequence it meets. Then binding is stabilized, and the 40S subunit is joined by a 60S subunit at the site characterized by the ribosome protection assay. When the leader sequence is long, a second 40S subunit could recognize the 5′ end before the first has progressed away from the initiation site, so there could be a queue of subunits proceeding to the initiation site.

The idea that ribosomes must start at the 5′ end is consistent with the monocistronic nature of eucaryotic mRNAs. There are some exceptional viral mRNAs that contain more than one coding region. But in these cases, only the coding region nearest the 5′ end is translated from the intact mRNA. The other(s) can be read only after a cleavage has occurred in the mRNA to generate a new 5′ end in the vicinity of the initiation codon. This supports the idea that internal initiation sites cannot be recognized directly.

LINKAGES BETWEEN PROTEIN SYNTHESIS AND LOCATION

The eucaryotic cell is a well-ordered structure, whose functions are exercised at specific locations. This general view applies in particular to protein synthesis. Polyribosomes can be divided into two classes: "free" and membrane-bound. These are engaged in the synthesis of different groups of proteins.

The "free" polyribosomes are somewhat misnamed, since in fact they are not able to diffuse freely through the cytoplasm, but are associated with the **cytoskeleton.** When cells are extracted with a nonionic detergent (Triton) in hypertonic buffer, most of the lipids and soluble proteins are removed, leaving a "cytoskeletal framework" anchored on a remnant nucleus. This framework is a complex network of fibers. The entire class of "free" polyribosomes is associated with the cytoskeleton, as seen visibly in the electron micrograph of **Figure 9.11,** or as detected by biochemical fractionation.

The polysomes tend to be clustered near the nucleus, where the mRNA enters the cytoplasm. Most of the products of translation are soluble proteins that rapidly diffuse away from the site of synthesis when they are released from the polysomes. Those proteins that are components of the cytoskeleton tend to be incorporated into it at points close to their sites of synthesis.

In contrast with the polysomes, monomeric ribosomes are free in the cell. Association with the cytoskeleton appears to be a function of the mRNA, which remains attached even after treatments that disaggregate the polysomes. The translation of mRNA may depend on its association with the cytoskeleton, as seen most clearly by the events involved in viral infection. Poliovirus infection inhibits host translation, and this is correlated with a rearrangement of the cytoskeleton that is accompanied by the release of cellular mRNAs. In the case of VSV (vesicular stomatis virus) infection, the viral mRNAs associate with the cytoskeleton shortly after their synthesis; they are translated only during this period. Then they are released from the cytoskeleton.

The existence of *in vitro* systems for the translation of mRNA shows that the act of protein synthesis can be disengaged from extraneous cellular structures.

Figure 9.11

The cytoskeleton is a fibrous network that is associated with polyribosomes.

The bar is 1 μm.
Photograph kindly provided by Sheldon Penman.

However, this does not diminish the importance of the observation that the natural course of events is for mRNA to be associated with the cytoskeleton during translation. This may be important in the metabolism of mRNA *in vivo* as well as being necessary for the proper location of newly synthesized cytoskeletal proteins.

The "free" class of polysomes essentially includes all those not associated with membranes; these polysomes are responsible for the synthesis of proteins that do not interact with membranes. Other proteins have specific locations within or outside the cell that depend on their ability to bind to or pass through a membrane. These proteins are synthesized by the membrane-bound polysomes.

Within the cell there are proteins sequestered into compartments, such as the mitochondrion or lysosome. Some proteins are membrane components, of which the most attention has been paid to those residing in the plasma membrane (which circumscribes the cytoplasm). Other proteins may be **secreted** from the cell into the surrounding milieu. How do these proteins find their ultimate destinations?

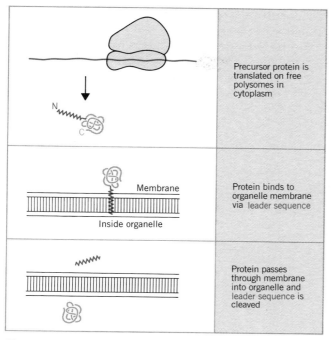

	Precursor protein is translated on free polysomes in cytoplasm
	Protein binds to organelle membrane via leader sequence
	Protein passes through membrane into organelle and leader sequence is cleaved

Figure 9.12

Leader sequences are used for proteins to recognize mitochondrial or chloroplast surfaces.

For many proteins that must be inserted in membranes, the sequence of the mature polypeptide is not itself sufficient to direct membrane insertion. Additional information is needed; and this most often takes the form of a **leader sequence** at the N-terminal end of the protein. The protein carrying this leader is called a **preprotein.** It is a transient precursor to the mature protein, since the leader is cleaved as part of the process of membrane insertion.

The **pre** sequence is distinct from the **pro** sequence that describes the additional regions present on proteins that exist as *stable* precursors. Some proteins may have both. For example, insulin is initially synthesized as **preproinsulin;** the **pre** sequence is cleaved during secretion, generating **proinsulin,** which is the substrate for processing to mature insulin.

The leader sequence plays different roles in different circumstances. For certain proteins synthesized within the cytoplasm, but destined to reside within the chloroplast or mitochondrion, the product of cyto-

plasmic protein synthesis is a precursor some 5000 daltons (roughly 45 amino acids) larger than the mature protein. This precursor is released from polysomes. If it is added to intact organelles *in vitro,* it can be incorporated into the compartment. As illustrated in **Figure 9.12,** this involves passage through the organelle membrane, during which the leader sequence is cleaved, probably by a protease located on the outside of the envelope. The leader sequence serves to provide information recognized by the organelle membrane and used to sequester the protein in a **posttranslational** process. Note that a cleavable leader is not the only acceptable form of such information; some mitochondrial proteins are recognized as such in their mature form, and may have an internal sequence that is able to ensure membrane passage without cleavage.

For proteins that are secreted through, or inserted into, other cellular membranes, the process of association most often starts during translation. The polyribosomes synthesizing these proteins are associated with the membrane of the endoplasmic reticulum. The preproteins are not released into the cytoplasm to form a precursor pool, but instead pass directly from the ribosome to the membrane. From the membrane, the proteins enter the Golgi apparatus, and then are directed to their ultimate destination, such as the lysosome or the plasma membrane.

A model for the mechanism of membrane insertion has been based on work with eucaryotic microsomal systems (containing ribosomes and endoplasmic reticulum). These systems are able to package *nascent* proteins into membranes; but they do not work with the addition of isolated preproteins. The **signal hypothesis** proposes that the leader characteristic of almost all secreted proteins constitutes a **signal sequence** whose presence distinguishes them from other proteins. With only rare exceptions, the N-terminus of secreted proteins consists of a cleavable leader of from 16 to 29 amino acids, which starts with two or three polar residues, but continues with a high content of hydrophobic amino acids; otherwise there is no noticeable conservation of sequence.

The signal sequence provides the means for ribosomes translating the mRNA to attach to the membrane. Some membrane receptor recognizes the signal sequence, perhaps by virtue of its hydrophobicity, and inserts the precursor protein directly into the membrane, probably as soon as the signal sequence and a few additional amino acids have been synthesized. A route to characterizing the protein receptors is provided by the discovery that salt-washed membranes cannot sponsor ribosomal attachment; but this ability can be recovered by adding the salt wash. The active component has been purified in the form of a complex of six proteins.

Figure 9.13 shows that as synthesis of the nascent polypeptide chain continues, there comes a point at which the protein is well inserted into the membrane, and the signal sequence can be cleaved. Then when the ribosomes complete translation, the protein is al-

Figure 9.13
The signal hypothesis proposes that ribosomes synthesizing secretory proteins are attached to the membrane via the leader sequence on the nascent polypeptide.

Membrane

Ribosome initiates protein synthesis on free mRNA

N-terminal leader on nascent polypeptide causes ribsomes to attach to membrane

Protein passes through membrane as translation continues and leader sequence is cleaved off

Protein has been secreted through membrane, ribosome has completed translation, and subunits are released into pool ready well on its way through the membrane. No free preprotein is ever generated.

Is the signal sequence entirely responsible for attaching ribosomes to the membrane? If it is, synthesis of all proteins must initiate in the same way, by binding ribosomes to free mRNA. Then, following the start of protein synthesis, those ribosomes translating mRNAs for secretory proteins will attach to the membrane through the signal sequence, while other polysomes remain free in the cytoplasm. This explains why it has always been impossible to isolate two pools of ribosomes, one free, one membrane-attached. However, it is not yet proven that the ribosome or mRNA is totally lacking in any responsibility for attachment. At all events, the mechanism by which proteins are synthesized in juxtaposition with the membrane, so that passage begins during translation, is called **cotranslational transfer.** It applies to a variety of secreted proteins, including some immunoglobulins and many hormones.

How does the protein pass through the membrane? The original idea was that the N-terminus would lead the way, as illustrated in Figure 9.13. However, it is possible that the N-terminus enters the membrane but does not pass through; in fact, the signal sequence could be anchored at the site of insertion, with the protein looping into the membrane. At all events, the critical feature is that the signal sequence can exercise its function only during translation, and not as part of a mature preprotein.

How are proteins that are secreted *through* the membrane distinguished from those that reside *within* it? The mechanism of insertion appears to be the same, via a signal sequence, but proteins that are not to pass through the membrane possess a second, internal **halt-transfer** signal. This may take the form of a cluster of hydrophobic amino acids adjacent to some ionic residues. The cluster serves as a "hook" that latches onto the membrane and stops the protein from passing right through.

There are some cases of proteins that do not have leader sequences, but that nonetheless are inserted into membranes. Probably some other sequence serves instead to promote cotranslational passage, although in this case the mechanism cannot involve cleavage.

Although originally conceived for eucaryotic cells, the signal hypothesis applies to bacteria. Mutations in N-terminal leaders may prevent secretion; they may be suppressed by mutations in other genes, including at least one ribosomal protein. Thus the ribosome itself is implicated in the mechanism of attachment to the membrane. This means that the signal sequence cannot have as its sole role the interaction with the membrane, but there must be other protein-protein interactions with the ribosome. As with eucaryotes, further signals may be necessary for proper location after the protein has entered the membrane. For example, the C-terminal region of the *E. coli* β-lactamase is necessary for the protein to leave the membrane to enter the periplasmic space on the other side.

There are some cases in bacteria in which secretion is posttranslational, the best known of which is the coat protein of phage M13. This is synthesized in the form of a procoat that can be inserted in the membrane. The procoat contains a leader sequence that is cleaved, and its possible role has been summarized in the **membrane trigger** hypothesis. The model argues that the role of the leader sequence is to alter the pathway for folding of the protein. With the leader, the protein exists in a water-soluble conformation that is appropriate for the cytoplasm in which it is synthesized. Loss of the leader converts it into a water-insoluble conformation that is appropriate for membrane insertion.

In at least some cases, the information that is used to establish the location of a protein is neither species- nor tissue-specific. For example, *X. laevis* oocytes translate injected rabbit globin mRNA to produce free globin protein, but translate rat albumin mRNA into a protein product that is incorporated into membrane vesicles. The signal on mammalian preproinsulin may be adequate to ensure its secretion when its mRNA is translated in the bacterium *E. coli*. There is a great deal that we have yet to learn about the basis for selectivity between membranes (in a eucaryotic cell, after all, there are nuclear, cytoplasmic, Golgi, mitochondrial, etc., membranes), but it seems probable that the use of a signal sequence is the major mechanism responsible for starting a protein on its route to or through a membrane.

FURTHER READING

Bacterial mRNAs have been reviewed mostly in the context of the individual genes that they represent; citations will be found in Chapters 14, 15, and 23. A principal source for general information on the structure and function of eucaryotic mRNA is **Lewin's** *Gene Expression, 2, Eucaryotic Chromosomes* (Wiley, New York, 1980), especially pp. 653–693. Capping has been reviewed by **Banerjee** (*Microbiol. Rev.* **44,** 175–205, 1980). The issue of protein location has been controversial, and the protagonists all have had their say recently. A critical paper in the series on the role of the cytoskeleton is from **Cervera, Dreyfuss,** & **Penman** (*Cell* **23,** 113–120, 1981). Reviews of the signal hypothesis are from **Blobel** et al. (*Symp. Soc. Exp. Biol.* **33,** 9–36, 1979), **Blobel** (*Proc. Nat. Acad. Sci. USA* **77,** 1496–1500, 1980), and **Davis** & **Tai** (*Nature* **283,** 433–438, 1980), with bacterial work reviewed by **Emr, Hall,** & **Silvahy** (*J. Cell. Biol.* **86,** 701–711, 1980). Alternative views have been considered by **Wickner** (*Ann. Rev. Biochem.* **48,** 23–45, 1979; and *Science* **210,** 861–868, 1980).

PART 3
PRODUCTION OF THE TEMPLATE

The code-script must itself be the operative factor bringing about the development [of the organism]. But . . . with the molecular picture of the gene it is no longer inconceivable that the miniature code should precisely correspond with a highly complicated and specified plan of development and should somehow contain the means to put it into operation.

Erwin Schrödinger, 1945

CHAPTER 10
RNA POLYMERASES:THE BASIC TRANSCRIPTION APPARATUS

How RNA came to be the intermediary between DNA and protein is a question only for evolutionary speculation. There is no way to discern the constituents of the prototypes for the living cell. It seems possible that in the first primitive cells there was no distinction between types of nucleic acid, so that what passed for the genome was involved directly in both replication and translation. At some point the separation of translation from the genome must have become advantageous, so that proteins were synthesized on messengers that were distinct from the genome itself. It is impossible to say how this related in time to the development of the other ribonucleic acid components involved in translation, but it is striking that RNA is present in the ribosome as well as constituting the tRNA adaptor. (It would not be surprising if the role of RNA in the ribosome at one time had been more prominent than is apparent today.)

All this speaks to the fact that RNA plays central roles in gene expression, not merely in constituting the messenger, but also in providing the means for its translation into protein. In a sense, these roles represent the various specific interests of an RNA conglomerate generally concerned with gene expression, but with several different and apparently independent functions. The production of each type of RNA has a common origin: transcription of DNA. In the case of mRNA, the product is an intermediate whose function requires translation. In the case of tRNA and rRNA, the transcriptional product is the active molecule. In procaryotes (and probably also in eucaryotes), transcription is the major stage at which gene expression is controlled. The first (and sometimes the only) step in control is the decision whether or not to transcribe the gene. In considering the various stages of transcription, we are therefore bound to keep in mind the opportunities that they offer for regulating gene activity.

Transcription is not the only means by which RNA can be synthesized. Viruses whose genomes are RNA specify enzymes that undertake a variety of reactions, but that share the central property of being able to synthesize RNA on a template itself consisting of RNA. This serves both to produce mRNA to support the process of infection (RNA transcription) and to perpetuate the virus itself (RNA replication). It happens in both procaryotic and eucaryotic cells. Yet a further reaction is possible in the case of the (eucaryotic) retroviruses, in which viral RNA serves as template for reverse transcription to produce a DNA product.

WHAT IS RNA POLYMERASE?

Transcription involves three distinct stages of reaction: initiation, elongation, and termination. The first and last encompass the interactions by which an enzyme binds to DNA to start RNA synthesis or dissociates from it at the cessation of synthesis. Between these stages, the central process of RNA synthesis involves the movement of the enzyme along DNA, producing an RNA chain whose sequence is determined by the template.

Originally defined simply by its ability to incorporate ribonucleoside triphosphates into ribonucleic acid under the direction of a DNA template, the enzyme **RNA polymerase** now can be seen as part of a more complex apparatus involved in transcription. The original enzymatic activity is the *minimum* component that can be described as RNA polymerase. It supervises the base pairing of the substrate ribonucleotides with DNA, and catalyzes the formation of phosphodiester bonds between them.

But ancillary activities may be needed both to initiate and to terminate the synthesis of RNA. The analogy with the division of labors between the ribosome and the protein synthesis factors is obvious. Sometimes it is difficult to decide whether a particular polypeptide that is involved in transcription at one of these stages should be considered a part of the "RNA polymerase" or an ancillary factor.

The products of transcription are discrete molecules that have predefined 5′ and 3′ ends. This means that transcription must initiate and terminate at particular sites on DNA. Initiation involves the formation of a complex between RNA polymerase and DNA at a site that surrounds the first base to be transcribed into the RNA. The entire sequence of DNA that is necessary for complex formation is called the **promoter.** This may include both the sequence that is stably bound in the initiation complex, and also other sequences in the vicinity (whose recognition is necessary, but that are not part of the stable binding site). The point at which the first nucleotide is incorporated is called the **startsite** or **startpoint.**

The activity required for initiation usually appears to be more complex than that required for elongation. All the components involved in elongation are necessary for initiation, but additional polypeptides also are involved. These additional polypeptides may or may not be recognized as subunits of the RNA polymerase. The distinction tends to rest on the generality of their involvement. If they constitute part of the usual enzyme preparation that happens to be involved only in the initiation step, but that is needed to recognize all promoters, they are likely to be classified as part of the enzyme. If they are needed only for the transcription of particular genes, they are likely to be classified as ancillary control factors.

Termination involves the ability to recognize the point at which no further bases should be added to the chain. The corresponding DNA site is often called the **terminator.** Two reactions are called for on the part of RNA polymerase. Its enzymatic activity must cease: no further phosphodiester bonds are synthesized. And it must dissociate from DNA and RNA, while the RNA is released.

Termination is less well defined than initiation. In bacteria there are different classes of terminator sites. At least some of them are not recognized as such by the form of the enzyme that is involved in elongation; additional protein factors are necessary. Again, there is variation in the degree to which these can be regarded as components of the enzyme. In eucaryotes, the issue of termination is cloudy, and it may prove to some degree to be a matter of taste how the term "RNA polymerase" is defined.

With bacterial enzymes, it is possible to begin to define the roles of individual polypeptides in the stages of transcription. With eucaryotes, the enzymes are less well purified, and the actual enzymatic activities have yet to be resolved from the crude preparations. So we shall take a broad view in seeking to define all the activities involved in transcribing RNA from DNA.

In this chapter, we shall deal with the mechanism of transcription and with the properties inherent in the preparations of RNA polymerase as commonly purified. In Chapter 11, we shall consider the nature of promoters. Then in Chapters 12 and 13, we shall come to the issue of the control of transcription via bacterial RNA polymerase. The stages of both initiation and termination are subject to regulation in both a host bacterium and in phage-infected cells, through the use of ancillary protein factors that modify or oppose intrinsic tendencies of the RNA polymerase enzyme.

SUBUNIT STRUCTURE OF BACTERIAL RNA POLYMERASE

A single type of RNA polymerase is responsible for all synthesis of mRNA, rRNA, and tRNA in bacteria. The total number of RNA polymerase molecules present in an *E. coli* cell may be about 7000. A large proportion of these are likely to be involved in transcription, with probably between 2000 and 5000 enzymes engaged in RNA synthesis at any one time, depending on the growth conditions. However, the major RNA polymerase does not provide the sole RNA-synthesizing capacity of the cell, since one of the proteins involved in replication of DNA (the *dnaG* product) is involved in synthesizing an RNA primer for DNA synthesis (see Chapter 33). But the function of this protein is restricted to the synthesis of RNA for use in replication.

The RNA polymerase that has been characterized best is that of *E. coli*, but its subunit structure is very similar in all other bacteria studied. The **complete enzyme** or **holoenzyme** has a molecular weight of about 480,000 daltons. It has the subunit constitution

2	α	40,000 each
1	β	155,000
1	β'	160,000
1	σ	85,000

The α, β, and β' subunits show rather constant molecular weights in different bacterial species; the σ varies more widely, from 44,000 to 92,000. The enzyme has a rather elongated structure with a maximum dimension of 15 nm.

The holoenzyme ($\alpha_2\beta\beta'\sigma$) can be separated biochemically into two components, the **core enzyme** ($\alpha_2\beta\beta'$) and the **sigma factor** (the σ polypeptide). The names reflect the fact that only the holoenzyme can initiate transcription; but then the sigma "factor" is released, leaving the core enzyme to undertake elongation. Thus the core enzyme has the ability to synthesize phosphodiester bonds with a DNA template, but cannot initiate transcription at the proper sites.

SIGMA FACTOR CONTROLS BINDING TO DNA

The function of the sigma factor is to ensure that RNA polymerase binds stably to DNA only at promoters, not at other sites.

The core enzyme itself has an affinity for DNA, in which electrostatic attraction between the basic protein and the acidic nucleic acid plays a major role. Probably this general ability to bind to any DNA, irrespective of its particular sequence, is a feature of all proteins that have specific binding sites on DNA (see Chapter 14). Any sequence of DNA that is bound by RNA polymerase in this reaction is described as a **loose binding site;** the enzyme●DNA complex is described as **closed,** because the DNA remains strictly in the double-stranded form. The association constant for forming a closed complex at a loose binding site is about 2×10^{11} M^{-1}; the half-life for dissociation of the enzyme from DNA is long, about 60 minutes.

Sigma factor introduces a major change in the affinity of RNA polymerase for DNA. The holoenzyme has a drastically *reduced* ability to recognize loose binding sites—that is, to bind to any general sequence of DNA. The association constant for the reaction is about 10^7 M^{-1}, and its half-life is less than 1 second. Thus sigma factor destabilizes the general binding ability by a factor of about 10,000, and the complexes are *very* short-lived.

But sigma factor also confers the ability to recognize *specific* binding sites. The holoenzyme binds to promoters very tightly, with an association constant of 10^{14} M^{-1} and a half-life of several hours. This association constant is an average value. Actually, there is (roughly) a hundredfold variation in the rate at which the enzyme binds to different promoters (determined by their particular DNA sequences), and this is a principal factor in determining how frequently initiation occurs at an accessible promoter.

As illustrated in **Figure 10.1,** the recognition of promoters by holoenzyme is very different from the reac-

Figure 10.1
Initiating transcription requires several steps, during which a closed binary complex is converted to an open form and then into a ternary complex.

tion of core enzyme with loose binding sites. The holoenzyme●promoter reaction starts with the formation of a similar closed (binary) complex. But then it is converted into an **open complex.** This involves the "melting" of a short region of DNA within the sequence bound by the enzyme. Typically the enzyme covers about 60 bp of DNA, within which 12–17 bp (according to different estimates; see later) are broken to generate a single-stranded region. The series of events leading to formation of the open complex is called **tight binding.**

Consistent with the need to disrupt the structure of DNA, the RNA polymerase forms open complexes and transcribes negatively supercoiled circular DNA more rapidly than linear DNA. The stress placed on a double helix by the supercoiling makes it easier to unwind the two strands. This may be important in initiating transcription at least at some promoters (see Chapter 11). Because its formation involves separation of the DNA strands, the holoenzyme●DNA open complex has decreased stability at low temperature or high salt, whereas the core enzyme●DNA closed complex has exactly the opposite dependence on these parameters.

The features of these types of binding suggest that RNA polymerase may find promoters on DNA by the process of trial and error illustrated in **Figure 10.2.** If any core enzyme is free in the cell, it would exist largely in the form of closed loose complexes, because the enzyme would enter into them rapidly and would leave them slowly. (We don't know for sure how much of the free RNA polymerase in the cell is in the form of core enzyme and how much exists as holoenzyme.) The holoenzyme associates with, and dissociates from, loose binding sites very rapidly. So it is likely to continue to make and break a series of closed complexes in an agitated manner until (by chance) it encounters a promoter. Then its recognition of the specific sequence will allow tight binding to occur by formation of an open complex.

Now the rate constant for binding promoters is very close to the limit imposed by the speed of diffusion. Since binding is so rapid, the dissociation of holoenzyme from loose binding sites should be the rate-limiting step. But this step actually is too slow, about 1 dissociation per second. RNA polymerase may therefore use another means to seek its binding sites, possibly the direct displacement of one bound sequence by another. This reaction could occur at about 1 displacement per millisecond. In other words, instead of moving about by leaving one binding site and diffusing to another, the enzyme is likely to take hold of one sequence of DNA, exchange it very rapidly for another, and continue in this promiscuous manner to exchange sequences until a promoter is obtained. Then the enzyme forms a stable, open complex, after which initiation occurs.

How does sigma factor influence the ability of the core enzyme to bind to DNA? Early experiments suggested that sigma does not itself bind directly to duplex DNA. However, it is possible that it may be able to bind to DNA that is supercoiled, although we have no information on whether it is able to recognize spe-

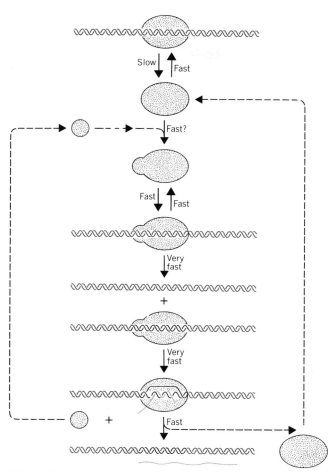

Figure 10.2
Sigma factor and core enzyme recycle at different points in transcription.

Core enzyme is released at termination; it will either bind to loose sites on DNA or will complex with sigma factor to form holoenzyme that can bind stably only at promoters. Sigma factor is released as soon as a ternary complex has formed at an initiation site.

THE SIGMA FACTOR CYCLE

Sigma factor is involved *only* in initiation. It is released from the core enzyme as soon as RNA synthesis has been initiated.

After the open (binary) complex has been formed, the next step is to incorporate the first two nucleotides, and then a phosphodiester bond forms between them. This creates a **ternary complex** of core polymerase●DNA●nascent RNA, as indicated in Figure 10.1. This reaction occurs so fast that the half-life of the binary complex at a tight binding site is only 0.2 second. Ternary complex formation completes the initiation reaction; and sigma factor is not part of the ternary complex.

We do not know the exact relationship between the dissociation of sigma factor and the synthesis of the phosphodiester bond. Sigma might be directly displaced by the (dinucleotide) nascent RNA; or its release might be caused by a change in the properties of the core enzyme that is induced by the nascent RNA. At all events, from this point on, transcription is accomplished by a ternary complex containing core enzyme.

The core enzyme in the ternary complex is very tightly bound to DNA. It is essentially "locked in" until elongation has been completed. When transcription terminates, the core enzyme is released from DNA as a free protein tetramer. As illustrated in Figure 10.2, it must then find another sigma factor in order to undertake a further cycle of transcription.

The existence of a cycle in which sigma factor and core enzyme come together only temporarily solves the dilemma of RNA polymerase in reconciling its needs for initiation with those for elongation. It is a dilemma because initiation requires tight binding *only* to particular sequences (promoters), while elongation requires tight binding to *all* sequences along which the enzyme must progress. The core enzyme has a high intrinsic affinity for DNA, which is increased by the presence of nascent RNA. But its affinity for loose binding sites remains too high to allow the enzyme to find promoters efficiently: the associations and dissociations involved in the trial and error of finding a tight binding site could take many hours. By reducing the stability of the loose complexes, sigma allows the process to occur much more rapidly; and by stabilizing the association at tight

cific nucleotide sequences. But we do know that when the holoenzyme forms a tight binding complex at a promoter, the sigma factor contacts DNA at a point in the region of initial melting, just before the startpoint. The addition of sigma factor may change the conformation of the core enzyme, so that it recognizes loose binding sites less effectively, but in such a way that the *combination* of core and sigma can specifically contact the tight binding sites.

binding sites, sigma factor drives the reaction irreversibly into the formation of open complexes. But then the holoenzyme would be paralyzed by its specific affinity for the promoter; so by releasing sigma, the enzyme reverts to a general affinity for all DNA, irrespective of sequence, that suits it to continue transcription.

THE CORE ENZYME SYNTHESIZES RNA

Only one strand of the DNA template is transcribed into RNA. To make this **template strand** available for base pairing with ribonucleotides, DNA must be unwound into its two strands. The core enzyme starts transcription at the separated DNA strands of an open promoter complex. As the enzyme moves along the template extending the RNA chain, a region of local unwinding moves with it.

Within the stretch of 60 bp or so that is associated with the enzyme, there will be only a shorter region in which the DNA strands are separated. The length of this unwound region has been hard to estimate. The susceptibility of the DNA in an open complex to a nuclease specific for single strands suggests a distance of about 12 bp (see Chapter 11). The extent of unwinding of DNA when RNA polymerase binds to a circular duplex molecule corresponds to about 17 separated base pairs.

As the DNA is unwound to provide the template, probably each of its strands enters a separate site in the enzyme structure. As **Figure 10.3** indicates, the template strand will be free just ahead of the point at which the ribonucleotide is being added to the RNA chain, and it will exist as a DNA-RNA hybrid in the region where RNA has just been synthesized. Thus the length of the hybrid region may be a little shorter than the stretch of unwound DNA. Probably the RNA-DNA hybrid is about 12 bp long.

As the enzyme leaves the area, the DNA duplex reforms, and the RNA is displaced as a free polynucleotide chain. Probably about the last 50 ribonucleotides added to the growing chain are complexed with DNA and/or enzyme at any moment. The area covered by the enzyme in which DNA is unwound and RNA is being synthesized is sometimes called the **growing point** (although this term is more usually used of DNA replication).

We still do not really understand the topology of the unwinding and rewinding, but the ability of purified RNA polymerase to transcribe double-stranded DNA *in vitro* implies that the reaction per se is accomplished by an intrinsic property of the enzyme. The formal possibility that the enzyme revolves around the axis of DNA as it proceeds seems unlikely; for this would involve movement of the entire RNA chain and its associated ribosomes. Another possibility is that the DNA revolves in the unwinding sense ahead of the enzyme, and revolves in the opposite sense behind it. This could require the assistance *in vivo* of other enzymatic activities to adjust the topology of the DNA accordingly.

All nucleic acids are synthesized from nucleoside 5′ triphosphate precursors. **Figure 10.4** shows the condensation reaction between the 5′ triphosphate group of the incoming nucleotide and the 3′-OH group of the last nucleotide to have been added to the chain.

The core enzyme must hold the two reacting groups in the proper apposition for phosphodiester bond formation; then, once they are covalently linked, it moves one base farther along the DNA template so that the reaction can be repeated. The reaction rate is fast, about 40 nucleotides/second at 37°C (see Chapter 9). As shown in Figure 10.3, the position of the last nucleotide added to the chain defines the **primer terminus site.** The location occupied by the incoming

Figure 10.3

RNA polymerase covers ~60 bp of DNA and has several active centers.

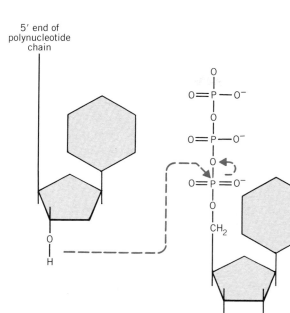

5′ end of
polynucleotide
chain

Figure 10.4
Phosphodiester bond formation involves a hydrophilic attack by the 3′-OH group of the last nucleotide of the chain on the 5′ triphosphate of the incoming nucleotide, with release of pyrophosphate.

The nucleic acid chain is said to grow from 5′ to 3′ because the 5′ end remains constant and each new nucleotide extends it one residue farther at the 3′ end.

nucleoside triphosphate is the **elongation nucleotide site.**

A difference in the source of the two reacting groups is seen at initiation, when the enzyme must directly bind two nucleoside triphosphates. The first base of an RNA chain is most commonly (but not always) a purine, and the enzyme has the ability *in vitro* to bind ATP and GTP to a site that is distinct from the elongation nucleotide site. This is known as the **initiation nucleotide site,** and it must largely overlap with the primer terminus site. The first nucleotide of the chain retains all its 5′ triphosphate residues.

All four nucleoside triphosphates can enter the elongation nucleotide site. The acceptability of an incoming precursor is judged by its base pairing with the template strand of DNA, an action apparently supervised by the enzyme. Probably the elongation nucleotide site has a structure that allows phosphodiester

bond formation to proceed only when the nucleotide is properly base paired with DNA. Presumably the nucleotide is expelled if its ability to base pair is deemed inadequate; then another can enter. (There is roughly a 25% chance of correct entry, which is much more favorable than the 3% chance that the correct aminoacyl-tRNA will enter the ribosome A site.)

Divalent cations are important in RNA polymerase function. Ribonucleotides enter the elongation nucleotide site as Mg^{2+}-NTP chelates. The core enzyme contains two atoms of zinc, which are probably involved in maintaining the nucleotide in the primer terminus/initiation nucleotide site in a conformation appropriate for reaction with the incoming nucleotide.

FUNCTIONS OF THE CORE SUBUNITS

Our knowledge of the topology of the core enzyme is really very primitive, and the best we can do at present is to make a diagrammatic representation of the sites defined by the various enzymatic functions, as illustrated in Figure 10.3. None of these sites has yet been physically located on one (or more) of the polypeptide subunits. However, there is some general information about the roles of individual subunits.

Two types of antibiotic both act on the β subunit, as defined by the location of mutations conferring resistance to them. The **rifamycins** (of which rifampicin is the most used) prevent initiation, acting prior to formation of the first phosphodiester bond, after which the ternary complex becomes resistant to inhibition. **Streptolydigins** inhibit chain elongation. The β subunit is also implicated as their target by reconstitution experiments, in which it proves to be the source of the β subunit that determines the response to the antibiotic. Also, the β subunit is the polypeptide labeled by certain affinity analogs of the nucleoside triphosphates. Together these results suggest that the β subunit may be involved in binding the nucleotide substrates.

Heparin is a polyanion that binds to the β′ subunit and inhibits transcription *in vitro*. The heparin competes with DNA for template binding prior to formation of a stable initiation complex. The β′ subunit is the most basic, which would fit with a role in template binding.

The α subunit exists in isolated form as a dimer that is able to bind to the β subunit; the $\alpha_2\beta$ complex then binds β' to form the core enzyme. When phage T4 infects *E. coli,* the α subunit is modified by ADP-ribosylation of an arginine. This is associated with a reduced affinity for the usual promoters that the holoenzyme recognizes. It is therefore possible that the α subunit plays a role in promoter recognition.

No one of the subunits can be singled out as the binding site for σ. All of them, directly or indirectly, appear to contribute to the conversion of core enzyme to holoenzyme.

PHAGE RNA POLYMERASES MAY BE "MINIMUM" ENZYMES

Why does bacterial RNA polymerase require a large, multimeric structure? The existence of much smaller RNA polymerases, comprising single polypeptide chains coded by certain phages, demonstrates that the apparatus required for RNA synthesis can be much smaller than that of the host enzyme.

These enzymes give some idea of the "minimum" apparatus necessary for transcription. They recognize a very few promoters on the phage DNA; they have no ability to change the promoters that they recognize. Thus they are limited to the intrinsic ability to recognize a very few specific DNA binding sequences and to synthesize RNA. How complex are they?

The RNA polymerases coded by the related phages T3 and T7 are single polypeptide chains of about 110,000 daltons each. They synthesize RNA very rapidly (at rates of about 200 nucleotides/second at 37°C). The T3 enzyme starts all RNA synthesis with formation of the dinucleotide pppGpG. This restriction suggests that the process of initiation is probably simpler than that of the *E. coli* enzyme.

By contrast, the enzyme present in the host bacterium is able to transcribe any one of many (say, about 1000) transcriptional units. Some of these units it is able to transcribe directly, without any further assistance, although the efficiency with which it does so is a characteristic of the affinity of the enzyme for the particular promoter. Many units, however, can be transcribed only in the presence of further protein factors. Sometimes these are specific for a single transcrip-

tional unit; sometimes they are involved in coordinating transcription from many units. Certain phages induce general changes in the affinity of the host RNA polymerase for promoters, so that it stops recognizing host genome sites and instead initiates at phage promoters. Thus the host enzyme requires the ability to interact with a variety of host and phage functions that modify its transcriptional activities. The complexity of the enzyme may therefore at least in part reflect the multiplicity of control signals to which it must respond.

COMPLEX EUCARYOTIC RNA POLYMERASES

The transcription apparatus of eucaryotic cells is more complex and less well defined than that of bacteria. There are three nuclear RNA polymerases, occupying different locations, each with a very large number of polypeptide components. Also, different RNA polymerase activities are found in mitochondria and chloroplasts.

The most prominent RNA-synthesizing activity is the enzyme RNA polymerase I, which is found in the nucleolus and is responsible for transcribing the genes for rRNA. This accounts for 50–70% of cellular RNA synthesis. The enzyme is stimulated more or less equally well by Mn^{2+} and Mg^{2+} ions.

The other major enzyme is RNA polymerase II, located in the nucleoplasm (the part of the nucleus excluding the nucleolus), representing 20–40% of cellular activity. This is responsible for the synthesis of heterogeneous nuclear RNA (hnRNA), the precursor to mRNA. The enzyme is stimulated much more effectively by Mn^{2+} ions than by Mg^{2+} ions.

A minor enzyme activity is RNA polymerase III, providing up to 10% of the cellular capacity for RNA synthesis. The enzyme is nucleoplasmic and is responsible for synthesis of many of the small nuclear RNAs and tRNAs. It is stimulated slightly more efficiently by Mn^{2+} than by Mg^{2+} ions.

The characteristic dependence on divalent cations was at one time a feature used to distinguish between the enzymes (which is why we have mentioned it). Now, however, the major distinction between the enzyme activities is drawn from their response to the bicyclic octapeptide **α-amanitin.** In animal, plant, and insect cells, the activity of RNA polymerase II is rapidly inhibited by low concentrations of α-amanitin (about 0.03 μg/ml). This property is therefore conserved rather widely. The activity of RNA polymerase III is inhibited in animal cells at much greater levels, about 20 μg/ml; but in insect and yeast cells it is highly resistant to inhibition. In all cells the RNA polymerase I enzyme is not inhibited. ("Not inhibited" means that more than, say, 500 μg/ml is required to produce an inhibitory effect.)

The crude enzyme activities all are large proteins, appearing as aggregates of 500,000 daltons or more. Their subunit compositions are complicated. Each enzyme has two large subunits, generally one about 200,000 daltons and one about 140,000 daltons. There are up to 10 smaller subunits, ranging in size from 10,000 to 90,000 daltons. We do not know whether any of the subunits found in the different enzymes are the same.

Because it is not yet possible to reconstitute active RNA polymerase from the subunits of any one of these enzymes, there is no evidence as to whether all of the protein subunits are integral parts of each enzyme. We do not know which subunits may represent catalytic activities and whether others may be involved in regulatory functions. It is possible that the enzyme preparations represent the basic transcription apparatus—essentially similar in all cells, and subject to regulation by further protein factors. Or the enzyme preparations may include such factors as well as a basic catalytic apparatus. The central issue of the constitution of the transcription apparatus, whether selection of genes to be transcribed resides in components isolated with it or instead with ancillary proteins yet to be identified, generally remains unresolved.

Isolated enzyme preparations can be used to synthesize RNA *in vitro,* and the reaction initiates at the appropriate promoter sites. Some heterologous systems have been developed. For example, an RNA polymerase II preparation from cultured human KB cells can transcribe the mouse globin gene. This means that neither tissue nor species-specific features are involved in the recognition per se of the promoter by

RNA polymerase. Thus there is some basic and conserved feature that is recognized as a promoter. Of course, this does not exclude the possibility that further protein factors or other sequences are involved in modulating the reaction in the natural situation.

In another system, RNA polymerase III of *X. laevis* can specifically transcribe the genes for 5S RNA only when provided with an additional protein factor, a 37,000 dalton polypeptide that is found complexed with the 5S RNA product in oocytes. This appears to be an ancillary factor necessary for the basic transcription apparatus to work with the 5S RNA genes, although not needed with other genes, since the RNA polymerase III preparations can transcribe tRNA genes without the factor (see Chapter 11).

The overall complexity of the eucaryotic transcription apparatus is therefore being defined only system by system, so at present all we can say is that there are three multimeric enzymes needed for RNA synthesis. Whether all their components are essential, and how many other proteins are needed, remains to be seen.

The RNA polymerase activities of mitochondria and chloroplasts appear to be smaller and distinct from the nuclear enzymes. Of course, the organelle genomes are much smaller, the resident polymerase needs to transcribe only a few genes, and the control of transcription is likely to be very much simpler. So these enzymes may be analogous to the phage enzymes that have a single fixed purpose and do not need the ability to respond to a more complex environment. None of these enzymes has yet been purified in sufficient quantity to use for an *in vitro* system.

FURTHER READING

A source for many reviews and original research articles is the volume edited by **Losick** & **Chamberlin,** *RNA Polymerase* (Cold Spring Harbor Laboratory, New York, 1976). Two chapters that cover the general matters dealt with here are written by **Chamberlin,** giving first a general overview (pp. 17–68) and then an account of interactions of the bacterial enzyme with its template (pp. 159–192).

CHAPTER 11
PROMOTERS:
THE SITES FOR
INITIATING TRANSCRIPTION

The key question in examining the interaction between an RNA polymerase and its promoter is how a protein can recognize a specific sequence of DNA. Does the enzyme have an active site that distinguishes the chemical structure of particular bases in the DNA double helix? How specific are its requirements? Promoters differ in their affinities for RNA polymerase, and this can be an important factor in controlling the frequency of initiation and thus the level of gene expression. How do changes in DNA sequence influence the ability to bind the enzyme?

Initiation of transcription is a crucial stage at which gene expression can be controlled. Often it is the primary, sometimes the only, step at which the decision is taken on whether or not to express a particular gene. What controls the ability of RNA polymerase to initiate transcription at a particular promoter? Some promoters are competent to support initiation by RNA polymerase, although extraneous protein factors may act to *prevent* the initiation process. In other cases, the polymerase itself is not adequate, and ancillary proteins are *necessary* for initiation to occur. The additional protein factors act by recognizing sequences of DNA that may be close to, or overlap with, the promoter itself. Thus while the promoter is defined in terms of the sequences that react with RNA polymerase, the

context within which it is located also may influence the ability to initiate transcription.

As sequences of DNA whose function is to be *recognized* by proteins, a promoter and any adjacent control sites differ from other sequences whose role is exerted by being transcribed or translated. The information for promoter function is provided directly by the DNA sequence: its structure is the signal. By contrast, expressed regions gain their meaning only after the information is transferred into the form of some other nucleic acid or protein.

A promoter is necessary for the transcription of a particular region of DNA physically contiguous with it. (In genetic terms, sequences of DNA that function without being expressed as diffusible molecules of RNA or protein are described as **cis-dominant** or **cis-acting** sites. This means that they are concerned only with the sequences lying on the same double helix of DNA on which they reside. They cannot influence homologous sequences lying on independent duplex DNA molecules. See Chapter 14.)

Once transcription has initiated at the startpoint, RNA polymerase continues along the template until a terminator sequence is reached. This action defines a **transcription unit** that extends from the promoter to the terminator. The critical feature of the transcription

174

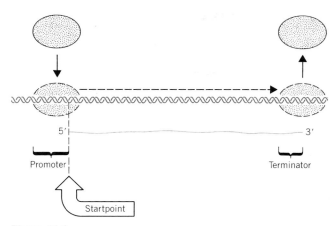

Figure 11.1

A transcription unit is a sequence of DNA transcribed into a single RNA, starting at the promoter and ending at the terminator.

unit, as depicted in **Figure 11.1,** is that it constitutes a stretch of DNA that is expressed via the production of a single RNA molecule. Depending on the system, it may include only one or several genes. Sometimes regions close to the promoter are described as **proximal,** while those farther from it are described as **distal.**

DEFINING THE STARTPOINT *IN VIVO* AND *IN VITRO*

The startpoint is the base pair on DNA that corresponds to the first nucleotide incorporated into the transcript by RNA polymerase (in terms of mechanism, this is the nucleotide that enters the initiation nucleotide site). Usually this is a single position, sometimes it is either of two adjacent base pairs, and occasionally it may involve any one of several adjacent positions. To define the startpoint, it is necessary to examine the 5′ end of the **primary transcript,** the immediate product of RNA polymerase before any changes occur to the RNA. Most information about startpoints is provided by *in vitro* studies, because of the difficulty of isolating the primary transcript *in vivo* before it is degraded (in procaryotes) or modified (in eucaryotes).

A common method for identifying the startpoint by hybridizing the transcript with its DNA template is illustrated in **Figure 11.2.** In principle, this consists of degrading all the DNA that cannot hybridize with the RNA, and then determining the sequence of the surviving DNA.

In bacteria, the 5′ end of the RNA is identified unequivocally by its retention of a triphosphate terminus. Although the instability of mRNA makes it difficult to isolate *in vivo,* for rRNA and tRNA it is often possible to block the processing of the primary transcript (which is a precursor to the mature RNA) and thus to identify the 5′ end. For all loci, the isolated genes can be used as templates for transcription *in vitro,* when the product can be isolated and the corresponding locations on DNA identified precisely. In those cases in which the same genes have been examined *in vivo* and *in vitro,* the same startpoints are used in each situation; this lends confidence to the use of *in vitro* systems to identify the startpoint when *in vivo* information is not available.

In eucaryotes, the precursors to mRNA are capped so rapidly after the initiation of transcription that it has been impossible to isolate molecules retaining the original 5′ triphosphate end *in vivo.* The capping reaction occurs only with the 5′ tri- or diphosphate as substrate, implying that the capping site must correspond with the startpoint. This view is supported by the development of systems for transcription *in vitro,* since in every case the first base of the transcript corresponds with the nucleotide found adjacent to the cap *in vivo.* On this assumption, startpoints for RNA polymerase II have been located by identifying the sequences on DNA that correspond to the start of the mRNA isolated *in vivo.*

For the products of RNA polymerase I (rRNA) and RNA polymerase III (small nuclear RNA and tRNA) that are not capped, it is possible to isolate RNA transcripts that still possess 5′ triphosphate ends. This allows the startpoints used *in vivo* to be identified unequivocally. (Some small nuclear RNAs are capped and so cannot be analyzed like this.)

Capping does not occur in organelles, and so in principle there is no impediment to identifying the startpoints used by all species of mitochondrial and chloroplast RNA products.

Figure 11.2

Startpoints can be determined by comparing the RNA product with the DNA template.

When the RNA is hybridized with the denatured strands of DNA, it forms an RNA-DNA hybrid with the strand that acted as its template; the other strand of DNA remains unpaired. Treatment with the enzyme S1 nuclease, which specifically degrades single-stranded DNA, destroys both the unpaired strand and regions of the template strand beyond the transcription unit. The RNA of the RNA-DNA hybrid can be removed. Then the sequence of the DNA can be determined or its length can be used to locate the position of the RNA on the template.

BINDING SITES FOR
E. COLI RNA POLYMERASE

The tight binding sites at which the RNA polymerase holoenzyme forms stable initiation complexes with DNA lie within promoters. These sequences of DNA can be recovered by the protocol summarized in **Figure 11.3,** in which RNA polymerase is bound *in vitro* to a template consisting of a particular transcription unit. Then digestion with the enzyme DNAase is used to degrade all the regions of DNA that are not protected by the RNA polymerase.

The fragments that are recovered vary from 41 to 44 bp in length. The polymerase actually can initiate transcription if left attached to the protected fragment;

this leads to the synthesis of a short RNA of 17–20 bases, terminated when the polymerase runs off the end of the fragment. This shows that the protected fragment corresponds to the sequence at which transcription is initiated, and locates the startpoint in the center of this binding site.

The sequences prior to the startpoint are described as **upstream;** those after the startpoint (that is, within the transcribed sequence) are **downstream.** It is conventional to write sequences so that transcription proceeds from left (upstream) to right (downstream). This corresponds to writing the mRNA in the usual 5′ to 3′ direction. Often the DNA sequence is written just to show the strand whose sequence is the same as the RNA. Base positions are numbered in both directions

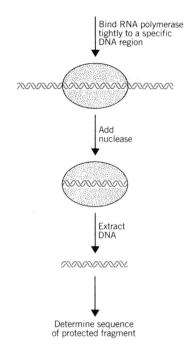

Bind RNA polymerase
tightly to a specific
DNA region

Add
nuclease

Extract
DNA

Determine sequence
of protected fragment

Figure 11.3

DNA binding sites for RNA polymerase can be recovered because they are protected by the enzyme against degradation by nuclease.

away from the startpoint, which is assigned the value +1, numbers then increasing downstream. Proceeding upstream, positions are numbered −1, increasing away from the startpoint. In these terms, the protected fragments extend from about −20 to +20.

A significant feature of the binding site is that, if the fragment is recovered, RNA polymerase cannot rebind to it. This means that, although the protected fragment is the sequence tightly bound by RNA polymerase, and is sufficient to support initiation, it is not adequate for initial binding. Other sequences beyond it must be necessary to recognize DNA. The promoter therefore includes these additional sequences as well as the tight binding site. These results leave open two possible types of structure for the promoter.

It may comprise a single binding site for the enzyme, but the additional sequences may be less tightly associated with it, or more exposed in some way, so that they are susceptible to added nuclease, while the tightly bound sequences are protected. In this case,

the discrepancy between the protected fragment and its inability to rebind simply reflects the geometry of the reaction between RNA polymerase and DNA.

Alternatively, the promoter may actually contain two types of sequence. The first could be recognized by the enzyme as a prerequisite for binding to the second. The transition from recognition to binding might involve simply a change in conformation—for example, a contraction along the DNA so that only the protected sequence is covered. Or it could represent a movement physically from a recognition site to a binding site.

The ability of RNA polymerase to recognize DNA can be defined further by a technique known as **footprinting.** A sequence of DNA bound to RNA polymerase (or any other protein) is partially digested with an **endonuclease.** This independently attacks individual phosphodiester bonds *within* a nucleic acid. Under the appropriate conditions, every phosphodiester bond that is in principle accessible to the nuclease is broken in some, but not in all, molecules of DNA. Only if RNA polymerase is blocking the access of the nuclease to DNA will a particular bond fail to be broken at all.

The positions that are broken are recognized by using DNA labeled on one strand at one end only. As **Figure 11.4** shows, following the nuclease treatment, the broken DNA fragments are recovered and electrophoresed on a gel that separates them according to their length. (The principle of using *partial* cleavage of each susceptible site followed by isolating an end-labeled fragment is the same developed for sequencing DNA, as described in Figure 3.4.) For every bond position that is susceptible, a band is found on the gel, corresponding to the distance from the labeled end to the site of breakage. For every position that is protected against cleavage by RNA polymerase, a band is missing. Each of the two strands of DNA can be analyzed separately by using it as the labeled strand. By combining the footprinting with DNA sequencing, the nucleotide sequence of the binding site can be determined as well as its position.

The results of such an analysis of one promoter (for the *lac* genes) show that the RNA polymerase binding site is more extensive than detected by the experiments to recover intact protected fragments. It extends for about 50 bp upstream from the startpoint

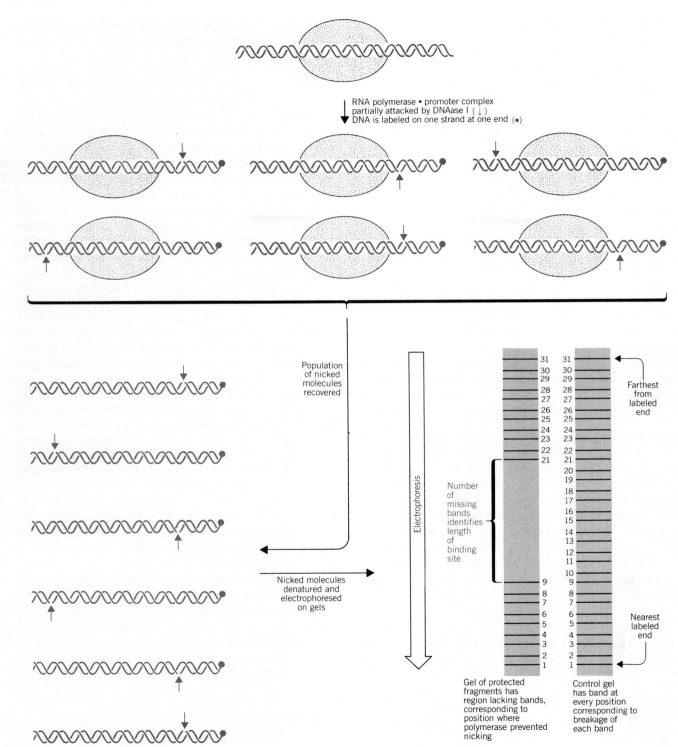

RNA polymerase • promoter complex
partially attacked by DNAase I (↓)
DNA is labeled on one strand at one end (●)

Population
of nicked
molecules
recovered

Electrophoresis

Nicked molecules
denatured and
electrophoresed
on gels

Number
of
missing
bands
identifies
length
of
binding
site

Farthest
from
labeled
end

Nearest
labeled
end

Gel of protected
fragments has
region lacking bands,
corresponding to
position where
polymerase prevented
nicking

Control gel
has band at
every position
corresponding to
breakage of
each band

and for about 20 bp downstream. The two strands are not equally well protected in all regions, especially at the ends of the binding site, which implies that the RNA polymerase has an asymmetric conformation when bound to the DNA. This accords with the need to transcribe only one strand of the DNA.

Somewhat similar results can be obtained with digestion experiments using an **exonuclease.** An exonucleolytic enzyme attaches to an end of DNA and degrades it continuously. If a protein is bound to the DNA, digestion stops when the nuclease encounters the protein. It is possible again to define the limits of the promoter in terms of the point on each side at which the exonuclease is blocked from proceeding. Once again, this shows that the sequence bound by RNA polymerase extends for about 44 bp upstream from the startpoint.

Together these results show that RNA polymerase binds asymmetrically to a region of DNA stretching from about 44–50 bp upstream to a point about 20 bp downstream. The terminal 25–30 bp of the binding site upstream must be more loosely associated with the enzyme. This region is bound sufficiently well to be protected against gentle endonucleolytic or exonucleolytic cleavage, but cannot withstand the stronger conditions that are used to retrieve intact fragments from the enzyme. These data argue in favor of a model in which there is a single binding site for RNA polymerase, in which (as we shall see) some points of contact are more important than others.

SEQUENCE HOMOLOGIES IN *E. COLI* PROMOTERS

An obvious way for a promoter to be constructed would be for an invariant sequence of DNA to provide the signal for binding RNA polymerase. Given the size of the bacterial genome, the minimum length that could provide an adequate signal is 12 bp. Any sequence shorter than this is likely to occur (just by chance) a sufficient number of additional times to provide false signals. The 12 bp sequence need not be contiguous; and, in fact, if a specified number of base pairs separates two constant shorter sequences, their combined length could be less than 12 bp, since the *distance* of separation itself provides a part of the signal (even if the intermediate *sequence* is itself irrelevent).

Attempts to identify the features in DNA that are necessary for RNA polymerase binding started with comparisons of the sequences of different promoters. Any essential nucleotide sequence should be present in all the promoters. Such a sequence is said to be **conserved.** However, a large number of promoters now have been sequenced in *E. coli* (more than 50), and a striking feature is the *lack* of any extensive con-

Figure 11.4
Footprinting identifies DNA binding sites for proteins.
The principle of the technique is that under conditions of partial digestion, every susceptible bond position in a free DNA is broken in one or another molecule, but the region covered by the DNA-binding protein is protected in all molecules that were complexed with protein. So two reactions are run in parallel: a control of pure DNA; and an experimental mixture containing the protein.

When the strands are separated and electrophoresed, a radioactive band is produced by each fragment that retains a labeled end. The position of the band corresponds to the number of bases in the fragment. The shortest fragments move the fastest, so distance from the labeled end is counted up from the bottom of the gel (see Figure 3.4). In the control, every bond is broken, generating a series of bands, one representing each base. In the figure, 31 bands can be counted. In the protected fragment, bonds cannot be broken in the region bound by the protein, so bands representing fragments of the corresponding sizes are not generated. In the figure, bands 10–20 are missing. This identifies a protein-binding site covering the region located 10–20 bases from the labeled end of the DNA.

Instead of using a single gel that analyzes DNA simply by length, both the control and experimental mixtures can be treated as shown in Figure 3.5 to generate four sequencing gels. Comparison of the two gel sets allows the sequence to be "read off" directly, identifying the nucleotide sequence of the binding site.

servation of sequence over the 60 bp associated with RNA polymerase. The sequence of much of the binding site may actually be irrelevant. But some short stretches within the promoter appear to be conserved. These are candidates for the signal recognized by RNA polymerase.

The startpoint is usually, but not always, a purine (with A found about twice as often as G). Presumably in the less common cases where a pyrimidine is found, the RNA polymerase is able to overcome its prejudice against inserting this base into the initiation nucleotide site. It is quite common for the startpoint to be the central base in the sequence CAT, but the conservation of this triplet is not high enough to regard it as an obligatory signal.

Just upstream from the startpoint a 6 bp region is recognizable in almost all promoters. This signal is not invariable, but is closely related to the sequence TATAAT. Sometimes it is known as the **Pribnow box.** Actually, it is quite rare for the sequence TATAAT itself to be present, but this is the **consensus** or **canonical sequence** that describes the base most often present at each position.

The identification of this sequence makes a general point. It is useful to define putative DNA recognition sites in terms of an idealized sequence that represents the predominant bases. The consensus is defined by aligning all the known examples so as to maximize their homology. For a sequence to be accepted as a consensus, each particular base must be reasonably predominant at its position—present in, say, more than 60% of the examples—and most of the actual sequences must be related to the consensus by only a small number of substitutions, say, 1 or 2.

The extent to which the base at each position of the Pribnow box is conserved varies somewhat. If bases that are found in as few as 40% of the promoters are included, the region can be extended farther. It can be summarized as

$$A_{46} \ T_{40} \ T_{40} \ T_{43} \ G_{43} \ N \ \mathbf{T_{89} \ A_{89} \ T_{50} \ A_{65} \ A_{65} \ T_{100}} \ G_{37}$$

Pribnow box

where the subscript denotes the percent occurrence of the most frequently found base; N indicates that there is no discernible preference for any base.

There is a fairly clear demarcation between the flanking sequences and the ends of the Pribnow box. If the frequency of occurrence indicates likely importance in binding RNA polymerase, we would expect the initial highly conserved TA and the final totally conserved T in the Pribnow box to be the most important bases. The bases between show intermediate conservation and presumably are involved; the next few positions upstream have a propensity toward A-T base pairs that might be helpful, although not essential.

The consensus sequence of the Pribnow box consists exclusively of A-T base pairs. This has aroused suspicion that its function may be concerned with the initial melting of DNA into single strands. The lower energy needed to disrupt the two H bonds of each A-T pair, compared with the three H bonds of a G-C pair, means that a stretch of A-T base pairs demands the minimum amount of energy for strand separation. Actually, most of the Pribnow boxes identified in bacterial promoters do have one G-C pair, so the melting reaction obviously can overcome more than the minimum requirement.

The usual assumption is that the function of a recognition site is likely to be fulfilled most effectively by the consensus sequence itself. However, the *context* of a base may be important, so (for example) two transversions of A-T to T-A pairs could compensate for each other and provide another sequence just as effective as the consensus. The efficiency of sequences related to the consensus therefore cannot be entirely predicted.

The center of the Pribnow box generally is close to 10 bp upstream from the startpoint. Sometimes it is therefore called the **−10 sequence.** The actual location of the hexamer varies from positions −11 to −5 to positions −14 to −8, more often the latter.

Similarities of sequence also occur at another location, centered about 35 bp upstream from the startpoint. This is called the **−35 sequence** and is sometimes known as the "recognition region," the name reflecting its proposed necessity for RNA polymerase recognition, in spite of its exclusion from the tightly bound protected region.

Again forming a consensus sequence by summing the most commonly occurring bases, the conserved −35 sequence is usually related fairly well to the ideal

$$T_{85} \ T_{83} \ G_{81} \ A_{61} \ C_{69} \ A_{52}$$

The distance between the −35 and −10 sites varies between 16 and 19 bp in known promoters. Although the actual sequence in the intervening region may be unimportant, these limits may be critical in holding the two sites at the appropriate separation to fit the geometry of RNA polymerase. Deletions that reduce the separation to 15 bp, or insertions that increase the distance to 20 bp, reduce the activity of the promoters in which they occur.

There are some promoters that lack the −35 sequence (as judged by the inability to discern any well enough related sequence). Usually, however, they have an adjacent site that is recognized by an ancillary protein that assists initiation, so possibly this reaction substitutes for the function of the −35 site. But since other promoters that are subject to the same assistance retain the −35 signal, the correlation is not exact. Possibly the −35 sequence is involved in establishing the efficiency with which RNA polymerase recognizes a particular promoter.

There are also a very few promoters that lack the −10 site. Thus neither of the conserved sequences is absolutely essential for promoter function. We do not yet know all the details of the recognition reaction, and so all that now can be said firmly is that the "typical" promoter can use the −35 and −10 sequences to be recognized by RNA polymerase. From their absences in other promoters, we realize that other means also can be used for recognition, but we do not know how many alternatives there are, nor exactly how they substitute for the absence of the consensus sequences.

UP AND DOWN PROMOTER MUTATIONS

Mutations in promoters affect the level of expression of the gene(s) they control, without altering the gene products themselves. Most are identified in the form of bacterial mutants that have lost, or have very much reduced, transcription of the adjacent genes. These are known as **down** mutations. Less often, mutants are found in which there is increased transcription from the promoter. These are called **up** mutations.

It is striking that both types of mutation can be produced by changes involving only a single base pair. The reaction of RNA polymerase with these sites in the promoter must be highly specific. Either the base present must be recognized directly by the enzyme or it must be essential for the process of initiation. (Of course, down mutations also can result from deletion of a more extensive stretch of the promoter.)

The information gained from studies *in vivo* simply identifies the overall nature of the change caused by the mutation. To analyze its effects at the molecular level requires characterization *in vitro* of the reaction between RNA polymerase and the mutant promoter. This can show at which stage the mutation influences the capacity of the promoter. Does it change the affinity of the promoter for RNA polymerase? Does it leave the enzyme able to bind but unable to initiate? Is the influence of an ancillary factor altered?

Figure 11.5 summarizes the locations of the point mutations that affect promoter function. Almost all of them fall within the two regions of homology at −35 and −10. The bases present at other positions in the area are clearly much less important or even irrelevant in most, but not all, promoters.

Within the Pribnow box centered at −10, the mutations confirm the importance of the three bases that are conserved. Most (18 out of 20) down mutations occur at one of these positions. Any substitution in the initial TA or final T acts as a down mutation. Almost all of these mutations result from the replacement of an A-T base pair by a G-C base pair. This reinforces the idea that the concentration of A-T base pairs is important. Yet the exceptional mutations in which there has been a transversion to replace an A-T pair with a T-A pair or vice versa show that the exclusion of G-C base pairs is not the sole determinant of function. For example, at the start of the box, one promoter has AA instead of the usual TA; but in another that has the TA, a substitution to change this to AA is recovered as a down mutation. So the bases must work together; and the context also is important in determining what is acceptable at each position.

The up mutation in this region occurs in the *lac* promoter, whose wild-type Pribnow box is TATGTT. The up mutation changes this to TATATT. So its effect is to increase homology with the consensus sequence. There must be some reason why this improvement in efficiency has not occurred spontaneously and been selected for in bacteria. (A very few other up mutations in the −10 region have been found; all occur in promoters whose homology with the Pribnow box is too poor to allow an unequivocal alignment, although probably they can be explained by an increase in homology with the consensus.)

Just to the left of the Pribnow box itself, still within the partially conserved −10 region, there are further examples of up or down mutations. Their occurrence suggests that in these particular promoters, these sites are contacted by RNA polymerase, and in such a way that the individual bases are important in these promoters, although they can be different in other promoters.

At the −35 site, most mutations represent a sub-stitution that introduces a base that must be unacceptable in the particular promoter, although the same base is acceptable in other promoters. Again, this argues for the importance of context effects, in which the RNA polymerase does not have an absolute requirement for the base present at any position, but does insist on some conjunction of bases present at particular locations. Some bases are never found at certain positions, and possibly their exclusion may be more important than the issue of which (other) base actually is present.

The effects of mutations at the −35 site are difficult

Figure 11.5
One face of the promoter contains the contact points for RNA.
The DNA sequence shows a typical promoter, with consensus sequences at −35 and −10, and a region of initial unwinding extending from within the Pribnow box to just past the startpoint.

Upper: sites at which modification prevents RNA polymerase binding are shown by arrows pointing toward the double helix. Sites at which the DNA is protected by RNA polymerase against modification are shown by the arrows pointing away from the double helix. Arrows pointing at bases indicate modifications of the base itself; arrows pointing between bases indicate modification of the connecting phosphodiester bond.

Center: sites at which mutations affect promoter function are indicated by small arrows (pointing down for down mutations and up for up mutations). The number of times a mutation is found at each position is given; this is a total for all promoters.

Lower: when the DNA is drawn as a double helix viewed from one side, as indicated diagrammatically, all the contact points lie on one face. Most lie on the partner strand (that is, not the template strand).

to interpret, because most occur in promoters that also require an ancillary protein factor to bind in this region. So it is possible that the mutation affects the ancillary protein rather than RNA polymerase itself (see later). From the more even distribution of mutations in this region, there seems to be less concentration on particular positions within or just adjacent to the -35 site, all of which is more or less equally involved in one promoter or another.

In all of this analysis, it is important to remember that "up" and "down" mutations are defined relative to the *usual* efficiency with which a promoter functions. This varies widely. So a change that is recognized as a down mutation in one promoter might not have been isolated in another (which in its wild-type state could even be less efficient than the mutant form of the first promoter). To determine the absolute effects of these mutations, it is therefore necessary to turn to studies *in vitro* that measure the affinity of RNA polymerase for different wild-type and mutant promoters under identical conditions. There is roughly a hundredfold variation in the rate at which RNA polymerase binds to different promoters *in vitro,* and this correlates with the frequency of transcription of the genes *in vivo.*

POINTS OF CONTACT IN THE PROMOTER

The points at which RNA polymerase contacts the promoter can be identified by treating RNA polymerase ●promoter complexes with reagents that modify particular bases. The presence of the enzyme may either increase or decrease the availability of a particular base (relative to a control consisting of the DNA by itself). Changes in sensitivity reflect the geometry of the complex and can be used to work out its overall conformation.

The reagent dimethylsulfate (DMS) reacts specifically with purines, to methylate G residues in the wide groove and A residues in the narrow groove of DNA. The methylation renders the purines labile to heat; so after heating, the base is lost from DNA, leaving an exposed position at which the phosphodiester backbone can be broken by alkali. (This reaction is used in determining the seqeuence of DNA, as described in Chapter 3.)

When DNA is substituted with bromouracil (BUdR)

in place of thymine, ultraviolet irradiation releases the bromine. This creates a free radical that then breaks the DNA chain.

The backbone of DNA is attacked directly by ethylnitrosourea, which converts the phosphodiester bond linking the two nucleotides to a phosphotriester (by an alkylation). The phosphotriesters are labile to alkali and provide positions at which the DNA chain can be broken.

The common feature of all these modifications is that they allow a breakage to be made at the corresponding bond in the polynucleotide chain. This site can be identified by the same approach used in the footprinting experiments (see Figure 11.4). DNA is labeled at the end of one strand, so that each breakage generates an electrophoretic band of the corresponding length. When the susceptibility of a complex of RNA polymerase bound at its promoter is compared with free DNA, some bands disappear. These identify sites at which the enzyme has protected the promoter against modification. Other bands may increase in intensity, identifying sites at which the DNA must be held in a conformation in which it is more exposed.

The reverse experiment can be performed with the methylating agent DMS. The DNA can be methylated *first;* and then it is bound to RNA polymerase. Actually, the DNA consists of a population of molecules, each methylated at some, but not all, of the susceptible sites. Those DNA molecules that cannot bind RNA polymerase are recovered and treated in the usual way to generate strand breakages whose positions can be identified. This locates points at which prior modification *prevents* RNA polymerase from binding to DNA. The same experiment can be done with the alkylation of the phosphodiester backbone.

The results of these experiments show that the regions at -35 and -10 contain most of the contact points for the enzyme. Within these regions, the same sets of positions tend both to prevent binding if previously modified, and to show increased or decreased susceptibility to modification after binding. **Figure 11.5** compares the points of contacts with the sites of mutation; although they do not coincide completely, they occur in the same limited region. At the two sites, the region of contact extends for 12–15 bp, somewhat longer than the maximally conserved region.

It is particularly interesting that the same *positions*

in different promoters may provide the contact points, even though a different base is present. This indicates that there may be a common mechanism for RNA polymerase binding although the reaction does not depend on the presence of particular bases at some of the points of contact. This model may explain why some points of contact are not sites of mutation. Also, not every mutation lies in a point of contact; could some influence the neighborhood without actually being touched by the enzyme?

RECOGNITION AND UNWINDING OF DNA

It is especially significant that the experiments with prior modification identify *only* sites in the same region that is protected by the enzyme against subsequent modification. These two experiments measure different things. The first identifies all those sites that the enzyme must recognize in order to bind to DNA. The second identifies all those sites that actually make contact in the binary complex. So if there were any positions involved in initial recognition, but not then needed to sustain stable binding, they would be revealed by sites at which prior modification prevented binding, but at which there was no protection in the complex. Since there are no such sites, and since the protected sites include all the recognition sites and also some additional positions, it seems that the enzyme first recognizes a set of sites that is necessary for it to "touch down," after which it extends its points of contact to further sites.

Another type of modification experiment allows the region of DNA that is unwound in the open binary complex to be identified directly. When the strands of DNA are separated, the unpaired adenines become susceptible to further methylation by DMS. If the DNA fragment is then recovered from the enzyme, the unwound region renatures, *except* at the methylated adenine residues, which now have methyl groups occupying the N1 position that is usually involved in base pairing. The methyl groups prevent the A bases from recognizing their T partners. This deficiency in pairing can be revealed by treatment with an enzyme, S1 nuclease, that specifically cleaves DNA at any unpaired site. As usual, the result of this breakage can be ana-

lyzed by determining the length of the end-labeled fragment that is generated.

The experiment generates a series of bands stretching from positions −9 to +3. This result shows that the region unwound for the initiation reaction includes the right end of the Pribnow box and extends just past the startpoint. Perhaps the unwinding is initiated earlier within the Pribnow box, but is not maintained there. (This measure for the extent of strand separation is less than the 17 bp estimated by the degree of unwinding; also see Chapter 10.)

When viewed in three dimensions, the points of contact upstream from the Pribnow box all lie on one face of DNA, as illustrated in Figure 11.5. These bases could be recognized in the initial formation of a closed binary complex. This would make it possible for the RNA polymerase●DNA recognition reaction to occur by contact between the enzyme and the one face of DNA, a simplifying model. As the DNA unwinding commences, further sites that originally lay on the other face of DNA might be recognized and bound.

Little is known about points of contact in the polymerase enzyme, but RNA polymerase bound at the promoter can be cross-linked to DNA in which bromouracil has been substituted for thymine. The results show that the σ subunit contacts the nontemplate strand at position −3, while the β subunit contacts position +3. This locates the sigma subunit in an appropriate position to be involved in promoter recognition and/or creating the initially unwound region. But we do not know whether it acts directly in these functions or has an indirect effect mediated via the core subunits.

The importance of strand separation in initiating transcription is emphasized by the effects of supercoiling on the template. As we have mentioned in Chapter 2, the introduction of negative supercoils into a closed molecule of DNA (one lacking free ends) assists the unwinding of the double helix. It has been known for quite some time that both procaryotic and eucaryotic RNA polymerases can initiate transcription more efficiently *in vitro* on supercoiled templates than on free DNA. The reason probably is that the supercoiled structure requires less free energy for the initial melting of DNA in the initiation complex.

The involvement of this effect in controlling promoter activity in bacteria is shown by the effects of interfering with enzymes that influence the degree of supercoiling. Among the relevant enzymes are DNA gyrase, which *introduces* negative supercoils, and topoisomerase I, which *relaxes* (removes) negative supercoils (see Chapter 32). The addition of inhibitors of DNA gyrase reduces transcription. The converse effect is seen with mutations that inactivate topoisomerase I, which increase transcription. Both effects are seen only at some promoters.

All of this suggests that the degree of negative supercoiling of DNA influences the efficiency of affected promoters. Inhibition of DNA gyrase reduces the expression of these transcription units by preventing the introduction of supercoils into DNA. At first it was thought that mutation of topoisomerase I increases expression by preventing the enzyme from removing supercoils. However, additional mutations may also be present, so the relationship between topoisomerase action and transcription is not yet clear.

Why are some promoters influenced by the extent of supercoiling while others are not? One possibility is that every promoter has a characteristic dependence on supercoiling, determined by its sequence. This would predict that some promoters have sequences that are easier to melt (and are therefore less dependent on supercoiling), while others have more difficult sequences (and thus have a greater need to be supercoiled). An alternative is that the location of the promoter might be important, if different regions of the bacterial chromosome have different degrees of supercoiling.

POSITIVE REGULATION AT THE PROMOTER

So far we have dealt with the promoter as a DNA sequence that is competent to bind RNA polymerase and then to initiate transcription. But there are some promoters at which RNA polymerase cannot initiate transcription without assistance from an ancillary protein. Such proteins are said to be **positive regulators,** because their presence is necessary to switch on the transcription unit. Several positive regulators are known, some coded by phages and some present within the host cell. At the outset, it has to be said that we do not properly understand the nature of the difference between promoters that can function per se and those that need a positive regulator.

Several loci in *E. coli* (and also in other bacteria) are subject to a positive control in which a high level of cyclic AMP is required for activity. The nucleotide functions by activating a protein, the **CAP** or **CRP** factor. (The battle over its names has not been resolved.) This protein is a dimer of about 45,000 daltons that is essential for transcription of those promoters under its control. We shall discuss its role as a coordinating factor more fully in Chapter 15.

The CAP factor acts by binding directly to DNA, and complexes of cyclic AMP•CAP•DNA can be isolated at each promoter at which it is necessary. In the *lac* operon, there are mutations in the binding site that prevent the promoter from responding to CAP *in vivo*, and these also prevent binding to DNA *in vitro*. At each promoter, the CAP binding site is upstream from the RNA polymerase site. Variations on a putative 11 bp consensus sequence can be identified in all the CAP binding sites. But the surprising feature is that these sites are found at rather different locations relative to the startpoint in several promoters. This makes it difficult to construct a uniform model for CAP action. Three examples encompass the range of locations.

In the *lac* operon, the region of DNA protected by CAP extends from about -72 to -52. It is possible that two dimers of CAP are bound. The binding pattern is consistent with the presence of CAP largely on one face of DNA, the same face that is bound by RNA polymerase. This would place the two proteins just about in reach of each other.

At the *gal* locus, however, the CAP binding site lies between -50 and -23. Sequences farther upstream are not necessary. It is likely that only a single CAP dimer is bound, probably more intimately in contact with RNA polymerase, because the CAP binding site extends well into the region generally protected by the RNA polymerase.

In the *ara* region, the binding site for a single CAP dimer is the farthest from the startpoint, at -107 to -78. Here the CAP cannot be in contact with RNA polymerase, because *another* regulatory protein binds in the region between the CAP and RNA polymerase binding sites.

One model for CAP action is that the main effect is

mediated by DNA-bound protein-protein interactions. Quite a modest interaction between two proteins would be sufficient to achieve a substantial increase in the ability of RNA polymerase to bind to a promoter. The geometry of the CAP-RNA polymerase interaction would have to be different in *lac* and in *gal,* while in *ara* there would need to be a triple-protein interaction.

Another possibility is that the effect of CAP is exercised entirely through its binding to DNA. Perhaps in the CAP-dependent promoters, RNA polymerase cannot accomplish the initial melting of DNA. This could be undertaken by CAP, so that a region of strand separation is propagated from its binding site. Relevant to this model is the observation that the mutations in topoisomerase I described in the last section release the dependence of these operons on CAP, allowing them to function without its assistance. It is possible that increased supercoiling (via its concomitant unwinding) provides an alternative to dependence on CAP.

The crystal structure of CAP identifies two separate structural domains. Probably the small C-terminal domain (~65 amino acids) binds DNA, because the larger N-terminal domain (~135 amino acids) has been shown to bind cyclic AMP. One model for the structure of the C-terminal domain suggests that it would be able to make the proper contacts only with DNA in a *left-handed* B-form. If CAP created a left-handed region near the promoter, this would generate an unwound region where the transition is made with the usual right-handed conformation. This (speculative) effect would be assisted by supercoiling.

The requirement for CAP varies in the different loci it controls. Just how this effect depends on the structure of the individual promoter is not clear, but one speculation is that it may be related to the -35 consensus sequence, whose absence is correlated with greater dependence on CAP. Perhaps the interaction of RNA polymerase at -35 and the interaction of CAP with DNA play related roles in initiation.

POSSIBLE CONSENSUS SEQUENCES FOR RNA POLYMERASE II

First attempts to infer the identities of eucaryotic promoters were made by comparing the nucleotide sequences upstream from the startpoints of various genes. Since RNA polymerase I transcribes only the genes for rRNA, all of which are identical, there is no way to distinguish the promoter. With RNA polymerase III, an insufficient number of genes from any one source have been characterized, but (as we shall see later) it turns out anyway that the promoter is not necessarily located in the immediate upstream region. With RNA polymerase II, a large variety of genes (more than 60) from various species have been sequenced, and the startpoints identified. Here (to recapitulate) a critical assumption is that the first base of the mRNA corresponds to the site of initiation, and is not generated by processing from a larger precursor. In a limited number of cases, this assumption has been validated by the demonstration that the site of initiation *in vitro* coincides with the start of the mRNA.

As with bacterial promoters, homologies in the region near the startpoint are rather restricted. There are three regions in which homologies of possible significance have been noted: the startpoint, a region centered at about -25, and a region centered at about -75.

At the startpoint, there is no extensive homology of sequence, but there is a tendency for the first base of mRNA to be A, flanked on either side by pyrimidines.

An almost ubiquitous sequence of 7 bp has a separation from the startpoint that varies from 19 to 27 bp. It has been found in mammals, birds, amphibians, and insects. From all known cases (irrespective of species), the consensus is

$$T_{82} \; A_{97} \; T_{93} \; A_{85} \; {A_{63} \atop T_{37}} \; A_{83} {A_{50} \atop T_{33}}$$

Often this is known as the **TATA** or **Hogness box.** The consensus consists entirely of A-T base pairs (at two positions the orientation is variable), and in only a minority of actual cases is a G-C pair present. The box tends to be surrounded by G-C-rich sequences, which could be a factor in its function. It is almost identical with the Pribnow box found in bacterial promoters; in fact, the only substantive difference is its location at -25 rather than -10. It cannot be absolutely essential for transcription, because there are a couple of cases in which it is entirely absent.

Farther upstream there is another sequence that has been conserved in some promoters, although it is absent from several. This sequence lies between about -70 and -80 and has the consensus

$$G G \, ^C_T \, C A A T C T$$

It is sometimes known as the **CAAT box.** We do not yet know whether it is a general feature of promoters or is specific for certain classes.

IN VITRO AND "IN VIVO" SYSTEMS

Attempts to define the promoters for eucaryotic RNA polymerases have taken advantage of the precedents established with bacterial RNA polymerase. There are two particular limitations with the eucaryotic systems. The first is that virtually no promoter mutations have been identified *in vivo,* so we start without any prior information on the location of the promoter. The second is that it has not yet been possible to characterize directly the sites that are bound by any one of the RNA polymerases. This is rendered difficult by the complexity of the enzyme preparations and the lack of information on exactly what constitutes the active structure of the enzyme. Of course, it is only a matter of time until a successful system is developed that will allow us to recover the DNA binding site from an initiation complex.

A less direct assay, but one that actually is more informative in some ways, is to define the promoter in terms of its ability to initiate transcription in a suitable test system. Three types of systems have been used.

The *in vitro* system takes the classic approach of purifying all the components and manipulating conditions until faithful initiation is seen. "Faithful" initiation is defined as the production of an RNA starting at the same site that is represented at the $5'$ end of mRNA. This has become possible fairly recently, and can now be achieved with each one of the three classes of RNA polymerase. The systems are in various stages of purification. Some consist of extracts of cells, containing the RNA polymerase, but not yet purified; others consist of the enzyme augmented by a cellular extract. In due course, all of these systems should be superseded by preparations in which all of the components have been defined; and it will be possible to compare the activities *in vitro* of RNA polymerases from different tissues or species.

The oocyte system follows the principles established for translation, and relies on injection of a suitable DNA template, this time into the nucleus of the *X. laevis* oocyte. The RNA that is produced can be recovered and analyzed. The main limitation of this system is that it is entirely restricted to the conditions that prevail in the oocyte.

In vivo systems do not (as the name might imply) necessarily involve characterizing the ability of a cell to transcribe one of its usual products. What is involved here is following the ability of a cultured cell line to transcribe a template that is introduced by transfection (a procedure that allows an exogenous DNA to enter a cell and survive; the events involved are discussed in Chapter 38). The system is genuinely *in vivo* in the sense that transcription is accomplished by the same apparatus that is responsible for expressing the cell's own genome; but it differs from the true situation in the sense that the template may comprise genes that would not usually be transcribed in the host cell.

The approach to characterizing the promoter is the same in all three systems. We seek to manipulate the template *in vitro* before it is submitted to the system for transcription. When a particular fragment of DNA can be used to initiate transcription, it is taken to include a functional promoter. Then the boundaries of the sequence constituting this promoter can be determined by reducing the length of the fragment from either end, until at some point it ceases to be active. The type of protocol is illustrated in **Figure 11.6.** The boundary upstream can be identified simply by progressively removing material from this end until promoter function is lost. For the boundary downstream, it is necessary to construct a DNA in which the shortened promoter is reconnected to the sequence to be transcribed (since otherwise there is no product to be assayed).

We have to take several precautions to avoid the introduction of extraneous effects. Because there may be topological demands for recognition of DNA by RNA polymerase, usually these experiments are done by placing the promoter-gene sequence next to a long constant upstream sequence of DNA. This avoids any "end-effects" in which promoter function is prevented

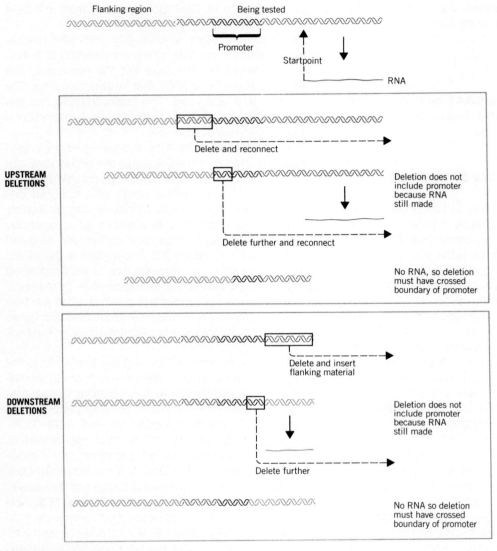

Figure 11.6
Promoter boundaries can be determined by the effects of deletions.
Deletions in the region being tested are made that extend farther into it from either end. When one deletion fails to prevent RNA synthesis but the next stops transcription, the boundary of the promoter must lie between them. (With the downstream mutations, the sequences comprising the transcription unit are altered.)

simply because it happens to be too near the terminus of the DNA fragment. Also, it ensures that the environment is the same in each case, in the form of the identical flanking sequence adjacent to the sequence that is to be tested for promoter function. In the *in vitro* systems, in which termination may not occur properly, the template may be cut at some distance from the promoter (usually about 500 bp downstream), to ensure that all polymerases "run off" in the same way, to generate an identifiable transcript.

Once the boundaries of the promoter have been defined, the importance of particular bases within it can be determined by introducing point mutations or other rearrangements in the sequence. As with the bacterial RNA polymerase, these can be characterized as *up* or *down* mutations. Some of these rearrangements may affect only the *rate* of initiation; others may influence the *site* at which initiation occurs, as seen in a change of the startpoint. To be sure that we are dealing with comparable products, in each case it is necessary to characterize the 5' end of the RNA, as described previously in Figure 11.2.

RNA POLYMERASE II FUNCTIONS ACCURATELY *IN VITRO*

Results with the *in vitro* systems are mostly fairly straightforward. Proceeding from the upstream direction, sequences can be progressively removed without any effect, until reaching a point located somewhere between -45 and -30. The exact limit varies with the system, but is always to the left of the TATA box. Once the limit is transgressed, transcription is reduced twentyfold or more. This is analogous to our concept of the bacterial promoter, which represents a short and well-defined sequence just upstream from the startpoint.

Proceeding from the downstream direction, the boundary of the promoter varies. In two well-characterized cases, it lies at about -10 or -12. This is well upstream from the startpoint, whose sequence must therefore be relatively less important. In such cases, when the startpoint itself has been deleted, initiation occurs at a point in the template that is the *same distance* from the promoter as the original startpoint, except that it may be adjusted by a base or two in order to find a purine (usually an A residue) with which to start. However, the efficiency of initiation may be somewhat reduced by the absence of the usual startpoint. In another case, the essential sequence extends to $+6$ and so includes the startpoint.

We do not know exactly what features of the region defined by these boundaries are necessary for recognition by RNA polymerase. The apparent failure always to include the startpoint could mean that the geometry of the complex may define where initiation occurs, presumably because the enzyme binds to the upstream site in such a way as to stretch downstream. The TATA box always lies within the region, and its importance is confirmed by the results of introducing two mutations into the (chick) conalbumin gene. Changing the T in the third position to a G provides a strong down mutation. It is not just the change in base-pair composition that is effective, because a change to an A has the same effect (this has simply reversed the orientation of the T-A pair). Thus the exact sequence of the TATA box may be important. The adequacy of this region in the *in vitro* assay is demonstrated by the ability of the sequence from -32 to -12 of the late adenovirus transcription unit to cause transcription to initiate about 30 bp downstream from wherever it is located in a bacterial plasmid.

All of these results were obtained with the same *in vitro* system, obtained from cultured mammalian cells. Is it a demonstration of evolutionary conservation that this is able to function not just with mammalian genes, but also with avian and insect genes? In the case of the insect gene fibroin, a homologous system from the silk gland of the moth also has been developed; and it demands exactly the same boundaries (-29 to $+6$). However, the silk gland extract transcribes the fibroin genes about three times better than it transcribes adenovirus; whereas the mammalian cell extract functions about three times better with adenovirus DNA, for which it provides a homologous system. The silk gland system also shows a specific increase in the efficiency with which it transcribes the homologous fibroin gene when the sequences upstream from -73 are retained. The existence of this effect only in the homologous system, together with the tendency of each *in vitro* system to work better with its homologous template, suggests that what we may see *in vitro* is part of a basic promoter function that does work across species barriers, but that represents only part of the full promoter function.

A DISCREPANCY BETWEEN THE *IN VIVO* AND *IN VITRO* BOUNDARIES

With *in vivo* or *in oocyte* systems, we see a strong dependence on sequences farther upstream from the

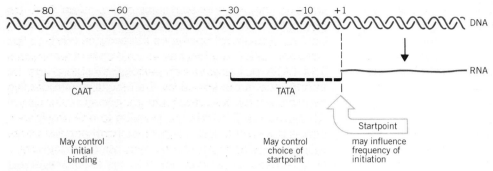

Figure 11.7
Promoters for RNA polymerase II contain separate sequence components with (possibly) different roles. The sequences between the components do not seem to be important.

TATA region. Their exact location varies, but in most cases they are about another 40 bp upstream from the TATA box. Deletion mapping experiments with the TK (thymidine kinase) gene of herpes virus show that a region between −100 and −60 controls the frequency of initiation. In its absence, initiation at the usual startpoint is reduced to 2% of the previous level. If the TATA box is removed, initiation continues to occur, but there is reduced accuracy in selecting the exact startpoint relative to the upstream region. Similarly, in mammalian globin genes, deleting a sequence of roughly 20 to 30 bp, centered about −70, causes a substantial drop in transcription. An upstream sequence also is important in a yeast gene. With a sea urchin histone gene, the sequence lies farther from the startpoint, between −139 and −111. These sequences are not active if the usual region preceding the startpoint is removed, but they increase the frequency of initiation there very considerably.

In vivo, the promoter therefore seems to have two separate components, as illustrated in **Figure 11.7.**

Usually upstream from −60 or thereabouts is an element that has a strong influence on the frequency of initiation, as seen by a drop typically of ten- to fiftyfold in transcription when it is deleted. This element contains the conserved sequence of the CAAT box. Probably it has a major effect on the binding of RNA polymerase. However, the residual transcription that occurs in its absence does initiate at the proper startpoint.

Close to the startpoint is a sequence element that

surrounds the conserved TATA box, and whose deletion introduces a more erratic quality into the choice of startpoint, although any overall reduction in transcription is relatively small. The role of this sequence could therefore be to align the RNA polymerase so that it initiates at the proper site. This is consistent with the fact that there are some promoters that naturally lack the TATA sequence, and whose mRNAs may start at more than one point instead of being restricted to the usual single startpoint.

How can a promoter consist of elements separated in this way, and altogether stretching over a greater distance from the startpoint than RNA polymerase could contact on DNA? Two models are illustrated in **Figure 11.8.** One possibility is to think in terms of models that suppose RNA polymerase initially contacts the site farther upstream and then moves to the site nearer the startpoint. Another view requires that we remember that *in vivo* DNA is unlikely to be stretched out in linear fashion; its compact organization could bring into juxtaposition sites that are separated on the duplex molecule (see Figure 29.9). Thus the binding site for RNA polymerase could consist of DNA sequences that are not contiguous, but that are held together in space by proteins that bind DNA. This implies further that the *spacing* (rather than the exact sequence) between these components of the promoter may be important.

In either type of model, we can see that the region farther upstream could be the most important for binding RNA polymerase, but need not necessarily be able to hold the enzyme in a configuration that allows it to

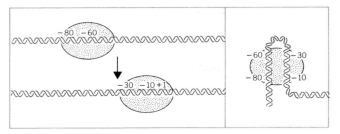

Figure 11.8
One model to reconcile the size of the promoter with the size of RNA polymerase supposes that the enzyme moves; another supposes that the DNA is more compactly organized.

recognize the exact startpoint. This could be the function of the region closer to the startpoint. Thus when the TATA box is deleted, RNA polymerase remains able to bind via the remaining upstream region, but its contacts in the region around the startpoint are less precise, allowing initiation to occur at more than one point. On the other hand, when the sequence farther upstream is absent, the ability of RNA polymerase to bind to the DNA is much reduced; but when it does so, the presence of the TATA box means that initiation is precise.

How are we to explain the discrepancy between the need *in vivo* for the sequence upstream from -60 and its apparent lack of importance *in vitro*? A major responsibility for this effect may lie with the differing efficiencies of transcription in the two circumstances. The *in vitro* system is relatively inefficient; less than 1% of the templates actually are transcribed. We do not know just what proportion of the templates is utilized *in vivo* or in oocyte, but it could well be very much higher. The *in vitro* level could even correspond to the residual expression that we see *in vivo* in the absence of sequences upstream from -60. At all events, the simpler structure of the template *in vitro,* as a DNA molecule not organized in the usual structure, may mean that the manner of recognition by RNA polymerase is different, apparently less efficient, perhaps because it uses only some parts of the promoter.

Even in considering the possibly more complex demands of the *in vivo* system, we have thought of the promoter essentially as an isolated region responsible for binding RNA polymerase. Some further results show that the situation may not be this straightforward. The

DNA of the virus SV40 contains two identical sequences of 72 bp each, repeated in tandem about 200 bp upstream from the usual startpoint of one transcription unit. They lie in a region that has an unusual nucleoprotein structure (see Chapter 30). Deletion mapping experiments show that the removal of both of these **72 bp repeats** greatly reduces transcription *in vivo*. Transcription is normal so long as one of these sequences remains. By this type of criterion, we might argue that this region constitutes an upstream component of the promoter.

But reconstruction experiments in which the 72 bp sequence is removed from the DNA and then is inserted elsewhere show that normal transcription can be sustained so long as it is present *anywhere* on the DNA molecule. In fact, if a β-globin gene is placed on a DNA molecule that contains the 72 bp sequence, its transcription is increased *in vivo* more than 200-fold, even when the 72 bp sequence is as much as 1400 bp upstream or 3300 bp downstream from the startpoint. And these are simply the limits that have been tested so far; we have yet to discover at what distance the 72 bp sequence fails to work.

This sequence is an **enhancer:** it is not part of the promoter, but is able to increase the frequency with which RNA polymerase initiates. What does it do? How can it work over such large distances? One possibility is that the enhancer changes the overall structure of the template—for example, by influencing the DNA-protein organization of chromatin, or by changing the density of supercoiling. Another idea is that it might be responsible for locating the template at a particular place within the cell—for example, attaching it to the nuclear matrix. Still possible, although perhaps less likely, is that it is directly involved in binding RNA polymerase (after which the enzyme would have to move to the promoter proper).

Several viral genomes have elements that bear a structural resemblance to the SV40 72 bp repeats, although there is no homology of sequence. The idea that these elements may fulfill a related function is supported by the demonstration that the analogous component of a retroviral genome can substitute for the enhancer sequences in SV40 DNA. It may be relevant that this part of the retroviral DNA is involved in activating cellular genes.

Enhancer sequences have not at present been

characterized as natural components of cellular transcription units. The interaction of the viral enhancer with a cellular promoter may vary, because not all promoters are susceptible to enhancement. Thus the α globin gene promoter is not enhanced by the presence of a 72 bp repeat from SV40.

The moral of these results is clear. We need to be careful in defining the components of the promoter. It is not enough just to show that a deletion of a particular sequence reduces transcription; we must also investigate the importance of the *location* of the deleted sequence. So what is a promoter? If we use a working definition that it constitutes a sequence or sequences of DNA that must be in a fixed location relative to the startpoint, the components around the CAAT and TATA boxes are included, but the enhancer is excluded.

From this definition, we see further that we cannot investigate the internal structure of the promoter simply by making deletions within it, because these could inactivate it by changing critical distances between components, even though the deleted sequences are themselves irrelevant. So we need to make small deletions that are replaced by other sequences of the same length. By this approach, we should be able to delineate those sequences in the general vicinity of the startpoint that are needed by RNA polymerase to initiate transcription.

A DOWNSTREAM PROMOTER FOR RNA POLYMERASE III

Until the promoter of the genes coding for 5S RNA in *X. laevis* was identified, an assumption in all attempts to identify promoter sequences was that the promoter would lie upstream from the startpoint. But in the 5S RNA genes, transcribed by RNA polymerase III, the promoter lies well *within* the transcription unit, more than 50 bases downstream from the startpoint. This location explains the lack of any evident consensus sequences in the upstream regions of genes transcribed by RNA polymerase III.

The 5S RNA gene can be transcribed when a plasmid carrying it is used as template for a nuclear extract obtained from *X. laevis* oocytes, the tissue in which this gene usually is expressed. The promoter was located by using plasmids in which deletions extended into the gene from either direction. The 5S RNA product continues to be synthesized when the entire sequence upstream from the gene is removed. This demonstrates that the promoter does not lie outside the gene.

When the deletions continue into the gene, a product very similar in size to the usual 5S RNA continues to be synthesized so long as the deletion ends before about base $+55$. The product consists of an RNA, the first part of which represents plasmid DNA; the second part represents the remaining segment of the usual 5S RNA gene sequence. But when the deletion extends past position $+55$, transcription does not occur. This means that the promoter lies downstream from position $+55$, but causes RNA polymerase III to initiate transcription at a more or less fixed distance away. In the wild-type gene, initiation occurs at a unique startpoint; when the usual startpoint is absent, initiation occurs at the purine base nearest to the position 55 bp upstream from the promoter.

When deletions extend into the gene from its distal end, transcription is unaffected so long as the first 80 bp remains intact. Once the deletion cuts into this region, transcription ceases. This places the downstream boundary of the promoter at about position $+80$.

So the promoter for 5S RNA transcription lies between positions $+55$ and $+80$ within the gene. A fragment containing this region is able to sponsor initiation of any DNA in which it is placed, at a position about 55 bp before its location. How does RNA polymerase initiate transcription upstream from its promoter? The most likely explanation is that the enzyme binds to the promoter, but is large enough simultaneously to contact regions 55 bp away. As with RNA polymerase II, the geometry of binding to the promoter must dictate the position of the startpoint, subject to the reservation that pyrimidines cannot be used for initiation, so the enzyme has some small flexibility in using the nearest purine to initiate. The difference between the enzymes, of course, is that RNA polymerase II reaches forward to the startpoint from its promoter, whereas RNA polymerase III reaches backward (assuming that the promoter is the binding site).

Are other promoters for RNA polymerase III similarly located within their transcription units? The bounda-

Figure 11.9

Three genes for RNA polymerase III all contain internal promoters (regions in red). The numbers indicate the first and last bases of the gene.

ries of some of them are indicated in **Figure 11.9.** In the VA1 gene of adenovirus, the promoter lies between +9 and +72, as defined by the loss of activity when deletions extend past either boundary. In the tRNA$_1$^Met gene of X. *laevis,* the promoter lies in two separate parts, both within the gene. Deletion mapping shows that both of the sequences from +8 to +30 and from +5l to +72 must be present for transcription. Any deletion between them that reduces their separation prevents initiation, but changes in the intermediate sequence, or insertions that increase the separation of the two components (at least up to 30 bp), have no effect. This promoter therefore consists of two separate regions of about 20 bp each, which must be separated by a minimum distance of 20 bp. Some other tRNA genes have comparable promoters. The sequences of these regions are highly conserved in eucaryotic tRNAs, a fact that had been interpreted solely in terms of tRNA function, but that now also may be attributed to the needs of the promoter.

The internal location of the promoter poses an important question. When a promoter lies outside the transcription unit itself, presumably it is able to evolve freely just to meet the needs of the enzyme. But the sequences needed for initiation of the RNA polymerase III transcription units are constrained also to meet the needs of rather different types of product, including tRNA, 5S RNA, VA RNA, and small nuclear RNAs. How then are they able also to provide whatever features are needed for recognition by RNA polymerase III?

One possible answer is that the enzyme in fact is able to recognize a wide variety of promoter sequences. This would not be entirely unprecedented given the very short consensus sequences found within the promoters recognized by other types of RNA polymerase. An alternative is that the enzyme does not itself recognize the promoters, but acts via an ancillary factor. In this case, there could be a different factor for each type of promoter. In fact, it turns out that RNA polymerase III can transcribe the 5S RNA genes only in the presence of an added factor, a 37,000-dalton protein that binds to the region from +45 to +96. This protein (several copies of which must bind to DNA to cover a region this large) serves a dual purpose: it also binds the 5S RNA product in the oocyte. The ancillary protein does not bind the DNA of the tRNA$_1$^Met gene. We may surmise that other factors of this sort remain to be characterized and will prove to be important in promoter selection.

Although the ability to transcribe these genes is conferred by the internal promoter, the startpoint does have some influence. Changes in the region immediately upstream from the startpoint can alter the efficiency of transcription. In fact, there is one striking case of two tRNA genes whose sequences are identical, but whose flanking upstream regions are different, and which accordingly are transcribed at very different frequencies. Thus the primary responsibility for recognizing the gene lies with the internal promoter, perhaps assisted by ancillary factors; but the responsibility for setting the frequency of initiation lies with the region at the startpoint, perhaps depending on an interaction with RNA polymerase III itself.

FURTHER READING

The interaction of bacterial RNA polymerase with its promoters has been well described by **Gilbert** in *RNA Polymerase* (Losick & Chamberlin, Eds., Cold Spring Harbor Laboratory, New York, 1976, pp. 193–206) and by **Siebenlist, Simpson,** & **Gilbert** (*Cell* **20,** 269–281, 1980). The analysis of consensus signals has been dealt with by **Rosenberg** & **Court** (*Ann. Rev. Genet.* **13,** 319–353, 1979). The involvement of supercoiling in promoter function has been reviewed by **Smith** (*Cell* **24,** 599–600, 1981). Some of the techniques and results in characterizing eucaryotic RNA polymerase II promoters have been discussed by **Corden** et al. (*Science* **209,** 1406–1414, 1981); some details of consensus sequences have been dealt with briefly by **Breathnach** & **Chambon** (*Ann. Rev. Biochem.* **50,** 349–383, 1981). The discrepancy between promoter function *in vivo* and *in vitro* was brought to light by **McKnight** et al. (*Cell* **25,** 385–398, 1981). The transcription of 5S genes has been reviewed by **Korn** (*Nature* **295,** 101–105, 1982). The nature of split promoters for tRNA genes has been analyzed by **Galli, Hofstetter,** & **Birnstiel** (*Nature* **294** 626–631, 1981) and reviewed by **Hall** et al. (*Cell* **29** 3–5, 1982). A thoughtful analysis of the effects of the SV40 enhancer has been provided by **Banerji, Rusconi,** & **Schaffner** (*Cell* **27,** 299–308, 1981); the existence of retroviral enhancers has been reported by **Levinson** et al. (*Nature* **295,** 568–572, 1981).

CHAPTER 12
GLOBAL SWITCHES IN INITIATION

Bacterial RNA polymerase is inviolate in a normally growing cell. The same basic enzyme transcribes all genes, and must therefore recognize a wide spectrum of promoters, whose associated genes are transcribed at different levels and on different occasions. The ability to initiate at particular promoters is controlled by a seemingly endless series of ancillary modifications to, or interference with, RNA polymerase. In none of these instances is any change made in the subunits of the enzyme itself. Presumably such a mechanism is not favored, because a change that allowed the RNA polymerase specifically to recognize one type of promoter might prevent it from finding another. This would reduce the flexibility of the enzyme.

Yet the division of labors between a core enzyme that undertakes chain elongation and a sigma factor that is involved in site selection immediately raises the question of whether there could be more than one type of sigma, each specific for a different class of promoters. This mechanism does not seem to be employed to control transcription in the bacterium per se, at least at all commonly. But in certain circumstances, a drastic change occurs in the life-style of the cell, when there is a switch from expression of its former functions to transcription of novel functions. Then there is less impediment to introducing permanent changes in the RNA polymerase itself.

There are two occasions on which substitutions of sigma factors are made, preventing expression of former genes, but promoting initiation of new genes. One is **sporulation,** when (certain types of) bacteria undergo a major morphological shift into a dormant form (the spore). The other occurs during **lytic infection** by a bacteriophage. When a phage enters the lytic pathway of infection, ultimately it destroys the cell in the process of reproducing itself. In all but the very simplest cases, the development of the phage involves shifts in the pattern of transcription. Yet there is only one well-authenticated case in which the change of transcriptional specificity involves substitution of the host sigma factor with one coded by the phage. (And this is in a bacterium of the sporulating kind.) Other mechanisms, often involving ancillary transcription factors, are used more commonly. It seems as though the introduction of changes in RNA polymerase itself is a mechanism employed only with reluctance, as a last resort. It is possible that the capacity to use substitutions in sigma factors has evolved in only a limited range of bacteria.

THE SPORULATION LIFE-STYLE

Sporulation is an alternative life-style available to some bacteria, such as *Bacillus subtilis.* At the end of the

vegetative phase of logarithmic growth, nutrients in the medium become depleted, and a series of morphological changes is initiated. This involves the replication of DNA and the segregation of a genome at one end of the cell, where eventually it is surrounded by the tough spore coat. The process takes about 8 hours.

Sporulation involves a drastic change in the biosynthetic activities of the bacterium, and a large number of genes are involved. The basic level of control lies at transcription. Synthesis of RNA continues throughout sporulation; its inhibition brings the process to a halt. A few of the genes that functioned in the vegetative phase are turned off during sporulation, but most continue to be expressed. In addition, a large number of genes specific for sporulation are expressed only during this period. At the end of the period of sporulation, about 40% of the mRNA in the bacterium is sporulation-specific, while 60% is similar to vegetative phase mRNA.

Genes that are necessary specifically for sporulation can be identified by spo^- mutations, which allow normal vegetative growth but block sporulation. These asporogenous mutants can be classified by the stage at which sporulation is blocked; they are described as *spo0* (sporulation cannot start at all), *spoI, spoII,* etc., as later stages are defective. Some of these constitute mutations in enzymes or structural products that are needed for spore development; but some identify genes concerned with regulating the transition in lifestyles.

The first indication that transcriptional specificity is altered in sporulating cells came from observations that some bacteriophages are able to infect vegetative cells, but cannot infect sporulating cells, because they fail to be transcribed. We are a long way from defining all the changes that occur during sporulation, or from knowing the range of mechanisms that are used, but some of the changes that occur at early times now are known to be caused by an alteration in the specificity of RNA polymerase at initiation.

The enzyme found in *B. subtilis* cells engaged in normal vegetative growth has the same structure as that of *E. coli,* $\alpha_2\beta\beta'\sigma$, where the sigma factor plays the same role and has a mass of 55,000 daltons. It is therefore described as σ^{55}. At least two further forms of the RNA polymerase can be found in sporulating

cells; they contain the same core enzyme as vegetative cells, but have different proteins in place of the vegetative σ^{55}. These changes in transcriptional specificity are summarized in **Figure 12.1.**

The form of RNA polymerase that may initiate sporulation was found by using a gene that is activated at the beginning of sporulation. This gene cannot be transcribed *in vitro* by the vegetative RNA polymerase. However, it can be transcribed by a form of the enzyme that has a protein of 37,000 daltons instead of σ^{55}. This new protein is called σ^{37} on the supposition that its role is to act as a new sigma factor able to redirect the attention of core enzyme to new genes.

What is responsible for the replacement of σ^{55} by σ^{37}? The σ^{37} protein can be found in vegetative cells, although then it is not associated with core enzyme. So probably some other protein, very likely a product of a *spo0* gene, is needed, perhaps to remove σ^{55} directly, or perhaps to modify core enzyme so that it has greater affinity for σ^{37}. Actually, mutation in any one of five *spo0* genes can block transcription of the early sporulation genes that are usually expressed via the $\alpha_2\beta\beta'\sigma^{37}$ enzyme, so the process of substituting σ^{37} in place of σ^{55} may be quite complex. Also, it may affect only part of the RNA polymerase population, since some vegetative enzyme may continue to be present during sporulation. The displaced σ^{55} is not destroyed, but can be recovered from extracts of sporulating cells.

Another form of the enzyme first appears in cells about 4 hours after sporulation has started. This contains a protein of 29,000 daltons associated with the usual core tetramer. The new protein, σ^{29}, allows the core enzyme to transcribe new genes *in vitro*. The σ^{29} protein cannot be synthesized by at least one *spo0* mutant, and so it is likely to be the product of an early sporulation gene that is needed for the transition into the next stage of sporulation. Again, it is mysterious how σ^{29} is able to replace σ^{37} or σ^{55} on the core enzyme.

RNA polymerase containing σ^{37} or σ^{29} actually remains able to transcribe some of the vegetative genes (and in the latter case, at least some of the earlier sporulation genes). So the three forms of RNA polymerase transcribe overlapping sets of genes; each recognizes some new genes and fails to recognize some old genes. We shall not know how this is ac-

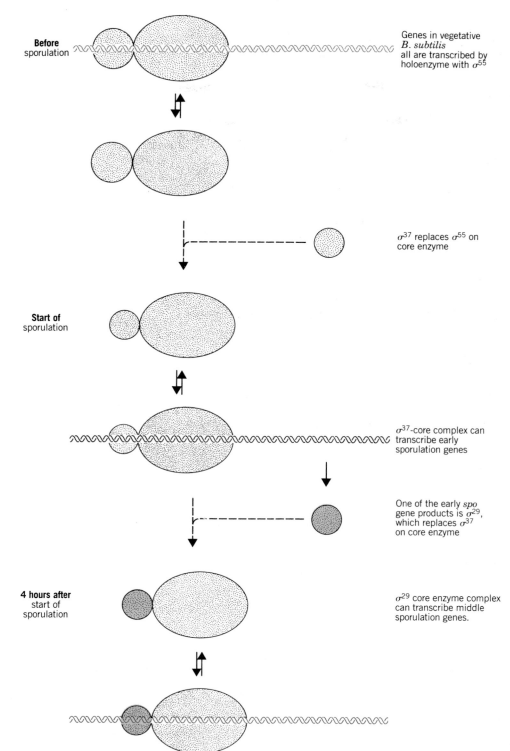

Genes in vegetative
B. subtilis
all are transcribed by
holoenzyme with σ^{55}

σ^{37} replaces σ^{55} on
core enzyme

σ^{37}-core complex can
transcribe early
sporulation genes

One of the early *spo*
gene products is σ^{29},
which replaces σ^{37}
on core enzyme

σ^{29} core enzyme complex
can transcribe middle
sporulation genes.

Before sporulation

Start of sporulation

4 hours after start of sporulation

Figure 12.1
**Sporulation involves successive
changes in the initiation specificity of *B. subtilis* RNA polymerase.**

complished until the promoters recognized by each form of the enzyme have been characterized.

ALTERNATIVE PHAGE SIGMA FACTORS

The phage SPO1 is a large, virulent invader of *B. subtilis*. The infective cycle passes through three stages of gene expression. Immediately upon infection the **early** genes of the phage are transcribed. After 4–5 minutes, the early genes cease transcription and the **middle** genes are transcribed. Then at 8–12 minutes, middle gene transcription is replaced by transcription of the **late** genes. What is responsible for these switches in transcription?

The early genes are transcribed by the holoenzyme of the host bacterium. They are essentially indistinguishable from host genes whose promoters have the intrinsic ability to be recognized by the RNA polymerase $\alpha_2\beta\beta'\sigma^{55}$. This enzyme has recognition properties very similar to, or identical with, those of the *E. coli* holoenzyme. The two RNA polymerases can recognize the same promoters *in vitro;* and early genes of SPO1 have hexameric sequences at -35 and -10 that fit with the consensus sequences described for *E. coli* in Chapter 11. (Although in the phage SPO1, T is replaced by hydroxymethyluracil, this does not seem to make any difference.)

The expression of phage genes is required for the transitions to middle and late gene transcription. Three regulatory phage genes, named *28, 33,* and *34,* control the course of transcription. Their functions are summarized in **Figure 12.2.** The pattern of regulation creates a cascade, in which the host enzyme transcribes an early gene whose product is needed to transcribe the middle genes; and then some of the middle gene products are needed to transcribe the late genes.

Mutants in the early gene *28* cannot transcribe the middle genes. The product of gene *28* (called gp28) is a protein of 26,000 daltons that replaces the sigma factor on the core enzyme. This creates an RNA polymerase that specifically transcribes the middle genes. When the enzyme is isolated from cells during the middle period of infection, or when gp28 is added to purified core enzyme, the polymerase becomes able to bind to the promoters of middle genes; it fails any

longer to bind to the early gene promoters. So the sole event required to make the transition from early to middle gene expression is the replacement of host sigma factor by phage gp28 (sometimes called σ^{gp28}). How gp28 displaces σ^{55}, and what happens to the host sigma polypeptide, is not known. Probably gp28 has a greater affinity for the core enzyme.

Two of the middle genes are involved in the next transition. Mutation in either gene *33* or *34* prevents the transcription of the late genes. The products of these genes are proteins of 13,000 and 24,000 daltons, respectively, that replace gp28 on the core polymerase. Again, we do not know how gp33 and gp34 exclude gp28 (or any residual host σ^{55}), but once they have bound to the core enzyme, it is able to initiate transcription only at the promoters for late genes.

The successive replacements of sigma factor have dual consequences. Each time this subunit is changed, the RNA polymerase becomes able to recognize a new class of genes, *and* no longer recognizes the previous class. These switches therefore constitute global changes in the activity of RNA polymerase. Probably all or virtually all of the core enzyme in the cell becomes associated with the sigma subunit of the moment, and the change is irreversible.

From the perspective of the enzyme, it is interesting that subunits of different sizes all can fulfill the same basic function of directing the attention of core enzyme to particular promoters. Taking sporulation and phage infection together, we know of (at least) five different sigma types with which the core can interact: σ^{55}, σ^{37}, σ^{29}, σ^{gp28}, and $\sigma^{gp33-34}$. Another possible factor, σ^{28}, has been found in vegetatively growing cells, which also contain a polypeptide, δ, that may be involved in enhancing proper promoter selection with various sigma factors. So the core enzyme has evolved the fundamental property of accepting direction from any one of many sigma factors.

EACH SIGMA MAY HAVE ITS OWN -35 AND -10 CONSENSUS SEQUENCES

What distinguishes the different classes of promoters? The sporulation genes do not display the usual consensus sequences that are found at -35 and -10 in the promoters of vegetative genes. Perhaps each set

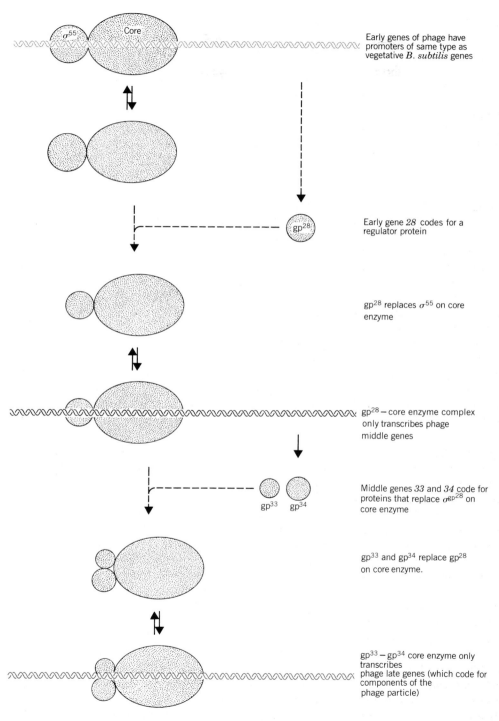

Early genes of phage have promoters of same type as vegetative *B. subtilis* genes

Early gene *28* codes for a regulator protein

gp²⁸ replaces σ⁵⁵ on core enzyme

gp²⁸ − core enzyme complex only transcribes phage middle genes

Middle genes *33* and *34* code for proteins that replace σᵍᵖ²⁸ on core enzyme

gp³³ and gp³⁴ replace gp²⁸ on core enzyme.

gp³³ − gp³⁴ core enzyme only transcribes phage late genes (which code for components of the phage particle)

Figure 12.2
Transcription of phage SPO1 genes is controlled by two successive substitutions of the sigma factor that change the initiation specificity.

of sporulation promoters has its own characteristic consensus. This also may be true of targets for the second vegetative factor, σ^{28}. And such a situation seems to be the case at least with the SPO1 middle genes, where sufficient sequences have been determined to calculate a consensus.

The RNA polymerase containing gp28 binds to sites on DNA that have the same size and location relative to the gene as those bound by host enzyme. The middle polymerase protects a region of DNA extending from positions $-40/-50$ to about $+20$.

At -10 there is a consensus sequence generally similar to the Pribnow box in its A-T-rich character, although all its base pairs are in one orientation as

$$T\ T\ T\ N\ T\ T\ T$$

compared with the vegetative consensus

$$T\ A\ T\ A\ A\ T$$

Probably the two -10 sequences share the same purpose.

But a difference is found in the -35 region. As with E. coli promoters, this is the only other part of the middle promoters that shows any sequence conservation. The homology at -35 takes the form of a hexameric sequence

$$A\ G\ G\ A\ G\ A$$

that is different from the vegetative consensus

$$T\ T\ G\ A\ C\ A$$

Either or both of the regions at -35 and -10 may be responsible for identifying promoters to the classes of RNA polymerase that should recognize them. Because the core enzyme can display (a minimum of) six different specificities in B. subtilis, it seems unlikely that each sigma factor could induce a particular conformation in the core enzyme that causes it to recognize a distinct set of promoters. It is easier to think in terms of sigma factors that themselves directly recognize features of the promoters, although this brings us back to the problem of how a small polypeptide can contact sites spanning up to 20 bp of DNA.

Apropos of the question of the responsibility for promoter recognition, it has been known for a long time that changes in transcriptional specificity occur during infection of E. coli by phage T4. These are mediated by changes in both sigma factor and the core subunits. Their effects have been hard to disentangle, but it is possible that both components may influence promoter selection. So we cannot conclude that sigma alone bears the burden of selecting promoters in every circumstance.

NEW PHAGE RNA POLYMERASES

Infection of E. coli with the phages T3 or T7 follows a course not dissimilar from that seen in SPO1 infection of B. subtilis. Immediately upon infection, the class I (early) genes are transcribed. Their transcription ceases, together with host protein synthesis, after about 6 minutes, when the class II and class III genes of the phage start to be transcribed. The class II genes are active until their transcription is halted at about 15 minutes after infection; the class III genes continue to be transcribed until lysis. The class III genes are transcribed more efficiently than the class II genes.

The switch from class I to class II/III gene expression again depends on a phage product. Mutants in gene 1 of the phage cannot make the transition. This gene codes for a protein of 107,000 daltons that is necessary for transcription of both class II and class III genes. But instead of modifying the host enzyme, this protein is a new, phage-specific RNA polymerase. We have mentioned in Chapter 10 that it may constitute a "minimum" RNA polymerase. It is indeed highly specific: the phages T3 and T7 have closely related organizations, but each codes for a gene 1 polypeptide that is almost entirely specific for its own genome.

The general strategy of campaign for these phages is therefore similar to that of SPO1, but the tactics by which it is achieved are somewhat different. The common feature is the existence of a class of genes transcribed by the host RNA polymerase, among which is a phage function that sponsors transcription of a further class or classes of phage genes. Since the phage T3/T7 product is an RNA polymerase in its own right, this transition does not prevent the transcription of host or early phage genes by the E. coli RNA polymerase. However, another early phage gene, 0.7, codes for a protein kinase that has a variety of targets, among which are the β and β' subunits of the host RNA polymerase. This may be responsible for preventing host and phage early gene transcription from continuing beyond the end of the early period.

Many of the promoters recognized by the phage T7 RNA polymerase have been sequenced. They are entirely different from the promoters recognized by the host enzyme. The T7 promoters are shorter than the bacterial promoters. They share a conserved sequence of 23 bp, stretching from -17 to $+6$. Within this region there is no exact equivalent to the Pribnow box, although the general constitution of the promoter is A-T rich, and there are two particular regions just of A-T bases, indicated in boldface.

$$\overset{-17}{\text{T A A T A}} \text{C G A C T C A C} \overset{-10}{\text{T A T A}} \overset{*}{\text{G}} \text{G G A G A} \overset{+6}{}$$

The nucleotide with the asterisk marks the startpoint $(+1)$. When the T7 RNA polymerase finds this sequence, it forms a complex in which the sequence CTATAG from -5 to $+1$ is melted.

The entire 23 bp sequence is absolutely conserved in all the class III promoters that have been sequenced. In each of the class II promoters, there are at least two base substitutions. Presumably this very limited variation is able to account for the differing efficiencies with which class II and class III promoters are recognized, but we do not understand how it is related to the earlier cessation of transcription at the class II promoters.

Only part of the conserved sequence may be necessary for initiation by T7 RNA polymerase. When the sequence upstream from -12 is replaced by other DNA, the class III promoter remains able to recognize the enzyme. This contrasts with the sequence of a T3 promoter, which is not recognized by T7 RNA polymerase, yet which shares the sequence from -9 to $+4$. It seems likely that the sequence that is shared between the T3 and T7 promoters plays the same role for each RNA polymerase, with just the question of

initial recognition being different and based on the sequences farther upstream.

When crosses are made between phages T3 and T7, recombinants in gene *1* can be obtained. These have the specificity in promoter selection that is characteristic of the phage providing the right end of the gene, corresponding to the C-terminal region of the protein. Probably this part of the RNA polymerase protein makes contact with the region of the promoter upstream from -9. From this type of experiment, we are beginning to approach a detailed definition of how the polymerase protein may recognize its promoters on DNA.

From the characteristics of these phage promoters, we see the same general phenomenon characteristic of bacterial promoters. Only a very short sequence of DNA need be conserved in order to confer specificity of recognition on the promoter. This sequence lies within, but is unlikely to constitute the whole of, the DNA binding site.

FURTHER READING

Most reviews on this subject approach from the perspective of the process; examples are sporulation treated by **Hoch** (*Adv. Genet.* **18,** 67–98, 1976) and phages T3/T7 by **Lewin** in *Gene Expression, 3, Plasmids and Phages* (Wiley, New York, 1977, pp. 682–723). The sequence of much of phage T7 determined by **Dunn** & **Studier** (*J. Mol. Biol.* **148,** 303–330, 1981) extends earlier views. Promoters have been considered in detail by **Rosa** (*J. Mol. Biol.* **147,** 199–204, 1981). More from the perspective of gene expression are the reviews of sporulation-induced changes by **Losick** & **Pero** in *RNA Polymerase* (Losick & Chamberlin, Eds., Cold Spring Harbor Laboratory, New York, 1976, pp. 227–246); and *Cell* (**25,** 582–584, 1981). An account of phage-induced RNA polymerases is provided by **Bautz** in *RNA Polymerase* (op. cit., pp. 273–284).

CHAPTER 13
TERMINATION AND
ANTITERMINATION

Once RNA polymerase has started transcription, the enzyme continues to move along the template, synthesizing RNA, until it meets a signal to cease transcription. At this point, the enzyme stops adding nucleotides to the growing RNA chain, releases the completed product, and dissociates from the DNA template. (We do not know in which order the last two events occur.) Termination requires that all hydrogen bonds holding the RNA-DNA hybrid together must be broken, after which the DNA duplex reforms.

The DNA sequence that provides the signal to cease transcription is called a **terminator.** Not very much is known about eucaryotic terminators, but terminators in bacteria and their phages have been characterized in detail. They vary widely in both their efficiencies of termination and their dependence on ancillary proteins, at least as seen *in vitro.*

At some terminators, the termination event can be *prevented* by specific ancillary factors that interact with RNA polymerase. The **antitermination** causes the enzyme to continue transcription past the terminator sequence. This is called **readthrough** (the same term that is used when ribosomes continue past a termination codon as the result of suppression).

Antitermination provides a sophisticated control mechanism. Different ancillary proteins (**antitermination factors**) allow RNA polymerase to bypass specific terminator sequences. Such interactions control the ability of the enzyme to read past a terminator into genes lying beyond. This mechanism for controlling gene expression is a feature of phage regulatory circuits.

In approaching the termination event, we must therefore regard it not simply as a mechanism for generating the 3' end of the RNA molecule, but as an opportunity to control gene expression. Thus the stages when RNA polymerase associates with DNA (initiation) or dissociates from it (termination) are both subject to specific control. There are interesting parallels between the systems employed in initiation and termination. Both require a breaking of hydrogen bonds (initial melting of DNA at initiation, RNA-DNA dissociation at termination), and both require additional proteins to interact with the core enzyme. In fact, they may be accomplished by forms of the enzyme that are alternatives to one another.

TERMINATORS MUST BE IDENTIFIED
IN VITRO

It is difficult to define the termination point of an RNA molecule that has been synthesized in the living cell. Whether in procaryotes or eucaryotes, it is always possible that the 3' end of the molecule has been generated by a cleavage of the primary transcript, and therefore does not represent the actual site at which RNA polymerase terminated. There is no marker for the 3' end comparable to the triphosphate that should identify the original 5' end.

The best identification of termination sites is provided by systems in which RNA polymerase terminates *in vitro.* Because the ability of the enzyme to terminate is strongly influenced by parameters such as the ionic strength, its termination at a particular point *in vitro* does not prove that this same point is a natural terminator. But now several cases are known in which the same 3' end can be identified on a particular (procaryotic) species of RNA, irrespective of whether it has been synthesized *in vivo* or *in vitro.* This is good evidence that we have found a natural termination sequence.

The termination site on DNA can be identified by comparing the template with the 3' terminal sequence of its RNA product. Many procaryotic terminators have been examined in this way. There are no similarities of sequence beyond the point at which the last base is added to the RNA. This suggests that the primary responsibility for termination lies with the sequences already transcribed by RNA polymerase. Thus the polymerase does not anticipate some forthcoming sequence of DNA and stop before transcribing it; it relies on scrutiny of the region that it is transcribing.

TERMINATION MAY OR MAY NOT NEED THE RHO FACTOR

When certain transcription units are used as templates for RNA synthesis *in vitro,* proper termination does not occur. The core enzyme may pause at the terminator, but then continues transcription, extending the RNA chain until some random event causes it to dissociate from the template. This produces a heterogeneous population of RNA products, due to the variable positions of their 3' ends.

Rho factor was discovered as a protein whose addition to an *E. coli in vitro* transcription system generates discrete RNAs, with unique 3' ends. This implies that the corresponding terminators in the DNA template can be recognized only when the rho factor

is present. This is called **rho-dependent termination.** To see the effects of rho factor *in vitro,* it is necessary to use conditions of relatively low ionic strength.

The need for rho factor *in vitro* is somewhat variable. Some terminators require relatively high concentrations of rho, while others function just as well at much lower levels. This effect is likely to reflect a genuine variation in the efficiency of termination *in vivo.*

Are there terminators that can function without the rho factor altogether? Some terminators seem able to cause the termination reaction to be completed when encountered by core enzyme alone *in vitro.* This is called **rho-independent termination.** However, we must treat this cautiously, because it is always possible that the rho factor actually is needed to recognize these signals in the bacterium. To see the effects of these terminators in the absence of rho factor *in vitro,* it is necessary to operate the transcription reaction at a relatively high ionic strength. This could create circumstances in which a natural (low) dependence on rho is obviated.

The *in vitro* results are necessarily indecisive because of the different circumstances under which rho-dependent and rho-independent termination must be studied. What information can we gain *in vivo?* There are mutants that are deficient in rho factor activity. Although their effects on different rho-dependent terminators may vary, they do not usually affect the termination event at the sites that are defined as rho-independent.

A hint that the two types of site genuinely represent some difference in the termination process is offered by their sequences. In neither case is there any extensive conservation of sequence in the region just preceding the point of termination. However, rho-independent termination always occurs at the end of a short stretch of U residues (as represented in the RNA). This feature is not found in rho-dependent sites, which entirely lack any sequence homology. But both types of terminator display a striking coincidence of secondary structure, generated by the presence of a **palindrome** just prior to the site of termination.

A DIGRESSION ABOUT INVERTED REPEATS

Palindromic sequences play an important role not only in termination, but also in other events involved with

Cruciform DNA

Rearrangement

Duplex DNA

Inverted repeats

Transcription

RNA hairpin

Figure 13.1

Inverted repeats could generate secondary structure in DNA or RNA.

The central duplex DNA shows the sequences of the inverted repeats. A cruciform would be generated if they rearranged so that each repeat pairs with the complement on its *own* strand of DNA instead of with the complement on the other strand.

If the upper DNA strand is transcribed, the RNA has the same sequence as the lower strand, and each repeat can base pair with its complement to generate a hairpin. This structure is exactly the same as the lower part of the cruciform. The inverted repeats themselves form the base-paired stem of the hairpin. The three bases that separate the two copies of the repeat appear as a single-stranded loop at the end of the stem.

controlling the function of mRNA (and possibly DNA). We shall therefore just briefly consider their nature before returning to the issue of termination itself.

A **palindrome** is a sequence of duplex DNA that is the same when either strand is read with a particular orientation. Thus the sequence

$$5' \quad GGTACC \quad 3'$$
$$3' \quad CCATGG \quad 5'$$

is palindromic, since when read 5' to 3', it is the same

(GGTACC) on either strand (that is, reading left to right on the upper strand, and reading right to left on the lower strand).

Formally this is called a region of **dyad symmetry,** in which the **axis of symmetry** identifies the central point about which the sequence is the same on either side (when read in opposite directions). By drawing a line through the axis of symmetry,

we can see that the same sequence $\frac{GGT}{CCA}$ is present on either side of the axis of symmetry, in inverse orientation, as indicated by the arrows. For this reason, the sequence $\frac{GGT}{CCA}$ is called an **inverted repeat.**

Now the two copies of an inverted repeat need not necessarily be contiguous. Consider the same repeat in a different situation

We still have an inverted repeat of the same triplet base-pair sequence, although the axis of symmetry has been shifted into the center of the three nonrepetitive base pairs that separate the two copies of the repeat.

An inverted repeat has a critical property that can be reflected in the secondary structure of DNA or RNA. The copies of the repeat on the *same* strand of DNA are *complementary* when read in the opposite direction. In the example above, if we read *away* from the axis of symmetry on the top strand, we get TGG going left and the complementary sequence ACC going right. These can base pair to form a **hairpin** in RNA or a **cruciform** in DNA, as illustrated in **Figure 13.1.**

Thus the presence of a palindromic sequence in DNA exactly corresponds to the ability of the RNA product to take up a base-paired secondary structure by reaction between the two copies of the inverted repeat. For example, such hairpin formation is exactly what we have described in Chapter 7 for the means by which the arms of the tRNA cloverleaf are constructed.

The conditions under which cruciforms may form *in vivo* are more controversial, because their formation requires DNA to abandon its continuous double helix to form the protruding loops. This is unlikely to occur spontaneously. But it could be promoted by supercoiling and rendered feasible by proteins that associate with the DNA.

CORE ENZYME PAUSES AT PALINDROMES

Palindromes are present in every procaryotic terminator. Some examples are given in **Figure 13.2.** The two rho-independent terminators represent extremes for the length of the hairpin loop that represents the palindrome in the RNA. In the first case, a distance of 54 bases is involved (including the loop as well as the base-paired stem); in the second, only 28 bases are involved. As in all rho-independent terminators, the hairpin is extremely rich in G-C pairs, and is followed immediately by the run of about six U residues. The hairpin of the rho-dependent terminator is about the same distance from the actual termination point, but is less G-C-rich and is not succeeded by a U run.

Although there are no homologies of primary sequence in the palindromic region, homology could exist at the level of secondary structure. What form might such a signal take? As we have mentioned, it seems unlikely that the DNA would generate a cruciform prior to transcription, although possibly such a structure could form when the strands are unwound for transcription. It seems more likely that the formation of a hairpin in the RNA product may be recognized by RNA polymerase. We know that the properties of the RNA itself are important in termination, since (for example) the substitution of ITP for GTP inhibits termination, probably because I forms less stable bonds with C than does G.

The palindromic sequence is implicated directly by the nature of point mutations that prevent termination. The mutations occur in a region of about 35 bp prior to the actual point of termination. This is somewhat similar in size to the sequence that is used to determine binding at the promoter. All of these mutations reside within the stem region of the hairpin. Most would disrupt base pairs, but some do not, so the stability of the stem cannot be the sole factor involved in its use in the terminator.

What is the effect of the hairpin on RNA polymer-

Figure 13.2
Terminator sequences include G-C-rich regions that form hairpins of varying lengths; these are followed by a run of U residues in rho-independent but not in rho-dependent sites.

ase? Probably all hairpins that form in the RNA product cause the polymerase to slow or pause in transcription. At a typical terminator, the polymerase pauses for about 60 seconds. (If continuing to transcribe, generally it would move about 2000 bp during 1 minute). However, the actual delay varies from case to case. In one extreme example, the RNA polymerase actually stops at the terminator, but although it does not continue transcription, it does not terminate either. The events that follow the pause depend on the type of terminator site.

At a rho-independent terminator, the polymerase completes the reaction by dissociating from the template and releasing the RNA. The string of U residues immediately following the hairpin probably is involved in this. The rU-dA RNA-DNA hybrid has an unusually weak base-paired structure, so it requires the least energy of any RNA-DNA hybrid to break the association between the two strands. Thus in rho-independent termination, the pausing of the polymerase at the

hairpin may provide an opportunity for the RNA-DNA hybrid to unravel from the weakly bonded rU-dA terminal region. Often the actual termination event takes place at any one of several positions toward or at the end of the U-run, as though the enzyme "stutters" during termination.

The importance of the stretch of U bases is confirmed by making deletions that remove some of them; then termination no longer occurs, although the RNA polymerase does still pause at the hairpin. The series of U bases, of course, corresponds to an A-T-rich region in DNA, so we see that A-T-rich regions may be important in termination as well as initiation.

HOW DOES RHO FACTOR WORK?

Rho factor functions solely at the stage of termination. It is a protein of about 55,000 daltons, probably active as a tetramer. It displays two activities *in vitro*. It functions catalytically as a termination factor; typically its maximum activity is displayed when it is present at about 10% of the concentration of the RNA polymerase. And it has an RNA-dependent NTPase activity, which is necessary for termination.

Does rho factor act via recognizing DNA, RNA, or RNA polymerase? Its NTPase activity requires the presence of a polyribonucleotide preferably of more than 50 or so residues. This suggests that rho binds RNA. A model for rho action supposes that it binds to the 5′ end of a nascent RNA chain, and then moves along the RNA, using the hydrolysis of NTP to provide the necessary energy.

Figure 13.3 shows that the pausing of RNA polymerase at a terminator site provides an opportunity for rho to catch up. Then the factor interacts with RNA polymerase to cause the release of RNA and the dissociation of the two proteins from nucleic acid. This model predicts that the critical feature of the terminator site is its ability to make RNA polymerase pause. The differences between the various terminators could be explained if they vary in the efficiency with which

Figure 13.3
Rho factor pursues RNA polymerase along the RNA and can cause termination when it catches the enzyme pausing at a hairpin in a terminator region.

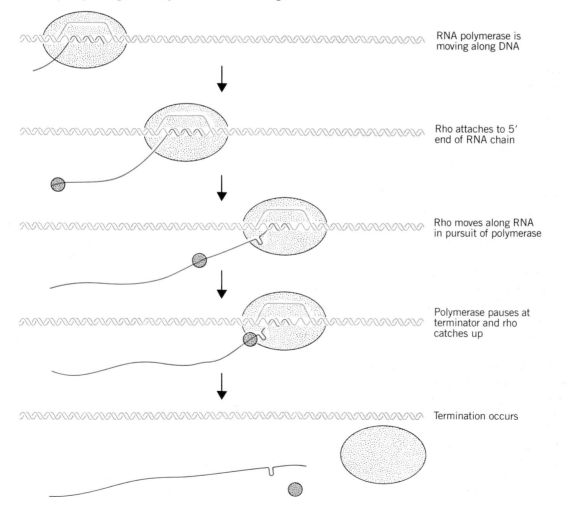

RNA polymerase is moving along DNA

Rho attaches to 5′ end of RNA chain

Rho moves along RNA in pursuit of polymerase

Polymerase pauses at terminator and rho catches up

Termination occurs

they cause a pause, for example, because they delay the polymerase for different characteristic periods. This could depend on the stability of the hairpin, determined by its length and G-C base-pair content. We do not know whether rho factor actually is able to cause termination at *any* point at which RNA polymerase has been halted or whether in addition it requires some sequence-specific information at the terminator.

The idea that rho moves along RNA leads to an important prediction about the relationship between transcription and translation. If rho is to be able to work, it must first have free access to the 5′ end, and then be able to move along the RNA. Either or both of these conditions may be prevented if ribosomes are translating an mRNA. Thus the ability of rho factor to

reach the terminator *while RNA polymerase is pausing there* may depend on what is happening in translation.

This may explain a phenomenon that has been puzzling for a long time. In some cases, a nonsense mutation in one gene of a transcriptional unit prevents the expression of subsequent genes in the unit. This is called **polarity** (see also Chapter 9). A common cause is the disappearance of the mRNA corresponding to the subsequent (distal) parts of the unit. There has been a long debate about whether the distal mRNA sequences disappear because they are synthesized but then rapidly degraded, or because the premature termination of translation prevents RNA synthesis from continuing.

Suppose that there may be rho-dependent termi-

Figure 13.4
The action of rho factor may create a link between transcription and translation when a rho-dependent terminator lies soon after a nonsense mutation.

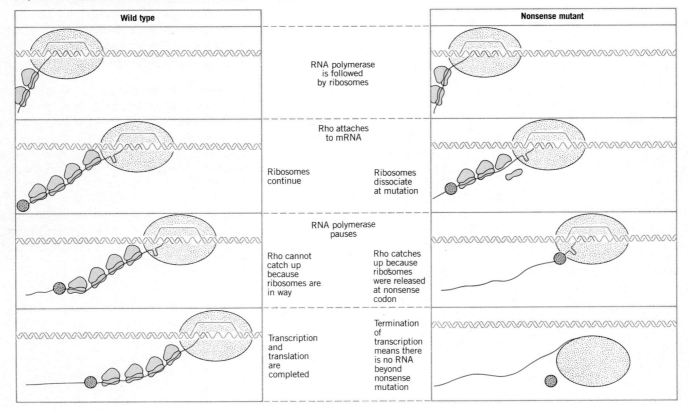

Wild type		Nonsense mutant
	RNA polymerase is followed by ribosomes	
	Rho attaches to mRNA	
Ribosomes continue		Ribosomes dissociate at mutation
	RNA polymerase pauses	
Rho cannot catch up because ribosomes are in way		Rho catches up because ribosomes were released at nonsense codon
Transcription and translation are completed		Termination of transcription means there is no RNA beyond nonsense mutation

nators *within* the transcription unit, that is, before the terminator that *usually* is used. The consequences are illustrated in **Figure 13.4.** These earlier terminators normally are not used, because the presence of the ribosomes delays rho in reaching the terminator; so rho never catches up with RNA polymerase while it is delayed there. But a nonsense mutation causes the ribosomes to be released from the mRNA, so that rho is free to move along the mRNA, and then it may catch up with the polymerase when it is at the terminator. This results in termination, with the release of RNA polymerase, so that the distal regions of the transcriptional unit are never expressed. (Why should there be internal terminators? Perhaps these are simply hairpins that mimic the usual rho-dependent terminator structure, and that are recognized as terminators only under the unusual conditions of polarity.)

MUTATIONS IN THE RHO FACTOR

The interactions between rho factor and other proteins, and some features of its action, have been revealed by the isolation of mutants of *E. coli* that have an altered rho factor. The gene coding for the factor now is known as *rho*. Because of the various effects of these mutations, different mutants originally were isolated by a variety of phenotypes and were given different names, of which the best known is *suA*. The *suA* mutations were named for their ability to suppress polarity.

The *rho* mutations show wide variations in their influence on termination. The basic nature of the effect is a failure to terminate. But the magnitude of the failure, as seen in the percent of readthrough *in vivo,* depends on the particular target locus that is examined. This is probably due to the leakiness of most *rho* mutations, which generally allow the production of a fairly substantial amount of rho activity. If different terminators require different levels of rho factor for termination, the amount of residual rho activity may allow some terminators to function (those active at low rho concentrations), while others fail to function and instead allow RNA polymerase to readthrough.

The ability of *suA* mutations to suppress polarity is explained if they reduce the probability that rho will act on the internal terminator that follows the nonsense codon. Then the termination of translation does not cause transcription also to terminate; and the regions of mRNA beyond the mutation can be translated by ribosomes that reattach farther along.

Some *rho* mutations can be suppressed by mutations in other genes. This provides an excellent way to identify proteins that interact with rho. The β subunit of RNA polymerase is implicated by two types of mutation. First, mutations in the *rpoB* gene coding for it can reduce termination at a rho-dependent site. Second, mutations in *rpoB* can restore the ability to terminate transcription at rho-dependent sites in *rho* mutant bacteria.

ANTITERMINATION IS A PHAGE-CONTROL MECHANISM

Termination provides an opportunity for certain phages to control the switch from early genes to the next stage of expression. The use of antitermination as a control mechanism is illustrated in **Figure 13.5,** which compares it with the use of new promoters.

We have seen in Chapter 12 that one mechanism for switching from early gene transcription to the next stage of expression is to replace the sigma factor of the host enzyme with another factor that redirects its specificity in initiation. An alternative is to synthesize a new phage RNA polymerase. In either case, the critical feature that distinguishes the new set of genes is their possession of *different promoters* from those originally recognized by host RNA polymerase. Thus the expression of the new set of genes does not depend on expression of the early genes once the critical sigma factor or new polymerase has been synthesized. The two sets of transcripts are independent. Early gene expression may cease when the switch to the next stage is made.

By contrast, the use of antitermination depends on a particular arrangement of genes. The early genes lie adjacent to the genes that are to be expressed next, but are separated from them by terminator sites. If termination is prevented at these sites, the RNA

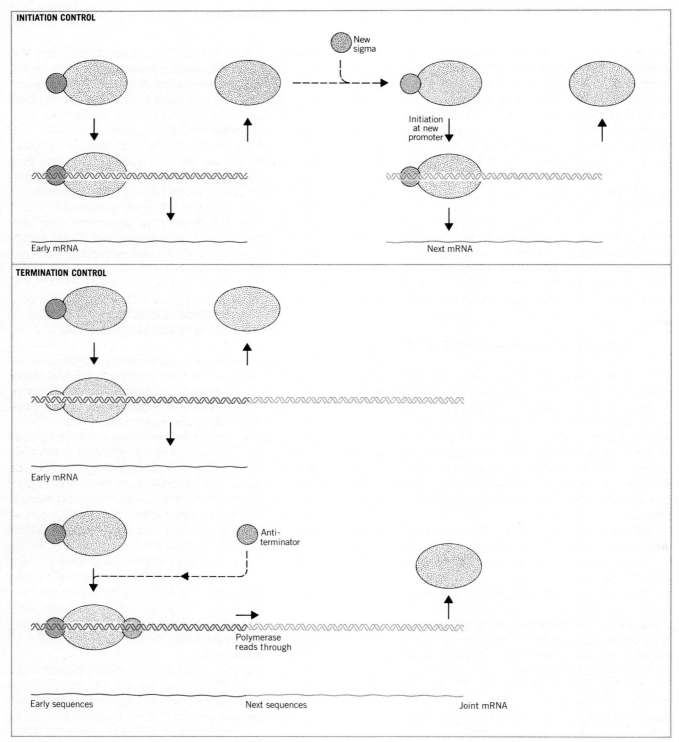

Figure 13.5
Switches in transcriptional specificity can be controlled at initiation or termination.

polymerase reads through into the genes on the other side. Thus in antitermination, the *same promoters* continue to be recognized by RNA polymerase. So the new genes are expressed only via extension of the RNA chain to form molecules that contain the early gene sequences at their 5′ end and the new gene sequences at the 3′ end. Since the two types of sequence remain linked, early gene expression inevitably continues.

The best-characterized example of antitermination is provided by phage lambda. The overall design of the control network is very similar to that of other phages in that the host RNA polymerase transcribes the early genes; and one of the early gene products is needed to transcribe the next set of phage genes. In lambda, the genes transcribed immediately by the host RNA polymerase are called the **immediate early** genes, and those expressed at the next stage are called the **delayed early** genes.

The regulator gene whose product controls the switch from immediate to delayed early expression was identified by mutations that prevent the transition. Lambda mutants in gene *N* can transcribe only the immediate early genes, and therefore do not proceed any further in the infective cycle. This effect is the same as that seen when gene *28* of phage SPO1 is mutated to prevent the production of σ^{gp28}. From the genetic point of view, the mechanisms of new initiation and antitermination are similar. Both are **positive controls** in which an early gene product must be made by the phage in order to express the next set of genes.

A map of the early region of phage lambda is shown in **Figure 13.6.** The immediate early genes are *N* and *cro,* and they are transcribed respectively to the left and to the right from the promoters indicated as P_L and P_R by the *E. coli* RNA polymerase. This means that different strands of the DNA are used as the template in the **leftward** and **rightward** transcription units. **Figure 13.7** shows that transcription by *E. coli* RNA polymerase itself stops at the ends of genes *N* and *cro* at the terminators t_{L1} and t_{R1}, respectively. Both terminators depend on rho; in fact, these were the terminators with which rho was originally identified.

The situation is changed by expression of the *N* gene. The product pN is an antitermination protein that allows the RNA polymerase to read past t_{L1} and t_{R1} into the delayed early genes on either side. Because pN is highly unstable, with a half-life of five minutes, continued expression of gene *N* is necessary to maintain transcription of the delayed early genes. This is no problem, since the *N* gene is part of the delayed early transcription unit, and its transcription must precede that of the delayed early genes.

Like other phages, still another control is needed to express the late genes that code for the components of the phage particle. This switch is regulated by gene *Q,* itself one of the delayed early genes. Its product, pQ, is another antitermination protein, one that specifically allows RNA polymerase initiating at another site, the late promoter $P_{R'}$, to readthrough a terminator that lies between it and the late genes. Thus by employing antitermination proteins with different specificities, a cascade for phage gene expression can be constructed.

Figure 13.6
Phage lambda has two early transcription units.
The map of lambda represents the two strands of DNA by parallel lines. The "upper" strand is transcribed toward the left; while the "lower" strand is transcribed toward the right. The promoters are indicated by the arrowheads. The terminators are indicated by the solid boxes. Genes *N* and *cro* are the immediate early functions, and are separated from the delayed early genes by the terminators.

Figure 13.7

Host RNA polymerase transcribes the immediate early genes of lambda.

The enzyme binds at the two promoters, P_L and P_R, and transcribes genes N and cro up to the terminators t_{L1} and t_{R2}, respectively. Termination requires the rho factor at both sites (this is not shown in the illustration). This transcription produces two immediate early mRNA species, coding for pN and Cro.

ANTITERMINATION DEPENDS ON SPECIFIC DNA SITES

The antitermination activity of pN is highly specific. It does not suppress termination generally at rho-dependent sites. For example, following the synthesis of pN, termination continues to occur as normal at the ends of genes of the host bacterium. However, the antitermination event is not determined by the terminators t_{L1} and t_{R1} because if a bacterial gene with a rho-dependent terminator is inserted into the lambda early transcription unit, pN causes antitermination. Thus antitermination occurs at *any* terminator that is encountered by an RNA polymerase that has traversed a lambda early transcription unit. This is summarized in **Table 13.1.**

There must be some site in each of these transcription units that is recognized by pN as a signal indicating that antitermination should occur at the next terminator to be encountered. In other words, the recognition site needed for antitermination lies at a different place from the terminator site at which the action eventually is accomplished. This conclusion estab-

lishes a general principle. When we know the site on DNA at which some protein exercises its effect, we cannot assume that this coincides with the DNA sequence that it initially recognizes. They may be far apart.

The recognition sites for pN are called *nut* (for *N* utilization). The sites responsible for determining leftward and rightward antitermination are described as

Table 13.1

Antitermination Affects All RNA Polymerases that Initiate at Lambda Early Promoters Irrespective of the Source of the Terminator

Type of Locus		What Happens in Lambda-Infected Cells
Promoter ⟶ Terminator		
Bacterial	Bacterial	Termination occurs
Early lambda	Early lambda	pN prevents termination
Early lambda	Bacterial	pN prevents termination

Figure 13.8
The pN protein binds to RNA polymerase as it passes a *nut* site and then allows it to read-through the terminators t_{L1} and t_{R1}, producing a joint mRNA that contains delayed early as well as immediate early gene sequences.

nutL and *nutR,* respectively, and we know that they must lie between P_L and t_{L1} on one side, and between P_R and t_{R1} on the other. Where exactly are they located? They can be mapped by obtaining hybrids of lambda with other phages in which particular regions of the genome are substituted, by obtaining deletions of lambda DNA that prevent antitermination, and also via the isolation of *nut⁻* point mutations that prevent antitermination.

Such experiments locate *nutL* between the start-point of P_L and the beginning of the *N* coding region; and *nutR* lies between the end of the *cro* gene and t_{R1}. As **Figure 13.8** demonstrates, this means that the two *nut* sites lie in very different positions relative to the organization of their transcription units. Whereas *nutL* is at the beginning, *nutR* is close to the terminator.

How does antitermination occur? When pN recognizes the *nut* site, it must act on RNA polymerase at that site to modify it in such a way that it no longer responds to rho factor at the terminator. The variable locations of the *nut* sites indicate that this event is

linked neither to initiation nor to termination, but can occur to RNA polymerase as it elongates the RNA chain past the *nut* site. Then the polymerase becomes a juggernaut that continues past the terminator, heedless of its signal.

The DNA sequences identify two sites of 17 bp, identical except for a single position, oriented in the same direction relative to transcription. The *nut* sequences include a 5 bp inverted repeat, the two copies of which are separated by a nonsymmetrical region, thus

A G C C C T G A A Pu A A G G G C A
T C G G G A C T T Py T T C C C G T

where the shaded box shows the inverted repeats, and the base pair Pu-Py indicates the single difference between *nutL* and *nutR*.

Is the ability of pN to recognize a short sequence within the transcription unit an example of a more widely used mechanism for antitermination? There are other phages, related to lambda, that have different *N* genes and different antitermination specificities. The region of the phage genome in which the *nut* sites lie has a different sequence in each of these phages, and presumably each phage has characteristic *nut* sites recognized specifically by its own pN. Each of these pN products must have the same general ability to interact with the transcription apparatus as an antitermination function, but has a different specificity for the sequence of DNA that activates the mechanism.

MORE SUBUNITS FOR RNA POLYMERASE?

The discovery of antitermination as a phage-control mechanism has led to the identification of further bacterial proteins that are components of the transcription apparatus. The bacterial functions with which pN interacts can be identified by isolating mutants of *E. coli* in which pN cannot succeed in antitermination. These mutants are refractory to infection by lambda, because the phage is restricted to expressing only its immediate early genes. Several of these mutations lie in the *rpoB* gene. This argues that pN interacts with the β subunit of the core enzyme.

Other *E. coli* mutations that prevent pN function lie in genes that do not code for components of RNA polymerase. They identify the *nus* loci, of which there are at least three: *nusA, nusB,* and *nusE.* (The term "*nus*" is an acronym for *N* utilization substance.) The *nus* loci must code for proteins that are part of the transcription apparatus, but that are not isolated with the RNA polymerase enzyme in its usual form. The *nusA* and *nusB* functions appear to be concerned solely with the termination of transcription, but *nusE* turns out to be identical with the locus *rpsJ,* coding for ribosomal protein S10. Probably the function of S10 in transcription is different from its function in the ribosome, although it is possible that it provides some link to couple the ribosome to transcription.

The *nusA* function is closely related to pN action; mutations in *N* can be obtained that overcome the block to antitermination that is imposed by *nusA* mutations. The idea that NusA is needed to mediate the action of pN is supported by the demonstration that the two proteins bind together *in vitro.* (NusA is one of two proteins that bind strongly to pN *in vitro.* The other is a polypeptide of 25,000 daltons of unknown genetic origin and function.) The pN protein is small (13,500 daltons), very basic, and rather asymmetric in shape. We do not yet have a satisfactory *in vitro* system in which to characterize its function.

The *nusA* product is an acidic protein of 69,000 daltons. It functions *in vitro* as a *termination factor* at the lambda sites t_{L2} and t_{R2}, which are classified as rho-independent sites, and lie farther along the genome from the rho-dependent sites t_{L1} and t_{R1}. Actually, what the *nusA* protein does is to cause RNA polymerase to pause at these terminators for a long time, about 15 minutes. Release of the nascent RNA chain requires the presence of rho factor, but at a very low concentration (which by itself is ineffectual in the absence of NusA protein). This suggests that the action of NusA is not the same as that of rho, but is complementary to it. Genetic experiments show that *nusA* and *nusB* mutations have similar properties (both suppress polarity, as might be expected from a deficiency in a termination factor), so probably they function in a similar capacity.

The NusA protein binds to the polymerase core enzyme, but does not bind to the holoenzyme. But when sigma factor is added to the $\alpha_2\beta\beta'$NusA complex, it displaces the NusA protein, thus reconstituting the $\alpha_2\beta\beta'\sigma$ holoenzyme. This suggests that RNA polymerase passes through the cycle illustrated in **Figure 13.9,** in which it exists in the alternate forms of an enzyme ready to initiate ($\alpha_2\beta\beta'\sigma$) and an enzyme ready to terminate ($\alpha_2\beta\beta'$NusA).

Initiation takes place by the steps we have already described in Chapter 10. Holoenzyme $\alpha_2\beta\beta'\sigma$ binds to a promoter and releases free sigma factor. This generates the core enzyme $\alpha_2\beta\beta'$ that is competent to undertake elongation. Then a NusA protein recognizes the core enzyme and binds to it to generate the $\alpha_2\beta\beta'$NusA complex. We do not know at what stage this happens; it might be sometime during elongation or could occur just at the terminator site. At any rate, while the $\alpha_2\beta\beta'$NusA protein is bound to DNA, the NusA component cannot be displaced. But when termination occurs, the enzyme is released in a state in which NusA either is automatically released from the core or can be displaced from it by sigma factor.

From the viewpoint of the core enzyme, life consists of an alternation between association with sigma for initiation and association with NusA for termination. The two factors are mutually incompatible associates

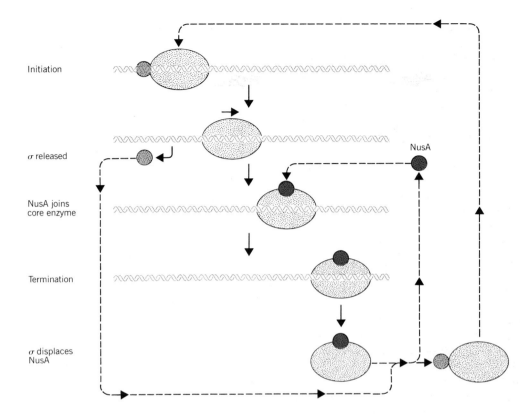

Initiation

σ released

NusA joins
core enzyme

NusA

Termination

σ displaces
NusA

Figure 13.9
RNA polymerase may alternate between initiation-competent and termination-competent forms as sigma and NusA alternately replace each other on the core enzyme.

of the core enzyme. There is no reason to regard one as any more a component of the enzyme than the other, so they seem really to represent alternative components of the enzyme. Thus the isolation of the core enzyme represents preparation of a completely minimal form of RNA polymerase, competent to engage in the basic function of RNA synthesis, but lacking subunits necessary for other functions. As we have remarked, it is a moot point where RNA polymerase ends and the wider transcription apparatus begins.

The NusA component seems to provide the link between pN function and the core enzyme. This makes sense if NusA is a termination subunit, whose incorporation into the enzyme confers specificity toward certain terminators. Thus pN may act as an antiterminator by preventing NusA from exercising a function that is necessary for termination. Immediately this raises the possibility that the host bacterium may contain other proteins, analogous to NusA (or to pN), whose interactions with the core enzyme control its recognition of termination sites.

This conclusion would explain a curious finding about NusA. It is identical with a protein previously identified as the L factor, which is *necessary* for transcription of some genes *in vitro*. This behavior is not at all what would be expected of a termination factor! But if it is the presence of NusA that allows these hypothetical bacterial functions to suppress termination at particular sites, NusA could be necessary to transcribe any gene lying beyond such a site. Alternatively, the NusA subunit itself might confer on the core enzyme the ability to read past some *t* sites while acknowledging others.

We do not know yet exactly how core enzyme, NusA protein, and rho factor cooperate at termination; but the identification of NusA as a termination subunit, and the existence of phage pN proteins of several specificities all presumably able to interact with NusA, imply that there could be a spectrum of other (unknown) functions that influence core enzyme or modify its ability. Also, this means that core enzyme usually may never terminate by itself; so the *in vitro* experiments on termination with core enzyme may reflect an aberrant ability. Thus the nature of the apparently rho-independent sites needs to be reinvestigated with *in vitro* systems that contain the other possible termination components.

DIFFICULTIES IN EUCARYOTES

Little is known about either the signals for termination or the process involved for most eucaryotic RNA polymerases. The major difficulty in analyzing transcripts is the lack of certainty about the actual site of termination. Although the RNA molecules identified *in vivo* have defined 3′ ends, how are we to know whether these were produced by a termination event or by cleavage within a longer original transcript?

For the products of RNA polymerase II, the problem is exacerbated by the processing that occurs at the 3′ end, in which a length of polyadenylic acid is added. In at least some cases, there is evidence that the 3′ terminus for polyadenylation is indeed generated by cleavage and not by termination. We therefore have some information about the signals needed to generate the end for addition of poly(A), but the 3′ region of the original primary transcript generally remains uncharacterized. There is a site in the DNA of the virus SV40 that causes termination; this sequence is like a rho-independent bacterial site, with a hairpin followed by a stretch of U bases.

For RNA polymerase I, the sole product is a large precursor that contains the sequences of the major rRNA. The precursor is subject to extensive processing, and it has not yet been proved that the 3′ end actually is generated by termination. However, the terminus does correspond to a run of U bases.

With RNA polymerase III, transcription *in vitro* produces molecules that have the same 5′ and 3′ ends as those synthesized *in vivo*. This suggests that proper initiation and termination have occurred. The system can be manipulated by introducing changes in the sequence of the template around the termination region.

When the 5S RNA genes of *X. laevis* are transcribed by the homologous enzyme *in vitro*, termination occurs within a run of four U bases. Termination usually occurs at the second, but there is heterogeneity, with some molecules ending in three or even four U bases. The same heterogeneity is seen in molecules synthesized *in vivo*, so it seems to be a bona fide feature of the termination reaction.

Just like the procaryotic terminators, the U run is embedded in a G-C-rich region. Although there are sequences of dyad symmetry present, they are not necessary for termination, since mutations of the template that abolish the symmetry do not prevent the normal completion of RNA synthesis. Nor are any sequences beyond the U run necessary, since all nucleotides to its right can be replaced by other sequences without any effect on termination.

The U run itself is not sufficient for termination, because regions of four successive U residues occur within transcription units read by RNA polymerase III. (However, there are no internal five-U runs, which fits with the greater efficiency that is seen in termination when the terminator is a U_5 rather than U_4 sequence.) The critical feature in termination must therefore be the recognition of a U_4 sequence in a context that is rich in G-C base pairs.

How does the termination reaction occur? It cannot rely on the weakness of the rU-dA RNA-DNA hybrid region that lies at the end of the transcript, because often only the first two U residues are transcribed before termination. Perhaps the G-C-rich region plays a role in slowing down the enzyme, but there does not seem to be any counterpart to the hairpin that may be involved in procaryotic termination. We do not know how the enzyme is able to respond so specifically to such a short signal.

FURTHER READING

The process of bacterial termination and its use as a control mechanism has been reviewed by **Roberts** in *RNA Polymerase*, (Losick & Chamberlin, Eds., Cold Spring Harbor Laboratory, New York, 1976, pp. 247–271) and by **Adhya** & **Gottesman** (*Ann. Rev. Biochem.* **47**, 967–996, 1978). The nature of the terminator signals has been reviewed in the context of consensus sequences by **Rosenberg** & **Court** (*Ann. Rev. Genet.* **13**, 319–353, 1979). The discovery of the more complex bacterial transcription apparatus has been put together by **Greenblatt** (*Cell* **24**, 8–9, 1981). Too little is known about eucaryotic termination for reviews yet to be available.

PART 4
CONTROL OF PROCARYOTIC GENE EXPRESSION

According to the strictly structural concept, the genome is considered as a mosaic of independent molecular blueprints for the building of individual cellular constituents. In the execution of these plans, however, coordination is evidently of absolute survival value. The discovery of regulator and operator genes, and of repressive regulation of the activity of structural genes, reveals that the genome contains not only a series of blueprints, but a coordinated program of protein synthesis and the means of controlling its execution.

Francois Jacob & Jacques Monod, 1961

CHAPTER 14
THE OPERON:
THE LACTOSE PARADIGM

Bacteria need to be able to respond swiftly to changes in their environment. Because of the conditions under which they live, capricious fluctuations in the supply of nutrients may occur at any time; survival depends on the ability to switch from metabolizing one substrate to another. Unicellular eucaryotes may share this subjection to an incessantly changing world; but more complex, multicellular organisms are restricted to a more constant set of metabolic pathways, and may not have the same need to respond to external circumstances.

Flexibility is therefore at a premium in the bacterial world. Yet economy also is demanded, since a bacterium that indulges in energetically expensive ways to meet the demands of the environment may be at a disadvantage. Certainly it would be expensive to produce unnecessarily all the enzymes for a metabolic pathway that cannot be used because the substrate is absent. So the bacterial compromise is to avoid synthesizing the enzymes of a pathway in the absence of their substrate; but to be ready at all times to produce the enzymes if the substrate should appear.

This line of reasoning explains the central features of the organization of bacterial genes. They are grouped in clusters, so that all the enzymes needed to make up a particular pathway are represented by genes adjacent to one another. The entire gene cluster may be transcribed into a single polycistronic mRNA that is sequentially translated by the ribosomes into each of the proteins. We have already alluded in previous chapters to the fact that few genes are individually transcribed in *E. coli;* most are part of larger transcription units. This form of organization allows the expression of all the genes in the unit to be coordinately regulated by controlling its transcription. This is accomplished by the interaction of a regulator protein with a control site that lies close to the promoter.

INDUCTION AND REPRESSION ARE CONTROLLED BY SMALL MOLECULES

It has been known since the beginning of this century that certain enzymes of yeast are present only when the cells are grown on the appropriate substrates. This effect now is known as enzyme **induction.** It has been characterized in bacteria, and in particular with *E. coli,* where the lactose system provides the paradigm for this sort of control mechanism.

When cells of *E. coli* are grown in the absence of a β-galactoside, they contain very few molecules of the enzyme β-galactosidase—say, fewer than five. The function of the enzyme is to break the β-galactoside into its component sugars. For example, lactose is cleaved into glucose + galactose, which are then further metabolized. There is no need for the enzyme in the absence of the substrate.

When a suitable substrate is added, the enzyme activity appears very rapidly in the bacteria, as the result of synthesis of new enzyme molecules. Within 2–3 minutes some enzyme is present, and soon there may be up to 5000 molecules of enzyme per bacterium. (Under suitable conditions, β-galactosidase can account for 5–10% of the total soluble protein of the bacterium.) If the substrate is removed from the medium, the synthesis of enzyme stops as rapidly as it originally started.

This type of rapid response to changes in nutrient supply is shown not only in dealing with the need to metabolize new substrates, but is used also to shut off endogenous synthesis of compounds that may suddenly appear in the medium. For example, tryptophan is an essential amino acid that E. coli synthesizes through the action of the enzyme tryptophan synthetase. But if tryptophan is provided in the medium on which the bacteria are growing, immediately the production of the enzyme is halted. This is called **repression.** It allows the bacterium to avoid devoting its resources to unnecessary synthetic activities.

Induction and repression represent the same phenomenon seen from different perspectives. In one case the bacterium adjusts its ability to use a given substrate for growth; in the other it adjusts its ability to synthesize a particular metabolic intermediate. The trigger for either type of adjustment is the small molecule that is the substrate for the enzyme, or the product of the enzyme activity, respectively. Small molecules that cause the production of enzymes able to metabolize them are called **inducers.** Those that prevent the production of enzymes able to synthesize them are called **corepressors.**

The ability to act as inducer or corepressor is highly specific. Only the substrate/product or a closely related molecule can serve. But in both cases, the activity of the small molecule does *not* depend on its interaction with the target enzyme. Thus there are molecules that adequately resemble the natural inducers for β-galactosidase, but that cannot be metabolized by the enzyme. The example par excellence is isopropylthiogalactoside (IPTG), one of several thiogalactosides with this property. Although it is not recognized by β-galactosidase, IPTG is a very efficient inducer.

Molecules that induce enzyme synthesis but are not metabolized are called **gratuitous inducers.** They are extremely useful because they remain in the cell in their original form. (A real inducer would be metabolized, and this interferes with study of the system.) The existence of gratuitous inducers reveals an important point. There must be some function of the system, distinct from the target enzyme, that recognizes the appropriate substrate; and its ability to recognize related potential substrates is different from that of the enzyme.

GENE CLUSTERS ARE COORDINATELY REGULATED

Induction or repression is not usually confined to a single enzyme. We have mentioned that often all the enzymes of a metabolic pathway are regulated together. And in addition to the enzymes actually involved in the pathway, other related activities may be included in the unit of coordinate control; an example is the protein responsible for transporting the small molecule substrate into the cell. The *lactose (lac)* genes provide an example of this type.

Three genes map in the cluster drawn in **Figure 14.1.** First, *lacZ* codes for the enzyme β-galactosidase, whose active form is a tetramer of about 500,000 daltons. Second, *lacY* codes for the β-galactoside permease, a 30,000 dalton membrane-bound protein constituent of the transport system. Third, *lacA* codes for β-galactoside transacetylase, an enzyme that transfers an acetyl group from acetyl-CoA to β-galactosides.

Mutations in either *lacZ* or *lacY* can create the *lac⁻* genotype, in which cells cannot utilize lactose. The *lacZ⁻* mutations abolish enzyme activity, directly preventing metabolism of lactose. The *lacY⁻* mutants cannot take up lactose from the medium. No defect is identifiable in *lacA⁻* cells, which is a puzzle. The significance of the acetylation reaction is uncertain, but it is possible that it may give an advantage when the bacteria grow in the presence of certain nonmetabolizable analogs of β-galactosides, because it results in detoxification and excretion.

Figure 14.1
The *lac* operon occupies about 6000 bp of DNA.

At the left the *lacI* gene has its own promoter and (presumably) terminator. The end of the *lacI* coding region is immediately adjacent to the *lacP* promoter, which is divided into a left half that binds CAP and a right half that binds RNA polymerase. The operator occupies the first 26 bp of the *lacZ* gene, which is extremely long and is followed by the *lacY* and *lacA* genes and (presumably) a terminator.

The cluster of three genes, *lacZYA,* is transcribed into a single mRNA from a promoter just upstream from *lacZ.* Their induction is controlled at the level of transcription. In the absence of an inducer, the gene cluster is not transcribed. When an inducer is added, transcription starts at the single promoter, *lacP,* and proceeds through the genes to a terminator located somewhere beyond *lacA.* This accomplishes a **coordinate regulation,** in which *all* the genes are expressed (or not expressed) in unison.

The mRNA is translated sequentially from its 5′ end, which explains why induction always causes the appearance of β-galactosidase, β-galactoside permease, and β-galactoside transacetylase, in that order. The expression of the three genes via a common mRNA explains why the relative amounts of the three enzymes always remain the same under varying conditions of induction. Induction essentially represents a switch that causes the genes to be expressed. Inducers may vary in their effectiveness, and other factors may influence the absolute level of transcription or translation, but the relationship between the three genes is predetermined by their organization.

The mRNA is extremely unstable, and decays with a half-life of only ~3 minutes. It is this feature that allows induction to be reversed so rapidly. Transcription ceases as soon as the inducer is removed; and in a very short time all the lactose mRNA has been destroyed, and the cell stops producing the enzymes.

A REGULATOR GENE CONTROLS THE STRUCTURAL GENES

The genes that code for the proteins required by the cell, for either enzymatic or structural functions, are called **structural genes.** The overwhelming majority of bacterial genes fall into this category, which therefore represents an enormous variety of protein structures and functions. Structural genes also include the genes coding for rRNA and tRNA. All types of structural gene tend to be organized into clusters that may be coordinately controlled, as in the example of the three *lac* structural genes, *lacZYA.*

What is the target for the small-molecule inducer? We know that it is distinct from the structural gene products. It is another protein, whose sole function is to control the expression of the structural genes. The gene that codes for this protein is called the **regulator gene.**

Regulator genes are responsible for controlling the expression of the structural gene clusters, usually via the synthesis of proteins that act to control transcription. The regulator proteins exercise this function by binding to particular sites on DNA.

We can distinguish between structural genes and regulator genes by the effects of mutations. A mutation in a structural gene simply deprives the cell of the particular protein for which that gene codes. But a mutation in a regulator gene influences the expression of all those structural genes that it controls. The nature of this influence reveals the type of regulation.

The *lac* genes are controlled by **negative regulation.** This means that they are transcribed *unless* they are turned *off* by the regulator protein. Thus a mutation that inactivates the regulator causes the genes to remain in the expressed condition. Since the function of the regulator is to *prevent* the expression of the structural genes, it is called a **repressor protein.**

Regulator proteins appear also in other guises. We have already seen that *positive* regulators are so called because they must be present in order for their target structural genes to be turned *on*. In the absence of the regulator, the genes therefore cannot be expressed. Examples are the control of initiation of transcription via the production of new sigma factors (Chapter 12) or the specific antitermination of transcription (Chapter 13).

THE CONTROL CIRCUIT OF THE OPERON

The concept that two different classes of genes might be distinguished by their functions was proposed by Jacob and Monod in 1961 in their classic formulation of the operon model. They defined an **operon** as a complete unit of gene expression, including structural genes, a regulator gene (or genes), and control elements (the sites at which the regulator proteins act). Each particular operon can be characterized by a network of interactions between regulator proteins and their target sites. This can be considered as a control circuit. The original model was constructed for the *lac* operon, which remains the best-characterized example, but similar arguments have been used to construct analogous circuits for other operons.

The basic control circuit of the lactose operon is illustrated in **Figure 14.2.** The cluster of three structural genes, *lacZYA*, forms a transcription unit that is conventionally drawn from left to right. Transcription starts at the promoter *lacP*, just to the left (upstream) of the first gene, *lacZ*. The regulator gene, *lacI*, lies a little farther to the left, and forms an independent transcription unit. It is transcribed from its own promoter into a monocistronic mRNA that codes for a repressor protein. The crucial features of the control circuit reside in the dual properties of this repressor: it can prevent transcription; and it can recognize the small-molecule inducer.

How does the repressor prevent transcription? It binds to a sequence of DNA called the **operator,** denoted *lacO*. This lies between the promoter *lacP* and the cluster of structural genes *lacZYA*. When the repressor binds at this site, its presence prevents RNA polymerase from initiating transcription at the promoter.

How does this interaction respond to the small-molecule inducer? The repressor protein has a very high affinity for the operator; in the absence of inducer, it binds there so that the adjacent structural genes cannot be transcribed. But the inducer can bind to the repressor to form a repressor•inducer complex that no longer associates with the operator. The key feature of this interaction is that the repressor protein has two binding sites, one for the inducer and one for the operator. When the inducer binds at its site, it changes the conformation of the protein in such a way as to influence the activity of the *other* site. This type of relationship is called **allosteric control.** The result in this case is that, when inducer is added, the repressor is converted to a form that leaves the operator. Then RNA polymerase can initiate transcription of the structural genes.

The lactose repressor is a tetramer of identical subunits of 38,000 daltons each. There are about 10 tetramers per cell (this means that there must be 40 translations of the *lacI* gene in each cell cycle). The regulator gene is transcribed at a rate that appears to be governed simply by the affinity of its promoter for RNA polymerase. (Mutations in the promoter can greatly increase the expression of the gene, a technique that was combined with amplification of the locus to obtain greater amounts of repressor than are normally found in the cell.)

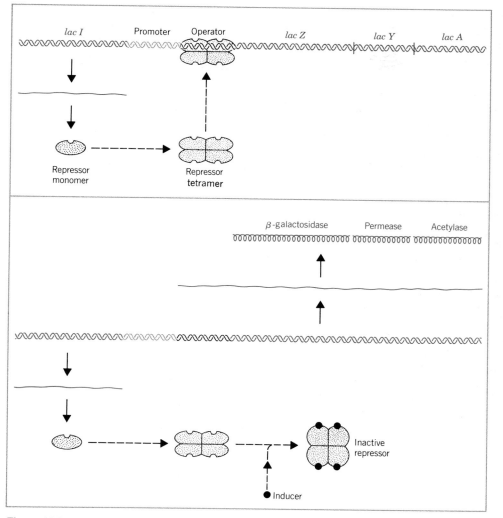

Figure 14.2
The *lac* operon is turned on by an inducer.

Upper. The *lacI* gene synthesizes a repressor whose tetramer binds to the operator and prevents transcription of the structural genes.

Lower. Addition of inducer converts the repressor into an inactive form that cannot bind to the operator; so transcription starts at the promoter and the three enzymes are synthesized.

CONSTITUTIVE MUTATIONS DEFINE REPRESSOR ACTION

The *lacI* regulator gene was originally identified by the isolation of mutations that affect the expression of *all three* structural genes and map outside them. Since the *lacI* mutations complement mutations in the structural genes, they can be taken to identify another gene coding for a diffusible product.

The *lacI⁻* genotype can arise from either point mutations or deletions. The latter indicates that it represents *loss* of the usual regulation. The *lacI⁻* mutants express the structural genes all the time, irrespective of whether the inducer is present or absent. This behavior is described as **constitutive** gene expression.

This conforms to our expectation for a negative control system. The *lacI*⁺ gene codes for a repressor protein that is able to turn off the transcription of the *lacZYA* cluster. Mutation of the gene to form the *lacI*[−] type allows the genes to be constitutively expressed because the repressor now is inactive.

We can confirm this conclusion by determining the

relationship between the *lacI*[−] constitutive mutant gene and the wild-type *lacI*⁺ gene when both are present in the same cell. This is accomplished by forming a **partial diploid,** when one copy of the operon is present on the bacterial genome itself, and the other is introduced via an independent self-replicating DNA molecule that carries only a few genes, including a copy of the operon or part of it. This additional DNA molecule is called a **plasmid.**

In cells in which one regulator gene is *lacI*⁺ and the

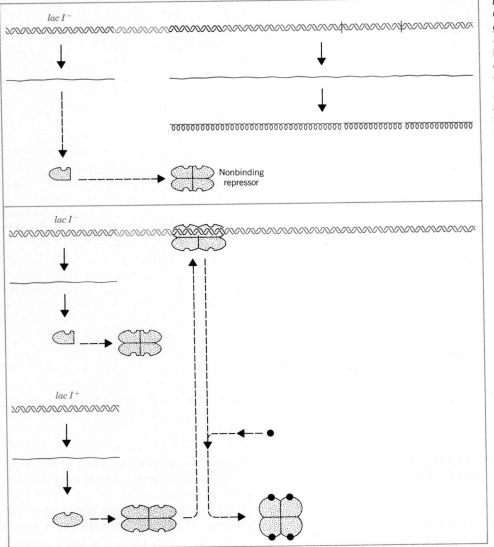

Figure 14.3
Constitutive mutants in the *lacI* gene are recessive.

Upper. A *lacI*[−] mutation causes an inactive repressor to be made. It cannot bind to the operator and so the operon is expressed.

Lower. The introduction of another *lacI*⁺ gene allows normal repressor to be made. This is able to act in *trans* to bind to the operator, being displaced from it when inducer is added.

lac I[−]

Nonbinding repressor

lac I[−]

lac I⁺

other is *lacI⁻*, normal regulation is restored. The structural genes again respond to normal regulation by being repressed in the absence of inducer. We can explain this in the way depicted in **Figure 14.3.** Constitutive mutations in the repressor abolish its ability to bind to the operator. The result is that transcription can initiate freely at the promoter. But the introduction of a second, wild-type regulator gene restores the presence of normal repressor. So once again the operon is turned off in the absence of inducer. In genetic parlance, the wild-type **inducibility** is dominant over the mutant constitutivity. This is the hallmark of negative control.

The *lacI* regulator gene lies just to the left of the cluster of structural genes. However, since it specifies a *trans*-diffusible product, it need not be located near the structural genes. Indeed, as we have seen, a *lacI* gene on an independent DNA molecule is able to control a *lacZYA* cluster on the bacterial chromosome (or the orientation can be reversed). In other operons in

E. coli, the regulator gene actually is located at some distance from the structural genes. However, we might speculate that an advantage to keeping a regulator near its structural genes is that it has a useful purpose only together with them, so there is no point risking chance separation.

THE OPERATOR IS *CIS*-DOMINANT

The operator was originally identified by constitutive mutations, denoted o^c, whose distinctive properties demonstrate that this region is not represented in a diffusible product. The structural genes contiguous with the o^c mutation are expressed constitutively. The reason is that the operator has suffered a mutation that prevents the repressor from binding to it. Thus the repressor is no longer able to prevent RNA polymerase from initiating transcription. So the operon is continuously expressed, as illustrated in **Figure 14.4.**

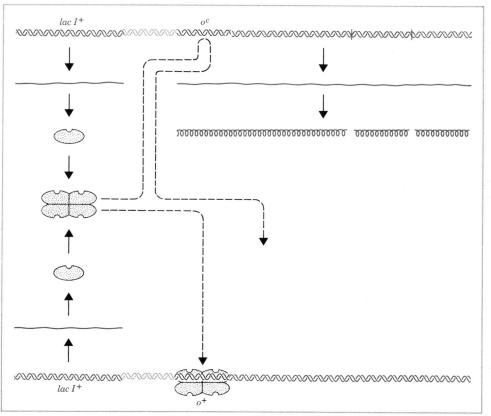

Figure 14.4
Operator-constitutive (o^c) mutations are *cis*-dominant.

The upper operon has a mutant operator that cannot bind repressor, so the adjacent structural genes are expressed constitutively. The lower operon has a wild-type operator that responds to repressor and is therefore expressed only on induction. The behavior of each set of structural genes is controlled solely by the contiguous operator, whose properties are not influenced by the other operator.

The operator can control *only* the *lac* genes that are adjacent to it. If a second *lac* operon is introduced into the bacterium on an independent molecule of DNA, it has its own operator. Neither operator is influenced by the other. Thus if one operon has a wild-type operator, it will be repressed under the usual conditions, while a second operon with an o^c mutation will be expressed in its characteristic fashion.

The way such mutations are tested in practice is to see whether two operons differing in their structural genes as well as in their operators can complement. Suppose we construct the diploid combination

$$o^c \; lacZ^+ \; lacY^- \; / \; o^+ \; lacZ^- \; lacY^+$$

There is one wild-type allele of each locus present in the cell, but we do not see wild-type gene expression. The constitutive operator causes the $lacZ^+$ and $lacY^-$ alleles to be expressed. But the $lacY^+$ gene in the other operon is not expressed under conditions of repression, because its operator remains wild type.

The ability of a site to control adjacent genes irrespective of the presence in the cell of other alleles of the site defines the phenomenon of **cis-dominance.** It indicates that the mutant site does not specify any product (usually protein, but in principle possibly also RNA) that can diffuse through the cell to exercise its effect at the other allele. Sometimes a *cis*-dominant site is referred to as **cis-acting,** to make the contrast with a **trans-acting** function, which, because it specifies a diffusible product, can act on all relevant sites in the cell, whether they are present on the same or different molecules of DNA (see Chapter 1).

The concept of *cis*-dominance applies to *any* sequence of DNA that functions by being recognized rather than by being converted into a diffusible product. Thus mutations in a promoter or terminator also are *cis*-acting.

A consequence of *cis*-dominance is that a mutation cannot be assigned to a complementation group. (For the ability to complement is characteristic only of genes expressed as diffusible products.) This means that when two *cis*-acting sites lie close together—for example, a promoter and an operator— we cannot classify the mutations by a complementation test. We are restricted to distinguishing them by their effects on the phenotype.

Finally, we should point out that *cis*-dominance is a characteristic of any site that is *physically contiguous* with the sequences it controls. Thus if a control site functions as part of a polycistronic mRNA, mutations in it will display *exactly the same pattern* of *cis*-dominance as they would if functioning in DNA. The critical feature is that the control site cannot be physically separated from the genes. From the genetic point of view, it does not matter whether the site and genes are together on DNA or on RNA.

UNINDUCIBLE MUTATIONS CAN OCCUR IN THE PROMOTER OR REPRESSOR

Mutants of the operon that are **uninducible** cannot be expressed at all. They fall into the same two types of genetic classes as the constitutive mutants. The dominance relationships of each type can be used in the same way to define the nature of the locus.

Promoter mutations, as we have mentioned, are *cis*-acting. Those that prevent RNA polymerase from binding at *lacP* render the operon nonfunctional because it cannot be transcribed.

Mutations in *lacI* that abolish the ability of repressor to bind the inducer can cause the same phenotype. The repressor is "locked in" to the active form that recognizes the operator and prevents transcription. The addition of inducer is to no avail.

Such mutants are described as *lacI^s*. **Figure 14.5** shows that they are *trans*-acting and dominant over wild type. This happens because the mutant repressor binds to all operators in the cell to prevent their transcription, and cannot be prized off, irrespective of the properties of any other repressor protein that may be present.

HOW DOES REPRESSOR BLOCK TRANSCRIPTION?

The repressor was originally isolated by purifying from extracts of *E. coli* the component able to bind the gratuitous inducer IPTG. (Because the amount of repressor in the cell is so small, in order to obtain enough material it was necessary to use a promoter up mu-

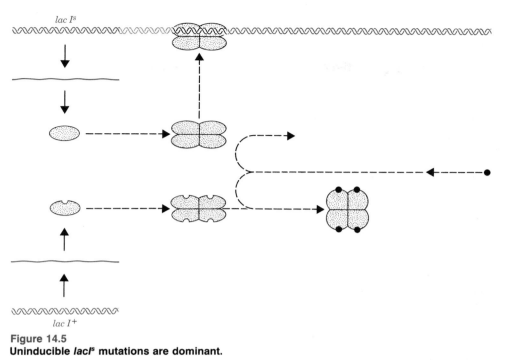

Figure 14.5
Uninducible *lacI^s* mutations are dominant.

The *lacI^s* gene produces a repressor that cannot respond to inducer; for example, because it has lost the inducer-binding site. Then it binds to the operator all the time, irrespective of the presence of inducer. This is dominant in *trans* because the introduction of a *lacI^+* gene does not make any difference; the uninducible repressor remains on the operator.

tation to increase *lacI* transcription, and to place this *lacI* locus on a DNA molecule present in many copies per cell. This results in an overall overproduction of 100–1000-fold.)

The binding of the purified repressor protein to DNA was first characterized by a **filter-binding assay.** This takes advantage of the ability of nitrocellulose filters to retain protein but not DNA. However, DNA complexed with protein is retained; so sequences of DNA that can bind to the repressor are specifically retained on the filter.

The repressor binds to DNA containing the sequence of the wild-type *lac* operator. The DNA must be in double-stranded form. The repressor does not bind if the DNA has been obtained from an *o^c* mutant. The addition of IPTG releases the repressor from *lacO* DNA *in vitro*. The *in vitro* reaction between repressor protein and *lacO* DNA therefore displays the charac-

teristics of control inferred *in vivo;* so it can be used to establish the basis for repression.

What is the precise location of the operator in relation to the promoter? Defined by the length of DNA protected by repressor *in vitro* against nuclease attack, *lacO* comprises a region of about 26 bp. It stretches from position −5 just upstream from the mRNA startpoint to position +21 within the transcription unit. Thus it may just overlap with the end of the promoter that extends from the Pribnow box into the transcription unit.

Does this overlap prevent repressor and RNA polymerase from binding simultaneously to *lac* DNA? Competition experiments show that the binding of either protein to the DNA prevents the binding of the other. This situation is illustrated in **Figure 14.6.** Similar results have been obtained in several other operons.

A common mechanism of repression is therefore for

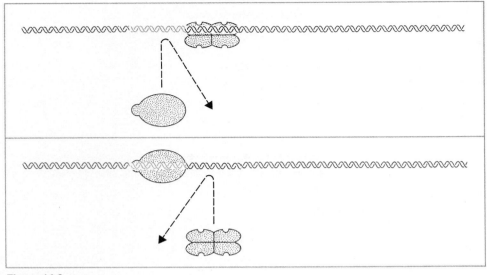

Figure 14.6
Repressor binding and RNA polymerase binding are mutually exclusive at the *lac* operator-promoter region.

the repressor to bind to an operator that at least partially overlaps the sequence that must be recognized by RNA polymerase. The overlap does not always occur in the same region of the promoter. For example, in phage lambda, the operator lies in the upstream region of the promoter (see Chapter 16). But the general rule is that when the repressor is bound at its operator, its presence denies the RNA polymerase access to its promoter, and therefore prevents transcription from being initiated.

CONTACTS IN THE OPERATOR

How does the repressor recognize the specific sequence of operator DNA? The interaction between the lactose repressor and operator is often taken as a paradigm for sequence-specific DNA-binding reactions. It is the best characterized from the aspects of both the DNA sequence and the protein structure.

The operator has a feature common to many recognition sites for proteins: it has an axis of dyad symmetry. The inverted repeats are not quite identical. As can be seen from **Figure 14.7,** there are three differ-

ences between them. (The prominent exception to the occurrence of symmetry in recognition sites, of course, is provided by the promoter, which must convey information about the direction of transcription.)

What use is made of the symmetry in recognizing the operator? The possibility that there is a rearrangement of DNA to form a cruciform has been excluded. When the repressor binds to the DNA, there is a change in the structure of the duplex equivalent to a rather small unwinding (but not enough to create free base pairs or a cruciform). Does the symmetry of the DNA sequence reflect a symmetry in the protein? This can be accomplished because the repressor is a tetramer of identical subunits, each of which must therefore have the same DNA-binding site.

The approaches used to define the points that the repressor contacts in the operator are the same as those that we have already discussed for the polymerase-promoter interaction (see Chapter 11). Constitutive point mutations identify individual base pairs that must be crucial; deletions of material on either side define the end points of the region. Experiments in which DNA bound to repressor is compared with unbound DNA for its susceptibility to methylation or

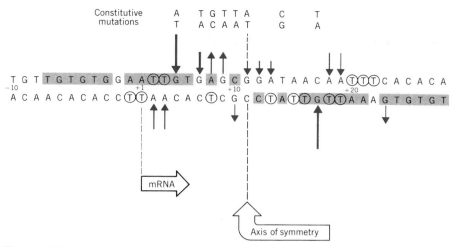

Figure 14.7
The *lac* operator has a symmetrical sequence.

The sequence is numbered relative to the startpoint for transcription at +1. The regions of dyad symmetry are indicated by the shaded blocks; the axis passes through the base pair at position +11. Substitutions that result in constitutive expression are shown by the base pairs above the sequence. Thymines at positions where the substitution of bromouracil permits cross-linking to the repressor are indicated by the circles. Arrows pointing into the sequence identify purines that are protected by repressor against methylation. Arrows pointing away from the sequence identify positions at which methylation is enhanced. The length of the arrow corresponds to the strength of the effect. The region of strong contacts lies from +1 to +23, but weaker contacts could be made beyond this.

UV cross-linking identify bases that are either protected or more susceptible when associated with the protein.

When the DNA bound to repressor is isolated by degrading unprotected sequences with nuclease, it does not extend over the entire region of symmetry. We have mentioned that it comprises the 26 bp region from −5 to +21. The area identified by constitutive mutations is even smaller. Within a central region extending over the 13 bp from +5 to +17, there are eight sites at which single base-pair substitutions so far have been shown to cause constitutivity. This emphasizes the same point made by the promoter mutations summarized earlier in Figure 11.5. A small number of essential specific contacts within a larger region can be responsible for sequence-specific association of DNA with protein.

The distribution of the sites of mutation suggests that the left side of the operator may be more susceptible to damage: it contains six mutations, compared

with the two on the right. The two mutations on the right are exact symmetrical counterparts of mutations found on the left. (At position +17, a T-A replaces C-G, while at position +5 A-T replaces G-C. Similarly, at position +14 a C-G replaces T-A, while at position +8 a G-C replaces A-T.) However, the magnitude of the constitutive effects is not identical for the members of each pair. Thus symmetry is clearly relevant to repressor-operator contacts, but nonetheless the repressor seems to bind to the left side more intimately.

The pattern of enhancement and protection of bases shows some features of symmetry within a general region of close contacts. Figure 14.7 shows that all but two of the thymine residues within the region from +1 to +22 can be cross-linked to repressor. Methylation experiments show that a large proportion of the purines (A and G) between +3 and +19 are protected quite strongly by repressor binding. A few display increased susceptibility, presumably due to the creation of "hydrophobic pockets" in the repressor-operator complex. The contacts between +1 and +6 are essentially symmetrical with those made between +21 and +16; contacts closer to the axis of symmetry are not symmetrical. Thus the repressor binds to the operator in such a way as to sit symmetrically about the outlying points of contact, while not lying symmetrically in the immediate vicinity of the axis of symmetry.

The central region that is crucial for repressor binding occupies roughly the first 20 bp downstream from the startpoint for transcription. But if we recall the points contacted by RNA polymerase at the promoter, Figure 11.5 shows that these tend to lie upstream from the startpoint. The major points of contact for the two proteins are therefore adjacent rather than overlapping. The reason for their mutual exclusion may be that actually it is important for each protein to stretch farther along the DNA, as suggested by the ability of repressor to protect the DNA up to −5 against nuclease digestion; or it may be that the presence of repressor provides sufficient steric hindrance to prevent the polymerase from settling onto its promoter.

THE INTERACTION OF REPRESSOR SUBUNITS

The repressor protein has two types of binding site that (as we have seen) interact to confer the ability to control gene expression in response to the environment. The repressor itself recognizes the specific nucleotide sequence that constitutes the operator. It also binds the small-molecule inducer, and as a result *loses* its ability to bind to the operator DNA. The two types of binding site can be identified within the repressor subunit by mutations in *lacI* that inactivate them. Their proper relationship *in vivo* depends on the multimeric structure of the repressor.

Monomeric subunits of the repressor associate at random in the cell to form the active protein tetramer. When two different alleles of the *lacI* gene are present, the subunits made by each can associate to form a heterotetramer, whose properties may differ from those of either homotetramer. This type of interaction between subunits is a characteristic feature of multimeric proteins and is described as **interallelic complementation** (see Chapter 1).

A particular type of interaction, called **negative complementation,** occurs with some repressor mutants. This is seen in the combination of *lacI^{-d}* with *lacI$^+$* genes. The *lacI^{-d}* mutation alone results in the production of a repressor that cannot bind the operator, and is therefore of the constitutive type representative of all *lacI$^-$* alleles. Because the *lacI$^-$* type of mutation inactivates the repressor, it is recessive to the wild type. However, the −*d* notation indicates that this variant of the negative type is *dominant* when paired with a wild-type allele.

The reason for the dominance is that the *lacI^{-d}* allele produces a "bad" subunit, which is not only itself unable to bind to operator DNA, but is also able as part of a tetramer to prevent any "good" subunits from binding. This demonstrates that it is indeed the repressor tetramer as a whole, rather than the individual monomer, that is needed to achieve repression. The poisoning effect also can be produced *in vitro* by mixing appropriate "good" and "bad" subunits.

The *lacI^{-d}* mutations identify the DNA-binding site of the repressor subunit. This explains their ability to prevent mixed tetramers from binding to the operator; a reduction in the number of binding sites must reduce the specific affinity too much. The map of the *lacI* gene shown in **Figure 14.8** shows that the *lacI^{-d}* mutations all are clustered at the extreme left end of the gene. This identifies the immediate N-terminal region of the protein as the DNA-binding site. Mutations of the recessive *lacI$^-$* type also occur elsewhere in the molecule, but could exert their effects on DNA binding indirectly.

The role of the N-terminal region in specifically binding DNA is shown also by its location as the site of occurrence of "tight binding" mutations. These increase the affinity of the repressor for the operator, sometimes so much that it cannot be released by inducer. They are rare.

Uninducible mutations of the *lacI^s* type render the repressor unresponsive to the inducer. This could be either because the protein has lost its inducer-binding site, or because it has become unable to transmit its effect to the DNA-binding site. Mutations of this class also include those that make the repressor subunits unable to aggregate to the tetrameric form. As can be seen from Figure 14.8, the *lacI^s* mutations occur in clusters that are rather regularly spaced along the gene.

Figure 14.8
Mutational analysis of the *lacI* gene identifies domains for different functions.

Distance along the gene is numbered in codons. The upper map shows that *lacI*⁻ mutations occur throughout the gene, but the *lacI*⁻ᵈ variety occurs only at the N-terminal end. The mutations that are *lacI*⁻ because the monomers cannot aggregate into tetramers are indicated in red. The lower map shows that *lacI*ˢ mutations do not start until residue 62, and tend to occur in clusters roughly between positions 150 and 300, generally separated by 26 residues or so.

The spacing may represent turns in the polypeptide chain. (A systematic study of the effects of amino acid replacements shows that about 60% of the positions are neutral, with no effect on repression.)

REPRESSOR AS A DNA-BINDING PROTEIN

The behavior of the isolated repressor protein *in vitro* shows directly that it possesses a DNA-binding site whose ability to remain attached to operator DNA is influenced by the structure of the rest of the molecule.

When the repressor is treated with trypsin, it is cleaved preferentially at amino acid 59. The C-terminal fragment of the protein, containing residues 60–360, is known as the **trypsin-resistant core.** It retains the ability to aggregate into a tetramer and to bind inducer; but it cannot bind the operator. The amino-terminal fragment consisting of residues 1–59 is known as the **long headpiece.** It may be cleaved again by trypsin at amino acid 51, generating a fragment of residues 1–51 called the **short headpiece.** Both the short and long headpieces retain the ability to bind to DNA; and when presented with operator DNA, they make the *same pattern of contacts* that is achieved by intact repressor (although they bind more weakly than the intact protein).

The ability of the headpieces to bind to the operator suggests that their structure per se is independent of the rest of the protein. This accords with models in which the DNA-binding site lies as an **arm** or **protrusion** of the N-terminal 50 amino acids from the body of the protein. The arm may be connected to the core by a **hinge** region, constructed from amino acids 50–80. The remainder of the protein, amino acids 81–360, is responsible for aggregating into the tetrameric structure and for binding inducer.

If the headpieces are protrusions that have the DNA-binding ability, what is the function of the tetrameric structure, and how does it contact DNA? One model suggests that the repressor has the shape of an elongated dumbbell or cylinder, about 115Å long, whose axis may lie across the DNA at a shallow angle of 15–20°. The cores of all four subunits are in contact in a central region; the headpiece binding sites lie at the extremities of the molecule, arranged in pairs.

It is possible that actually only two of the four subunits contact DNA, leaving the other two DNA-binding sites unoccupied. Although the core of the subunit does not recognize the operator, it does have an ability to bind to DNA in a nonsequence-specific manner, so it is possible that it too makes contacts (although these are not pronounced enough to influence the pattern of protection of DNA relative to the headpiece alone).

This model therefore pictures the repressor-DNA complex as a tetrameric protein symmetrically disposed on the DNA, and with only a small proportion (about 20%) of the polypeptide chain intimately contacting DNA. This last feature is shared by another quite different model, in which the axis of the repressor is perpendicular to the axis of DNA, so that only the two headpieces at one end are bound to the operator.

GETTING OFF DNA

A repressor tetramer is bound tightly to the operator. An inducer comes along and binds to the repressor. How does the protein get off the DNA?

Various inducers cause characteristic reductions in the affinity of the repressor for the operator *in vitro*. These changes correlate with the effectiveness of the inducers *in vivo*. This means that the mechanism of induction relies on a reduction in the attraction between operator and repressor. How is this accomplished?

The rate at which the repressor dissociates from the operator is rather slow, typically displaying a half-life that varies *in vitro* from 10–20 minutes, depending on the conditions. But when IPTG is added, there is an immediate reduction in the stability of the complex, as seen by a drastic reduction in its half-life.

This result distinguishes between the two models for repressor action illustrated in **Figure 14.9.** The equilibrium model (left) calls for repressor bound to DNA to be in rapid equilibrium with free repressor; inducer would bind to the free form of repressor, and thus unbalance the equilibrium by preventing reassociation with DNA. But the rate of dissociation of the repressor from the operator (as measured in the absence of inducer) is *much* too slow to be compatible with this model. This means that instead the inducer must bind *directly* to repressor protein complexed with the operator. As indicated in the model on the right, inducer binding must produce a change in the properties of the repressor that makes it let go of the operator.

Binding of the repressor-IPTG complex to the operator can be studied by using greater concentrations of the protein in the methylation protection/enhancement assay. The large amount compensates for the low affinity of the repressor-IPTG complex for the operator. The complex makes exactly the same pattern of contacts with DNA as does free repressor or head-

Figure 14.9
Does the inducer bind to free repressor to upset an equilibrium (left) or directly to repressor bound at the operator (right)?

piece. An analogous result is obtained with mutant repressors whose affinity for operator DNA is increased; they too make the same pattern of contacts.

Overall, a range of forms of the repressor whose affinities for the operator span seven orders of magnitude all make the same contacts with DNA. Changes in the affinity of the repressor for DNA must therefore occur by influencing the general conformation of the headpiece in binding DNA, not by making or breaking one or a few individual bonds. By changing the conformation of the *entire* DNA-binding region, many or all of the bonds with DNA must be simultaneously weakened. How does this happen? We know from a variety of techniques that binding of inducer causes an immediate conformational change in the repressor protein. Binding of two molecules of inducer to the repressor tetramer is adequate to release repression. But we do not yet know how the overall conformational change is related to the affinity of repressor for the operator.

The most likely model is that a change in conformation is transmitted from the core via the hinge to the headpiece. **Figure 14.10** represents such a model diagrammatically. The conformation of the DNA-binding site is altered from a state in which it exactly fits the DNA sequence to a state in which the headpiece is held in a slightly different register that cannot contact the DNA tightly enough.

Figure 14.10

Repressor displacement involves a general change in structure that weakens all its contacts with the operator.

This is illustrated by showing a change in the angle between the headpiece and core when the repressor binds inducer.

Left. The headpiece is specifically bound to the operator, lying along the DNA to make specific contacts.

Right. Inducer changes the conformation so that the headpiece can no longer lie exactly on the DNA. Only a monomer subunit is shown for clarity.

STORING SURPLUS REPRESSOR

Probably all proteins that have a great affinity for a specific sequence also possess a low affinity for any (random) DNA sequence. A large number of low-affinity sites will compete just as well for a repressor tetramer as a small number of high-affinity sites. There is only one high-affinity site in the *E. coli* genome: the operator. The remainder of the DNA can be considered to provide low-affinity binding sites. How is the repressor partitioned between the operator and the rest of DNA? What happens to the repressor when it has bound inducer and dissociated from the operator?

To achieve specificity for the operator within the context of the genome, the repressor's specific binding affinity must be very much greater than its random affinity. In fact, the repressor binds about 4×10^6 times better to operator DNA than to any random DNA sequence of the same length. (This is the ratio of the association constants of the repressor for *lacO* versus other DNA.)

Now the operator represents 26 bp in a genome of 4.2×10^6 bp, so that the excess of nonoperator DNA over operator DNA is about 2×10^5. Put another way, there are 2×10^5 low-affinity binding sites for the repressor, compared with the one high-affinity bonding site.

The relative preference of the repressor for the operator DNA *within the bacterial genome* is given by dividing the preference per se for the high-affinity site (4×10^6) by the number of low-affinity sites (2×10^5). The answer indicates that the repressor has about a twentyfold better probability of binding the operator than any other site.

This explains why mutations that reduce the affinity of the operator for the repressor by as little as twenty- or thirtyfold have sufficient effect to be constitutive. Within the genome, the mutant operators can be overwhelmed by the preponderance of random sites. Their specific affinity for repressor relative to these sites is simply no longer great enough to offset their minority status.

The consequence of these affinities is that in an uninduced cell, one tetramer of repressor will usually be bound to the operator. All or almost all of the remaining tetramers will be bound at random to other regions of DNA, as illustrated in **Figure 14.11.** There are likely to be very few or no repressor tetramers free within the cell.

When inducer binds to the repressor, the affinity for the operator is reduced by about 10^3-fold. The affinity for general DNA sequences remains unaltered. Thus in an induced cell, probably all or virtually all the repressor tetramers will be "stored" on random DNA sites. The effect of induction is therefore not to generate free repressor, but to change its distribution on DNA.

Several important biological conclusions follow from

the storage of repressor on DNA. Most directly, the ability to bind to the operator very rapidly is not consistent with the time that would be required for multiple dissociations from, and reassociations with, DNA. The discrepancy excludes random-hit mechanisms for finding the operator, suggesting that the repressor may be able to move directly from a random site on DNA to the operator. Movement could be accomplished either by sliding along the DNA very rapidly (which seems improbable because there would be obstacles in the form of other bound proteins) or by direct displacement from site to site. The latter is the same mechanism that we have already considered for RNA polymerase (see Chapter 10). A displacement reaction might be aided by the presence of more binding

sites per tetramer (four) than are actually used to contact DNA at any one time (two).

The ability of a large number of nonspecific sites to overpower a small number of high-affinity sites also implies that the binding of repressor to the operator will be very sensitive to both the total DNA concentration and the total repressor concentration in the cell. The difference in expression of the lactose operon between its induced and repressed states is 10^3-fold. In other words, even when inducer is absent, there is a basal level of expression that is about 0.1% of the induced level. This would be reduced if there were more repressor protein present, increased if there were less. Thus it could be impossible to establish tight repression if there were fewer repressors than the 10

Figure 14.11
Repressor loves DNA.

Virtually all the repressor in the cell is bound to DNA. In the uninduced state shown in (**1**), a tetramer is present at the operator. When induction causes it to leave, it becomes bound at another, random site on DNA (**2**). The other tetramers also bind inducer, but this does not affect their general affinity for DNA. Repressor may move from one site on DNA to another by direct displacement, as illustrated in (**3**). So when inducer is removed, the operator is contacted by a repressor tetramer that is bound at a random site and moves to restore the uninduced state in (**4**).

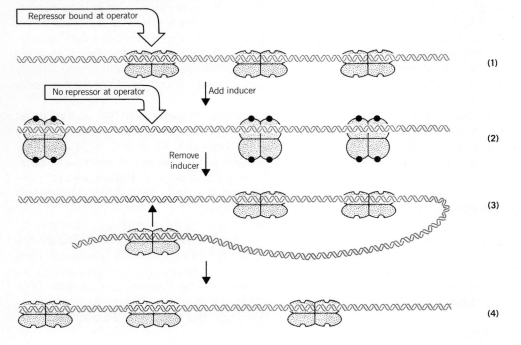

found per cell; and it might become difficult to induce the operon if there were too many.

A PARADOX OF INDUCTION

The lactose operon contains the structural gene *(lacZ)* coding for the β-galactosidase activity needed to metabolize the sugar; it also includes the gene *(lacY)* that codes for the protein needed to transport the substrate into the cell. But if the operon is repressed, how does the inducer enter the cell to start the process of induction by binding to repressor protein?

The basal level of expression of the operon may be crucial in ensuring that there is always a minimal amount of the protein present in the cell, enough to start the process off. This turns out to be necessary not just for uptake of the inducer (some of which could enter anyway via another uptake system), but also to start the metabolic pathway.

Although lactose is the substrate for β-galactosidase, it is not an inducer of the operon. In fact, it has a slight effect as an antiinducer (it binds to the repressor to *increase* its affinity for the operator). So how is induction to be accomplished?

The natural inducer of the operon has been obtained by isolating repressor from induced cells. The protein is bound to **allolactose.** The predominant reaction of β-galactosidase when presented with lactose is to break it to glucose and galactose. But a side reaction transfers the galactose moiety to certain acceptor molecules, producing (principally) allolactose and galactobiose. Since the acceptor that is used to form allolactose is in fact glucose, the reaction ends up as a molecular rearrangement.

The convoluted pathway for induction *in vivo* must therefore involve the uptake of a little lactose, some of which is converted to allolactose by the enzyme. This allolactose then induces the operon. The strong inducing effect of the small amount of allolactose that continues to be made is sufficient to outweigh the weak antiinducing effect of the lactose substrate.

FURTHER READING

The best source of recent reviews of the function of the *lac* operon is **Miller** & **Reznikoff's** (Eds.) *The Operon* (Cold Spring Harbor Laboratory, New York, 1978). Chapters of special interest include **Beckwith** (pp. 11–30) on the organization of the system, **Zabin** & **Fowler** (pp. 89–122) on the protein products, **Miller** (pp. 31–88) on the *lacI* gene, **Beyreuther** (pp. 123–154) and **Weber** & **Geisler** (pp. 155–176) on the repressor protein, and **Barkley** & **Bourgeois** (pp. 177–220) on repressor binding to DNA and inducers.

CHAPTER 15
CONTROL CIRCUITS:
A PANOPLY OF OPERONS

Sometimes it seems that every conceivable mechanism is used for controlling gene expression in one situation or another. But through this variety of control mechanisms, there runs a common thread. Regulation results from the interaction of a regulator protein with a sequence of nucleic acid, often one that has dyad symmetry. In this chapter, we shall look at some further examples of these interactions from the perspective of their linkage to form circuits that control individual operons. Similar mechanisms occur in slightly different guises in many systems, and the intricate pattern of connections varies from system to system.

The operon represents an arrangement that permits *coordinate* regulation of a group of structural genes. But, superimposed on this, there may be further controls that create *differences* in the extent to which individual genes are expressed.

DISTINGUISHING POSITIVE AND NEGATIVE CONTROL

Positive and negative control systems are defined by the response of the operon when no regulator protein is present. The characteristics of the two types of control system are mirror images.

Genes under negative control are expressed *unless* they are switched off by interaction of a repressor protein with their control site. Any action that interferes with gene expression can provide a negative control, but there is a uniformity in these mechanisms: a repressor protein either binds to DNA to prevent RNA polymerase from initiating transcription, or binds to mRNA to prevent a ribosome from initiating translation.

Negative control provides a fail-safe mechanism: if the regulator protein is inactivated, the system functions and so the cell is not deprived of these enzymes. It is easy to see how this might evolve. Originally the system functioned constitutively, but then cells able to interfere specifically with its expression acquired a selective advantage by virtue of their increased efficiency.

For genes under positive control, expression is possible *only* when an active regulator protein is present. How is this to be achieved? There is a variety of positive controls. Those that act at the initiation of transcription in particular operons are exact counterparts of the negative control systems. But instead of *interfering* with initiation, the regulator protein is *essential* for it. It interacts with DNA and with RNA polymerase to *assist* the initiation event. Such a protein may be called an **apoinducer.** Other positive controls provide for the global substitution of RNA polymerase subunits involved in setting initiation specificity (Chapter 12), or act as antitermination factors in transcription (Chapter 13).

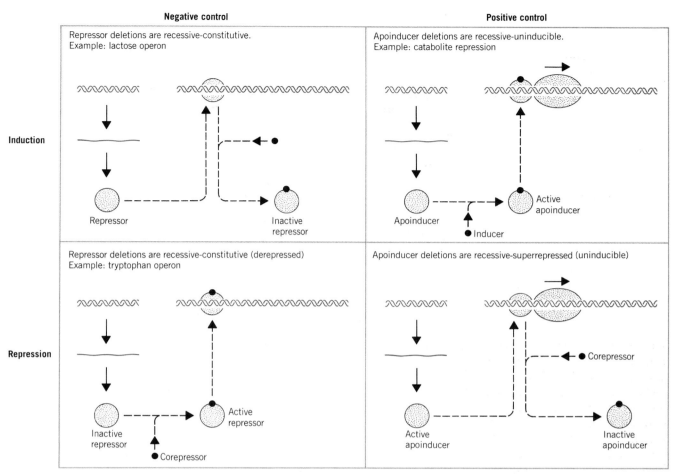

	Negative control	Positive control

Negative control

Repressor deletions are recessive-constitutive.
Example: lactose operon

Positive control

Apoinducer deletions are recessive-uninducible.
Example: catabolite repression

Induction

Repressor

Inactive
repressor

Apoinducer

Inducer

Active
apoinducer

Repressor deletions are recessive-constitutive (derepressed)
Example: tryptophan operon

Apoinducer deletions are recessive-superrepressed (uninducible)

Repression

Inactive
repressor

Corepressor

Active
repressor

Active
apoinducer

Corepressor

Inactive
apoinducer

Figure 15.1
Control circuits are versatile and can be designed to allow positive or negative control of induction or repression.

It is more difficult to see how positive control evolved, since the cell must have had the ability to express the regulated genes even before any control existed. Presumably some component of the control system must have changed its role. Perhaps originally it was used as a regular part of the apparatus for gene expression; then later it became restrained so as to act only in certain circumstances.

Operons are defined as **inducible** or **repressible** by the nature of their response to the small molecule that regulates their expression. Just as it may be advantageous for a bacterium to induce a set of enzymes only after addition of the inducer substrate that they metabolize, so also may it be useful to repress the enzymes that synthesize some compound if it is provided in adequate amounts by the medium. Thus inducible operons function only in the *presence* of the small-molecule inducer. Repressible operons function only in the *absence* of the small-molecule **corepressor** (so called to distinguish it from the repressor protein). The terminology used for repressible systems describes the active state of the operon as **derepressed;** this has the same meaning as *induced*. The condition in which a (mutant) operon cannot be derepressed is sometimes called **superrepressed;** this is the exact counterpart of *uninducible*.

Either positive or negative control could be used to achieve either induction or repression by utilizing appropriate interactions between the regulator protein and the small-molecule inducer or corepressor. **Figure 15.1** summarizes four simple types of control circuit. Induction is achieved when an inducer inactivates a repressor protein or activates an apoinducer protein. Repression is accomplished when a co-repressor activates a repressor protein or inactivates an apoinducer protein.

The genetic consequences of inactivating the regulator protein can be used to discriminate between negative and positive control systems. Inactivation of a repressor protein creates the recessive constitutivity (or derepression) typical of negative control systems. On the other hand, a mutation that inactivates an apoinducer causes the recessive uninducibility or superrepression that typifies positive control systems.

THE TRYPTOPHAN OPERON IS REPRESSIBLE

The five structural genes of the tryptophan operon are arranged in a contiguous series, coding for the three enzymes that convert chorismic acid to tryptophan by the pathway given in **Figure 15.2.** Transcription starts at a promoter at the left end of the cluster. Adjacent to it is the operator that binds the repressor protein coded by the unlinked gene *trpR*. Between the operator and the coding region of the first gene there is a leader sequence. Transcription of the structural genes is terminated at a rho-independent site, *trpt,* 36 bp beyond the end of the last coding region. About 250 bp later there is another terminator, *trpt',* that is rho-dependent. Essentially the same operon is present in *E. coli* and in *S. typhimurium.*

Both the activity and the synthesis of the tryptophan enzymes are controlled by the level of tryptophan in the cell. The classic feedback loop of **end-product inhibition** applies to the enzymes: the catalytic activities of the first enzyme of the pathway are inhibited by tryptophan, the ultimate product. This means that when the cell has sufficient tryptophan, it is able to cut off the synthesis of further molecules of the amino acid by inhibiting the beginning of the pathway.

Figure 15.2
The *trp* operon consists of five contiguous structural genes preceded by a control region.

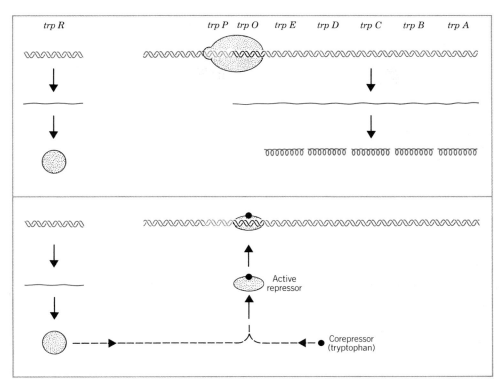

Figure 15.3

Transcription of the *trp* operon is controlled by a repressor.

Upper. In the absence of tryptophan, the repressor is inactive and RNA polymerase initiates transcription.

Lower. The addition of tryptophan activates the repressor so that it binds at the operator and prevents transcription.

Tryptophan also functions as the corepressor that activates the repressor protein. This is the classic mechanism for repression, one of the examples given in Figure 15.1, and the circuit is illustrated in more detail in **Figure 15.3.** In conditions when the supply of tryptophan is plentiful, the operon is repressed because the repressor protein•corepressor complex is bound at the operator. When tryptophan is in short supply, there is no corepressor to activate the repressor protein, which therefore has reduced specificity for the operator and is stored elsewhere on DNA.

The *trp* repressor functions as a tetramer of four identical subunits of about 12,000 daltons each. There are about 20 molecules of tetramer per bacterium. The repressor can bind to the operator *in vitro* only in the presence of tryptophan. Thus the only significant difference from the interactions in the *lac* operon is that the presence of tryptophan *increases* the affinity of the repressor protein for the operator.

The sequence of the *trp* promoter-operator region is given in **Figure 15.4.** RNA polymerase binds to a region extending from about −40 to +18, within which

Figure 15.4

The operator for the *trp* genes lies within the promoter.

Transcription initiates at +1. The usual hexamers related to the consensus sequences are present around −10 and −35. There is a 10 bp repeat within the operator (with one nonmatching position). All the o^c mutations lie within the dyad symmetry, but outside the −10 consensus sequence with which it overlaps.

it makes contacts analogous to those we described in Chapter 11. The promoter has the usual consensus-like sequences at -35 and -10.

The operator lies entirely within the promoter. The points contacted by the repressor mostly lie symmetrically on the two DNA strands and occupy the region from positions -23 to -3. The operator has a region of dyad symmetry, which includes the consensus sequence of the promoter at -10.

Although the exact relationship of the operator and promoter sequences is different from that of the *lac* operon, the general mechanism of repression is the same: the binding of RNA polymerase and the binding of repressor are mutually exclusive.

A MODIFICATION
OF COORDINATE REGULATION

Deprivation of repressor causes roughly a seventyfold increase in the frequency of initiation events at the *trp* promoter. Even under repressing conditions, the structural genes continue to be expressed at a low **basal** or **repressed level.** This means that the efficiency of repression at the operator is much lower than that seen in the *lac* operon (where the basal level is only about 1/1000 of the induced level).

Under conditions of repression, the basal level of the last three polypeptides of the operon is about five times greater than that of the *trpED* product. The noncoordinate expression of *trpCBA* is caused by the internal initiation of transcription at a promoter designated *trpP2,* which lies toward the end of the *trpD* gene. *TrpP2* is an unregulated promoter with a low affinity for RNA polymerase; so transcription is initiated at a low constitutive frequency. The product is an mRNA for *trpCBA,* whose translation increases the basal level of its products.

A rationale for additional expression of the last three genes may lie in the amino acid composition of the proteins. Unlike the other proteins, *trpB* and *trpC* products contain tryptophan. In conditions of really severe starvation for tryptophan, it might therefore be impossible to synthesize these proteins. Their absence would block any recovery via endogenous synthesis. But if the cell has an adequate basal level of these enzymes, their activity could be sufficient to start the process.

THE TRYPTOPHAN OPERON IS
CONTROLLED BY ATTENUATION

It has been known almost since studies of the tryptophan operon began that in addition to the promoter-operator complex there is another site involved in regulation. Its existence was first revealed by the observation that deletions of a sequence between the operator and the *trpE* coding region can increase the expression of the structural genes. This effect is independent of repression: both the basal and the derepressed levels of transcription are increased. Thus what is involved here concerns events that occur *after* RNA polymerase has set out from the promoter (irrespective of the conditions prevailing at initiation).

The regulator site is called the **attenuator.** It lies within the transcribed leader sequence of 162 nucleotides that precedes the initiation codon for the *trpE* gene. The attenuator is a barrier to transcription. It consists of a rho-independent termination site like those we have discussed in Chapter 13. A short G-C-rich palindrome is followed by eight successive U residues. RNA polymerase terminates there, either *in vivo* or *in vitro,* to produce a 140-base transcript.

The termination event at this site responds to the level of tryptophan, as illustrated in **Figure 15.5.** In the presence of adequate amounts of tryptophan, termination is efficient. But in the absence of tryptophan, RNA polymerases can continue into the structural genes. Such regulation of termination of transcription is called **attenuation.**

We see that events at the attenuator work in the same direction of control as those at the operator. When tryptophan is present, the operon is repressed; and most of the RNA polymerases that escape from the promoter then terminate at the attenuator. When tryptophan is removed, RNA polymerase has free access to the promoter, and also is no longer compelled to terminate prematurely.

Attenuation has an eight- to tenfold effect on transcription. Under conditions of efficient termination, about 10% of the RNA polymerases that encounter the terminator can get past. Under conditions of attenuation, probably almost all of the RNA polymerases can continue transcription. Together with the seventyfold increase in initiation of transcription that results from the release of repression, this allows roughly a 600-fold range of regulation of the operon.

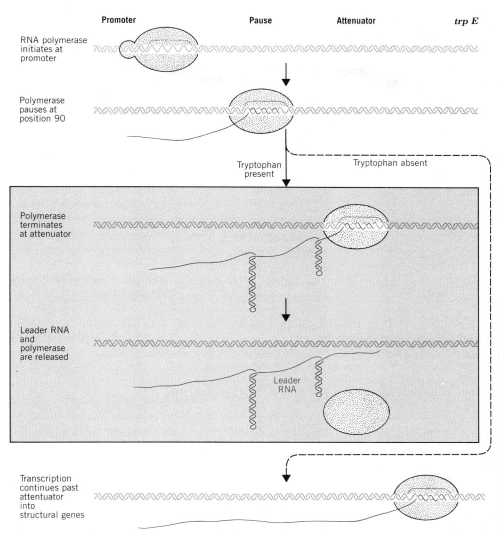

Figure 15.5

An attenuator controls the progression of RNA polymerase into the *trp* genes.

After initiating at the promoter (whether under basal expression or derepression), RNA polymerase proceeds to position 90, where it pauses. When it reaches the attenuator, in the presence of tryptophan there is a (roughly) 90% probability of termination to release the 140 base leader RNA. In the absence of tryptophan, the polymerase continues into the structural genes (*trpE* starts at +163).

ALTERNATIVE SECONDARY STRUCTURES CONTROL ATTENUATION

How can termination of transcription at the attenuator respond to the level of tryptophan? The sequence of the leader region suggests a mechanism.

Figure 15.6 shows that it contains a ribosome binding site whose AUG codon is followed by a coding region of 13 codons. Is this sequence translated into a **leader peptide?** Although no product has been detected *in vivo*, probably this is because it is unstable. We know that the ribosome binding site is functional, because when it is fused to a structural gene it sponsors effective translation.

What is the function of the leader peptide? It contains two tryptophan residues in immediate succession. Tryptophan is a rare amino acid in *E. coli* proteins, so this seems unlikely to be mere coincidence. Its significance is indicated by the (old) observation that mutations in the gene for Trp-tRNA synthetase prevent full repression of the operon. The effect of these mutations is to deprive the cell of charged Trp-tRNA, making it unable to translate these codons. This mimics what happens when the cell runs out of tryptophan.

Under these conditions, ribosomes presumably initiate translation of the leader peptide, but stop when

Leader peptide

Met Lys Ala Ile Phe Val Leu Lys Gly Trp Trp Arg Thr Ser

pppAAGUUCACGUAAAAAGGGUAUCGACAAUGAAAGCAAUUUUCGUACUGAAAGGUUGGUGGCGCACUUCCUGA

AACGGGCAGUGUAUUCACCAUGCGUAAAGCAAUCAGAUACCCAGCCCGCCUAAUGAGCGGGCUUUUUUUU

Attenuator sequence   Leader termination

GAACAAAAUUAGAGAAUAACAAUGCAAACACAAAAACCG ———

Met Gln Thr Gln Lys Pro . . .

trpE

Figure 15.6
The leader sequence of the *trp* genes codes for a short peptide.

they reach the Trp codons. The sequence of the mRNA suggests that this **ribosome stalling** may in turn influence termination at the attenuator.

The leader sequence can be written in alternative base-paired structures. The ability of the ribosome to proceed through the leader region may control transitions between these structures. The structure determines whether the mRNA can provide the features needed for termination.

Figure 15.7 draws these structures. In the first, region **1** pairs with region **2**; and region **3** pairs with region **4**. The pairing of regions **3** and **4** generates the hairpin that precedes the U_8 sequence: this is the essential signal for termination. Probably the RNA would take up this structure in lieu of any outside intervention.

A different structure is formed if region **1** is prevented from pairing with region **2**. In this case, region

Figure 15.7
The *trp* leader region can exist in alternative base-paired conformations.

In the center is a diagrammatic representation showing the four regions that can base pair. Region **1** is complementary to region **2**, which is complementary to region **3**, which is complementary to region **4**. On the left is the conformation produced when region **1** pairs with region **2**, and region **3** pairs with region **4**. On the right is the conformation when region **2** pairs with region **3**, leaving regions **1** and **4** unpaired.

Figure 15.8
The conformation of *trp* mRNA is controlled by the position of the ribosome; a difference of just a few bases in its position on mRNA determines whether regions 3 and 4 can pair to form the terminator hairpin.

2 is free to pair with region **3.** Then region **4** has no available pairing partner; so it is compelled to remain single-stranded. Thus the terminator hairpin cannot be formed.

Figure 15.8 shows that the position of the ribosome

can determine which structure is formed, in such a way that termination is attenuated only in the absence of tryptophan.

When tryptophan is present, ribosomes are able to synthesize the leader peptide. They will continue along the leader section of the mRNA to the UGA codon, which lies between regions **1** and **2.** As shown in the figure, by progressing to this point, the ribosomes extend over region **2** and prevent it from base pairing. The result is that region **3** is available to base pair with region **4,** generating the terminator hairpin. Under these conditions, therefore, RNA polymerase terminates at the attenuator.

When there is no tryptophan, ribosomes stall at the Trp codons. These are part of region **1.** Thus region **1** is sequestered within the ribosome and cannot base pair with region **2.** If this happens even while the mRNA itself is being synthesized, regions **2** and **3** will be base paired before region **4** has been transcribed. This compels region **4** to remain in a single-stranded form. In the absence of the terminator hairpin, RNA polymerase continues transcription past the attenuator.

Starvation for other amino acids does not have this result, because the positions at which the ribosome stalls leave regions **1** and **2** able to base pair, so that regions **3** and **4** in turn can base pair to form the terminator hairpin. The example of glycine starvation is shown in the figure. The only exception is arginine, whose codon immediately follows the Trp codons. Ribosome stalling at this position also attenuates termination, although less effectively than stalling at the Trp codons.

This model is supported by the properties of mutations that influence events at the attenuator. The most common are mutations that cause a deficiency in termination, increasing expression of the structural genes. These mutations are similar to those found in other terminators. One class eliminates base pairs in the double-stranded region **3:4,** thus reducing the stability of the terminator hairpin. Another type lies in the stretch of U residues, with similar effects.

Some mutations in the leader region increase termination at the attenuator. This may prevent starvation for tryptophan from relieving termination. One of these mutations destabilizes the pairing between regions **2** and **3.** Thus region **3** remains available to pair with region **4** to form the terminator hairpin, even when cells are starved for tryptophan. Another mutation changes the AUG initiation codon of the leader peptide, so that translation is prevented. The ensuing increase in termination demonstrates that transcription

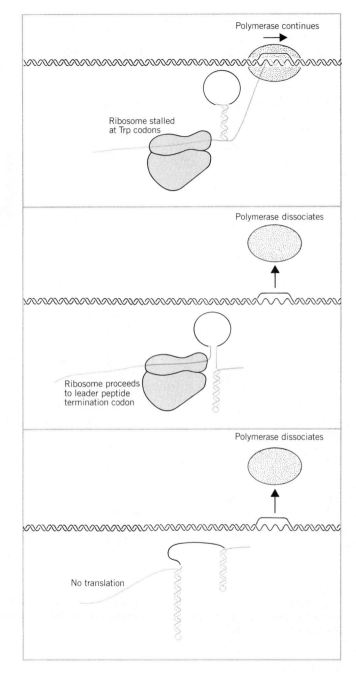

Figure 15.9
The alternatives for RNA polymerase at the attenuator depend on the location of the ribosome.

through the attenuator depends on the ability to translate the leader region.

Attenuation therefore provides a mechanism to sense the inadequacy of the supply of Trp-tRNA. Thus it responds directly to the need of the cell for tryptophan in protein synthesis. **Figure 15.9** summarizes the status quo when either termination or readthrough is occurring at the attenuator.

The critical aspect of the sensing mechanism is the position of the Trp codons in the leader peptide relative to the location of the attenuator. A feature that may ensure that all the events occur with the right timing is the presence of a site that causes the RNA polymerase to pause at base 90 along the leader. This may be necessary to give time for ribosomes to bind to establish the conformation of regions **1** and **2** before regions **3** and **4** are synthesized. Even so, it seems likely that the nature of the attenuation mechanism leaves it more susceptible to chance variation than the repressor•operator interaction, which probably explains why quantitatively its effect is less.

THE GENERALITY OF ATTENUATION

How widespread is the use of attenuation as a control mechanism for bacterial operons? It is used in at least six operons that code for enzymes concerned with the biosynthesis of amino acids. So a feedback from the level of the amino acid available for protein synthesis to the production of the enzymes may be common.

The histidine operon contains a cluster of nine structural genes representing the enzymes that synthesize histidine from PRPP. Deprivation of histidine causes a tenfold increase in its transcription, which has an inverse relationship with the amount of His-tRNA in the cell. The operon can be derepressed by mutating any one of several genes so that the production of functional His-tRNA is inhibited.

This situation is explained by the presence of an attenuator within the leader region that lies between the promoter and the first structural gene. Just as in the tryptophan operon, the attenuator comprises a terminator sequence that is preceded by a leader peptide. The leader peptide contains seven successive codons for histidine, as shown in the sequence in **Figure 15.10.** Alternative secondary structures for the leader region of the mRNA can be written so that a ribosome stalled at the histidine codons prevents the terminator hairpin from forming.

Cis-acting mutations in the leader region influence transcription of the operon. Some cause constitutive expression; others render the operon uninducible. Their properties are explained by the changes they cause in the DNA sequence. A constitutive mutation is caused by deleting the attenuator; so termination never occurs and RNA polymerase always reads through into the structural genes. Uninducible mutations may act either

Figure 15.10
Leader peptide sequences for amino acid biosynthetic operons contain multiple codons for the amino acid(s) that regulate the operon.

Operon	Leader peptide sequence	Regulatory amino acids
his	Met-Thr-Arg-Val-Gln-Phe-Lys-His-His-His-His-His-His-His-Pro-Asp	His
pheA	Met-Lys-His-Ile-Pro-Phe-Phe-Phe-Ala-Phe-Phe-Phe-Thr-Phe-Pro	Phe
leu	Met-Ser-His-Ile-Val-Arg-Phe-Thr-Gly-Leu-Leu-Leu-Leu-Asn-Ala-Phe-Ile-Val-Arg-Gly-Arg-Pro-Val-Gly-Gly-Ile-Gln-His	Leu
thr	Met-Lys-Arg-Ile-Ser-Thr-Thr-Ile-Thr-Thr-Thr-Ile-Thr-Ile-Thr-Thr-Gly-Asn-Gly-Ala-Gly	Thr Ile
ilv	Met-Thr-Ala-Leu-Leu-Arg-Val-Ile-Ser-Leu-Val-Val-Ile-Ser-Val-Val-Val-Ile-Ile -Ile-Pro-Pro-Cys-Gly-Ala-Ala-Leu-Gly-Arg-Gly-Lys-Ala	Leu, Val, Ile

on the mRNA secondary structure or via the leader peptide. A mutation that prevents the alternative hairpin from forming ensures that the terminator hairpin *always* forms, so that the polymerase is unable to transcribe the structural genes. Another mutation converts the CAA codon representing glutamine at position 5 of the leader peptide into a UAA ochre codon. The termination of translation at this point effectively creates a "no translation" situation similar to that shown in Figure 15.9, in which the terminator hairpin always forms.

The similarities between the tryptophan and histidine attenuation mechanisms are pronounced. In each case, it is deprivation of the aminoacyl-tRNA that directly prevents termination of transcription and so causes the structural genes to be transcribed. A difference is that the tryptophan operon also has a repressor-operator interaction; whereas in the histidine operon, attenuation provides the sole control. (The mutations in the *his* leader region were originally named *hisO*, because they were thought to identify an operator; this makes the point that the nature of such mutations cannot be determined until the molecular mechanism of control is investigated biochemically.)

Several other operons exhibit the same general features. Their leader peptide sequences are summarized in **Figure 15.10;** in each case stalling of the ribosome at the codons representing the regulator amino acid(s) can cause the mRNA to take up a secondary structure in which a terminator hairpin cannot form. The *thr* and *ilv* operons are subject to **multivalent repression.** Each is derepressed by starvation for more than one amino acid, two in the case of *thr,* three in the case of *ilv.* The sequence of the leader peptide shows how this is accomplished: codons for the various amino acids that regulate the operon are interspersed in such a way that ribosome stalling at any one of them is able to prevent formation of the terminator hairpin.

Several of the leaders are much longer than the *trp* leader, and the positions at which stalling can occur to cause derepression are more extensive. Actually, it is possible that in these cases stalling of more than one ribosome in the leader region may be necessary to achieve maximum derepression. Also, there may be several loops of secondary structure involved, but the general principle remains the same in each case.

Formally, the mechanism of attenuation can be described as one in which the ribosome functions as the equivalent of a positive regulator protein. Its binding at the site(s) of stalling is necessary to prevent termination. It is rendered unable to bind at this site (inactivated) by the binding of certain aminoacyl-tRNA(s), which therefore provide the equivalent of the corepressor.

REPRESSION CAN OCCUR AT MULTIPLE LOCI

The *trp* repressor protein has more responsibilities than just to act on the operator at the *trp* locus. It also represses transcription at two further loci.

The *aroH* gene codes for one of the three enzymes that catalyze the initial reaction in the common pathway of aromatic amino acid biosynthesis. Its expression is repressed by tryptophan, via the activation of the *trp* repressor. (The other enzymes are coded by the genes *aroF* and *aroG,* which are repressed by other regulators.) The effect of these controls is to extend the regulatory network, beyond the pathways for completing synthesis of the various aromatic amino acids, to the preceding steps at which the initial compounds are made.

The *trpR* regulator gene is repressed by its own product, the *trp* repressor. Thus the repressor protein acts to reduce its own synthesis. This circuit is an example of **autogenous control.** We shall see that such circuits are quite common in regulatory and other genes, and may be either negative or positive. Negative autogenous control is the most common; in this case, the protein inhibits its own synthesis, so that the level in the cell is autoregulatory. When the level becomes too high, production of further repressor is prevented, because the protein inhibits transcription of its own gene. When the level of repressor drops, the protein fails to inhibit its own synthesis, so the level is restored by resuming transcription. (In positive autogenous control, the protein assists its own synthesis; as we shall see in Chapter 16, this type of connection serves to provide an on/off switch.)

A related operator sequence, extending over 21 bp, is present at each of the three loci at which the *trp* repressor acts. The conservation of sequence is in-

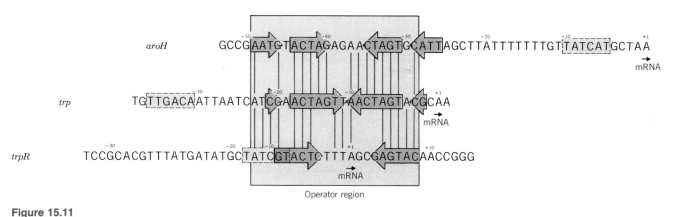

Operator region

Figure 15.11
The *trp* repressor recognizes operators at three loci.

Vertical lines show the conserved bases in the operators. Arrows indicate the regions of dyad symmetry. Boxes indicate promoter consensus sequences (with good fit in *aroH* and *trp,* poor fit in *trpR*). There is no -35 sequence in *aroH* and no -10 sequence in *trp.*

dicated in **Figure 15.11.** Each operator contains appreciable (but not identical) dyad symmetry. Presumably the features present at all three operators include the important points of contact for *trp* repressor. This explains how one repressor protein acts on several loci: each locus has a copy of a specific DNA-binding sequence recognized by the repressor (just as each promoter has specific sequences in common with other promoters).

A notable feature of the dispersed operators is their presence at different locations relative to the startpoint in each locus. Figure 15.11 shows that in the *trp* operon the operator occupies positions -23 to -3, while in *trpR* it lies between positions -12 and $+9$, but in the *aroH* locus it lies farther upstream, between -49 and -29. We have already seen that different repressors bind at sites located at various positions within or adjacent to the promoters: by revealing that the *same* repressor is effective at differently positioned operators, the *trp* situation reinforces the conclusion that repression is a matter of blocking access to the promoter.

THE ARABINOSE OPERON HAS DUAL CONTROL

Up to this point, we have dealt with systems that can be characterized formally as examples of either nega-

tive control or positive control. But the arabinose system fits neither category. It is regulated by a protein that interacts with a control site at the promoter, but this interaction displays features of both negative and positive control.

Several genes are controlled as part of this system, whose organization is summarized in **Figure 15.12.** The principal gene cluster codes for three enzymes that convert L-arabinose to D-xylulose-5-phosphate. These are the products of the genes *araBAD,* which are transcribed into a polycistronic mRNA in that order. Two unlinked genes, *araE* and *araF,* coding for membrane and periplasmic proteins, respectively, that are involved in arabinose uptake, are subject to the same regulation as *araBAD.* Sometimes a system that includes dispersed genes under common control is referred to as a **regulon.**

The regulator gene is *araC,* and it lies just beyond the promoter for *araBAD.* It is transcribed in the opposite direction from the *araBAD* cluster. Thus the region between *araBAD* and *araC* must contain two promoters in opposite orientation. They are said to initiate **divergent transcription.**

Both uninducible and constitutive mutations have been found in *araC.* Deletions or nonsense mutations create the uninducible *araC⁻* type, which is recessive to *araC⁺.* This is the relationship expected of a positive control system, in which *araC* codes for a protein that is active only in the presence of arabinose, and

Figure 15.12

The *ara* operon consists of a regulator gene *araC* separated from the structural gene cluster *araBAD* by a control region (and the unlinked genes *araF* and *araE* also are controlled by *araC*).

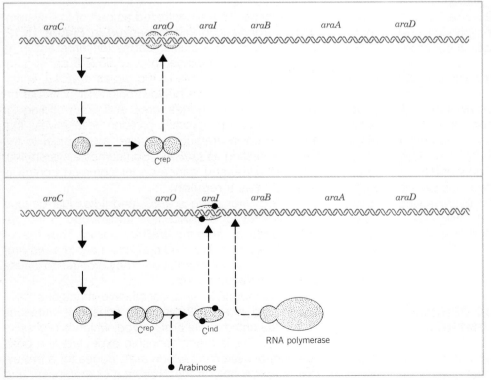

Figure 15.13

The *ara* operon is both negatively and positively controlled by C protein.

Upper. In the repressed state, C^{rep} binds to *araD* and prevents transcription.

Lower. In the induced state, C^{ind} binds to *araI* and assists the initiation of transcription.

that is needed to allow RNA polymerase to initiate transcription at the promoter. But the *araC*^{c} constitutive mutations also are recessive to *araC*$^+$. This is what would be expected of a negative control system, in which *araC* codes for a repressor protein that is inactivated by arabinose (comparable to the *lac* operon).

How is this genetic conflict to be resolved? A model is drawn in **Figure 15.13.** The *araC* gene codes for a repressor protein, originally called P1, now called C^{rep}. This binds to a control site to prevent transcription of the *ara* genes. Arabinose functions as a small-molecule inducer that binds to the repressor. So far this is typical of negative control. But arabinose binding does not merely inactivate the repressor's ability to bind to the operator. Instead, it actually converts the repressor into an inducer protein, originally called P2, now called C^{ind}, whose binding at the promoter is necessary to switch on transcription. This is characteristic of positive control.

THE CROWDED *ara* REGULATORY REGION

The two forms of the C protein must have different binding sites in the *ara* control region. These have been identified by *cis*-acting mutations. The operator, *araO*, is defined in the usual way as the site at which C^{rep} binds. The initiator, *araI*, is defined as the site involved in the initiation of transcription. It includes the binding site for C^{ind}. The approximate positions of these sites are given in Figure 15.13.

The *ara* control region extends between the startpoints of the two divergent transcription units *(araC* and *araBAD).* Their distance apart is only 147 bp. The two promoters themselves occupy an appreciable part of this region. In addition to the binding sites for C^{rep} and C^{ind}, there must also be at least one binding site for the regulator protein CAP (see Chapter 11), since both transcription units require CAP•cyclic AMP. Our present view of the disposition of these various regulatory elements is illustrated in **Figure 15.14.**

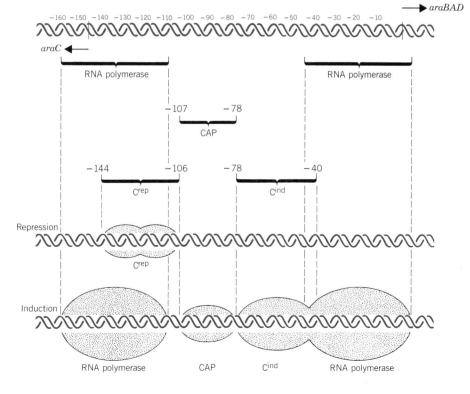

Figure 15.14

Induction of the *ara* operon involves multiple DNA-binding events.

The upper diagram shows a map of the *ara* control region with the binding sites for C protein, CAP, and RNA polymerase (assumed). Under conditions of repression, only C^{rep} is bound as shown in the center. Under conditions of induction, CAP, C^{ind}, and RNA polymerase reside at adjacent sites next to *araBAD*, while another RNA polymerase binds to the *araC* promoter.

Purification of the C protein has allowed the DNA sequence to which each form binds to be identified directly. The protein is a dimer of identical subunits of 33,000 daltons. It can be obtained in the C^{rep} form by binding the antiinducer D-fucose; it exists in the C^{ind} form when it is bound to arabinose. Footprinting experiments show that in either form it covers about 38 bp of DNA.

In the form of C^{rep}, it binds only to a region located between -44 and -106 relative to the *araBAD* startpoint. In the form of C^{ind}, it retains some affinity for this site, but binds mostly to the region from -78 to -40. Thus the conversion from one form of the protein to another is accompanied by its switch in its site of binding from a 38 bp region centered 125 bp upstream to another site centered 59 bp upstream. By analogy with the properties of other DNA-binding proteins, this movement could be accomplished by a displacement reaction.

The CAP protein (when activated by cyclic AMP) has two binding sites in the control region (see also later in this chapter). The major site lies between -107 and -78, as shown in the figure. [Another site of less affinity (not shown) lies between -146 and -121; it is not clear whether it is actually used *in vivo*.] Remember that CAP is needed to activate *both* transcription units; this may be accomplished from the one central binding site.

The promoters have not been directly characterized by binding RNA polymerase, but it is safe to assume that each extends about 50 bp upstream from the startpoint.

How do these various binding sites interact to produce the characteristic dual control of *araBAD*?

Expression of the operon requires three proteins to be bound simultaneously, probably contacting each other as depicted in Figure 15.14. We know that both C^{ind} and CAP exert their effects via the promoter, because of the properties of two types of constitutive mutations. The *araI^c* type allows expression in the absence of C^{ind}; these lie in the -35 region. The *araX^c* type allows expression in the absence of both C^{ind} and CAP; these lie in the -10 region. So usually the ability of RNA polymerase to initiate at the *araBAD* promoter depends on the simultaneous presence at their binding sites of C^{ind} and CAP, but this requirement can be released by mutation. What we do not understand is

how these three proteins interact. The order of binding sites suggests that CAP interacts with C^{ind} as a prerequisite for the interaction of C^{ind} with RNA polymerase. This contrasts with the direct interaction between CAP and RNA polymerase that seems to occur at other loci.

How the *araBAD* unit is repressed is not clear. We do not understand how C^{rep} binding so far upstream could influence the promoter. If this is the site at which negative control is exerted, the mechanism must be different from other repressors. It may be that one effect of C^{rep} is to prevent CAP from binding at its site. If this is correct, the effect of converting C^{rep} into C^{ind} is to flip the protein from one side of the CAP site (at which it inhibits CAP binding) to the other side (at which it assists it).

The *araC* gene is subject to autogenous repression by C^{rep}. The binding site for C^{rep} overlaps with the promoter for *araC* in the traditional way. In the repressed state, there is a low-level basal transcription of *araC* that produces C^{rep}. This forms an autogenous circuit that maintains just enough C^{rep} to keep both *araC* and *araBAD* repressed. Then the addition of arabinose to convert C^{rep} to C^{ind} has two effects. It relieves the repression of *araC*, so that more C protein is synthesized. And the C^{ind} activates the *araBAD* cluster. The entire circuit therefore is designed with maximum economy, using the appearance of arabinose to increase the amount of regulator protein as well as to constitute the on/off switch. So induction changes the situation from the production of the minimum adequate amount of C^{rep} to the high output of C^{ind}.

A DOUBLE PROMOTER FOR THE GALACTOSE OPERON

A set of three enzymes coded by the *galactose* operon is responsible for metabolizing galactose to glucose-1-phosphate. The genes *galKTE* are transcribed from either of two startpoints into a polycistronic mRNA that is translated sequentially into the three enzymes. Transcription is subject to classic negative regulation: a repressor protein coded by the unlinked *galR* gene is inactivated by the addition of galactose. The presence of activated CAP is necessary for transcription.

Figure 15.15
The *gal* operon has alternative startpoints.

Figure 15.15 displays the *gal* control sequence. The uses of its alternative startpoints are determined by the conditions of initiation. In the presence of active CAP, initiation occurs at a point indicated as +1 in the usual way. (All base positions are defined relative to this, for both promoters.) This startpoint is taken to identify promoter *galP1*, which has a Pribnow box at −12 to −16 and lacks a −35 consensus sequence. In the absence of CAP, the operon still can be transcribed, but initiation occurs upstream at position −5. This corresponds to promoter *galP2*, which has a Pribnow box at positions −17 to −11.

So there are two overlapping promoters, with a difference between their startpoints of just 5 bp. We do not know whether they are independent, in the sense that an RNA polymerase makes an entirely different set of contacts for each, or whether there is a single binding site with the same contacts made farther upstream and an adjustment made around the startpoint. The important general principle established by the existence of *galP1* and *galP2* is that expression of a set of structural genes may be initiated in more than one way, and therefore is subject to independent control systems.

The CAP binding site is rather close to the startpoint, within the usual distance covered by RNA polymerase. When CAP is bound there, it seems certain to interact directly with RNA polymerase, controlling its conformation so that it uses the downstream promoter, *galP1*.

The operator is adjacent to the CAP binding site, but farther upstream. Repressor binding inhibits the use of both *galP1* and *galP2*. Given the distance of the operator from the startpoint, and the interposition

of the CAP binding site, we do not understand how the repressor works. It is possible that it just manages to exercise steric hindrance to prevent RNA polymerase from binding, and this so far seems the only explanation for its inhibition of *galP2*. It could prevent use of *galP1* indirectly, by preventing the binding of CAP. This is the same model we have considered for repression of *araBAD*.

From comparing the regulation of several operons, we can see that the same general set of mechanisms is employed. The initiation of transcription is controlled by proteins that bind to specific sites on DNA, at which their presence may either enhance or inhibit binding of other proteins at overlapping or adjacent sites. In each case the ultimate result is to control the ability of RNA polymerase to bind to form a stable initiation complex. But the way these interactions are put together may vary greatly, so that even two simple interactions, such as repressor-operator binding and CAP-promoter binding, work together somewhat differently in (for example) the *gal* and *lac* operons.

CATABOLITE REPRESSION IMPOSES PREFERENCE FOR GLUCOSE

When glucose is available, it is used in preference to other sugars. Thus when *E. coli* finds (for example) both glucose and lactose in the medium, it metabolizes the glucose and represses the use of lactose. This is accomplished by preventing expression of the genes of the *lactose* operon, an effect called **catabolite repression.** The same effect is found with other operons, including galactose and arabinose. Thus catabolite repression represents a general coordinating system that exercises a preference for glucose by inhibiting the expression of the operons that code for the enzymes of alternative metabolic pathways.

Catabolite repression is set in train by the ability of glucose to reduce the level of cyclic AMP (cAMP) in the cell. We do not know exactly how this is accomplished; it may involve the rate of synthesis. At all events, expression of the catabolite-regulated operons shows an inverse relationship with the level of cyclic AMP.

Two types of mutation abolish catabolite repression. The first, cya^-, introduces a defect in the enzyme **adenylate cyclase** that synthesizes cyclic AMP. (The reaction uses ATP as substrate and consists of making a $3'-5'$ link via phosphodiester bonds; the result is a monophosphate joined to *both* of the positions at which phosphates usually can be found.) The second type of mutation, known alternatively as cap^- or crp^-, identifies the regulator protein that acts directly on the target operons. The protein is known as CAP (for catabolite activator protein) or CRP (for cyclic AMP receptor protein).

As we have already seen within the context of the individual operons, CAP is a positive control factor that is necessary for the initiation of transcription at dependent promoters. The protein is active only in the presence of cyclic AMP, which behaves as the classic small-molecule inducer (see Figure 15.1). Thus reducing the level of cyclic AMP renders the protein unable to bind to the control region, which in turn prevents RNA polymerase from initiating transcription. So the effect of glucose in reducing cyclic AMP levels is to deprive the relevant operons of a control factor necessary for their expression.

We have already mentioned that the action of CAP is puzzling, because in each target operon it binds to a site at a different location relative to the startpoint. (These are at -72 to -52 in *lac*, -50 to -23 in *gal*, and -107 to -78 in *araBAD*.) The alternative models for its action that we have discussed in Chapter 11 are for it to interact with adjacent proteins (RNA polymerase in *lac* and *gal*, C^{ind} in *ara*) or to exert a structural effect on DNA that may be transmitted a varying distance downstream within the promoter.

So we have not yet demonstrated that the dependence of these various operons on CAP necessarily reflects a common mode of regulation at the level of individual protein-protein or protein-DNA interactions. However, it accomplishes the same purpose: to turn off alternative metabolic pathways when they become unnecessary because the cell has an adequate supply of glucose. Again, this makes the point that coordinate control, of either negative or positive type, can extend over dispersed loci by repetition of binding sites for the regulator protein.

AUTOGENOUS CONTROL OF RIBOSOMAL PROTEIN TRANSLATION

About 70 or so proteins constitute the apparatus for gene expression. The ribosomal proteins are the major component, together with the ancillary proteins involved in protein synthesis. The subunits of RNA polymerase and its accessory factors make up the remainder. Coordinate controls ensure that these proteins are synthesized in amounts appropriate for the growth conditions: when bacteria grow more rapidly, they devote a greater proportion of their efforts to the production of the apparatus for gene expression (also see later). An array of mechanisms is used to control the expression of the genes coding for this apparatus; and probably we have not yet identified all of the regulatory systems that are employed.

Almost every one of these proteins is represented only by a single gene in *E. coli*. The genes coding for ribosomal proteins, protein-synthesis factors, and RNA polymerase subunits all are intermingled and organized into a small number of operons.

The organization of the operons that have been characterized so far is summarized in **Table 15.1.** About half of the genes for ribosomal proteins (often abbreviated as **r-proteins**) map in four operons that lie close together. These are known as *str*, *spc*, *S10*, and α (each named simply for the first one of its functions to have been identified). The *rif* and *L11* operons lie together at a different location.

Each operon contains a mélange of functions. The *str* operon has genes for small subunit ribosomal proteins as well as for EF-Tu and EF-G. The *spc* and *S10* operons both have genes interspersed for both small and large ribosomal subunit proteins. The α operon has genes for proteins of both ribosomal subunits as well as for the α subunit of RNA polymerase. The *rif* locus has genes for large subunit ribosomal proteins and for the β and β' subunits of RNA polymerase.

In most cases, there is no evident correlation with the functions of the proteins. Certainly the 30S subunit proteins are not coded in any pattern related to their assembly into the subunit. The group of 50S subunit proteins coded in the *rif/L11* cluster does show a correlation with the assembly map, in which L7/L12, L10, and L11 may be related. So in this case the proteins could assemble into some subparticle as they are synthesized.

All except one of the ribosomal proteins are needed in equimolar amounts, which must be coordinated with the level of rRNA. (There could also be variations for ribosomal proteins that have extracurricular activities; for example, S10 is implicated in the antitermination function of pN.) The dispersion of genes whose products must be equimolar, and their intermingling with genes whose products are needed in different amounts, pose some interesting problems for coordinate regulation.

The exceptional ribosomal protein is L7/L12, which is present in (probably) four copies per ribosome. Another exception is EF-Tu, which is present in amounts roughly equimolar with aminoacyl-tRNA—that is, about 10 times greater than the ribosomes. (This is the one case in which there is more than one gene, so the need for extra synthesis is divided between the two genes *tufA* and *tufB*.) Another difference occurs between ribosomes and RNA polymerase, which is present in somewhat smaller amounts. So some mechanism must increase the synthesis of L7/L12 and EF-Tu, and decrease the synthesis of RNA polymerase subunits, relative to a constant level of ribosomal proteins.

A feature common to all of the operons described in Table 15.1 is autogenous regulation of some of the genes by one of the products. Usually the regulatory

Table 15.1
Genes for Ribosomal Proteins, Protein Synthesis Factors, and RNA Polymerase Subunits Are Interspersed in a Few Operons

Operon	Genes and Proteins (in order from promoter)	Regulator
str	*rpsL-* *rpsG-* *fusA-* *tufA* S12 S7 EF-G EF-Tu	S7
spc	*rplN-* *rplX-* *rplE-* *rpsN-* *rpsH-* *rplF-* *rplR-* *rpsE-* *rpl0-* *rpmD* L14 L24 L5 S14 S8 L6 L18 S5 L15 L30	S8
S10	*rpsJ-* *rplC-* *rplB-* *rplD-* *rplW-* *rplS-* *rplV-* *rpsC-* *rpsQ-* *rplP-* *rpmC* S10 L3 L2 L4 L23 S19 L22 S3 S17 L16 L29	L4
α	*rpsM-* *rpsK-* *rpsD-* *rpoA-* *rplQ* S13 S11 S4 α L17	S4
L11	*rplK-* *rplA* L11 L1	L1
rif	*rplJ-* *rplL-* *rpoB-* *rpoC* L10 L7/ β β' L12	L10

The overall organization of these operons identifies two major gene clusters. One contains the operons *str*-space-*S10*-*spc*-α (where "space" = 14,000 bp); the other contains the adjacent operons *L11-rif*. Each operon is written so that it has its promoter at the left end. The regulator protein is indicated at the right, and the proteins subject to regulation are underlined (in red). Genes whose inclusion in the regulation is uncertain are indicated by broken lines.

protein inhibits expression of a contiguous set of genes within the operon, always including its own gene.

The exact pattern of inhibition varies from operon to operon. Thus S7 inhibits synthesis of itself and EF-G, but not of S12 or EF-Tu. Similarly, S8 inhibits expression of its entire operon except for the first two genes. On the other hand, L4 and L1 probably inhibit the expression of all genes in their respective operons. In the operons containing RNA polymerase subunit genes, S4 inhibits the expression of all the other small ribosomal subunit genes; and similarly L10 inhibits its own production, but not that of L7/L12 or the RNA polymerase subunits.

In each case, it is the accumulation of the protein that inhibits further synthesis of itself and of whatever other genes are involved. The effect often is caused at the level of translation of the polycistronic mRNA, and in several cases can be reproduced *in vitro*. Thus an excess of free ribosomal protein triggers a repression of translation.

Each of the regulators is a ribosomal protein that is among those that bind directly to rRNA. Its effect on translation is a result of its ability also to bind to its own mRNA. In some cases, the binding site has been characterized. In the example of the L11 operon, protein L1 binds to a site in the vicinity of the initiation codon for the first gene. This action probably inhibits ribosome binding. The inhibition affects translation of both genes of the operon, probably because there are only three nucleotides between the L11 and L1 genes. Thus it is plausible to suppose that the genes can only be translated sequentially.

The use of r-proteins that bind rRNA to establish autogenous regulation immediately suggests that this may serve as a mechanism to link r-protein synthesis to rRNA synthesis. A generalized model is depicted in **Figure 15.16.** Let us suppose that the binding sites for the autogenous regulator r-proteins on rRNA are much stronger than those on the mRNAs. Then so long as any free rRNA is available, the newly synthesized r-proteins will associate with it to start ribosome assembly. There will be no free r-protein available to

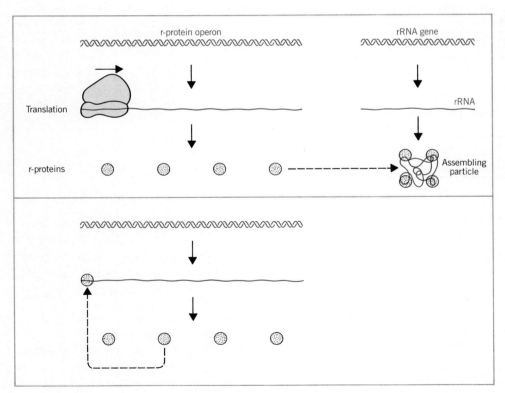

Figure 15.16
Translation of the r-protein operons is autogenously controlled and responds to the level of rRNA.

Upper. When rRNA is being produced, ribosomal proteins associate with it, so there is no free pool, and translation of the r-protein mRNA continues.

Lower. When all rRNA has been assembled into ribosomes and none is free, the r-proteins accumulate. One of them binds to the r-protein mRNA and prevents further translation.

bind to the mRNA, so its translation will continue. But as soon as the synthesis of rRNA slows or stops, free r-proteins begin to accumulate. Then these are available to bind their mRNAs, repressing further translation. This circuit ensures that each r-protein operon responds in the same way to the level of rRNA.

Two objectives are accomplished by this regulation of the r-protein operons. First, the level of r-proteins is established in correspondence with the growth conditions of the cell. By controlling the level of rRNA, the cell controls its production of all ribosomal components. Second, the other proteins coded by these operons are synthesized at their own rates, divorced from the translation of the r-protein genes. The β subunits of RNA polymerase may be subject to their own autogenous regulation. Both EF-Tu and L7/L12 are translated with increased efficiency. So within the control circuits of these operons, there is provision for allowing disparate rates of synthesis for coordinately regulated proteins.

Autogenous control of translation is not the only mechanism that is employed to allow variation to occur within the operon. Some operons contain additional, internal promoters; attenuation may be employed; processing of mRNA can be influenced.

AUTOGENOUS CONTROL AND MACROMOLECULAR ASSEMBLIES

Autogenous regulation may turn out to be a common type of control among proteins that are incorporated into macromolecular assemblies. The advantage is evident. The assembled particle itself may be unsuitable as a regulator, because it is too large, too numerous, or too restricted in its location. But the need for synthesis of its components may be reflected in the pool of free precursor subunits. If the assembly pathway is blocked for any reason, free subunits accumulate and shut off the unnecessary synthesis of further components.

Actually, the first example of autogenous regulation was provided by the gene 32 protein of phage T4. This plays a central role in genetic recombination, DNA repair, and replication, in which its function is exercised by virtue of its ability to bind to single-stranded DNA. An indication that p32 synthesis is regulated

autogenously was provided by the fact that nonsense mutations cause the inactive protein to be overproduced. This means that when the function of the protein is prevented, more of it is made. This effect occurs at the level of translation; the gene 32 mRNA is stable, and remains so irrespective of the behavior of the protein product.

When single-stranded DNA is present in the phage-infected cell, it will in effect sequester the p32 protein. However, in the absence of single-stranded DNA, or at least in conditions in which there is a surplus of p32, the protein prevents translation of its own mRNA. This repression shows the same cooperative features as its binding to DNA, in which binding of one protein molecule permits another to bind more easily. So we may suspect that it is mediated by p32 binding to mRNA to prevent initiation of transcription. Probably this occurs at a long single-stranded region near the ribosome binding site.

Another system in which autogenous regulation occurs is presented by eucaryotic cells. Tubulin is the monomer from which microtubules, a major filamentous system of all eucaryotic cells, are synthesized. The production of tubulin mRNA is controlled by the free tubulin pool. When this pool reaches a certain concentration, the production of further tubulin mRNA is prevented. Again, the principle is the same: tubulin sequestered into its macromolecular assembly plays no part in regulation, but the level of the free precursor pool determines whether further monomers are added to it.

HARD TIMES PROVOKE THE STRINGENT RESPONSE

When bacteria find themselves in such poor growth conditions that they lack a sufficient supply of amino acids to sustain protein synthesis, they shut down a wide range of activities. This is called the **stringent response.** We can view it as a mechanism for surviving hard times: the bacterium husbands its resources by engaging in only the minimum of activities until nutrient conditions improve, when it reverses the response and again engages its full range of metabolic activities.

The stringent response causes a massive (ten- to

twentyfold) reduction in the synthesis of stable (ribosomal and transfer) RNA species. This alone is sufficient to reduce the total amount of RNA synthesis to only 5–10% of its previous level. The synthesis of only some mRNAs is reduced; overall this causes roughly a threefold reduction in mRNA synthesis. The rate of protein degradation is increased. A large number of metabolic adjustments occur, as seen in reduced synthesis of nucleotides, carbohydrates, lipids, etc.

Deprivation of any one amino acid, or mutation to inactivate any aminoacyl-tRNA synthetase, is sufficient to initiate the stringent response. The trigger that sets the entire series of events in train is the presence of uncharged tRNA in the A site of the ribosome. Under normal conditions, of course, only aminoacyl-tRNA is placed in the A site by EF-Tu (see Chapter 6). But when there is no aminoacyl-tRNA available to respond to a particular codon, the uncharged tRNA becomes able to gain entry. Of course, this blocks any further progress by the ribosome, and the activities that follow are described as an **idling reaction.**

When cells are starved for amino acids, they accumulate two unusual nucleotides, originally designated **magic spots I** and **II.** Now these are known by their structures as **ppGpp** (guanosine tetraphosphate, with diphosphates attached to both 5′ and 3′ positions) and **pppGpp** (guanosine pentaphosphate, with a 5′ triphosphate group and a 3′ diphosphate). In all strains of *E. coli,* the stringent response is accompanied by the accumulation of ppGpp; the pppGpp may or may not be produced. These nucleotides are typical small-molecule effectors whose activity is mediated by the ability to bind to protein(s) to alter the conformation.

The components that are involved in producing ppGpp and pppGpp have been identified by mutations that eliminate the stringent response. These produce **relaxed mutants,** in which starvation for amino acids does not cause any reduction in stable RNA synthesis or in any of the other reactions comprising the stringent response.

The most common site of relaxed mutation lies in the gene *relA,* which codes for a protein called the **stringent factor.** This factor is associated with the ribosomes, although the amount is rather low—say, less than 1 molecule for every 200 ribosomes. So per-

Figure 15.17
Stringent factor catalyzes the synthesis of pppGpp and ppGpp; ribosomal proteins can dephosphorylate pppGpp to ppGpp.

haps only a minority of the ribosomes are able to make the stringent response.

Ribosomes obtained from stringent bacteria can synthesize ppGpp and pppGpp *in vitro,* provided that the A site is occupied by an uncharged tRNA *specifically responding to the codon.* Ribosomes extracted from relaxed mutants cannot perform this reaction; but they are able to do so if the stringent factor is added.

Figure 15.17 shows the pathways for synthesis of the unusual guanine nucleotides. The stringent factor is an enzyme that catalyzes the synthetic reaction in which ATP is used to donate a pyrophosphate group to the 3′ position of either 5′ GTP or GDP. The former is used as substrate more frequently. However, pppGpp may be converted to ppGpp by several enzymes; among those able to perform this dephosphorylation are the translation factors EF-Tu and EF-G. The production of ppGpp via pppGpp is the most common route.

What is involved in the idling reaction? Mutations in another locus able to cause the relaxed type turn out to lie in the 50S subunit protein L11. This is located in the vicinity of the A and P sites. The stringent response could involve the TψC region of tRNA, since mutations in tRNA, especially involving this region, can influence the reaction. It is possible that this region of tRNA (located at the hinge of the L tertiary structure) could be involved in transmitting a signal that the right tRNA is paired with the codon, thus allowing the ribosome to proceed with peptide transfer. The same pathway could be part of the route for establishing the idling reaction. The presence of a properly paired uncharged tRNA in the A site may trigger a conformational change in the ribosome; but because no amino acid is present on the tRNA, the idling reaction occurs instead of polypeptide transfer from the peptidyl-tRNA.

What does ppGpp do? It may be an effector for controlling several reactions, including the inhibition of transcription. Many conflicting effects have been reported, among which two stand out. There is a specific inhibition of the *initiation* of transcription at the promoters of operons coding for rRNA. And the addition of ppGpp slows the *elongation* phase of transcription of many or most templates *in vitro*. The cause is increased pausing by the enzyme. This effect is responsible for the general reduction in transcription efficiency when ppGpp is added *in vitro*. We do not yet know the specificity of such inhibition, but it will not be surprising if there is variation in magnitude from operon to operon in such a way that particular operons are more inhibited.

It is interesting that unusual nucleotides are used in at least two control systems with a general coordinating function. Both appear to be specific to bacteria. Although cyclic AMP is the classic "second messenger" in eucaryotic cells, there is no evidence that it

plays any role in the specific control of initiation of transcription. Similarly, an extensive search for ppGpp in eucaryotes has failed to detect any.

FURTHER READING

The circuitry of several operons has been reviewed in **Miller** & **Reznikoff's** (Eds.) *The Operon* (Cold Spring Harbor Laboratory, New York, 1978). This includes **Platt** on the *trp* operon (pp. 263–302), **Lee** on the *ara* operon (pp. 389–410), and **De Crombrugghe** & **Pastan** on cyclic AMP and the *gal* operon (pp. 303–324). Work on attenuation of the *trp* operon has been brought further up to date by **Yanofsky** (*Nature* **289,** 751–758, 1981) and **Platt** (*Cell* **24,** 10–23, 1981). The functions of this operon have been summarized in terms of its sequence by **Yanofsky** et al. (*Nuc. Acids Res.* **9,** 6647–6668, 1981). An argument about the mode of control in *ara* can be followed through **Ogden** et al. (*Proc. Nat. Acad. Sci.* **77,** 3346–3350, 1980) and **Lee** et al. (*Proc. Nat. Acad. Sci.* **78,** 752–756, 1981). The model for autogenous control of r-protein synthesis was presented by **Nomura** et al. (*Proc. Nat. Acad. Sci.* **77,** 7084–7088, 1980). Stringent control has been generally reviewed by **Gallant** (*Ann. Rev. Genet.* **13,** 393–415, 1979).

CHAPTER 16
LYTIC CASCADES AND
LYSOGENIC REPRESSION

Some phages have only a single strategy for survival. On infecting a susceptible host, they subvert its functions to the purpose of producing a large number of progeny phage particles. As the result of this **lytic infection,** the host bacterium dies. In the typical lytic cycle, the phage DNA (or RNA) enters the host bacterium, its genes are transcribed in a set order, the phage genetic material is replicated, and the protein components of the phage particle are produced. Finally, the host bacterium is broken open **(lysed)** to release the assembled progeny particles.

Other phages have a dual existence. They are able to perpetuate themselves via the same sort of lytic cycle in what amounts to an open strategy for producing as many copies of the phage as rapidly as possible. But they also have an alternative, closet existence, in which the genome is present in the bacterium in a latent form known as **prophage.** This is called **lysogeny.** In lysogenic bacteria, the prophage is **integrated** into the bacterial genome, and it is inherited as any other part of it, in the same way as bacterial genes. By virtue of its possession of a prophage, a lysogenic bacterium has **immunity** against infection by further phage particles of the same type, so usually there is only one copy of a prophage of any particular type in a bacterial genome.

Transitions between the lysogenic and lytic modes of existence may occur in either direction. When a phage produced by a lytic cycle enters a new bacterial host cell, it may either repeat the lytic cycle or enter the lysogenic state. The choice depends on the conditions of infection and the genotypes of phage and bacterium. A prophage may be freed from the restrictions of lysogeny by the process called **induction,** in which it is **excised** from the bacterial genome, to generate a free phage DNA that then proceeds through the lytic pathway.

Another type of existence within bacteria is represented by **plasmids.** These are autonomous elements whose genomes exist in the cell as **extrachromosomal** units. Plasmids are self-replicating circular molecules of DNA that are maintained in the cell in a stable and characteristic number of copies; that is, the number remains constant from generation to generation.

Some of these elements also have alternative lifestyles. They can exist either in the autonomous extrachromosomal state; or they can be inserted into the bacterial chromosome, and then be carried as part of it like any other sequence. These are called **episomes.** (As a note of caution, since their original definition, the terms "plasmid" and "episome" have sometimes been used loosely as though interchangeable, so they do not always carry the precise connotations described here.)

Like lysogenic phages, plasmids and episomes maintain a selfish possession of their bacterium and often make it impossible for another element of the same type to become established. This also is called **immunity.** (The basis for plasmid immunity is different from lysogenic immunity; see Chapter 31.)

THE LYTIC CYCLE HAS DISTINCT STAGES

The purpose of lytic development is to produce as many as possible progeny phage particles by infection of a bacterial host cell. Infection starts with a single phage particle that attaches to a bacterium. **Icosahedral** phages have a **head** that contains the DNA and a **tail** that attaches to the bacterium. The DNA is injected into the bacterium; the particle is left outside. **Filamentous** phages have a rodlike structure, and their interaction with the bacterium is less well understood.

Phage genomes of necessity are small. Indeed, as with all viruses, the principal restriction is the need to package the nucleic acid within its protein coat. This dictates many of the viral strategies for reproduction. Typically they take over the apparatus of the host cell, which, instead of replicating and expressing bacterial genes, replicates and expresses phage genes.

Usually the phage genes include functions that ensure preferential replication of phage DNA. These functions may be concerned with the initiation of replication or may even provide a new DNA polymerase. Some change is always made in the capacity of the host cell to engage in transcription. This may involve replacing the RNA polymerase or modifying its capacity for initiation or termination. The result is always the same: phage mRNAs are preferentially transcribed. So far as protein synthesis is concerned, usually the phage is content to use the host apparatus, redirecting its activities principally by replacing bacterial mRNA with phage mRNA.

Lytic development is accomplished by a pathway in which the phage functions are expressed in a particular order. This ensures that the right amount of each component is present at the appropriate time.

The cycle can be divided into the two general parts illustrated in **Figure 16.1. Early infection** describes the period from entry of the DNA to the start of its replication. **Late infection** defines the period from the start of replication to the final step of lysing the bacterial cell to release progeny phage particles.

In the usual order of battle, the early phase is devoted to the production of enzymes involved in the reproduction of DNA. These include the enzymes concerned with DNA synthesis, recombination, and sometimes modification. Their activities cause a **pool** of phage genomes to accumulate. In this pool, genomes are continually replicating and recombining, so that the events even during one lytic cycle concern a population of phage genomes.

When the late phase begins, the protein components of the phage particle are synthesized. Often a large number of proteins may be needed to make up the head and tail structures, so the largest part of the phage genome consists of late functions. In addition to these, there may also be assembly proteins, whose presence is necessary to construct the particle, but which are not themselves part of it. By the time these components are assembling into heads and tails, replication of DNA has reached its maximum rate. The genomes then are inserted into the empty protein heads, tails are added, and the cell is lysed.

LYTIC DEVELOPMENT IS CONTROLLED BY A CASCADE

Many phages display an organization in which the genetic map closely reflects the sequence of lytic development. The concept of the operon is taken to somewhat of an extreme, in which the genes coding for proteins with related functions are clustered to allow their control with the maximum economy. This allows the pathway of lytic development to be controlled with a small number of regulatory switches.

To arrange for the expression of phage genes in a particular order, the lytic cycle is under positive control. Each group of phage genes can be expressed only when an appropriate signal is given.

Usually only a few phage genes can be expressed by the transcription apparatus of the host cell. Their

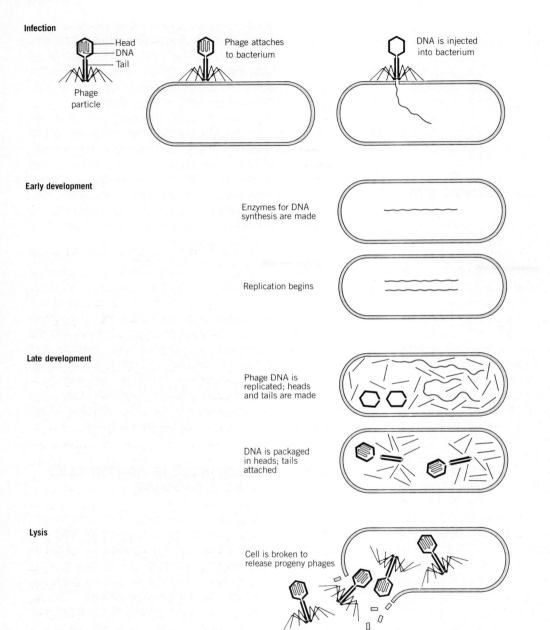

Figure 16.1
Lytic development takes place by producing more phage genomes and protein particles that are assembled into progeny phages.

promoters are indistinguishable from those of host genes. The name of this class of genes depends on the phage. In most cases, they are known as the **early genes.** In phage lambda, they are given the evocative description of **immediate early.** Irrespective of the name, they constitute only a preliminary, representing just the initial part of the early period. Sometimes they are exclusively occupied with the transition to the next period. At all events, one of these genes always codes for a protein that is necessary for transcription of the next class of genes.

This second class of genes is known variously as the **delayed early** or **middle** group. As the name suggests, its expression starts in the early period, typically as soon as the regulator protein is available. Depending on the nature of the control circuit, the initial set of early genes may or may not continue to be expressed at this stage (see Figure 13.5). Often the expression of host genes is reduced. Together the two sets of early genes account for all the phage functions except those needed to assemble the particle coat itself and to lyse the cell.

When the replication of phage DNA begins as the result of early gene expression, it is time for the **late genes** to be expressed. Their transcription at this stage usually is provided for by embedding a further regulator gene within the previous (delayed early or middle) set of genes. This regulator may be another antitermination factor (as in lambda) or it may be another sigma factor (as in SPO1).

The use of these successive controls, with a regulator gene in each set necessary for expression of the next gene set, creates a cascade in which groups of genes are turned on (and sometimes off) at particular times. The detailed construction of each phage cascade is different, but the results are similar, as the following sections show.

FUNCTIONAL CLUSTERING IN PHAGES T7 AND T4

The genome of phage T7 has three classes of genes, each constituting a group of adjacent loci. As **Figure 16.2** shows, the class I genes are the immediate early type, expressed by host RNA polymerase as soon as the phage DNA enters the cell. Among the products of these genes are enzymes that interfere with host gene expression (and therefore with their own expression) and a phage RNA polymerase. The phage enzyme is responsible for expressing the class II genes (concerned principally with DNA synthesis functions) and the class III genes (concerned with assembling

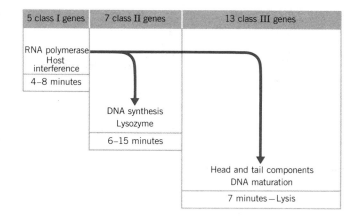

Figure 16.2
Phage T7 contains three classes of genes that are expressed sequentially. The genome is about 38,000 bp.

the mature phage particle). We do not yet understand the basis for the difference in expression between the class II and class III genes (see Chapter 12).

Phage T4 has one of the larger genomes, organized with extensive functional grouping of genes. **Figure 16.3** presents the genetic map. Genes that are numbered are **essential:** a mutation in any one of these prevents successful completion of the lytic cycle. Genes indicated by three-letter abbreviations are **nonessential,** at least under the usual conditions of infection. We do not really understand the inclusion of so many nonessential genes, but presumably they confer a selective advantage. (In most phage genomes, most or all genes are essential.)

There are three phases of gene expression, and a summary of the functions of the genes expressed in each is given in **Figure 16.4.** The early genes are transcribed by host RNA polymerase. At a slightly later point, the quasi-late genes are transcribed, but we do not yet understand how this is controlled (their name reflects a mode of expression that does not fit with the categories we have described for other phages). Together the early and quasi-late genes account for virtually all of the phage functions concerned with the synthesis of DNA, modifying cell structure, or transcribing and translating phage genes.

The two essential genes in the "transcription" category fulfill a regulatory function: their products are

Figure 16.3
The map of phage T4 is circular and shows extensive clustering of related functions in the 165,000 bp genome.

necessary for late gene expression. In phage T4, there is a mechanical link between replication and late gene expression. Only actively replicating DNA can be used as template for late gene transcription. This occurs via the modification of the host RNA polymerase so that it requires some feature present only in replicating DNA

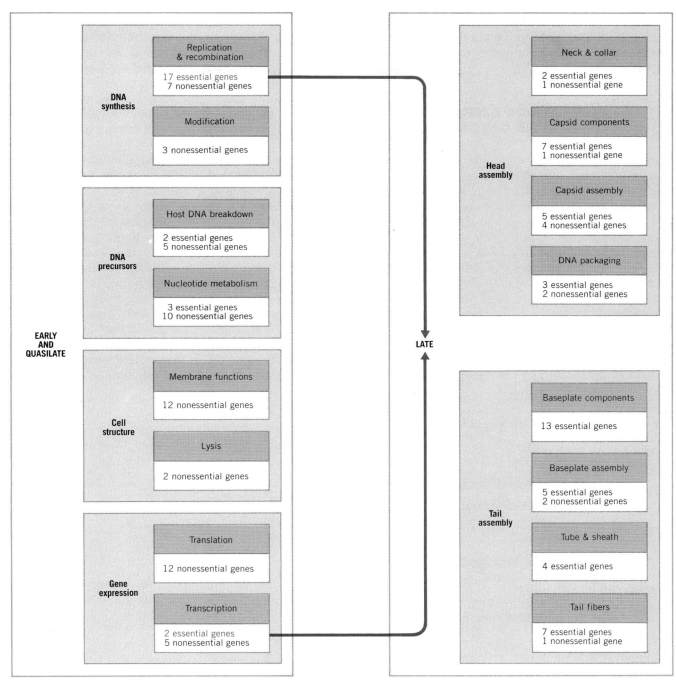

Figure 16.4
The phage T4 lytic cascade falls into parts: early and quasi-late functions are concerned with DNA synthesis and gene expression; late functions are concerned with particle assembly.

(probably nicks). This ensures that the phage protein components (of which there are a very large number) are produced in amounts appropriate for the pool of available DNA.

HOW LAMBDA EXERCISES ITS LYTIC CASCADE

One of the most intricate cascade circuits is provided by phage lambda. Actually, the cascade for lytic development itself is straightforward, with two regulators controlling the successive stages of development. But this circuit for the lytic cycle is interlocked with the circuit for establishing lysogeny, as summarized in **Figure 16.5.** When lambda DNA is introduced into a new host cell, the lytic and lysogenic pathways start off the same way. Both require expression of the immediate early and delayed early genes. But then they diverge, lytic development following if the late genes are expressed, lysogeny ensuing if the synthesis of the necessary repressor is established.

Lambda has only two immediate early genes, transcribed independently by host RNA polymerase. One of these is *cro,* which has dual functions in preventing synthesis of the lysogenic repressor (a necessary action if the lytic cycle is to proceed) and also in turning off expression of the immediate early genes (which are not needed later in the lytic cycle). The other immediate early gene is *N,* which, as we have seen in Chapter 13, codes for an antitermination factor that allows transcription to proceed into the delayed early genes.

The delayed early genes include two replication genes (needed for lytic infection) and seven recombination genes (some involved in recombination during lytic infection, two necessary to integrate lambda DNA into the bacterial chromosome for lysogeny). The *cII/cIII* pair of regulators is needed to start up the synthesis of the lysogenic repressor. The *Q* regulator is an antitermination factor that allows host RNA polymerase to proceed into the late genes. So the delayed early genes serve two masters: some are needed for the phage to enter lysogeny, the others ensure that the lytic cycle follows its proper order.

To disentangle these two pathways, let us first just consider the lytic cycle. **Figure 16.6** gives the map of lambda DNA. A group of genes concerned with reg-

Figure 16.5

The lambda lytic cascade is interlocked with the circuitry for lysogeny.

Figure 16.6

The lambda map shows clustering of related functions. The genome is about 46,500 bp.

ulation is surrounded by genes needed for recombination and replication. Within the regulation group are the immediate early genes, *N* and *cro*. They are transcribed from different strands of DNA—*N* toward the left, and *cro* toward the right. In the presence of the pN antitermination factor, transcription continues to the left of *N* into the recombination genes, and to the right of *cro* into the replication genes (see also Figures 13.7 and 13.8). This is illustrated in **Figure 16.7,** which shows the true state of lambda DNA during infection.

The map shown in Figure 16.6 gives the organization of the lambda DNA as it is found in the phage particle. But shortly after infection, the ends of the DNA join to form a circle. Figure 16.7 demonstrates that this welds the late genes into a single group, containing the lysis genes *S–R* from the right end of the linear DNA, and the head and tail genes *A–J* from the left end.

The late genes are expressed as a single transcription unit, starting from a promoter P_R, that lies between *Q* and *S*. In the absence of the product of gene *Q* (which is the last gene in the rightward delayed early unit), this transcription is constitutive but terminates at a site t_{R3} located close to the promoter $P_{R'}$. The transcript is known as 6S RNA and is 194 bases long. (Probably some of the RNA polymerases transcribing the delayed early genes manage to escape into the late genes, but in insufficient amounts to produce

enough of the head and tail components to sustain lytic development.) When pQ becomes available, however, it suppresses termination at t_{R3}, and the 6S RNA is extended, with the result that the late genes are heavily expressed.

Late gene transcription does not seem to terminate at any specific point, but continues through all the late genes into the region beyond. A similar event happens with the leftward delayed early transcription, which continues past the recombination functions. Transcription in each direction is probably terminated before the polymerases could clash.

LYSOGENY IS MAINTAINED BY AN AUTOGENOUS CIRCUIT

Looking at the lambda lytic cascade, we can see that the entire program is set in train by initiating transcription at the two promoters P_L and P_R for the immediate early genes *N* and *cro*. Because lambda uses antitermination to proceed to the next stage of (delayed early) expression, the same two promoters continue to be used right through the early period.

The expanded map of the regulatory region in **Figure 16.6** shows that the promoters P_L and P_R lie on either side of the *cI* gene. Associated with each promoter is an operator (O_L, O_R) at which a repressor

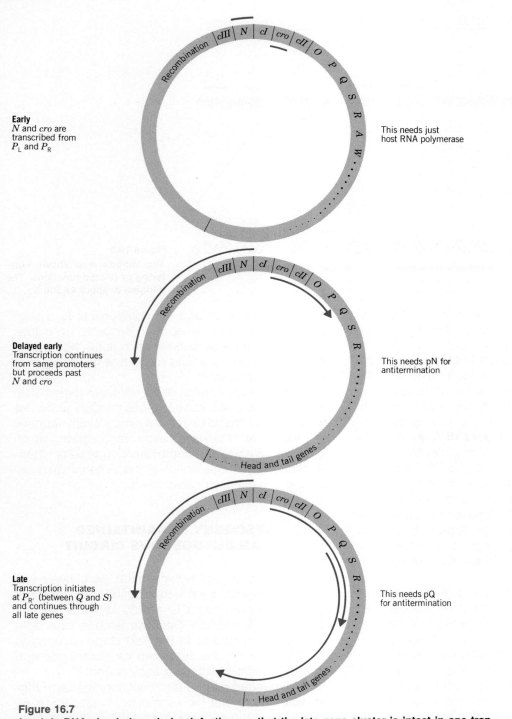

Early
N and cro are transcribed from P_L and P_R

This needs just host RNA polymerase

Delayed early
Transcription continues from same promoters but proceeds past N and cro

This needs pN for antitermination

Late
Transcription initiates at $P_{R'}$ (between Q and S) and continues through all late genes

This needs pQ for antitermination

Figure 16.7
Lambda DNA circularizes during infection, so that the late gene cluster is intact in one transcription unit.

protein binds to prevent RNA polymerase from initiating transcription. The sequence of each operator overlaps with the promoter that it controls; so often these are described as the P_L/O_L and P_R/O_R control regions.

Because of the sequential nature of the lytic cascade, the control regions provide a pressure point at which entry to the entire cycle can be controlled. By denying RNA polymerase access to these promoters, the repressor prevents the phage genome from initiating the lytic cycle. The operators are revealed as the targets for repressor action by **virulent** mutations that alter the DNA sequence that the repressor recognizes. These mutations prevent the repressor from binding at O_L or O_R, with the result that the phage inevitably proceeds into the lytic pathway. Virulent mutations in phages are the equivalent of operator-constitutive mutations in bacterial operons.

The repressor protein is coded by the *cI* gene. Mutants in this gene cannot maintain lysogeny, but are fated always to enter the lytic cycle. This gives rise to the name of the gene, which reflects the **clear plaque** phenotype of the resulting infection. When a bacterial culture is infected with a phage, the cells are lysed to generate regions that can be seen on a culture plate as a **plaque.** With wild-type phages, this is a small area of clearing that is cloudy, because it contains some cells that have established lysogeny instead of being lysed. The effect of a *cI⁻* mutation is to prevent lysogeny, so that the plaques contain only lysed cells, and are therefore clear in appearance.

The *cI* gene is transcribed from a promoter P_M, sometimes also called P_{RM}, that lies at its right end (see Figure 16.6). (The subscript "M" indicates maintenance of repressor.) Transcription is terminated at the left end of the gene. The mRNA actually starts with the AUG codon that is used to initiate translation; because of the absence of the usual ribosome binding site, the mRNA is translated somewhat inefficiently (see Chapter 9), producing only a low level of repressor protein.

The repressor binds independently to the two operators. It has a single function at O_L, but has dual functions at O_R. These are illustrated in **Figure 16.8.**

At O_L the repressor has the same sort of effect that we have already discussed for several other systems: it prevents RNA polymerase from initiating transcrip-

tion at P_L. This stops the expression of gene *N*. Since P_L is used for all leftward early gene transcription, this action prevents expression of the entire leftward early transcription unit. Thus the lytic cycle is stymied before it can begin.

At O_R, repressor binding prevents the use of P_R. Thus *cro* and the other rightward early genes cannot be expressed. (We shall see later why it is important to prevent the expression of *cro* when lysogeny is being maintained.)

But the presence of repressor at O_R also has another effect. The promoter for repressor synthesis, P_M, is adjacent to the rightward operator O_R. It turns out that RNA polymerase can initiate at P_M *only* when repressor is bound at O_R. In this action, the repressor is acting as a positive regulator protein that is necessary for transcription of the *cI* gene. Since it is in fact the product of this gene, there is an autogenous circuit, in which the presence of repressor is necessary to support its own continued synthesis.

The nature of this control circuit explains the biological features of lysogenic existence. Lysogeny is stable because the control circuit ensures that, so long as the level of repressor is adequate, there is continued expression of the *cI* gene. The result is that O_L and O_R remain occupied indefinitely. By repressing the entire lytic cascade, this action maintains the prophage in its inert form.

The presence of repressor explains the phenomenon of immunity. If a second lambda phage DNA enters a lysogenic cell, repressor protein synthesized from the resident prophage genome will immediately bind to O_L and O_R in the new genome. This prevents the second phage from entering the lytic cycle.

Induction of the prophage into the lytic cycle occurs if the lysogenic circuit is broken. This happens when the repressor present in the cell is inactivated. The autogenous nature of the circuit implies that this should be a very sensitive response. First, the absence of repressor allows RNA polymerase to bind at P_L and P_R, starting the lytic cycle as shown in Figure 16.8. Second, because the presence of repressor is necessary for its own synthesis, the expression of the *cI* gene stops as soon as the existing repressor is destroyed. This prevents any more repressor from being synthesized to replace the molecules that have been damaged. So the lytic cycle can start without any un-

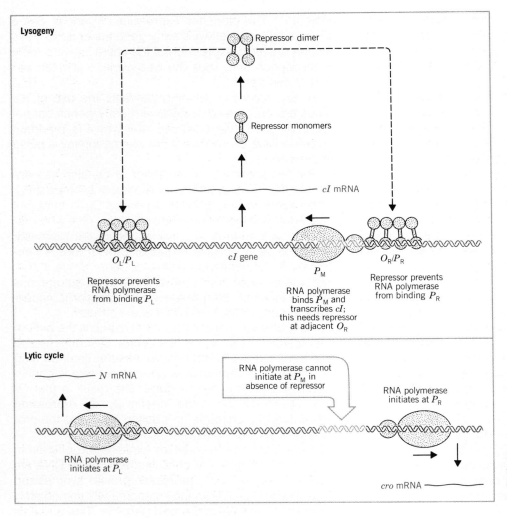

Figure 16.8
Lysogeny is maintained by an autogenous circuit (upper). If this circuit is interrupted, the lytic cycle starts (lower).

due interference from the circuit that maintains lysogeny.

The region of the left and right operators, the *cI* gene coding for the repressor and the *cro* gene that codes for another regulator protein, determines the immunity of the phage. That is to say, any phage that possesses this region has the same type of immunity, because it specifies both the repressor protein and the sites on which the repressor acts. Accordingly, this is called the **immunity region.** When we say that a lysogenic phage confers immunity to any other phage of the same type, we mean more precisely that the immunity is to any other phage that has the same immunity region (irrespective of differences in other regions).

REPRESSOR IS A DIMER WITH DISTINCT DOMAINS

The lambda repressor functions as a multimeric protein with distinct domains. The repressor subunit is a polypeptide of 27,000 daltons that dimerizes in solution. The N-terminal domain, residues 1–92, provides the operator-binding site. The C-terminal domain, residues 132–236, is responsible for dimer formation. The two regions are joined by a connector of 40 residues.

When the isolated protein is digested with a protease, each domain is released as a fragment that is relatively resistant to further degradation.

Each domain can exercise its function independently of the other. The C-terminal proteolytic fragment can form oligomers. The N-terminal fragment can bind the operators, although with a lower affinity than the intact repressor. Thus the information for specifically contacting DNA is contained within the N-terminal domain, but (as also with the *lac* repressor) the efficiency of the process is enhanced by the attachment of the C-terminal domain. Probably this happens because the ability of the intact repressor to dimerize ensures that when it binds to DNA its two N-terminal domains each contact DNA simultaneously.

The dimeric structure of the repressor is crucial in maintaining lysogeny. The induction of a lysogenic

Figure 16.9

Repressor forms dimers that bind to the operator. The affinity of the N-terminal domains for DNA is controlled by the dimerization of the c-terminal domains.

Lysogenic repression	
Dimers of repressor bind to DNA	Carboxyterminal domains form dimer — Connector — Aminoterminal domains bind operator
Induction	
Dimeric carboxyterminal domains are released from DNA-bound aminoterminal domains	Cleavage by protease
Lytic cycle starts	
Individual aminoterminal domains dissociate from DNA	

prophage to enter the lytic cycle is caused by cleavage of the repressor subunit in the connector region, between residues 111 and 112. (In a sense, this is the counterpart to the allosteric change in conformation that results when a small-molecule inducer inactivates the repressor of a bacterial operon, a capacity not possessed by the lysogenic repressor.)

The cleavage releases the dimeric C-terminal domains from the N-terminal domains that are bound to the operator. As illustrated in **Figure 16.9**, the N-terminal domains then no longer have sufficient affinity to remain on the operator. So they dissociate from DNA, allowing lytic infection to start.

Thus the balance between lysogeny and the lytic cycle depends on the concentration of repressor. Intact repressor is present in the cell at a concentration sufficient to ensure that the operators are occupied. But if the repressor is cleaved, this concentration is inadequate, because of the lower affinity of the separate N-terminal domain for the operator. Too high a concentration of repressor would make it impossible to induce the lytic cycle in this way; too low a level, of course, would make it impossible to maintain lysogeny.

REPRESSOR BINDS COOPERATIVELY AT EACH OPERATOR

Each operator contains three repressor-binding sites. As can be seen from **Figure 16.10**, each binding site is a sequence of 17 bp displaying partial symmetry about an axis through the central base pair. No two of the six individual repressor-binding sites are identical in sequence, but they all conform with a consensus sequence. The binding sites within each operator are separated by spacers of 3–7 bp that are rich in A-T base pairs. The sites at each operator are numbered so that O_R consists of the series of binding sites $O_R1–O_R2–O_R3$, while O_L consists of the series $O_L1–O_L2–O_L3$. In each case, site 1 lies closest to the startpoint for transcription in the promoter, and sites 2 and 3 lie farther upstream.

Faced with this triplication of binding sites at each operator, how does repressor decide where to start binding? At each operator, site 1 has a greater affinity (roughly tenfold) than the other sites for the repressor. The result is that the repressor always binds first to

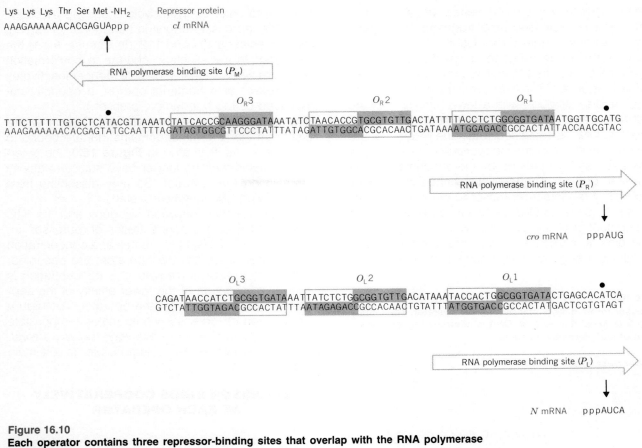

Figure 16.10
Each operator contains three repressor-binding sites that overlap with the RNA polymerase binding site.

The orientation of O_L has been reversed from usual to facilitate comparison with O_R.

O_L1 and O_R1. Thus low concentrations of repressor *in vitro* protect a fragment of about 25 bp at each operator against nuclease digestion. The fragment corresponds to the 17 bp of site 1 plus a few adjacent nucleotides. When the concentration of repressor is increased, a fragment of 50 bp is protected next, corresponding to O_L1-O_L2 or O_R1-O_R2. Greater concentrations of repressor are needed to fill site 3, but when this is achieved, an entire operator fragment of 80 bp is protected.

At each binding site, the pattern of points contacted by repressor suggests that it binds symmetrically, so that each N-terminal domain of the dimer contacts a similar set of bases. Thus the individual N-terminal region contacts a half-binding site. The side of the

binding site that the repressor contacts is indicated by the shading in Figure 16.10. In the double-helical structure of DNA, the points of contact lie primarily along the major groove of DNA.

Repressor binds to each operator in a cooperative manner. First, a repressor dimer binds to site 1, because of its greater affinity. However, the presence of a dimer at site 1 greatly increases the affinity with which a second dimer can bind to site 2. When both sites 1 and 2 are occupied, this interaction does *not* extend farther, to site 3. This means that at the concentrations of repressor usually found in a lysogen, both sites 1 and 2 are filled at each operator, but site 3 is not occupied.

If site 1 is inactive (because of mutation), then repressor binds cooperatively to sites 2 and 3. That is, binding at site 2 assists another dimer to bind at site 3. This interaction occurs directly between repressor dimers and not via conformational change in DNA. Probably the connector region of the first repressor orients the C-terminal regions of the dimer in such a way that they contact the C-terminal regions of just the second dimer (which therefore lies in the wrong direction to offer assistance to the third).

From the sequences shown in Figure 16.10, we can see that O_L1 and O_R1 lie more or less in the center of the RNA polymerase binding sites of P_L and P_R, respectively. Occupancy of O_L1–O_L2 and O_R1–O_R2 thus physically blocks access of RNA polymerase to the corresponding promoter.

A different relationship is shown between O_R and the promoter P_M for transcription of cI. The RNA polymerase binding site is just about adjacent to O_R2. This explains how repressor autogenously regulates its own synthesis. When two dimers are bound at O_R1–O_R2, the dimer at O_R2 interacts with RNA polymerase, probably through a protein-protein interaction. Unlike the interaction between repressors, this effect resides in the amino terminal domain, which can stimulate use of P_M even as an independent fragment. This and other interactions at O_R/P_R are illustrated in **Figure 16.11.**

Figure 16.11

Interactions between repressor dimers and with RNA polymerase control the use of the three promoters in the lambda right control region.

(a) In the usual lysogenic situation, repressor dimers interact at O_R2 and O_R1, blocking expression from P_R, but promoting use of P_M.

(b) A dimer bound at O_R1 interacts with a dimer bound at O_R2, but binding at O_R3 is independent. With all three sites filled, neither P_R nor P_M is used.

(c) If O_R1 is mutated, then dimers bound at O_R2 and O_R3 can interact. Neither P_R nor P_M can be used in this case.

Note that since it is O_R2 that must be filled for the repressor dimer to come close enough to assist the RNA polymerase at P_M, the cooperative dependence on a dimer initially binding at O_R1 may be important in ensuring that the autogenous loop stays in existence. In mutants that lack O_R1 activity, a single repressor dimer bound at O_R2 can inhibit transcription from P_R but stimulate it from P_M. The O_R2 site is only 1 bp closer to the startpoint for P_R than to that for P_M, yet there is a drastic difference in effect!

But what happens if a repressor dimer binds to O_R3? This site overlaps with the RNA polymerase binding site at P_M. Thus if the repressor concentration becomes great enough to cause occupancy of O_R3, the transcription of cl is prevented. This leads in due course to a reduction in repressor concentration; O_R3 then becomes empty, and the autogenous loop can start up again because O_R2 remains occupied. This provides a mechanism for preventing the concentration of repressor from becoming too great. (However, repressor concentration in lysogens usually is not high enough to trigger this effect, whose importance *in vivo* is therefore unclear.) In the formal sense, the repressor is an autogenous regulator of its own expression that functions positively at low concentrations and negatively at high concentrations.

Virulent mutations have been found in sites 1 and 2 of both O_L and O_R. Every one of these changes the base composition. The mutations vary in their degree of virulence, according to the extent to which they reduce the affinity of the binding site for repressor, and also depending on the relationship of the afflicted site to the promoter. Consistent with the conclusion that O_R3 and O_L3 usually are not occupied, virulent mutations are not found in these sites.

HOW IS REPRESSOR SYNTHESIS ESTABLISHED?

The control circuit for maintaining lysogeny presents a paradox. The presence of repressor protein is necessary for its own synthesis. This explains how the lysogenic condition is perpetuated. But how is the synthesis of repressor established in the first place?

When a lambda DNA enters a new host cell, the bacterial RNA polymerase cannot transcribe cl, because there is no repressor present to aid its binding at P_M. But this same absence of repressor means that P_R and P_L are available. So the first event when lambda DNA infects a bacterium is for genes N and cro to be transcribed. Then pN allows transcription to be extended further. This allows clll (and other genes) to be transcribed on the left, while cll (and other genes) are transcribed on the right (see Figure 16.6).

The cll and clll genes share with cl the property that mutations in them cause clear plaques. But there is a difference. Mutants that are cl⁻ can neither establish nor maintain lysogeny. Mutants that are cll⁻ or clll⁻ have some difficulty in establishing lysogeny, but once established, they are able to maintain it by the same autogenous circuit we have already discussed. This implicates the cll and clll genes as regulators whose products are needed for an alternative system for repressor synthesis. The system is needed only to *initiate* the expression of cl in order to circumvent the inability of the autogenous circuit to engage in de novo synthesis.

Between the cro and cll genes there is another promoter, called P_E, sometimes also known as P_{RE}. (The "E" indicates its use for repressor-establishment.) This promoter can be recognized by RNA polymerase only in the presence of the products of the cll and clll genes, which are therefore positive regulator proteins. Their action is illustrated in **Figure 16.12.**

The P_E promoter has an unusual sequence that lacks both the normal consensus sequences. The deficiency may explain its dependence on cll/clll. The importance of the -35 and -10 regions in recognition by RNA polymerase, however, is indicated by the existence of cY mutations. These have effects similar to those of cll⁻ and clll⁻ in preventing the establishment of lysogeny; but they are *cis*-acting instead of *trans*-acting. They fall into two groups, located around the -35 and -10 positions of P_E. So they are promoter mutations. It is thought that cll/clll may act at the -35 site to allow RNA polymerase to bind at the -10 site.

The P_E transcript contains the **antisense** strand of the cro gene, which usually is transcribed in the opposite direction (i.e., from P_R). The antisense cro se-

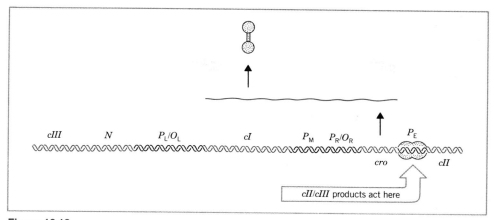

Figure 16.12

Repressor synthesis is established by the action of cII/cIII products at P_E to initiate transcription that extends from the antisense strand of *cro* through the *cI* gene.

quence is not translated on the P_E transcript, but the *cI* coding region is very efficiently translated (in contrast with the weak translation of the P_M transcript mentioned earlier). In fact, repressor is synthesized about seven or eight times more effectively via expression from P_E than from P_M.

For a long time it seemed puzzling that the *cro* gene could simultaneously be transcribed in opposite directions. We really don't know what happens if two RNA polymerases meet when traveling toward each other. The sensible presumption is that a tangle would develop, blocking transcription, but it is possible that the polymerases are able to pass in some way we do not understand. One explanation may be that, because the P_E promoter is stronger, its use suppresses the use of P_R; this would explain an effect of *cII/cIII* in reducing expression of the rightward early genes.

Now we can see how lysogeny is established on a new infection. **Figure 16.13** recapitulates the initial stages of transcribing *N* and *cro* and extending this transcription via pN action into *cIII* and *cII*. Then the presence of the cII/cIII products allows P_E to be used for transcription extending through *cI*. Repressor protein is synthesized in high amounts from this transcript. Immediately it binds to O_L and O_R.

By directly inhibiting any further transcription from P_L and P_R, the binding turns off the expression of all

phage genes. This halts the synthesis of cII and cIII proteins, which are unstable, so they decay rapidly, with the result that P_E can no longer be used. Thus the synthesis of repressor via the establishment circuit is brought to a halt.

But repressor now is present at O_R. This switches on the maintenance circuit for expression from P_M. The result is that repressor continues to be synthesized, although now at the lower level typical of P_M function. So the establishment circuit starts off repressor synthesis at a high level, and then repressor is able to turn off all other functions, while turning on the maintenance circuit, which functions at the low level adequate to sustain lysogeny.

We shall not now deal in detail with the other functions needed to establish lysogeny, but we can just briefly remark that the infecting lambda DNA must be inserted into the bacterial genome (see Chapter 35). The insertion requires the product of gene *int*, which is expressed from its own promoter P_I, at which cII/cIII products are necessary. The sequence of P_I shows homology with P_E. The functions necessary for establishing the lysogenic control circuit are therefore under the same control as the function needed physically to manipulate the DNA. Thus the accomplishment of lysogeny is under a control that ensures all the necessary events occur with the same timing.

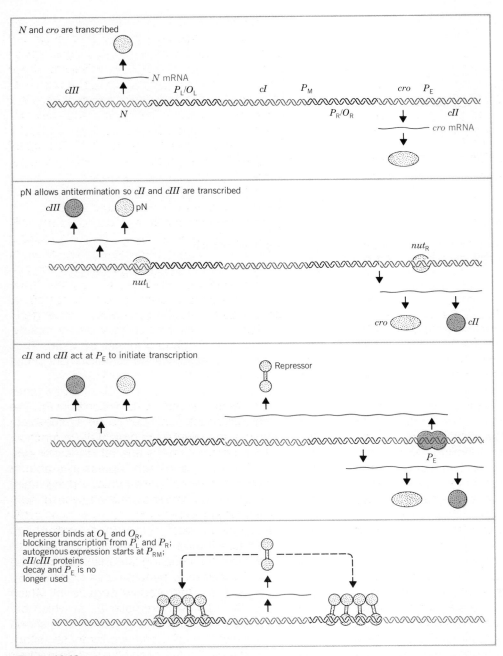

Figure 16.13
A cascade is needed to establish lysogeny, but then this circuit is switched off and replaced by the autogenous repressor-maintenance circuit.

ANTIREPRESSOR IS NEEDED FOR LYTIC INFECTION

We started this chapter by saying that lambda has the alternatives of becoming lysogenic or starting a lytic infection. We have seen that lysogeny is established by a program that leads to the start of an autogenous maintenance circuit that inhibits the entire lytic cascade by applying pressure at two points. The program for establishing lysogeny actually proceeds through some of the same events that we described earlier in terms of the lytic cascade (expression of delayed early genes via expression of N is needed). We now face a problem. How does the phage enter the lytic cycle?

What we have left out of this account so far is the role of gene *cro,* which codes for another regulator protein, sometimes known as the **antirepressor.** This regulator is responsible for preventing the synthesis of the repressor protein; this action shuts off the possibility of establishing lysogeny. Mutants of the cro^- type usually establish lysogeny rather than entering the lytic pathway, because they lack the ability to switch events away from the expression of repressor.

The Cro product is a small protein (only 9000 daltons) that forms a dimer that acts within the immunity region. It has two effects. It prevents the synthesis of repressor via the maintenance circuit; that is, it prevents transcription via P_M. It also inhibits the expression of early genes from both P_L and P_R. This means that, when a phage enters the lytic pathway, Cro has responsibility both for preventing the synthesis of repressor and for turning down the expression of the early genes.

To achieve its function, Cro binds to the same sites as the repressor. It also binds symmetrically at these sites, making contacts similar to those made by repressor. How can two proteins have the same sites of action, yet have such opposite effects? The answer lies in the different affinities that each protein has for the individual binding sites within the operators. Let us just consider O_R, where more is known, and where Cro exerts both its effects. The series of events is illustrated in **Figure 16.14.**

Cro has an affinity for O_R3 that is greater than its affinity for O_R2 or O_R1. So it binds first to O_R3. This

inhibits RNA polymerase from binding to P_M. Thus the first action that Cro accomplishes is to prevent the maintenance circuit for lysogeny from coming into play.

Then Cro binds to O_R2 or O_R1. Its affinity for these sites is similar, and there is no cooperative effect. Its presence at either site is sufficient to prevent RNA polymerase from using P_R. This in turn stops the production of the early functions (including Cro itself). Because the cII/cIII products are unstable, any use of P_E is brought to a halt. So the two actions of Cro together block *all* production of repressor.

So far as the lytic cycle is concerned, Cro turns down (although it does not completely eliminate) the expression of the early genes. Its incomplete effect is explained by its affinity for O_R1 and O_R2, which is about eight times lower than that of repressor. This effect of Cro does not occur until the early genes have become more or less superfluous, because pQ is present; by this time, the phage has started late gene expression, and is concentrating on the production of progeny phage particles.

A DELICATE BALANCE: LYSOGENY VERSUS LYSIS

The programs for the lysogenic and lytic pathways are so intimately related that it is impossible to predict the fate of any individual phage genome that enters a new host bacterium. Will the antagonism between repressor and Cro be resolved by establishing the autogenous maintenance circuit shown in Figure 16.13, or by turning off repressor synthesis and entering the late stage of development shown in Figure 16.14? The same pathway is followed in both cases right up to the brink of decision. Both involve the expression of the immediate early genes and extension into the delayed early genes. The difference between them comes down to the question of whether repressor or Cro will obtain occupancy of the two operators.

The early phase during which the decision is taken is limited in duration in either case. No matter which pathway the phage follows, expression of all early genes will be prevented as P_L and P_R are repressed; and, as a consequence of the disappearance of cII and cIII, production of repressor via P_E will cease. The critical

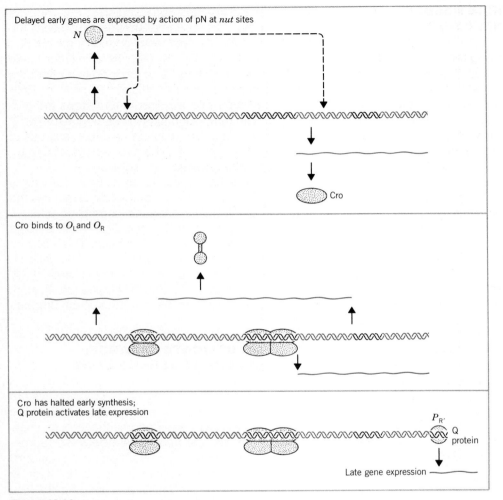

Figure 16.14
The lytic cascade requires Cro protein, which directly prevents repressor maintenance via P_M, as well as turning off delayed early gene expression, indirectly preventing repressor establishment via P_E.

question is therefore whether the cessation of transcription from P_E is followed by activation of P_M with the establishment of lysogeny, or whether P_M fails to become active and the pQ regulator commits the phage to lytic development.

The initial event in establishing lysogeny is the binding of repressor at O_L1 and O_R1. Binding at the first sites will rapidly be succeeded by cooperative binding of another repressor dimer at O_L2 and O_R2. This shuts off the synthesis of Cro and starts up the synthesis of repressor via P_M.

The initial event in entering the lytic cycle is the binding of Cro at O_R3. This stops the lysogenic-maintenance circuit from starting up at P_M. Then Cro must bind to O_R1 or O_R2, and to O_L1 or O_L2, to turn down early gene expression. By halting production of cII and cIII, this action leads to the cessation of repressor synthesis via P_E. The shutoff of repressor establishment occurs when the unstable cII and cIII proteins decay.

Thus there is a tangle of events all competing in the same region. RNA polymerase, repressor, and Cro protein bind to sites in the left and right control regions in mutually assisting and exclusive ways. Which regulator triumphs may depend on a sensitive quantitative and temporal balance. Fluctuations in the timing and frequencies of transcription and translation in individual cells may determine which pathway is followed in each case.

FURTHER READING

The basis for this view of lambda is analyzed in detail in **Lewin's** *Gene Expression, 3, Plasmids and Phages* (John Wiley, New York, 1978, pp. 274–535), which also deals with phage T4 (pp. 536–681), phages T3 and T7 (pp. 682–723), and other phages. Two chapters in *The Operon* (Miller & Reznikoff, Eds., Cold Spring Harbor Laboratory, New York, 1978) deal with lambda: **Rosenberg** et al. (pp. 345–371) on regulatory circuits, and **Ptashne** (pp. 325–343) on the action of repressor. The interplay between repressor and Cro has been reviewed by **Ptashne** et al. (*Cell* **19**, 1–11, 1980).

PART 5
CONSTITUTION OF THE EUCARYOTIC GENOME

The present day genetics concept visualizes the appearance of an organism as a result of an interaction of the whole set of genes the organism possesses and the environment in which it develops. A change in any of the genes, called a mutation, is liable to upset the balance of that system and show up on the organism as a character, usually as an abnormality, in some respect poorer than the wild type. . . . Studies with deficiencies, *viz.,* material where one or several genes are missing, show that the majority of deficiencies are lethal to the organism when present in a homozygous condition. This suggests that the presence of at least the majority of genes is essential in order that an organism may live. Moreover, the work with *D. melanogaster* deficiencies indicates that many of them are cell-lethal, *viz.,* that even a small group of cells located among normal tissues but containing a homozygous deficiency cannot exist. This suggests that genes are active in every cell and that, probably, the majority of them perform there a function highly important in the vital processes of the cell.

Milislav Demerec, 1935

CHAPTER 17
EUCARYOTIC GENOMES:
A CONTINUUM OF SEQUENCES

The "eucaryotic genome" should be a somewhat nebulous concept. The essence of being eucaryotic is that the major part of the genome is sequestered in the nucleus and is safeguarded by the nuclear membrane from exposure to the cytoplasm. At the gross level, the total amount of nuclear DNA varies enormously, the number of chromosomes into which it is segregated is extremely variable, the relatively miniscule part of the genome contained in organelles also shows wide characteristic variations, and there can be major differences in the types of sequences that constitute the nuclear genome.

Yet there are features to be found in eucaryotic genomes that are absent from procaryotic genomes. The integrity of individual genes can be interrupted, there can be multiple and (sometimes) identical copies of particular sequences, and there may be large blocks of DNA that do not code for protein. Because of the division between nucleus and cytoplasm, the arrangements for gene expression must necessarily be different in eucaryotes from those in procaryotes.

Do these features represent any underlying uniformity in the organization of eucaryotic DNA? We should remember that the eucaryotic kingdom is extremely broad, and at present we have detailed information about the genetic organization of only a few types of species. In describing the characteristics of eucaryotic DNA, we are dealing with features that may be represented to widely varying degrees in different individual genomes. We are therefore concerned with establishing the options available for the organization of eucaryotic DNA, rather than with defining a hypothetical "typical" pattern.

THE C-VALUE PARADOX DESCRIBES VARIATIONS IN GENOME SIZE

The total amount of DNA in the (haploid) genome is a characteristic of each living species known as its **C value.** It may be measured chemically in picograms (pg) of DNA or can be determined by the technique of DNA reassociation (see later), when it is generally given in base pairs or daltons. The equivalence is 1 pg $= 0.965 \times 10^9$ bp $= 6.1 \times 10^{11}$ daltons. There is enormous variation in the range of C values, from as little as a mere 10^4 bp for a mycoplasma to as much as $\sim 10^{11}$ bp for some plants and amphibians.

The **C value paradox** takes its name from our inability to explain the content of a genome in terms of its anticipated functions. There are two aspects to the paradox. First, there are huge variations in C values between certain species whose apparent complexity does not vary correspondingly; there can be rather substantial variations even between certain closely related species.

Figure 17.1 summarizes the range of C values found in different evolutionary phyla. There is some increase in the minimum genome size that is found in each phylum as the complexity increases. This is not absolute, but we see that unicellular eucaryotes (whose life-styles can somewhat resemble the procaryotic) get by with genomes that are small, although larger than those of the bacteria. Being eucaryotic per se does not imply a vast increase in genome size; for example, the yeast *S. cerevisiae* has a genome size of about 2.3×10^7 bp, only five times greater than that of the bacterium *E. coli.*

A further modest increase in genome size (just over twofold) is adequate to support the slime mold *D. discoideum,* able to live in either unicellular or multicellular modes. Another increase in complexity is necessary to produce the first fully multicellular organisms. For example, the nematode worm *C. elegans* has a DNA content of 8×10^7 bp. Then any close relationship between complexity of the organism and content of DNA becomes obscure, although it is necessary to have a genome of more than 10^8 bp to make an insect, of more than 4×10^8 bp to assemble an echinoderm, of more than 8×10^8 bp to produce a bird or amphibian, and of more than 2×10^9 bp to develop a mammal.

In some cases, the spread of genome sizes is quite small. For example, birds, reptiles, and mammals all show little variation within the phylum, with a range of genome sizes in each case about twofold. But in other cases there is quite a wide range of values, often more than tenfold. This reflects some surprising discrep-

Figure 17.1
DNA content of the haploid genome is not closely related to the morphological complexity of the species. Genome size does increase through the procaryotes and the protostomes, but extends over a wide range for the deuterostomes. The range of DNA values within a phylum is indicated by the shaded area.

ancies between genome size and complexity of the organism.

An extraordinary range of C values is found in amphibians, where the smallest genomes are just below 10^9 bp while the largest are almost 10^{11} bp. It is hard to believe that this could reflect a 100-fold variation in the number of genes needed to specify different amphibians. To reinforce this skepticism, there are some cases where rather closely related species show surprising variations in total genome size. For example, two amphibian species whose overall morphologies are very similar may have a difference of, say, 10 in their relative amounts of DNA. Again, it seems unlikely that there could be a tenfold difference in the gene number. Yet if the gene number is roughly similar, most of the DNA in the species with the larger genome cannot be concerned with coding for protein: what can be its function?

This leads us to the second aspect of the C-value paradox: the apparent absolute excess of DNA compared with the amount that could be expected to code for proteins. Indeed, this is often referred to as the problem of excess eucaryotic DNA. We now know that some of the excess is accounted for because genes are much larger than the sequences needed to code for proteins (principally because of the intervening sequences that may break up a coding region into different segments.) However, this does not seem sufficient to resolve the problem. Let us consider a couple of examples.

Even if we suppose that the average mammalian gene is as large as 10,000 bp (which actually is longer than almost all known interrupted genes), the number of genes in the mammalian genome would be 300,000. Is this plausible?

Probably not; for although gene numbers are not known with much precision, there are two reasons why it seems likely that they are to be counted in tens of thousands rather than in hundreds of thousands.

Some direct evidence is gained by counting the number of genes via estimates of the number of different mRNAs (see Chapter 18). While this cannot be done for every cell type to generate a sum total for the organism, it seems that the number of genes expressed in a given cell type is roughly about 10,000. Most of these are common to most or all cells of the organism. So this value is probably within a factor of (say) 2–4 of the total expressed gene number.

A less direct line of evidence is provided by attempts to estimate the number of essential genes by identifying all the loci that can be mutated. Much of this work has been performed with *D. melanogaster,* where we can infer that the total essential gene number is likely to be about 5000. With a reasonable estimate for the size of the insect gene as 2000 bp, this corresponds to a total length of 10^7 bp, just 10 times less than the amount of available DNA.

Of course, the genes identified by mutation are those in which damage has visible or lethal effects. Perhaps only some genes fall into this class. This would imply that at least a large proportion, possibly even the majority, of genes are concerned with specifying proteins that are not essential for the survival of the organism (at least in the sense that mutational damage does not cause any detectable effect).

So we are left with several key questions. What proportion of the DNA of the genome actually is concerned with representing proteins in the sense that it lies in a gene, either in the coding region itself or in an intervening or transcribed flanking sequence? Of the number of genes, how many are essential and how many dispensable? What is the function (if any) of DNA that does not reside in genes? What effect does a large change in total size have on the operation of the genome, as in the case of the related amphibians?

REASSOCIATION KINETICS DEPEND ON SEQUENCE COMPLEXITY

The reassociation of two complementary sequences of DNA occurs by base pairing in a reversal of the process of denaturation by which they were separated (see Figure 2.15). We have already discussed many experiments that rely on the extension of this technique to isolate individual DNA or RNA sequences by their ability to hybridize with a particular probe. Now we shall see that the kinetics of the reassociation reaction reflect the variety of sequences that are present; and the reaction can be used to quantitate genes and their RNA products. When performed in solution, such reactions are described as **liquid hybridization.**

Renaturation of DNA depends on random collision of the complementary strands; it therefore follows second-order kinetics. This means that the rate of reaction is governed by the equation

$$\frac{dC}{dt} = -kC^2 \tag{1}$$

where C is the concentration of DNA that is single-stranded at time t, and k is a reassociation rate constant.

By integrating this equation between the limits of the initial concentration of DNA, C_0 at time $t = 0$, and the concentration C that remains single-stranded after time t, we can describe the progress of the reaction as

$$\frac{C}{C_0} = \frac{1}{1 + k \cdot C_0 t} \tag{2}$$

Thus when the reaction is half complete, at time $t_{1/2}$,

$$\frac{C}{C_0} = \frac{1}{2} = \frac{1}{1 + k \cdot C_0 t_{1/2}} \tag{3}$$

so that

$$C_0 t_{1/2} = \frac{1}{k} \tag{4}$$

The reassociation of any particular DNA can be described by the rate constant k (in units of liter-nucleotide-moles^{-1}-sec^{-1}), or in the form of its reciprocal $C_0 t_{1/2}$ (given in nucleotide-moles × sec/liter).

This equation shows that the parameter controlling the reassociation reaction is the product of DNA concentration (C_0) and time of incubation (t). This is usually described simply as the **Cot.** Similarly, the value required for half-reassociation is called the **Cot$_{1/2}$.** A greater Cot$_{1/2}$ implies a slower reaction, since it is the product of the concentration and time required to proceed halfway.

The reassociation of DNA usually is followed in the form of a **Cot curve,** which plots the fraction of DNA that has reassociated ($1 - C/C_0$) against the log of the Cot. **Figure 17.2** gives Cot curves for several genomes. The form of each curve is similar, with renaturation occurring over about a 100-fold range of Cot values between the points of 10% reaction and 90% reaction. But the Cot required in each case is very different, and can be characterized by the Cot$_{1/2}$ (taken by the measured Cot at 50% reassociation).

The Cot$_{1/2}$ is directly related to the amount of DNA in the genome. This reflects a situation in which, as the genome becomes more complex, there is a smaller number of copies of any particular sequence within a given mass of DNA. For example, if the C_0 of DNA is 12 pg, it will contain 3000 copies of each sequence in a bacterial genome whose size is 0.004 pg, but it will contain only 4 copies of each sequence present in a eucaryotic genome of size 3 pg. Thus the same *absolute* concentration of DNA measured in moles of nucleotides per liter (that is, the C_0) will provide a concentration of each eucaryotic sequence that is 750 times (3000/4) lower than that of each bacterial sequence. Since the rate of reassociation depends on the concentration of complementary sequences, for the eucaryotic sequences to be present at the same *relative* concentration as the bacterial, it is necessary to have 750 times more DNA (or to incubate the same amount of DNA for 750 times longer). Thus the Cot$_{1/2}$ of the eucaryotic reaction is 750 times the Cot$_{1/2}$ of the bacterial reaction.

The Cot$_{1/2}$ of a reaction therefore indicates the *total length* of *different* sequences that are present. This is described as the **complexity.** It is usually given in base pairs, but can be expressed in daltons or any other mass unit.

The renaturation of the DNA of any genome (or part

Figure 17.2
Rate of reassociation is inversely proportional to the length of the reassociating DNA.

of a genome) should display a $Cot_{1/2}$ that is proportional to its complexity. Thus the complexity of any DNA can be determined by comparing its $Cot_{1/2}$ with that of a standard DNA of known complexity. Usually *E. coli* DNA is used as a standard. Its complexity is taken to be identical with the length of the genome (implying that every sequence in the *E. coli* genome of 4.2×10^6 bp is unique). Thus we can write

$$\frac{Cot_{1/2} \text{ (DNA of any genome)}}{Cot_{1/2} \text{ (E. coli DNA)}}$$
$$= \frac{\text{Complexity of any genome}}{4.2 \times 10^6 \text{ bp}} \quad (5)$$

EUCARYOTIC GENOMES CONTAIN SEVERAL SEQUENCE COMPONENTS

When the DNA of a eucaryotic genome is characterized by reassociation kinetics, usually the reaction occurs over a range of Cot values spanning up to eight orders of magnitude. This is much broader than the 100-fold range expected from equation **2** and shown for the examples of Figure 17.2. The reason is that the equation applies to a single **kinetically pure** reassociating component. A genome actually may have several of these components, each reassociating with its own characteristic kinetics.

Figure 17.3 shows the reassociation of a (hypothetical) eucaryotic genome, starting at a Cot of 10^{-4} and terminating by a Cot of 10^4. The reaction falls into three distinct phases, outlined by the shaded boxes. Actually, there is a plateau separating the first two phases, but the second and third overlap slightly. Each of these phases represents a different kinetic component of the genome.

The first fraction to reassociate is called the **fast component.** In this case, it represents 25% of the total DNA, renaturing between Cot values of 10^{-4} and about 2×10^{-2}, with a $Cot_{1/2}$ value of 0.0013.

The next fraction is called the **intermediate component.** This represents 30% of the DNA. It renatures between Cot values of about 0.2 and 100, with a $Cot_{1/2}$ value of 1.9.

The last fraction to renature is called the **slow component.** This is 45% of the total DNA; it extends over a Cot range from <100 to about 10,000, with a $Cot_{1/2}$ of 630.

To calculate the complexities of these fractions, each must be treated as an independent kinetic component whose reassociation is compared with a standard DNA. The slow component represents 45% of the total DNA, so its concentration in the reassociation reaction is 0.45 of the measured C_0 (which refers to the total amount of DNA included). Thus the $Cot_{1/2}$ applying to the slow fraction alone is $0.45 \times 630 = 283$. Thus if this DNA were isolated as a pure component free of the other fractions, it would renature with a $Cot_{1/2}$ of 283. Suppose that under these conditions, *E. coli* DNA reassociates with a $Cot_{1/2}$ of 4.0. Comparing these two values, we see from equation **5** that the complexity of this fraction is 3.0×10^8 bp. Treating the other components in a similar way shows that the intermediate component has a complexity of 6×10^5 bp, and the fast component has a complexity of only 340 bp. This places on a quantitative basis our statement that the faster a component reassociates, the lower its complexity.

Reversing the argument, if we took three DNA preparations, each containing a unique sequence of the appropriate length (340 bp, 6×10^5 bp, and 3×10^8 bp, respectively) and mixed them in the proportions

Figure 17.3
The reassociation kinetics of eucaryotic DNA show three types of component (indicated by the shaded areas).

	Fast Component	Intermediate Component	Slow Component
Percent of genome	25	30	45
$Cot_{1/2}$	0.0013	1.9	630
Complexity, bp	340	6.0×10^5	3.0×10^8
Repetition frequency	500,000	350	1

25:30:45, each would renature as though it were a single component described by equation **2;** together the mixture would display the same kinetics as those determined for the whole genome of Figure 17.3.

NONREPETITIVE DNA COMPLEXITY CAN ESTIMATE GENOME SIZE

The complexity of the slow component corresponds with its physical size. Suppose that the genome reassociating in Figure 17.3 has a chemically determined haploid DNA content of 7.0×10^8 bp. Then 45% of it is 3.15×10^8 bp, which is only marginally greater than the value of 3.0×10^8 bp measured by the kinetics of reassociation. In fact, given the errors of measurement inherent in both techniques, we can say that the complexity of the slow component is the same whether measured chemically or kinetically. The two values are referred to as the **chemical complexity** and **kinetic complexity.**

The coincidence of these values means that the slow component comprises sequences that are unique in the genome: upon denaturation, each single-stranded sequence is able to renature *only* with the corresponding complementary sequence. This part of the genome is the sole component of procaryotic DNA (as witnessed by the examples of Figure 17.2 and implied by the use of *E. coli* DNA as a standard); and it is usually a major component in eucaryotes. It is called **nonrepetitive DNA.**

We can use the kinetic complexity of nonrepetitive DNA to estimate the complexity of the genome. This just reverses the calculation that we performed earlier. For the example of Figure 17.3, the complexity of nonrepetitive DNA is 3.0×10^8 bp. If this fraction is unique and represents 45% of the genome, the whole genome should have a size of $3.0 \times 10^8 \div 0.45 = 6.6 \times 10^8$ bp. This provides an independent assessment for genome size that we can compare with the chemical complexity, which in this case is 7.0×10^8 bp.

Virtually every eucaryotic genome has a nonrepetitive DNA component. The sole exceptions are some plants that were generated by a recent polyploidization, where as a result there is more than one copy of every sequence.

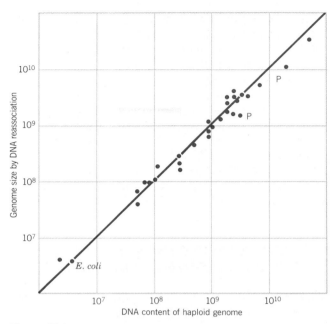

Figure 17.4
There is a good correlation between the kinetic complexity and chemical complexity of eucaryotic genomes, with the exception of polyploid genomes (indicated by *P*).

Figure 17.4 plots the relationship between genome size, as determined by reassociation kinetics of nonrepetitive DNA, and the haploid DNA content, as determined by chemical analysis. There is good agreement. This verifies the underlying assumption that the nonrepetitive DNA component does indeed consist of individual sequences present in only one copy per genome.

This is an important point in light of the C value paradox. The presence of nonrepetitive DNA implies that larger genomes are not generated simply by increasing the number of copies of the same sequences present in smaller genomes. In this case, larger genomes would behave as though polyploid; as we have already mentioned, this results in the absence of nonrepetitive DNA. So the reassociation data exclude models that would explain the C value paradox in terms of simple increases in the number of copies of each gene present in the haploid genome. This means that we must account for the differences in genome size on the basis that larger genomes genuinely contain greater diversity of sequences.

EUCARYOTIC GENOMES CONTAIN REPETITIVE SEQUENCES

What is the nature of the components that renature more rapidly than the nonrepetitive (slow) DNA? In the example of Figure 17.3, the intermediate component occupies 30% of the genome, which, according to the chemical complexity, should correspond to a total amount of material of $0.3 \times 7 \times 10^8 = 2.1 \times 10^8$ bp. But the kinetic complexity of this component is only 6×10^5 bp.

Thus the unique length of DNA that corresponds to the $Cot_{1/2}$ for reassociation is much shorter than the total length of the DNA chemically occupied by the component in the genome. In other words, the intermediate component behaves as though consisting of a sequence of 6×10^5 bp that is present in 350 copies in every genome (because $350 \times 6 \times 10^5 = 2.1 \times 10^8$). Following denaturation, the single strands generated from any one of these copies are able to renature with their complements from *any one* of the 350 copies. This effectively raises the concentration of reacting sequences in the reassociation reaction, explaining why the component renatures at a lower $Cot_{1/2}$.

Sequences that are present in more than one copy in each genome are described as **repetitive DNA**. The number of copies present per genome is called the **repetition frequency (f)**. Formally, the repetition frequency is defined by the ratio

$$f = \frac{\text{Chemical complexity}}{\text{Kinetic complexity}} \qquad (6)$$

Together, the complexity and repetition frequency describe the properties of the sequence components of the genome. For example, if a given genome consists of 1 sequence that is A bp long, 10 copies of a sequence that is B bp long, and 100 copies of a sequence that is C bp long, the complexity is $A + B + C$. The repetition frequencies of sequences A, B, and C are 1, 10, and 100, respectively. Sequence A is nonrepetitive; sequences B and C are repetitive.

The repetitive DNA fraction includes all components whose repetition frequency is significantly greater than 1. For practical purposes, repetition frequencies between 1 and 2 are taken to indicate sequences that belong in the nonrepetitive component. (It is impossible for a sequence to be represented a nonintegral number of times in the haploid genome!) In the example of Figure 17.3, the application of equation **6** to the nonrepetitive DNA gives a value of $3.15 \times 10^8 / 3.0 \times 10^8 = 1.05$. The usual range of values for nonrepetitive DNA in real genomes is from about 0.8 to 1.6.

There may be several components in the repetitive DNA of any genome, but two general types are distinguished by the range of Cot values within which reassociation occurs. These correspond approximately to the intermediate and fast components of Figure 17.3. The intermediate fraction, reassociating usually in a range between a Cot of 10^{-2} and that of nonrepetitive DNA, is said to consist of **moderately repetitive DNA**. The fast fraction, reassociating before a Cot of 10^{-2} is reached, is said to consist of **highly repetitive DNA**.

Highly repetitive DNA takes its name from the very large number of copies of the basic reassociating sequence present. Generally this sequence is rather short. In the example of Figure 17.3, the fast component consists of a sequence only 340 bp long, present in 500,000 copies per genome. Because of the short length of the reassociating unit, sometimes this is also referred to as **simple sequence DNA.**

Rather than proceed through the series of calculations we have described, the repetition frequency of any component in a reassociation curve can be determined directly by comparison with the nonrepetitive DNA component. The inverse ratio of their $Cot_{1/2}$ values gives the relationship between the repetition frequencies of any two components in the same reassociation curve. So if we assume that nonrepetitive DNA really is unique (its f is defined to be 1.0), we can write an alternative equation for the frequency of any repetitive component.

$$f = \frac{Cot_{1/2} \text{ of nonrepetitive DNA component}}{Cot_{1/2} \text{ of repetitive DNA component}} \qquad (7)$$

As an example, in the case of Figure 17.3, the intermediate component has an average value of $f =$

630/1.9 = 332. (This is slightly different from the value calculated via equation **6,** because that corresponds to allowing the *f* for nonrepetitive DNA to be 1.05.)

MODERATELY REPETITIVE DNA CONSISTS OF MANY DIFFERENT SEQUENCES

A repetitive DNA component represents only an average behavior that is useful for describing its sequences. The relevant parameters do not necessarily represent the properties of any particular sequence. The moderately repetitive component of Figure 17.3 includes a total length of 6×10^5 bp of DNA, repeated about 350 times per genome. But this does not correspond to a single, identifiable, continuous length of DNA. Instead, it is made up of a variety of individual sequences, each much shorter, whose total length together comes to 6×10^5 bp. These individual sequences may be dispersed about the genome. Their average repetition is 350, but some will be present in more copies than this and some in less.

When a eucaryotic genome is analyzed by reassociation kinetics, the individual sequence components are rarely so well separated as shown in Figure 17.3. In fact, they often overlap too extensively to be distinguished by eye. An example is drawn in **Figure 17.5.** The genome of the sea urchin *S. purpuratus* renatures over eight orders of magnitude, and points of inflection can be seen that distinguish the various

components, but the proportion and $Cot_{1/2}$ of each is not immediately evident.

The procedure followed to identify individual components is to use a computer program that attempts to fit individual curves (for hypothetical kinetically pure components) to each region of the reassociation curve. Together the individual curves overlap to give a sum that corresponds to the behavior actually observed for the genome. As we noted earlier, if DNA preparations of the appropriate complexities were mixed in the specified proportions, they would give the same curve as that obtained experimentally.

An example of such a resolution is shown in Figure 17.5, where the three individual curves represent components occupying 19%, 27%, and 50% of the total DNA (in order of reassociation). In this case, the first two components both would be classified as moderately repetitive DNA, with $Cot_{1/2}$ values of 0.53 and 8.3; these correspond to components of complexity 1.0×10^6 bp repeated 160 times, and of complexity 2.3×10^7 repeated 10 times. The third component in the curve is nonrepetitive DNA.

The usual approach to making computer fits is to seek the minimum number of individual kinetic components whose sum gives a good enough approximation to the total curve. But this need not necessarily reflect the biological construction of the genome. Often it is possible to achieve an equally good resolution with a greater number of components. For example, the data shown in Figure 17.5 can be interpreted just

Figure 17.5

Several components can be discerned in the reassociation kinetics of *S. purpuratus* DNA.

The experimental curve is drawn through the actual data points; the curves without data points show the three minimum ideal components that could account for the experimental curve.

as well with three moderately repetitive components, occupying 10%, 27%, and 25% of the genome, with repetition frequencies of 8000, 240, and 19. This analysis also reduces the size of the nonrepetitive DNA component.

The real situation may therefore comprise a continuum of repetitive components reassociating over a range from about 10 times to more than 20,000 times that of the nonrepetitive component. The kinetic components introduced to solve the curve do not correspond to discrete repetitive fractions of the genome, but are just useful approximations for descriptive purposes.

For most eucaryotic genomes, two or three components are sufficient to describe the range of repetitive sequences. The proportion of the genome occupied by repetitive DNA varies widely. For lower eucaryotes, most of the DNA is nonrepetitive, with only 10–20% falling into one or more moderately repetitive components. In animal cells, up to half of the DNA often is occupied by moderately and highly repetitive components. In plants and amphibians, the repetitive proportion may be even greater, sometimes representing the majority of the genome, with the moderately and highly repetitive components accounting for up to 80%. The length of the nonrepetitive DNA component tends to increase with overall genome size, but is not tightly related to it.

This offers a modest relief from the C value paradox, in the sense that the range of complexities of nonrepetitive DNA is less than the range of genome sizes. If we assume that genes are contained within the nonrepetitive DNA, it remains true that we have to explain rather a substantial variation in complexities that does not seem to accord with the morphological features of the organism, but the range is reduced. For example, a fruit fly has a nonrepetitive DNA complexity of 1×10^8 bp and a housefly has one of 2.8×10^8 bp, so the difference is only threefold, compared with a discrepancy in genome sizes of sixfold. Seen in these terms, the mammalian nonrepetitive DNA complexity tends to be about 9×10^8 bp, the frog *X. laevis* is 1.6×10^9 bp, and plants and amphibians extend to about 9×10^9 bp.

Finally, we should emphasize that there is no single feature that characterizes all eucaryotic genomes. We have mentioned that in plants that have become poly-ploid very recently, there may be no nonrepetitive DNA, so that the slowest reassociating fraction has a repetition frequency of 2–3 or more. (True, this is probably a temporary stage in evolution, which usually may lead to divergence to regenerate nonrepetitive DNA.) In crab genomes, there may be no moderately repetitive DNA at all—just highly repetitive and nonrepetitive DNA. In lower eucaryotes, there may be no highly repetitive DNA.

MEMBERS OF REPETITIVE SEQUENCE FAMILIES ARE RELATED BUT NOT IDENTICAL

The different components of eucaryotic DNA can be isolated by separating the duplex molecules according to the Cot needed for their renaturation. The procedure is to allow the renaturation reaction to proceed only to a particular Cot value. At this point, the DNA that has renatured is separated by virtue of its double-stranded structure; the remainder remains single-stranded. Usually the duplex and single-stranded molecules are separated by using hydroxyapatite columns, which preferentially retain duplex DNA.

Renatured nonrepetitive DNA behaves very much like the original preparation of DNA before its denaturation. When subjected to a denaturation procedure, it melts sharply at a T_m only very slightly below that of the original native DNA. This shows that strand reassociation has been accurate: each unique sequence has annealed with its exact complement.

Quite different behavior is shown by renatured repetitive DNA, which tends to melt gradually over rather a wide temperature range, as shown in **Figure 17.6.** This means that it does not consist of exactly paired molecules. Instead, there must be appreciable mispairing in the renatured double-stranded molecules. In fact, the more mispairing in a particular molecule, the fewer hydrogen bonds there are to break in melting, and thus the lower the T_m. The breadth of the melting curve shows that renatured repetitive DNA contains a spectrum of sequences, ranging from those that have been formed by reassociation between sequences that are only partially complementary, to those formed by reassociation between sequences that are very nearly or even exactly complementary. How can this happen?

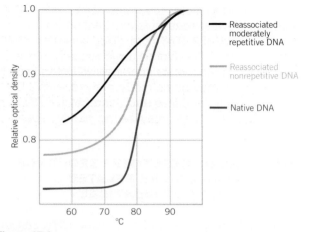

Figure 17.6
The denaturation of reassociated nonrepetitive DNA takes place over a narrow temperature range close to that of native DNA, but reassociated moderately repetitive DNA melts over a wide temperature range.

Repetitive DNA components consist of families of sequences that are not exactly the same, but that are related. The members of each family consist of a set of nucleotide sequences that are sufficiently similar to renature with one another. The differences are the result of base substitutions, insertions, and deletions, all creating points within the related sequences at which the complementary strands cannot base pair. The proportion of these changes establishes the relationship between any two sequences. When two closely related members of the family renature, they form a duplex with high T_m; when two more distantly related members associate, they form a duplex with a lower T_m.

The ability of related but not identical complementary sequences to recognize each other can be controlled by the **stringency** of the conditions imposed for reassociation. A higher stringency is imposed by (for example) an increase in temperature, which requires the formation of a greater number of base pairs in a given length of DNA. So by performing the hybridization reaction at high temperatures, reassociation may be restricted to rather closely related members of a family; at lower temperatures, more distantly related members may anneal. Note also that, as the stringency of hybridization increases, the proportion

of the genome characterized as nonrepetitive also increases.

Analysis of the members of repetitive families shows that there is wide variety in their construction. In some cases, well-defined families can be identified, whose members are quite closely related. Such families remain intact even when the stringency of hybridization is increased. In other cases, there is a continuum of variously related sequences, so that the size of the family decreases continuously as the stringency is increased. The measured size of such repetitive families is arbitrary, since it is determined by the particular hybridization conditions that are used. These results tell us that eucaryotic genomes contain sequences related to one another, most likely derived by divergence from some common ancestral sequence that was multiplied, but the extent of such relationships varies enormously.

The reassociation of two sequences that are related but not identical occurs more slowly than the reaction between identical sequences. This means that a greater Cot is required for the reassociation of related sequences. So the Cot$_{1/2}$ values observed for repetitive fractions may be higher than really corresponds to the complexity. Thus most repetitive components have lower complexities and greater repetition frequencies than implied by their reassociation kinetics. We shall see that the effect is especially marked for highly repetitive DNA (in Chapter 24).

MODERATELY REPETITIVE DNA IS INTERSPERSED WITH NONREPETITIVE DNA

In most genomes, neither moderately repetitive nor nonrepetitive DNA takes the form of a long continuous sequence uninterrupted by sequences of other components. Instead, both components are present in the form of individual sequences interspersed with one another. In many cases, there is some degree of regularity in the type of interspersion.

The most common type is **short period interspersion,** in which the moderately repetitive DNA is present as individual sequences of average length about 300 bp. These alternate with nonrepetitive sequences whose average length varies from 800 to 1500 bp in various genomes. This type of pattern may occupy about half of the total genome, and usually it accounts for most of both the moderately repetitive and nonrepetitive DNA.

Some genomes display a **long period interspersion,** in which there still is an alternation of moderately repetitive and nonrepetitive DNA components, but each type of sequence is very much longer. In neither type of interspersion is there any clear significance to attribute to the pattern.

FURTHER READING

A principal source for the information discussed in this chapter is **Lewin's** *Gene Expression, 2, Eucaryotic Chromosomes* (Wiley, New York, 1980, pp. 479–530). The reassociation technique was introduced by **Britten** & **Kohne** (*Science* **161,** 529–540, 1968), and subsequent results were reviewed by **Britten** & **Davidson** (*Quart. Rev. Biol.* **46,** 111–133, 1971) and again by **Davidson** & **Britten** (*Quart. Rev. Biol.* **48,** 565–613, 1973).

CHAPTER 18
STRUCTURAL GENES:
AS REPRESENTED IN mRNA

Mendelian genetics for simple traits imply that there is only one copy of each determining factor in the haploid genome. This can be mapped to a particular locus; and the simplest assumption is that each such locus is occupied by a DNA sequence representing a single protein, as exemplified by the definition of the gene in Chapter 2. This is the classic view of a structural gene: a unique component of the genome, the only sequence coding for its protein product, and therefore identifiable by mutations that impede the protein function.

But there is also another class of genes in the form of multiple sequences all coding for the same protein (or sometimes for closely related proteins). These may be difficult or even impossible to identify by mutation, because inactivation of any one copy does not impede the activity of the remainder. Genetic data are therefore biased toward characterizing the first type of gene, so we must turn to direct analysis of the DNA of the genome to determine the numbers and proportions of unique and repeated genes.

For the purpose of identifying and characterizing structural genes, mRNA provides the ideal intermediate. On the one hand, the protein to which it corresponds can be determined by translating the mRNA. On the other hand, the gene from which it is derived can be obtained by hybridizing the mRNA with the genomic DNA. An individual mRNA provides a handle, as it were, for proceeding back from its protein to the gene. A population of mRNAs, as defined by the spectrum of sequences found in the polysomes, defines the entire set of genes that is expressed in a cell or tissue. Thus the constitution of the mRNA reveals both the nature and number of structural genes. By means of nucleic acid hybridization, we can come to grips with some central questions. How many copies of each gene are there? How many genes are expressed in a particular cell type? How much overlap is there between the sets of genes whose expression defines different cell types?

ARE STRUCTURAL GENES
UNIQUE OR REPETITIVE?

We can envisage three possibilities for the organization of the eucaryotic genome in terms of its component structural genes.

A structural gene may be unique, consisting of a DNA sequence representing its particular protein and unrelated to any other sequences in the genome. Such a sequence, of course, would be part of the nonrepetitive DNA component.

Alternatively, a structural gene may be unique in the

sense that it is indeed the only DNA sequence coding for its exact protein product, but there could be other sequences in the genome, related to it and coding for related proteins. This type of situation occurs when a coding sequence is duplicated and the copies subsequently diverge to generate different proteins. Depending on the relationship between the DNA sequences, the number of copies, and the not entirely predictable effects of mismatching during reassociation, such sequences could appear in either nonrepetitive DNA or moderately repetitive DNA.

Still another situation is that structural genes may be repeated: there may be more than one copy of a sequence coding for a particular protein. Because of the degeneracy of the genetic code, DNA sequences coding for the same polypeptide need not actually be identical. However, any sequence present in a reasonable number of copies (more than two or three) should behave as part of the moderately repetitive sequence component. (Of course, such sequences could constitute only a part of this component, because we know that much of the moderately repetitive DNA consists of individual sequences that are too short to code for proteins and are interspersed with nonrepetitive DNA.)

The kinetics of hybridization can be used to determine how many copies there are of each gene corresponding either to members of the mRNA population or to individual mRNAs. More precisely, the question asked by such analysis is: which sequence component of the genome—nonrepetitive or repetitive—is represented in mRNA?

Given the limitations of these experiments, structural genes that are found to be part of nonrepetitive DNA need not necessarily be unique. They could be present in a small number of copies, but this should certainly be less than, say, three or four. To determine the precise number, it is necessary to isolate the DNA corresponding to individual mRNAs. This means proceeding from the kinetics of hybridization to the use of the techniques described in the next chapter.

For structural genes that lie in repetitive DNA, the repetition frequency of the DNA sequence gives only a crude estimate of the gene number. To determine the number of genes and the relationships between them, again it is necessary to isolate the individual members of the family. The properties of repetitive

DNA as a whole correspond to families that consist of related but not identical sequences, but this does not exclude the presence within this component of some families all of whose members are identical or very closely related.

MOST STRUCTURAL GENES LIE IN NONREPETITIVE DNA

The genome sequence components represented in mRNA can be determined by using the RNA as a **tracer** in a reassociation experiment. A very small amount of radioactively labeled RNA (or DNA) is included together with a much larger amount of cellular DNA. Because the cellular DNA is present in great excess, the amount withdrawn into a hybrid with the tracer RNA does not change the concentration of single-stranded DNA sequences. The reaction is governed by the reassociation of the complementary cellular DNA strands (as though the tracer RNA were not present). This is described as a **DNA-driven** reaction. The reassociation of the whole DNA is followed by the usual means (change in optical density or retention on hydroxyapatite). The reassociation of the tracer is followed by the entry of its radioactivity into duplex form.

The tracer RNA (or DNA) participates in the reaction as though it were just another member of the sequence component from which it was transcribed. The component is identified by comparing the radioactive tracer curve with the reassociation curve for whole DNA. Thus the Cot values at which the labeled RNA hybridizes can be taken to identify the repetition frequencies of the corresponding genomic sequences. When an individual mRNA molecule is used, it hybridizes with a single $Cot_{1/2}$ value determined by the repetition frequency of the gene or genes representing it. For a unique gene, this will be the same as nonrepetitive DNA; while for a repeated gene, there will be a corresponding decrease in the $Cot_{1/2}$. When a population of mRNA molecules is used, each mRNA hybridizes with a characteristic $Cot_{1/2}$, so that overall the curve is the sum of the individual components. This can be resolved in the same way as the reassociation curve of the genomic DNA itself.

With a population of mRNAs, a typical result resembles that shown in **Figure 18.1.** A small proportion of

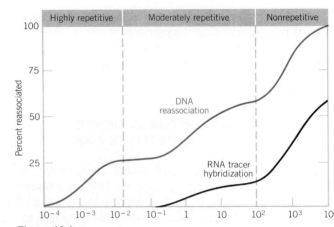

Figure 18.1
The hybridization of an mRNA tracer preparation in a reassociation curve shows that most mRNA sequences are derived from nonrepetitive DNA, the remainder from moderately repetitive DNA, and none from highly repetitive DNA.

the RNA, generally 10% or less, hybridizes with a Cot$_{1/2}$ corresponding to moderately repetitive sequences. The major component hybridizes with a Cot$_{1/2}$ identical with or very close to that of nonrepetitive DNA. Usually this represents up to 50% of the total RNA.

What is the relationship between the mRNA sequences that hybridize with nonrepetitive DNA and those that hybridize with repetitive DNA? They can be separated into different classes (by retrieving the RNA that hybridizes first). This shows that independent molecules are involved: one class represents genes that lie in nonrepetitive DNA, and the other corresponds to genes that lie in repetitive DNA.

The main problem in interpreting these results is to decide on the nature of the sequences that do not hybridize at all, which often correspond to up to half of the total RNA. Probably most of this material is derived from the nonrepetitive sequences. One reason why they fail to hybridize may be that they are mRNAs present in very large amounts, large enough to interfere with the assumption that their quantity in the tracer is small relative to the DNA. In other words, if the DNA excess is not sufficient in a DNA-driven reaction, some of the RNA may fail to hybridize.

From these results, it is clear that most of the mRNA,

perhaps up to 80%, is derived from sequences that reassociate with the nonrepetitive DNA component. Because of the difficulties in detecting very low degrees of repetition (especially when the repeated copies are related rather than identical, so that their reassociation is slowed by mispairing), these genes are not proven to be absolutely unique, but at least should be present in fewer than three or four copies per genome. Only a small proportion of genes are openly present in repetitive DNA; whether these multiple copies are identical or related is not revealed by the hybridization kinetics. Note that because these experiments are conducted in terms of the *total mass of mRNA,* and different genes are expressed as different amounts of mRNA, this technique does not prove that 80% of the *genes* lie in the nonrepetitive DNA component, although this class does appear to comprise a majority.

The hybridization kinetics of individual mRNAs have been determined for several genes, most notably the α and β globins of several mammalian species. Usually the results suggest that there are one or two copies of each gene. Actually, this underestimates the number of related sequences in the genome, since experiments to isolate the genes directly often identify further sequences related to them (see Chaper 21).

HOW MANY NONREPETITIVE GENES ARE EXPRESSED?

The number of nonrepetitive DNA sequences represented in RNA can be determined directly in terms of the proportion of the DNA that is able to hybridize with RNA. When a small amount of single-stranded DNA is hybridized with a large amount of RNA, all the sequences in the DNA that are complementary to the RNA should react to form an RNA-DNA hybrid. This is called a **saturation experiment,** because the critical feature is that the excess of RNA should be great enough to ensure that every available complementary DNA sequence actually is hybridized. In this **RNA-driven** reaction, the controlling parameter is the product of RNA concentration and time, known as the **Rot.** This is exactly analogous to the use of Cot values to describe DNA-driven reactions.

The usual way to follow a saturation hybridization is

to plot the percent of hybridizing nonrepetitive DNA against the Rot value. A curve of this sort is shown in **Figure 18.2.** In this example, the reaction is complete by a Rot value of about 300, but it is necessary to extend it further to be sure that a plateau has been reached. No further DNA is hybridized as the Rot is increased up to 1200, which demonstrates that all the available DNA has indeed been hybridized.

From the proportion of hybridized DNA, we can calculate the number of genes represented in the mRNA population used to drive the reaction. At saturation, 1.35% of the available nonrepetitive DNA is hybridized. Because only one strand of DNA is transcribed into RNA, only half of the DNA in principle is potentially able to hybridize with RNA (the other half is identical in sequence with it). Thus 2.70% of the total sequences of nonrepetitive DNA are represented in the mRNA.

The nonrepetitive DNA itself represents 75% of a genome of 8.1×10^8 bp. Thus the complexity of DNA represented in the RNA population is $0.027 \times 0.75 \times 8.1 \times 10^8 = 1.7 \times 10^7$ bp. The RNA population was a preparation of mRNA taken from the polysomes, of average length 2000 bases. Thus the total number of different messengers is $1.7 \times 10^7 / 2000 = 8500$. This corresponds to the total number of genes expressed in the tissue from which the mRNA was taken, the gastrula embryo of the sea urchin.

Similar experiments have been performed for quite a large number of cases. For a lower eucaryote—for example, yeast—the total number of expressed genes is relatively low, roughly 4000. For somatic tissues of higher eucaryotes, the number usually is between 10,000 and 15,000. This is similar for plants and for vertebrates. The total amount of DNA represented in mRNA typically is therefore a very small proportion of the genome, of the order of 1–2%. The only substantial exception to this type of value is presented by mammalian brain, where much larger numbers of genes appear to be expressed, although the exact quantitation is not certain.

This type of experiment can be performed only for nonrepetitive DNA, in which each DNA sequence can be hybridized only by the mRNA originally derived from it. Because of the presence of multiple copies of identical or related sequences in moderately repetitive

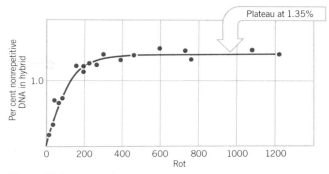

Figure 18.2

Hybridizing an excess of mRNA with nonrepetitive DNA until saturation is reached shows that only a small proportion of the DNA is represented in the mRNA.

The experiment is performed by using purified nonrepetitive DNA that carries a radioactive label; so the proportion of radioactivity entering the hybrid is measured. The data in this figure show the reaction between nonrepetitive DNA of the sea urchin *S. purpuratus* and the mRNA extracted from polysomes of the gastrula embryo.

DNA, an RNA derived from this component might be able to hybridize other genomic sequences in addition to the particular sequence from which it was originally transcribed. At saturation, therefore, a large number of additional DNA sequences could be contained in the hybrid, thus overestimating the number of expressed genes. In fact, to be sure that this has not happened, the DNA that has been saturated with RNA usually is recovered (by degrading the RNA) and then used in a reassociation experiment to show that its $Cot_{1/2}$ does indeed correspond to that expected of nonrepetitive DNA. This is called a **playback experiment.** So the number of expressed genes estimated by this technique refers strictly to the majority whose sequences lie in nonrepetitive DNA.

ESTIMATING GENE NUMBERS BY THE KINETICS OF RNA-DRIVEN REACTIONS

Because RNA is in excess over the DNA, in an RNA-driven reaction the RNA concentration remains essentially unchanged by the small amount drawn into hybrid form during the reaction. This means that a saturation analysis follows first-order kinetics, as described by the equation

$$\frac{R}{R_o} = e^{-k \cdot R_o t} \qquad \textbf{(8)}$$

so that when the reaction is half complete, and $R/R_o = 0.5$,

$$k = \frac{ln2}{Rot_{1/2}} \qquad \textbf{(9)}$$

However, the $Rot_{1/2}$ displayed by an RNA-driven saturation reaction cannot be used to determine the complexity of the RNA population. The reason is that different RNA sequences are present at different concentrations, depending on the characteristic levels at which their genes are expressed. This means that the R_o as measured by the mass of RNA is not applicable for any particular sequence. In fact, usually quite a large proportion of the mass of mRNA is provided by just a few sequences. These very quickly saturate their complements in DNA, leaving the reaction to be driven by the remaining sequences. Thus the real mass of RNA driving the reaction is much less than the measured R_o. This means that the measured $Rot_{1/2}$ is much too high; it includes the large mass of RNA that does not participate in the reaction because its complementary sequences are very quickly saturated.

Another technique has been developed to allow the kinetics of an RNA-driven hybridization reaction to be used to determine complexity. The protocol is illustrated in **Figure 18.3.** It makes use of a labeled "cDNA" prepared by reverse transcription of the mRNA. We shall discuss this technique in Chapter 19; all we need to know for now is that the cDNA consists of single-stranded DNA that is complementary to the mRNA.

The hybridization consists of a reaction between excess mRNA and the labeled cDNA previously prepared from it. Each mRNA sequence should be represented in the cDNA population with a frequency corresponding to its proportion of the RNA. Since all of the cDNA sequences have been derived from the mRNA, all of the labeled cDNA should be driven into the hybrid form. For a single component, the $Rot_{1/2}$ of reaction is proportional to complexity in the same way that $Cot_{1/2}$ is determined by complexity in a reassociation reaction. Thus the complexity of an unknown RNA population may be determined by comparing its $Rot_{1/2}$ with the $Rot_{1/2}$ of a standard reaction.

Figure 18.3
The hybridization of excess mRNA × cDNA is used to determine mRNA complexity.

For example, the reaction of excess globin mRNA with its cDNA is often used as a standard. Then we can write

$$\frac{\text{Complexity of any RNA}}{\text{Complexity of globin mRNA}}$$

$$= \frac{Rot_{1/2} \text{ of any RNA}}{Rot_{1/2} \text{ globin mRNA}} \qquad \textbf{(10)}$$

This is exactly analogous with equation **5** of Chapter 17 for determining the complexity of unknown preparations by comparing the $Cot_{1/2}$ values of unknown and standard DNAs. In the Rot analysis also, the calculated complexity gives the total length of different sequences without making any implication about organization in terms of individual sequences.

Just as with a DNA reassociation curve, a single component hybridizes over about two decades of Rot values, and a reaction extending over a greater range must be resolved by computer curve-fitting into individual components. Again this represents what may really be a continuous spectrum of sequences. To determine the complexity of each component, it is treated as though it were the only one. This is done by converting the measured $Rot_{1/2}$ value to the $Rot_{1/2}$ value that would be shown if the component were pure; the observed $Rot_{1/2}$ is multiplied by the proportion of the reaction occupied by the particular component. Essentially this corrects the R_o so that it includes only the mass of RNA present in this component.

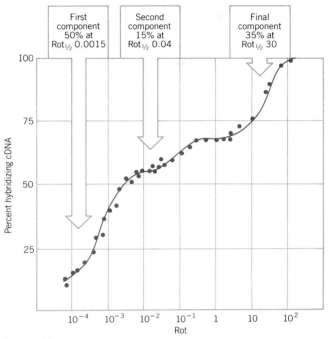

First component 50% at Rot$_{1/2}$ 0.0015

Second component 15% at Rot$_{1/2}$ 0.04

Final component 35% at Rot$_{1/2}$ 30

Figure 18.4

Hybridization between excess mRNA and cDNA identifies several components, each characterized by the Rot$_{1/2}$ of reaction.

The reaction is expressed relative to the final amount of cDNA hybridized, but usually this is close to 100%, almost always more than 90%. These data show the reaction using mRNA derived from the chick oviduct.

An example of an excess mRNA × cDNA reaction that generates three fairly clear components is given in **Figure 18.4.** The control for this reaction was the hybridization between purified ovalbumin mRNA (2000 bases long) and its cDNA, which had a Rot$_{1/2}$ of 0.0008. This means that a Rot$_{1/2}$ of 0.0004 is demanded for each 1000 bases of complexity.

The first component has an observed Rot$_{1/2}$ of 0.0015 and represents 50% of the total reaction. Thus the Rot$_{1/2}$ applicable to the purified component should be 0.5 × 0.0015 = 0.00075, almost exactly the same as the control Rot$_{1/2}$ for ovalbumin. This suggests that the first component is in fact just ovalbumin mRNA! This fits well with the known properties of the oviduct tissue, where ovalbumin is indeed the predominant product of gene expression, with its RNA corresponding to about half of the messenger mass.

The next component has an observed Rot$_{1/2}$ of 0.04, and provides 15% of the reaction. Thus its purified Rot$_{1/2}$ should be 0.15 × 0.04 = 0.006. Its complexity is therefore 0.006/0.0004 × 1000 = 15,000 bases. This would correspond to 7–8 mRNA species of average length 2000 bases.

The last component has a Rot$_{1/2}$ of 30 and provides 35% of the reaction; so its purified Rot$_{1/2}$ of 0.35 × 0.30 = 10.5 corresponds to a complexity of 10.5/0.0004 × 1000 = 26,000,000 bases. This corresponds to about 13,000 mRNA species of average length 2000 bases.

From this analysis, we can see that about half of the mass of mRNA in the cell represents a single mRNA, about 15% of the mass is provided by a mere 7–8 mRNAs, and about 35% of the mass is divided into the large number of 13,000 mRNA species. It is therefore obvious that the mRNAs present in the various components must be present in very different amounts.

GENES ARE EXPRESSED AT WIDELY VARYING LEVELS

The average number of molecules of each mRNA per cell is called its **abundance** or **representation.** It can be calculated quite simply if the total mass of RNA in the cell is known. For each component, total mass = abundance × complexity, so that as a general equation,

$$\text{Abundance} = \frac{\text{gm of mRNA in cell} \times \text{fraction in component} \times 6 \times 10^{23}}{\text{Complexity of component in daltons}} \quad \textbf{(11)}$$

The equation is usually expressed in this form, since total mRNA is measured (chemically) in picograms, while complexity is determined (by hybridization) in bases or daltons.

In the example shown in Figure 18.4, there are 0.275 pg mRNA per cell. This corresponds to 100,000 copies of the first component (ovalbumin mRNA), 4000 copies of each of the 7–8 mRNAs in the second component, but only about 5 copies of each of the 13,000 mRNAs that constitute the last component.

The oviduct is a somewhat extreme case, with so much of the mRNA represented in only one mRNA species, but in most cells there is a large part of the mass that contains very few mRNAs. This **abundant** component typically consists of fewer than 100 different mRNAs present in more than 1000 or even 10,000 copies per cell. It often corresponds to a major part of the mass, approaching 50% of the total mRNA. The abundant component may be resolved into two components, as seen in the example of the oviduct, or may seem to constitute one component; but for mRNAs that are frequently expressed, probably each gene has a characteristic and different level.

In all cases, around half of the mass of the mRNA consists of a large number of sequences, of the order of 10,000, each represented by only a small number of copies in the mRNA—say, less than 10. This is the **scarce** or **complex** mRNA class. (It is this class that drives a saturation reaction.)

The total numbers of expressed genes as estimated by the saturation technique or the kinetic technique usually are in fairly good agreement. The kinetic technique provides a lower estimate (because some sequences that are very scarce may fail to react). The saturation technique provides a higher estimate (because there are always likely to be some sequences included that are present in more than one copy per genome). Thus for chick oviduct the kinetic technique identifies about 13,000 mRNAs, while the saturation technique corresponds to 15,000.

OVERLAPS BETWEEN mRNA POPULATIONS

Many somatic tissues of higher eucaryotes have an expressed gene number in the range of 10,000 to 20,000. How much overlap is there between the genes expressed in different tissues? For example, the expressed gene number of chick liver is in the range of 11,000 to 17,000, compared with the value we have just quoted for oviduct of 13,000 to 15,000. How many of these two sets of genes are identical? How many are specific for each tissue?

We see immediately that there are likely to be substantial differences among the genes expressed in the abundant class. Ovalbumin, for example, is synthesized only in the oviduct, not at all in the liver. This means that 50% of the mass of mRNA in the oviduct is specific to that tissue. Taking a more general view, in many cases the abundant mRNAs represent genes whose proteins are a major product of the cell type. In terms of the mass of mRNA and of protein synthesized, there may therefore be appreciable differences between cell types. But the abundant mRNAs invariably represent only a small proportion of the expressed genes. In terms of the total number of genes of the organism, and of the number of changes in transcription that must be made between different cell types, we need therefore to know the extent of overlap between the genes represented in the scarce mRNA classes of different cell phenotypes.

There is no really sensitive technique for measuring overlaps between mRNA populations en masse, but some reasonable estimates have been obtained by adapting the methods used to determine the population complexity.

Additive saturation experiments show how many sequences are different between two populations. **Figure 18.5** shows the example of chick liver and oviduct. With mRNA derived from each individual tissue, 2.05% and 1.80% of nonrepetitive DNA is saturated in the two cases, respectively. If the two mRNA populations comprised entirely different sequences, together they should saturate 2.05 + 1.80 = 3.85% of the nonrepetitive DNA. But the actual value is 2.4%, only slightly greater than that of liver alone. Thus about 75% of the sequences expressed in liver and oviduct are the same (though since this is a saturation experiment, the data do not show whether they are present in the same or different abundances in the two tissues). In other words, about 12,000 genes are expressed in both liver and oviduct, about 5000 additional genes are expressed only in liver, and about 3000 additional genes are expressed only in oviduct.

Another way to estimate the extent of overlap is to hybridize the mRNA of some tissue with nonrepetitive DNA. Then the DNA that reacts is isolated to constitute the **mDNA** preparation, while the DNA that does not react constitutes the **null DNA** preparation. These DNA preparations are then hybridized separately with an excess of mRNA from some other tissue. The proportion of the mDNA preparation that reacts identifies the proportion of genes expressed in the second as well as in the first tissue. The amount of the null DNA prep-

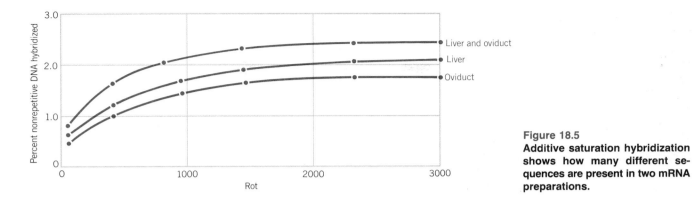

Figure 18.5
Additive saturation hybridization shows how many different sequences are present in two mRNA preparations.

aration that reacts identifies the number of genes expressed in the second, but not in the first, preparation.

Heterologous experiments also can be performed by using excess mRNA of one source to drive a reaction with cDNA representing the mRNA of another source. These experiments show that the sequences that are abundant in one tissue tend not to be abundant in another tissue, although sometimes they are present in the scarce mRNA class. Between mouse liver and kidney, about 90% of the scarce mRNAs are identical, leaving a difference between the tissues of only 1000–2000 in terms of the number of expressed genes. The general result obtained in several comparisons of this sort is that only about 10% of the mRNA sequences of a cell are unique to it. The majority of sequences are common to many, perhaps even all, cell types.

This suggests that the common set of expressed genes, numbering perhaps about 10,000 in a mammal, may comprise functions that are needed in all cell types. Sometimes this type of function is referred to as a **housekeeping** or **constitutive** activity. It contrasts with the activities represented by specialized functions (such as ovalbumin or globin) needed only for particular cell phenotypes. These are sometimes called **luxury** genes. Of course, if we take into account all the various cell phenotypes of the organism, there may certainly be as many luxury functions as housekeeping functions. Still, the total number of genes (at least as represented in nonrepetitive DNA) is unlikely to be more than, say, 2–3 times the number of housekeeping genes—within the range of 20,000 to 40,000.

Some specialized cell types can function with a relatively small number of expressed genes. A detailed examination of expressed genes in the sea urchin *S. purpuratus* has revealed that the oocyte contains about 18,000 mRNA sequences. During embryogenesis, the number of expressed genes falls to about 13,000 at blastula, 8500 at gastrula, and 7000 at pluteus. But in some adult tissues—tubefoot, intestine, coelomocyte—the number of expressed genes falls in the range of only 2500 to 3000. Thus these cells can be maintained with much less than the housekeeping number of 10,000 that has been implied for mammalian cells.

An especially significant feature of this series of cells is that there is very extensive overlap in the expressed sequences, so that genetic development occurs by a progressive reduction in the expressed gene number. Experiments with the mDNA/null DNA assay have shown that most of the genes that are expressed at a later stage have also been expressed at an earlier, embryonic stage. It is even possible that all the structural genes of the organism are expressed in the oocyte, which would imply that the sea urchin genome consists of fewer than 20,000 genes, whose coding sequences correspond to only 3% of the total haploid DNA.

FURTHER READING

A detailed review of the estimates of gene number is presented in **Lewin's** *Gene Expression,* **2,** *Eucaryotic Genomes* (Wiley, New York, 1980, pp. 694–727). An earlier view of these experiments is summarized in **Lewin** (*Cell* **4,** 77–93, 1975).

CHAPTER 19
DEALING WITH DNA

What we might think of as the technology for dealing with DNA has advanced dramatically in the past few years. Among the remarkable range of manipulations to which we can now subject DNA are procedures for reconstructing DNA molecules by joining sequences from one source to other sequences of quite different origin. The product is often described as **recombinant DNA** and the techniques (more colloquially) as **genetic engineering.** They are equally applicable to procaryotes and eucaryotes, although the power of this approach is especially evident with eucaryotic genomes. With these techniques, we can isolate and characterize genes that otherwise would not be accessible.

At the heart of these techniques lies the ability of restriction enzymes to cleave DNA at particular, rather short nucleotide sequences. We have already discussed in Chapter 3 the use of restriction enzymes to make physical maps of DNA and to generate fragments whose nucleotide sequences can be determined directly. Thus when the DNA corresponding to a particular gene (for example) is available, it is now a relatively routine, not to say mundane, task to determine its sequence. Using such information, together with several chemical and biochemical techniques, we can break and rejoin DNA molecules at virtually any desired site.

In this chapter, we shall be concerned with the first steps in characterizing particular genes. How do we isolate the RNA from the cytoplasm or the DNA from the genome that represents a particular gene? Having identified this nucleic acid, how do we proceed to obtain a sufficient amount of material to characterize? Can we use this to produce the gene product *in vitro* or *in vivo*? Can we introduce changes in this sequence that will influence its expression and help reveal the nature of regulatory signals?

ANY DNA SEQUENCE CAN BE CLONED IN BACTERIA

The **cloning** of a fragment of DNA allows indefinite amounts to be produced from even a single original molecule. (A **clone** is defined as a large number of cells or molecules all identical with an original ancestral cell or molecule.) The property that makes cloning of DNA possible is the ability of bacterial plasmids and phages to continue their usual life-style after additional sequences of DNA have been incorporated into their genomes.

The insertion generates a **hybrid** or **chimeric** plasmid or phage, consisting in part of the authentic DNA of the original genome and in part of the additional "foreign" sequences. These chimeric elements replicate in bacteria in exactly the same way as the original plasmid or phage and so can be obtained in large amounts. Copies of the original foreign fragment then can be retrieved from the progeny. Since (with some rare exceptions) the properties of the chimeric species are unaffected by the particular foreign sequences that are involved, virtually any sequence of DNA can be cloned in this way. Because the phage or plasmid is used to "carry" the foreign DNA as an inert part of the genome, it is often referred to as the **cloning vector.**

A critical feature of any cloning vector is that it should possess a site at which foreign DNA can be inserted without disrupting any essential function. The simplest way is to find a restriction enzyme that makes only a single break at such a site in the vector DNA. Then it is necessary to be able to *select* the chimeric genome from the original vector, since otherwise it may be difficult to isolate the small proportion of chimeric genomes from the starting material.

Plasmids are small autonomous elements that replicate in bacteria independently of the bacterial chromosome (see Chapter 31). Their genomes are always circular duplexes of DNA, with a replication control system that maintains them in the bacterium at a characteristic level. There are two general types. **Single-copy** plasmids are maintained at 1 plasmid DNA per host chromosome. **Multicopy** plasmids are present in greater amounts, typically about 10–20 genomes per cell. Some plasmids are under **relaxed replication control,** which means that they accumulate in very large amounts (~1000 genomes per cell) when the bacteria stop growing. These often are used to provide cloning vectors, because of the increased yield of material.

Cleavage at one site in a circular plasmid DNA converts the genome into a linear molecule, as depicted in **Figure 19.1.** Then the two ends can be joined to the ends of a linear foreign DNA, regenerating a circular chimeric plasmid. There is no formal limit on the length of foreign DNA that can be inserted. The chimeric plasmid can be perpetuated indefinitely in bacteria. It may be isolated by virtue of its size or circularity—for example, by gel electrophoresis.

Many plasmids carry genes that specify resistance to antibiotics. This is useful in designing cloning sys-

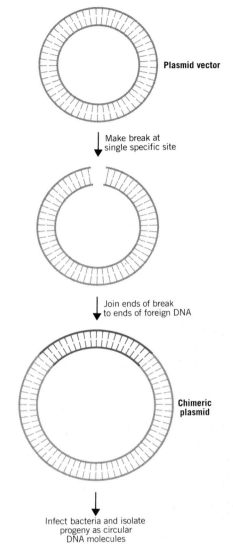

Plasmid vector

Make break at single specific site

Join ends of break to ends of foreign DNA

Chimeric plasmid

Infect bacteria and isolate progeny as circular DNA molecules

Figure 19.1
Plasmid vectors can be used to clone any fragment of DNA that is inserted at an appropriate site.

tems. A common procedure is to use a plasmid that has two genes each specifying resistance to a different antibiotic. One of these is used simply to identify bacteria that carry the plasmid, by virtue of the resistance that the cells acquire to the antibiotic. The other is used to distinguish chimeric plasmids from parental vectors. If the site used to insert foreign DNA lies within this gene, the chimeric plasmid *loses* the antibiotic resistance. Thus a parental vector can be selected by its resistance to both antibiotics; whereas a chimeric plasmid can be selected by its retention of resistance to one antibiotic, but sensitivity to the other.

Often natural plasmids have some, but not all, of the desired properties of a cloning vector. Several series of developments have been undertaken to engineer better vectors from the natural starting material. This may involve the introduction of changes in the replication control system or the addition of genes determining resistance to particular antibiotics. One of the standard cloning vectors most widely used at present is **pBR322,** which was derived by several sequential alterations of earlier cloning vectors. It is a multicopy plasmid that has genes for resistance to tetracycline and ampicillin; several restriction enzymes have unique cleavage sites at useful locations.

Phages provide another type of vector system. Usually the phage is a linear DNA molecule, so that a single restriction break generates two fragments. These are joined together with the foreign DNA to generate a chimeric phage as shown in **Figure 19.2.** The procedure for isolating chimeric phage genomes is to allow the phage to proceed through the lytic cycle to produce particles. This has the advantage of conveniently isolating the chimeric genomes. However, it does impose a limit on the length of foreign DNA that can be cloned, because the capacity of the phage head prevents genomes that are too long from being packaged into progeny particles.

To ameliorate this problem, often a fragment of the vector that does not carry any essential phage genes is *replaced* by the foreign DNA, instead of just inserting additional material. This has been taken to a fine art with phage lambda, where a new vector has been created by manipulations of the DNA to produce a shorter genome (lacking nonessential genes) that has only a single restriction site for the enzyme EcoRI. This cloning vector itself actually is *too short* to be packaged into the phage head, which has a minimum as well as maximum length requirement. Thus it is *necessary* for a foreign DNA fragment to be inserted into the cleaved parental vector in order to generate a phage that can be perpetuated as progeny particles. This

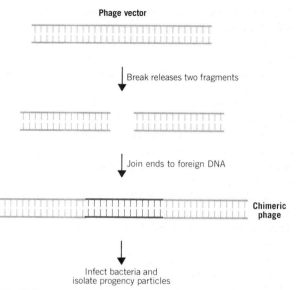

Phage vector

Break releases two fragments

Join ends to foreign DNA

Chimeric phage

Infect bacteria and isolate progeny particles

Figure 19.2
Phage vectors can be used to clone a foreign DNA that is inserted into a nonessential region of the genome.

creates an automatic selective system for obtaining chimeric phage genomes.

The utility of this type of vector has been increased by the development of systems for packaging the DNA into the phage particle *in vitro*. An attempt to combine some of the advantages of plasmids and phages has led to the construction of **cosmids.** These are plasmids into which have been inserted the particular DNA sequences (*cos* sites) that are recognized for packaging lambda DNA into its particle. These vectors still can be perpetuated in bacteria in the plasmid form, but can be purified by packaging *in vitro* into phages. They are still subject to the length limitation imposed by the particle head, but it is not necessary for the genome to carry the genes for lytic development.

We have dealt with cloning vectors in the context of using bacterial hosts. Sometimes it is useful to use a eucaryotic host. The only known authentic eucaryotic plasmids reside in yeast. By reconstruction of DNA, some "dual-purpose" or "shuttle" plasmids have been obtained that have the necessary sequences for surviving in either *E. coli* or *S. cerevisiae*. Thus the one vector can be used with either host.

CONSTRUCTING THE CHIMERIC DNA

To join a foreign DNA fragment to a cloning vector requires a reaction between the ends of the fragment and the vector. This can be accomplished by generating complementary sequences on the fragment and vector, so that they can recombine into a chimeric DNA when mixed together.

The most common method is to use restriction enzymes that make **staggered cuts** to generate short, complementary single-stranded **sticky ends.** The classic example is the enzyme EcoRI, which cleaves each of the two strands of duplex DNA at a different point. These sites lie on either side of a short palindromic sequence that is part of the site recognized by the enzyme.

In a sequence of DNA in which *N* indicates bases not involved in recognition, the EcoRI enzyme cuts its recognition site (highlighted in color) at the bonds indicated by the vertical arrows.

$$\downarrow$$

5′ . . . pNpNpNpGpApApTpTpCpNpNpN . . . 3′
3′ . . . NpNpNpCpCpTpApApGpNpNpNp . . . 5′

$$\uparrow$$

The DNA fragments on either side of the target site fall apart, and because of the stagger of individual cutting sites, they have protruding single-stranded regions that are complementary—the sticky ends indicated by the color.

pApApTpTpCpNpNpN . . . . 3′
GpNpNpNp . . . 5′

and

5′ . . . pNpNpNpG
3′ . . . NpNpNpCpTpTpApAp

Their complementarity allows the sticky ends to anneal by base pairing. When two different molecules are cleaved with EcoRI, the *same* sticky ends are generated on both. This enables one to anneal with the other, as illustrated in the protocol of **Figure 19.3.** The procedure generates a chimeric plasmid that is intact except for the lack of covalent bonds between the vector and the foreign DNA. The missing bonds are sealed by the action of the enzyme DNA ligase *in vitro*. This technique for recombining two DNA molecules has both pros and cons.

An advantage is that the chimeric plasmid possesses regenerated EcoRI sites at either end of the inserted DNA. Thus the foreign DNA fragment can be retrieved rather easily from the cloned copies of the chimeric vector by cleaving with EcoRI.

A disadvantage is that any EcoRI sticky end can anneal with any other EcoRI sticky end. Thus some vectors reform by direct reaction of their ends, without gaining an insert, while others may gain an insert of several foreign fragments joined end to end. It is therefore necessary to select chimeric plasmids that have gained only a single insert.

A problem in relying exclusively on restriction enzymes to generate ends for the joining reaction is that their recognition sites may not lie at convenient points (especially in the foreign DNA sequence). Another method allows any DNA end to be used for recombination with a plasmid.

The plasmid is cleaved as before with an enzyme that recognizes a single site in a suitable location, but this need not necessarily generate staggered ends. To both the 3′ ends of the plasmid DNA, poly(dA) is added, by using the enzyme **terminal transferase** with the precursor dATP. In the same way, poly(dT) is added to the 3′ ends of the foreign DNA molecule that is to be inserted; these ends can have been generated in any convenient manner. As **Figure 19.4** demonstrates, the poly(dA) on the plasmid can anneal *only* with the poly(dT) on the insert fragment. Thus only one reaction is possible: the insertion of a single foreign fragment into the vector.

A drawback of this technique is that it is not so easy to retrieve the inserted fragment from the cloned chimeric plasmid, because the recognition site for the restriction enzyme in the original vector has been abolished by the insertion of the foreign DNA. The inserted sequence is flanked on either side by the poly(dA): poly(dT) paired region. The technique also

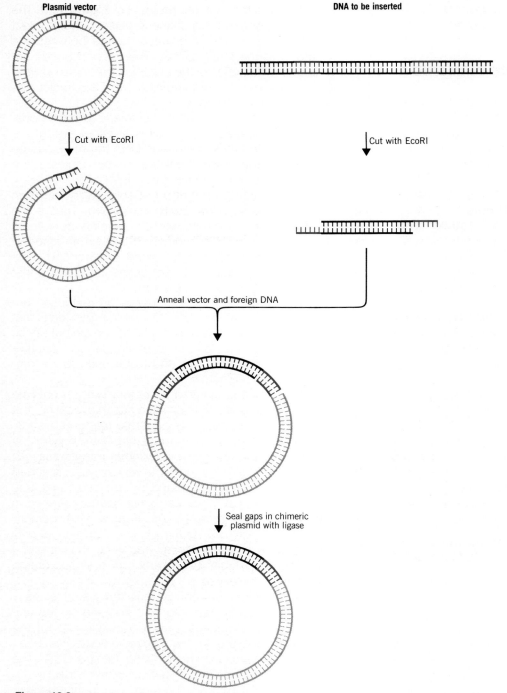

Figure 19.3
Any DNA sequence that lies between EcoRI cleavage sites can be inserted into a plasmid vector that has an EcoRI cleavage site by cutting and annealing the two DNA molecules.

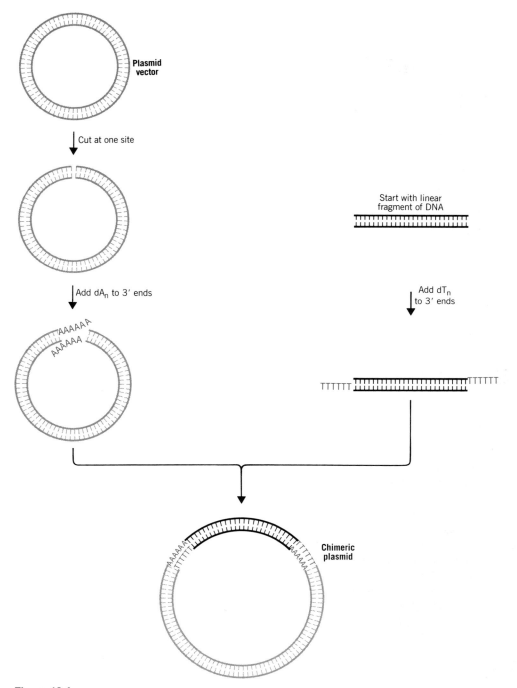

Figure 19.4
The poly(dA:dT) tailing technique allows any two DNA molecules to be joined by adding poly(dA) to the 3′ ends of one and adding poly(dT) to the 3′ ends of the other.

can be used with poly(dC) and poly(dG), which has the advantage that, if the plasmid has been cleaved with PstI, the poly(dC):poly(dG) joining regenerates the PstI cleavage site on either side of the insert.

Another useful technique is **blunt-end ligation.** This relies on the ability of the T4 DNA ligase to join together two DNA molecules that have blunt ends, that is, they lack any protruding single strands. (This reaction is in addition to the activity of joining broken bonds within a duplex). Blunt-end ligation directly joins the fragments generated by cleaving DNA with restriction enzymes that cut across both DNA strands at the same position. The great advantage of this technique is that any pair of ends may be joined together, irrespective of sequence. This is especially useful when we want to join two defined sequences without introducing any additional material between them. A problem inherent in this technique is that there is no control over which pairs of blunt ends are joined together, so it is necessary first to perform the reaction and then to isolate the desired products from among the other products.

There are numerous variations of these methods. One technique uses short DNA duplexes ("linkers") that contain the EcoRI (or some equivalent) recognition palindrome. These linkers can be synthesized chemically, and then are added covalently to the ends of a plasmid or an insert by blunt-end ligation. This allows the inserted DNA to be retrieved by cleavage with EcoRI, but without imposing any restrictions on the original choice of sites to generate the ends. With sufficient manipulation, it is now possible to insert any foreign DNA fragment into any particular vector site, and to arrange for retrieval of the fragment when necessary.

When a foreign DNA fragment is inserted into a plasmid, it can be connected in either orientation, that is, with either of the ends of the foreign DNA joined to either of the ends of the plasmid. This does not matter when the purpose of cloning is simply to amplify the inserted sequence. However, it may be important when the experiment is designed to obtain expression of the foreign DNA. In this case, populations of plasmids carrying the plasmid in either orientation may be obtained via random insertion, after which they are characterized by restriction mapping to identify the desired class. Or the experiment may be designed so as to permit insertion in one orientation only. For ex-

ample, each of the DNAs, vector and insert, may be cleaved with two restriction enzymes that make different sticky ends, to generate the type of pattern where each DNA has the sequence

$$\text{End 1} \underline{\hspace{4cm}} \text{End 2}$$

Now if only the two end-1 sequences can anneal together, and only the two end-2 sequences can anneal, the insertion can take place only in one orientation, generating the chimeric plasmid

$$
\begin{array}{ccc}
\text{End 1} & \text{——Insert——} & \text{End 2} \\
\| & & \| \\
\text{End 1} & \text{——Plasmid——} & \text{End 2}
\end{array}
$$

The inserted sequence has replaced lost material that was between the sites at which end-1 and end-2 were generated in the original plasmid.

COPYING mRNA INTO DNA

For the purpose of obtaining a DNA sequence that represents a particular protein, the ideal approach is to start with the mRNA, which is, after all, the template used to produce the protein *in vivo.* The existence of reverse transcription makes it possible to synthesize a duplex DNA from an mRNA. This is especially easy for mRNAs that carry a poly(A) tail at the 3′ end, as illustrated in **Figure 19.5**

First, a **primer** is annealed to the poly(dA). This is a short sequence of oligo(dT), and its purpose is to provide a free 3′ end that can be used for extension by the enzyme reverse transcriptase. (Like other enzymes that synthesize DNA, reverse transcriptase cannot initiate formation of a polynucleotide chain without a priming end.) The enzyme engages in the usual 5′–3′ synthesis, adding deoxynucleotides one at a time, as directed by complementary base pairing with the mRNA template.

The product of the reaction is a hybrid molecule, consisting of a template RNA strand base paired with the complementary DNA strand. The only practical problem is the propensity *in vitro* of reverse transcriptase to stop at various points before it has reached the 5′ end of the mRNA. In this case, the resulting reverse transcript falls short of representing the entire mRNA, because it lacks some of the sequences com-

5' mRNA 3'
AAAAA$_n$

↓ Anneal primer

AAAAA$_n$
TTTTT$_n$

End of cDNA
hooks round

↓ Make DNA copy with
reverse transcriptase

AAAAA$_n$
TTTTT$_n$

↓ Treat with alkali
to remove RNA

TTTTT$_n$

↓ Use DNA polymerase to extend
hook into complementary strand

TTTTT$_n$

↓ Treat with S1 nuclease
to cleave hook

TTTTT$_n$

This is duplex DNA copy of original mRNA

Figure 19.5
mRNA can be copied into double-stranded DNA.

plementary to the 5' end. However, by judicious adjustment of the experimental conditions, often it is possible to persuade reverse transcriptase to proceed all the way.

A useful reaction tends to occur at the end of the reverse transcription. Upon reaching the end of the mRNA, the enzyme may cause the reverse transcript to "loop back" on itself, by using the last few bases of the reverse transcript as a template for synthesis of a complement. That is, the end of the complementary DNA is used to direct synthesis of a short sequence that is identical with the mRNA and displaces it. This creates a short hairpin, usually 10–20 bp long, at this end of the reverse transcript.

At this juncture, the original mRNA is degraded by treatment with alkali (a procedure that does not affect DNA). This produces a single-stranded DNA that is complementary to the mRNA; it is called **cDNA.** The hairpin at its 3' end provides a natural primer for the next step, the use of the enzyme DNA polymerase I of *E. coli* to convert the single-stranded cDNA into a duplex DNA via synthesis of the complementary strand. In this reaction, the enzyme uses the cDNA as tem-

plate for synthesis of a sequence identical with the original mRNA. The product is a duplex molecule with a hairpin at one end. The hairpin is cut by the enzyme S1 nuclease (which specifically degrades single-stranded DNA) to generate a conventional DNA duplex.

The duplex DNA can be cloned to generate large amounts of a synthetic gene that represents the mRNA sequence in the form of duplex DNA. This is called a **cDNA clone.** (From this terminology, a somewhat looser use of the term "cDNA" has emerged, being taken to describe the duplex insert and not just the original single-stranded reverse transcript.)

ISOLATING SPECIFIC GENES FROM THE GENOME

One of the principal uses of cloning technology is to isolate individual genes directly from the genetic material. Any particular gene represents only a very small part of a eucaryotic genome. For example, in a typical mammal, the size of the genome is about 10^9 bp, so that a single gene of (say) 5000 bp represents only 0.00005% of the total nuclear DNA. To identify such a tiny proportion, we need to have a very specific **probe** that reacts *only* with the particular sequence in which we are interested, to pick it out from the vast excess of other sequences. The usual technique is to use a highly labeled radioactive probe of RNA or DNA, whose hybridization with the gene is assayed by autoradiography.

The crux of the matter is to obtain the mRNA that represents a particular protein. This can be quite difficult when the product is rare. There are several techniques for isolating an mRNA via the properties of its product, including the sensitive detection of products after injecting the RNA into a *Xenopus* oocyte. A powerful technique that isolates cDNA complementary to an mRNA is **hybrid-arrested translation;** this relies on the ability of the cDNA to *prevent* translation of the mRNA by hybridizing to it, with the result that the product disappears from an *in vitro* translation system.

Originally, mRNA itself was used as a probe for gene isolation. This suffers from several practical disadvantages. The mRNA is never completely pure and may be hard to obtain in sufficient amounts. In fact, the technique works well enough for genes whose mRNAs are abundant, but cannot be used for genes that are expressed at low levels. Another difficulty is that often the mRNA cannot be obtained with a sufficiently intense radioactive label.

The use of mRNA as a probe has therefore been superseded by the use of a cloned cDNA copy of the mRNA. Large amounts of purified material can be obtained by the techniques we have already described. This material can be labeled *in vitro* to achieve a very high specific radioactivity.

The first step toward identifying the gene corresponding to a particular probe is to break the DNA of the genome into fragments of a manageable size. It is desirable to obtain the gene in as few fragments as possible (ideally only one). Usually the maximum genome lengths that can be handled are in the range of 15–20 kb. Sometimes it is not possible to obtain a gene in the form of a single fragment, and then its structure must be determined by piecing together the information gained from its various fragments.

Two techniques can be used to fragment a genome. One is to make a restriction digest. Then every fragment ends in a site that was recognized by that particular enzyme. An alternative is to shear the DNA mechanically, by introducing breaks in it at a frequency that is controlled by the conditions of shearing. Then the sites of breakage are entirely random.

When the DNA of a entire genome is digested with a restriction enzyme, the frequency of breakage is controlled by the length of the sequence recognized by the enzyme. The longer the sequence, the less often it occurs by chance. For example, the probability that a particular 4 bp sequence will occur is $0.25^4 = 1/256$, so that an enzyme with this short recognition sequence will cleave DNA rather frequently. The frequency declines to 1/1000 for a 5 bp sequence and to 1/4000 for a 6 bp sequence. (This calculation assumes that each base is equally well represented in DNA, which usually is not the case. The frequency of cutting can be decreased by using an enzyme whose target sequence contains base pairs that are less predominant in the base composition of the DNA, and vice versa.)

The distribution of sites recognized by an enzyme becomes a matter of chance taken over the genome as a whole. So a restriction digest of eucaryotic DNA generates a continuum of fragments. When electrophoresed on a gel, these fragments form a smear in which no distinct bands are evident (with the exception of some repeated sequences, which are discussed in Chapter 24).

When a specific probe is available, however, it is possible to detect the corresponding sequences in the restriction smear. The procedure is illustrated in **Figure 19.6.** The key feature is the denaturation of

Figure 19.6
Southern blotting allows DNA fragments corresponding to a particular probe to be isolated directly from a digest of the DNA of the genome.

Cut genome into random fragments of DNA

Electrophorese on agarose gel

Smear of DNA without distinct bands

"Blot" onto nitrocellulose

Hybridize with radioactive probe for particular sequence

Autoradiography identifies reacting sequences

Obtain equivalent DNA fragments for characterization from an original agarose gel

DNA to give single-stranded fragments that are transferred from the agarose gel to a nitrocellulose filter on which they become immobilized. The procedure used to transfer the DNA is somewhat akin to blotting, and this is used colloquially as a description of the procedure. When performed with DNA, it is known as **Southern blotting** (named for the inventor of the procedure).

The agarose gel is placed on a filter paper that has been soaked in a concentrated salt solution. Then the nitrocellulose filter is placed on top of the gel, and some dry filter paper is placed in contact with the nitrocellulose. The salt solution is attracted to the dry filter paper. To get there, it must pass through the agarose gel and then through the nitrocellulose filter. The DNA is carried along with it, but becomes trapped in the nitrocellulose.

When immobilized on the nitrocellulose, the DNA can be hybridized *in situ* with a radioactive probe. Only those fragments complementary to a particular probe will hybridize with it. Because the probe is radioactive, the hybridization can be visualized by autoradiography. Each complementary sequence gives rise to a labeled band at a position determined by the size of the DNA fragment.

The technique can also be performed with RNA. To blot RNA from agarose onto a medium suitable for hybridization, some changes in the technique are necessary. The procedure then is known as **Northern blotting.**

SHOTGUN CLONING TO FORM LIBRARIES

There are practical difficulties in directly isolating genomic fragments corresponding to a probe, and a powerful technique for isolating a particular gene is to reverse the order of events and to clone the genome first. Then the clone(s) containing a particular sequence is selected. Vectors carrying DNA derived from the genome itself are called **genomic** or **chromosomal DNA clones** (in distinction to the cDNA clones that are representations of the mRNA).

Cloning an entire genome (as opposed to specific fragments) is often called a **shotgun experiment.** It is done by breaking the whole genome into fragments of a manageable size. Then the fragments are inserted into a cloning vector to generate a population of chimeric vectors. A set of cloned fragments of this sort is called a **genome library.** Once a library has been obtained with either a phage or plasmid vector (more often a phage, because it is easier to store the necessary large numbers of chimeric DNAs), it can be perpetuated indefinitely, and is readily retrieved whenever a new probe is available to seek out some particular fragment.

There are technical difficulties in cloning the fragments generated by mechanical shear; it is easier to insert restriction fragments into the vector. However, one problem is that restriction sites may occur at inconvenient locations—for example, in the middle of a gene that is to be cloned. One way to avoid this is to use more than one restriction enzyme, that is, to repeat the experiment with different enzymes whose recognition sites lie at different locations. But this is time consuming, and when a long sequence is involved, it may be difficult to find an enzyme that does not cleave within it.

To make a useful library by restriction digests, therefore, a trick is employed to control the frequency of cutting so as to obtain longer fragments. An enzyme with a short (4 bp) recognition sequence is used under conditions that generate a *partial* digest. Any particular target site is cleaved only occasionally, so that the presence of a target site in some sequence does not prevent it from being obtained without breakage. The infrequent cleavage at each site, together with the frequent distribution of sites, means that the fragment distribution approaches an entirely random cleavage of the genome. But each fragment ends in the same sequence, which can contain a sticky end and is therefore useful for cloning.

The number of random fragments that must be cloned to ensure a high probability that *every* sequence of the genome is represented in at least one chimeric plasmid decreases with the fragment size and increases with the genome size and the desired probability. For a probability level of 99%, 1500 cloned fragments are needed with *E. coli*, but the necessary library size increases to 4600 with yeast, 48,000 with *D. melanogaster*, and 800,000 with mammals. Libraries of cloned fragments reaching this probability level have been established in all these cases.

How is a particular genomic clone to be selected

from the library? One technique is called **colony hybridization.** This is illustrated in **Figure 19.7.** Bacterial colonies carrying chimeric vectors are lysed on nitrocellulose filters. Then their DNA is denatured *in situ* and fixed on the filter. The filter is hybridized with a radioactively labeled probe that represents the desired sequence (usually a cloned cDNA). Any colonies in which it occurs are visualized as dark spots by autoradiography. Then the corresponding chimeric vectors can be recovered from the original reference library.

To react with a probe, a genomic clone needs only to carry some of the probe's sequence (usually the minimum is reckoned to be about 50 bp). Indeed, when a eucaryotic gene is large, it is likely to be fragmented during the formation of a library, and the various fragments will appear in different clones that react with the probe. None of the genomic clones may have the entire sequence represented in the probe. In this case, it is necessary to reconstruct the original genomic sequence by taking advantage of the overlaps that occur between the individual fragments, all of which are likely to have different ends.

A critical assumption underlying this approach is that a cloned eucaryotic sequence will be perpetuated with complete fidelity in bacteria. This is borne out by the data available. There are some exceptional cases in which deletions have occurred during passage through the bacterial host.

Figure 19.7
Colony hybridization allows a chimeric plasmid carrying a particular sequence to be selected by its complementarity with a radioactively labeled probe.

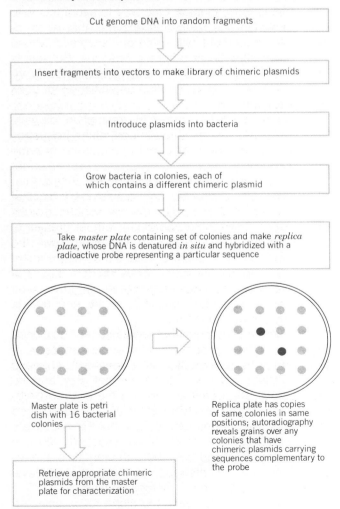

Master plate is petri dish with 16 bacterial colonies

Replica plate has copies of same colonies in same positions; autoradiography reveals grains over any colonies that have chimeric plasmids carrying sequences complementary to the probe

EUCARYOTIC GENES CAN BE TRANSLATED IN BACTERIA

Because the genetic code is universal, a given coding sequence should have the same meaning in all situations. (The sole exception lies with mitochondrial genes, where there are differences in the genetic code, as described in Chapter 4). Thus when an intact coding sequence for a eucaryotic protein is carried on a chimeric vector, it should be possible to transcribe the sequence into an mRNA that can be translated within the bacterial host. The only caveats are that modifications made to the protein in its natural habitat may not occur, and, of course, there is always the risk that the polypeptide chain may be unstable in the bacterial environment. However, in an appropriate environment, it should be possible to translate any eucaryotic coding sequence into the corresponding polypeptide.

What conditions must be fulfilled to obtain effective expression of eucaryotic genes in bacteria? Because promoters differ in sequence between eucaryotes and procaryotes, the gene to be expressed must be placed under the control of a bacterial promoter. This can be done by incorporating the eucaryotic coding sequence into an existing transcription unit, but usually this manipulation is not sufficient to ensure efficient translation. Generally the eucaryotic insert is therefore placed adjacent to a promoter from which it is transcribed as a monocistronic mRNA. This can be done by finding an appropriate site for insertion within the cloning vector; a widely used example is the promoter for the ampicillin-resistance gene of the vector pBR322 (whose usual product is the enzyme β-lactamase). Another procedure is to insert a bona fide bacterial promoter into the cloning vector for the purpose of expressing the foreign gene. A popular choice is the *lac* promoter of *E. coli,* usually using a mutant version that has an increased frequency of use.

To obtain translation of the resulting mRNA, there must be a proper ribosome binding site in juxtaposition to the AUG initiation codon. Cloning vectors have been constructed in which there is a restriction site for inserting foreign DNA just next to a ribosome binding site. Usually this is arranged so that the inserted DNA contributes the AUG codon. **Figure 19.8** shows that any foreign DNA sequence (procaryotic or eucaryotic) that starts with an AUG codon can be placed at this site and will be transcribed and translated in bacteria. This produces a protein that corresponds precisely to the inserted coding region.

Another technique is to insert a coding sequence shortly after the start of a bona fide protein. This produces a hybrid protein, with an N-terminal region consisting of the bacterial protein whose sequence has been interrupted, and the remainder of the protein corresponding to the foreign coding region. The N-terminal bacterial sequence can be quite short and may be useful. For example, it can provide a signal sequence that ensures secretion of the hybrid protein into the medium; and it may serve to protect the hybrid protein from degradation in the bacterium.

What source should be used to provide the eucaryotic coding sequence? Because many eucaryotic genes are interrupted, the coding sequence in a genomic DNA clone may not form a continuous region. In such cases, the genomic DNA cannot be

Figure 19.8
Any eucaryotic coding sequence can be translated by insertion into a suitable site in a cloning vector.

translated in bacteria. The coding sequence is therefore best obtained from a cDNA clone prepared from the mRNA template, in which the coding sequence is uninterrupted. It is not necessary for this to include any 5′ and 3′ nontranslated regions; the coding sequence from the initial AUG to the termination codon will suffice.

Sometimes we wish to characterize a bacterial promoter, but the products of its natural genes may be difficult to assay. This can be overcome by the reverse type of manipulation, in which the readily assayable *lac* genes are connected to the promoter. Then its activity can be followed by the standard assays for β-galactosidase. A vector that accomplishes this is a variant of phage Mu, called the **Mud phage.**

FURTHER READING

The topics of this chapter are dealt with in detail in **Lewin's** *Gene Expression,* **2,** *Eucaryotic Chromosomes* (Wiley, New York, 1980, pp. 761–785). An earlier review is by **Sinsheimer** (*Ann. Rev. Biochem.* **46,** 415–438, 1977).

CHAPTER 20
STRUCTURAL GENES:
INTERNAL ORGANIZATION

For a long time there were somewhat uneasy suspicions that there might be something unusual about eucaryotic genes, that they might differ in some fundamental way from bacterial genes. The roots of this idea lay in the apparent discrepancies of DNA content summarized by the C value paradox. This was reinforced by measurements of structural gene complexity implying that only a small part of the genome can be accounted for by its representation in the mRNA template (see Chapter 18).

One way to escape from this situation is to suppose that a large proportion (even a majority) of DNA sequences are not part of the structural genes. The amount of such DNA, and what functions it may (or may not) have, remains to be established.

Another idea is to suppose that the structural gene—or at least the transcription unit—is *much larger than the sequence represented in mRNA.* This is supported indirectly by observations that the size of the RNA found in the nucleus (of higher eucaryotes) is much larger than that of mRNA. A large part of the additional length must be removed during processing of the RNA for transport to the cytoplasm.

Originally, the implication was taken to be that the transcription unit might include extensive sequences on one or the other side of the region represented in mRNA (most likely upstream from it), perhaps involved in regulation. There could also be extensive nontranscribed regions needed for regulation, so that the unit of gene expression might be rather large relative to the size of the mRNA.

Indeed, it turns out that many eucaryotic genes are much longer than their mRNAs. But this occurs because there are *interruptions* that separate different parts of the coding region in DNA. This explains at least part of the size discrepancy between mRNA and nuclear RNA; these **intervening sequences** are included in the primary transcript, but are removed from it during maturation into mRNA. Although comprising part of the structural gene, the intervening sequences do not have any coding function, and it is not clear whether they serve any structural purpose other than to be removed during gene expression.

At this point, we should recapitulate the terminolgy that we mentioned in Chapter 3 for describing the relationship between a gene and its RNA product. An interrupted gene consists of an alternating series of **exons** and **introns.** The exons are the sequences that are present in the RNA (this can be mRNA, rRNA, or tRNA, so the exons may or may not have a protein-coding function). The introns are the intervening sequences that are removed from the primary transcript and are therefore absent from the mature RNA. A gene naturally must start and end with exons (corresponding to the 5' and 3' ends of the RNA), but there may be any number of introns within it. Often the exons and introns are numbered or assigned letters in order along the gene.

The interruption of coding regions is unique to the eucaryotes, but is not found with all eucaryotic genes. Indeed, an insufficient number of genes has been characterized yet to determine any average relationship between the size of the gene and that of the mRNA. So we do not yet know to what extent the presence of introns helps to resolve the C value paradox, but it is unlikely to resolve it completely.

Proceeding from the structure of the gene itself, the next question is to determine its content. How much material on either side of a gene is involved in its function? How far is it to the next gene, and how much of the genome comprises sequences that lie between transcription units? Do adjacent genes tend to be related and is there any mechanism for regulating regions of the chromosome longer than single genes?

THE DISCOVERY
OF INTERRUPTED GENES

The existence of interrupted genes was revealed by results obtained in experiments designed to identify the DNA of the gene corresponding to a particular mRNA. When this work began, its purpose was not to compare the two nucleic acid sequences per se, but to locate the genomic DNA sequence corresponding to the mRNA. This in turn allows the flanking sequences—including putative regulatory regions—to be identified and characterized. Thus the intention of this approach was to proceed back from the mRNA to the gene, to allow the recovery of intact transcription units, including promoters and other elements not necessarily represented in the mRNA. The aim of characterizing the context within which a structural gene resides still remains an object of these experiments, one that has been achieved in several cases, as we shall see in Chapter 21.

Originally it was assumed that an mRNA would have the same sequence as the DNA from which it is transcribed. As a control to show that an isolated genomic DNA sequence did indeed correspond with the mRNA used to isolate it, their sequences were compared,

either by electron microscopy or by restriction mapping. But what this revealed was a discrepancy between the sequences in the form of additional regions present in the genomic DNA but absent from mRNA.

VISUALIZING INTERRUPTED GENES
BY ELECTRON MICROSCOPY

Electron microscopy can be used to visualize RNA-DNA hybrids in two ways. In both the object of the exercise is to distinguish RNA-DNA hybrids from unpaired DNA regions. Then the length and relative position of the hybrid region identify the location of the DNA sequence representing the mRNA.

An RNA can be hybridized with a single-stranded DNA, in which case the complementary regions form a double-stranded hybrid while the other regions re-

Figure 20.1
Single-stranded DNA is hybridized with RNA to form a structure in which the hybrid region can be visualized by electron microscopy. The exact length of duplex nucleic acid varies with the conditions, but roughly 1 μm = 3250 bp.

In this and the subsequent figures, the upper part of the figure shows the events that are involved in terms of the individual strands of DNA and RNA, and the lower part shows what is actually seen by electron microscopy, where thicker and thinner regions have to be interpreted in terms of double- and single-stranded regions.

Figure 20.2
Under the conditions for R-loop formation, RNA hybridizes with DNA by displacing one strand of the double helix.

Figure 20.3
When RNA hybridizes with two regions of single-stranded DNA that are separated by an intervening sequence, the interruption protrudes as a nonhybridized, single-stranded loop of DNA from the RNA-DNA hybrid.

main single-stranded. The double-stranded region is thicker than the single-stranded, although sometimes it is difficult to distinguish the exact junctions of the two regions. **Figure 20.1** shows that this technique visualizes a continuous DNA sequence representing an RNA as a thicker region within the thinner single strand of DNA.

In the technique of **R-loop mapping,** an RNA is hybridized with a double-stranded DNA under conditions in which an RNA-DNA hybrid is more stable than the original DNA duplex. This allows the RNA to displace one DNA strand from the duplex in the region in which it can pair with its complement. **Figure 20.2** shows that the displaced strand forms a loop flanked by the undisturbed double-stranded DNA. This gives rise to the name of the technique.

The critical feature to be noted in both Figures 20.1 and 20.2 is that when an RNA corresponds to a continuous sequence of DNA, the two molecules are colinear. There is a single, continuous hybrid region within the DNA.

But when a gene possesses an internal sequence that is *not* represented in mRNA, this part of the DNA

cannot pair with the RNA. Yet the RNA can pair to form a hybrid with the sequences on either side of the interruption. This results in the formation of a characteristic type of structure in each method for electron microscopic visualization.

When RNA is hybridized with single-stranded DNA, the intervening sequence is extruded as a single-stranded loop of DNA from within the region of the RNA-DNA hybrid. This is illustrated in **Figure 20.3.** The length and relative position of the loop within the hybrid define the size and location of the intron within the transcription unit.

When the technique of R-loop mapping is used, the RNA displaces one strand of the DNA to form an RNA-DNA hybrid with the regions on either side of the intervening sequence. But the intervening sequence itself remains inviolate in its original double-stranded form. This produces the structure illustrated in **Figure 20.4,** in which the two RNA-coding regions are brought together in the hybrid, as seen by the two displaced single-stranded DNA loops. At the junction of these

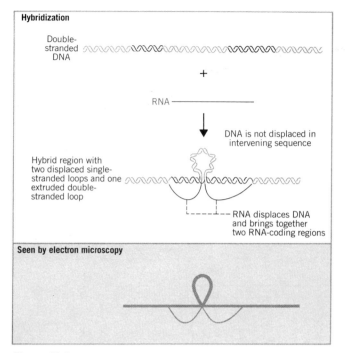

Figure 20.4
When RNA is hybridized with an interrupted gene by R-loop mapping, the intervening sequence cannot hybridize and remains as double-stranded DNA. It is extruded as a thick duplex loop from the RNA-DNA hybrid region. The DNA sequences displaced by RNA form single-stranded loops that identify the individual RNA-coding segments.

loops, a double-stranded loop of DNA is extruded, corresponding to the intervening sequence. An example in which a single intron is visualized in the mouse minor β-globin gene is shown in **Figure 20.5.** (There is also a second intron in this gene that is too small to be visualized by electron microscopy; see later.)

The first results of electron microscopic visualization were obtained with the viral genomes of adenovirus and SV40. In each case, when late mRNAs were hybridized with the viral DNA, the main part of the mRNA formed a hybrid with one region of the viral DNA, but the 5′ terminal part of the mRNA hybridized with a sequence or sequences far upstream in the viral DNA.

The situation in adenovirus is depicted in **Figure 20.6,** which demonstrates that the 5′ end of the late mRNA actually is an amalgam of three very short RNA sequences that represent widely separated sites in

Figure 20.5
R-loop mapping of the mouse minor β-globin gene identifies a long intervening sequence. The electron micrograph is shown on top and an interpretive drawing below it.
Photograph kindly provided by Philip Leder.

the viral genome. In this instance, the tripartite 5′ region corresponds to a nontranslated leader sequence, while the coding region is provided by the main body of the mRNA. **Figure 20.7** visualizes this organization by electron microscopy. As is evident from this hybrid, when there is a series of intervening sequences, these structures can become tangled and difficult to interpret.

The resolution of electron microscopic mapping is good to within about 50–100 bp, depending on the

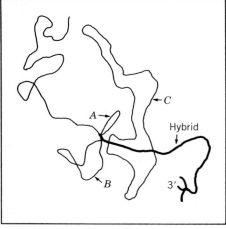

Figure 20.6
Late mRNA of adenovirus is constructed via the removal of three intervening sequences.

The upper part of the figure shows the structure of the viral genome, with the intervening sequences, *A, B,* and *C,* that are removed from the RNA. The lower diagram demonstrates the result of hybridizing the mRNA with the single-stranded DNA, to generate a hybrid region that extrudes three single-stranded DNA loops.

Figure 20.7
One of the original electron micrographs of an RNA hybridization with single-stranded adenovirus DNA identifying the mosaic structure of the late transcription unit.

Photograph kindly provided by Philip Sharp.

conditions. Segments smaller than this can be missed on an electron microscopic map.

RESTRICTION MAPPING OF INTERRUPTED GENES

The restriction maps of both mRNA and genomic DNA are most readily obtained via the use of cloned sequences. In the case of mRNA, of course, the use of a cloned cDNA is necessary to provide the DNA substrate. For the gene, it is possible as an alternative to

use direct cleavage of genomic DNA. Whatever the source, when a gene contains a sequence that is not present within the corresponding mRNA (cDNA), this is revealed in one of two ways. **Figure 20.8** shows that if the intervening sequence does not have a recognition site for whatever restriction enzyme is being employed, there is a change in the length of the corresponding fragment. The presence of the intron means that a greater distance separates the two restriction sites on either side of it in the gene than in the mRNA. Thus the length of this fragment is increased in the digest of the gene.

Figure 20.8

The presence of an intervening sequence that lies between two cleavage sites increases the size of the corresponding restriction fragment.

The figure compares a gene and cDNA that have two cleavage sites for a restriction enzyme. This generates three fragments, labeled A, B, and C in order of decreasing size. The intervening sequence lies between the two cleavage sites that generate A, so just the size of this fragment is increased in the gene.

Figure 20.9

The presence of an intervening sequence creates new restriction fragments when it contains a cleavage site for a particular enzyme.

An additional cleavage site in the gene increases the number of cuts to three from the two found with cDNA. Thus fragment A from cDNA is replaced by A1 and A2 from genomic DNA. The combined size of A1 + A2 exceeds that of A by the length of the intervening sequence.

On the other hand, if the intervening sequence possesses a recognition site for the enzyme, an extra cut is made within it. Since this cut occurs only in the gene and not in the mRNA sequence, cleavage of the gene generates two fragments corresponding to the one generated by the message. This is illustrated in **Figure 20.9.**

When a map is constructed by using several restriction enzymes, a mixture of these two results is obtained. For some enzymes, cleavages will not occur in an intervening sequence; then the lengths of the corresponding fragments are increased. For other enzymes, additional fragments will be generated from the same intervening sequence. Putting this together, when an entire restriction map is constructed for a gene that has an intervening sequence, what we see is that the map at each end of the gene corresponds with the map at each end of the message sequence.

But at some point within the gene, the maps diverge, and an additional region is present within the gene, containing a series of recognition sites not represented in the message. The resolution of restriction mapping is adequate to detect gene segments down to about 20–30 bp.

Figure 20.10 compares the restriction maps of a β-globin gene and mRNA. Two introns can be recognized. The larger is the same as the intron seen by electron microscopy in Figure 20.5. The smaller shows up only on the restriction map.

A useful procedure for analyzing an interrupted gene is to characterize the hybrid formed between the mRNA and DNA, as summarized in **Figure 20.11.** In addition to the RNA-DNA hybrid regions, the molecule contains

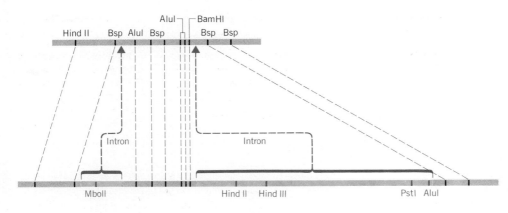

Figure 20.10
Comparison of the restriction maps of cDNA and genomic DNA for mouse β globin shows that the gene has two additional regions not present in the cDNA. The other regions can be aligned perfectly between cDNA and gene.

single-stranded flanking DNA and the extruded single-stranded intron. The molecule can be treated in two ways.

The flanking DNA can be removed by the enzyme exonuclease VII. Then the single-stranded DNA corresponding to the length of the gene itself can be isolated by degrading the RNA with alkali.

The individual regions of the gene can be isolated by treatment with S1 nuclease, an enzyme that specifically cleaves any single-stranded DNA, including even very short single-stranded gaps within a duplex structure. Thus this treatment removes the flanking regions and the intron. The product is an RNA-DNA hybrid, of the length of the mRNA, with a gap in the covalent integrity of the DNA strand. Treatment with alkali removes the RNA, releasing the individual fragments of DNA. These identify the individual exons of the gene.

In the form illustrated in the figure, the technique is often known as Berk-Sharp mapping, named for its authors. In more generalized form, it is sometimes called S1 nuclease mapping. The property of this enzyme for breaking all bonds in unpaired DNA is very useful. Often it is used to locate the end of an RNA sequence on DNA. For example, DNA is hybridized with RNA and then S1 nuclease is used to remove the flanking regions. The DNA then is terminated *exactly* at the ends of the RNA molecule (as we have mentioned previously in Figure 11.2). By comparing the sequence of this DNA fragment with the sequence of the original DNA, the exact startpoint for the mRNA can be located.

CHARACTERIZING GENOMIC DNA FRAGMENTS

When genomic DNA is cleaved with a restriction enzyme that is known not to cut within a particular message sequence, cuts will be made on either side of the gene. These cuts per se should generate a single fragment. What are we to conclude when two (or more) fragments that hybridize with the mRNA actually are released in such an experiment?

One possibility is that there is more than one copy of the gene representing the mRNA. (The copies need not necessarily be identical, just closely related enough to hybridize with one another.) In this case, the sequences on either side of each copy could be different, so that the restriction sites lie at different chance locations with regard to the gene. Then each copy will be released in a restriction fragment of different size.

The alternative is that the gene contains an intervening sequence (or sequences) that has a recognition site for the enzyme. In fact, an additional fragment will be generated for every intron in which there is such a site. (The reverse result, that a single fragment is obtained, does not prove the absence of an intervening sequence, because the intron need not possess a site for the enzyme. Nor does it exclude the possibility that there is a small number of tandem copies.)

How are the alternatives of multiple gene copies and interrupted gene structures to be resolved? The usual approach is to use probes derived from **subclones** that represent only specific regions of the mRNA. These are obtained by cleaving the cloned

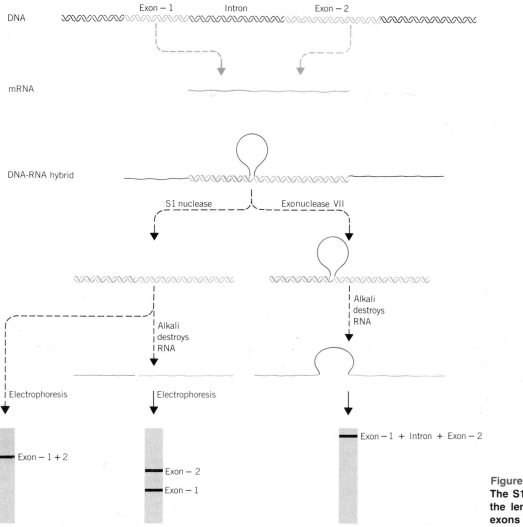

Figure 20.11
The S1 mapping protocol allows the lengths of both introns and exons to be measured.

cDNA into 5′ and 3′ regions with a suitable restriction enzyme (that is, one that cuts at an appropriate place within the message sequence). Then these end-specific probes are recloned separately into new plasmids.

Figure 20.12 shows that if each of two fragments represents a different copy of the gene, each will have its complete sequence; so both fragments will hybridize with both 5′ and 3′ probes.

Figure 20.13 shows that if two genomic fragments represent different parts of the gene that have been separated by cleavage within an intervening sequence, one will hybridize with the 5′ probe of cDNA; the other will hybridize with the 3′ probe.

When an interrupted gene is obtained in fragments, how are we to be certain of identifying every intron and every exon? The structure is deduced by obtaining *overlapping* fragments that correspond to different regions. This is a general principle of restriction mapping and DNA sequencing (as described in Chapter 3), and is necessary to exclude the possibility that part of the gene has been missed.

Suppose, for example, that within the gene there is an intron carrying two restriction sites for some enzyme, as illustrated in **Figure 20.14.** When these are cleaved, the fragments to the left (*A*) and to the right (*B*) each consist in part of the sequence represented in mRNA and in part of the intron. The presence of the

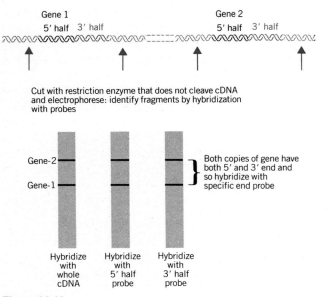

Figure 20.12

Multiple genes produce multiple intact copies when the genome is cleaved with a restriction enzyme that does not cut within the cDNA.

Two identical genes, each consisting of a 5′ half and a 3′ half, give different fragments because the restriction sites lie at different relative positions in the flanking DNA. The fragments are identified by blotting a gel of the electrophoresed DNA and hybridizing with a labeled cDNA probe. Since each fragment has both ends of the gene, each hybridizes with both the specific 5′ and 3′ cDNA probes.

Figure 20.13

A single gene with an intron(s) can produce multiple fragments when the genome is cleaved with a restriction enzyme that does not cut within the cDNA.

If the enzyme cuts within the intron, two fragments are released that can hybridize with a labeled cDNA probe. But when 5′ and 3′-specific probes are used, each fragment hybridizes with only one probe.

first component allows each to hybridize with the message probe. But the central fragment (C) consists *only* of intron sequences; because it lacks any message sequence, it does not hybridize with a probe representing the message. So the sequence will not be detected in a digest made with this enzyme. This type of occurrence is avoided by using several enzymes, extending the map in such a way that the *junction between each pair of adjacent fragments also is represented intact in some other fragment.* This may involve using regions of the intervening sequence as probes.

Ultimately a comparison of the nucleotide sequences of the gene and cDNA clone shows exactly where and how large the intervening sequences are. Resolution at the sequence level is necessary before we can be sure that all the segments of the gene have been identified. Indeed, there have been several cases

where the genomic sequence presented a surprise in the form of an additional short coding segment. When a coding segment is shorter than about 50 bp, it may fail to hybridize with the cDNA probe, and can therefore pass unnoticed within the intervening sequences on either side.

GENES COME IN ALL SHAPES AND SIZES

No particular rhyme or reason yet has been discerned in the very varied structures that have been found for eucaryotic genes. Some genes are uninterrupted, so that the sequence in the genome is colinear with that of the mRNA. More genes at present appear to have interruptions, but the intervening sequences vary enormously in both number and size.

Some critical features are common to all interrupted genes. The order of the parts of an interrupted gene is the same in the genome as in its mature RNA prod-

Figure 20.14
When genomic fragments are identified by hybridization with cDNA or mRNA, regions of introns that lie between two restriction sites may be missed.

uct. Genes are thus split rather than dispersed. Introns of nuclear genes generally have termination codons in all reading frames, and so cannot code for any polypeptides. An interrupted gene retains the same structure in all tissues, including the germ line and somatic tissues in which it is or is not expressed. Thus the presence of an intervening sequence is an invariant feature.

All classes of genes may be interrupted: nuclear genes coding for proteins, nucleolar genes coding for rRNA, and genes coding for tRNA. Interruptions also are found in mitochondrial genes in yeast and in chloroplast genes. Interrupted genes do not appear to be excluded from any class of eucaryotes, although their proportion does vary. They may, for example, be more predominant in vertebrates than in fungal nuclear structural genes.

Intervening sequences must be removed before a gene can be translated (or before an interrupted rRNA or tRNA can fulfill its function). They are present in the primary transcript, but absent from the mature RNA product. The process of their removal is called **RNA splicing.** Essentially this involves a controlled deletion of each intervening sequence from the primary transcript, with the ends of the RNA on either side being joined together to form a covalently intact RNA. This is illustrated in **Figure 20.15.** We shall discuss the mechanism of RNA splicing in Chapter 26.

Some interrupted genes possess only one or a few introns. The globin genes provide an extensively studied example (see Chapter 21). The two general types of globin gene, α and β, share a common type of structure in which two introns occur at constant positions relative to the coding sequence. The consistency of the organization of mammalian β-globin genes is evident from the summary in **Figure 20.16.** In the 146 amino acid chain of mouse β globin, the introns are located between codon positions 30/31 and between 105/106.

Interruptions occur at the homologous positions in all known active globin genes, including those of several mammals, birds, and a frog. The first intron is always fairly short, and the second is usually longer, but the actual lengths can vary. Most of the variation in overall lengths between different globin genes re-

Figure 20.15
An interrupted gene consists of alternating exons and introns, and the introns are removed by splicing of the RNA transcript, generating an mRNA consisting only of exon sequences.

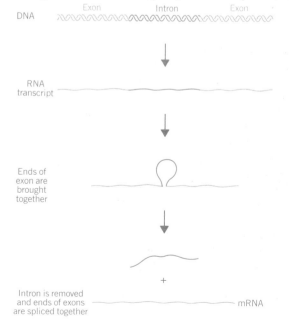

	Exon 1	Intron 1	Exon 2	Intron 2	Exon 3
Length in bp	142-145	116-130	222	573-904	216-255
Represents	5′ nontranslated + amino acids 1-30		Amino acids 31-104		Amino acids 105-end + 3′ nontranslated

Figure 20.16
All functional globin genes have an interrupted structure with three exons. The lengths indicated in the figure apply to the mammalian β-globin genes.

sults from the variation in the second intron. In the mouse, the second intron in the α-globin gene is only 150 bp long, so the overall length of the gene is 850 bp, compared with the 1382 bp of the major β-globin gene. Thus the variation in length of the genes is much greater than the range of lengths of the mRNAs (α-globin mRNA is 585 bases, β-globin mRNA is 620 bases). The central exon of each type of gene is constant in length, so the variation in the mRNAs reflects the limited variation in the lengths of the first and third exons; mostly this represents fluctuations in the length of the nontranslated segments (the leader and trailer).

Other genes may have more introns. Two examples are shown in **Figure 20.17.** The ovalbumin and con-albumin genes of the chicken have 8 exons and 17 exons, respectively. This makes the total lengths of the genes 7500 and 10,000 bp, compared with their respective mRNAs of 1859 and 2500 bases.

(In this and other cases, we define the length of the structural gene as the distance in the genome between points corresponding to the 5′ and 3′ terminal bases of mRNA. We know that transcription probably does start at the 5′ end of the mRNA; but it is not certain whether it terminates at the 3′ end or whether this end is generated by cleavage from a longer precursor. In the latter case, the length of the gene would be greater, including additional sequences downstream.)

Genes that are either much longer or more highly mosaic than these are known. The present extremes in these parameters are mouse DHFR and chicken α collagen.

The gene for mouse DHFR is organized into 6 exons that correspond to the 2000-base mRNA. But these extend over more than 31,000 bp of DNA. Here the introns are exceedingly long.

The gene for chicken proα2 collagen is split into more than 50 exons, each rather short. The lengths of characterized exons vary from 45 to 249 bp. Some of the introns are short, comparable in length to the exons; but some are much longer, so that the total length of exons—about 5000 bp—is spread out over 40,000 bp of the genome.

In some eucaryotic structural genes, therefore, the exons actually comprise only a small proportion of the total length of the gene. Together with their often rather

Figure 20.17
Scale maps of two chicken genes show that the exons tend to be rather small units (of the order of 100 bp) connected by much longer introns (although some exons are also long).

Ovalbumin   8 exons: total length 1859   structural gene 7500 bp

Conalbumin   17 exons: total length 2500   structural gene 10,000 bp

1000 bp

Introns
Exons

short lengths, this gives a picture of the gene as comprising a series of brief exons spread out over a large area of the genome, which (presumably) fills only the function of providing this structural gene. In such cases, the map visualizes the exons as interruptions in the introns, rather than vice versa. This is clear from Figure 20.17. Many genes must be painstakingly pieced together, as it were, by joining many short exons that reside considerable distances apart.

From the data now available, we can form no firm conclusion on whether a eucaryotic genome is likely to consist of a greater proportion of exons or introns, but clearly the mosaic form of the gene goes some way toward explaining the minor proportion of DNA that is represented in mRNA. Just how far remains to be seen. We have already mentioned our suspicion that it will not solve the problem of finding functions for all the DNA.

INTRONS IN GENES CODING FOR rRNA and tRNA

Introns occur in the gene for large rRNA in several cases. Actually, interrupted genes were first discovered in *D. melanogaster,* where about two thirds of the genes for 28S rRNA have an interruption of up to 5000 bp in length. (Genes for rRNA are always present in multiple, usually identical, copies; see Chapter 24.) But since the other third are not interrupted, and since the interrupted genes do not seem to be used to synthesize rRNA, this left open the possibility that they constitute damaged, nonfunctional genes. Indeed, this intron may represent insertions in previously uninterrupted genes (see Chapter 37).

In some lower eucaryotes, however, *all* of the copies of the large rRNA gene are interrupted. This includes *Tetrahymena pigmentosa* (a ciliate) and *Physarum polycephalum* (a slime mold). Their situation is therefore analogous to that of the nonrepetitive nuclear structural genes: the intron must be removed by splicing to make an active gene product. Yet in other lower eucaryotes, such as *S. cerevisiae* (yeast) and *D. discoideum* (a slime mold), the rRNA genes all are uninterrupted. In the mitochondrion of *S. cerevisiae* and *N. crassa* (a fungus), and in the chloroplast of *C. reinhardii* (an alga), again the gene for large rRNA is in-

terrupted; but it is uninterrupted in the mitochondria and chloroplasts of other species.

There seems to be no systematic relationship to be discerned in the distribution of interrupted and uninterrupted rRNA genes. However, in all cases but one, there is only a single intron. It is always located at a position about two thirds of the way along the gene. (The rRNA molecules vary in length from about 3000 to 4000 bases.) Could this be a homologous position in these very different organisms? There are no interruptions in the genes for small rRNAs.

In two cases, there is a polymorphism in the presence of the intron: some strains of the organism possess an interrupted rRNA type of gene; other strains possess the uninterrupted type. This occurs in the nucleus of *T. pigmentosa* and in the mitochondrion of *S. cerevisiae.* In neither case does there appear to be any difference in the ability to utilize the gene. The coexistence of the two forms of the gene implies that the intron is neither essential nor deleterious.

Introns have been found in tRNA genes of the yeast nucleus. An interesting feature is that the intron is always located at the same relative location, at the start of the anticodon loop. This allows the interrupted tRNA precursor to take up a conformation in which the anticodon is base paired with part of the intron. This may be relevant to the mechanism of splicing, and is illustrated later in Figure 26.1.

The length of the intron varies quite widely, being as short as 14 or as long as 46 bp in different tRNA genes. Yet again it cannot be an essential feature of the yeast nuclear tRNA gene per se, because some tRNA genes are not interrupted. In families of tRNA genes, some members of a family may be interrupted, while other, related genes are continuous.

INTRONS ARE NONREPETITIVE AND EVOLVE RAPIDLY

By using restriction fragments that correspond to specific regions of a structural gene, it is possible to test whether they are related to other sequences in the genome. The cloned restriction fragment can either be used as a tracer in hybridization with whole genome DNA to determine its repetition frequency, or it can be used as a probe to detect corresponding sequences in a restriction digest. By obtaining appropriate restriction fragments, regions of either exons or introns can be used.

Often it turns out that the exons of a gene are related to those of another gene; when an exonic probe is used, it detects fragments that are part of another gene or genes. This implies that the two genes have evolved by duplication of some common ancestral gene, whose duplicates then accumulated differences.

Usually the introns are not related to other sequences. And when two genes have related exons, the relationship between their introns is likely to be more distant than the relationship between the exons. This means that, while the exons are under the constraint of needing to code for a particular sequence of amino acids, the introns are not, and therefore can accumulate more mutations.

So in the original sense of considering nonrepetitive DNA to comprise sequences that are unique in the genome, introns come closest to representing this feature. In asking whether structural genes are nonrepetitive, we see therefore that the answer is ambiguous: the entire length of the gene may be unique as such, but its exons sometimes are related to those of other genes. So at least for some genes, the exons constitute slightly repetitive sequences embedded in a unique context of introns.

Some detailed comparisons between corresponding exons and introns have been made in the globin genes, and we shall discuss these in more detail in Chapter 21. The overall degree of divergence between the exons is related to the differences in the proteins. It is caused mostly by base substitutions. However, many of these do not affect codon meanings, either because they lie in third-base positions or because they are in nontranslated regions. The pattern of homology tends to extend past the exon-intron boundaries for a short distance into the intron.

In the second intron, there is extensive divergence in sequence, as shown by the almost twofold variation in its length among the β-globin genes. Its pattern of divergence involves both changes in size (due to deletions and insertions) and base substitutions. In some pairwise combinations of β-globin genes, there has been so much variation in the second intron that corresponding sequences cannot even be recognized. In other words, introns evolve much more rap-

idly than the exons. These changes imply that the intron does not have a sequence-specific function (or at least not one that is the same in different species). Whether its presence is at all necessary for gene function is impossible to say, except to note that no functional β-globin genes have been found in which it is absent.

EXON-INTRON JUNCTIONS HAVE A CONSENSUS SEQUENCE

By comparing the nucleotide sequence of mRNA with that of the structural gene, the junctions between exons and introns can be assigned. Two features (or lack thereof) are significant. First, there is no extensive homology or complementarity between the two ends of an intron. This excludes any possibility that extensive secondary structure could form to link them together as a preliminary step in splicing. Second, the junctions do prove to have a well-conserved, though rather short, consensus sequence; its ubiquity argues that it may be involved in splicing in the nucleus (see Chapter 26).

In many cases, it is not possible to assign a unique site for the exon-intron junction solely on the basis of the mRNA/gene comparison. The difficulty is that a short sequence, usually between 1 and 4 bp, is repeated at each end of the intron. **Figure 20.18** shows an example from mouse (or rabbit) β globin, in which any one of four pairs of sites could mark the ends of the intron.

In spite of this difficulty, it is possible to assign a specific end to each intron by relying on the homology of exon-intron junctions. They can all be aligned in

Figure 20.18
Repetition of the trinucleotide AGG at both ends of the second intron in mouse β globin allows several possible frames for splicing.

	gly	arg	leu
Messenger sequence	GGC	AGG	CTG

Genome sequence	GGCAGGTG_____TTT AGGCTG
Possible splicing frames	GGC_____AGGCTG
	GGCA_____GGCTG
	GGCAG_____GCTG
	GGCAGG_____ CTG

such a way as to relate closely to the consensus sequence

Exon |←————————— Intron —————————→| Exon
A_{64} G_{73} G_{100} T_{100} A_{62} A_{68} G_{84} T_{63} $\cdots$ $6Py_{74-87}$ N C_{65} A_{100} G_{100} N

where the arrows mark the putative ends of the intron. (In this as in other cases, we write just the sequence of the DNA strand that is identical with the RNA product.)

The subscripts indicate the percent of junctions at which the specified base (or type of base) is found. The really high conservation is found only immediately within the intron at the presumed junctions. This identifies the essential ends

$$GT \cdots \quad \cdots AG$$

Because the intron defined in this way starts with the dinucleotide GT and ends with the dinucleotide AG, this is often described as the **GT-AG rule.**

Note that these two junctions are different in sequence, and so define the ends of the intron directionally. They are named proceeding from left to right along the intron, that is, as the **left** and **right** splicing sites. Sometimes they are called the **donor** and **acceptor** sites, respectively, but we shall eschew this description because it implies a mechanism for which we have no evidence.

This consensus sequence is found in nuclear structural genes of many eucaryotes. So far as we know, it is common to all higher eucaryotes, and this implies that there may be a common mechanism for splicing the introns out of RNA. However, the consensus does not apply to the introns of mitochondria and chloroplasts, nor to the yeast tRNA genes; at least in these cases, a fundamental difference in splicing mechanism may apply (see Chapter 26).

ONE GENE'S INTRON CAN BE ANOTHER GENE'S EXON

So far we have dealt with the organization of interrupted genes in terms of an invariant alternating pattern of exons and introns. Although this is scarcely the relationship anticipated from the original definition of the cistron, it is not at odds with it. The practical consequence of the mosaic structure is that mutations in a protein should map in clusters that correspond to the exons, rather than in a continuous segment of DNA. None of the mutations in each cluster will complement either with one another or with the mutations in other clusters. In other words, a mosaic gene is expressed only via the splicing together of exons carried by one molecule of RNA, thus conforming to the original concept of the *cis/trans* test (see Chapter 2).

Will mutations in the introns be detected by impairment of gene function? We know from the high rates at which introns evolve that their sequences are under less constraint, which argues that many, perhaps most, changes in an intron will pass unnoticed. After all, these sequences are removed during expression of the gene. There are some examples of mutations in introns of nuclear genes that affect the ability of the RNA to be recognized by the splicing apparatus. Usually these change the consensus sequences, preventing proper removal of the introns (see Chapter 26). Such mutations are *cis*-acting and thus appear as additional members of the single complementation group.

In contrast with the common situation in which a gene is mosaic but otherwise conforms to the traditional view, there are some cases in which the fixed relationship between gene and protein does not hold. Here more than one mRNA sequence can be derived from a single stretch of DNA. The meaning of a particular region of DNA therefore is not invariant, but depends on the pathway selected for its expression. In this case, the pattern of complementation between mutations cannot be expected to conform entirely to the traditional pattern.

A tissue-specific change that occurs in the expression of mouse amylase mRNA is summarized in **Figure 20.19.** The same amylase protein is synthesized in the liver and the salivary gland of the mouse, and is coded by a single gene. The mRNAs present in each tissue are identical in their coding region, but differ in the first part of the 5′ nontranslated leader. The leader is longer in liver and different in sequence from that of salivary gland.

The amylase coding sequence starts about 50 bp within exon 2, and is formed by joining exon 2 to exon 3 and the subsequent exons. This sequence may be preceded by either of two exons. In liver, the first 161 bases of the mRNA are coded by exon L, which lies

Figure 20.19
Alternative splicing generates two amylase mRNAs from one mouse gene.

about 4500 bp upstream. In salivary gland, the first 50 bases of the mRNA are coded by exon S, which lies about 7300 bp upstream. Thus exons S and L provide alternative initial sequences for the amylase mRNA. The sequence of exon L is in fact part of the long intron that is removed by splicing in the salivary gland.

Another change in a processing pathway for mRNA maturation has been characterized in the expression of immunoglobulin genes. This involves the substitution of one exon by another at the 3′ end of the transcription unit. This changes the C-terminal amino acid

sequence of the protein, with the result that a region responsible for membrane attachment is replaced by a region that allows secretion. This is discussed in Chapter 39.

In late adenovirus expression, the same leader sequence can be spliced to any one of several different coding sequences. **Figure 20.20** presents a diagrammatic version of these events. There is a single transcription unit whose expression is initiated at a single point. The first part of the transcription unit carries three sequences that are spliced together to form a

Figure 20.20
The same leader can be joined to different messenger bodies during late adenovirus expression. Only three coding exons are shown, but actually there are five groups of messenger bodies defined by common 3′ ends, and these include at least nine individual different coding regions.

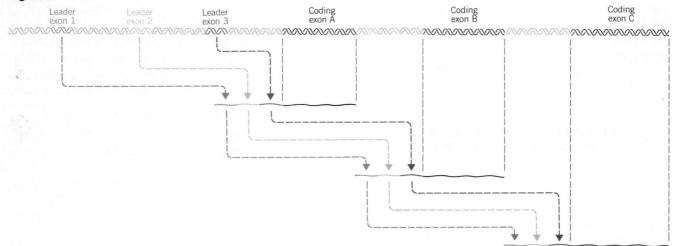

nontranslated leader. (This is the situation depicted in Figures 20.6 and 20.7.) The components of the leader are quite short and are indicated as leader exons 1, 2, and 3. Downstream from these sequences, there are several coding regions, each representing a late viral protein. These are indicated as coding exons A, B, C, etc. In each primary transcript, the tripartite leader formed by the first two splicing events then can be spliced to *one* of the coding regions.

The 3' end of each coding exon is generated by cleaving the primary transcript and polyadenylating the terminus. The sequences farther downstream then are discarded. The 5' end of the coding exon then is spliced to the tripartite leader. For the first coding exon, this poses no particular problem: an intron is removed in the usual way. But for coding exon B, the "intron" comprises the entire region between its 5' end and leader exon 3. This includes coding exon A. Similarly, when coding exon C is spliced to the leader, both coding exons A and B are part of the "intron" that is removed. So none of these coding exons is inevitably used to direct protein synthesis in a particular transcript molecule: in fact, only one can be utilized per RNA molecule, and the others are discarded if downstream or spliced out if upstream from the chosen exon. The generation of the 3' end of the nuclear RNA is probably responsible for determining which exons are allowed to be spliced to the tripartite leader.

Another situation occurs in the small DNA viruses SV40 and polyoma, where alternative splicing patterns allow the simultaneous production of proteins whose amino acid sequences overlap. This happens during both early and late transcription.

The early transcription of SV40 leads to the synthesis of two proteins, **T** (called "big T") and **t** (called "little t"). These have the identical N-terminal amino acid sequence, but differ in the C-terminus. They are coded by the same stretch of DNA, as illustrated in **Figure 20.21.**

The coding region for T antigen is divided into two exons. The first exon carries the 5' nontranslated region and the N-terminal part of the protein; the second exon carries the remainder of the protein and the 3' nontranslated region. The intron between is 347 bp long, and displays the usual consensus sequences at the exon-intron junctions.

The mRNA for t antigen also has a single intron. This is part of the same intron removed in T-antigen formation, but consists of only its last 67 bases. Thus the right splicing junction is identical in the two cases, and the difference lies in the choice of the left junction that is spliced to it. The left junction used in t-mRNA formation also follows the consensus rules.

Translation of the T and t antigens starts at the same AUG codon in their common initial sequence. The amino acid sequences diverge at the point where the T sequence is spliced to its second exon. The result is to produce two proteins, sharing the same N-terminal half, but having a different C-terminal half. The t antigen actually is coded entirely by the first exon of the

Figure 20.21
Alternative splicing patterns generate the T and t antigens of SV40 from the same DNA sequence.

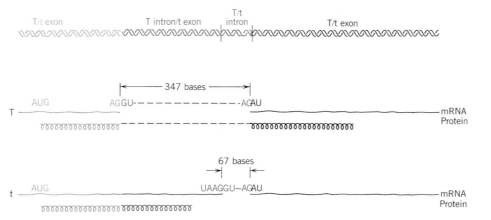

t mRNA. The termination codon lies immediately before the end of the exon. (Although the second T exon is present in t mRNA, it is not translated because protein synthesis terminates at the t-UAA codon; this codon does not terminate T synthesis, because it is removed as part of the T intron.) In these circumstances, the definition of introns and exons does not have the clear significance attributable in cellular genes.

AN INTRON THAT MAY CODE FOR A REGULATOR PROTEIN

Two interrupted genes in the yeast mitochondrion display unprecedented genetic properties that are so much at odds with the traditional view of the gene that they could be explained only by virtue of the molecular analysis of DNA. The organization of these genes is illustrated in **Figure 20.22.**

The locus *oxi3* codes for subunit 1 of cytochrome oxidase. The gene has a coding region of 1530 bp, extending over about 10,000 bp of DNA. It is organized into eight exons, two of which are very short.

The gene coding for cytochrome b is known variously as *box* and *cob*. It exists in two versions. Some strains of yeast have a "long" gene, in which the coding region of 1155 bp extends over about 6400 bp, organized into six exons (numbered B1 to B6), some of which are very short. Other strains have a "short" gene, about half the length, in which the first four exons of the "long" gene all form a single, continuous exon. In other words, the first three introns of the gene all may be present or all may be absent. Both forms of the gene are expressed equally well. The genetic analysis that we shall discuss has been performed with the long gene. The introns are identified according to their numbering in the long gene, I1 to I5.

When RNA molecules corresponding to the *box* gene are isolated, they fall into a discrete series of sizes. The shortest species is about 3000 bases long and provides the mRNA. This is longer than the coding region and includes 5′ and 3′ nontranslated regions

Figure 20.22
Two genes of the yeast mitochondrion have mosaic structures.
The gene *oxi3* has 8 exons that code for subunit 1 of cytochrome oxidase, (probably) 510 amino acids long. The "short" and "long" versions of the *box* gene code for the same cytochrome-b protein of 385 amino acids. The clusters of mutations are indicated as *box* followed by a number that reflects time of isolation (not map position). Their locations on the map indicate the exons or introns to which they are assigned. Both genes as depicted correspond only to the protein-coding regions of the mRNA and do not include 5′ and 3′ nontranslated regions.

that are not shown on the map in Figure 20.22. Among the larger RNA molecules, each size possesses one or more of the introns. The largest molecule, of 8500 bases, is probably the intact primary transcript. The others are intermediates in which some introns have been spliced out while others remain.

Mutations preventing the synthesis of cytochrome-b can be mapped into a series of clusters. Each cluster covers a short genetic distance, of the order of 1% recombination. The clusters are separated from each other by greater distances, in the range of 2–9% recombination. Each cluster is indicated as *box* followed by a number. The clusters fall into three classes.

Mutations in four of the clusters affect the cytochrome-b protein directly. All of these mutants synthesize normal mRNA. The effects of the mutations are mediated at the level of translation, by causing missense or nonsense readings of the message. None of these mutations complements any other, in either the same or a different cluster. By this criterion, they all lie in the same gene. These clusters identify some of the exons; the correspondence is *box4* = B1, *box8* = B3, *box1* = B4, *box6* = B6 (see Figure 20.22). Mutations have not been found in the other two exons, perhaps because of their small sizes (B2 = 14 bp, B5 = 51 bp). The properties of these groups of mutations correspond exactly with those predicted for an interrupted gene. In fact, this is the only case known so far in which the genetics of a mosaic gene can be studied.

Mutations in the *box9* and *box2* clusters also fail to complement with mutations of the other groups. By this genetic criterion, they are therefore indistinguishable from the exonic mutations. But their biochemical properties are different, as indicated by a failure to synthesize the normal mRNA. Sequence analysis of the DNA shows that both these clusters lie in I4. The *box9* mutations lie in a sequence of 8 bp about 350 bp downstream from the junction with B4. The *box2* mutations lie toward the other end of the intron, about 25 bp upstream from the junction with B5. Both groups of mutations prevent the splicing together of B4 and B5 by removal of I4. Their existence reveals two important general points. First, mutations in specific sites can prevent recognition of particular splicing junctions; and these sites may be quite distant from the actual junctions themselves. Second, such mutations cannot be distinguished genetically from mutations in the protein itself. (The same lack of discrimination applies to mutations in promoters or operators and their structural genes, the classic *cis*-acting type; see Chapter 14.)

Three other groups of mutations, *box3, box10,* and *box7,* each represent a small cluster of mutations located within an intron. Their behavior in complementation tests reveals that these mutations are of a novel type. The mutations within each of these three clusters behave as a complementation group. But although they cannot complement other mutations in the cluster, they *can* complement mutations in any other cluster, either of this type or of the exon type.

Formally, this means that each of these three clusters codes for a *trans*-acting, diffusible product that is distinct from the cytochrome-b protein, but nonetheless is necessary for its synthesis. In practical terms, this means that within three of the introns there are sequences whose function is to code for some product that has an independent existence and plays some regulatory role in cytochrome-b production.

The nature of this role is suggested by another property of these mutations. They all block the production of cytochrome-b mRNA by causing precursor RNAs to accumulate. These precursors are: 7500 bases for *box3* mutants, 7100 bases for *box10,* 3500 for *box7.* This suggests that each cluster of mutations may block the maturation of the RNA at a specific stage, by inactivating a diffusible product that probably is needed to remove a particular intron. (We know that the inhibition cannot be caused simply by mutations in the splicing junctions, because these would be *cis*-acting like *box2* and *box9,* rather than *trans*-acting.)

The nature of this function is suggested by the sequence of the gene. The salient features are shown in diagrammatic form in **Figure 20.23.** The first exon (B1) carries the N-terminal 139 codons (417 bp) of cytochrome-b. An intron of 765 bp (I1), blocked in all reading frames, separates the first exon from the second, very short exon (B2,5 codons). This exon is followed by the long second intron (I2). The significant feature about this intron is that *the first 840 bp represent an open reading frame in exact register with the reading frame of the preceding exon.*

Figure 20.23

Successive splices in the box gene could generate mRNAs coding for overlapping proteins. Removal of only the first intron generates the mRNA for RNA maturase; removal of the second intron then gives the coding sequence for cytochrome b.

The *box3* mutations all lie in this region, and those that have been sequenced turn out to create nonsense codons in the open reading frame. This part of the intron may therefore have a coding function; and the protein may provide the diffusible function involved in splicing. It has been called the **RNA maturase.** (The remainder of the intron is blocked in all reading frames.)

What is the structure of the RNA maturase? Indirect evidence suggests that its translation is not initiated within the intron, but occurs via readthrough from the second exon. This could occur if the first splicing intermediate shown in Figure 20.23 is translated. The RNA is generated just by removing the first intron. This connects the first and second exons, creating a reading frame that continues into the second intron. Its total length is 424 codons; and its translation would produce a protein with the 144 N-terminal amino acids of cytochrome-b and 279 amino acids coded by the intron.

If the RNA maturase coded by this sequence is needed specifically to splice out the second intron, the enzyme action creates the exquisitely sensitive negative feedback loop illustrated in **Figure 20.24.** The removal of the second intron to join the first two exons to the third exon disrupts the sequence coding for the maturase. Thus the splicing activity of RNA maturase leads to loss of the capacity to synthesize the enzyme.

So an equilibrium will be achieved between the levels of the two forms of the RNA and the RNA maturase protein (which is indeed present in exceedingly small amounts and therefore has been very difficult to characterize).

Is this a general model for other introns in the mitochondrion? We lack detailed evidence, but in both the *box* and *oxi3* genes the predominant form taken by introns is to display an open reading frame in exact register with the preceding exon. This cannot be a coincidence. It makes it seem likely that these reading frames are translated.

Unlike nuclear structural genes of yeast, mitochondrial exon-intron junctions do not follow the GT-AG rule; and they do not display any other consensus sequence. So an RNA maturase could be specific for a particular intron or introns. The probable function of the *box3* RNA maturase is to recognize only the ends of the second intron, so that the flanking exons can be spliced together. In another case, there is an indication that an intron-coded function may be involved in two splicing events; since *box7* mutants (in the fourth intron of the *box* gene) are defective not only in the splicing of the *box* mRNA, but also of *oxi3* mRNA. The *box7* mutations create nonsense codons in the open reading frame of the intron.

Is any evolutionary relationship evident between different intron coding functions? There are some related

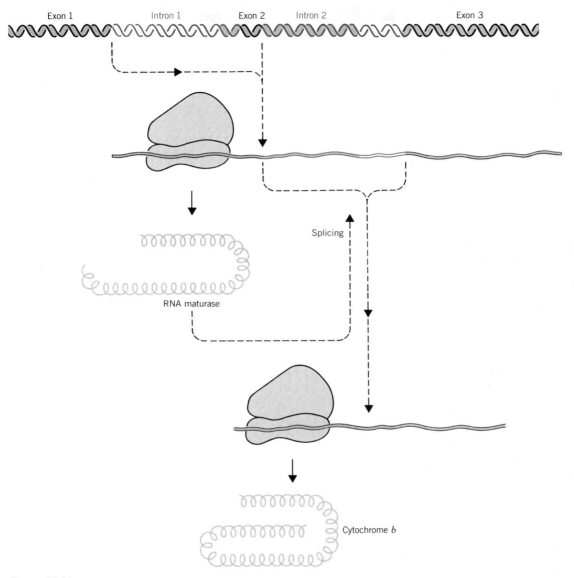

Figure 20.24
If the RNA maturase specifically splices out the intron that codes for it, a negative feedback loop will control the amounts of the protein and the two mRNAs.
Shaded regions on the DNA and RNA indicate open reading frames; regions in outline are blocked in all reading frames.

sequences among the intron open reading frames, but we do not yet know whether these are significant. In all of the intron coding regions, the pattern of codon usage is somewhat different from that found in the exons. While it is not clear what advantage accrues to the mitochondrion from maintaining such an intricate system for expressing its interrupted genes, we could speculate that at some time there was an independent protein involved in splicing the second *box* intron. Its gene might later have been translocated into the intron, giving rise to the present arrangement. Of course, it is still true that the sole purpose of the intron coding function appears to be to remove the sequence that codes for it from the mRNA.

HOW HAVE INTERRUPTED GENES EVOLVED?

What was the original form of genes that today are interrupted? Did the ancestral protein-coding units consist of uninterrupted sequences of DNA, into which introns were subsequently inserted during their evolution to the present form? Or did these genes initially arise as interrupted structures, which since have been maintained in this form? Another form of this question is to ask whether the difference between eucaryotic and procaryotic genes is to be accounted for by the acquisition of introns in the eucaryotes or by the loss of introns from the procaryotes.

Could the mosaic structure be a remnant of an ancient approach to the reconstruction of genes to make novel proteins? Suppose that an early cell had a number of separate protein-coding sequences. One aspect of its evolution is likely to have been the reorganization and juxtaposition of different polypeptide units to build up new proteins. If the protein-coding unit must be a continuous series of codons, every such reconstruction would require a precise recombination of DNA to place the two protein-coding units in register, end to end in the same reading frame. Furthermore, if this combination should not be successful, the cell has been damaged, because it has lost the original protein-coding units. But if an approximate recombination of DNA could place the two protein-coding units within the same transcription unit, splicing patterns could be tried out to combine the two proteins into a single polypeptide chain. And if these combinations are not successful, the original protein-coding units remain available for further trials. Essentially, this allows the cell to try out controlled deletions in RNA without suffering the damaging instability that could occur from applying this procedure to DNA.

If current proteins evolved by the combination of originally separate ancestral proteins, this is likely to have occurred sequentially over some period of time, with one exon added at a time to existing proteins. If this model is realistic, we can ask whether the different functions from which these genes were pieced together are discernible in their present structure. In other words, can we identify particular functions of current proteins with individual exons?

In some cases, there is indeed a clear relationship between the structures of the gene and protein. The example par excellence is provided by the immunoglobulin proteins, which are coded by genes in which every exon corresponds exactly with a known functional domain of the protein (see Chapter 39).

There are other instances in which some of the exons can be identified with particular functions. Often the first exon, coding for the N-terminal region of the polypeptide, specifies the region that provides the signal sequence involved in membrane secretion. An example is insulin.

Sometimes the evolution of a gene may involve the duplication of exons, creating an internally repetitious sequence in the protein. For example, in chicken collagen a 54 bp exon appears to have been multiplied several times. (In a *D. melanogaster* collagen, however, there are just two large exons.)

In some instances, exons can be identified that code for segments that clearly extend across the junction between two different domains of the protein. Examples are found in collagen. In the ADH gene of *D. melanogaster,* there is a curious duality, with some exons corresponding exactly to protein domains, while others divide a domain into two parts.

There are some interesting cases in which homologous genes differ in structure. This may provide information about their evolution. An example is insulin. In rodents, there are two genes for insulin, whereas other mammals have only one gene. The single insulin gene has two introns. Of the two insulin genes in the rat, one has both introns, whereas the other has only one intron. This implies that the ancestral insulin gene had two introns. Its duplication in rodents was followed by the precise removal of one intron.

A fascinating case of evolutionary conservation is presented by the globins, all of whose genes have three exons. The two introns are located at constant positions relative to the coding sequence. The central exon appears to represent the heme-binding domain of the globin chain. The active protein is a tetramer containing two globin chains of the α type and two of the β type.

Another perspective on this structure is provided by the existence of two other types of protein that are related to globin. Myoglobin is a monomeric oxygen-binding protein of animals, whose amino acid sequence suggests a common (though ancient) origin with the globin subunits. Leghemoglobins are oxygen-binding proteins present in the legume class of plants; like myoglobin, they are monomeric. They too may have a common origin with the other heme-binding proteins.

Leghemoglobin genes contain three introns, the first and last of which occur at points in the coding sequence that are homologous to the locations of the two introns in the globin genes. This remarkable similarity suggests an exceedingly ancient origin for the heme-binding proteins in the form of a split gene.

The central intron of leghemoglobin separates two exons that together code for the sequence corresponding to the single central exon in globin. Could the central exon of the globin gene have been derived by a fusion of two central exons in the ancestral gene, bringing together the sequences coding for two parts of the protein chain that together form the heme-binding structure? It may be interesting to determine the structure of the myoglobin gene.

The equation of at least some exons with protein domains supports the idea that this has been a fundamental relationship in the evolution of genes. It is clear that the duplication and juxtaposition of exons may have played an important role in evolution. We cannot trace the actual events involved in the evolution of every gene; there are several relationships between exons and protein domains that do not conform to a simple equation, but these could be accounted for if events such as exon fusions have modified the ancestral structure during the evolution of nuclear genes. There are some situations, however, in which we are faced with more widespread discrepancies in gene structures.

The genes of yeast and mammalian mitochondria code for virtually identical mitochondrial proteins, in spite of a striking difference in gene organization. Vertebrate mitochondrial genomes are very small, with an extremely compact organization of continuous genes (see Chapter 22), whereas yeast mitochondrial genomes are larger and have some complex interrupted genes. Which is the ancestral form?

Is the presence of introns in the yeast mitochondrion at odds with the view that mitochondria are the result of an ancient endosymbiotic event in which a bacterial prototype was inserted into eucaryotic cytoplasm? This is not a problem if we assume that the insertion predated the loss of introns from the procaryotic line, although then we have to assume that the ability to splice out introns was lost independently during the evolution of bacteria and of vertebrate mitochondria.

Polymorphisms seem common in genes for rRNA and tRNA, where alternative forms can often be found, with and without introns. In the case of the tRNAs, where all the molecules conform to the same general structure, it seems unlikely that evolution brought together the two regions of the gene. After all, the different regions are involved in the base pairing that gives significance to the structure. So here it may be that the introns were inserted into continuous genes.

FURTHER READING

An extensive discussion of structures of interrupted genes is given in **Lewin's** *Gene Expression, 2, Eucaryotic Chromosomes* (Wiley, New York, 1980, pp. 790–847). Some further thoughts are given in *Cell* (**22,** 645–646, 1980). The first major review of gene structure since the discovery of interrupted genes is provided by **Breathnach** & **Chambon** (Ann. Rev. Biochem. **50,** 349–383, 1981).

PART 6
CLUSTERS OF
RELATED SEQUENCES

In the genetic programme, therefore, is written the result of all past reproductions, the collection of successes, since all traces of failures have disappeared. The genetic message, the programme of the present-day organism, therefore, resembles a text without an author, that a proof-reader has been correcting for more than two billion years, continually improving, refining and completing it, gradually eliminating all imperfections.

François Jacob, 1973

CHAPTER 21
STRUCTURAL GENES:
EXTERNAL RELATIONSHIPS

It is beginning to seem that few genes are alone in the eucaryotic genome. Although most genes behave as members of nonrepetitive DNA in liquid hybridization analysis, more detailed characterization of their sequences by isolation of genomic fragments usually reveals the presence of other sequences related to their exons. Sometimes these other sequences are functional genes, less or more distantly separated in evolution; sometimes they are nonfunctional **pseudogenes,** relics of evolution, descended from genes that once must have been functional, but that now have become inactive and accumulated many mutational changes.

Considering the interrupted, sometimes very extensively spread out structure of eucaryotic genes, we can picture the eucaryotic genome as a sea of introns (mostly but not exclusively unique in sequence), in which islands of exons (sometimes very short) are strung out in individual archipelagoes that constitute genes. Comparing the corresponding exons in related genes, we see a relationship that emphasizes the importance in evolution of gene duplication as a mechanism for generating new genes. One copy can evolve via mutation, while the other retains its original function. Thus the history of a gene is likely to encompass a series of events in which first its component exons were brought together to form a coding region; later

the entire series of exons and introns constituting the gene may have been duplicated. The duplication will have been followed by lesser divergence in exon sequences and greater divergence in intron sequences. From the sequences of related genes, we can therefore attempt to reconstruct the history of their evolution.

A set of genes descended by duplication and variation from some ancestral gene is called a **gene family.** Its members may be clustered together or dispersed on different chromosomes (or a combination of both). **Gene clusters** range from extremes where a duplication has generated two adjacent related genes to cases where hundreds of identical genes lie in a tandem array. Situations where related genes are dispersed at different locations must have arisen by translocation at some time after a duplication event.

The members of a structural gene family have related or identical functions, although they may be expressed at different times or in different cell types. In cases where there is an extensive tandem repetition of a gene or genes, the product is needed in unusually large amounts. Examples are the genes for rRNA or histone proteins (see Chapter 23). Sometimes the significance of a lesser repetition is evident, in providing slightly different proteins for particular circumstances. For example, different globin proteins are provided for use in embryonic and adult red blood cells, while different actins are utilized in muscle and in nonmuscle cells. Sometimes no significance is discernible in a repetition; for example, there is no difference apparent in the expression or function of the duplicate insulin genes of rodents (and we know that a single insulin gene indeed is adequate in other mammals).

THE MULTIPLE TYPES OF GLOBIN PROTEINS

The transport of oxygen through the bloodstream is a function central to the animal kingdom, coded by an ancient gene family. The major constituent of the red blood cell is the globin tetramer, associated with its heme (iron-binding) group in the form of hemoglobin. Genes coding for the globin polypeptide chains have been analyzed in several species. All functional globin genes have the same general structure, divided into three exons as shown previously in Figure 20.16. We have concluded that all globin genes are derived from a single ancestral gene (see Chapter 20). We may therefore learn a great deal about the general mechanisms involved in the evolution of gene systems by tracing the development of individual globin genes within and between species.

In adult cells, the globin tetramer consists of two identical α chains and two identical β chains. The α and β globin genes are coded by independent genetic loci whose expression must be coordinated to ensure equivalent production of both polypeptides. This system therefore provides an example of the need for simultaneous control of dispersed genes to generate a particular cell phenotype.

Embryonic blood cells contain hemoglobin tetramers that are different from the adult form. Each tetramer contains two identical α-like chains and two identical β-like chains, each of which is related to the adult polypeptide and is later replaced by it. This is an example of developmental control, in which different genes are successively switched on and off to provide alternative products that fulfill the same function at different times.

The details of the relationship between embryonic and adult hemoglobins vary with the organism. The human pathway consists of three stages (which still

are not completely delineated). The hemoglobins found at different stages of development are

Embryonic (up to 8 weeks)	$\zeta_2\epsilon_2$ and $\zeta_2\gamma_2$ and $\alpha_2\epsilon_2$
Fetal	$\alpha_2\gamma_2$
Adult	$\alpha_2\delta_2$ and $\alpha_2\beta_2$

Zeta and alpha are the two α-like chains. Epsilon, gamma, delta, and beta are the β-like chains. In terms of developmental events, the simplest pathway is to suppose that zeta is the first α-like chain to be expressed, but it is soon replaced by alpha itself. While in the β-pathway, epsilon and gamma are expressed first, with delta and beta replacing them later. In adults, the $\alpha_2\beta_2$ form provides 97% of the hemoglobin, $\alpha_2\delta_2$ is about 2%, and about 1% is provided by persistence of the fetal form $\alpha_2\gamma_2$.

GLOBIN GENES ARE ORGANIZED IN CLUSTERS

The division of globin chains into α-like and β-like is supported by the structure of the protein and by the organization of the genes. The entire β-like family is defined by the ability to combine directly with the α chain. The zeta chain is defined as α-like by its ability to combine with the embryonic β-like chains. Each type of globin turns out to be coded by genes organized into a single cluster. In man, the α cluster lies on chromosome 11, and the β cluster lies on chromosome 16.

Each cluster was discovered by the ability of genomic fragments to hybridize with probes representing adult globin mRNA. Extension of this analysis along the DNA led to the discovery of further genes, some representing known or unknown embryonic globins, others comprising nonfunctional pseudogenes.

The structures of the two clusters in the human genome are illustrated in **Figure 21.1.** The organization of these genes in other higher primates is very similar. For example, the β clusters of gorilla and baboon are virtually indistinguishable from that of man. All of the genes are transcribed from left to right, and they lie in the order of developmental expression: 5'—embryonic—fetal (in β cluster)—adult—3'.

Stretching over 50 kb, the β cluster contains five

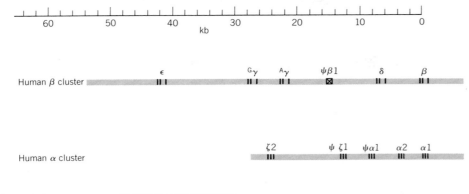

III Interrupted gene
(exons are dark)

⊠ Gene of unknown structure

Figure 21.1
The human α-like and β-like globin gene families each are organized into a single cluster that includes functional genes and pseudogenes (ψ).

functional genes—epsilon, two gamma, delta, and beta—and one pseudogene. The existence of all of the functional genes was known previously. The two gamma genes differ in their coding sequence in only one amino acid; the G variant has glycine at position 136, where the A variant has alanine.

The more compact α cluster extends over 28 kb and includes one active zeta gene, one zeta pseudogene, two alpha genes, and one alpha pseudogene. The existence of two α genes, coding for the identical protein, was suspected previously on the basis of genetic

evidence. Two (or more) identical genes present on the same chromosome are described as **nonallelic** copies.

A similar general organization is found in other vertebrate globin gene clusters, but details of the types and numbers of genes may vary. Some examples of β-globin clusters are illustrated in **Figure 21.2.**

Four β-like genes have been identified in the rabbit: two embryonic, one pseudo, and one adult. Again, the functional genes lie in order of expression, 5' to 3'.

In the goat, there are seven β-like globin genes (not

Figure 21.2
Clusters of β-globin genes and pseudogenes are found in other vertebrates.

shown in the figure). Two are embryonic, one fetal, two adult, and two pseudo.

In the mouse, there must be at least nine β-like globin genes. Seven genes so far have been identified. These include the two adult β genes, the major being expressed at greater levels than the minor. The genes β*h1* and β*h0* have sequences that about equally well resemble the known adult and embryonic genes; so these could provide a counterpart to the human β-like fetal globin genes (although the protein products have not yet been detected). The gene ε*y3* is closely related to the known embryonic proteins, and seems likely to code for an unknown β-like embryonic chain. Genes have not yet been identified for the two known embryonic β-like globins, ε*y2* and ε*z*. In addition to the five identified and two unidentified functional genes, there are two other genes, β*h3* and β*h2,* whose sequences do not seem to fit known mouse globin types, and which may be pseudogenes.

Clusters can be smaller and need not include pseudogenes. In the chicken, the β-globin cluster is less than 14 kb in length, and seems to include only four functional β-like genes. The outside two are embryonic and the inside two are adult, so there is no correlation between time of expression and position of the gene.

The characterization of these gene clusters makes an important general point. There may be more members of a gene family, both functional and nonfunctional, than we would suspect on the basis of protein analysis. The extra functional genes may represent duplicates that code for identical polypeptides; or they may be related to known proteins, although different from them, and presumably expressed only briefly or in low amounts. The pseudogenes show varying relationships with the functional genes, and can be analyzed only at the level of the DNA sequence. Taking these features into account, it is hard to be certain that all the members of a cluster have been identified until flanking regions have been analyzed well beyond the terminal members.

Are *all* globin genes confined to the clusters? In each genome, all the *active* globin genes lie in the α and β clusters, sometimes together with pseudogenes. In the mouse genome, some additional pseudogenes have been found that are not part of the clusters. The α cluster consists of one embryonic and two

adult genes. Two identified pseudogenes each lie on different chromosomes from the cluster. There may also be others. These pseudogenes must have originated by translocation events that moved a gene away from the cluster.

With regard to the question of how much DNA is needed to code for a particular function, we see that coding for the β-like globins requires a range of from about 14 kb to 50 kb in different mammals and a bird. Needless to say, this is much greater than we would expect just from scrutinizing the known β-globin proteins. It does not seem likely that any other proteins are coded within the region of the β cluster, but we do not yet know how much of the noncoding DNA (both flanking and in introns) serves functions necessary for β-globin gene expression. And until many more genes or gene clusters coding for particular proteins have been identified, we shall not be able to tell whether the complexity of the globin clusters is unusual or typical.

UNEQUAL CROSSING-OVER REARRANGES GENE CLUSTERS

There are frequent opportunities for rearrangement in a cluster of related or identical genes. We can see the results by comparing the mammalian β clusters included in Figures 21.1 and 21.2. Although the clusters serve the same function, and all have the same general organization, each is different in size, there is variation in the total number and the types of β-globin genes, and the numbers and structures of pseudogenes are different. All of these changes must have occurred since the mammalian radiation, about 85 million years ago (this was the last time in evolution common to all the mammals). The comparison makes the general point that gene duplication, rearrangement, and variation may be as important a factor in evolution as the slow accumulation of point mutations in individual genes to allow selection of improved proteins. What types of mechanism are responsible for reorganization of the genes?

A gene cluster can expand or contract by **unequal crossing-over,** when recombination occurs between nonallelic genes, as illustrated in **Figure 21.3.** Usually, recombination occurs (as described in Chapter 1) be-

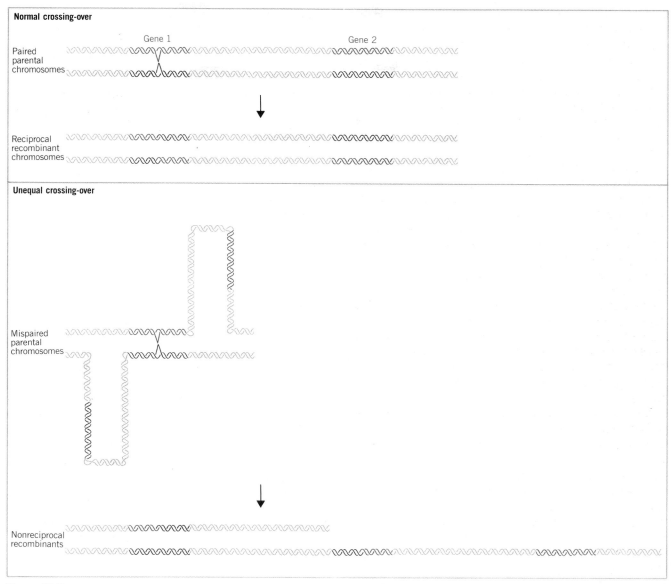

Figure 21.3
Gene number can be changed by unequal crossing-over.

If gene 1 of one chromosome pairs with gene 2 of the other chromosome, the other gene copies are excluded from pairing, as indicated by the extruded loops. Recombination between the mispaired genes produces one chromosome with a single (recombinant) copy of the gene and one chromosome with three copies of the gene (one from each parent and one recombinant).

tween corresponding sequences of DNA held in exact alignment between the two homologous chromosomes. However, when there are two copies of a gene on each chromosome, occasionally a misalignment may occur to allow pairing between them. (This requires some of the adjacent regions to go unpaired.)

When a recombination event occurs between the mispaired gene copies, it generates **nonreciprocal recombinant chromosomes,** one of which has a duplication of the gene and one a deletion. The duplication includes all the material between the points of unequal recombination; the deletion extends between the same points. The first recombinant therefore has an increase in the number of gene copies from 2 to 3, while the second has a decrease from 2 to 1.

In this example, we have treated the noncorresponding gene copies 1 and 2 as though they were entirely homologous. However, unequal crossing-over also can occur when the adjacent genes are well related (although presumably the probability of establishing mispairing is less than when they are identical). A more serious obstacle to unequal crossing-over is presented by the interrupted structure of the genes. In a case such as the globins, the corresponding exons of adjacent gene copies are likely to be well enough related to support pairing; but the sequences of the introns may have diverged quite appreciably. The restriction of pairing to the exons considerably reduces the continuous length of DNA that can be involved and may correspondingly lower the chance of unequal crossing-over. Indeed, one idea is that divergence between introns may enhance the stability of gene clusters by hindering the occurrence of unequal crossing-over.

MANY α-THALASSEMIAS RESULT FROM UNEQUAL CROSSING-OVER

The occurrence of unequal crossing-over in the human globin gene clusters is revealed by the nature of certain thalassemias. A **thalassemia** may result from any mutation that prevents synthesis of either α or β globin. Depending on the severity, the thalassemias are classified as α^0 and β^0 (no detectable synthesis of one type of chain) or α^+ and β^+ (reduced levels of synthesis). Thalassemias have been identified through patients suffering from anemias (which tend to be concentrated in certain human populations, such as the Mediterranean). Many of the zero type result from deletions of part of the relevant globin gene cluster. In at least some cases, the ends of the deletion lie in regions that are homologous, which is exactly what would be expected if it had been generated by unequal crossing-over.

Two general types of deletion are found among the α-thalassemias. The **α-thal-1** eliminates both the α genes. The **α-thal-2** eliminates only one of the two α genes. The names refer to the individual chromosome carrying the deletion. Depending on the diploid combination of thalassemic chromosomes, an individual may have any number of α chains from zero to four. There are few differences from the wild type (four α genes) in individuals with three or two α genes. With only one α gene, the excess β chains form the unusual tetramer β_4, which causes **HbH disease.** The complete absence of α genes results in **hydrops fetalis,** which is fatal at or before birth.

Figure 21.4 summarizes the deletions that cause these α-thalassemias. The α-thal-1 deletions all are long, varying in the location of the left end, and with the positions of the right ends unknown at present. The α-thal-2 deletions all are short. The L form removes 4.2 kb of DNA including the α2 gene. It probably results from unequal crossing-over, because the ends of the deletion lie in regions that are homologous, just to the right of the ψα1 and α2 genes, respectively. The R form results from the removal of exactly 3.7 kb

Figure 21.4
Thalassemias result from various deletions in the α-globin gene cluster.

of DNA, the precise distance between the α1 and α2 genes. It appears to have been generated by unequal crossing-over between the α1 and α2 genes themselves. This is precisely the situation depicted in Figure 21.3.

The same unequal crossing-over that generated the thalassemic chromosome should also have generated a chromosome with three α genes. Individuals with such chromosomes have been identified in several populations. In some, the frequency of the triple α locus is about the same as that of the single α locus; in others, the triple α genes are much *less* common than single α genes. This suggests that unknown selective factors may operate in different populations to adjust the gene levels.

At all events, variations in the number of α genes are found relatively frequently, which argues that unequal crossing-over in the cluster must be fairly common. It appears, in fact, to be somewhat more common than in the β cluster. One possible reason is that the introns in α genes are much shorter, and therefore present less impediment to mispairing between nonhomologous genes.

NEW GENES GENERATED IN β-THALASSEMIAS

The β-thalassemias are interesting in several respects, especially because many of them may represent defects in regulation. This provides a rare opportunity to dissect the expression of a gene family.

The β⁰-thalassemias vary widely in type. Some of this class have no detectable deletion, and so the complete absence of β chains must be due to mutations that affect transcription or translation. In one case, mutation has generated a nonsense codon at position 17, so the defect is in translation. In another, the defect lies at an earlier stage of gene expression; nuclear RNA is produced, but there is no mRNA in the cytoplasm. The cause is a mutation that interferes with the splicing of mRNA (see Chapter 26).

The β⁰ deletions are summarized in **Figure 21.5.** In some (rare) cases, only the β gene is affected. In these, there is a deletion of 600 bp, extending from the second intron through the 3′ flanking regions. In the other cases, more than one gene of the cluster is affected.

The **Hb Lepore** type provided the classic evidence for deletion via unequal crossing-over between linked genes. The β and δ genes differ only about 7% in sequence. Unequal recombination deletes the material between the genes, thus fusing them together (see Figure 21.3). The fused gene produces a single β-like chain that consists of the N-terminal sequence of δ joined to the C-terminal sequence of β.

Several types of Hb Lepore now are known, the difference between them lying in the point of transition from the δ to the β sequences. Thus when the δ and β genes pair for unequal crossing-over, the exact point of recombination determines the position at which the switch from δ to β sequence occurs in the amino acid chain.

The reciprocal of this event has been found in the form of **Hb anti-Lepore,** which is produced by a gene

Figure 21.5
Deletions in the β-globin gene cluster cause several types of thalassemia.

that has the N-terminal part of β and the C-terminal part of δ. This fusion gene should lie between normal δ and β genes.

Evidence that unequal crossing-over can occur between more distantly related genes is provided by the identification of **Hb Kenya,** another fused hemoglobin. This contains the N-terminal sequence of the $^A\gamma$ gene and the C-terminal sequence of the β gene. The fusion must have resulted from unequal crossing-over between $^A\gamma$ and β, which differ about 20% in sequence.

The absence of both δ and β can result in either of two phenotypes. In **HPFH** (hereditary persistence of fetal hemoglobin), there are no clinical symptoms; the disease is ameliorated because the synthesis of fetal hemoglobin ($\alpha_2\gamma_2$) continues after the time in development at which it would usually have been turned off. In **δβ thalassemia,** there are anemic symptoms, because, although γ gene expression may continue in adult life, it is less effective than in HPFH.

A variety of deletions prevent synthesis of both δ and β, and these are named in Figure 21.5 according to the class of phenotype. One idea about the difference between the HPFH and δβ thalassemias is that deletions causing HPFH might remove some regulator site—located between $^A\gamma$ and δ—usually responsible for turning off the expression of the γ genes. A similar switch could also exist in the α cluster, because the deletion of both α genes leads to continued expression of the ζ gene, which usually is turned off in adult life.

GENE CLUSTERS SUFFER CONTINUAL REORGANIZATION

From the differences between the globin gene clusters of various mammals, we may infer that duplication followed (sometimes) by variation has been an important feature in the evolution of each cluster. The human thalassemic deletions demonstrate that unequal crossing-over continues to occur as a current event in both globin gene clusters. Each such event generates a duplication as well as the deletion, and we must account for the fate of both recombinant loci in the population. Deletions can also occur (in principle) by recombination between homologous sequences lying on the *same* chromosome. This does not generate a corresponding duplication.

It is difficult to estimate the natural frequency of these events, because selective forces rapidly adjust the levels of the variant clusters in the population. There may be a rough correlation between the likelihood of an unequal crossing-over and the relationship of the genes—the more closely related (including both exons and introns), the greater the chance of mispairing. (However, some unequal recombinations do not involve the genes themselves, but rely on repetitive sequences nearby.)

Generally a contraction in gene number is likely to be deleterious and selected against. However, in some populations, there may be a balancing advantage that maintains the deleted form at a low frequency.

What is the result of an expansion? The only examples that have been characterized are the triple α locus and the anti-Lepore. Individuals who possess 5 α-globin genes (one normal locus and one triple) do not display any change in hemoglobin synthesis. However, it is possible that a further increase (to the triple homozygote) could be deleterious, by unbalancing globin synthesis through the production of excess α chains. Individuals with anti-Lepore have the fusion βδ gene as an addition to the normal β and δ genes. It is possible that the additional chain is deleterious because it interferes with the assembly of normal hemoglobin.

So these particular changes in the number of genes are unlikely to have a selective advantage that will cause them to spread through the population. However, the structures of the present human clusters show several duplications that attest to the importance of such mechanisms. The *functional* sequences include two α genes coding the same protein, fairly well-related β and δ genes, and two almost identical γ genes. These comparatively recent independent duplications have survived in the population, not to mention the more distant duplications that originally generated the various types of gene. Other duplications may have given rise to pseudogenes or have been lost.

Two striking examples have occurred in which attempts to propagate clusters by cloning in *E. coli* have led to the frequent generation of deletions in the bacterium. One gene of the four-member chicken β clus-

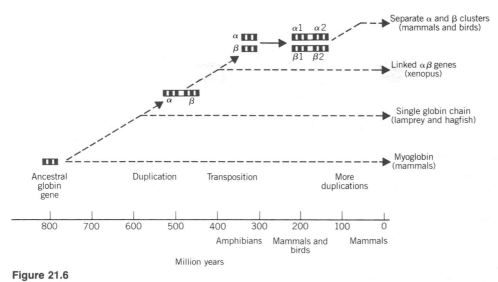

Figure 21.6
All globin genes have evolved by a series of duplications, transpositions, and mutations from a single ancestral gene.

ter was hard to isolate, because it was usually deleted during cloning, probably by recombination involving homologous sequences on either side. In the human α cluster, two types of deletion that occur during propagation in *E. coli* correspond exactly with the two types of α-thal-2 deletion shown in Figure 21.4. The occurrence of exchanges in bacteria that have consequences identical to those observed naturally in the human population suggests that these events are an inevitable consequence of the mechanisms involved in DNA recombination. Thus we may expect continual duplication and deletion to be a feature of all gene clusters.

AN EVOLUTIONARY TREE FOR GLOBIN GENES

From the organization of globin genes in a variety of species, we should be able eventually to trace the evolution of present globin genes from a single ancestral globin gene. Our present view of the evolutionary descent is pictured in **Figure 21.6.**

Since there are separate clusters for α and β globins in both birds and mammals, the α and β genes must have been physically separated before the mammals and birds diverged from their common ancestor, an event that occurred probably about 270 million years ago.

The preceding stage of evolution is represented by the state of the globin genes in the frog *X. laevis,* where the α and β genes are linked, lying only about 8 kb apart. (The entire doublet is present in two non-identical copies, probably due to an ancestral tetra-ploidization.) The amphibians separated from the mammalian/avian line about 350 million years ago, so we can place the separation of the α- and β-globin genes—presumably by a transposition—somewhere around the period of early vertebrate evolution.

For the preceding stage, we have evidence of protein chains, but not yet of genes. Some "primitive" fish have only a single type of globin chain, so they must have diverged from the line of evolution before the ancestral globin gene was duplicated to give rise to the α and β variants. This appears to have occurred roughly 500 million years ago, during the evolution of the bony fish.

We can trace the globin gene even further back, since the amino acid sequence of the single chain of mammalian myoglobin argues that it diverged from the globin line of descent about 800 million years ago. We do not yet have any evidence about its gene. Representing an even earlier separation, the leghemoglobin gene of plants is related to the globin gene (see Chapter 20).

SEQUENCE DIVERGENCE INDICATES EVOLUTIONARY SEPARATION

Most changes in protein sequence occur by small mutations that accumulate slowly with time. Point mutations and small insertions and deletions occur by chance, probably with more or less equal probability in all regions of the genome, except for hotspots at which the frequency is much increased. Many mutations that change the amino acid sequence are *deleterious* and will be eliminated fairly rapidly by natural selection (the rate of removal depending on the extent of the damaging effect). Fewer mutations will be advantageous, but these may spread through the population, eventually replacing the former sequence. When a new variant replaces the previous version of the gene, it is said to have been **fixed** in the population. A highly contentious issue is what proportion of mutational changes in an amino acid sequence may be **neutral,** that is, without any effect on the function of the protein, and able therefore to accrue as the result of **random drift and fixation.**

The rate at which mutational changes accumulate is a characteristic of each protein, presumably depending at least in part on its flexibility with regard to change. Within a species, a protein evolves by mutational substitution, followed by elimination or fixation within the single breeding pool. The presence in the population of two (or more) allelic variants is called a **polymorphism.** This can be a stable situation, in which case either there is no advantage to each form or its advantages and disadvantages cancel out. Or the polymorphism may be transient, as must be the case while one variant is replacing another. When we scrutinize the gene pool of a species, what we see are the variants that have survived.

When a species separates into two new species, each now constitutes an independent pool for evolution. By comparing the corresponding proteins in two species, we see the differences that have accumulated between them *since the time when their ancestors ceased to interbreed.* Some proteins are highly **conserved,** showing little or no change from species to species. This indicates that almost any change is deleterious and therefore is selected against.

The difference between two proteins is expressed as their **divergence,** the percent of residues that are different. When the corresponding protein is examined in a range of species, the divergence in each pairwise comparison is (more or less) proportional to the time since they separated. This provides an **evolutionary clock** that measures the accumulation of mutations at an apparently even rate during the evolution of a given protein.

The rate at which divergence has developed can be measured as the percent difference per million years, or as its reciprocal, the unit evolutionary period (UEP), which is the time in millions of years that it takes for 1% divergence to develop. Once the clock has been established by pairwise comparisons between species (remembering the practical difficulties in establishing the actual time of speciation), it can be applied to related genes *within* a species. From their divergence, we can therefore calculate the time since the duplication in which they originated.

TWO TYPES OF DIVERGENCE IN DNA

The divergence between nucleic acid sequences can be different from that between the corresponding proteins. The source of this difference is the representation of each amino acid in a three-base codon, in which often the third base has no effect on the meaning. We must therefore divide the available nucleotides into potential **replacement sites** and **silent sites.** At replacement sites, a mutation alters the amino acid that is coded. The effect of the mutation (deleterious, neutral, or advantageous) depends on the result of the amino acid replacement. At silent sites, mutation only substitutes one synonym codon for another, so that there is no change in the protein. Usually the replacement sites account for 75% of a coding sequence and the silent sites provide 25%.

Although silent mutations are neutral with regard to

the protein, they could affect gene expression via the sequence change in RNA. Potential effects are changes in the secondary structure of the RNA, which might influence transcription, processing, or translation. Another possibility is that the change in synonym codons calls for a different tRNA to respond; this could influence the efficiency of translation. And, of course, in addition to the coding sequences of a gene, there are nontranslated regions. Here again, mutations are potentially neutral, apart from their effects on either secondary structure or (usually rather short) regulatory signals.

The mutations in replacement sites should correspond with the amino acid divergence, which essentially is a count of the percent of changes. Actually, the measured divergence underestimates the differences that have occurred during evolution, because of the occurrence of multiple events at one codon. Usually a correction is made for this. Without the correction, a nucleic acid divergence of 0.45% at replacement sites corresponds to an amino acid divergence of 1% (assuming that the average number of replacement sites per codon is 2.25).

To take the example of the human β- and δ-globin chains, there are 10 differences in 146 residues, a divergence of 6.9%. The DNA sequence has 31 changes in 441 residues. However, these changes are distributed very differently in the replacement and silent sites. There are 11 changes in the 330 replacement sites, but 20 changes in only 111 silent sites. This gives (corrected) rates of divergence of 3.7% in the replacement sites and 32% in the silent sites, almost an order of magnitude in difference.

The striking difference in the divergence of replacement and silent sites immediately demonstrates the existence of much greater constraints on nucleotide positions that influence protein constitution than on those that do not. This makes it unlikely that more than a small proportion of the amino acid changes are neutral. If we take the rate of mutation at silent sites to indicate the underlying rate of mutational fixation (this is to assume that there is no selection at all at the silent sites), then over the period since the β and δ genes diverged, there should have been changes at 32% of the 330 replacement sites, a total of 105. All but 11 of these have been eliminated, which means that roughly 90% of the mutations did not survive.

USING THE CLOCK TO FOLLOW THE GLOBIN GENES

By comparing the sequences of homologous genes in different species, the rate of divergence at both replacement and silent sites can be determined.

In pairwise comparisons, there is an average divergence of 10% in the replacement sites of either the α- or β-globin genes of mammals that have been separate since the mammalian radiation occurred about 85 million years ago. This corresponds to a replacement divergence rate of 0.12% per million years, or to a UEP of 8.5.

This rate is steady when the comparison is extended to genes that diverged in the more distant past. For example, the average replacement divergence between corresponding mammalian and chicken globin genes is 23%. Relative to a separation roughly 270 million years ago, this gives a rate of 0.09% per million years.

To go further back, we can compare the α with the β-globin genes within a species. These have been diverging since they separated ≥ 500 million years ago (see Figure 21.6). They have an average replacement divergence of about 50%, which gives a rate of 0.1% per million years.

These data are plotted in **Figure 21.7,** which shows that replacement divergence in the globin genes has a clock with an average rate of about 0.096% per million years or a UEP of 10.4. Considering the uncertainties in estimating the points at which the species diverged, the results lend good support to the idea that there is a linear clock.

The data on silent site divergence are much less clear. In every case, it is evident that the silent site divergence is much greater than the replacement site divergence, by a factor that varies from 2 to 10. But the spread of silent site divergences in pairwise comparisons is so great that it is hard to know whether a clock is applicable.

From Figure 21.7 it is clear that the rate is not linear with regard to time. *If we assume that there must be zero divergence at zero years of separation,* we see that the rate of silent site divergence is much greater for about the first 100 million years of separation. Then it becomes slower. One interpretation is that a fraction of roughly half of the silent sites is rapidly (within 100

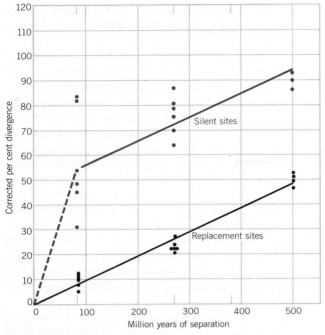

Figure 21.7
Divergence of DNA sequences depends on evolutionary separation. Each point on the graph represents a pairwise comparison.

Figure 21.8
Replacement site divergences between pairs of β-globin genes allow the history of the human cluster to be reconstructed.

This tree accounts for the separation of *classes* of globin genes. Duplications of individual genes are of unknown origin, as indicated by the circles. The time of the α-ζ divergence is not known.

million years) saturated by mutations; this fraction behaves as neutral sites. The other fraction accumulates mutations more slowly, at a rate approximately the same as that of the replacement sites; this fraction identifies sites that are silent with regard to the protein, but that come under selective pressure for some other reason. At all events, these uncertainties make the silent sites unreliable as an evolutionary clock.

Now we can reverse the calculation of divergence rates to estimate the times since genes within a species have been separate. The difference between the human β and δ genes is 3.7% for replacement sites. At a UEP of 10.4, these genes must have diverged 10.4 × 3.7 ≈ 40 million years ago—about the time of separation of the lines leading to New World monkeys, Old World monkeys, great apes, and humans. All of these higher primates have both β and δ genes, which suggests that the gene divergence commenced just before this point in evolution.

Proceeding further back, the divergence between

the replacement sites of γ and ε genes is 10%, which corresponds to a time of separation about 100 million years ago. The separation between embryonic and fetal globin genes therefore may just have preceded or may have accompanied the mammalian radiation.

The β and γ genes have a replacement divergence of 18%, which corresponds to a separation approaching 200 million years ago. Thus the divergence between adult globin genes and embryonic/fetal genes must have occurred some time after the separation of the mammals from the birds (that is, during the evolution of the reptiles preceding the mammalian radiation).

We can therefore construct the evolutionary tree for the human globin gene clusters depicted in **Figure 21.8.** The features that evolved before the mammalian radiation should be found in all mammals. The features that evolved afterward must have evolved independently in different mammalian lines. In each species there have been comparatively recent changes in the structures of the clusters, since we see differences in gene number (one β globin in human, two in mouse) or type (we are not yet sure whether there are separate embryonic and fetal β-like globins in rabbit and mouse).

MECHANISMS FOR MAINTAINING ACTIVE SEQUENCES

The duplication of a gene is likely to result in an immediate relaxation of the evolutionary pressure on its sequence. Now that there are two identical copies, a change in the sequence of either one of them will not deprive the organism of a functional protein, since the original amino acid sequence continues to be coded by the other copy. Thus the selective pressure on the two genes is diffused, until one of them mutates sufficiently away from its original function to refocus all the selective pressure on the other. It seems possible that, immediately following a gene duplication, changes accumulate more rapidly in one of the copies, leading eventually to a new function (or to its disuse in the form of a pseudogene). If a new function develops, the gene then may evolve at the same, slower rate characteristic of the original function. This is likely to be the sort of mechanism responsible for the separation of functions between embryonic and adult globin genes.

Yet there are some instances where duplicated genes retain the same function, coding for the identical or nearly identical proteins. Identical proteins are coded by the two human α-globin genes, and there is only a single amino acid difference between the two γ-globin proteins. Is this not paradoxical in view of the expectation that duplication should be followed by variation? In these cases, we can probably exclude the only simple explanation, that both genes are needed and remain under selective pressure because of the necessity to make enough protein. So how is selective pressure maintained on both genes?

Two general types of mechanism have been proposed. They share the principle that nonallelic genes are not independently inherited, but that both must be continually regenerated from *one* of the copies of a preceding generation. When a mutation occurs in one copy, either it is by chance eliminated (because the sequence of the other copy takes over), or it is spread to both duplicates (because the mutant copy becomes the dominant version). Spreading exposes a mutation to selection. The result is that the two genes evolve together as though only a single locus existed. This is called **coincidental** or **concerted evolution** (occasionally **coevolution**).

One mechanism supposes that the sequences of the nonallelic genes are directly compared with one another and homogenized by enzymes that recognize any differences. This can be done by exchanging single strands between them, to form genes one of whose strands derives from one copy, one from the other copy. Any differences show as improperly paired bases, which may be the subject of attention from enzymes able to excise and replace a base, so that only A-T and G-C pairs survive. This type of event is called **gene conversion** and is associated with genetic recombination as described in Chapter 35.

Another mechanism is for both copies of the gene physically to be regenerated from one copy. This could occur by unequal crossing-over. Following the sort of event depicted in Figure 21.3, for example, the chromosome carrying a triple locus could suffer deletion of one of the genes. Of the two remaining genes, 1½ represent the sequence of one of the original copies; only ½ of the sequence of the other original copy has survived. Any mutation in the first region now exists in both genes and is subject to selective pressure. This type of event is called **crossover fixation** and is discussed in the context of satellite DNA evolution in Chapter 24.

We should be able to ascertain the scope of these mechanisms by comparing the sequences of duplicate genes. If they are subject to concerted evolution, we should not see the accumulation of silent site substitutions between them (because the homogenization process applies to these as well as to the replacement sites). We know that the extent of the maintenance mechanism need not extend beyond the gene itself, since there are cases of duplicate genes whose flanking sequences on either side are entirely different. Indeed, we may see abrupt boundaries that mark the ends of the sequences that were homogenized. Also, we must remember that the existence of such mechanisms can invalidate the determination of the history of such genes via their divergence, *because the divergence reflects only the time since the last homogenization/regeneration event, not the original duplication.*

PSEUDOGENES ARE DEAD ENDS OF EVOLUTION

Pseudogenes are defined as such by their possession of sequences that are related to those of the functional genes, but that cannot be translated into a functional protein.

Some pseudogenes have the same general structure as functional genes, with sequences corresponding to exons and introns in the usual locations. They are rendered inactive by mutations that prevent any or all of the stages of gene expression. The changes can take the form of abolishing the signals for initiating transcription, preventing splicing at the exon-intron junctions, or terminating translation prematurely. Usually, a pseudogene has several deleterious mutations, presumably because once it ceased to be active, there was no impediment to the accumulation of further mutations. Pseudogenes of this type have been found in many systems, including globin, immunoglobulins, histocompatibility antigens, and so on.

A typical example is the rabbit pseudogene, $\psi\beta2$, which has the usual organization of exons and introns, and is related most closely to the functional gene $\beta1$. But there is a deletion of one base pair at codon 20 of $\psi\beta2$, causing a frameshift that would lead to termination shortly after. Several point mutations have changed later codons representing amino acids that are highly conserved in the β globins. Neither of the two introns any longer possesses boundary sequences that conform adequately to the GT-AG rule, so probably the introns could not be spliced out even if the gene were transcribed. However, there are no transcripts corresponding to the gene, possibly because there have been changes in the 5' flanking region.

Since this list of defects includes mutations potentially preventing each stage of gene expression, we have no means of telling which event originally inactivated this gene. However, from the divergence between the pseudogene and the functional gene sequences, we can estimate when the pseudogene originated and when its mutations started to accumulate.

If the pseudogene had become inactive as soon as it was generated by duplication from $\beta1$, we should expect both replacement site and silent site divergence rates to be the same. (There is no reason for them to be different if the gene is not translated.) But actually there are fewer replacement site substitutions than silent site substitutions. This suggests that at first there was selection against replacement site substitution. From the relative extents of substitution in the two types of site, we can calculate that $\psi\beta2$ diverged from $\beta1$ about 55 million years ago, remained a functional gene for 22 million years, but has been a pseudogene for the last 33 million years.

Similar calculations can be made for other pseudogenes. Some also appear to have been active for some time before becoming pseudogenes. Some appear to have been inactive from the very time of their original generation. The general point made by the structures of these pseudogenes is that each has evolved independently during the development of the globin gene cluster in each species. This reinforces the conclusion that the creation of new genes, followed by their acceptance as functional duplicates, variation to become new functional genes, or inactivation as pseudogenes, is a continuing process in the gene cluster.

The mouse $\psi\alpha3$ gene has an interesting property: it precisely lacks both introns. Its sequence can be aligned (allowing for accumulated mutations) with the α-globin mRNA. The apparent time of inactivation coincides with the original duplication, which suggests that the original inactivating event may have been associated with the loss of introns. Without introns, the gene may have been unable to function, and thus will have begun immediately to collect mutations. The main question is how the introns could have been lost so precisely. There is no reason why the systems for reconstructing DNA should recognize the exon-intron boundaries, which argues for the involvement at some level of the mRNA itself. This could take the form of insertion of a reverse transcript into the genome (perhaps carried by a retrovirus for which such action is normal) or of pairing between the mRNA and one strand of DNA at some stage of a recombination process (see Chapter 38).

If pseudogenes are evolutionary dead ends, simply an unwanted accompaniment to the rearrangement of functional genes, why are they still present in the genome? Do they fulfill any function or are they entirely without purpose, in which case there should be no selective pressure for their retention? It is necessary to remember that we see those genes that have survived in present populations. In past times, any number of other pseudogenes may have been eliminated. This elimination could occur by deletion of the sequence as a sudden event or by the accretion of mutations to the point where the pseudogene can no longer be recognized as a member of its original sequence family (probably the ultimate fate of any pseudogene that is not suddenly eliminated).

Even relics of evolution can be duplicated. In the β-globin genes of the goat, there are two adult species, β^A and β^C. Each of these has a pseudogene a few kilobases upstream from it (called $\psi\beta^Z$ and $\psi\beta^X$, respectively). The two pseudogenes are better related to each other than to the adult β-globin genes; in particular, they share several inactivating mutations. Also, the two adult β-globin genes are better related to each other than to the pseudogenes. This implies that an original $\psi\beta$-β structure was itself duplicated, giving two functional β genes (which diverged further into the β^A and β^C genes) and two nonfunctional genes (which diverged into the current pseudogenes). The general implication to read into this is that the mechanisms responsible for gene duplication, deletion, and rearrangement act on all sequences that are recognized as members of the cluster, whether or not they are functional. It is left to selection to discriminate among the products.

GENE FAMILIES ARE COMMON FOR ABUNDANT PROTEINS

From experiments to count the number of genes by isolating the genomic fragments corresponding to a particular mRNA or cDNA probe, we know that in several cases there are multiple sequences in the genome that correspond to the probe. Of course, until the nucleotide sequence of each of these is determined, we do not know whether all of them are active or some are pseudogenes. Usually, it is assumed that most of the genes are likely to be active.

By this criterion, there are multiple genes for interferon, actin, and tubulin. Within the gene families, there may be more repetition of some subtypes of gene than others. For example, there seem to be at least eight genes relating to the α (leucocyte) type of interferon in man, but only one for β (fibroblast) interferon.

The multiple genes may be clustered or dispersed. In the case of interferon, there is evidence for clustering. In the example of the actin and tubulin genes of *D. melanogaster,* the copies seem to lie at different chromosomal locations, so they are effectively unlinked.

Is the preservation of structure a usual feature of multiple gene copies? The interferon genes seem to have a common structure, entirely lacking introns. But the actin genes have interrupted structures that vary widely. In these genes, the protein-coding regions are highly homologous, but there is little or no relationship between the flanking or even the nontranslated regions within a species. For example, in *D. melanogaster* there are actin genes with introns in different positions. None of these is the same as the single actin gene found in yeast. There are also differences with the actin genes of the sea urchin, at least some of which are clustered. So if all of these actin genes are derived from a common ancestor, there has been considerable change in the pattern of exons and introns. One possibility is that the ancestral gene had more introns, different introns having been lost in individual copies or species. Of course, this would argue for a more frequent rate of change than we have seen in the globin genes.

The only general conclusions that we can draw from these somewhat restricted results is that functional genes may be interrupted or uninterrupted and that changes in the pattern of interruptions need not affect gene activity. Genes coding for the same or related proteins need not be located in tandem, but may have become dispersed as individual genes or perhaps as small clusters.

FURTHER READING

A review of the structure and function of human hemoglobins is **Maniatis et al.** (*Ann. Rev. Genet.* **14,** 145–178, 1980). The general characteristics of gene and protein evolution are reviewed by **Wilson et al.** (*Ann. Rev. Biochem.* **46,** 573–639, 1977). An original report of a pseudogene lacking introns was made by **Nishioka et al.** (*Proc. Nat. Acad. Sci. USA* **77,** 2806–2809, 1980). Pseudogenes have been encapsulated in review by **Little** (*Cell* **28,** 683–684, 1982). The basis for thalassemic defects has been brought up to date by **Weatherall** & **Clegg** (*Cell* **29,** 7–9, 1982).

CHAPTER 22
GENOMES SEQUESTERED
IN ORGANELLES

No counterpart to the operon has been found in higher eucaryotic genomes, but there are situations in which a defined region of DNA contains a set of genes all of which have the task of contributing to the same aspect of the phenotype. Mitochondria and chloroplasts both possess DNA genomes that code for all of the RNA species and for some of the proteins involved in the functions of the organelle. Although the organelle genome is not a cluster in the sense of comprising a tandem repetition of genes derived by duplication and variation from a common ancestor, we can regard it as a cluster in the sense that all the genes code for proteins whose purpose is related.

In some ways, the organization of an organelle genome may be more akin to the construction of a viral genome than to a cellular genome, in coding for some, but not all, of the functions needed to perpetuate the organelle. The others are coded in the nucleus, expressed via the cytoplasmic protein synthetic apparatus, and imported into the organelle. The relationship is symbiotic rather than parasitic, since the organelle is a permanent component of the cell and not a virus passing through. In effect, the organelle genome comprises a length of DNA that has been physically sequestered in a defined part of the cell,

and is accordingly subject to its own form of expression and regulation.

ORGANELLE GENES SHOW
NONMENDELIAN INHERITANCE

The first evidence for the presence of genes outside the nucleus was provided by nonMendelian inheritance in plants (observed in the early years of this century, just after the rediscovery of Mendelian inheritance). Two characteristics distinguish nonMendelian from Mendelian inheritance.

First, there are abnormal segregation ratios when a cross is made between mutant and wild type. The extreme form is uniparental inheritance, when the genotype of only one parent is inherited and that of the other parent is permanently lost. In less extreme examples, the progeny of one parental genotype exceed the other. Usually, it is the mother whose genotype is preferentially inherited. This is sometimes described as **maternal inheritance.** The important point is that the genotype contributed by the parent of one particular sex predominates, in contrast with the behavior of Mendelian genetics when reciprocal crosses

353

show the contributions of both parents to be equally inherited (see Chapter 1).

The second feature is that in those cases where both genotypes survive, the wild and mutant phenotypes segregate during somatic growth of the plant. Thus in a single heterozygous plant, some tissues may have one parental phenotype, while other tissues display the alternative parental phenotype. This contrasts with the stability in all somatic cells of the entire nuclear genotype present in the zygote, as defined by Mendelian genetics.

Traditional genetic techniques could not be applied to the inheritance of these markers, so all that could be said from these results was that whatever event is responsible does not originate with the chromosomes. Each gene had to be dealt with independently, since no recombination between markers was observed; so it was impossible to construct linkage maps. There was a debate over whether genetic or epigenetic events were involved. (**Epigenetic changes** are those that influence the phenotype, but do not comprise alterations of genotype; for example, stable changes in gene expression.)

When it became clear that true genetic events were involved, the term **cytoplasmic inheritance** became popular to describe the location of the pertinent genes. However, we shall not use this description, since it is important to be able to distinguish between events in the general cytoplasm and those in specific organelles. A better general term for genes not residing within the nucleus is **extranuclear,** leaving to be resolved in each particular case the question of which organelle (if any) carries the gene(s). This leaves us able to use the term **cytoplasmic protein synthesis** to describe the final stage of expression of *nuclear* genes; while organelle genes are transcribed and translated in the *same* organelle compartment in which they reside.

The presence of DNA in mitochondria and chloroplasts has been demonstrated directly by its isolation from purified preparations of the organelles. The features of organelle inheritance are explained by the location and properties of the DNA.

The representation of organelle genetic information in a zygote depends on the relative contributions of each parent or on differential ability of the information to survive (or on both). This contrasts with nuclear genetic information, which derives equally from each parent.

Maternal inheritance in higher animals can be predicted by supposing that the mitochondria are contributed probably entirely by the ovum and not at all by the sperm. Thus the mitochondrial genes are derived exclusively from the mother; and in males they are discarded each generation.

A different cause for uniparental inheritance is found in the alga *Chlamydomonas reinhardii,* in which gametes of equal size fuse to form the zygote. Both parents contribute chloroplast DNA to a zygote. But the chloroplast DNA provided by one parent is degraded shortly after zygote formation, leaving the DNA of the other parent entirely in possession of the organelle.

Given the existence of these two mechanisms, it is not possible in other situations (for example, in higher plants) to predict the reason for the predominance of one parental genotype. Even in those cases where the parents contribute different amounts of cytoplasm to the zygote, so that one is likely to provide more copies of the organelle genome than the other, it is possible that selective mechanisms operate to alter the balance further.

The association of somatic segregation with nonMendelian inheritance suggests that there are multiple copies of the organelle genes in the zygote, and that these may be distributed into different cells during somatic division. We do not know what rules govern the segregation of organelle DNA, in particular whether the parental origin has any influence on the process.

The question of which organelle carries a particular genetic marker is not trivial. Since the only organelle shown to possess DNA in higher animal cells is the mitochondrion, probably it is the sole extranuclear residence of genetic material. But in plants and in some unicellular eucaryotes, both chloroplasts and mitochondria are present. Again these are the only organelles shown to possess DNA; and it is likely (although not proven) that any marker displaying nonMendelian inheritance resides in one or the other compartment. Yet it may not be evident in which of these compartments a particular gene resides. In some cases, the location is clear from the nature of the effect that a mutation has on the phenotype; but in other instances, the defect takes a more general nature not clearly attributable to either organelle.

ORGANELLE GENOMES ARE CIRCULAR DNA MOLECULES

Almost all organelle genomes so far characterized take the form of a single molecule of DNA of unique sequence. This can usually be recovered as a circular duplex, although sometimes breakage during isolation is so frequent that most of the material is recovered in the form of linear fragments. There are some exceptions in ciliate protozoans where mitochondrial DNA is a linear molecule. Usually, there are several copies of the genome in the individual organelle. Since there are multiple organelles per cell, there may be a large number of organelle genomes per cell. Thus although the organelle genome itself is unique, it constitutes a repetitive sequence relative to any nonrepetitive nuclear sequence.

Chloroplast genomes are relatively large, usually about 140 kb in higher plants, and up to 200 kb in lower eucaryotes. This is comparable to the size of a large bacteriophage, for example, T4 at about 110 kb. **Table 22.1** shows that in two lower eucaryotes, there are many of the circular DNA molecules in each chloroplast. The total amount of DNA in the chloroplasts forms a proportion of total cellular DNA as high as several percent. Unfortunately, we do not have comparable figures for higher plants, but here the amount of nuclear DNA is very much greater, and so the proportion of mass present in the chloroplast must be rather small.

Mitochondrial genomes show a variation in total size approaching an order of magnitude, as can be seen from **Table 22.2.** Mammals have small mitochondrial genomes, small enough to have been completely sequenced in man, mouse, and cow; all are about 16.5 kb. Data on the number of organelles per cell are available only for cultured cell lines, but the number is large (several hundred). The total amount of mito-

Table 22.1
Chloroplast DNA Shows a Narrow Range of Complexities

Species	Complexity of ct DNA, kb	No. Genomes per Organelle	No. Organelles per Cell	Ratio ct DNA to Total DNA
C. reinhardii	195	70-100	1	14%
E. gracilis	135	40	15	3%
Higher plants	140	Unknown	Unknown	Unknown

Table 22.2
Mitochondrial DNA Varies Widely in Complexity

Species	Complexity of mt DNA, kb	No. Genomes per Organelle	No. Organelles per Cell	Ratio mt DNA to Total DNA
Mouse (L cell)	16.2	2	500	0.2%
Human (HeLa)	16.6	10	800	1%
D. melano-gaster	18.4	Unknown	Unknown	Unknown
X. laevis	18.4	Unknown	Unknown	Unknown
S. cerevisiae	84	4	22	18%
P. sativum	110	Unknown	Unknown	Unknown

chondrial DNA relative to nuclear DNA is small, less than 1%. There seems to be more than one genome per mitochondrion, but fewer than ten. The size of the mitochondrial genome is slightly larger in the fruit fly and frog, where we lack information about the number of genomes and organelles.

In yeast, the mitochondrial genome is much larger. In *S. cerevisiae,* the actual size varies quite widely among different strains, but is about 84 kb. There are about 22 mitochondria per cell by conventional measurement, which suggests that there are some 4 genomes per organelle. In growing cells, the proportion of mitochondrial DNA can be as high as 18%. (Other yeasts may have much larger mitochondrial genomes.)

Plants themselves show a wide range of variation in mitochondrial DNA size. There is some evidence that the plant mitochondrial genome does not conform to the pattern of a single molecule of DNA, but may consist of more than one, varying in size. The total size of the plant mitochondrial genome may be roughly comparable to that of the chloroplast genome, so that in plants the total amount of information carried outside the nucleus may be relatively large.

ORGANELLES EXPRESS THEIR OWN GENES

Both of the organelles concerned with energy conversion run a mixed economy. Most of their constituent proteins are imported from the surrounding cytoplasm, where their synthesis represents the final stage of expression of nuclear genes. But each organelle also engages in its own protein synthesis. In the case of mitochondria, this effort is devoted to the production of a small number of proteins, each of which is a component of an oligomeric aggregate that includes some protein subunits imported from the cytoplasm, as illustrated in **Figure 22.1.** In the case of chloroplasts, the number of proteins produced by endogenous gene expression may be greater.

The apparatus for organelle protein synthesis is itself of mixed origin. Most or all of its protein components are transported into the organelle from the surrounding cytoplasm. But in both mitochondria and

chloroplasts there appears to be no passage permitted for nucleic acids in either direction through the organelle membrane. Thus among the products of the organelle itself are all of the RNA components of the protein synthetic apparatus.

The origin of the apparatus for gene expression in yeast mitochondria is summarized in **Table 22.3.** It is generally similar in the mitochondria of other species and in chloroplasts. In the yeast mitochondrion, eight mRNAs can be identified, and they are transcribed by an RNA polymerase that (presumably) is coded within the nucleus. Only the mRNAs that have been transcribed from mitochondrial genes can be translated in the organelle; and, conversely, this is the only compartment in which they can be expressed.

The mitochondrial ribosome consists of the usual larger and smaller subunits, each with a single rRNA coded within the mitochondrion, and with a large number of proteins imported from the cytoplasm. However, the ribosome structures are diverse in different species (see Chapter 8).The set of tRNAs is smaller than needed in the cytoplasm, as described in Chapter 7. The aminoacyl-tRNA synthetases responsible for charging the tRNAs are brought in from the cytoplasm. All of the protein components of the synthetic apparatus probably are unique to the mitochondrion; that is, they are coded by nuclear genes different from those coding for the comparable functions for gene expression in the nucleus and surrounding cytoplasm.

Corresponding with the eight mitochondrial mRNAs, eight proteins can be identified as the products of mitochondrial synthesis. The synthetic origin of a protein can be determined *in vivo* by virtue of its susceptibility to drugs that preferentially inhibit either cytoplasmic or organelle protein synthesis. Chloroplast and mitochondrial protein synthesis usually is sensitive to antibiotics that inhibit bacterial protein synthesis—for example, erythromycin and chloramphenicol. These do not affect cytoplasmic protein synthesis, which, however, is susceptible to cycloheximide; organelle protein synthesis remains resistant to cycloheximide. The origins of organelle proteins can be determined *in vitro* by characterizing the products made in isolated organelle preparations.

The complexes containing the proteins synthesized

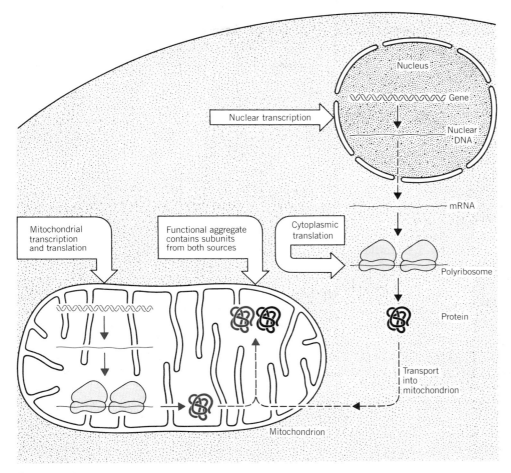

Figure 22.1
Mitochondrial protein aggregates are assembled from the products of expression of nuclear genes and mitochondrial genes.

Table 22.3
The Mitochondrial Apparatus for Protein Synthesis Is Assembled from RNA Synthesized in the Mitochondrion and Proteins Imported from the Cytoplasm

Component	RNA Synthesized in Mitochondrion	Proteins Synthesized in Cytoplasm
Template	~8 mRNAs	RNA polymerase
Ribosome	2 rRNAs	~75 ribosomal proteins
Adaptor	22 tRNAs	22 aminoacyl-tRNA synthetases

Table 22.4
Mitochondrial Protein Complexes Are Assembled from Proteins Synthesized in Different Cell Compartments

Protein Complex in Yeast	Subunits Synthesized in Mitochondrion	Subunits Synthesized in Cytoplasm	Subunits of Uncertain Origin
Oligomycin-sensitive ATPase (340,000 daltons)	ATPase 9 (membrane factor)	ATPase 2, 3, 4, 7 (F1 ATPase)	ATPase 1, 5, 6, 8
Cytochrome c oxidase (137,000 daltons)	CO 1, 2, 3	CO 4, 5, 6, 7	
Cytochrome bc₁ complex (160,000 daltons)	Cytochrome b apoprotein	6 subunits	

in the yeast mitochondrion are summarized in **Table 22.4.** The ATPase consists of two units, a membrane factor of ≥2 subunits coded by the mitochondrial genome, and the soluble F1 ATPase of ~5 subunits synthesized in the cytoplasm. The cytochrome c oxidase similarly consists of subunits from both sources. The cytochrome bc₁ complex consists of a single protein from the mitochondrion associated with six subunits from the cytoplasm. The ribosome small subunit contains one protein (Var1) coded in the mitochondrion. Mutations identifying almost all of the mitochondrial genes have been isolated.

Nuclear mutations that abolish each of these complexes also have been found. In both organelle and nuclear mutants, the protein complex fails to assemble properly, so that several subunits appear to be absent. This suggests that the assembly of each of these complexes into its usual membrane-associated form is an intricate process in which the absence of one component may prevent the assembly of the others. This has the practical consequence of making it difficult to equate nuclear mutations with individual subunits. Also, there are some nuclear mutations that prevent complex assembly, but seem to leave the individual subunits unaffected. This may mean that there are some nuclear-coded products that are necessary for assembly, but that do not actually comprise part of the final aggregate.

If an individual nuclear mutation can prevent the synthesis of a particular protein component, the simplest explanation is that it has occurred in a unique structural gene. This implies that there is a numerical discrepancy between the representation of organelle-coded and nuclear-coded components, since there are many more copies of the mitochondrial genome (see Table 22.2). Presumably the nuclear-coded genes are expressed more efficiently. The discrepancy has been confirmed directly in the case of a chloroplast function, the RuBP carboxylase of maize, in which the large subunit is coded by an organelle gene present in many copies per cell, while the small subunit is coded by a nuclear gene whose DNA is nonrepetitive.

Is the particular division of labors displayed by the yeast mitochondrion and nucleus typical of other species? The overall constitution of the mixed protein aggregates is similar at least in mammalian and other fungal mitochondria. Genes coding for the same functions are present in yeast and mammalian mitochondrial genomes. Our general expectation is therefore that there will prove to have been a substantial conservation of the coding functions found in the mitochondrial genomes of different species.

One exception to this rule has been found. The smallest subunit of the ATPase is coded by the mitochondrial genome in yeast and in *Aspergillus nidulans.* But in another fungus, *Neurospora crassa,* the corresponding protein appears to be synthesized in the cytoplasm. Nor is any gene corresponding to it present in mammalian mitochondrial DNA. If this subunit is coded by a nuclear gene in these cases, at some time there must have been a transfer of the gene between mitochondrial and nuclear genomes.

THE LARGE MITOCHONDRIAL GENOME OF YEAST

The fivefold discrepancy in size between the *S. cerevisiae* (84 kb) and mammalian (16 kb) mitochondrial genomes alone alerts us to the fact that there must be a great difference in their genetic organization in spite of their common function. We have already mentioned that there is no evidence for any substantial increase in yeast in the number of endogenously synthesized products concerned with mitochondrial enzymatic functions. Does the additional genetic material in yeast mitochondria represent other proteins, perhaps concerned with regulation, or is it unexpressed?

The existence of genetic recomination in yeast mitochondria (see later) has made it possible to construct a genetic map, which has been correlated with the physical map by the use of naturally occurring deletions in the mitochondrial genome (called petites). The extent of each gene has been determined by mapping the corresponding mRNA.

The map shown in **Figure 22.2** accounts for the major RNA and protein products of the yeast mitochondrion (excluding the tRNA molecules, of which there are about 22, and which have not yet been completely mapped). The most notable feature is the dispersion of loci on the map.

This genome has the distinction of providing the

Figure 22.2
The yeast mitochondrial genome contains both interrupted and uninterrupted protein-coding genes, rRNA genes, and tRNA genes (positions not indicated). Arrows indicate direction of transcription.

oli genes code for subunits of oligomycin-sensitive ATPase

oxi genes code for subunits of cytochrome *c* oxidase (CO)

box codes for cytochrome *b*
par unknown functions
var small ribosome subunit protein

extremely rare situation where the genes for rRNA are separated. The gene for 15S rRNA is uninterrupted and lies about 25,000 bp away from the gene for 21S rRNA. In some strains of yeast the latter gene has a single intron (as shown on the map); in other strains it is uninterrupted.

The two most prominent loci are the mosaic genes *box* (coding for cytochrome *b*) and *oxi3* (coding for subunit 1 of cytochrome oxidase), whose organization we have already discussed in Chapter 20. Together these two genes are almost as long as the entire mitochondrial genome in mammals! Many of the long introns in these genes have open reading frames in register with the preceding exon; at least in some cases, there is evidence for translation of the intron. This adds several proteins, presumably all synthesized in low amounts, to the complement of the yeast mitochondrion.

The remaining genes appear to be uninterrupted. They correspond to the other two mitochondrially coded subunits of cytochrome oxidase, to the subunit(s) of

the ATPase, and (in the case of *var1*) to a mitochondrial ribosomal protein.

Quite a large proportion of the yeast mitochondrial genome, about 25%, consists of short stretches rich in A-T base pairs, which probably lack any coding function. However, this still leaves a considerable amount of genetic material to be accounted for, and it will be surprising if some further genes are not discovered in the regions that have been left open on the map. Even given this likelihood, though, we can conclude that the total number of yeast mitochondrial genes is unlikely to exceed about 20.

THE COMPACT MITOCHONDRIAL GENOME OF MAMMALS

A very different picture is seen in the mammalian mitochondrial DNA. Its organization is extremely compact; there are no introns, some genes actually overlap, and almost every single base pair can be assigned

Figure 22.3
The sequence of human mitochondrial DNA identifies 22 tRNA genes, 2 rRNA genes, and 13 potential protein-coding regions, some of which can be identified with known products.

to a gene. With the exception of the D-loop, a region concerned with the initiation of DNA replication, no more than 87 of the 16,569 bp of the human mitochondrial genome can be regarded as lying in intercistronic regions.

The complete nucleotide sequences of the human and murine mitochondrial genomes show extensive homology in organization. The map of the human genome is summarized in **Figure 22.3.** There are 13 potential protein-coding regions. These include cytochrome *b,* the usual three subunits of cytochrome oxidase, and one of the subunits of ATPase. The remaining reading frames are indicated as URF (unidentified reading frame). The functions of these (presumed) proteins are unknown at present.

The organization of the genes has some striking features that reflect the peculiarities of gene expression in the mammalian mitochondrion. Most genes are expressed in the same direction; and tRNA genes lie between the genes coding for rRNA or protein.

In many cases, there is no separation at all between genes. The last base of one gene is adjacent to the first base of the next gene. In some cases, there is an overlap, most commonly of just one base, so that the last base of one gene serves as the first base of the next gene. Five of the reading frames lack a termination codon; these end with U or UA, and so an ochre codon is created by polyadenylation of the transcript (mitochondrial mRNA acquires a short sequence of poly (A) at its 3′ end). In three cases, the termination codon appears to be AGA or AGG, usually used as arginine in the genetic code (see Chapter 4).

Almost all of the genes are expressed in the clockwise direction, as indicated by the arrows on the map. Of the 15 protein-coding or rRNA-coding regions, 14 are transcribed in the same direction. (Atypical features are presented by three of the URF genes. URF4L and URFA6L have an unusually large overlap with the adjacent genes. URF6 is different from the main set of genes only by its direction of expression.) Of the 22 tRNA genes, 14 are expressed in the clockwise direction and 8 are read counterclockwise.

In only one case are two clockwise coding regions found in the form of contiguous genes (ATPase 6 and CO3). In every other case, at least one tRNA gene separates adjacent coding regions. The mRNA species corresponding to all the protein-coding regions

have been identified; and in each case, an initiation codon lies within six bases of the start. The codon used for initiation apparently can be AUG, AUA, or AUU. The mRNAs terminate with nonsense codons or incomplete nonsense codons. So nontranslated 5′ and 3′ regions are virtually absent from human mitochondrial mRNAs.

The punctuation of the rRNA- and protein-coding regions by the tRNA genes does not leave room for promoters comparable to those found in eucaryotic nuclei or in bacteria. There appears to be a single promoter for clockwise transcription, located in the region of the D-loop. This suggests the model illustrated in **Figure 22.4.** Transcription starts just before the tRNA gene in front of the 12S rRNA gene, and continues almost all the way around the circle, to terminate in the D-loop. The strand that is transcribed is called the **H strand** (because mitochondrial DNA can be separated into heavy and light single strands on the basis of their density).

The significance of the alternation of tRNA genes with rRNA- and protein-coding genes is that the tRNAs

Figure 22.4
Human mitochondrial DNA is transcribed into a single transcript, from which tRNAs are cleaved to release the rRNAs and mRNAs.

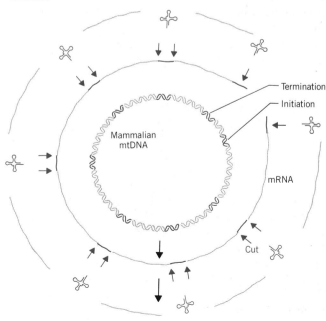

indicate sites of cleavage. By cleaving the primary transcript on either side of each tRNA gene, all except two of the genes (ATPase 6 and CO3) give rise to monocistronic products. The rRNA molecules appear to be synthesized in greater amounts than the mRNAs. This could be caused by premature termination of some proportion of the transcripts at some point after the two rRNA genes have been transcribed.

This mitochondrial DNA therefore presents the closest analogy to a bacterial operon that has been found in the eucaryotes. It is transcribed from a single promoter region, but individual tRNAs, rRNAs, and mRNAs are released from the transcript. Processing of the transcript therefore becomes the central event in gene expression. (There are also bacterial operons in which the separation of tRNA from rRNA or protein-coding regions must be accomplished posttranscription; see Chapter 23.)

How are the tRNA genes coded by the L strand expressed? Probably a similar mechanism applies, with a giant transcript produced for the L strand, from which the tRNA sequences and perhaps some mRNA sequences are conserved, while the rest is degraded. The origin for L-strand (counterclockwise) transcription also lies in the region of the D-loop. Very large L-strand transcripts can be isolated, so that their processing probably takes place some time subsequently to transcription. Large H-strand transcripts cannot be found, which suggests that they are processed more or less immediately after the transcription of each tRNA junction.

RECOMBINATION OCCURS IN (SOME) ORGANELLE DNAs

The prerequisite for recombination is the coexistence of genomes of both parents, a condition that is prevented in situations where absolute uniparental inheritance prevails. But the existence of instances where both organelle genomes survive immediately raises the question of whether they can interact. Is complementation possible between different mutations; can recombination between genomes occur? Complementation requires intermingling of the products of gene expression; recombination requires physical exposure of the two genomes to each other. Both demands are somewhat at odds with the static view of the individual organelle that might be inferred from the description of its endogenous gene expression as an event strictly confined to its own compartment. Yet in at least one species, recombination can occur between chloroplast DNAs; and in another, both complementation and recombination can occur with mitochondrial genomes.

In the alga *C. reinhardii,* the chloroplast DNA usually is inherited uniparentally. However, under the unusual circumstance when the genes of the usually predominant parent are irradiated with ultraviolet, the chloroplast genes of the other parent also survive. In the resulting biparental zygotes, recombination occurs between the chloroplast genomes, and this can be used to construct a genetic map.

In the yeast *S. cerevisiae,* recombination between mitochondrial markers is the norm. The nature of the recombination event is unknown, but a cross between two parents whose mitochondrial DNAs have gross differences generates progeny possessing physically recombinant DNAs. This demonstrates directly that the mitochondrial DNA of one parent comes into contact with the mitochondrial DNA provided by the other parent.

How do the parental organelle DNAs reach each other? The most likely explanation is that organelles can undergo fusion, to generate a single compartment containing genomes from both parents. This concept has been taken further in the case of yeast, where it is possible that instead of many individual mitochondria, there is a large branched structure (or structures) corresponding to many mitochondria. Perhaps there is a continual state of flux, with making and breaking of membranes. This raises the possibility that the presence of a genome is related to some unit smaller than the organelle itself; then the total number of genomes per organelle would be determined by the number of these constituent subunits.

In yeast it takes some hours after the production of a zygote before recombination occurs. But in the interim period, complementation can be detected between different mutations. This provides a unique opportunity to define the genetic organization of the organelle genome, since mutations can be both mapped and assigned to complementation groups. (A similar approach also has proved successful with the *C. reinhardii* chloroplast.)

REARRANGEMENTS OF YEAST MITOCHONDRIAL DNA

The yeast mitochondrial genome has a surprising fluidity of structure. We have already mentioned that different forms of some genes are found in different strains of *S. cerevisiae*. Indeed, the total size of the mitochondrial genome can vary substantially, by about 10,000 bp between strains. Extensive deletions of mitochondrial DNA can occur, as found in the existence of **petite** strains of *S. cerevisiae* (and certain other yeasts). These may arise spontaneously or may be induced by certain treatments, such as addition of ethidium bromide.

All petite mutations lack mitochondrial function. Their existence is possible because the alternative life-styles of yeast allow it to survive aerobically (when respiration is essential) or anaerobically (when it is dispensable). Thus mitochondrial mutations behave as conditional lethals for which aerobic growth provides the nonpermissive condition, while anaerobic growth provides the permissive condition under which the organism survives. Loss of mitochondrial function thus forces adoption of the anaerobic life-style. Of course, such a choice is not possible in (for example) animal cells, where loss of mitochondrial function is lethal. (*C. reinhardii* offers an analogous genetic situation with the chloroplast, since photosynthesis is dispensable in the presence of acetate.)

Petite mutations of *S. cerevisiae* fall into three types. **Nuclear petites** have Mendelian (that is, nuclear) mutations that abolish mitochondrial activity. **Neutral (*rho⁰*) petites** have an extreme situation in which all mitochondrial DNA is absent. This is a recessive genotype. **Suppressive (*rho⁻*) petites** have grossly abnormal mitochondrial DNA. The *rho⁻* mitochondria contain circles of DNA that are much smaller than the usual genome, containing only a small part of the usual complexity, ranging from 0.2% to 36% in different cases.

Thus more of the mitochondrial genome is deleted than is retained.

The sequence retained in a *rho⁻* petite often is amplified, to generate a large number of copies. Amplification occurs either by increase of the ploidy (the number of DNA molecules) or by formation of multimeric DNA molecules, each containing many copies of the sequence. The retained sequence need not necessarily represent a formerly contiguous region of the mitochondrial genome, but can have different regions juxtaposed. Sometimes further rearrangements of the sequence can occur (in this case, the petite is unstable).

The DNA retained in a petite strain can recombine with the DNA of another petite or of a wild-type strain (called *grande,* by comparison). This allows the genetic markers present in each petite to be correlated with the sequence of DNA that has been retained. In fact, it is the use of such petites for this deletion mapping that allows construction of the mitochondrial genetic map shown in Figure 22.2

Any sequence of mitochondrial DNA can be retained in a petite. This implies that all segments of the DNA can be independently replicated. We do not know how this facility is provided and whether it relies on any feature equivalent to an origin for replication (see Chapter 31).

When a petite strain of yeast is crossed with a wild-type strain, in a certain proportion of the progeny, only the petite genotype is found. The wild-type mitochondrial DNA has disappeared. This is called **suppression;** and the characteristic proportion of progeny in which it occurs gives the degree of **suppressiveness.**

The cause of suppression in highly suppressive petites is preferential replication of the petite mitochondrial DNA. This probably happens because the petite DNA possesses an increased concentration of origin-like sequences, called *rep* regions, relative to wild-type DNA. Thus we see that yeast mitochondrial DNA does contain a specific sequence(s) that can be used to initiate replication. But, unlike its counterpart in other genomes, this sequence is dispensable, because petite DNA lacking it can survive (although it is less suppressive). The cause of suppression therefore is not necessarily the same in less suppressive petites.

The general point made by these studies of petite and wild-type mitochondrial DNA is that there is extensive flexibility in the organization of the genome, with regard both to its expression and replication. Perhaps the difference from other mitochondrial DNAs lies in the ability of yeast altogether to dispense with mitochondrial functions. It will be interesting to characterize the enzyme systems that act on this DNA.

FURTHER READING

A general review of chloroplast and mitochondrial inheritance is provided by **Gillham** in *Organelle Heredity* (Raven Press, 1978). The topics of this chapter are reviewed in **Lewin's** *Gene Expression, 2, Eucaryotic Chromosomes* (Wiley, New York, 1980, pp. 583–637). The sequence of the human mitochondrial genome was analyzed by **Anderson et al.** (*Nature* **290,** 457–465, 1981).

CHAPTER 23
IDENTITY AND VARIATION
IN TANDEM GENE CLUSTERS

From the perspective of the individual gene, all extremes are found in genome organization. A gene may exist in a location where it is the only sequence of its kind, although it is likely to have relatives elsewhere. It may be a member of a small cluster of genes descended from a common ancestor and retaining related functions (as in the globin system). Other gene clusters may be larger, comprising an extensive repetition of genes that are either identical or very closely related. Genes (or other sequences) that are found as multiple copies adjacent to each other are said to be **tandemly** repeated. One type of tandem gene cluster codes for histone proteins, another for rRNAs. In both cases, the extensive repetition may reflect the need for large amounts of the product.

The close tandem clustering of genes into a defined region whose only purpose is to code for a particular function or functions suggests a way to describe the organization of sequences. Each tandem cluster can be defined in terms of a **repeating unit.** This is the basic entity whose multiple end-to-end repetitions make up the cluster. The copies of the repeating unit may be identical, or they may show variations.

Usually this repeating unit is longer than the sequence that is transcribed; and so it may be divided into two regions. The **transcription unit** is the dis-

tance between the first and last bases transcribed into RNA. The **nontranscribed spacer** is the remaining part of the repeating unit, and comprises the region lying between adjacent transcription units. This form of organization is illustrated in **Figure 23.1.**

The restriction map of a tandem gene cluster has a characteristic feature. It is circular. When the repeating units are identical, each generates exactly the same pattern of restriction cuts. Adjacent repeats share their terminal fragments. In the example shown in the figure, fragment A is adjacent to fragment B, which is adjacent to fragment C, which in turn is adjacent to fragment A, thus generating the circular map.

This pattern is broken only at the two ends of the entire cluster, shown in the figure by fragments X and Y. But usually these are not detected, because each is unique, compared with the multiple repetition of the internal fragments. For example, if there were 100 copies of the repeating unit, fragments X and Y would constitute only 1% of the material present in fragments A, B, and C.

The nontranscribed spacer is usually shorter than the transcription unit. We can define the character of the spacer because of the clarity with which regions can be delineated within a tandem gene cluster. The spacer is simply a part of the repeating unit that is not

Linear organization of cluster

Figure 23.1
A tandem gene cluster may comprise an alternation of transcription unit and nontranscribed spacer and generates a circular restriction map.

expressed in RNA, but whose functions (if any) remain concerned with the sole purpose of the cluster: production of the RNA coded by the transcription unit.

This view may prompt us to ask whether the somewhat longer sequences between individual genes in (for example) the globin gene cluster should be regarded in the same light. However, there is an important difference in the two situations. Every intergenic region in a globin gene cluster is different, whereas the nontranscribed spacers of a tandem gene cluster are repeated along with the genes.

THE REPETITIVE NATURE OF HISTONE GENES

Histones are nuclear structural proteins par excellence. In (probably) all eucaryotes, the chromosomes contain five types of histone protein: H1 present in half-molar; and H2A, H2B, H3, and H4 present in equimolar amounts (see Chapter 29). The total mass of the histones is roughly equivalent to the mass of the DNA itself. In dividing somatic cells, histone synthesis occurs pari passu with DNA replication; as soon as a stretch of DNA is replicated, histones are available to associate with it. This means that a large amount of histone protein must be synthesized during the short period while DNA is synthesized. The need for the somatic cell to synthesize a genome's worth of histones in a relatively short time may be the basic cause of histone gene repetition.

There is no particular relationship between genome size and histone gene number, which implies that the efficiency of expression of histone genes must vary in different species. But the repetition of histone genes is a common aspect of the provision for production of histones. Usually, there is the same number of copies of each histone gene. In the chicken the repetition frequency is about 10, in mammals it is about 20. This increases to roughly 40 in *X. laevis* and to about 100 in *D. melanogaster*. In several sea urchin species, there are in the range of 300–600 copies of each histone gene.

In these latter organisms, the initial nuclear divisions in embryogenesis occur very rapidly indeed. The large gene number in sea urchins may be needed to allow de novo histone gene expression to keep up with DNA synthesis. In amphibians, histone gene expression precedes the start of embryogenesis, and a large amount of histone mRNA is accumulated in the oocyte. Also, some histone proteins may be stored. This is an exception to the rule that histone genes are expressed only during the phase of DNA synthesis.

On the basis of those species characterized so far, a split is seen in the form of organization between genomes with a relatively low repetition of histone genes

(<50) and those with a high repetition (>100). Associated with low repetition, there are various relationships between histone genes. Associated with high repetition, there is a more rigid organization in which all the histone genes are organized into a single type of tandemly repeated cluster. In all these cases, the histone genes have the same structure; at present, it appears to be a rule that all functional histone genes lack introns.

A VARIETY OF TANDEM GENE CLUSTERS FOR HISTONES

The first histone gene clusters to be characterized were those of the sea urchin. Three (not closely related) species of sea urchin—*P. miliaris, S. purpuratus,* and *L. pictus*—all display the same form of organization. All five histone genes are part of a single repeating unit that constitutes a single tandem gene cluster in the genome. The genes are separated by nontranscribed spacers. They are all transcribed in the same direction, although each independently gives rise to its own mRNA. The predominant organization of these sea urchins is summarized in **Figure 23.2,** which shows that the main difference between species lies in the nontranscribed spacers. These differ not only in length but also in sequence. There is no cross-hybridization between corresponding spacers in different species, although the genes may be quite well related.

This situation poses a paradox. Presumably all of these clusters have evolved by duplication of the entire repeating unit, whose common organization suggests that it should have existed before speciation of the sea urchins. Yet selective forces have acted so as to preserve the functions of the genes, while allowing the intermingled spacers entirely to diverge *between* species although remaining constant *within* each. This implies that some corrective mechanism must act on the cluster within each species.

Actually, each of the five spacers is not perfectly preserved. Some variation is seen in the lengths of the corresponding spacers of individual repeating units, as detected by (slight) heterogeneity in the lengths of the appropriate restriction fragments. These variations between repeating units are referred to as **microheterogeneity.**

There is also variation in the sequences of the genes. In the sea urchins, different varieties of histones are synthesized in early embryogenesis (during the phase of rapid nuclear division) and in later development. The gene clusters that have been characterized code for the predominant early type. We do not yet know how the later variants are organized. Are gene clusters turned on and off as units; or are genes controlled individually within clusters containing more than one type of variant?

In one sea urchin, two types of repeating unit specifying the early embryonic genes have been characterized. They differ by virtue of divergence in their spacers. Each type of unit appears to be tandemly repeated; the two types of cluster are not intermingled. This implies that whatever mechanisms are responsible for maintaining homogeneity must be able to act on subregions of the cluster and need not extend throughout it.

The principle that all five genes are organized into a single repeating unit is followed also in *D. melanogaster,* although here the organization of the repeating unit is different. The genes lie in a different order and are transcribed in different directions. Two transloca-

Figure 23.2
Repeating units of histone gene clusters vary in organization.

tions would be needed for conversion between the sea urchin and fruit fly types of cluster. Spacer variants also are present in the *D. melanogaster* cluster.

An intriguing variation on this form of organization is found in the newt *N. viridescens,* in which there is a single type of repeating unit (different from both sea urchin and fruit fly). However, the individual repeating units do not lie adjacent to one another, but are separated by up to 10–50 kb of highly repetitive DNA.

A more variable form of organization is found in some cases. In *X. laevis,* the histone genes are organized in a cluster, but there is heterogeneity in their organization, since the neighbors of a given type of gene may vary. If there are tandem repeats, there must be more than one class, each containing genes in a different order. In chickens, there is a cluster of histone genes, but the genes lie in a variety of orders, and there are no tandem repeats. Proceeding to mammals, there is again evidence that the genes may not be systematically organized into repeating units, but may lie in smaller groups or even as individual genes. Although much more data will be needed to evaluate these patterns of histone gene organization, at present they appear to be somewhat intermediate between the small gene cluster (such as globin) and the tandem gene cluster.

There is an interesting contrast between the structures of individual histone genes and their overall organization. All histone genes share the same general uninterrupted structure; and corresponding genes in different species code for well-related proteins. Yet there is appreciable flexibility in the relationships be-

tween the various classes of genes, which vary from rigorous tandem clustering to an apparently erratic ordering. All of this reinforces the general conclusions suggested by the example of the globin gene cluster. Individual genes may be tailored to their function by particular mutations in an otherwise conserved sequence, but there is continual generation of new copies and subsequent rearrangement of the overall organization of the cluster.

GENES FOR rRNA AND tRNA ARE REPEATED

Many genes represent each of the RNA components of the protein synthetic apparatus. In a sense, the structural genes coding for mRNA form the largest class, although since there are so many different mRNAs, each gene is represented in only a relatively small number of messenger molecules. To match the large number of mRNAs that need translation, the cell contains many ribosomes and tRNA molecules.

Ribosomal RNA is by far the predominant product of transcription, constituting some 80–90% of the total mass of cellular RNA in both eucaryotes and procaryotes. This proportion is the result of its size, since there are several tRNA molecules per ribosome. Both rRNA and tRNA are represented by multiple genes, and the numbers found in the haploid genome are summarized in **Table 23.1** for the best-investigated cases.

The number of major rRNA genes varies from 7 in *E. coli,* to between 100 and 200 in lower eucaryotes,

Table 23.1
There Are Multiple rRNA and tRna Genes in All Genomes

Species	Number of 18S/ 28S Genes	Ratio of rDNA to Total DNA	No. 5S Genes	No. tRNA Genes
E. coli	7	1.0%	7	60
S. cerevisiae	140	5.5%	140	250
D. discoideum	180	17.0%	180	?
D. melanogaster				
(X)	250	1.3%	165	850
(Y)	150			
Human	280	0.4%	2000	1300
X. laevis	450	0.18%	24000	1150

to several hundred in higher eucaryotes. In virtually every case, the genes for the large and small rRNA are found as a tandem pair. (The sole exception is the yeast mitochondrion.) In bacteria and in some lower eucaryotes, the gene for 5S RNA is part of the same unit, so the total number of 5S genes is the same as that of the major rRNAs. In bacteria, the 5S gene is cotranscribed with the major rRNA genes; in eucaryotes it is transcribed independently. In higher eucaryotes, the genes for 5S RNA are separately organized into their own clusters, and their number exceeds that of the major rRNA genes.

The lack of any detectable variation in the sequences of the rRNA molecules implies that all the copies of each gene must be identical, or at least must have differences below the level of detection in rRNA (about 1%). In bacteria, the multiple 16S-23S rRNA gene pairs are dispersed. In most eucaryotic nuclei, the rRNA genes are contained in a tandem cluster or clusters. These regions are sometimes known as **rDNA.** (In some cases, the proportion of rDNA in the total DNA, together with its atypical base composition, is great enough to allow its isolation as a separate fraction directly from sheared genomic DNA.)

In the case of tRNA, the exact number of genes is hard to determine, because the extensive secondary structure of the molecule creates technical difficulties in the hybridization reaction. Probably our present estimates are therefore less than the actual number.

A TANDEM REPEATING UNIT CONTAINS BOTH rRNA GENES

The pair of major rRNAs is transcribed as a single precursor in both bacteria and eucaryotic nuclei. Following transcription, the precursor is cleaved to release the individual rRNA molecules by the pathways discussed in more detail in Chapter 25. The salient features of the organization of the genes are described in **Figure 23.3.** The transcript starts with a 5′ leader, followed by the sequence of the small rRNA, a region called the **transcribed spacer,** and finally the sequence of the large rRNA toward the 3′ end of the molecule.

The transcription unit is therefore longer than the combined length of the mature rRNAs. **Table 23.2** summarizes some examples of the relationship between the primary transcript and the rRNAs.

The transcription unit is shortest in bacteria, where the rRNA sequences constitute 80% of its total length of 6000 bases. No particularly systematic pattern is seen in a range of eucaryotes in which the transcript length varies from about 7000 to 8000 bases, with 70–80% of the sequence representing the rRNAs. The precursor is longest in mammals (where it is known as 45S RNA, according to its rate of sedimentation). Here the mature rRNA sequences occupy only just

Figure 23.3

A single transcription unit contains the sequence for both the small and large rRNA molecules.

Table 23.2

Each rRNA Precursor is Longer than the Two Mature rRNAs

Species	Length of Transcript	Length of rRNA		rRNA Percent of Transcript
		Large	Small	
E. coli	5600	2904	1542	80
S. cerevisiae	7200	3750	2000	80
D. melanogaster	7750	4100	2000	78
X. laevis	7875	4475	1925	79
N. tabacum	7900	3700	1900	71
G. domesticus	11250	4626	1800	57
M. musculus	13400	5000	1950	52

more than 50% of the length of the complete transcript, so that the leader and transcribed spacer must be much increased in length.

What happens to the nonribosomal parts of the RNA precursor? The leader and the transcribed spacer regions are discarded during the maturation of rRNA, presumably to be degraded to nucleotides for reutilization. However, the transcribed spacer contains the sequences of small RNA molecules that are released from it by cleavage. In mammals and in *Xenopus,* there is a short sequence that forms the 5.8S RNA, a small molecule that hydrogen bonds with the 28S rRNA in the ribosome. In bacteria, as we shall see later, tRNA sequences may lie in the transcribed spacer and (sometimes) also at the 3′ end of the primary transcript.

The alternation of transcription unit and nontranscribed spacer in the tandem repeats of the eucaryotic nucleus can be seen directly in electron micrographs. The example shown in **Figure 23.4** is taken from the newt *N. viridescens,* in which each transcription unit is intensively expressed, so that many RNA polymerases are simultaneously engaged in transcription on each repeating unit. The polymerases are so closely packed that the increase in the length of their products forms a characteristic matrix moving along the transcription unit.

The nontranscribed spacer varies widely in length between and (sometimes) within species. **Table 23.3** summarizes the situation.

In yeast there is a short nontranscribed spacer, relatively constant in length. In *D. melanogaster,* there is almost a twofold variation in the length of the nontranscribed spacer between different copies of the repeating unit. A similar situation is seen in *X. laevis.* In each of these cases, all of the repeating units are present as a single tandem cluster on one particular chromosome. (In the example of *D. melanogaster,* this

Figure 23.4

Transcription of rDNA clusters generates a series of matrices, each corresponding to one transcription unit and separated from the next by the nontranscribed spacer.

Photograph kindly provided by Oscar Miller.

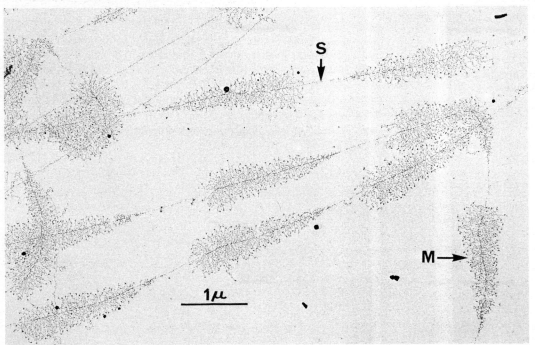

Table 23.3
Length of the Tandem Repeating Unit of rDNA Clusters Is Variable

Species	Length of Repeating Unit	Length of Nontranscribed Spacer	Length of Transcript
S. cerevisiae	8950	1750	7200
D. melanogaster	11500–14200	3750–6450	7750
X. laevis	10500–13500	2300–5300	7875
M. musculus	44000	30000	13400

These data refer to uninterrupted genes. The range would be greater if the interrupted 28S genes of
D. melanogaster were included.

happens to be the sex chromosome. The cluster on the X chromosome is larger than that on the Y chromosome, so that female flies have more copies of the rRNA genes than male flies. However, not all of the genes are active, as we shall see later.)

In mammals the repeating unit is very much larger, comprising the transcription unit of about 13,000 bp and a nontranscribed spacer of roughly 30,000 bp. Usually, the genes lie in several dispersed clusters—in the case of human and mouse residing on five and six chromosomes, respectively. One interesting question is how the corrective mechanisms that presumably function within a single cluster to ensure constancy of rRNA sequence are able to work when there are several clusters.

The region of the nucleus where rRNA synthesis occurs has a characteristic appearance, with a core of fibrillar nature surrounded by a granular cortex. The fibrillar core is the region at which the rRNA is transcribed from the DNA template; and the granular cortex is formed by the ribonucleoprotein particles into which the rRNA is assembled. The whole area is called the **nucleolus.**

Particular chromosomal regions can be seen to be associated with a nucleolus. These are called **nucleolar organizers.** Each nucleolar organizer corresponds to a cluster of tandemly repeated rRNA genes, so in a diploid nucleus there are twice as many nucleolar organizers as there are tandem clusters in the haploid genome. The concentration of the tandemly repeated rRNA genes, together with their very intensive tran-

scription, is responsible for creating the characteristic morphology of the nucleoli.

SOME rRNA GENES ARE EXTRACHROMOSOMAL

Two situations occur in which rRNA transcription units do not lie on the chromosome but instead are present as independent **extrachromosomal** molecules.

A stable situation is found in certain lower eucaryotes, in which the genes responsible for producing rRNA are carried by extrachromosomal DNA molecules, present in many copies per nucleus. The example of D. discoideum is shown in **Figure 23.5.** The extrachromosomal molecule is a large palindromic dimer, containing two transcription units in opposite orientation, separated by about 20,000 bp. Each extrachromosomal dimer also contains two 5S genes, although these are transcribed independently from the major transcription unit. We know very little about the arrangements for inheritance of this molecule.

Figure 23.5
The rRNA genes in D. discoideum lie on an extrachromosomal palindrome.

Chromosomal rDNA   Spacers   vary in length

                transcription units   all the same length

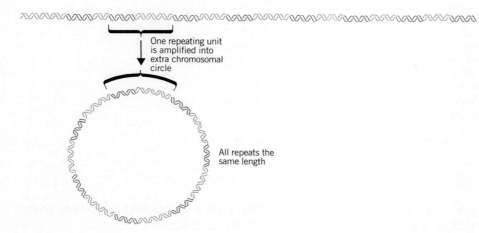

One repeating unit is amplified into extra chromosomal circle

All repeats the same length

Figure 23.6
Each extrachromosomal rDNA in the X. laevis oocyte is a circle generated by end-to-end amplification of a single repeating unit of chromosomal rDNA.

A transient state occurs during the development of some animals. During oogenesis in many species, a large number of *additional* rRNA genes are suddenly generated. The process has been characterized most fully in *X. laevis.* Circular extrachromosomal molecules of DNA each containing many repeating units are produced by **amplification** of the chromosomal repeating units. Unlike the chromosomal genes, in which adjacent repeating units can have spacers of different length, on the extrachromosomal circles the adjacent units all have spacers of the *same* length. This suggests that each amplified circle is produced by making multiple end-to-end copies from a single chromosomal repeating unit, as illustrated in **Figure 23.6.** The additional genes are needed to achieve sufficient production of rRNA in the oocyte, and are not a feature of somatic cells, which rely on the chromosomal genes. (The mechanism of this amplification is a form of replication called the rolling circle, which is discussed in Chapter 31 and illustrated in Figure 31.14.)

ABOUT NONTRANSCRIBED SPACERS AND PROMOTERS

The variation in length of the nontranscribed spacer in a single gene cluster contrasts with the conservation of sequence of the transcription unit. In spite of this variation, the sequences of longer nontranscribed spacers remain homologous with those of the shorter nontranscribed spacers. This implies that each nontranscribed spacer is *internally repetitious,* so that the variation in length results from changes in the number of repeats of some subunit.

The general nature of the nontranscribed spacer is illustrated by the example of **Figure 23.7,** which shows its organization in *X. laevis.* Regions that are fixed in length alternate with regions that may vary. Each of the three repetitious regions comprises a variable number of repeats of a rather short sequence (less than 100 bp). This accounts for the overall variation in length.

One of the fixed regions (*A*) is unique in sequence and length. The others are the short constant sequences called **Bam islands.** (This description takes its name from their isolation via the use of the Bam restriction enzyme.) They are puzzling, because the last half of each Bam island is very similar to the sequence that immediately precedes the start of the transcription unit, which is likely to contain the promoter. This similarity therefore raises the question of why transcription does not start at the Bam islands as well as at the proper promoter.

Another situation in which there are seemingly identical promoters, some of which initiate transcription and others of which do not, is provided in *D. melan-*

Figure 23.7

The nontranscribed spacer of *X. laevis* rDNA has an internally repetitive structure that explains its variation in length.

ogaster. Here about two thirds of the genes for the large rRNA contain an insertion. There are two types of insertion, each variable in length but present at a characteristic location (see Chapter 37). None of the insertions can be characterized as introns, since the genes carrying them are not expressed. But their 5′ flanking regions appear no different from those of the active uninterrupted genes. In some way that we do not yet understand, the presence of the insertion is associated with the failure of transcription to initiate at the beginning of the gene, 6800 bp away.

5S GENES AND PSEUDOGENES ARE INTERSPERSED

In most eucaryotes, the organization of 5S genes is distinct from that of the major rRNA genes. The total number of 5S genes may be greater, their location(s) different, and they are transcribed by RNA polymerase III instead of by RNA polymerase I (the nucleolar enzyme). We really have little idea whether and how the transcription of the 5S genes and the major rRNA genes is coordinated.

The 5S genes usually are organized as a cluster or clusters of tandem repeats. They have been analyzed in detail in *X. laevis*, in which most or all of the chromosomes have a cluster of 5S genes located at or very close to the telomere (the end).

Several types of 5S RNA sequence are found in *X. laevis*, falling into two principal types. One is found exclusively in oocytes; the other is synthesized in somatic cells. The same division of types is found in the related frog *X. borealis*. In fact, the corresponding types of 5S RNA in the two species are more closely related

to each other than either is to the other type in its own species. The oocyte and somatic 5S RNAs of *X. laevis* differ in six bases.

About 20,000 genes code for 5S RNA in *X. laevis*, but the vast majority represent the oocyte type. The reason for this large number may be to provide sufficient 5S genes to keep up with the expression of the amplified major rRNA genes during oogenesis. The genes for 5S RNA are not amplified.

The major 5S oocyte gene cluster takes the familiar form of an alternation between genes and nontranscribed spacers. However, the repeating unit is internally repetitive, presumably due to a past duplication. Its structure is shown in **Figure 23.8.**

The spacer has an A-T-rich sequence that is based on varying numbers of repeats of a 15 bp sequence and some variants closely related to it. The length of each nontranscribed spacer therefore varies according to the number of tandem copies of this repeat. The average length of the A-T-rich sequence is about 400 bp.

The remaining 300 bp of the repeating unit is G-C-rich. It starts with a 49 bp sequence, which is followed by the 120 bp sequence of the 5S gene itself. Next

Figure 23.8

The repeating unit of the major oocyte 5S gene cluster of *X. laevis* contains a gene and a pseudogene.

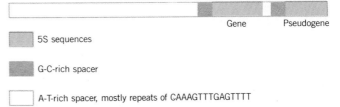

come about 1½ of the A-T-rich repeating units, succeeded by another copy of the 49 bp G-C-rich spacer sequence. Then there is a repeat of the first 101 bp of the gene; as an incomplete sequence that is not represented in RNA, it constitutes a pseudogene. The region including the pseudogene represents a repeat of the sequence around the active gene from -73 to $+101$. A surprising feature is that the cluster as a whole must contain an equal number of genes and pseudogenes.

The promoter for the transcription of 5S RNA lies entirely within the gene (see Chapter 11). There are only nine point mutations to change the sequence of the pseudogene from that of the gene. The pseudogene therefore contains the region of the gene that provides the internal promoter, with a few changes. We need to know whether these changes are sufficient to account for the failure of the pseudogene to give rise to an RNA product *in vivo*.

Not all the oocyte 5S genes are identical. A minor component of oocyte 5S RNA, called trace 5S RNA, displays some difference in sequence. This also is coded by a tandem array of genes and spacers; but the repeat length is only 350 bp, and the spacer is unrelated to that of the major cluster. Again, we see that a gene may undergo rather small changes in sequence, while the spacer changes completely. In a similar way, a 5S gene cluster in *X. borealis* has a spacer sequence unrelated to either of those characterized in *X. laevis*.

AN EVOLUTIONARY DILEMMA

The same problem is encountered by all clusters of identical genes. How is constancy of sequence maintained when there are so many copies of each gene? In more practical terms, the question becomes how selection can be imposed to prevent the accumulation of deleterious mutations.

When there are many copies of a gene, the immediate effects of mutation in any one copy can only be very slight. The consequences of the individual mutation are diluted by the large number of copies of the gene that retain the wild-type sequence. This suggests that an appreciable proportion of mutant copies could accumulate before the effect becomes strong enough to be eliminated by evolution. Lethality becomes quantitative, a conclusion that is reinforced by the observation that half of the units of the rDNA cluster of *X. laevis* or *D. melanogaster* can be deleted without ill effect. So how are these units prevented from gradually accumulating deleterious mutations? And what chance is there for the rare favorable mutation to display its advantages in the cluster?

The inevitable conclusion seems to be that some mechanism must exist for allowing particular variants to be scrutinized by evolution. Two types of mechanism have been proposed.

The **sudden correction** model supposes that every so often the entire gene cluster is replaced by a new set of copies that is derived from one or at least from only a very few of the copies present in the previous generation. To impose sufficient selective forces, "every so often" need be only every few generations; but in practical terms, any mechanism must be constructed on a regularly recurring basis—that is, every generation. The small number of copies that gives rise to the new cluster is a "master" set on which selection acts; the copies that are discarded are irrelevant to selection. The master copies could constitute a particular set or might be chosen at random.

The difficulty with this model is that it predicts regeneration of the cluster from a few *repeating units*. This means that both the transcription units and spacers should be regenerated, so that the entire repeating unit should be homogeneous in sequence (or at least, should show no more variation than the small number of master copies). But in fact the spacers show a continual range of variation, while only the transcription units are constant.

The **crossover fixation** model supposes that the entire cluster is subject to continual rearrangement by the mechanism of unequal crossing-over. Essentially this applies to the tandem cluster on a grand scale the mechanisms we have already discussed for globin genes on a more restricted and occasional basis. We shall explore their consequences in more detail in Chapter 24, but it is immediately evident that tandem clustering provides frequent opportunities for "mispairing" of genes whose sequences are actually the same, but that lie in different positions in their clusters.

By continually expanding and contracting the number of units via unequal crossing-over, it is possible for all the units in one cluster to be derived from rather a small number of those in an ancestral cluster. The variable lengths of the spacers are consistent with the idea that unequal crossing-over events take place in spacers that are internally mispaired. This can explain the homogeneity of the genes compared with the variability of the spacers. For the genes are subject to selection when individual repeating units are amplified within the cluster and therefore exposed to selection; but the spacers are irrelevant and can accumulate changes.

BACTERIAL rRNA GENES FORM MIXED OPERONS WITH tRNA GENES

The seven operons coding for rRNA in *E. coli* are named *rrnA-G*. They are not closely linked on the chromosome; and we have no information on how sequence homogeneity is maintained. Because the number of genes is small, it should be possible to determine their exact sequences, and thus to see directly whether all the rRNA genes are identical or have minor variations not detectable in the rRNA population.

Each of these operons is transcribed into a single RNA precursor from which the mature rRNA molecules are released by cleavage. In bacteria in which the cleavage reaction has been blocked by mutation of the responsible enzyme, the precursor accumulates

as an RNA sedimenting at about 30S. The overall organization of all the *rrn* operons is the same, containing all three rRNA molecules in the order 16S-23S-5S, as illustrated in **Figure 23.9**. The two major rRNAs lies in the same order that is found in the eucaryotic transcription unit; the presence of the 5S RNA in the same transcription unit is unique to the procaryotes.

All the *rrn* operons have a dual promoter structure. The first promoter, *P1,* lies about 300 bp upstream from the start of the 16S rRNA sequence. Probably it is the principal promoter; it uses ATP or GTP for initiation, and has the usual consensus sequences. Up to the first 150 bp of the transcription unit may be different in the various *rrn* operons. Within this region, about 110 bp from *P1*, there is a second promoter, *P2*, which may have the less usual starting nucleotide of CTP, and whose consensus sequences may be less evident. The 5′ end of the individual precursor to 16S rRNA (which is slightly longer than the mature rRNA) lies after the start of the sequence that is common for all *rrn* operons.

Between the sequences for 16S rRNA and 23S rRNA, there is a transcribed spacer region of 400–500 bp. Within this region, however, lies a sequence or sequences coding for tRNA. As summarized in the generalized diagram of Figure 23.9, in four of the *rrn* operons the transcribed spacer contains a single tRNA sequence, that of tRNA$_2^{Glu}$. In the other three *rrn* operons, the transcribed spacer contains the sequences

Figure 23.9
The *rrn* operons contain genes for both rRNA and tRNA.
Each functional sequence is separated from the next by a transcribed spacer region; the lengths of the leader and trailer depend on which promoters and terminators are used. Each RNA product must be released from the transcript by cuts on either side.

of two tRNAs (tRNA$_1^{Ile}$ and tRNA$_{1B}^{Ala}$). Thus on the basis of the "spacer tRNA" sequences, we can distinguish two types of rRNA operon. The tRNAs are released from the precursor RNA by cleavage; the remaining regions of the spacer presumably are degraded.

Yet another variation is found in the *rrn* operons. In some, the last sequence coded in the precursor is the 5S RNA. In others, there may be some additional tRNA sequences. For example, in *rrnC* there are two genes, tRNA^{Trp} and tRNA$_1^{Asp}$, located at the 3' end. (The presence of the tRNA^{Trp} gene makes this particular *rrn* operon indispensable for *E. coli,* since this is the only copy of the gene, and thus provides sole capacity for the utilization of tryptophan in protein synthesis.) There is no correlation between spacer tRNA sequences and distal tRNAs.

tRNA GENES MAY LIE IN CLUSTERS

We have only an incomplete picture of the organization of tRNA genes in either bacteria or eucaryotes, but there seems to be a tendency for tRNA genes to lie in clusters.

In eucaryotes, the scanty information available is provided by cases where genomic sequences corresponding to a particular tRNA probe have been cloned. Often there turn out to be sequences close by that code for other tRNAs. In contrast to tandem gene clusters, the tRNAs coded in the same region all may be different. We do not know whether the clustered tRNA genes are transcribed independently or form a common unit of gene expression.

A similar situation is seen in *E. coli,* in which tRNA genes form clusters, each of which is concerned with several amino acid acceptors. At least some of these form operons, transcribed from a single promoter into one long precursor RNA. Phage T4 contains a cluster of eight tRNA genes expressed in a similar manner. This contrasts with the situation in mammalian mitochondria, in which tRNA genes punctuate the sequence of structural genes (see Chapter 22).

In every known case, a common principle applies. The tRNA is formed as part of a longer precursor RNA from which it is released by cleavage and other processing reactions. These are discussed in Chapter 25.

Two of the tRNA gene clusters in *E. coli* have been characterized in detail, and both prove to contain other sequences as well as the tRNAs. These two clusters are named for their possession of tyrosine tRNA genes. The *tyrU* cluster includes the genes for four tRNAs, tRNA$_4^{Thr}$, tRNA$_2^{Tyr}$, tRNA$_2^{Gly}$ and tRNA$_3^{Thr}$. The *tyrT* cluster contains two identical genes coding for tRNA$_1^{Tyr}$.

In each cluster, the single promoter is located before the first tRNA gene, as shown in the example of

Figure 23.10
The *tyrU* operon codes for 4 tRNAs and the protein EF-Tu.

Figure 23.10. Beyond the last tRNA gene, there is a sequence coding for protein. In the *tyrU* cluster this is the gene *tufB* (one of two coding for the protein synthesis factor EF-Tu). In the *tyrT* cluster, the coding sequence represents protein P, a protamine-like polypeptide (protamines are very basic proteins associated with DNA in some spermatozoa).

Presumably the mRNA for each of these proteins is generated by cleavage of the primary transcript between the last tRNA sequence and the start of the structural gene. There is a short common sequence in the two operons that might be used for this purpose. Unlike other bacterial operons, in which ribosomes probably attach to the mRNA almost as soon as the RNA polymerase has left the promoter, in this case the mRNA is likely first to be generated by cleavage from the preceding tRNA sequences.

FURTHER READING

The elucidation of histone gene structure, and changing views on its degree of conservation, can be traced back through a series of reviews, from **Hentschel** & **Birnstiel** (*Cell* **25,** 301–313, 1981) to **Kedes** (*Ann. Rev. Biochem.* **48,** 837–870, 1979). The structure of eucaryotic rDNA clusters is dealt with in **Lewin's** *Gene Expression, 2, Eucaryotic Genomes* (Wiley, New York, 1980, pp. 875–906).

CHAPTER 24
ORGANIZATION OF
SIMPLE SEQUENCE DNA

Within the eucaryotic genome there are many sequences that do not code for protein (or RNA). Very likely constituting a majority of the DNA, these sequences are under different evolutionary constraints from those imposed by the need to represent a series of amino acids. Some of them, of course, are part of transcriptional units—for example, the nontranslated flanking regions in the mRNA or the introns that are removed during maturation of mRNA.

Others may play roles as signals that are recognized by proteins, including elements such as promoters for transcription, origins for DNA replication, sites for folding the chromosome, points for attaching the kinetochore, and other cellular functions. Very few of these have been characterized to the degree that a particular function can be associated with a particular sequence. Really the only one for which any features are known in the eucaryotic genome is the promoter for transcription. If we include viral genomes, origins for DNA replication also are known. From the little information that is available, a reasonable idea is that control signals of this type are likely to be relatively short in sequence.

Examining this problem from the reverse perspective, we can inquire what we know about short sequences that are repeated in identical or related copies in the genome. Of course, this amounts to a definition

of repetitive DNA as characterized by the DNA reassociation kinetics described in Chapter 17. To recapitulate, the moderately repetitive component consists of a variety of sequence families, present in varying degrees of repetition and with differing internal relationships. The members of these families often are interspersed in a more or less regular way with longer stretches of nonrepetitive DNA. By contrast, the highly repetitive component generally consists of very short sequences repeated many times in tandem in large clusters. Because of the short repeating unit, this is sometimes described as "simple sequence DNA."

In no case do we have any information about the function of the sequence family. In this chapter, we shall therefore describe the known properties of the various types of repetitive DNA, some of which yield some interesting information about the mechanisms deployed in their evolution.

THE Alu FAMILY

In at least some mammalian genomes, an appreciable proportion of the interspersed moderately repetitive component comprises members of a single family. Its short length and high degree of repetition thus make it comparable to the simple sequence DNA, except that the individual members of the family are dispersed around the genome instead of being confined to tandem clusters.

In the human genome, a large part of the moderately repetitive DNA exists as sequences of about 300 bp that are interspersed with nonrepetitive DNA. The duplex DNA corresponding to the moderately repetitive sequence component can be isolated by renaturation at intermediate Cot, followed by degradation of the adjacent regions of nonrepetitive DNA that remain unpaired. At least half of the renatured duplex material is cleaved by the restriction enzyme AluI at a single site. This site is located at a point 170 bp along the sequence.

The cleaved sequences all may be members of a single family, known as the **Alu family** after the means of its identification. There are roughly 300,000 members in the haploid genome (equivalent to one member for every 6 kb of DNA). The individual Alu sequences are widely dispersed; at least one copy of the sequence is present in virtually every clone in a library carrying genomic DNA of average length 18 kb. A related sequence family is present in the mouse (where the 50,000 members are called the B1 family), in the Chinese hamster (where it is called the Alu-equivalent family), and in other mammals.

The individual members of the Alu family are related rather than identical. It is likely that we will be able to derive a consensus sequence for the family in each species. Ten members of the human family have a consensus that reflects a 130 bp tandem duplication, with an unrelated sequence of 31 kb inserted in the right half of the dimer. A short but different sequence is repeated at each end of individual copies of Alu, which is reminiscent of the termini of transposable elements (see Chapter 38). The individual sequences have an average homology with the consensus sequence of 87%. The mouse B1 repeating unit is 130 bp long, corresponding to a monomer of the human unit. It has 70–80% homology with the human sequence.

A variety of properties have been found for the family, and its ubiquity has prompted many suggestions on its function, but it is not yet possible to discern its true role.

Part of the Alu sequence is a 14 bp region that is almost identical with a sequence present at the origin of replication in the papova viruses (such as SV40)

and in hepatitis B virus. This raises the possibility that the Alu family could be connected with origins of replication for the eucaryotic genome, although the number of members of the family argues against this, since there are about ten times more Alu sequences than the expected number of replication origins (see Chapter 31).

Another feature is that at least some members of the family can be transcribed *in vitro* into a small RNA by the enzyme RNA polymerase III (which is responsible for transcribing small nuclear RNAs, tRNAs, and 5S RNA). In the Chinese hamster, some (although not all) members of the Alu-equivalent family appear to be transcribed *in vitro*. Transcription units of this sort are found in the vicinity of other transcription units. For example, there are two such units in the region of the human β-globin gene cluster.

Several small RNAs are components of eucaryotic cells (see Chapter 27). The 7S RNA sequence appears to be conserved in higher eucaryotes and also is incorporated into some retrovirus virions. Its sequence has been determined in human and rat cells, where it takes the same general form, corresponding to the insertion of a sequence of 140 bases into an Alu monomeric sequence at a point 100 bases from the 5′ end. Thus the RNA is homologous with the Alu sequence for its first 100 bases and last 40 bases. The central inserted sequence is the same in man and the rat, arguing that the evolution of the coding sequence from an Alu member occurred before the mammalian radiation.

Members of the Alu family may be included within structural gene transcription units, as seen by their presence in long nuclear RNA. The presence of multiple copies of the Alu sequence in a single nuclear molecule can generate secondary structure. In fact, the presence of Alu family members in the form of the inverted repeats described in the next section is responsible for most of the secondary structure found in mammalian nuclear RNA (see Chapter 27).

INVERTED REPEATS RENATURE INSTANTLY

Most of the DNA of any genome renatures in a bimolecular reaction according to the second-order kinetics described in Chapter 17. These apply to the nonrepetitive, moderately, and highly repetitive components. But usually there is also another component that reassociates more rapidly than highly repetitive DNA. This fraction is obtained in duplex form even after the shortest periods allowed for reassociation. It is, of course, an exaggeration to say that it reassociates "instantly," but the reaction is certainly extremely rapid, occurring in a Cot range below 10^{-4} or 10^{-5}.

The fraction is sometimes described as the **zero-time binding DNA,** because it is found in duplex form essentially at time 0 in the reassociation reaction. Usually it is removed before the kinetics of the other components are determined. (This means that the DNA is denatured and then the duplex molecules are removed by hydroxyapatite chromatography, after which the single strands are incubated to follow the second-order reassociation reaction.)

The reassociation of the zero-time binding fraction

is **unimolecular,** implying that only a single molecule of DNA is involved in the reaction. This means that the pair of reassociating components both must lie on a single strand of DNA. Thus the reaction is an intramolecular reassociation of sequences that lie in reverse orientation close together. These are the **inverted repeats** or **palindromes** that we have already described in Chapter 13. For this reason, the fraction is sometimes graphically referred to as **foldback** or **snapback DNA.**

Two types of palindrome can be distinguished, as illustrated in **Figure 24.1.** When the inverted repeats are contiguous, their reassociation forms a hairpin. When they are separated by some other sequence, they reassociate to form a duplex stem with a single strand loop.

The inverted repeats seem to be widely dispersed throughout the genome. They include members of all sequence classes, both nonrepetitive and repetitive. Little is known about their function.

Figure 24.1

Denaturation and renaturation form hairpins from adjacent inverted repeats and form stems and loops from separated inverted repeats.

They are present in the long transcription units for nuclear RNA, where they give rise to much of its characteristic secondary structure. Many of the inverted repeats found in human nuclear RNA are members of the Alu family described in the previous section. Actually, considering the Alu family as a whole, about one third of the members occur as directly adjacent inverted repeats. The remainder occur either individually or in the company of a more distantly located inverted repeat. Altogether the Alu family accounts for half or more of the inverted repeats in the human genome.

HIGHLY REPETITIVE DNA FORMS SATELLITES

Highly repetitive DNA takes the form of very short sequences repeated many times in tandem. This sequence component therefore forms blocks of material, each block consisting of very long arrays of multiple copies of the highly repetitive sequence. In some cases, the multiple copies are identical; in others, they are related.

The tandem repetition of a short sequence often creates a fraction that has distinctive physical properties. This feature may be used to isolate it. In some cases, the repetitive sequence has a base composition that is different from the genome average, which may allow it to form a separate fraction by virtue of its buoyant density.

The buoyant density of a duplex DNA depends on its G-C content according to the empirical formula

$$\rho = 1.660 + 0.00098 \, (\%\text{G-C}) \text{ g-cm}^{-3} \quad (12)$$

Buoyant density usually is determined by centrifuging DNA through a **density gradient** of CsCl. The DNA forms a band at the position corresponding to its own density. Fractions of DNA differing in buoyant density by more than about 0.005 g-cm^{-3} can be separated on a gradient. The discrimination corresponds to a difference in G-C content of about 5%.

When a eucaryotic DNA is centrifuged on a density gradient, most of the genome forms a continuum of fragments that appears as a rather broad peak cen-

Main   Satellite
band
1.701   1.690

Optical absorbance

◄── Buoyant density

Figure 24.2
Mouse DNA is separated into a main band and a satellite by centrifugation through a density gradient of CsC1.

tered on the buoyant density corresponding to the average G-C content of the genome. This is called the **main band.** Sometimes an additional, smaller peak (or peaks) is seen at a different value. This material is called **satellite DNA.**

A good example is mouse DNA, shown in **Figure 24.2.** The graph is a quantitative scan of the bands formed when mouse DNA is centrifuged through a CsCl density gradient. The main band contains 92% of the genome and is centered on a buoyant density of 1.701 g-cm^{-3} (corresponding to its average G-C of 42%, typical for a mammal). The smaller peak represents 8% of the genome and has a distinct buoyant density of 1.690 g-cm^{-3}. It contains the mouse satellite DNA, whose G-C content is much lower than any other part of the genome (it is roughly 30%).

Satellites are present in many eucaryotic genomes. They may be either heavier or lighter than the main band; but it is uncommon for them to represent more than 5% of the total DNA. The resolution of satellite DNA is often much improved by using centrifugation through gradients of Cs_2O_4 containing silver ions, or certain other reagents, including various dyes. Sometimes a single satellite is more clearly separated from the main band; sometimes multiple satellite sequences can be resolved.

The behavior of satellite DNA on density gradients is anomalous more often than not. Thus when the actual base composition of a satellite is determined, it is often different from what had been predicted from its buoyant density. The reason is that ρ is a function not just of base composition, but of the constitution in terms of nearest neighbor pairs. For simple sequences, these are likely to deviate from the random pairwise relationships needed to obey equation **12.** Also, satellite DNA may be methylated, which changes its density.

Most of the highly repetitive DNA of a genome often can be isolated in the form of satellites. Where there is some highly repetitive DNA component that does not separate as a satellite, on isolation its properties often prove to be similar to those of satellite DNA. That is to say that it consists of multiple tandem repeats with anomalous centrifugation. Material isolated in this manner is sometimes referred to as a **cryptic satellite.** Together the cryptic and apparent satellites usually account for all the tandemly repeated blocks of highly repetitive DNA. When a genome has more than one type of highly repetitive DNA, each exists in its own satellite block (although sometimes different blocks are adjacent).

SATELLITE DNA OFTEN LIES IN HETEROCHROMATIN

Where are the blocks of highly repetitive DNA located? An extension of nucleic acid hybridization techniques allows the location of satellite sequences to be determined directly in the chromosome complement. In the technique of **in situ** or **cytological hybridization,** the DNA of the chromosomes is denatured by treating cells that have been squashed on a cover slip. Then a solution containing a radioactively labeled DNA or RNA probe is added. The probe hybridizes with its

Figure 24.3
Cytological hybridization shows that mouse satellite DNA is located at the centromeres.

Photograph kindly provided by Mary Lou Pardue and Joe Gall.

complements in the denatured genome. The location of the sites of hybridization can be determined by autoradiography.

Labeled satellite DNAs often are confined to the heterochromatin present around the centromeres of mitotic chromosomes. **Heterochromatin** is the term used to describe regions of chromosomes that are permanently tightly coiled up and inert, in contrast with the **euchromatin** that represents most of the genome (see Chapter 28).

An example of the localization of satellite DNA for the mouse chromosomal complement is shown in **Figure 24.3.** In this case, the ends of the chromosomes are labeled, because this is where the centromeres are located in the *M. musculus* chromosomes.

The identification of satellite DNA with heterochromatin implies that highly repetitive DNA is not expressed in the form of RNA or protein (the last is evident also from its simple sequence). Attempts to detect RNA products corresponding to satellite DNA have confirmed that the highly repetitive sequences usually are not transcribed.

The centromeric location of satellite DNA suggests that it may have some structural function in the chromosome. Since the centromeres are the regions where the kinetochores are formed at mitosis and meiosis for providing chromosome movement, this function could be connected with the process of chromosome segregation. But that is all we know about its role, which

apart from this general suggestion remains quite mysterious.

ARTHROPOD SATELLITES HAVE VERY SHORT IDENTICAL REPEATS

In the arthropods, as typified by insects and crabs, each satellite DNA appears to be rather homogeneous in sequence constitution. Usually, a single very short repeating unit accounts for more than 90% of the satellite. This makes it relatively straightforward to determine the sequence, which has been done in several cases.

In *Drosophila virilis,* there are three major satellites and also a cryptic satellite, together representing a large proportion of the genome, about 40%. The sequences of the satellites are summarized in **Table 24.1.** The three major satellites have closely related sequences. A single base substitution is sufficient to generate either satellite II or III from the sequence of satellite I. The cryptic satellite can be derived from satellite II by 2 base changes.

The satellite I sequence is present in other species of *Drosophila* related to *virilis,* and so may have preceded speciation. The sequences of satellites II and III seem to be specific to *D. virilis,* and so may have evolved from satellite I after speciation.

The main feature of these satellites is their very short

Table 24.1
Satellite DNAs of *D. virilis*

Satellite	Predominant Sequence	No. Copies	Percent of DNA
I	A C A A A C T T G T T T G A	1.1×10^7	25
II	A T A A A C T T A T T T G A	3.6×10^6	8
III	A C A A A T T T G T T T A A	3.6×10^6	8
Cryptic	A A T A T A G T T A T A T C		

More than 95% of each satellite consists of a tandem repetition of the predominant sequence. Note that satellites II and III have exactly the same base composition (1 G-C pair out of 7 = 14% G-C), but have buoyant densities of 1.688 and 1.671 g-cm^{-3}, respectively.

repeating unit: only 7 bp. Similar satellites are found in other species. For example, *D. melanogaster* has a variety of satellites, several of which have very short repeating units (5, 7, 10, or 12 bp). Comparable satellites are found in the crabs.

The close sequence relationship found among the *D. virilis* satellites is not necessarily a feature of other genomes, where the satellites may have unrelated sequences. It is clear that each satellite has arisen by a lateral amplification of a very short sequence. This sequence may represent a variant of a previously existing satellite, or could have some other origin. At all events, it seems likely that satellites may continually be generated and lost from genomes. Thus it is difficult to ascertain evolutionary relationships, since a current satellite could have evolved from some previous satellite that has since been lost. The important feature of these satellites is that they represent very long stretches of DNA of very low sequence complexity and within which constancy of sequence can be maintained.

One feature of many of these satellites is a pronounced asymmetry in the orientation of base pairs on the two strands. In the example of the *D. virilis* satellites shown in Table 24.1, in each of the major satellites one of the strands is much richer in T and G bases. This increases its buoyant density, so that upon denaturation this **heavy strand** (**H**) can be separated from the complementary **light strand** (**L**). This can be useful in sequencing the satellite.

MAMMALIAN SATELLITES CONSIST OF HIERARCHICAL REPEATS

In the mammals, as typified by various rodents, the sequences comprising each satellite show appreciable divergence between tandem repeats. Common short sequences can be recognized by their preponderance among the oligonucleotide fragments released by chemical or enzymatic treatment. However, the predominant short sequence usually accounts for only a small minority of the copies. The other short sequences are related to the predominant sequence by a variety of substitutions, deletions, and insertions.

But a series of these variants of the short unit can constitute a longer repeating unit that is itself repeated in tandem with some variation. Thus mammalian satellite DNAs are constructed from a hierarchy of repeating units. These longer repeating units constitute the sequences that renature in reassociation analysis. They can also be recognized by digestion with restriction enzymes.

Some uncertainty in evaluating the complexity of a repetitive DNA is introduced by the effect of mismatching on the reassociation of related but not identical repeating units. This effect is particularly pronounced with satellite DNA, and it means that the length of the reassociating unit can be assessed only within a factor of two or so.

When any satellite DNA is digested with an enzyme that has a recognition site in its repeating unit, one fragment will be obtained for every repeating unit in which the site occurs. In fact, when the DNA of a eucaryotic genome is digested with a restriction enzyme, most of it gives a general smear, due to the random distribution of cleavage sites. But satellite DNA generates sharp bands, because a large number of fragments of identical or almost identical size are created by cleavage at restriction sites that lie a regular distance apart.

Direct sequence determination of satellite DNA can be difficult. Using the discrete bands generated by restriction cleavage, we can attempt to obtain a sequence directly. However, if there is appreciable divergence between individual repeating units, there will be different nucleotides present at the same position in different repeats, so that the sequencing gels will be obscure. If the divergence is not too great—say, within about 2%—it may be possible to determine directly the average repeating sequence.

Individual segments of the satellite can be inserted into plasmids for cloning. A difficulty is that the satellite sequences sometimes tend to be excised from the chimeric plasmid by recombination in the bacterial host. However, when the cloning succeeds, it is possible to determine the sequence of the cloned segment unambiguously. While this gives the actual sequence of a repeating unit or units, we should need to have many individual such sequences to reconstruct the type of divergence typical of the satellite as a whole.

By either sequencing approach, the information that

Figure 24.4
The repeating unit of mouse satellite DNA contains two half-repeats, whose homologies are indicated by the bars.

can be gained is limited to the distance that can be analyzed on one set of sequence gels. The repetition of divergent tandem copies makes it impossible to reconstruct longer sequences by obtaining overlaps between individual restriction fragments.

The satellite DNA of the mouse *M. musculus* is cleaved by the enzymes EcoRII or Sau96I into a series of bands, including a predominant monomeric fragment whose electrophoretic mobility corresponds to a length of 230–240 bp. Using this fragment, direct sequencing identifies a single nucleotide sequence of 234 bp. This sequence must be repeated with few variations throughout the 60–70% of the satellite that is cleaved into the monomeric band. We shall now analyze this sequence in terms of its successively smaller constituent repeating units.

Figure 24.4 depicts the sequence in terms of two half-repeats. By writing the 234 bp sequence so that the first 117 bp are aligned with the second 117 bp, we see that the two halves are quite well related. They differ at 22 positions, corresponding to 19% diver-

gence. This means that the current 234 bp repeating unit must have been generated at some time in the past by duplicating a 117 bp repeating unit, after which differences accumulated between the duplicates.

Within the 117 bp unit, we can recognize two further subunits. Each of these is a quarter-repeat relative to the whole satellite. The four quarter-repeats are aligned in **Figure 24.5.** The upper two lines represent the first half-repeat of Figure 24.4; the lower two lines represent the second half-repeat. We see that the divergence between the four quarter-repeats has increased to 23 out of 58 positions, or 40%. Actually, the first three quarter-repeats are somewhat better related, and a large proportion of the divergence is due to changes in the fourth quarter-repeat.

Looking within the quarter-repeats, we find that each consists of two related subunits (one-eighth-repeats), shown as the α and β sequences in **Figure 24.6.** The α sequences all have an insertion of a C and the β sequences all have an insertion of a trinucleotide, relative to a common consensus sequence. This sug-

Figure 24.5
The alignment of quarter-repeats shows that there are homologies between the first and second half of each half-repeat.

```
           10          20          30          40          50
GGACCTGGAATATGGCGAGAAAACTGAAAATCACGGAAAATGAGAAATACACACTTTA

GGACGTGAAATATGGCGAGᴳAAAACTGAAAAAGGTGGAAAATTᵀAGAAATGTCCACTGTA
   60       70       80       90       100      110

GGACGTGGAATATGGCAAGAAAACTGAAAATCATGGAAAATGAGAAACATCCACTTGA
   120      130      140      150      160      170

CGACTTGAAAAATGACGAAATCACTAAAAAACGTGAAAAATGAGAAATGCACACTGAA
   180      190      200      210      220      230
```

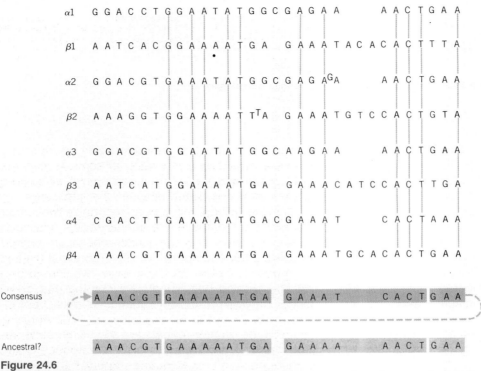

Figure 24.6
The alignment of eighth-repeats shows that each quarter-repeat consists of an α and a β half. The consensus sequence gives the most common base at each position. The ancestral sequence shows three 9 bp units, very closely related to the consensus sequence, that could be predecessor to the α and β units.

gests that the quarter-repeat originated by the duplication of a sequence like the consensus sequence, after which changes occurred to generate the components we now see as α and β. Further changes then took place between tandemly repeated αβ sequences to generate the individual quarter- and half-repeats that exist today. Among the one-eighth-repeats, the present divergence is 19/31 = 61%.

Within the consensus sequence, we can recognize three related 9 bp sequences. (Remember that the satellite sequence is continuous, so that for the purpose of deducing the consensus sequence, we can treat it as a circular permutation, as indicated by joining the last GAA triplet to the first 6 bp.) If in one of the repeats we take the next most frequent base at

two positions instead of the most frequent, we obtain three well-related 9 bp sequences.

$$\text{G A A A A A C G T}$$
$$\text{G A A A A A T G A}$$
$$\text{G A A A A A A C T}$$

This is as good a guess as we can make for the original constitution of the satellite.

The consensus sequence is analyzed directly in **Figure 24.7,** which demonstrates that the current satellite sequence can be treated as derivatives of a 9 bp sequence. The overall consensus sequence is GAAA$^{AA}_{T}$A$^{TGA}_{ACT}$, which is effectively an amalgam of the three 9 bp repeats of Figure 24.6.

1	2	3	4	5	6	7	8	9	
			G	G	A	C	C	T	
G	G	A	A	T	A	T	G	G	C
G	A	G	A	A	A	A	C	T	
G	A	A	A	T	C	A	C		
G	G	A	A	A	T	G	A		
G	A	A	A	T*	C	A	C	T	
T	T	A	G	G	A	C	G	T	
G	A	A	A	T	A	T	G	G	C
G	A	G	^{GA}A	A	A	A	C	T	
G	A	A	A	A	G	G	T		
G	G	A	A	A	T	T^{T}	A		
G	A	A	A	T*	C	A	C	T	
G	T	A	G	G	A	C	G	T	
G	G	A	A	T	A	T	G	G	C
A	A	G	A	A	A	A	C	T	
G	A	A	A	T	C	A	T		
G	G	A	A	A	T	G	A		
G	A	A	A	C*	C	A	C	T	
T	G	A	C	G	A	C	T	T	
G	A	A	A	A	T	G	A	C	
G	A	A	A	T	C	A	C	T	
A	A	A	A	A	C	G	T		
G	A	A	A	A	T	G	A		
G	A	A	A	T*	C	A	C	T	
G	A	A							

1	2	3	4	5	6	7	8	9
G_{20}	A_{16}	A_{21}	A_{20}	A_{12}	A_{17}	T_8	G_{11}	A_5
				T_7	C_5	A_8	C_9	T_{15}
						C_7		

* indicates inserted triplet in β sequence

C in position 10 is extra base in α sequence

Figure 24.7
The existence of an overall consensus sequence is shown by writing the satellite sequence in terms of a 9 bp repeat.

RECONSTRUCTING THE STAGES OF MOUSE SATELLITE DNA EVOLUTION

From the sequence of the satellite DNA, we can both reconstruct its evolution and account for its present properties. A model showing the possible steps involved in its evolution is given in **Figure 24.8.**

The general principle of this model is that at various times a group of repeating units may be suddenly amplified laterally to generate a large number of identical tandem copies. An event of this sort is called a **saltatory replication.** Then the identity of the copies is lost as mutations accumulate in them. At some subsequent time, a group of these copies again may be subject to a saltatory replication. The extent of divergence among the copies that are amplified at each time will depend on the period that has passed since the last saltatory replication imposed identity on the satellite. The satellite can evolve by a series of these saltatory replications, alternating with accumulation of mutations.

Suppose that the present satellite originated with the tandem repetition of a sequence that was either GAAAAATGT or something closely related to it. (This sequence might have been part of a satellite or could have some quite different origin: its existence is as far back as we can pursue the satellite.) All the original 9 bp units were identical, but with time mutations created differences between them. Then three adjacent units with the ancestral sequence hypothesized in Figure 24.6 were amplified, giving a tandem repeat of 27 bp.

Mutations occurred in this unit, including cases in which one unit gained an additional C while its neighbor gained a triplet insertion (of unknown sequence). This pair of repeats, now together 58 bp long, was subject to saltatory replication and gave a satellite that we can describe as $(\alpha\beta)_n$.

Figure 24.8
**The evolution of mouse satellite
DNA can be explained by an alternation of saltatory replications and accumulation of mutations.**

Once again, the satellite accumulated point mutations, deletions, and insertions to create divergence
between its repeating units. Two adjacent αβ pairs
were utilized in the next saltatory replication, giving a
116 bp repeating unit ($\alpha_c\beta_c\alpha_d\beta_d$). After further mutations, two of these adjacent units were amplified to
give the present satellite.

VARIATIONS IN THE PRESENT
REPEATING UNIT

The average sequence of the Sau monomeric fragment of the mouse satellite DNA explains its proper-

ties. The longest repeating unit of 234 bp is identified
by the restriction cleavage. The unit of reassociation
between single strands of denatured satellite DNA is
probably the 117 bp half-repeat, because the 234 bp
fragments can anneal both in register and in half-register (in the latter case, the first half-repeat of one strand
renatures with the second half-repeat of the other). In
the oligonucleotide digest, the most common fragments, accounting for 4% of the total amount of DNA,
are GA_5TGA, GA_4TGA, and GA_4CTGA, all of which
can be found in the 234 bp unit and are related to the
proposed ancestral units.

So far, we have treated the present satellite as though
it consisted of identical copies of the 234 bp repeating

unit. Although this unit does appear to account for the majority of the satellite, variants of it also are present. Some of these are scattered at random throughout the satellite; others are clustered.

The existence of variants is implied by our description of the starting material for the sequence analysis as the "monomeric" fragment. When the satellite is digested by an enzyme that has one cleavage site in the 234 bp sequence, it also generates dimers, trimers, and tetramers relative to the 234 bp length. These arise when a repeating unit has lost the enzyme cleavage site as the result of mutation.

The monomeric 234 bp unit is generated when two adjacent repeats each have the recognition site. A dimer occurs when one unit has lost the site, a trimer is generated when two adjacent units have lost the site, and so on. With the enzymes EcoRII or Sau96I, most of the satellite is cleaved into a member of this repeating series, as shown in the example of **Figure 24.9.** The declining number of dimers, trimers, etc., shows that there is a random distribution of the repeats in which the enzyme's recognition site has been eliminated by mutation.

Figure 24.9
Digestion of mouse satellite DNA with the restriction enzyme EcoRII identifies a series of repeating units (1, 2, 3) that are multimers of 234 bp and also a minor series (½, 1½, 2½) that includes half-repeats (see text later). The band at the far left is a fraction resistant to digestion.

Some restriction enzymes show a different type of behavior with the satellite DNA. They continue to generate the same series of bands. But they cleave only a small proportion of the DNA, say 5–10%. This implies that a particular region of the satellite contains a concentration of the repeating units that have this particular restriction site. Presumably the series of repeats in this domain all are derived from an ancestral variant that possessed this recognition site (although in the usual way, some members since have lost it by mutation).

THE CONSEQUENCES OF UNEQUAL CROSSING OVER

In a region of nonrepetitive DNA, recombination occurs between precisely matching points on the two homologous chromosomes, generating the reciprocal recombinants we have discussed in general terms in Chapter 1, and that we shall discuss in molecular detail in Chapter 35. The basis for this precision is the ability of two duplex DNA sequences to align exactly. We know that unequal recombination can occur when there are multiple copies of genes whose exons are related, even though their flanking and intervening sequences may differ (see Chapter 21). This happens because of the mispairing between corresponding exons in *nonallelic* genes. Imagine how much more frequently misalignment must occur in a tandem cluster of identical or nearly identical repeats. Except at the very ends of the cluster, the close relationship between successive repeats may make it impossible even to define the exactly corresponding repeats.

Consider a sequence consisting of a repeating unit "ab" with ends "x" and "y". If we represent one chromosome in black and the other in red, the exact alignment between "allelic" sequences would be

xababababababababababababababababababy

xababababababababababababababababababy

But probably *any* sequence ab in one chromosome could pair with *any* sequence ab in the other chromosome. In a misalignment such as

xabababababababababababababababababy

xababababababababababababababababababababy

the region of pairing is no less stable than in the perfectly aligned pair, although of course it is shorter. We do not know very much about how pairing is initiated prior to recombination, but very likely it starts between short corresponding regions and then spreads along. If it starts within satellite DNA, it is more likely than not to involve repeating units that do not have exactly corresponding locations in their clusters.

Now suppose that a recombination event occurs within the unevenly paired region. The recombinants will have different numbers of repeating units. In one case, the cluster has become longer; in the other, it has become shorter,

xabababababababababababababababababy
×
xabababababababa bababababababababababy
↓
xababababababababababababab ababababababababababy
+
xababababababab ababababababy

where "×" indicates the site of the crossover.

If this type of event is common, clusters of tandem repeats will be subject to continual expansion and contraction. Unfortunately, we do not yet have much data on the extent to which satellite DNA clusters vary in size between different individual genomes in a species.

Unequal recombination has another consequence when there is internal repetition in the repeating unit. In the example just given, the two clusters are misaligned with respect to the positions of the repeating units within each cluster, but they are aligned **in register,** as seen by the correspondence between individual ab repeats and ab repeats. But suppose that the "a" and "b" components of the repeating unit are themselves sufficiently well related to pair. Then the two clusters can align in **half-register,** with the "a" sequence of one aligned with the "b" sequence of the other. How frequently this occurs will depend on the closeness of the relationship between the two halves of the repeating unit. (We have already mentioned that reassociation between the denatured satellite DNA strands *in vitro* commonly occurs in the half-register.)

When a recombination event occurs, it changes the length of the repeating units that are involved in the reaction.

xababababababababababab**ab**ababababy
×
xababababababababab**ab**ababababy
↓
xababababababababab**aab**ababababy
+
xabababababababab**b**babababy

In the upper recombinant cluster, an "ab" unit has been replaced by an "aab" unit. In the lower cluster, the "ab" unit has been replaced by a "b" unit.

This type of event explains a feature of the restriction digest of mouse satellite DNA. In addition to the integral repeating series in Figure 24.9, there is a fainter series of bands at lengths of ½, 1½, 2½, and 3½ repeating units. Suppose that in the preceding example, "ab" represents the 234 bp repeat of mouse satellite DNA, generated by cleavage at a site in the "b" segment. The "a" and "b" segments correspond to the 117 bp half-repeats.

Then in the upper recombinant cluster, the "aab" unit generates a fragment of 1½ times the usual repeating length. And in the lower recombinant cluster, the "b" unit generates a fragment of half of the usual length. (The longer fragments in the half-repeat series are generated in the same way as longer fragments in the integral series, when some repeating units have lost the restriction site by mutation.)

Turning the argument the other way around, the identification of the half-repeat series on the gel shows us that in mouse satellite DNA, the 234 bp repeating unit consists of two half-repeats well enough related to pair sometimes for recombination. Also visible in Figure 24.9 are some rather faint bands corresponding to ¼- and ¾-spacings. These will be generated in the same way as the ½-spacings, when recombination occurs between clusters aligned in a quarter-register. The decreased relationship between quarter-repeats compared with half-repeats explains the reduction in frequency of the ¼- and ¾-bands compared with the ½-bands.

CROSSOVER FIXATION COULD MAINTAIN IDENTICAL REPEATS

A general presumption about satellite DNA is that its sequence is not under high selective pressure (if in-

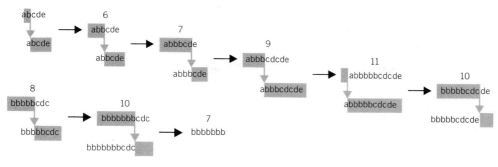

Figure 24.10
Unequal recombination allows one particular repeating unit to occupy the entire cluster.
The numbers indicate the length of the repeating unit at each stage.

deed under any at all). Unlike a sequence that codes for protein, where mutation may inactivate the product, satellite DNA seems likely to serve whatever functions it has probably by its presence instead of by virtue of its exact sequence. This idea fits perfectly well with the structure of satellites whose repeating units are related rather than identical. Mutations have accumulated since the satellite was last rendered uniform in sequence. But how are we to explain the presence in arthropods of satellites most of whose repeating units remain identical? Even if the sequence were important, there is still the difficulty of how selection could be imposed on so many copies.

Unequal recombination has been proposed as an alternative model to saltatory replication to explain the evolution of satellite DNA. The basic idea is that unequal crossing-over occurs frequently at random sites. The result of a series of random events is to cause one repeating unit to take over the entire satellite. This process is called **crossover fixation.**

The spread of a particular repeating unit through the satellite is illustrated in **Figure 24.10.** Suppose that a satellite consists initially of a sequence *abcde,* where each letter represents a repeating unit. The different repeating units are closely enough related to each other to mispair for recombination. Then by a series of unequal recombinations, the size of the repetitive region may increase or decrease, and also one unit may spread through to replace all the others.

This model actually predicts that *any* sequence of DNA that is not under selective pressure will be taken over by a series of identical tandem repeats generated in this way. The critical assumption is that the process of crossover fixation is fairly rapid relative to mutation, so that new mutations either are eliminated (their repeats are lost) or come to take over the entire cluster (as shown for the spread of variant "b" in Figure 24.10).

We may wonder whether the existence of domains that contain particular repeating units (as in mouse satellite DNA) could represent a partial spreading of the unit from its origin. For example, we can see that in the intermediate stages of the spreading illustrated in Figure 24.10, there are clusters that have a domain consisting of a series of "b" variants.

FURTHER READING

The idea of saltatory replication was introduced by **Southern** (*J. Mol. Biol.* **94,** 51–70, 1975). The implications of unequal crossing-over were developed by **Smith** (*Science* **191,** 528–535, 1976). The organization of inverted and tandem repeats is reviewed in **Lewin's** *Gene Expression,* **2,** *Eucaryotic Chromosomes* (Wiley, New York, 1980, pp. 531–569). Alu sequences have been reviewed by **Schmid** & **Jelinek** (*Science* **216,** 1065–1070, 1982).

PART 7
REACHING MATURITY: RNA PROCESSING

If hnRNA molecules much larger than mRNA act as messenger precursors, it is tempting to speculate that at least some repetitive sequences act as elements that control processing. . . . One approach to the critical problem of defining the relationship between mRNA and genome, that is how structural genes are expressed, may be to define the structure of the precursor hnRNAs comprising the primary transcripts, so that the unit of transcription can be compared both with the unit of translation and with the organization of the genome.

Benjamin Lewin, 1975

CHAPTER 25
CUTTING AND TRIMMING STABLE RNA

The "processing" of the mature molecule from a precursor is a crucial stage in the production of many forms of RNA. All stable RNAs in bacteria are synthesized as primary transcripts from which the mature rRNA or tRNA must be released. A similar pathway is followed for eucaryotic rRNA and tRNA. By contrast, most bacterial mRNAs are primary transcripts, so only in the exceptional case is processing necessary for translation. In eucaryotes, however, the production of an mRNA from an interrupted gene is the most labor-intensive of all RNA processing, sometimes involving a very large number of splicing reactions in which the exons are connected via removal of the introns (see Chapter 26).

The processing reactions all are highly specific, producing mature RNA molecules with unique 5' and 3' ends. We must therefore ask what features of the RNA substrates are recognized by the processing enzymes to accomplish such precise maturation.

These reactions are the responsibility of particular ribonucleases: for they depend on the enzymatic ability to cleave phosphodiester bonds in RNA. As with all nucleases, they fall into the general classes of **exonucleases** and **endonucleases.**

The endonucleases cut individual bonds *within* RNA molecules, generating discrete fragments. They are involved in cutting reactions, when the mature sequence is separated from the flanking sequence.

Exonucleases remove residues one at a time from the end of the molecule, generating mononucleotides. Their attitude toward the substrate may be **random** or **processive.** A randomly acting enzyme removes a base from one RNA molecule and then dissociates; for its next catalytic event, the enzyme may attack either the same or a different RNA substrate molecule. Processive action means that the enzyme stays with one substrate molecule, removing further bases until its mission is accomplished.

Exonucleases are involved in trimming reactions. From the perspective of the RNA substrate, the extra residues are whittled away, base by base. All the characterized exonucleases proceed along the nucleic acid chain from the 3' end. This means that additional material at the 3' end can be trimmed, but additional material at the 5' end can be released only by cutting.

As well as participating in specific maturation reactions, ribonucleases accomplish the degradation of superfluous RNA. "Superfluous" molecules include mature species that turn over, as well as the extraneous material that is discarded during the cutting of precursors. In particular, mRNA whose time has come must be degraded (its time comes rapidly in bacteria, more slowly in eucaryotes, but in very few cases is mRNA stable enough to survive for a protracted period). We might expect that the ribonucleases engaged in this turnover should be less specific than those involved in particular maturation pathways, since their role is to degrade the RNA completely rather than to generate specific termini. Both endoribonucleases and exoribonucleases are found in this class. In a sense, of course, the degradation of an RNA to mononucleotides is its final maturation, returning its components for reutilization via the appropriate metabolic pathways.

Ribonucleases in fact vary widely in their specificities. None is known that has any multibase sequence requirement in its substrate. Thus there is no counterpart to the restriction nucleases that attack specific sequences in DNA. Many ribonucleases have some small specificity in the sense that they cleave phosphodiester bonds at particular individual bases. But any specificity greater than this seems to be accomplished by relying on the conformation of the RNA substrate.

Substrate recognition may take the form of seeking a specific feature of secondary structure (such as a hairpin of certain size), or examining the overall secondary/tertiary structure (as seems to apply in tRNA processing).

The only situation in which there is any hope of gaining sufficient information to define the entire cellular set of ribonucleases is in *E. coli.* Here not only can the enzymes be defined biochemically, but their effects *in vivo* can also be determined by selecting bacteria in which a particular enzyme has been mutated. Indeed, both lines of evidence often are needed to distinguish enzymes or to show that two apparently different activities reside in the same enzyme. (In time, yeast also may become approachable in these terms.)

We do not yet know precisely how many ribonucleases there are in *E. coli,* but the current count is less than 10. It seems likely that combinations of a small number of ribonuclease activities provide the necessary range of reactions to process rRNA and tRNA. Although mRNA is actively degraded, the process has been intractable to analysis so far, and we do not yet have any detailed information about the relevant enzymatic activities.

PHOSPHODIESTER BONDS CAN BE CLEAVED ON EITHER SIDE

Nucleases vary in their enzymatic activities as well as in their substrate specificities. When the phosphodiester bond linking two nucleotides is cleaved, it can in principle be cut on either side of the phosphate group. The consequences are illustrated in **Figure 25.1.** Cleavage on one side generates 3'-hydroxyl and 5'-phosphate termini. Cleavage on the other side generates 3'-phosphate and 5'-hydroxyl termini.

Two ribonucleases of practical importance, although of uncharacterized physiological function, are ribonuclease T1 and pancreatic ribonuclease. Both are endoribonucleases with rather simple sequence requirements. Ribonuclease T1 is obtained from *Aspergillus oryzae* (but is common in other fungi and in bacteria). It hydrolyzes the bond on the 3' side of a G residue. Thus it cleaves a sequence such as

5'. . .pNpNpNpNpNpNpNpNpNpNpGpNpNpN. . .3'

to generate the products

5'. . .pNpNpNpNpNpNpNpNpNpNpGp   3'

                     5'HO-NpNpN. . .3'

Note that the new ends have 3'-phosphate and 5'-hydroxyl termini. The same types of ends are generated by pancreatic ribonuclease, which cleaves on the 3' side of either pyrimidine.

Before the advent of the current methods for nucleic acid sequencing, these enzymes were used to digest RNA into oligonucleotides from which the sequence of the molecule could be pieced together. Given their low specificity, presumably the enzymes are concerned with the degradation of RNA *in vivo.* This may explain the type of ends that they produce.

In the natural state, all RNA molecules have a polarity in which the 3' end bears a hydroxyl group and the 5' end is phosphorylated (it may have a triphosphate or monophosphate or even a cap, depending on whether and what changes have been introduced at the original 5' terminus). Enzymes concerned with RNA processing generate the same type of termini. That is, all the cutting and trimming enzymes that we discuss in this chapter generate 5'-phosphate and 3'-hydroxyl ends.

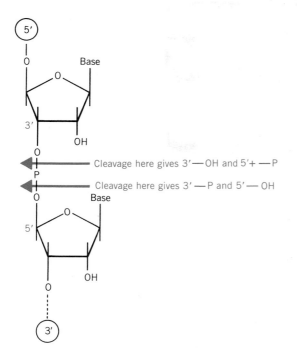

Figure 25.1
The nature of the new termini is determined by which side of the phosphodiester bond is cleaved.

RNAase III RELEASES THE PHAGE T7 EARLY mRNAs

Bacterial mRNAs almost all are translated in the same form in which they are transcribed. Most are polycistronic, some are monocistronic. An exceptional situation, however, is presented by the early genes of phage T7. These are transcribed into a polycistronic mRNA. But then the molecule is cleaved at several sites to release mRNAs for the individual genes. (A similar cleavage may also occur with late RNA, but is less well characterized.)

This form of expression is found in phage T3 as well as in T7. Since the two phage DNA sequences are quite divergent, the retention of a common pathway suggests that it offers some distinct selective advantage. But we do not know why this unique mRNA processing pathway should be favored by these two phages.

We have mentioned in Chapter 9 that cleavage is used to generate individual mRNAs from certain poly-cistronic RNAs that are transcribed from eucaryotic viruses. In this case, we understand the purpose, which is to generate monocistronic mRNAs suitable for translation, but we do not know anything about the mechanism.

The transcription of the early region of phage T7 is illustrated in **Figure 25.2.** Initiation occurs at one of a group of three promoters (*A1, A2, A3*) and continues to a terminator (*t*) about 7000 bp later. There are six genes within this region, and cuts are made at inter-cistronic boundaries to release five individual mRNAs (four monocistronic and one bicistronic).

The enzyme responsible for making the cuts is **RNAase III.** When T7 DNA is transcribed *in vitro* by *E. coli* RNA polymerase, the product is a single large RNA representing the entire early region. This can be cleaved by purified RNAase III into the individual mRNAs found *in vivo*. Bacterial mutations (called *rnc⁻*) that inactivate RNAase III block the processing reaction *in vivo*. Thus in the *rnc⁻* mutant cells, T7 DNA is transcribed into the same large precursor that is generated *in vitro*. Although remaining uncleaved in the mutant bacterial host, this precursor can be translated *in vivo* as a polycistronic mRNA, giving the usual early proteins and thus sustaining infection.

The successful expression of T7 in *rnc⁻* mutants adds to our puzzlement about the need for the cleavage reaction. However, it turns out that cleavage is necessary for translation of the bicistronic *1.1/1.2* mRNA. Phage mutants with an altered pattern of cleavage that extends the *1.1/1.2* mRNA for an extra 29 bases fail to translate either coding region. The reason is that the longer mRNA takes up a secondary structure that prevents initiation at the *1.1* coding sequence (and failure to translate the *1.1* coding sequence in turn prevents translation of *1.2*). This effect was revealed by attempts to grow T7 on a mutant *E. coli* strain in which expression of gene *1.2* is necessary for phage growth (usually it is nonessential). In this circumstance, therefore, and perhaps in others in the natural habitat, cleavage may be a necessary reaction.

Another instance in which RNAase III may affect translation is presented by an mRNA of phage lambda that carries the sequence of gene *int*. This mRNA usually is not translated (Int protein is synthesized from another mRNA). Mutations in a site called *sib*, located

Figure 25.2
The T7 early region is transcribed into a single RNA that is cleaved into five individual mRNAs by RNAase III.

Figure 25.3
A typical cleavage site for RNAase III in T7 RNA lies within a short hairpin that contains an unpaired bubble.

downstream from the *int* coding sequence, allow Int protein to be translated from the usually inactive mRNA. The inhibitory effect of the wild-type *sib* sequence is called **retroregulation.** The *sib* sequence loses its ability to prevent translation in *rnc*⁻ hosts. This suggests that *sib* may be a target site for RNAase III; when the mRNA is cleaved here, either it acquires a conformation that precludes initiation, or it is degraded. Thus mutations in either *sib* or *rnc* prevent cleavage and allow translation to occur.

What is the nature of the sites recognized by RNAase III? The targets have been identified by determining the sequences at the 5′ and 3′ ends of the mRNA molecules generated by cleavage. These sequences can be aligned with the known DNA sequence of phage T7 to reveal the complete environment of the cleavage site.

The enzyme RNAase III was originally discovered as an endoribonuclease (then of unknown function) that is active with double-stranded RNA. The duplex structure is an important feature of the sites recognized by the enzyme.

Three of the T7 early cleavage sites show quite strong sequence homology, but the other two are different. The longest homology between all five is a triplet sequence close to the site of cleavage. However, all the sites conform to the same general structure, a duplex hairpin containing an unpaired "bubble." An example is drawn in **Figure 25.3.** RNAase III cleaves at a single point at four of the sites, and at two points in one site, within or near to the bubble. The only information we have about the dependence of cleavage on the RNA structure is that mutations that reduce the stability of the hairpin may decrease or eliminate cutting by the enzyme.

RNAase III SEPARATES rRNAs FROM THEIR PRECURSOR

In the host bacterium, RNAase III is involved in the processing of rRNAs. The 16S, 23S, and 5S rRNAs all are synthesized as part of a common primary transcript (see Figure 23.9). In *rnc*⁻ cells, the mature rRNAs do not appear; instead, a 30S precursor containing them accumulates. This precursor can be cleaved by RNAase III *in vitro* to generate the molecules that serve as individual precursors to the mature RNA species. In the case of the 16S and 23S rRNAs, these are known as p16 and p23. Each precursor is slightly longer than the rRNA found in the ribosome, due to the presence of additional sequences at both the 5′ and 3′ termini.

A common mechanism generates the ends of both the p16 and p23 RNAs. In each case, the regions in the primary transcript that contain the ends are complementary. They can therefore base pair to generate a duplex structure that contains both the 5′- and 3′-terminal regions of the product. The terminal regions around the p16 sequence can form a duplex stem of 26 bp. The sequences at the ends of the p23 region can base pair to form a longer (although imperfect) duplex. A remarkable feature of this reaction is that such a large distance separates the complementary sequences, about 1600 nucleotides for the p16 and about 2900 nucleotides for the p23.

Figure 25.4 shows that in each case RNAase III cleaves on both sides of the double-stranded stem, thus simultaneously generating the 5′ and 3′ ends of the p16 or p23 molecules. This action of RNAase III differs in two respects from that most commonly seen with the T7 early RNA. The sequence that is recognized is a continuous stem, without any prominent bubble. And two cuts are made rather than one (this also happens in an atypical T7 site).

There is no extensive sequence homology between either the p16 and p23 RNAase III sites or between either of these and any of the T7 sites. The features required by RNAase III for specific cutting of RNA therefore are mysterious, except for the common demand for a double-stranded structure either surrounding or including the actual site of cleavage. The enzyme could depend on a combination of primary sequence and secondary structure.

The creation of the recognition site for RNAase III by base pairing between regions at the 5′ and 3′ ends of each precursor sequence means that each processing reaction can occur only after both ends of the target sequence (p16 or p23) have been synthesized. It is likely that *in vivo* the RNA is associated with some ribosomal proteins before this happens. However, we know this is not a prerequisite for the reaction, because the isolated 30S RNA can be cleaved *in vitro* by RNAase III. Thus the ability of the 5′- and 3′-terminal regions to base pair is maintained in either

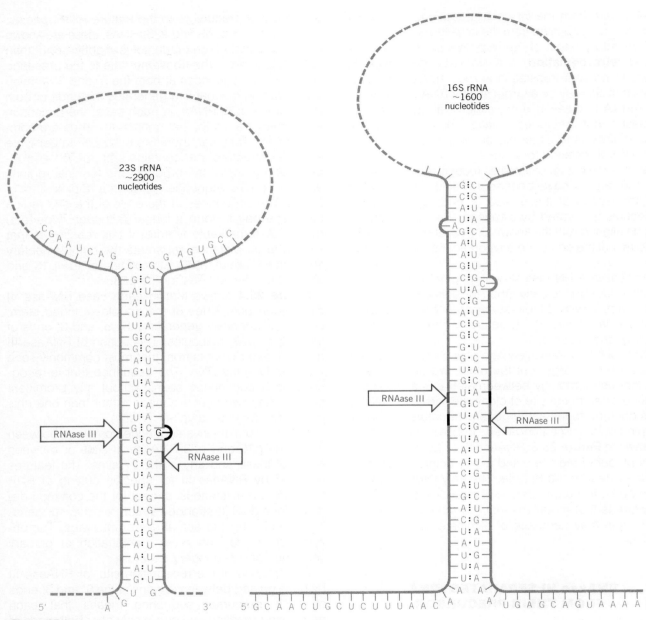

Figure 25.4
Regions containing the 5′ and 3′ ends of the p16 and p23 RNAs base pair to form duplex stems, each of which is cleaved at two points by RNAase III.

the absence or the presence of the ribosomal proteins.

Since the cleavages accomplished by RNAase III do not release the actual 5′ or 3′ termini of either 16S or 23S rRNA, further processing reactions are necessary. Not very much is known about these reactions,

except that their occurrence is contingent on the prior action of RNAase III. The tRNA sequences that also are present in the *rrn* transcription units are processed by RNAase P and other tRNA-processing enzymes (see later).

CLEAVAGE SITES IN THE PATHWAY FOR EUCARYOTIC rRNA RELEASE

The maturation of eucaryotic rRNA can be followed in some detail because of the relatively slow rate at which the primary transcript matures through intermediate stages to give the final rRNA molecules. The major processing intermediates can be isolated from a variety of cells, among which the pathway in mammals is well characterized.

The mammalian primary transcript is a 45S RNA containing the sequences of both the 18S and 28S rRNAs, but almost twice their combined length (see Table 23.2). As seen in HeLa (human) cells, it contains

~110 methyl groups, added during or immediately after its transcription. Almost all of these are attached to ribose moieties, in a variety of oligonucleotide sequences. The methyl groups are *conserved* during processing of the precursor into mature rRNAs. There are ~39 of the original methyl groups on mature 18S rRNA, plus another 4 that are added later in the cytoplasm. There are ~74 methyl groups (all original) on the 28S rRNA. This suggests that methylation is used to distinguish the regions of the primary transcript that mature into rRNA.

More than one pathway has been found for maturation, as seen from the sizes of the intermediate RNA molecules. But all the pathways can be reconciled by supposing that a small number of cleavage sites can be utilized in varying orders. Two mammalian pathways, originally characterized in HeLa (human) and L (mouse) cells, are illustrated in **Figure 25.5**. In each case, there are cleavage sites on the 5′ side of the 18S gene, within the spacer between the 18S and 5.8S

Figure 25.5
Mature rRNAs are released from the 45S mammalian precursor by cleavages within each spacer region.

gene, and in the spacer between the 5.8S and 28S sequences. (The 5.8S sequence becomes a small RNA that associates with the 28S rRNA by base pairing.) The difference between the two pathways shown in the figure lies just in the order with which the cleavage sites are utilized. Other variations also are possible. In some cases, more than one pathway is found in a given cell type.

The figure shows the minimum number of likely cleavage sites used to generate all the 5′ and 3′ ends. We do not know whether these cleavages actually generate the mature ends, or whether cleavage releases individual precursors that then are cut or trimmed further. Nor do we have any information about the ribonuclease activities responsible for the process. We do know, however, that the 45S RNA associates with proteins immediately upon its synthesis, so that the processing pathway applies to a ribonucleoprotein rather than to free RNA.

tRNAs ARE CUT AND TRIMMED BY SEVERAL ENZYMES

Processing is a necessary step in the production of all known tRNAs of bacteria and eucaryotic cytoplasm. In eucaryotes, newly synthesized tRNAs sediment at 4.5S or more, corresponding to molecules of roughly 100 nucleotides or so, compared with the mature size of 4S tRNA at 70–80 nucleotides. In bacteria, all tRNAs are synthesized in the form of precursors with additional sequences at both the 5′ and 3′ ends. In some cases, a single primary transcript may contain several tRNA sequences, up to about six, which must be separated. So far as we know, the tRNA part of the precursor usually takes up the same secondary structure that is displayed by mature tRNA, and this may be an important feature in its recognition by ribonucleases during maturation. Mutations in the tRNA moiety may prevent processing; the result is that defective mature tRNAs do not accumulate.

A single enzyme seems to be responsible for generating the 5′ end of all tRNA molecules in E. coli. This is **RNAase P,** an endoribonuclease that cleaves at the junction of either the 5′ leader sequence or the intercistronic sequence (when a tRNA is part of a polycistronic transcript). The primary sequences in which the

cleavage occurs are different in all the precursors. Mutations in tRNAs that prevent processing by RNAase P occur in several regions of the molecule. This suggests that RNAase P must recognize a common feature of tRNA tertiary structure, maintained in the precursor, and relying on different parts of the tRNA.

Ribonclease P is an unusual enzyme. It has both protein and RNA components. The active enzyme can be dissociated into the protein and RNA, neither of which retains the endonucleolytic activity. But activity can be restored by mixing the two components. The exact structure of the enzyme has not yet been pinned down, but current indications are that the RNA component is of appreciable size, say ~360 bases ($\approx$120,000 daltons), and therefore represents a greater part of the mass than the protein component, which is only ~20,000 daltons.

The RNA component is essential for catalytic activity per se. The obvious speculation is that its role is to recognize the tRNA via a base-pairing mechanism. This interaction could be responsible for orienting the protein component in the right position to cleave its substrate specifically. Potential counterparts to RNAase P have been found in eucaryotes.

Two genes in E. coli have been identified in which mutations abolish RNAase P activity: rnpA and rnpB. All we know about them is that they lie well apart. Could one code for the protein and the other for the RNA component?

In these mutants, a large number of precursor tRNA molecules accumulate. Some of the polycistronic precursors are cleaved into individual monocistronic precursors, although none has an authentic 5′ terminus. This cleavage indicates that there is another enzyme able to cleave in the intercistronic regions. The enzyme responsible may be RNAase III.

Why are some but not other precursors cleaved to monocistronic precursors in the absence of RNAase P? Even if the same set of enzymes are involved in cleavage of all tRNA precursors, they may act in a different order in different precursors. This is borne out by studies of individual precursors, in which analogous events may occur at different stages. The compulsion for a particular order of events may be that only one cleavage site is accessible in the original transcript, but its cleavage then changes the structure to expose the site for the next enzyme.

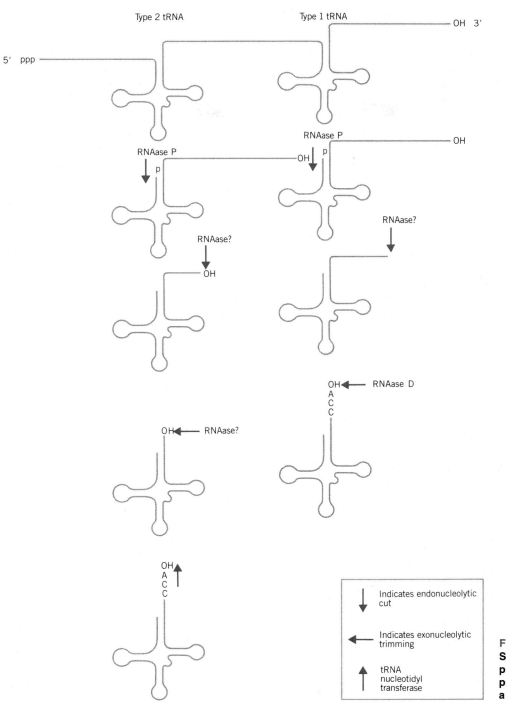

Figure 25.6
Several enzymes are needed to process a (hypothetical) dimeric precursor containing a type I and a type II tRNA.

The other steps in tRNA maturation are not so well established. A summary of the necessary events is given in **Figure 25.6.**

Endonucleolytic cleavage probably occurs also on the 3' side of tRNA sequences. We do not yet know which enzyme(s) is responsible.

The actual 3′ terminus is generated by an exonucleolytic activity. An enzyme called **RNAase D** removes bases one at a time from the 3′ end (possibly of the original precursor, possibly generated by cleavage) until it arrives at the -CCA terminus common to all mature tRNAs. The enzyme seems to recognize the overall tRNA structure rather than the CCA sequence itself as a signal to stop, because it is able to remove an additional CCA sequence added to the authentic 3′- terminal sequence. RNAase D is a randomly acting endonuclease.

Another enzyme with exonucleolytic activity is **RNAase II.** This was once thought to be involved in tRNA processing, but it is able to degrade tRNA sequences completely. So probably it is concerned with general degradation rather than with specific maturation. Its action is processive.

In bacteria, there are two types of tRNA precursor, as distinguished by their 3′ sequences. Type I molecules have a CCA triplet that is the demarcation between the mature tRNA sequence and the additional 3′ material. But type II molecules (coded by certain phages) have no CCA sequence. After the additional 3′ nucleotides have been removed from the precursor, the CCA must be added. We do not know how the 3′ terminus for CCA addition is generated in the type II tRNAs, whether it is a function of RNAase D like the type I tRNAs, or whether a different enzyme is involved. In eucaryotes, probably all tRNAs are of type II.

The addition of CCA is the function of the enzyme **tRNA nucleotidyl transferase,** which adds the triplet (or part of it) to any tRNA sequence lacking it. Mutants of *E. coli* with reduced levels of this enzyme (called cca^-) grow slowly, which suggests that it is essential for tRNA biosynthesis. The terminal A residue of all bacterial tRNAs turns over constantly, so this enzyme is needed also in a repair function as well as for the initial production of type II tRNAs.

FURTHER READING

There are few reviews that disentangle the enzymes actually involved in tRNA and rRNA processing from the various activities that have been detected. **Gegenheimer** & **Apirion** have attempted to sort out the genetics and biochemistry of bacterial enzymes (*Microbiol. Rev.* **45,** 502–541, 1981). A report on RNAase III that also covers earlier ground is from **Bram, Young,** & **Steitz** (*Cell* **19,** 393–401, 1980). The processing of tRNAs has been reviewed by **Altman** in *Transfer RNA* (MIT Press, Cambridge, Mass. 1978, pp. 48–77) and updated in *Cell* (**23,** 3–4, 1981).

CHAPTER 26
MECHANISMS OF RNA SPLICING

The discovery that interrupted genes are transcribed into precursor RNAs from which the exons are recovered via splicing raises a series of questions about gene expression. Viewing RNA splicing simply as one of the mechanisms involved in gene expression, we want to know at what point it occurs vis à vis the other modifications of RNA (capping and polyadenylation). Are introns excised from a precursor in a particular order? What is responsible for the remarkable precision of the process? Does it occur at a particular location in the nucleus; is it connected with other events, for example, nucleocytoplasmic transport? Do such connections create differences in the pathways for expression of interrupted and uninterrupted genes? Is the maturation of RNA (via splicing or other mechanisms) used to *regulate* gene expression by discriminating among the available precursors?

RNA splicing involves breakage of the phosphodiester bonds at exon-intron boundaries and the formation of a bond between the ends of the exon. This could occur either as two successive, independent reactions, or as one coordinated transfer reaction. The breakage of phosphodiester bonds is common to other RNA processing reactions, but their creation in RNA is unique to the splicing reaction. Whatever the enzymatic mechanism(s), a critical feature is that the reacting groups must be brought into juxtaposition for bond formation.

There appears to be no unique mechanism for RNA splicing. We can identify at least four distinct types of splicing system. The removal of introns from yeast nuclear tRNA precursors involves enzymatic activities whose dealings with the substrate seem to resemble those of the tRNA processing enzymes, since a critical feature is the conformation of the tRNA precursor. Excision of the intron from a ciliate rRNA precursor occurs as an autonomous property of the RNA. The removal of introns from yeast mitochondrial RNAs probably lacks a single mechanism, but at least in some cases requires the intercession of a protein coded by the intron that is to be removed (see Chapter 20). Introns are removed from the nuclear RNAs of higher eucaryotes by a system that apparently recognizes nothing more specific than the short consensus sequences conserved at exon-intron boundaries in a wide range of species. So if there was a single evolutionary origin for the interrupted gene, considerable divergence since has occurred in the means of its reconstitution in RNA.

YEAST tRNA SPLICING INVOLVES CUTTING AND REJOINING

Several nuclear tRNA genes in yeast are interrupted. In each case, there is a single intron, located just one nucleotide beyond the 3′ side of the anticodon. The introns vary in length from 14 to 46 bp. Those in related tRNA genes are related in sequence, but the introns in tRNA genes representing different amino acids are unrelated. However, all the introns include a sequence that is complementary to the anticodon of the tRNA. This creates an alternative conformation for the anticodon arm in which the anticodon is base paired to form an extension of the usual arm. An example is drawn in **Figure 26.1** Only the anticodon arm is affected—the rest of the molecule retains its usual structure.

An opportunity to study the splicing reaction is provided by the existence of a temperature-sensitive mutant of yeast that fails to remove the introns and accumulates the interrupted precursors in the nucleus. Flanking sequences can be processed as usual. We do not know the molecular basis for the defect; but it is specific to interrupted genes, because the tRNAs coded by uninterrupted genes can mature and be transported to the cytoplasm.

This is very useful, because it allows the tRNA precursors to be isolated in a form in which their only difference in sequence from the mature tRNA is the presence of the intron. (The precursors may also lack some base modifications.) Then these molecules can be used as substrates for a cell-free system extracted

from wild-type cells. The splicing of the precursor can be followed quite simply by virtue of the resulting size reduction. This is seen by the change in position of the band on gel electrophoresis, as illustrated in **Figure 26.2.** The reduction in size can be accounted for by the appearance of a band representing the intron.

The cell-free extract can be purified by assaying the ability to splice the tRNA. The *in vitro* reaction requires ATP and can be split into two steps by omitting the nucleotide. The first step involves phosphodiester bond cleavage, taking the form of an atypical nuclease reaction. The second step involves bond formation; this is called a **ligation** reaction, and the responsible enzyme activity is described as an **RNA ligase.** We do not yet know whether the nuclease and ligase activities are separate or reside in one enzyme. The two stages of reaction are depicted in **Figure 26.3.**

Without ATP, the nuclease activity cleaves the precursor at both ends of the intron. This releases two "half-tRNA" molecules and a linear intervening sequence. These intermediates have unique ends. They have the unusual termini of 5′-hydroxyl and 3′-phosphate. (All other known RNA processing enzymes

Figure 26.1
The intron in yeast tRNA^Phe base pairs with the anticodon to change the structure of the anticodon arm.

Figure 26.2
Splicing of yeast tRNA *in vitro* can be followed by assaying the RNA precursor and products by gel electrophoresis.

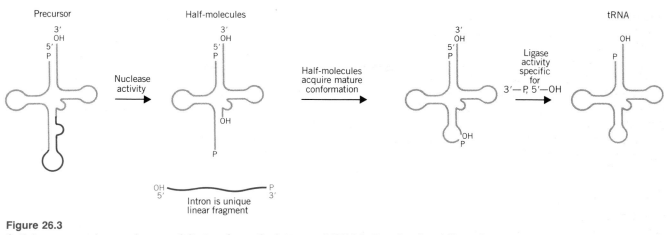

Figure 26.3

Splicing occurs via a nuclease activity to release the intron and tRNA half-molecules, followed by a ligase activity to link the half-molecules covalently.

cleaves on the other side of the phosphate bond, as mentioned in Chapter 25.)

When ATP is added, the second reaction occurs. The two "half-tRNAs" are able to base pair to form a tRNA-like structure. The RNA ligase activity links the two halves covalently by making a phosphodiester bond. The presence of the 3'-phosphate is necessary for the ligase activity.

An RNA ligase activity isolated from wheat germ also has an unusual substrate specificity. The enzyme joins an RNA end with a 2', 3' cyclic phosphate group to an RNA end with a 5'-phosphate group. For this reaction to be involved in splicing, two additional events must occur between the cleavage and ligation reactions. An additional phosphate group must be added to one of the RNA ends generated by the cleavage reaction so that both ends are phosphorylated. This is a conventional kinase reaction. And the phosphate on the 3' end must be linked also to the 2'-OH position. The cyclization might occur independently or might be part of the cleavage reaction that generated the free end.

The ligation reaction is

$$\text{Sugar}\!\!\bigg\langle\genfrac{}{}{0pt}{}{2'}{3'}\bigg\rangle\text{p} + \text{p-5'-sugar} \rightarrow \text{sugar-3'p-5'-sugar}$$

Thus the product of this splicing event has a 2'-phosphate group at the position where the 3'-5' phospho-

diester bond links the two RNA ends. The additional phosphate is presumably removed by a phosphatase, but in the interim could be useful in marking the site where the ligation occurred.

The yeast tRNA precursors also can be spliced in an extract obtained from the germinal vesicle (nucleus) of *Xenopus* oocytes. This shows that the reaction is not species-specific. *Xenopus* must have enzymes able to recognize the introns in the yeast tRNAs.

The features of the precursor that are recognized in splicing can be identified by introducing changes into the RNA molecule, either via manipulation *in vitro* of the cloned gene, or by mutation *in vivo*. The irrelevance of the exact sequence and size of the intron is indicated by the natural variation between genes. It is confirmed by an experiment in which an extra 21 bp segment was inserted into the intron; this extends the length of the additional arm present in the precursor, but has no effect on splicing.

An advantage of the tRNA system is that mutants can be obtained rather readily. For example, there is a yeast variant in which an interrupted tRNA^{Tyr} gene has been mutated to an ochre suppressor. From these mutant cells, we can select further variants that are *ineffective* in suppression. Some of these lack the suppressor tRNA because mutations have prevented processing of the precursor. The series of three tRNA structures drawn in **Figure 26.4** consists of the wild-type precursor, the suppressor tRNA precursor (which

Figure 26.4
Three variants of yeast tRNA^Tyr have different conformations in the anticodon region of the interrupted precursor.

is spliced normally), and a mutant that is spliced seven times less efficiently.

The secondary structure of the intronic region is different in each case. The suppressor has a shorter base-paired stem, and one end of the intron is base paired instead of single-stranded. The *a122* mutation actually lies at the splicing junction and creates a large unpaired bubble in the duplex stem; the other end of the intron is base paired instead of lying in the unpaired loop. So the splicing sites are recognized whether they lie in base-paired or in single-stranded regions, and even a change at the boundary reduces only the efficiency and does not alter the sites used in splicing.

This begins to seem reminiscent of the conclusions we advanced in Chapters 7 and 25 about tRNA structure and expression. Specific bases do not seem consistently to be implicated in recognition reactions, which instead rely on the conformation of the molecule as a whole. In particular, it may be important to recognize in the precursor those structural features that are conserved in the mature tRNA. This is consistent with the observation that mature tRNA competitively inhibits the splicing of precursors, presumably by being recognized by the enzyme(s).

THE EXTRAORDINARY SPLICING OF *TETRAHYMENA* rRNA

The gene for 26S rRNA in (some strains of) *Tetrahymena thermophila* is interrupted by a single intron. The situation of this gene is unusual. *Tetrahymena* has a life-style in which the diploid genome is maintained in the **micronucleus,** but the genes are amplified in the **macronucleus.** The amplification generates a large number of extrachromosomal linear molecules, each comprising a palindromic dimer of rDNA. The two transcription units are identical, and each is expressed via a 35S precursor RNA, in which the 26S sequence lies at the 3' end.

When isolated macronuclei are incubated, the intron can be excised from the precursor. It accumulates as a linear fragment of 400 bases. With longer incubation periods, this molecule is converted to a circular RNA. These events are summarized in **Figure 26.5.** Can

Figure 26.5
Splicing of the *Tetrahymena* 35S rRNA precursor can be followed by gel electrophoresis.

The 35S precursor RNA forms a rather broad band. The removal of the intron is revealed by the appearance of a rapidly moving small band. (No change is seen in the 35S RNA band because of its breadth and the smallness of the reduction in size. No free exons are seen.) When the intron becomes circular, it electrophoreses more slowly, as seen by a higher band.

these features be reproduced by incubating the isolated precursor RNA in an *in vitro* system?

An intriguing result was obtained when the 35S precursor RNA was isolated. In the presence of only a monovalent cation, a divalent cation, and GTP, the splicing reaction occurred. This means that any enzyme responsible for the splicing activity must have been purified together with the precursor RNA. But the procedure used to isolate the precursor was to extract the RNA with phenol, treat it with proteases (which should degrade any proteins), and use chemical treatments that also should destroy any proteins that are present. Such treatments usually would be considered to constitute strong evidence against the survival of any protein.

So there are two possibilities, either of which a priori would be thought to be rather unlikely. The splicing activity may reside in an enzyme that is very tightly bound indeed to the RNA (perhaps through covalent bonds) and that retains its enzymatic activity through this series of severe treatments. Or there might be no splicing enzyme, and the reaction is an intrinsic property of the RNA molecule itself. This would imply that the RNA acquires some particular conformation that

brings the relevant bonds into juxtaposition for an autonomous bond breakage and reunion reaction. The possession of splicing ability by precursor RNA transcribed from a copy of the gene cloned in *E. coli* shows that in fact the property is intrinsic to the RNA (although still the reaction could be assisted by a protein *in vivo*).

The reaction itself has some unusual features. The demand for a guanosine nucleotide is unique; no other base can be substituted. But a triphosphate is not needed; GTP, GDP, GMP, and guanosine itself all can be used. This means that although a G residue is involved, there is no net energy requirement. The guanosine nucleotide must have a 3'-hydroxyl group.

When the guanosine nucleotide carries a radioactive label, the radioactivity enters the excised linear intron fragment. It turns out that the G is linked to the 5' end of the intron by a normal phosphodiester bond. This happens to every fragment released by splicing. When the splicing reaction is allowed to continue *in vitro,* the linear intron becomes a circle. This involves joining of the 3' end to a site 15 bases from the 5' end (the terminal 15 base segment, including the additional G, is released).

Figure 26.6
The molecules formed by the *Tetrahymena* splicing reaction can be explained by phosphoester transfer reactions.

The splicing reaction formally may be equivalent to a phosphoester transferase activity, as illustrated in **Figure 26.6.** The guanine nucleotide is a cofactor that provides a free 3'-hydroxyl group to which the 5' end of the intron is transferred. This reaction could be followed by a second, similar reaction, in which the 3'-hydroxyl created at the end of the first exon is used to attack the second exon. These two transfers seem to be connected. Because no free exons have been observed, their ligation appears to be part of the same reaction that releases the intron. The cyclization of the linear intron fragment could be explained by another phosphoester transfer, in which the bond 15 bases from the end is broken and the released 5' end is transferred to the 3'-hydroxyl group of the intron molecule. (There is a parallel for this type of enzymatic activity, involved in transferring 3'-OH to 5'-P without net input of energy, in the DNA nicking-closing enzymes discussed in Chapter 32.)

NUCLEAR RNA SPLICING FOLLOWS PREFERRED PATHWAYS

Highly mosaic genes present some interesting problems for gene expression. A large number of introns must be removed; and the exons must be connected in the correct sequence. The reaction does not proceed sequentially along the precursor, which means that individual pairs of intron ends must be distinguished within the precursor RNA.

Until the development of blotting techniques, it was

all but impossible to identify nuclear precursors to particular mRNAs, because of the exceedingly small amount of nuclear RNA and its inevitable contamination with mRNA from the cytoplasm. But with Northern blotting, it is possible to fractionate nuclear RNA on a gel, transfer the material to a special paper, and use a radioactively labeled probe to hybridize with some particular sequence. Then autoradiography reveals bands whose positions identify the RNA molecules containing the hybridizing sequence. (The corresponding technique of Southern blotting with DNA is illustrated in Figure 19.6.) The probe used to identify the bands of RNA can be a cDNA representing the mRNA or a cloned sequence corresponding to some particular part of the gene (for example, an exon or an intron).

When nuclear RNA from chick oviduct is analyzed by using a probe against ovomucoid or ovalbumin mRNA, in each case a discrete series of bands is obtained. This itself suggests that splicing may occur via definite pathways. (If the seven introns of either gene were removed in an entirely random order, there would be more than 300 precursors with different combinations of introns, and we should not see discrete bands.)

The largest band corresponds to the size of the gene and is probably the primary transcript. The smallest is the size of the mRNA. In between, each band represents a particular precursor(s) from which some but not other of the introns have been removed. **Figure 26.7** shows a Northern blot analysis of the precursors to ovomucoid mRNA, in which the most prominent intermediates are *b* (lacks introns E and F), *d–e* (lack introns E, F, and G), and *g* (contains only intron C).

Further information about the pathway can be gained by electron microscopic analysis of individual nuclear RNA molecules. Each molecule is hybridized with the DNA of the intact gene and then examined to see which introns have been removed. Any intron that has been lost shows up as a loop in the DNA. The results of such an analysis are summarized in **Table 26.1**. They show that there does not seem to be an *obligatory* pathway, since intermediates can be found in which different combinations of introns have been removed. However, there is evidence for a *preferred* pathway or pathways.

When only one intron has been lost, it is virtually

Figure 26.7
Northern blotting of nuclear RNA with an ovomucoid probe identifies discrete precursors to mRNA.

always E or F. But either can be lost first. When two introns have been lost, E and F are again the most frequent, but there are other combinations. Intron C is never or very rarely lost at one of the first three splicing steps. From this pattern, we can see that there is a preferred pathway in which F and E are lost first, G and D are lost next, B tends to be removed next, and A and C are excised last. But clearly there are other pathways, since (for example), there are some mole-

Table 26.1
Introns Are Removed from Ovomucoid Nuclear RNA in a Preferred Order

Total No. Introns Removed	Frequency of Loss of Individual Introns						
	A	B	C	D	E	F	G
1	5	0	0	0	30	60	5
2	20	20	0	2	60	60	25
3	5	5	5	30	100	95	60
4	10	25	35	95	90	90	55
5	40	75	65	85	100	75	60
6	55	100	80	90	100	100	80

cules in which D or G is lost last. A caveat in interpreting these results is that we do not have proof that all these intermediates actually lead to mature mRNA, but this is a reasonable assumption.

The general conclusion suggested by this analysis is that the conformation of the RNA may influence the accessibility of the splicing junctions. As particular introns are removed, the conformation changes, and new pairs of splicing sites become available. But the ability of the precursor to remove its introns in more than one order suggests that either the molecule may take up alternative conformations or more than one pair of splicing junctions might be available in any particular conformation. More options must become available as the number of introns increases; and when we consider larger genes, it is more difficult to see how provision is made for the imposition of specific secondary structure.

One point emerging from this type of analysis is that all the intermediates are polyadenylated. This implies that usually transcription is completed, and poly(A) is added to at least an appreciable proportion of the molecules before any splicing occurs. In this case, splicing succeeds transcription and is not concomitant with it. However, this sequence of events is not obligatory, because splicing can continue when polyadenylation is inhibited (see Chapter 27).

We have assumed so far that each intron is excised in a single step. But this need not necessarily be so. In the mouse β-globin gene, a 15S precursor can be identified that retains only the longer intron. This may mean that the shorter intron always is removed first. But then the longer intron may be removed in two discrete stages, for an intermediate can be recovered that retains only part of it. In seeking to identify the sites involved in splicing, therefore, we must remember that although they must include the exon-intron junctions, other sites between also could be utilized in intermediate ''hops.''

What happens to the excised intron fragments? Hybridization experiments with the nuclear RNA population show that the concentration of ovalbumin intron sequences is about tenfold lower than the concentration of exon sequences. Probably this means that the released intron fragments are degraded rather rapidly. This also makes the point that the precursors identified on Northern blots or in the electron microscopic analysis represent the *steady-state* population of the nucleus. Any intermediates that are more rapidly processed will be present in relatively reduced amounts. So the precursors that we identify could be biased by an increased concentration of any intermediates that are processed slowly. To elucidate the splicing pathway in detail, we will need to obtain *pulse-chase* data, where a radioactive label is briefly incorporated into the primary precursor and then followed through the stages of maturation.

SPLICING JUNCTIONS MAY BE INTERCHANGEABLE

To hone in on the molecular events involved in splicing, we must consider the nature of the **splicing junctions,** the sequences immediately surrounding the sites of breakage and reunion. As we have mentioned in Chapter 20, these are named by their positions in the intron. The left splicing junction (sometimes called the donor) is at the left end of the intron; the right junction (sometimes called the acceptor) is at the other end.

Does recognition of the correct pair of left and right splicing junctions depend on the individual sequences or is it imposed by some external feature of the interaction with the splicing apparatus? In other words, is each correct pair in some way marked by a unique feature of sequence or structure, or are all left junctions functionally equivalent and all right junctions similarly indistinguishable?

From the sequences at and around the splicing junctions, we know that there is no complementarity between corresponding left and right junctions, which excludes the idea that they might be brought together directly by base pairing. This leaves the possibility that individual pairs of junction sequences might be recognized specifically by proteins or by RNA sequences. Models of this sort may apply in the yeast mitochondrion, where only a few pairs of junctions are involved, as described in Chapter 20, but seem harder to visualize for the much larger numbers in the nucleus.

An associated question is whether there is any tissue specificity in RNA splicing. Is the mechanism of splicing the same for all genes, irrespective of circumstance, so that recognition of the proper junctions depends only on the RNA and the common splicing apparatus? Or is the capacity to splice a particular nuclear RNA present only in the cells in which it is usually expressed? These issues can be explored by studying the splicing of authentic RNA precursors in novel cellular situations, and by constructing synthetic genes with new combinations of splicing junctions.

The same techniques that we described in Chapter 19 for cloning foreign sequences in bacterial phages or plasmids can be used to insert genes into eucaryotic viruses. By making appropriate restriction cleavages and then rejoining the fragments, globin, insulin, and other genes from a variety of sources have been inserted into eucaryotic viral vectors. One of the most

common vectors is the monkey virus SV40, which can be perpetuated in the cells of several mammals. The new genes can be inserted into either the early or late transcription unit of the virus. They are expressed via the formation of a precursor RNA that usually contains viral sequences at the 5′ terminus followed by the foreign gene. Such RNAs can be spliced normally, as indicated in the example of **Figure 26.8.**

This provides two important pieces of information. First, correct splicing does not depend on the integrity of the natural precursor RNA, because a foreign gene (or sometimes just part of it) can be properly processed in the context of viral sequences. Second, the information carried in the sequence is adequate to support

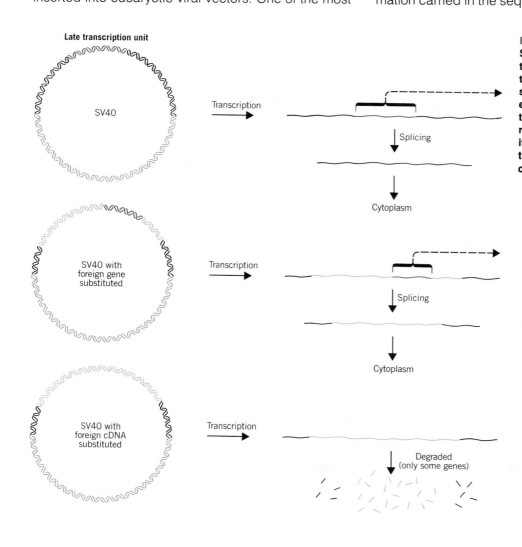

Figure 26.8
SV40 vectors carrying foreign interrupted genes in place of the late transcription unit are transcribed, spliced normally, and the product exported to the cytoplasm. When the foreign insert is an uninterrupted cDNA representing the gene, it is transcribed but in some cases the product is degraded in the nucleus (see text later).

l and *r* indicate left and right splicing junctions

Figure 26.9

Splicing occurs normally in a transcript of a unit constructed by linking the exons of two different genes.

splicing in cells derived from a different tissue and/or species from the usual place of expression. The signals involved in splicing must therefore be highly conserved.

The reconstruction approach has been taken a step

further by producing synthetic "genes" in which an exon from one authentic gene is linked to an exon from a different gene. This is illustrated diagrammatically in **Figure 26.9.** In an actual experiment, the first exon of the early SV40 transcription unit was linked to the third

Figure 26.10

Exons *L* and *C* are spliced together when the left junction of intron 2 is deleted but not when it is intact.

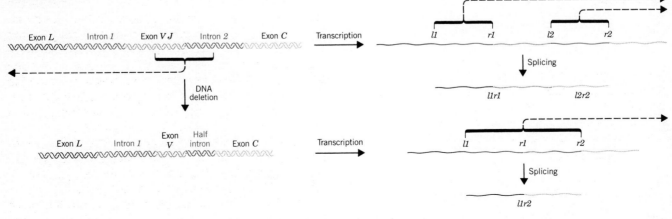

l and *r* indicate left and right splicing junctions.

exon of mouse β globin. The hybrid intron was spliced out perfectly. Thus the left junction of an SV40 intron (*l1* in the figure) can be spliced to the right junction of a mouse β-globin intron (*r2* in the figure). This implies that in principle any left splicing junction may be able to react with any right splicing junction.

The paradox between this conclusion and the usual circumstances in which splicing occurs only between the left and right junctions of the *same* intron is highlighted by the situation summarized in **Figure 26.10.** This compares the expression of a wild-type immunoglobulin gene with that of a mutant in which a deletion has removed the region including the junction between exon *VJ* and intron *2*. (The structure and expression of immunoglobulin genes are discussed in detail in Chapter 39.)

In the wild type, the left junction of each intron is spliced together only with the right junction of that intron. There is no reaction between the left junction of intron *1* and the right junction of intron *2*. But in the deletion, the right junction of intron *1* is ignored, and the left junction of this intron is spliced to the right junction of intron *2*. This means that these two junctions have an inherent ability to recognize each other for splicing, although usually they do not do so when the left junction of intron *2* is present.

The moral of these experiments is that probably all left splicing junctions look similar to the splicing apparatus, and all right splicing junctions look similar to it. We may infer that the significant properties of each type of junction should comprise common features that can be discerned in all the sequences at exon-intron boundaries.

MUTATIONS IN CONSENSUS SEQUENCES CAN AFFECT SPLICING

We have already seen that all introns in nuclear structural genes that code for proteins can be aligned by the GT···AG rule to show homology with a short consensus sequence at the exon-intron boundaries. As described in Chapter 20, the consensus is

Left junction   Right junction
↓   ↓
Exon···A G G T A A G T···Intron··· 6Py N C A G···Exon

The arrows identify bonds whose splicing would con-

vert the precursor to the mature RNA, but these have not yet been confirmed as the actual sites of splicing by examining the ends of the intermediates. (In many cases, a sequence repetition means that the splicing sites cannot be identified unequivocally on the basis of sequence alone; in the example of the consensus itself, the arrows could be moved one or two bases to the left at each junction.)

From scrutinizing the sequences of splicing junctions, these short consensus sequences are the only common features that are apparent. They describe the exon-intron boundaries of an extremely wide range of species, from animals to yeast. But there is no more extensive conservation to be discerned even when we consider particular groups of introns (such as those in a species, those in a gene family, etc.).

Can such short sequences provide for the specificity of splicing? We know that at least in some cases, no extensive sequence is needed at the boundary. For example, a deletion in SV40 that removes almost all sequences upstream from a left consensus sequence does not prevent its use in splicing. We need further experiments of this sort to see whether "splicing junction" is synonomous with "consensus sequence." We do know that the junctions do not always function autonomously. A short deletion in SV40 some 200 bp distant from a left splicing junction may prevent its use (presumably via the introduction of conformational changes in the RNA). Similarly, *cis*-acting mutations that prevent splicing of an intron of the *box* gene of yeast mitochondrial DNA may be located some distance from the splicing junction (see Chapter 20).

Our implication of the consensus sequences as the sites that are recognized in splicing is really faute de mieux. But their importance is shown directly by the existence of some point mutations that affect consensus sequences and influence splicing. These have been found in the form of human thalassemias, in which the production of either α- or β-globin protein is prevented. (There are many causes of thalassemia, some of which we have reviewed in Chapter 21.)

In an α-thalassemia, the production of mRNA from the α2 gene is prevented by a mutation that deletes a pentanucleotide sequence from the left junction of intron *1*. No stable RNA product is found.

A β⁰-thalassemia, in which there is no production of β globin, is produced by a point mutation in a consensus sequence. This changes the GT at the left junction of intron *2* into an AT. Presumably this prevents the production of mRNA.

In β$^+$-thalassemia, there is a large reduction in the *amount* of β-globin mRNA and protein synthesis. But the mRNA and protein that are made appear to be normal. Some β$^+$-thalassemic genes have been cloned to identify the basis for the defect.

In one case, there is a single point mutation, located 21 bp before the right end of the small (first) intron. This changes the wild-type sequence from a G into an A, creating the mutant sequence

Wild type
G   Intron/exon
↓   ↓
C C T A T T A G T C T A T T T T C C C A C C C T T A G G C T
Mutant

The sequence surrounding the mutant site is almost identical with the sequence that surrounds the splicing junctions—the mutation creates homology at 6 out of 7 positions. (And the base that is different is at a position whose member varies at authentic splicing junctions in other β-like globin genes.) Note that the 7 base sequence extends to the right of the splicing site, a region not included in the consensus sequence at the right junctions.

The result of this resemblance is that splicing takes place at the site of mutation instead of at the proper junction. This happens both in human patients with thalassemia, and when a cloned copy of the gene is expressed in either monkey or human (tissue culture) cells. Up to 90% of the precursor RNA molecules are spliced at the mutant instead of at the authentic splicing site. The small amount of splicing at the proper site generates the active β-globin mRNA that is present in the patients. We do not know why the mutant site is used more efficiently; it could be because it lies closer to the left junction, or because of some feature of the conformation of the RNA.

The overall efficiency of splicing is reduced with the mutant RNA, and there is also an accumulation in the nucleus of entirely unspliced precursor RNA. This implies that failure to splice intron *1* may cause a retardation in the splicing of intron *2* (which is consistent with reports that intron *1* is spliced out first; see ear-

lier). This type of effect could be mediated by secondary structure.

What happens to the misspliced RNA? This is not entirely clear and may depend on circumstance. In the thalassemic individual, it does not accumulate in the cytoplasm of reticulocytes. This implies either that the precursor or the misspliced product is unstable or that there is a deficiency in transporting it from nucleus to cytoplasm. But in tissue culture cells, the misspliced RNA makes its way to the cytoplasm as effectively as the normal mRNA. We do not know whether it enters the polysomes, but if it does it would be translated into an aberrant and prematurely terminated protein. (The use of the mutant splicing junction introduces a nonsense codon into the reading frame.)

Is splicing *necessary* for the production of mRNA from interrupted genes? (We know that any requirement can be bypassed by mRNA from naturally uninterrupted genes.) The use of SV40 vectors allows direct comparison of the expression of interrupted and uninterrupted versions of the same gene. An interrupted gene is provided by a genomic fragment; an uninterrupted gene is created by the cDNA that represents the mRNA. In these cases, when a transcription unit lacks any splicing junctions, it is transcribed normally. But in some cases (for example, globin), the resulting RNA is unstable, as illustrated earlier in **Figure 26.8**. This suggests that the transcripts of certain interrupted nuclear genes may have either of two fates: to be spliced and then transported to the cytoplasm; or to be degraded in the nucleus. This effect is not seen for other genes, which can be expressed equally well in either interrupted or uninterrupted form.

IS snRNA INVOLVED IN SPLICING?

The changes in splicing caused by mutations that create or abolish consensus sequences show that these sequences are involved in the recognition of exon-intron boundaries. But still this does not reveal just how the sequences are recognized. We can think of two general types of system. An enzyme could itself specifically recognize the two consensus sequences. Or an RNA might base pair with them to create a secondary structure that is recognized by an enzyme.

This would make use of conformation in the same way as other RNA processing enzymes (although in this case, the base pairing need not necessarily be autonomous, and might be influenced by proteins).

What could be the source of such an RNA? It could be a sequence located elsewhere in the same transcript, although this seems unlikely since none has been noticed in any of the nuclear sequences. Or there could be an independent RNA that base pairs with the unspliced precursor to bring the intron ends together.

Both the nucleus and cytoplasm of eucaryotic cells contain quite a large number of discrete small RNA species. These range in size from 100 to 300 bases, and they vary in abundance from about 10^5 to 10^6 molecules per cell. Some are synthesized by RNA polymerase III, others by RNA polymerase II, among which some of the products are capped like mRNAs.

Those that are restricted to the nucleus are called **small nuclear RNAs (snRNA)**; those found in the cytoplasm are called **small cytoplasmic RNAs (scRNA)**. In their natural state, they exist as ribonucleoprotein particles (snRNP and scRNP). Colloquially, these are sometimes known as *snurps* and *scyrps*. Some sequence relationships are found among the various snRNAs; and the reactions of the snRNPs with antibodies suggest that some of them may share common protein subunits.

The 5'-terminal sequence of one of these species, U1 snRNA, is complementary to the consensus sequences at the splicing junctions. The U1 snRNA is present, with only modest variations of sequence, in animal, bird, and insect cells. The probable secondary structure of the human U1 snRNA is drawn in **Figure 26.11.** Note that the 5'-terminal 11 nucleotides are sin-

Figure 26.11
Human U1 snRNA has a secondary structure in which the 5'-terminal end is single stranded. Other species have similar sequences and structures.

gle-stranded and therefore in principle available for reaction.

A model for the base pairing of this region with the splicing junctions is illustrated in **Figure 26.12.** The upper part depicts the overall structure of the intron vis à vis the U1 snRNA. The lower part shows the pattern of base pairing between the U1 snRNA sequence and the sequences at the ends of a typical intron. The exact pattern of pairing is difficult to predict, because the sequence UCC in the U1 snRNA could pair with the AGG found at either of the exon-intron junctions.

The base pairing is not extensive, and we do not know whether it would by itself be adequate to bring the splicing junctions together. It is shown at its best in the figure. In actual junctions, there seem always to be 4–6 bp able to form at the left junction site, but there may be only 2 instead of 3 at the right junction site. In considering the role this might play in splicing, we must remember that the specificity of the base-pairing reaction may depend on the environment. Thus

Figure 26.12
The 5′ end of U1 snRNA could base pair with the left splicing junction and (perhaps) also with the right splicing junction.

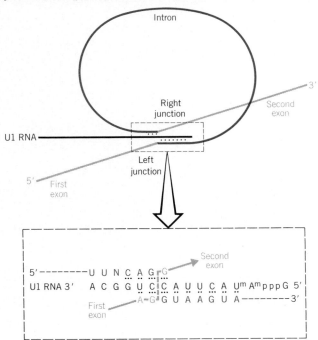

we would not a priori predict that the triplet codon-anticodon reaction would bring sufficient specificity to translation, but it does so within the ribosomal environment. Could the specificity of the U1 snRNA•nuclear RNA interaction similarly rest on the environment provided by protein components?

In lieu of an *in vitro* system for nuclear splicing from which the components can be purified, how are we to find out whether U1 snRNA is in fact involved in splicing? Some indirect evidence is provided by antibodies obtained from patients with the autoimmune disease **lupus erythematosus.** Such patients have a variety of reactions in which their antibodies react against normal components of the human cell, but among them it is common to find antibodies directed against nucleic acids. In some cases, the target is one or several of the snRNP types. The reaction is directed against the ribonucleoprotein particle, and probably recognizes a protein subunit or subunits.

Nuclei isolated from HeLa cells infected with adenovirus are able to continue splicing the viral RNAs. When these nuclei are incubated with the anti-snRNP antibodies, the splicing reaction is halted. The inhibition seems to be specific to splicing, and does not represent a general interference with gene expression, because other stages (such as transcription or polyadenylation) continue as normal. Only antibodies that react with U1 snRNP are effective; other snRNPs are not involved.

Some other evidence for a possible connection between U1 snRNA and nuclear RNA is provided by experiments with psoralens. These are cross-linking reagents that act specifically on base-paired regions of nucleic acids. The cross-links can be reversed by irradiation with ultraviolet light. When nuclear RNP preparations are treated with psoralen, and then the links are reversed with ultraviolet, some U1 snRNA is released. This implies that it was trapped with the nuclear RNA because it was base paired to it. Only a very small proportion of the U1 snRNA in the nucleus participates in this reaction.

Is U1 snRNA the only possible RNA assistant in splicing or are there others? Some viruses code for small RNA molecules. One of these is adenovirus, which produces the VAI RNA. It has complementarity to the splicing junctions of some adenovirus genes; it could be involved in their recognition. If small, diffusible RNAs are involved in creating structures that are recognized for splicing, the production of specific RNA assistants could be needed to accomplish particular splicing events.

We still are left with a major unanswered question. None of the available data explains how the recognition of splicing junctions is restricted so that only the left and right junctions of the same intron are spliced. Models that invoke the conformation of the RNA are rendered less likely by the apparently normal splicing of hybrid genes. One possibility is that the splicing apparatus acts in a processive manner. Having recognized a junction, the enzyme is compelled to scan the RNA in the appropriate direction until it meets a junction of the other type. This would restrict splicing to adjacent junctions. It could explain those cases in which the junctions are uniquely defined and interact directly (although some introns are very long indeed). It does not explain the use of multiple steps in splicing an intron, nor does it admit the possibility of alternative splicing patterns, such as those seen in SV40 where a common left junction may be spliced to more than one right junction. (Of course, in all of this we are assuming that splicing usually is accurate; but we have no data on the issue of how often erroneous splices may actually occur.)

FURTHER READING

There have been several short reviews in *Cell* devoted to the issue of splicing. General mechanisms involving consensus sequences are addressed by **Lewin** (**22,** 324–326, 1980) and **Sharp** (**23,** 643–646, 1981). The structures and functions of snRNAs have been addressed by **Zieve** (**25,** 296–297, 1981) and by **Lerner** & **Steitz** (**25,** 298–300, 1981). The differences with organelle splicing were emphasized by **Lewin** (**22,** 645–646,1980). The development of the auto-splicing *in vitro* system from *Tetrahymena* was reported by **Cech et al.** (**27,** 487–496, 1981). An earlier review on general mechanisms, including yeast tRNA, is by **Abelson** (*Ann. Rev. Biochem.* **48,** 1035–1069, 1979).

CHAPTER 27
CONTROL OF RNA PROCESSING

Protein-coding structural genes in the nucleus of a higher eucaryotic cell are transcribed in the nucleoplasm. But the nucleoplasmic RNA is not like mRNA. Its average size is much larger, it is very unstable, and it has a much greater sequence complexity. Taking its name from its very broad size distribution, it is called **heterogeneous nuclear RNA(hnRNA)**.

Before the discovery of interrupted genes, there was a heated debate as to whether mRNA is derived from hnRNA via a size reduction, or whether the hnRNA might serve some other purpose altogether. Now some of the discrepancies between hnRNA and mRNA can be reconciled by invoking the reduction in size and complexity that occurs via RNA splicing. Many protein-coding structural genes are much longer than their mRNAs, and the primary transcript should be at least as long as the known gene (it could be longer if it included additional flanking sequences).

But we shall see that its inclusion of introns as well as exons does not entirely explain the high instability of the hnRNA. On average, only a small proportion of the transcripts of each gene actually yield an mRNA product. And in some cases, there are transcription units whose products are found in the nucleus, but that do not seem at all to give rise to cytoplasmic mRNA. This prompts the question of whether the expression of some genes might be controlled at the level of hnRNA, by virtue of deciding whether or not the processing pathway should produce mRNA from a particular nuclear RNA.

hnRNA IS LARGE AND UNSTABLE

The transcription of genes in the nucleoplasm represents only a small part of the total RNA-synthesizing activity of the cell. The intensive transcription of rRNA genes in the nucleolus is responsible for most (up to 90%) of the synthesis of RNA. In fact, this makes it difficult to follow the synthesis of hnRNA en masse, because its small proportion is hard to distinguish from the rRNA. However, this can be accomplished by scrutinizing only part of the hnRNA population or by using special circumstances.

One option is to examine the very largest hnRNA fractions (which are distinct from rRNA precursors because they sediment more rapidly than 45S). Or it is possible to inhibit rRNA synthesis preferentially with actinomycin. Another technique is to follow just for a short time the fate of a pulse (i.e., very brief) radioactive label, which enters hnRNA more rapidly than rRNA. None of these conditions is entirely satisfactory, but they are adequate to outline the general features of nucleoplasmic transcription.

Two prominent features are the size and instability of hnRNA. Most of a radioactive label incorporated into hnRNA actually turns over in the nucleus. It never even enters the cytoplasm. The lifetime of hnRNA varies somewhat in different cell types. There may be only one, or there may be two, components with regard to the decay rate; but the half-lives generally lie in the range of from a few minutes to about an hour or so.

The nuclear turnover demonstrates that *most of the mass of hnRNA is synthesized and degraded entirely within the nucleus.* This does not make any implication about the fate of *individual* molecules or parts of molecules, so from these data per se we do not know whether the nuclear-degraded material represents only introns or includes entire molecules of hnRNA.

One of the difficulties in determining the true size of hnRNA is its tendency to aggregate. There is some secondary structure within the hnRNA, a large part of which represents reactions between inverted repeats of the Alu family or its equivalent (see Chapter 24). But there are also *inter*molecular reactions; and these probably are responsible for forming the very rapidly sedimenting material.

The size distribution of hnRNA is compared with the average size of mRNA in **Figure 27.1.** The histogram measures the lengths of hnRNA molecules after they were spread for electron microscopy under denaturing conditions (which should prevent aggregation between molecules). The distribution probably underestimates the lengths of the primary transcripts, for two reasons. Any breakage in a long molecule during its isolation will generate two shorter molecules. And some of the molecules could be partially processed intermediates (remember the series of ovomucoid precursors in Figure 26.7, all of which are present in the steady-state hnRNA population).

But even with these caveats, it is clear that most hnRNA is much longer than mRNA. Based on the analysis of pulse labels incorporated into nascent transcripts, it seems now that the size distribution of (mammalian) transcription units has an average in the range of 8000–10,000 bp. The overall distribution extends from ~2000 to ~14,000 bp. So the average hnRNA molecule is 4–5 times larger than the average mRNA of 1800–2000 bases.

The range of lengths suggests at once that hnRNA

Figure 27.1
hnRNA has a broad size distribution extending from about the size of mRNA to more than 14,000 bases.

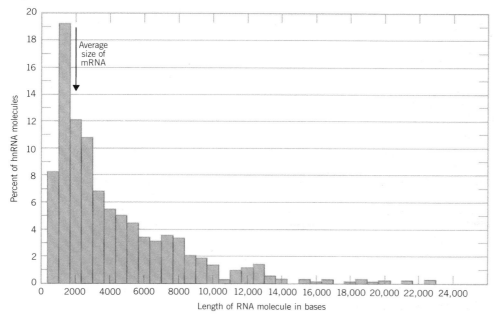

is rather long to exist in the form of an extended RNA chain within the nucleus. In fact, it is found in the form of a ribonucleoprotein (hnRNP), in which the hnRNA is associated with several proteins. They appear to comprise a more complex set than those associated with mRNA. The mass of protein in the hnRNP is about four times the mass of the hnRNA.

The hnRNP is associated with the **nuclear matrix.** This is a dense fibrillar network that lies on the nuclear side of the membrane. It forms a sort of inner shell around the nucleus and also extends into the interior.

Figure 27.2
The nuclear matrix consists of a fibrillar network.

A HeLa cell was first subjected to gentle extraction with a nonionic detergent to remove membraneous material, and then treatment with DNAase and high salt concentration was used to remove chromatin and some of the cytoskeleton (the fibers of the surrounding cytoplasm). The small ring structures at the surface of the nuclear matrix may be the nuclear pores. The large dark objects are the nucleoli.

Photograph kindly provided by David Capco and Sheldon Penman.

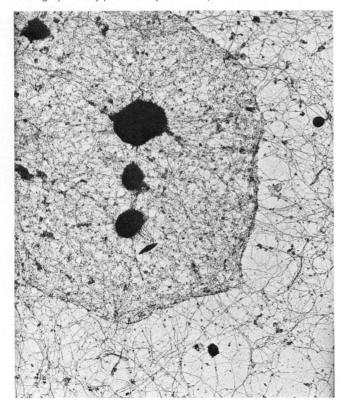

Figure 27.2 is an electron micrograph showing the nuclear matrix within the remnants of the surrounding cytoplasm. The structure of the fibrils is not yet understood, although we do know that they are composed of a large number of proteins.

The chromatin itself is intermittently attached to the nuclear matrix, and it is likely that the primary transcripts individually become attached soon after or even during their transcription. In fact, their processing may take place on the matrix. We do not know whether this location is essential in the sense that processing can occur *only* here. But in thinking about the events involved in RNA maturation, we should remember that the process may be complex topographically and is not accomplished simply by a set of "soluble" enzymes floating around the interior of the nucleus.

We have virtually no information on how mature mRNA is exported to the cytoplasm. It is possible that its emigration could be linked spatially to processing. As for the route of export, around the nucleus is a series of "pores" that could provide a route for RNA transport, but we do not know whether in fact they are used for this purpose.

mRNA IS DERIVED FROM hnRNA

The ends of some hnRNA molecules carry the same modifications that are seen in mRNA. The 5' end gains a (partially) methylated cap; the 3' end is polyadenylated. The similarities of these end modifications suggest that hnRNA and mRNA may have a precursor-product relationship. [Also, among the proteins bound to the hnRNP, the major 74,000-dalton species associated with poly(A) may be the same as that present in mRNA.] The paradox originally presented by the apparent identity of the extremities of hnRNA and mRNA was that the hnRNA is so much larger. It was not apparent how both ends of the molecule could find their way into mRNA. Now we see that this is resolved by reducing the size by removing the introns.

Pulse-chase experiments can be used to follow the fate of the modified ends. This is illustrated diagrammatically in **Figure 27.3.** A radioactively labeled precursor is added for a very brief period of time. This is the pulse label. Then it is removed and replaced by an unlabeled precursor. This is the chase. The amount

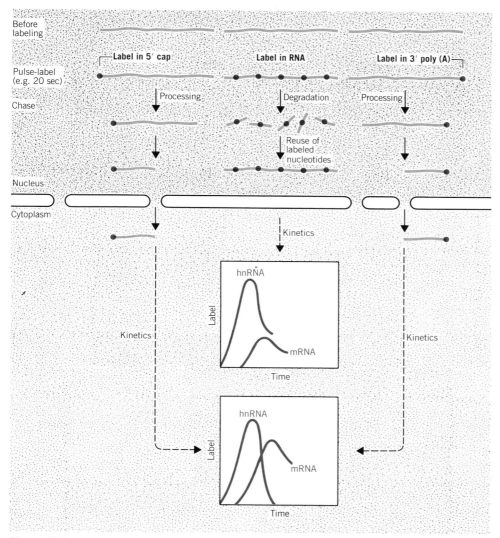

Figure 27.3
Radioactive labels in either 5′ caps or 3′ poly(A) first appear in hnRNA and then are chased into mRNA, but most of a label in the body of hnRNA is released by degradation and then reused.

of the label in hnRNA and in mRNA is measured at intervals of time.

For a short time after the pulse, the radioactive label accumulates in the hnRNA. Because the hnRNA molecules are unstable, it then declines rapidly. Different results are obtained depending on whether the label was incorporated into one of the ends or into the body of the RNA.

Because the end-modifications are stable, they can be chased from hnRNA into mRNA. In fact, the principal route for the labeled ends to leave the hnRNA population is via the processing of the RNA, followed by its departure for the cytoplasm. So the decline of the curve representing the amount of label in hnRNA is accompanied by an increase in the curve representing the amount of label in mRNA. The coincidence of the decline in hnRNA and rise in mRNA demonstrates that the ends of the mRNA are derived from the ends of the hnRNA. For these parts of the molecule there is a conventional precursor-product relationship between hnRNA and mRNA.

Probably about 20% of the hnRNA is capped following its transcription. At least a large proportion of the caps can be chased from hnRNA into mRNA. Although the data do not prove that *all* the 5' caps in hnRNA give rise to mRNA, this relationship seems plausible. Capping occurs very rapidly after the start of transcription, as witnessed by our inability to characterize original 5'-triphosphate ends on primary transcripts.

At the other end of the molecule, the length of the nuclear poly(A) is slightly greater than that found in mRNA. A few bases are removed at the stage of nucleocytoplasmic passage. There has been some discussion about whether *all* of a label in poly(A) can be chased from the nucleus to the cytoplasm. The answer is not unequivocal—it may vary in different cells and in different circumstances—but it is clear that at least the great majority of poly(A) tails added to hnRNA do find their way into mRNA.

When labeled nucleotides are incorporated into the internal regions of hnRNA, their main fate is to be released from the molecule when it is degraded. Then they are reutilized in the synthesis of further RNA, entering both rRNA and hnRNA. The result is a generalized decline in the label in hnRNA. A small proportion of the label does enter mRNA in the cytoplasm, but because of its minority status, the appearance in mRNA does not coincide with the decline in hnRNA in a way that demonstrates a conventional precursor-product relationship.

Early attempts to pulse-chase nucleotides from the body of hnRNA into mRNA thus were clouded by the nuclear turnover of the majority of the label. But improvements in the technique later made it possible to show that mRNA indeed is derived from hnRNA. From such experiments, it is possible to calculate the proportion of the *mass* of the hnRNA that is converted to mRNA. This is about 5% in mammalian cells. Given the fivefold discrepancy in average size, it corresponds to a conversion of (roughly) 25% of the hnRNA molecules.

This means that reduction in size of the transcript via loss of introns can account for the fate of only about a quarter of the hnRNA population. The rest consists of molecules that turn over *entirely* within the nucleus: no part of these molecules is used to provide cytoplasmic RNA. [The only caveat is that this analysis relies on the isolation of the poly(A)$^+$ mRNA; so some further part of the hnRNA population might give rise to poly(A)$^-$ mRNA. But although this would increase the proportion that consists of precursors, it would not change the basic conclusion.]

THE IMPORTANCE OF POLYADENYLATION

Poly(A) is added to the 3' ends of hnRNA molecules after transcription, by an enzyme, poly(A) polymerase, that uses ATP as substrate. The addition of poly(A) to hnRNA can be prevented by the analog **3'-deoxyadenosine,** also known as **cordycepin.** This is a specific inhibitor of polyadenylation; it does not stop the transcription of hnRNA. But its addition prevents the appearance of mRNA in the cytoplasm. This shows that polyadenylation is *necessary* for the maturation of mRNA from hnRNA. [This is true, of course, of the poly(A)$^+$ mRNA; the poly(A)$^-$ mRNA can bypass this requirement.]

Only some of the hnRNA is polyadenylated, about 30% in mammalian cells, compared with a value of about 70% for mRNA. There is often a rough correlation between the proportion of hnRNA molecules that is polyadenylated and the proportion that gives rise to cytoplasmic poly(A)$^+$ mRNA molecules. This suggests that polyadenylation could provide a signal that a particular hnRNA molecule is to be processed. However, although necessary, polyadenylation is unlikely itself to be sufficient; because, in this case, there would always be strict conservation of the poly(A) between nucleus and cytoplasm.

One of the important unresolved questions is the origin of the 3' ends to which the poly(A) is added. We know that in the case of some viruses, the ends are generated by cleavage of longer precursors, as illustrated in **Figure 27.4.** A common feature of both cellular and viral mRNAs is the presence of a sequence AAUAAA in the region from 11 to 30 nucleotides upstream from the site of poly(A) addition. When this sequence is deleted from the SV40 late transcription unit, cleavage and polyadenylation no longer occur at the usual site. The AAUAAA sequence is necessary, but not by itself sufficient, for the reaction.

The ubiquity of this signal provides a hint that the same mechanism—cleavage and polyadenylation—may be employed in cellular transcription units. There is virtually no information about the mechanism of termination by RNA polymerase II, so at present we do not know in what proportion of cases the 3′ end used for polyadenylation is generated by termination vis à vis cleavage.

Polyadenylation usually precedes RNA splicing. In the case of cellular transcripts, both the full length and the various intermediate RNA precursor molecules are polyadenylated. In the case of the processing of mRNAs from the late transcript of adenovirus, kinetic studies show that the cleavage to generate a 3′ end for polyadenylation occurs *before* the tripartite leader is spliced to a messenger body. (This is the system illustrated in Figure 20.20.)

Figure 27.4
The sequence AAUAAA is necessary for cleavage and polyadenylation of SV40 mRNA.

The order of events could mean that splicing is simply a slow step relative to transcription and polyadenylation; or it could result from some dependence of splicing on the polyadenylation event. However, splicing continues to occur in nuclei in which the synthesis of poly(A) is prevented by the addition of cordycepin (3′-deoxyadenosine). This implies that the relationship is not causal mechanistically [although in those cases where alternative splicing junctions are used, cleavage of the 3′ end may control the *choice* of splicing junction (see, for example, Chapter 39)].

hnRNA IS MORE COMPLEX THAN mRNA

Given the difference in size between hnRNA and mRNA, we expect hnRNA to have a greater sequence complexity. For those transcription units in which a primary transcript is spliced to yield a smaller mRNA, the sequence complexity of hnRNA should include both exons and introns, while the sequence complexity of mRNA should include only the introns. Unfortunately, it is difficult to measure the sequence complexity of hnRNA with as much precision as that of mRNA. First, we lack a functional test to isolate hnRNA free of other RNAs. Second, the use of hybridization kinetics is restricted by the presence in hnRNA of repetitive sequences that are interspersed with the nonrepetitive sequences forming the majority of mRNAs.

From measurements of the ability of hnRNA to saturate nonrepetitive DNA, its complexity falls in a range that varies from as little as four to as much as ten times greater than that of mRNA. Hybridization experiments confirm that the hnRNA population does indeed include all the sequences that are present in cytoplasmic mRNA. What is the nature of the additional sequences?

This question can be asked in another form. Do the hnRNA molecules that turn over entirely within the nucleus comprise transcripts of the *same* or of *different* genes as those that give rise to mRNA. The consequences are illustrated in **Figure 27.5.**

If the same genes are involved, in the typical mammalian cell the average gene is transcribed into hnRNA molecules, 75% of which are entirely degraded, and

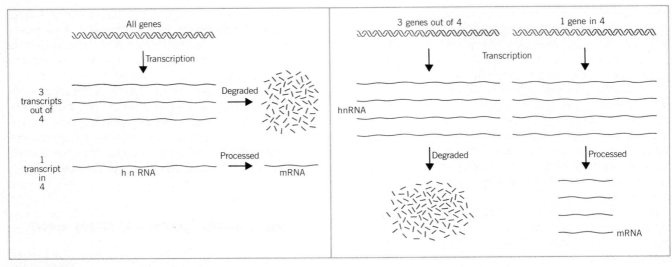

Figure 27.5
**Is processing of all hnRNAs wasteful as shown in the first part, or are some transcripts effi-
ciently processed while others do not give rise at all to mRNA as shown in the second part?**

25% of which donate 20% of their length (the exons) to become mRNA. Thus processing is extremely wasteful. In this case, the difference in complexities between hnRNA and mRNA is accounted for entirely by the loss of introns. In the example of mammalian cells, the difference between hnRNA and mRNA complexities is about fourfold, very similar to the size discrepancy. The data are not precise enough to say whether *all* the extra sequences in hnRNA can be accounted for by introns, but certainly it seems likely that this is the major cause.

If different genes give rise to hnRNA molecules that are conserved or that are degraded entirely, 75% of the transcription units must represent hnRNA molecules all copies of which are degraded: none of these donates mRNA to the cytoplasm. (The strict conclusion is that none gives rise to mRNA in this particular cell type; it could do so in another.) The other 25% of the genes give transcripts all copies of which mature to mRNA (conserving 20% of their length). In this case, the complexity of hnRNA should be much greater than that of mRNA, not only because the molecules are longer, but also because there is an entirely inde-

pendent set of transcripts from those that yield mRNA. In some sea urchin tissues, where the difference in nuclear and cytoplasmic RNA complexities is tenfold, the discrepancy seems too great to be accounted for just in terms of loss of introns.

Of course, these two models are extremes, and no doubt the situation in actual nuclei lies somewhere between. It is clear, however, that either the processing of expressed genes is inefficient or some transcription units are expressed only in nuclear RNA.

IS THERE CONTROL AFTER TRANSCRIPTION?

The small proportion of the genome represented in hnRNA (say, less than 5% in mammalian cells) makes it evident that the control of transcription is a primary event in gene expression. We know that certain genes are transcribed only in the cells in which they are expressed. Two examples are the globins (which are transcribed only at the appropriate stages of erythroid development) and ovalbumin (which is transcribed only when chick oviduct is stimulated by estrogens). Of course, these are genes that are expressed very actively, and their particular mode of control may not be the same as other genes that are expressed at lower levels.

Quantitative changes in gene expression can occur between the stages of transcription and translation. Some sequences are present at rather different abundances in nuclear RNA and in cytoplasmic mRNA. This means that the level of transcription is not the sole factor responsible for establishing the level of mRNA. Changes in stability or in the efficiency of processing could intervene to alter the relative abundance of an RNA sequence between nucleus and cytoplasm.

A situation in which general quantitative changes occur is provided by the transition that cells undergo in culture from the resting (that is, nondividing) to the growing state (when they are actively proceeding through the cell cycle).

Growing cells have more RNA than resting cells, and in particular they have an increased proportion of mRNA relative to rRNA. But there are no changes in the rate of transcription of hnRNA, in the proportion of molecules that are polyadenylated, or in the stability of the mRNA. What happens is that a greater proportion of the poly(A)$^+$ hnRNA is converted into poly(A)$^+$ mRNA in growing cells.

The major part of the change is quantitative rather than qualitative. Some new sequences are expressed,

Figure 27.6
The same nuclear transcripts are degraded in the adult intestine but used to give mRNA in the embryonic blastula of the sea urchin.

~25,000 hnRNAs   ~3,000 mRNAs   ~20,000 hnRNAs   ~13,000 mRNA

Sequences found in intestine mRNA
Sequences found in blastula mRNA
Nuclear sequences not represented in mRNA

but for the most part, the change represents an increase in the efficiency with which a constant set of sequences in hnRNA is converted into mRNA. Put the other way, in resting cells there is a decrease in the efficiency of conversion. In the light of our previous question about the nature of the nuclear-restricted sequences, this makes it clear that, at least in resting cells, some of them do represent surplus copies of expressed sequences.

A situation in which control over the selection of sequences is exercised at the level of nuclear RNA is found in the sea urchin. In *S. purpuratus,* the total complexities of the nuclear RNA of several adult and embryonic cell types are in the same general range (175,000 to 225,000 kb, consisting of molecules of average length 8800 bases). This bears no relationship at all to the complexities of the mRNAs, which vary from 6000 to 26,000 kb, consisting of molecules of average length 2000 bases.

The sequences present in the two tissues are compared in **Figure 27.6.** The blastula mRNA is much more complex. It contains some sequences that are shared with intestine mRNA, but the majority of blastula mRNAs are unique to the embryonic state. Most (~80%) of these sequences are present in the nuclear RNA of the intestine as well as the blastula.

This means that the intestine nucleus contains RNA precursors that are confined to the nucleus instead of being used to produce mRNA. But in the blastula, *these same sequences* are used to give rise to mRNAs. So in this case, many genes are transcribed in tissues irrespective of whether they are finally expressed. The transcripts are selected for processing and nucleocytoplasmic transport only in the appropriate tissues. Only a minority of genes seem to be controlled at the level of transcription.

MODELS FOR CONTROLLING GENE EXPRESSION

It is too early to form any general view of the frequency with which different levels of control are used to provide the stage at which gene expression is regulated. The concept of the "level of control" implies that gene expression is not necessarily an automatic process once it has begun. It could in principle be regulated in a gene-specific way at any one of several sequential steps. We can distinguish (at least) four potential control points, forming the series

Activation of gene structure
↓
Initiation of transcription
↓
Processing transcript and transport to cytoplasm
↓
Translation of mRNA

The existence of the first step is implied by the discovery that genes themselves may exist in either of two structural conditions. Relative to the state of most of the genome, genes are found in an "active" state in the cells in which they are transcribable. The nature of this state, which is detected by changes in the association of proteins with DNA, is discussed in Chapter 30. The change of structure is distinct from and precedes the transcription event (possibly by several cell generations). This suggests that acquisition of the "active" structure must be the first step in gene expression.

The transcription of a gene in the active state is controlled at the stage of initiation, that is, by the interaction of RNA polymerase with its promoter. This is now becoming susceptible to analysis in the *in vitro* systems (see Chapter 11). There is at present no evidence for control at subsequent stages of transcription in eucaryotic cells, for example, via antitermination mechanisms.

The primary transcript always must be modified by capping, and usually also by polyadenylation. Introns must be spliced out from the transcripts of interrupted genes. The mature RNA must be exported from the nucleus to the cytoplasm. It is not yet possible to distinguish between these steps as points of control, so regulation of gene expression by selection of sequences at the level of nuclear RNA might involve any or all of these stages. We have no idea at all in molecular terms of what events may be involved in controlling gene expression at the level of nuclear RNA, either quantitatively or qualitatively.

Finally, the translation of the mRNA in the cytoplasm can be specifically controlled. There is little evidence for the employment of this mechanism in adult somatic cells, but it does occur in some embryonic situations, as described in Chapter 9. The mechanism is presumed to involve the blocking of initiation of translation of some mRNAs by specific protein factors.

We must look beyond the question of the use and type of mechanism employed at each potential level of control to the general issue of how gene expression is coordinated between different loci. The genes that are expressed to generate any particular cell phenotype may lie in a large number of different locations. Each gene function probably is required to contribute to a characteristic spectrum of cell phenotypes. How does each cell activate the right combination of genes to produce its phenotype?

Eucaryotic gene expression generally is regarded as being under positive control: genes are inactive unless they are turned on. This is consistent with the need for a gene to change from the usual inactive state to the "active" structure before it can be expressed. The subsequent stages of control might in principle be subject to either positive or negative control. Models for controlling eucaryotic gene expression usually assume positive control, but could easily be converted to negative control.

There are two general types of solution to the problem of expressing overlapping combinations of genes in different cell phenotypes. Either the genes could be repeated, or single genes could be subject to multiple controls.

An extreme of gene repetition would be to have one copy of the gene for each cell type in which it must be expressed. Each copy would be under a different control, specific for the cell type in which it is expressed. Irrespective of whether the genes specify identical or related proteins, there is no problem with the application of selective pressure if each gene is essential in a particular time or place. Certainly some genes (maybe most genes) are present in more than one copy per genome. We do not yet know whether these copies usually are under different controls, or whether they usually represent repetitions all of which are expressed at the same time and/or place.

If a single gene is expressed in different circumstances in combination with various other genes, it is likely to have multiple control elements. The situation usually is considered by analogy with the operon model. Relatively short sequences of DNA may be used as recognition sites to control the activity of contiguous structural genes. We need not specify *how* they con-

Figure 27.7
The basic components of eucaryotic gene control could comprise a sensor site-integrator gene that regulates a receptor site-structural gene.

trol gene activity: their action could involve a structural change prior to transcription, it could concern initiation at the promoter, or it could be exercised in the nuclear RNA transcript.

A formal network of this sort has been promulgated by Britten and Davidson from the basic model illustrated in **Figure 27.7.** The structural gene is positively controlled by an adjacent **receptor site.** This responds to an **activator** molecule, considered in the model to be RNA, but it could alternatively be protein. The activator is synthesized by an *integrator gene,* which (of course) is the direct equivalent of a bacterial regulator gene. The only difference is that the integrator gene is itself subject to external control. The controlling element, adjacent to it, is called a **sensor site.** It is supposed to respond to functions that control gene expression, for example, hormones. The model is shown for control prior to transcription, but easily can be adapted to other levels.

A single activator can control many genes if each of its target loci possesses a copy of the appropriate receptor. There is ample precedent for such repetition in dispersed bacterial loci that are under common control (see Chapter 15). All the genes possessing the same receptor comprise a **set** that is expressed in response to the activator. The set is somewhat equivalent to an operon since all its genes always are expressed together. It is therefore likely to include genes specifying proteins with related functions, such as the enzymes of a metabolic pathway.

A structural gene can be expressed in different circumstances if it possesses more than one receptor. Each receptor is able to activate the gene by virtue of its reaction with the activator molecule. By placing a

Figure 27.8
Each structural gene can be activated by more than one receptor and therefore belongs to more than one set.

series of receptors adjacent to a structural gene, it can be included in several different sets, as depicted in **Figure 27.8.** By appropriate repetition of receptors, each of the three structural genes (*x, y, z*) belongs to two of the three partially overlapping sets (*a, b, c*).

Different sets of genes can be controlled together by supposing that a sensor may control several integrators. This allows a single stimulus to one sensor to activate several sets of genes. The genes all controlled by one sensor are called a **battery.** A set can

be included in more than one battery if the integrator elements are repeated as illustrated in **Figure 27.9.**

This hierarchy of elements allows any particular gene to be combined for expression with any other gene with economy of control. The model requires the integrator and receptor sequences to be repeated, which therefore defines them as members of the repetitive DNA component. Indeed, an impetus for the construction of the model was to find a role for repetitive sequences in the genome.

Figure 27.9
Integrators can be repeated and each sensor may control several integrators, so that a signal to one sensor may activate several sets of genes, and each set can be included in more than one battery.

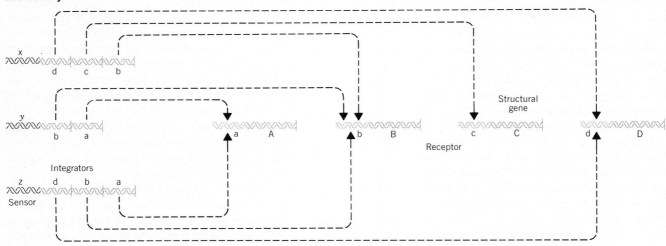

THE POTENTIAL OF CELLULAR POLYPROTEINS

The classic example of the operon accomplishes co-ordinate control of a group of genes by placing them in a single unit of transcription. When dispersed genes must be coordinately controlled, the same regulator element may be repeated at each locus, a mechanism of proven worth in bacterial genomes (see Chapter 15) and of hypothetical value in eucaryotic genomes (see previous section). When expression of a gene is required in more than one set of circumstances, one option is to provide it with alternative regulator elements, each able independently to sponsor its expression; again, this is found in bacterial genomes and may be postulated for eucaryotes. Another option is to repeat the structural gene itself, providing each copy with a different regulator element. (In a sense, this is the antithesis of the Britten-Davidson model.) Some of the members of eucaryotic gene families might fall into this category. (Repeated genes also may code for nonidentical, although closely related, proteins. Divergence between the proteins might accompany the evolution of differences in regulation, so that each protein best fits the circumstances of its employment.)

Some of these options are displayed dramatically by the synthesis of **polyproteins.** The name was first used to describe the multifunctional products of retroviral RNA genomes. Several viral proteins are synthesized in the form of a common precursor, which is cleaved at specific points to release the individual functional proteins. This situation formally complies with the apparent rule that eucaryotic translation systems only handle monocistronic messengers, but allows several proteins to be synthesized as the result of a single transcription event.

Several forms of cellular polyproteins since have been discovered. In some cases, the polyprotein contains a series of identical or related polypeptides. Thus an enkephalin precursor contains six copies of Met-enkephalin (the peptide Tyr-Gly-Gly-Phe-Met) and also one copy of Leu-enkephalin (Tyr-Gly-Gly-Phe-Leu). The polyprotein here functions principally as an amplification mechanism, so that each cycle of transcription and translation produces several copies of the polypeptide product.

The more complex example of the gene for pro-opiomelanocortin (POMC) is summarized in **Figure 27.10.** Genes from several mammals have a similar structure, in which a single intron is present in the signal part of the coding region, and several protein functions are coded by a single exon. In each species, a single gene has been characterized, although other (not necessarily identical) copies may also be present. The single polyprotein product of 32,000 daltons carries several known polypeptide hormones and also has other regions whose sequences suggest that they too may be of biological importance. All of the actual or potential individual products have dipeptide borders consisting of pairs of basic amino acids that may provide trypsin-like cleavage points. The use of these

ƍ indicates Lys-Lys or Lys-Arg potential cleavage sites

Figure 27.10
The POMC gene is translated into a polyprotein whose products depend on tissue-specific processing.

cleavage sites is different in two tissues in which the gene is expressed.

In the anterior lobe of the pituitary, the POMC protein first is cleaved once, releasing the N-terminal fragment and β-lipotropin (the C-terminal 70–90 amino acids). Then the N-terminal fragment is cleaved to release the ACTH hormone (39 amino acids long). In this tissue, the reaction proceeds no further.

In the intermediate lobe, ACTH is cleaved again, releasing α—melanotropin (α-MSH), which consists of the 13 N-terminal amino acids. The β-lipotropin also is cleaved once, releasing the analgesic β-endorphin, consisting of the 31 C-terminal amino acids.

In addition to these known hormones, there are two other sequences with potential MSH activity; they are denoted β-MSH and γ-MSH, but we do not yet know if and when they are generated.

What is the purpose of this form of organization? We still do not understand all the functions of these hormones, although we do know that they are produced as part of the response to a painful stress. The function of the sequence overlaps between the alternative hormones will become clear only when we understand their biological activities better.

These events establish the principle that a single gene may give rise to different (although overlapping) products by virtue of variation in the pathway for protein processing (presumably regulated by the processing enzymes of the relevant tissues). A further possibility is that repetitive copies of genes for polyproteins could allow the *same protein* to be synthesized in the company of alternative companions. The principle is illustrated in **Figure 27.11.** Two polyprotein

genes respectively contain the sequences of the individual proteins A, B, C and A, B, D. Thus activation of either gene leads to synthesis of A and B; but in one case this is accompanied by production of C, in the other case by production of D. This situation is formally equivalent to repetition of structural genes in polycistronic transcription units, although the products become separated at the stage of protein processing instead of at translation.

A family of genes coding for ELH, a neuropeptide of the marine mollusc *Aplysia,* may fall into this category. *Aplysia* DNA contains about 5 genes for the ELH hormone (a polypeptide of 36 amino acids). Although the full set of genes has yet to be characterized, several different mRNA species coding for ELH have been identified. Each of these is translated into a different polyprotein. The polyproteins all have an ELH sequence; but the other sequences represent different products, which may include other polypeptide hormones. Different mRNAs may be synthesized in different tissues, allowing the release in each case of a specific combination of hormones. This mechanism therefore allows overlapping combinations of hormones to be synthesized by activating appropriate members of a repetitive gene family.

It is notable that these examples of cellular polyproteins involve the generation of small products (generally less than 40 amino acids each). Does this reflect some difficulty in synthesizing small proteins de novo, which has led to the necessity for cleavage from longer precursors (and thus opened the way for evolution of alternative products)? It will be interesting to characterize the genes that code for multiple, related poly-

Figure 27.11

Are there genes coding for polyproteins consisting of overlapping products?

proteins. Will they prove to have been assembled by exon shuffling of the sort described in Chapter 20, or has the entire process evolved at some subsequent stage of expression of repeated genes whose ancestors did not have these multiple functions?

FURTHER READING

The properties of nuclear RNA are discussed in detail in **Lewin's** *Gene Expression,* **2,** *Eucaryotic Genomes* (Wiley, New York, 1980, pp. 728–760). The developing relationship between hnRNA and mRNA was previously reviewed in *Cell* (**4,** 11–20 & 77–94, 1975). More recent information on the signals for polyadenylation and its relationship with splicing is to be found in the original papers from **Fitzgerald** & **Shenk** (*Cell* **24,** 251–260, 1981) and **Zeevi, Nevins** & **Darnell** (*Cell* **26,** 39–46, 1981). Their model for gene control was published by **Britten** & **Davidson** in *Science* (**165,** 349–357, 1969). Mammalian polyproteins have been reviewed by **Herbert** & **Uhler** (*Cell* **30,** 1–2, 1982). The case of *Aplysia* was discovered by **Scheller et al.** (*Cell* **28,** 707–719, 1982).

PART 8
THE PACKAGING OF DNA

Do bacteria contain a nucleus? This question has agitated many workers; but at the present moment it appears no more meaningful than the everlasting controversies about the authorship of Shakespeare's plays which was finally resolved by the recognition that another fellow with the same name must have been the author. The important property of the nucleus of animal and plant cells is not so much its appearance or shape as the orderly changes, connected with definite biological tasks, which it can be seen to undergo.

Erwin Chargaff, 1947

CHAPTER 28
ABOUT GENOMES
AND CHROMOSOMES

A common problem is presented by the packaging of DNA into phages and viruses, into bacterial cells and eucaryotic nuclei. The length of the DNA as an extended molecule would vastly exceed the dimensions of the compartment that contains it. The DNA (or in the case of some viruses, the RNA) must therefore be compressed exceedingly tightly to fit into the space available. The condensation of the nucleic acid depends on the presence of proteins to which it binds. Usually these proteins are basic, and their positive charges neutralize the negative charges of the nucleic acid. We should like to know the specificity of this packaging: is the DNA folded into a *particular* pattern, or is it different in each individual copy?

The magnitude of the discrepancy between the length of the nucleic acid and the size of its compartment is evident from the examples summarized in **Table 28.1.** For bacteriophages and for eucaryotic viruses, whether they are long and thin (filamentous) or approximately spherical (icosahedral), a large amount of nucleic acid, whether DNA or RNA, whether single-stranded or double-stranded, effectively fills the container.

For bacteria or for eucaryotic cell compartments, the discrepancy is even greater. The DNA is contained in a compact area that does not occupy the entire compartment. This is seen as the **nucleoid** in bacteria and as the mass of **chromatin** in eucaryotic nuclei at interphase. Among the examples given in the table, the problem culminates in the need to fit 1.8 m of human DNA into a nucleus whose diameter may be as little as 6 μm (6×10^{-6} m). During division of the eucaryotic cell, the genetic material becomes even more tightly packaged, in the form of individually recognizable mitotic **chromosomes.**

The overall compression of the DNA can be described by the **packing ratio,** the ratio of the length of DNA to the length of the unit that contains it. For example, the smallest human chromosome contains about 4.6×10^7 bp of DNA (about 10 times the genome size of the bacterium *E. coli*). This is equivalent to 14,000 μm ($= 1.4$ cm) of extended DNA. Measured in the condensed mitotic state of metaphase, the chromosome is about 2 μm long. Thus the packing ratio of DNA in the metaphase chromosome is about 7000.

Packing ratios cannot be established with this certitude for the more amorphous overall structures of the bacterial nucleoid or eucaryotic chromatin. However, the usual reckoning is that mitotic chromosomes are likely to be 5–10 times more tightly packaged than interphase chromatin, which therefore has a typical ratio of 1000–2000.

A general principle is evident in the organization of

Table 28.1
The Length of Nucleic Acid Is Much Greater than the Dimensions of the Surrounding Compartment

Compartment	Shape	Dimensions	Type of Nucleic Acid	Length
Tobacco mosaic virus	Filamentous	$0.008 \times 0.3 \ \mu m$	1 single-stranded RNA	2 μm (= 6.4 kb)
Phage fd	Filamentous	$0.006 \times 0.85 \ \mu m$	1 single-stranded DNA	2 μm (= 6 kb)
Adenovirus capsid	Icosahedral	Diameter 0.07 μm	1 double-stranded DNA	11 μm (= 35 kb)
Phage T4 capsid	Icosahedral	$0.065 \times 0.10 \ \mu m$	1 double-stranded DNA	55 μm (= 170 kb)
E. coli bacterium	Cylindrical	$1.7 \times 0.65 \ \mu m$	1 double-stranded DNA	1300 μm (= 1.3 mm = 4.2×10^3 kb)
HeLa (human) mitochondrion	Oblate spheroidal	$3.0 \times 0.5 \ \mu m$	~10 identical molecules of double-stranded DNA	50 μm (5 μm = 16 kb each)
HeLa nucleus	Spheroidal	Diameter 6 μm	46 chromosomes of double-stranded DNA	1.8 m (= 6×10^6 kb)

all genetic material. It exists as a compact mass, in a delineated area; and its various activities, such as replication and transcription, must be accomplished within these confines.

CONDENSING VIRAL GENOMES INTO THEIR COATS

From the perspective of packaging the *individual* sequence, there is an important difference between a cellular genome and a virus. The cellular genome is essentially indefinite in size; the number and location of individual sequences can be changed by duplication, deletion, and rearrangement. Thus it requires a *generalized* method for packaging its DNA, insensitive to the total content or distribution of sequences. By contrast, two restrictions define the needs of a virus. The amount of nucleic acid to be packaged is *predetermined* by the size of the genome. And it must all fit within a coat assembled from a protein or proteins coded by the viral genes.

A virus particle is deceptively simple in its superficial appearance. The nucleic acid genome is contained within a **capsid.** This is a symmetrical or quasi-symmetrical structure assembled from one or from only a few proteins. Attached to the capsid, or incorporated into it, may be other structures, assembled from distinct proteins, and necessary for infection of the host cell.

Viewed in these terms, the tolerances for constructing the virus particle are rather tight. The internal volume of the capsid is rarely much greater than the volume of the nucleic acid it must hold. Simple calculations suggest that the difference is usually less than twofold, and often the internal volume is barely larger than the nucleic acid.

In its most extreme form, the restriction that the capsid must be assembled from virally coded proteins means that the entire shell is constructed from a single type of subunit. The rules for assembly of identical subunits into closed structures restrict the capsid to one of two types. The protein subunits may stack sequentially in a helical array to form a **filamentous** or rodlike shape. Or they may form a pseudospherical shell; this type of structure actually is a polyhedron with **icosahedral symmetry.** It now turns out that some viral capsids are assembled from more than a single type of protein subunit, but although this extends the exact type of structure that can be formed, viral capsids still all conform to the general classes of quasi-crystalline filaments or icosahedrons.

There are two types of solution to the problem of how to construct a capsid that contains nucleic acid. The protein shell can be assembled around the nucleic acid, condensing the DNA or RNA by protein-nucleic acid interactions during the process of assembly. Or the capsid can be constructed from its component(s) in the form of an empty shell, after which the nucleic acid must be inserted, being condensed as it enters.

Assembly of the capsid around the genome occurs

in the case of single-stranded RNA viruses and phages that are filamentous. The best characterized example is TMV (tobacco mosaic virus). Assembly starts at a duplex hairpin that lies within the RNA sequence. From this **nucleation center,** it proceeds bidirectionally along the RNA, until reaching the ends. The unit of the capsid is a two-layer disk, each layer containing 17 identical protein subunits. The disk forms a circular structure, which is converted to the helical form by interaction with the RNA. The RNA is coiled in a helical array on the inside of the protein shell, as illustrated in **Figure 28.1.**

The entire length of this RNA genome exists in a structure determined by its interaction with the protein shell. The same general type of responsibility for internal organization is presented also by the single-stranded RNA genome of turnip yellow mosaic virus, which has a spherical capsid. Here again, the position of the RNA within the capsid is determined directly by its binding to the proteins of the shell.

The events involved in the assembly of spherical capsids of DNA viruses are best characterized for the phages lambda and T4. In both cases, an empty head shell is assembled by the interactions between a small set of proteins. Then the duplex genome is inserted into the head, a process that is accompanied by a structural change in the capsid.

Figure 28.2 summarizes the assembly of lambda. It starts with a small head shell that contains a protein "core." This is converted to an empty headshell of more distinct shape. Then DNA packaging begins, the head shell expands in size though remaining the same shape, and finally the full head is sealed by the addition of the tail.

For both lambda and T4, the DNA that is to be inserted into the empty head takes the form of **concatemeric** molecules. These are multiple genomes joined end to end, as depicted in **Figure 28.3.** Each phage has its own mechanism for recognizing the proper amount of DNA to insert.

The ends of the lambda genome are marked by sequences called *cos* sites. Cleavage occurs at the left *cos* site (as defined on the usual map) to generate a free end that is inserted into the capsid. The insertion of DNA continues until the right *cos* site is encountered, when it is cleaved to generate the other end. The end that goes into the capsid last during assembly comes out first when a new host cell is infected.

Any DNA contained between two *cos* sites can be packaged. This is the basis of the "cosmid" cloning technique described in Chapter 19. However, the distance between the *cos* sites can be varied only slightly from the usual length of lambda DNA. Packaging does not occur if the distance is either too great or too small. This demonstrates that there must be *enough* DNA to complete the packaging reaction, as well as showing that there is room in the head only for a very little extra DNA (about 15%).

With phage T4, insertion starts at a *random* point in the concatemeric precursor. It continues until a genome's worth of DNA (a "headful") has been inserted. This implies the existence of some mechanism for measuring the amount of DNA. [Actually, the amount that is inserted is slightly greater than the length of the unit genome, with the result that there is a **terminal redundancy** corresponding to the additional length. In the terms of Figure 28.3, the first virion might contain the DNA from *A* to *A*, the next from *B* to *B*, and so on, so that each genome has a (different) letter repeated at each end.]

Now a double-stranded DNA considered over short distances is a fairly rigid rod. Yet it must be folded into a very compact structure to fit within the capsid. Little is known about how this is accomplished, except that there are "internal proteins" in the capsid, together with the DNA. One possibility is that they pro-

Figure 28.1
A helical path for TMV RNA is created by the stacking of protein subunits in the virion.

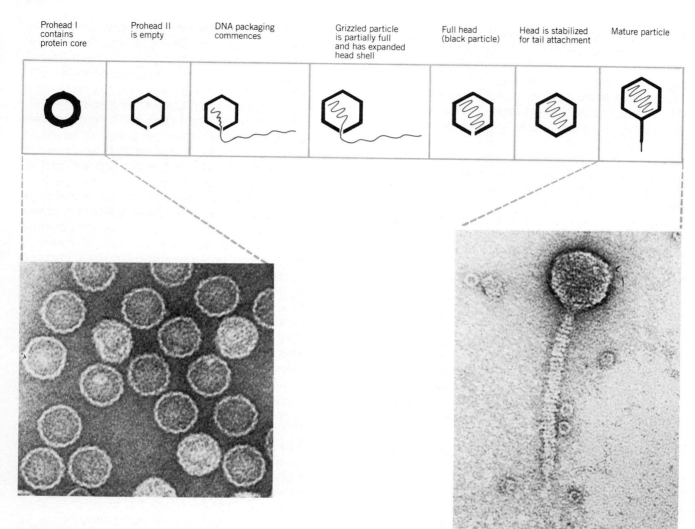

| Prohead I contains protein core | Prohead II is empty | DNA packaging commences | Grizzled particle is partially full and has expanded head shell | Full head (black particle) | Head is stabilized for tail attachment | Mature particle |

Figure 28.2
The empty head of phage lambda changes shape and expands when it becomes filled with DNA.

The electron micrographs show the particles at the start and end of the maturation pathway.
Photographs kindly provided by A. F. Howatson.

Figure 28.3
Concatemeric DNA consists of a tandem series of phage genomes.

Concatemer

ABC————XYZABC————XYZABC————XYZABC————XYZ

Individual genome

ABC————XYZ

vide some sort of "scaffolding" onto which the DNA condenses. (This would be a counterpart to the use of the proteins of the shell in the plant viruses mentioned earlier.) How specific is this packaging? It cannot depend on particular sequences, because deletions, insertions, and substitutions all fail to interfere with the assembly process.

These varying mechanisms of virus assembly all accomplish the same end: packaging a single DNA or

RNA molecule into the capsid. However, there are some viruses whose genomes consist of multiple nucleic acid molecules. Reovirus contains ten double-stranded RNA segments, all of which must be packaged into the capsid. Nothing is known about how the assembly process selects one copy of each different molecule in order to collect a complete set of genetic information.

Some plant viruses are multipartite: their genomes consist of segments each of which is packaged into a *different* capsid. An example is alfalfa mosaic virus, which has four different single-stranded RNAs, each packaged independently into a coat comprising the same protein subunit. A successful infection depends on the entry of one of each type into the cell.

The four components of the virus exist as particles of different sizes. This means that the same capsid protein can package each RNA into its own characteristic particle. This is a departure from the packaging of a unique length of nucleic acid into a capsid of fixed shape. However, in the case of viruses whose capsids have only one authentic form, the assembly pathway often displays a side effect. This is the formation of some aberrant (although inactive) **monster** particles in which the head is longer than usual. Thus the capsid protein(s) has an intrinsic ability to assemble into a particular type of structure, but the exact size and shape is not invariant. Often there are other, **assembly proteins** that are not themselves part of the head shell, but that play a role in limiting its options so that it assembles along the desired pathway. Cellular genomes also may employ proteins whose functions are to direct the assembly of others (see Chapter 29).

THE BACTERIAL GENOME IS A FOLDED NUCLEOID

Although bacteria do not display structures with the distinct morphological features of eucaryotic chromosomes, their genomes nonetheless are organized into definite bodies. The genetic material can be seen as a fairly compact clump or series of clumps that occupies (very roughly) about one third of the volume of the cell. **Figure 28.4** displays a thin section through a bacterium in which this **nucleoid** is evident.

In bacteria that have partially replicated their DNA,

the nucleoid may contain more than one genome's worth of DNA. By the time of cell division, the material has separated into two nucleoids that are partitioned into the daughter cells. The segregation mechanism probably involves the attachment of the bacterial genome to the membrane. This mechanism provides a counterpart to the mitotic segregation of eucaryotic chromosomes. Instead of being pulled apart on a spindle, the sequences of a specific locus on each bacterial genome may be attached to a site on the membrane. The two membrane sites move apart as the cell grows; and when division occurs, each lies in a different compartment, taking its nucleoid with it (see Chapter 31).

The nucleoid of *E. coli* cells has been isolated directly in the form of a very rapidly sedimenting complex, consisting of about 80% DNA by mass. (The analogous complexes in eucaryotes have about 50% DNA by mass, as described in Chapter 29.) In bacteria that contain more than a single copy of the genome (because DNA replication has occurred), the complex is proportionately increased in size.

The complex can be unfolded to a less compact

Figure 28.4
A thin section shows the bacterial nucleoid as a compact mass in the center of the cell.
Photograph kindly provided by Jack Griffith.

state by treatment with reagents that act on RNA or protein. This suggests that its structure is stabilized by these components. The possible role of proteins is evident. The role that RNA might have has been quite refractory to analysis. Attempts to isolate any specific RNA involved in a structural capacity all have failed. So have attempts to implicate nascent RNA chains (that is, the incomplete products of currently proceeding transcription).

When *E. coli* cells are lysed, fibers are released in the form of loops attached to the broken envelope of the cell. As can be seen from **Figure 28.5,** the DNA of these loops is not found in the extended form of a free duplex, but is folded into a more compact shape, presumably by virtue of its association with proteins. The rapidly sedimenting complex may consist of a compact organization of these loops.

Several DNA-binding proteins with at least a superficial resemblance to eucaryotic chromosomal proteins have been isolated in *E. coli.* Their properties are summarized in **Table 28.2.**

Two of the proteins, HU and H, are identified as fairly abundant components of the cell whose amino acid compositions resemble the histones that bind to eucaryotic DNA (see Chapter 29). The proteins are small, highly basic, and bind strongly to DNA. The physiological role of this binding, however, remains to be determined.

A potential DNA-binding protein has been identified by virtue of the presence of its gene in a tRNA operon (see Chapter 23). This is protein P. It has an amino acid composition resembling the protamines that bind

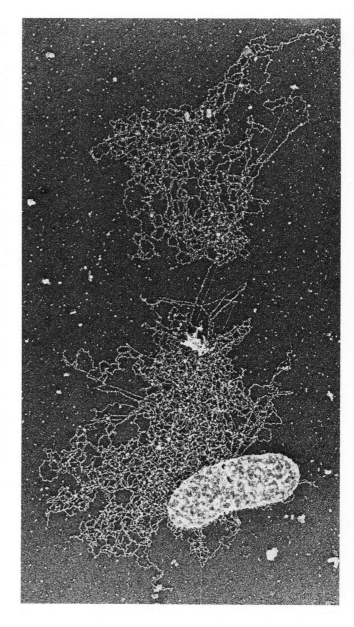

Figure 28.5
The nucleoid spills out of a lysed *E. coli* cell in the form of loops of a fiber.

Photograph kindly provided by Jack Griffith.

Table 28.2
***E. coli* Contains Several DNA-Binding Proteins**

Protein	Composition	Content/Cell	Eucaryotic Relatives	Locus
HU	2 nonidentical subunits of 9000 daltons	40,000 dimers	Histone H2B	Not known
H	2 identical subunits of 28,000 daltons	30,000 dimers	Histone H2A	Not known
HLP1	Monomer of 17,000 daltons	20,000 copies	Not known	*firA*
P	Subunit of 3000 daltons	Not known	Protamines	Not known

to DNA in certain sperm. Its quantity and functions are not known.

The one protein firmly implicated in interacting with the genetic apparatus is HLP1. This is coded by the locus *firA*, in which mutations can abolish the resistance to rifampicin conferred by some mutations in the β subunit of RNA polymerase. The properties of the mutants do not show whether the protein interacts with RNA polymerase or with DNA; HLP1 is small, basic, and can bind DNA like the other proteins.

THE NUCLEOID CONTAINS MANY SUPERCOILED LOOPS

The DNA of the compact body isolated *in vitro* behaves as a closed duplex structure, as judged by its response to ethidium bromide. This small molecule intercalates between base pairs to generate *positive* superhelical turns in "closed" circular DNA molecules, that is, molecules in which both strands have covalent integrity. In "open" circular molecules, which contain a nick in one strand, or with linear molecules, the DNA can rotate freely in response to the intercalation, thus relieving the tension.

In a natural closed DNA that is *negatively* supercoiled, the intercalation of ethidium bromide first removes the negative supercoils and then introduces positive supercoils. The amount of ethidium bromide needed to achieve zero supercoiling is a measure of the original density of negative supercoils.

Some nicks occur in the compact nucleoid during its isolation; they can also be generated by limited treatment with DNAase. But this does not abolish the ability of ethidium bromide to introduce positive supercoils. This capacity of the genome to retain its response to ethidium bromide in the face of nicking means that it must have many independent **domains;** the supercoiling in each domain is not affected by events in the other domains.

This autonomy suggests that the structure of the bacterial chromosome may have the general organization depicted diagrammatically in **Figure 28.6.** Each domain consists of a loop of DNA, the ends of which are secured in some (unknown) way that does not allow rotational events to propagate from one domain to another. There are about 100 such domains per

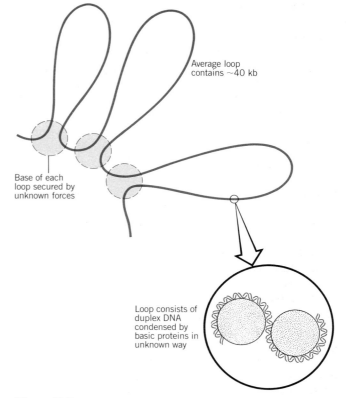

Average loop contains ~40 kb

Base of each loop secured by unknown forces

Loop consists of duplex DNA condensed by basic proteins in unknown way

Figure 28.6
The bacterial genome consists of a large number of loops of duplex DNA (in the form of a fiber), each secured at the base to form an independent structural domain.

genome, so that each consists of about 40 kb (13 μm) of DNA, organized into some more compact fiber whose structure has yet to be characterized.

The existence of separate domains could permit different degrees of supercoiling to be maintained in different regions of the genome. This is a pertinent factor in considering the different susceptibilities of particular bacterial promoters in response to supercoiling (see Chapter 11).

Now when DNA is supercoiled, there can be two results. If the DNA remains free, its path is **unrestrained,** and negative supercoils generate a state of torsional tension that can be relieved by unwinding the double helix, as described in Chapter 2. The DNA may be in a dynamic equilibrium between the states of tension and unwinding (see Chapter 32). However, the supercoiling can be **restrained** if proteins are bound to the DNA to hold it in a particular three-dimensional configuration. In this case, the supercoils are represented by the path the DNA follows in its fixed association with the proteins; and the energy of interaction between the proteins and the supercoiled DNA influences the stability of the double helix. For example, if negatively supercoiled DNA is bound by a protein relatively specific for single-stranded DNA, there could be a permanent local denaturation of DNA in the area of binding. On the other hand, the binding of histone proteins (to form a nucleosome) stabilizes the double helix of a negatively supercoiled DNA (see Chapter 29).

Are the supercoils in *E. coli* DNA mostly restrained *in vivo* or is the double helix subject to the torsional tension characteristic of free DNA? Measurements of supercoiling *in vitro* encounter the difficulty that restraining proteins might have been lost during isolation. However, the situation *in vivo* can be investigated by use of the cross-linking reagent psoralen, which binds more readily to DNA when it is under torsional tension. The reaction of psoralen with *E. coli* DNA *in vivo* corresponds to an average density of one negative superhelical turn for every 200 bp. This result demonstrates that supercoils *do* create torsional tension *in vivo*. Most of the natural supercoils in *E. coli* DNA may be unrestrained.

Many of the important features of the structure of the compact nucleoid remain to be established. What is the specificity with which domains are constructed—do the same sequences always lie at the same relative locations, or can the contents of individual domains shift? How is the integrity of the domain maintained? What structural roles are played by the DNA-binding proteins? Biochemical analysis by itself may be unable to answer these questions fully, but if it is possible to devise suitable selective techniques, the properties of structural mutants should lead to a molecular analysis of nucleoid construction.

THE CONTRAST BETWEEN INTERPHASE CHROMATIN AND MITOTIC CHROMOSOMES

Individual **chromosomes** come into the limelight for only a brief period, during the act of cell division. Only

Figure 28.7
The sister chromatids of a mitotic pair each consist of a fiber (~ 30 nm in diameter) compactly folded into the chromosome.
Photograph kindly provided by E. J. DuPraw.

then can each be seen as a compact unit, with a packing ratio of up to 10,000. **Figure 28.7** is an electron micrograph of a sister chromatid pair, captured at metaphase. (The sister chromatids are daughter chromosomes produced by the previous replication event, still joined together at this stage of mitosis, as described in Chapter 1.) Each consists of a fiber with a diameter of about 30 nm and a nubbly appearance, folded into the chromosomal structure.

During most of the life cycle of the eucaryotic cell, however, its genetic material occupies an area of the nucleus in which individual chromosomes cannot be distinguished. The structure of this interphase **chromatin** does not change visibly between divisions. For example, usually no disruption is evident during the period of replication, when the amount of chromatin doubles. Chromatin is fibrillar, although the overall configuration of the fiber in space is hard to discern in detail. The fiber itself, however, is similar to (and could be identical with) that found in the mitotic chromosomes.

Chromatin can be divided into two types of material. These are visible in **Figure 28.8.** In most regions, the fibers are much less densely packed than they are in the chromosome. This material is called **euchromatin.** Its overall packing ratio is probably about 1000–2000 during interphase, assuming that it is additionally condensed by some 5–10-fold at mitosis.

Some regions of chromatin are very densely packed with fibers, displaying a condition comparable to that of the chromosome at mitosis. This material is called **heterochromatin.** It passes through the cell cycle with relatively little change in its degree of condensation. Often the various heterochromatic regions aggregate into a densely staining **chromocenter.**

The same fibers run continuously between euchromatin and heterochromatin, which implies that these states represent different degrees of condensation of the genetic material. In the same way, euchromatic regions exist in different states of condensation during interphase and during mitosis. The important conclusion to be drawn from these features is that the genetic material is organized in a manner that both permits alternative states to be maintained side by side in chromatin, and allows cyclical changes to occur in the packaging of euchromatin between interphase and division.

There is a correlation between the structural condition of the genetic material and its transcriptional activity. Mitotic chromosomes are transcriptionally inert; cells virtually cease transcription during the process of division. Heterochromatin also is transcriptionally inactive. There are two classes of heterochromatin,

Figure 28.8
A thin section through a nucleus stained with Feulgen shows heterochromatin as compact regions clustered near the nucleolus and nuclear membrane.

Photograph kindly provided by Edmund Puvion.

each containing a different type of sequence, and in both cases the lack of transcription is unequivocal.

Constitutive heterochromatin consists of particular regions that usually are *never* expressed. This includes satellite DNA sequences. **Facultative heterochromatin** takes the form of entire chromosomes that are inactive in one cell lineage, although they can be expressed in other circumstances. The example par excellence is the mammalian X chromosome, one copy of which (selected at random) is entirely inactive in a given female. (This compensates for the presence of two X chromosomes, compared with the one present in males.) The inactive X chromosome is perpetuated in a heterochromatic state; the active X chromosome is part of the euchromatin. Here it is possible to see a correlation between transcriptional activity and structural organization when the *identical DNA sequences* are involved in both states.

Figure 28.9
G-banding generates a characteristic lateral series of bands in each member of the chromosome set.

This example of the human complement was provided by N. Davidson.

A very tight condensation of the genetic material is thus associated with (perhaps is responsible for) its inactivity. Note, however, that the reverse is not true. Active genes are indeed contained within euchromatin; but only a small minority of the sequences in euchromatin are transcribed at any time. Thus location in euchromatin is *necessary* for gene expression, but is not *sufficient* for it. We may wonder whether the gross changes seen between euchromatin and heterochromatin are mimicked in a lesser manner by changes in the structure of euchromatin, to give transcribed regions a less condensed structure than that of nontranscribed regions.

When isolated *in vitro,* both interphase chromatin and mitotic chromosomes possess large loops of the constituent fiber, apparently secured at the base to form independent domains analogous to those found in the bacterial nucleoid. Their structure is discussed in Chapter 29.

Because of the diffuse state of chromatin, we cannot at present determine the specificity of its organization. But we can ask how ordered is the structure of the chromosome? Do particular sequences always lie at particular sites, or is the folding of the fiber into the overall structure a more random event? Some years ago, the development of the **G-banding** technique showed that each chromosome has a different and reproducible ultrastructure. When subjected to certain treatments and then stained with the chemical dye Giemsa, chromosomes generate a series of bands. An example of the human set is presented in **Figure 28.9.**

Until the development of this technique, chromosomes could be distinguished only by their overall size and the relative location of the centromere (see later). Now each chromosome can be identified by its characteristic banding pattern. This pattern is reproducible enough to allow translocations from one chromosome to another to be identified by comparison with the original diploid set.

The banding technique is of enormous practical use, but the mechanism of banding remains a mystery. All that is certain is that the dye stains untreated chromosomes more or less uniformly. So the generation of bands depends on a variety of treatments that change the response of the chromosome (presumably by extracting the component that binds the stain from the nonbanded regions). But the variety of effective treatments is so great that no common cause yet has been discerned. These results imply that there is some definite long-range structure, but its basis is unknown.

The simplest model for chromosome structure is to suppose that each chromosome contains a single, very long duplex of DNA. This view is consistent with the continuity of the 30 nm fiber throughout the chromosome. Early experiments showed that the fiber contains a single compressed duplex of DNA, and this conclusion is confirmed in detail by experiments that are discussed in Chapter 29. The pattern of chromosome replication resembles that of the individual DNA molecule: it is semiconservative (see Figure 2.10 and Chapter 31). This is unlikely to be the case if a chromosome carries many independent molecules of DNA. Thus in accounting for interphase chromatin and mitotic chromosome structure, we have to explain the packaging of a single, exceedingly long molecule of DNA into a form in which it can be transcribed and replicated, and can become cyclically more and less compressed.

THE EUCARYOTIC CHROMOSOME AS A SEGREGATION DEVICE

During mitosis, the sister chromatids move to opposite poles of the cell (as illustrated in Figure 1.5). This movement depends on the attachment of the chromosome to microtubules connected to or around the centrioles at the poles. The chromosomal site of attachment is identified by a constricted region called the **centromere.** This constriction is clear in the photograph of Figure 28.7, which shows the sister chromatids at the metaphase stage of mitosis, when they are still attached to one another in the centromeric region.

The term "centromere" historically has been used in both the functional and structural sense to describe the feature of the chromosome responsible for its movement. The centromere is pulled toward the pole during mitosis, and the attached chromosome is dragged along behind, as it were. From this perspective, the chromosome is a device for attaching a large number of genes to the apparatus for division.

The centromere is essential for segregation, as shown

by the results of chromosome breakage. A single break generates one piece that retains the centromere, and another, an **acentric fragment,** that lacks it. The acentric fragment does not become attached to the mitotic spindle; and as a result it may fail to be included in either of the daughter nuclei. (Note that there can be *only* one centromere per chromosome. When translocations generate chromosomes with more than one centromere, aberrant structures form at mitosis, since the two centromeres on the *same* sister chromatid can be pulled toward different poles, leading to chromosome breakage.)

The regions flanking the centromere often are rich in satellite DNA sequences and may contain a considerable amount of constitutive heterochromatin. In contrast with its state in interphase chromatin, the heterochromatin is not immediately evident in mitotic chromosomes, but can be visualized by a technique called **C-banding.** In the example of **Figure 28.10,** all the centromeres show as darkly staining regions. This

Figure 28.10
C-banding generates intense staining at the centromeres of all chromosomes.

Photograph kindly provided by N. Davidson.

property (and also the characteristic features of interphase heterochromatin) do not rest just on the peculiarities of satellite DNA, but must depend on proteins specifically present in this material. Constitutitive heterochromatin cannot be identified around *every* known centromere, and so generally it is felt that it is unlikely to be directly associated with the division mechanism.

What is the feature of the centromere that is responsible for segregation? Within the centromeric region, a darkly staining fibrous object of diameter or length around 400 nm can be seen. This is called the **kinetochore.** This term describes the visible body that appears to be directly attached to the microtubules. Usually it is assumed that a specific sequence of DNA in some way defines the site at which the kinetochore should be established. The structure of the kinetochore varies, and has been difficult to analyze in relation to its function. However, there have been claims that it contains DNA, which would imply that it may form *directly* on the responsible chromosomal sequence.

If a centromeric sequence of DNA is responsible for segregation, any molecule of DNA possessing this sequence should move properly at cell division, while any DNA lacking it will fail to segregate. This prediction has been used to isolate centromeric DNA in the yeast, *S. cerevisiae.* Yeast chromosomes do not display visible kinetochores comparable to those of higher eucaryotes, but otherwise divide at mitosis and segregate at meiosis by the same mechanisms.

Recent genetic engineering has produced plasmids of yeast that have the ability to be replicated like chromosomal sequences (see Chapter 31). However, they are unstable at mitosis and meiosis, disappearing from a majority of the cells because they segregate erratically. Fragments of chromosomal DNA have been isolated by virtue of their ability to confer mitotic stability on these plasmids. In one case, the DNA fragment contains genetic markers known to map near the centromere, which confirms its derivation.

By reducing the sizes of the fragments that are incorporated into the plasmid, the minimum length necessary for mitotic centromeric function can be identified. At the present, there are two such fragments, *CEN3* of 627 bp and *CEN11* of 900 bp. (The number identifies the chromosome of origin.) It is interesting that these small fragments are adequate at mitosis, but appear not to support proper segregation at meiosis. It is possible that chromosome pairing and segregation at the first division of meiosis require a greater length of DNA than is needed for simple mitotic division.

The sequences of both the *CEN3* and *CEN11* fragments have been determined. There are no extensive regions of homology, but there are some rather short homologies and also an A-T-rich region of the same length. The conserved sequence is given in **Figure 28.11.** It seems significant that the distance between the short homologous regions is the same in the two centromeres. Neither centromeric sequence has any open reading frames. There are some similarities between this sequence and certain satellite DNA sequences in higher eucaryotes. The availability of these fragments should allow deletions and other changes to be made to determine directly what features are necessary for centromeric function.

Another feature that is essential in all chromosomes is the **telomere.** In some way that we do not yet un-

Figure 28.11
Four conserved regions can be identified by the sequence homologies between *CEN3* and *CEN11* of yeast.

derstand, this "seals" the end. It has been known for a long time that the end must be a special structure, because chromosome ends generated by breakage are "sticky" and tend to react with other chromosomes, whereas natural ends are stable.

Again, this problem has been brought to the molecular level by using yeast. The plasmids that survive in yeast all are circular DNA molecules. Linear plasmids are unstable (because they are degraded). Could an authentic telomeric DNA sequence confer stability on a linear plasmid?

A region from the end of a known natural linear DNA molecule—the extrachromosomal rDNA of *Tetrahymena*—is able to render the yeast plasmid stable in linear form. This putative telomeric sequence consists of a repetitive sequence with a series of nicks. These nicks are perpetuated at the same sites in yeast, a remarkable interspecies conservation. Each end of the linear DNA may be a hairpin.

The *Tetrahymena* sequence can be replaced by fragments from yeast DNA. These fragments have related sequences. At the present, they are available in the form of segments about 4 kb in length. These are likely to prove to be the natural telomeres of yeast chromosomes. The same type of analysis performed with the centromeric DNA, using size reduction and inducing mutations, should reveal how they function.

THE EXTENDED STATE
OF LAMPBRUSH CHROMOSOMES

It would be extremely useful to visualize gene expression in its natural state, to see what structural changes are associated with transcription. But the nature of the material restricts such analysis to only some unusual circumstances.

The compression of DNA in chromatin, coupled with the difficulty of identifying particular genes within it, makes it impossible to visualize the transcription of individual active genes. (However, they do display some distinctive biochemical properties that can be analyzed *in vitro,* as described in Chapter 30.)

Mitotic chromosomes are inert in gene expression, and, in any case, are so compact as to preclude the identification of individual loci. The distinct regions that

are rendered discernible by the G-banding technique, as in the example of Figure 28.9, each contain of the order of 10^7 bp of DNA, which could include many hundreds of genes.

Lateral differentiation of structure is evident in many chromosomes when they first appear for meiosis. At this stage, the chromosomes resemble a series of beads on a string. The beads are densely staining granules, properly known as **chromomeres.** However, usually there is little gene expression at meiosis, and it is not practical to use this material to identify individual genes.

Gene expression can be visualized directly in two situations, in which the chromosomes are found in a highly extended form that allows individual loci (or groups of loci) to be distinguished. These are the lampbrush chromosomes (best characterized in certain amphibians), and the polytene chromosomes (best characterized in some insects).

The lampbrush chromosomes are formed during an unusually extended meiosis, which can last up to several months! During this period, the chromosomes are held in a stretched-out form in which they can be visualized in the light microscope. Later during meiosis, the chromosomes revert to their usual compact size. So the extended state essentially proffers an unfolded version of the normal condition of the chromosome.

The lampbrush chromosomes are meiotic bivalents, so that each consists of two pairs of sister chromatids. **Figure 28.12** shows an example in which the sister chromatid pairs have mostly separated and are now held together only by chiasmata (the sites of crossing-over). Each sister chromatid pair forms a series of ellipsoidal chromomeres, about 1–2 μm in diameter, which are connected by a very fine thread. This thread contains the two sister duplexes of DNA and runs continuously along the chromosome, through the chromomeres. The lengths of the individual lampbrush chromosomes in the newt *Notophthalmus viridescens* range from 400 to 800 μm, compared with the range of 15–20 μm seen later in meiosis. So the lampbrush chromosomes are about 30 times less tightly packed. The total length of the entire lampbrush chromosome set may be 5–6 mm, organized into about 5000 chromomeres.

The lampbrush chromosomes take their name from

Figure 28.12
A lampbrush chromosome is a meiotic bivalent in which the two pairs of sister chromatids are held together at chiasmata (indicated by arrows).

Photograph kindly provided by Joe Gall.

Figure 28.13
A lampbrush chromosome loop is surrounded by a matrix of ribonucleoprotein.

Photograph kindly provided by Oscar Miller.

the lateral loops that extrude from the chromomeres at certain positions. (These resemble a lampbrush, a now-extinct object.) The loops extend in pairs, one from each sister chromatid. The loops are continuous with the axial thread, which suggests that they represent chromosomal material that has been extruded from its more compact organization in the chromomere.

The loops are surrounded by a matrix of ribonucleoproteins. These contain nascent RNA chains. Often a transcription unit can be defined by the increase in the length of the RNP moving around the loop. An example is shown in **Figure 28.13.**

So the loop is an extruded segment of DNA that is being actively transcribed. In some cases, loops corresponding to particular genes have been identified. Then the structure of the transcribed gene, and the nature of the product, can be scrutinized *in situ.*

POLYTENY FORMS GIANT CHROMOSOMES

The interphase nuclei of some tissues of the larvae of Dipteran flies contain chromosomes that are greatly enlarged relative to their usual condition. They possess both increased diameter and greater length. **Figure 28.14** is an example of a chromosome set from the salivary gland of *D. melanogaster*. They are called **polytene chromosomes.**

Each chromosome consists of a visible series of **bands** (more properly, but rarely, described as chromomeres). There is an approximately tenfold range in the sizes of the bands, the largest having a breadth of ~0.5 μm and the smallest ~0.05 μm. (The smallest can be distinguished only under the electron microscope.) The bands contain most of the mass of DNA and stain intensely with appropriate reagents. The regions between them stain more lightly and are called **interbands.** There are about 5000 bands in the *D. melanogaster* set.

The centromeres of all four chromosomes of *D. melanogaster* aggregate to form a chromocenter that consists largely of heterochromatin (and in the male includes the entire Y chromosome). Allowing for this, about 75% of the haploid DNA set is organized into the band-interband alternation. The length of the chromosome set is about 2000 μm, and 75% of the DNA is 1.3×10^8 bp, which would extend for about 40,000

Figure 28.14
The polytene chromosomes of *D. melanogaster* form an alternating series of bands and interbands.

Photograph kindly provided by José Bonner.

μm, so the average packing ratio is ~20. This demonstrates vividly how extended the genetic material is relative to the usual states of interphase chromatin or mitotic chromosomes.

What is the structure of these giant chromosomes? Each is produced by the successive replications of a synapsed diploid pair. The replicas do not separate, but remain attached to each other in their extended state. At the start of the process, each synapsed pair has a DNA content of 2C (where C represents the DNA content of the individual chromosome). Then this doubles up to 9 times, at its maximum giving a content of 1024C. The number of doublings is different in the various tissues of the *D. melanogaster* larva and, of course, in other Dipteran flies.

Each chromosome can be visualized as a large number of parallel fibers running longitudinally, tightly condensed in the bands, less condensed in the interbands. Probably each fiber represents a single (C) haploid chromosome content. This gives rise to the name "polyteny." The degree of polyteny is the number of haploid chromosomes contained in the giant chromosome.

The banding pattern is characteristic for each strain of *Drosophila*. The constancy in the number, and the linear arrangement of the bands, was first noted in the 1930s, when it was realized that the bands form a **cytological map** of the chromosomes. Rearrangements—such as deletions, inversions, or duplications—result in alterations of the order of bands. The linear array of bands can be equated with the linear array of genes, as first realized by Bridges (whose original map of the chromosomes still remains the basis for current work, with some modifications). Thus genetic rearrangements, as seen in a linkage map, can be correlated with structural rearrangements of the cytological map. Ultimately, a particular mutation can be located in a particular band.

The positions of particular genes on the cytological map can be determined directly by the technique of *in situ* or **cytological hybridization** that we have described in Chapter 24 in the context of highly repetitive DNA. The protocol is recapitulated in **Figure 28.15.** A radioactive probe representing a gene (most often a labeled cDNA clone derived from the mRNA) is hybridized with the denatured DNA of the polytene chromosomes *in situ*. By the superimposition of grains at a particular band or bands, autoradiography identifies the position or positions of the corresponding genes. An example is shown in **Figure 28.16.** With this type

Figure 28.15

Individual bands containing particular genes can be identified by *in situ* hybridization.

Squashed target cells

Microscope slide

Freeze in dry ice
Wash with ethanol
Dip in agar solution
Denature DNA
Add radioactive probe
Wash off unreacted probe
Autoradiography

Black areas identify silver grains corresponding to sites where the probe hybridized

Target cell

Figure 28.16

A magnified view of bands 87A and 87C shows their *in situ* hybridization with labeled RNA extracted from heat-shocked cells.

Photograph kindly provided by José Bonner.

of technique at hand, it is therefore possible to determine directly the band within which a particular sequence lies.

TRANSCRIPTION DISRUPTS THE CHROMOSOME STRUCTURE

One of the intriguing features of the polytene chromosomes is that active sites can be visualized. Some of the bands are found at times in an expanded or **puffed** state, in which chromosomal material is extruded from the axis. An example of some very large puffs (called Balbiani rings) is shown in **Figure 28.17.**

What is the nature of the puff? It consists of a region in which the chromosome fibers unwind from their usual state of packing in the band. The fibers remain continuous with those in the chromosome axis. Puffs usually emanate from single bands, although when they are very large, as typified by the Balbiani rings, the swelling may be so extensive as to obscure the underlying array of bands.

The pattern of puffs is related to gene expression. During larval development, puffs appear and regress in a definite, tissue-specific pattern. A characteristic pattern of puffs is found in each tissue at any given time. Puffs are induced by the hormone ecdysone that controls *Drosophila* development. Some puffs are induced directly by the hormone; others are induced indirectly by the products of earlier puffs.

The puffs are *sites where RNA is being synthesized.* The amount of synthesis determines the size of the puff. The sites of puffing differ from the ordinary bands in the accumulation of additional proteins. At the present, the characterization of these proteins is only rather primitive, but in due course it should be possible to establish whether they are a cause or a consequence of the puffing. From this, it should be possible to determine the nature of the molecular events that are responsible for the expansion of material.

So a puff represents a band containing a gene that is being transcribed actively. Is the band a unit of gene expression? To this somewhat vexed question, there is at present no unequivocal answer.

Genetic experiments have concentrated on correlating the number of lethal loci with the bands. A **lethal locus** is defined by mutations that prevent the orga-

Figure 28.17
Chromosome IV of *C. tentans* has three Balbiani rings in the salivary gland.
Photograph kindly provided by Bertil Daneholt.

nism from surviving. Thus it identifies an essential gene. The number of lethal loci appears to be close to the number of bands; and in defined areas of the chromosomes, individual bands can be equated with lethal loci on virtually a 1:1 basis. This would argue that there are about 5000 lethal loci in *D. melanogaster*.

Now the average band contains about 25,000 bp of haploid DNA. There is no way to predict a priori how many genes this may contain: it could be several packed quite closely, or only one with large introns or extensive flanking material. However, the analysis of DNA representing individual genes shows that in several cases there is more than one transcription unit within a single band. If these bands should prove to be typical of the majority, the total number of genes could be several times 5000. We do not yet know whether the different transcription units in such bands are subject to related controls. It does seem plausible that sufficiently intense expression of any one gene within a band may lead to the appearance of a puff.

There is at present, therefore, somewhat of a discrepancy between the genetic and biochemical data. The first argues for the probable existence of only one essential gene per band, the second argues for the presence of several genes per band. There is a possible reconciliation that many genes may be nonessential, at least as defined by the criterion of the lethal locus. This leaves the puzzle of why lethal loci should be distributed regularly at approximately one per band.

The features displayed by lampbrush and polytene chromosomes suggest a general conclusion. In order to be transcribed, the genetic material is dispersed from its usual more tightly packed state. The question to keep in mind is whether this dispersion at the gross level of the chromosome mimics the events that occur at the molecular level within the mass of ordinary interphase euchromatin.

FURTHER READING

Phage morphogenesis has been dealt with in detail in **Lewin's** *Gene Expression, 3, Plasmids and Phages* (Wiley, New York, 1978, pp. 496–535, 642–673). The unusual chromosomes discussed in this chapter are the subject of pp. 455–475 of *Gene Expression, 2, Eucaryotic Chromosomes.* The closest approach to the centromere was reported by **Fitzgerald-Hayes, Clarke,** & **Carbon** (*Cell,* **29,** 235–244, 1982).

CHAPTER 29
NUCLEOSOME PARTICLES AND
THE STRUCTURE OF CHROMATIN

One of the most striking advances of the past decade has been the elucidation of the general structure of chromatin. Before this, we knew only that chromatin contains DNA, RNA, and protein in a compact form of organization in which the majority of DNA sequences are structurally inaccessible and functionally inactive. The reason for their inaccessibility was assumed to lie in binding by proteins—the DNA was often described as "covered" by proteins—but there was no information at all about the nucleoprotein structure at the level of the DNA-protein interaction.

Enormous progress has been made since the critical insight in 1974 that chromatin is constructed from a subunit that has the same type of design in all eucaryotes. This consists of about 200 bp of DNA organized by an octamer of small, basic proteins (histones) into a beadlike structure. It turns out that DNA lies on the surface of this particle, while the protein components constitute the interior.

At the level of the fundamental DNA-protein interaction, we can define the structure of the particle in terms of the path of the DNA and the interactions between the nucleic acid and the proteins and among the individual proteins. We can ask whether the conformation of the particle can vary, and how such variations might be related to the functions of chromatin.

What happens to the particle when chromatin is transcribed or replicated? These processes may require other proteins (nonhistones) to modulate the behavior of the particle.

Replication requires reproduction of the series of particles. The protein content in addition to the DNA content of chromatin must be doubled. As well as asking how the particle is assembled, we must inquire what happens to other proteins present in chromatin. Since replication is likely to disrupt the structure of chromatin, it both poses a problem for maintaining regions with specific structure and offers an opportunity to change the structure.

The high overall packing ratio of the genetic material immediately suggests that DNA cannot be directly packaged into the final structure of chromatin. There must be *hierarchies* of organization.

The first level is the winding of the DNA into the beadlike particles, with a packing ratio of about 6. These particles are an invariant component of euchromatin, heterochromatin, and chromosomes.

The second level of organization is the coiling of the series of beads into a helical array to constitute the ~30 nm fiber that is found in both interphase chromatin and mitotic chromosomes (see Figure 28.7). In chromatin this brings the packing ratio of DNA to about

456

40. It is not certain whether the fiber is unaffected or has an increased packing ratio in mitotic chromosomes.

The final packing ratio is determined by the third level of organization, the packaging of the fiber itself. This gives an overall packing ratio of ≥1000 in euchromatin, cyclically interchangeable with the packing into mitotic chromosomes to achieve an overall ratio of ≤10,000. This too is likely to be a function modulated by nonhistone proteins (and also possibly involving chemical modification of histones), as is the difference between euchromatin and heterochromatin.

THE PROTEIN COMPONENTS OF CHROMATIN

The relative proportions of the components of chromatin vary with the method of isolation and the source. The mass of protein is always greater than the mass of DNA, by up to twofold. The proteins are divided into two types: the histones and the nonhistones. The mass of RNA is less than 10% of the mass of DNA. Much of the RNA consists of nascent chains that are still associated with the template DNA.

The **histones** are the most basic proteins in chromatin and account for just about the same mass as that of the DNA. Five classes of histones were originally characterized by the relative proportions of each type of basic amino acid, as expressed in the lysine/arginine ratio. A relic of this description still lingers in the names of the histones. The same classes can be recognized in virtually all eucaryotes. Their properties are summarized in **Table 29.1.**

The histones of four classes interact directly with DNA to form the first-level series of particles in chromatin. The need to preserve this fundamental reaction explains the conservation of histone types through eucaryotic evolution. The fifth class of histones is concerned with the interactions between particles. The constancy of the histones suggests that the histone-DNA interactions, histone-histone interactions, and histone-nonhistone interactions may be generally similar in different species; so we may be able to deduce general mechanisms for the formation of both the primary particle and the folding of a series of particles into a higher-order structure.

The histones of the first four classes have an appreciable content of acidic amino acids, as well as the basic amino acids. These proteins are therefore highly charged. The basic/acidic ratios lie in the range of 1.4 to 2.5. These histones fall into two groups.

The **arginine-rich** histones are the two species **H3** and **H4.** These are among the most conserved proteins known in evolution. They even have identical sequences in species as far distant as the cow and the pea. Only a few amino acid replacements have been found in the H3 and H4 sequences of other species. The conservation of the entire sequence shows that all of it must be important in the function of the protein; and by implication, this function must be identical in greatly different species, which lends support to the concept of a common basis for chromatin structure.

The **slightly lysine-rich** histones comprise two proteins. They are called **H2A** and **H2B** (but, in contrast

Table 29.1
Histones are Highly Basic Proteins

Histone	Basic Amino Acids		Acidic Amino Acids	Basic/Acidic Ratio	Molecular Weight
	Lys	Arg			
H1	29%	1%	5%	5.4	23,000
H2A	11%	9%	15%	1.4	13,960
H2B	16%	6%	13%	1.7	13,774
H3	10%	13%	13%	1.8	15,342
H4	11%	14%	10%	2.5	11,282

These data refer to mammalian (bovine) histones; others are similar.

with the implication of the nomenclature, they are independent proteins, not related variants). The same two types can be recognized in various eucaryotes, but show appreciable interspecific variation in sequence.

The fifth class is the **very lysine-rich,** which consists of several rather closely related proteins, with overlapping amino acid sequences. These are the **H1** histones (also there is a variant in avian red blood cells called **H5**). They show appreciable variation between tissues and between species (and this class apparently is absent from yeast). Although these are the most basic histones, they can be extracted readily from chromatin with quite dilute (0.5 M) salt solution.

As their disphonious name suggests, the **nonhistones** include all the other proteins of chromatin. They are therefore presumed to be more variable between tissues and species—although hard evidence still is lacking on the extent of the variability—and they comprise a relatively smaller proportion of the mass than the histones. They also comprise a much larger number of proteins, so that any individual protein is present in amounts much smaller than any histone.

Within the class of nonhistone proteins, there may be functions concerned with gene expression and with higher-order structure. Thus RNA polymerase may be considered to be a prominent nonhistone. The HMG (high-mobility group) proteins comprise a discrete and well-defined subclass of nonhistones. A major problem in working with other nonhistones is their contamination with other nuclear proteins.

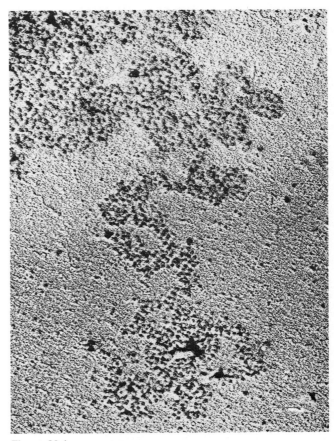

Figure 29.1
Chromatin spilling out of lysed nuclei consists of a compactly organized series of particles. The bar is 100 nm.

Photograph kindly provided by Pierre Chambon.

CHROMATIN CONTAINS DISCRETE PARTICLES

When interphase nuclei are suspended in a solution of low ionic strength, they swell and rupture to release fibers of chromatin. **Figure 29.1** shows a lysed nucleus in which fibers are streaming out. In some regions, the fibers consist of tightly packed material, but in regions that have become stretched out, they can be seen to consist of discrete particles. These are called **nucleosomes.** In especially extended regions, individual nucleosomes are connected by a fine thread; this is a free duplex of DNA. So a continuous duplex thread of DNA runs through the series of particles.

Individual nucleosomes can be obtained by treating chromatin with the enzyme **micrococcal nuclease,** an endonuclease that cuts the DNA thread at the junction between nucleosomes. First, this releases groups of particles; finally, it releases single nucleosomes. The monomeric nucleosomes can be seen clearly in **Figure 29.2** as compact particles (whose actual shape turns out to be disklike; see later). They sediment at a rate of about 11S, corresponding to a total mass in the range of 250,000–300,000 daltons. The protein/DNA ratio is about 1.25. Dimers, trimers, etc., have corresponding properties when viewed by electron microscopy or characterized biochemically.

Figure 29.2
Individual nucleosomes are released by digestion of chromatin with micrococcal nuclease. The bar is 100 nm.

Photograph kindly provided by Pierre Chambon.

The monomeric nucleosome contains ~200 bp of DNA associated with a histone octamer. This octamer consists of two copies each of H2A, H2B, H3, and H4. Sometimes these are known as the **core histones.**

Their association is illustrated diagrammatically in **Figure 29.3.**

This model explains the stoichiometry of the core histones in chromatin, in which H2A, H2B, H3, and H4 are present in equimolar amounts, with 1 molecule of each per ~100 bp of DNA. The direct experimental evidence for their association into an octamer consists of cross-linking data, in which the four core histones can be cross-linked preferentially to form a series of bands that extend up to the size of an octamer.

Associated with each nucleosome is a single molecule of H1. It is present in chromatin overall with a stoichiometry of half that of the other histones. Usually it is present in (up to) molar amounts when multimeric nucleosomes are isolated, but the amount is reduced in preparations of monomers. All of the H1 can be removed without affecting the structure of the nucleosome, which shows that it has an external location.

Including the H1, the combined mass of the protein and DNA should be about 262,000 daltons per nucleosome, with roughly equal amounts of DNA and protein. This is within the range of experimentally determined values. Any additional protein content probably represents small amounts of nonhistone proteins that are associated with the nucleosomes.

Figure 29.3
The nucleosome consists of approximately equal masses of DNA and histones (if HI is included).

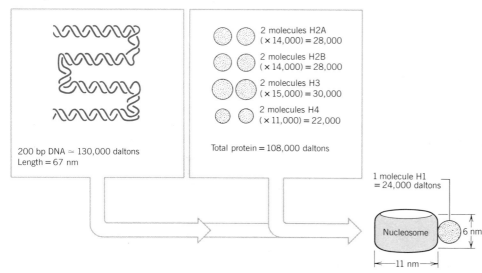

200 bp DNA ≃ 130,000 daltons
Length = 67 nm

2 molecules H2A
(× 14,000) = 28,000

2 molecules H2B
(× 14,000) = 28,000

2 molecules H3
(× 15,000) = 30,000

2 molecules H4
(× 11,000) = 22,000

Total protein = 108,000 daltons

1 molecule H1
= 24,000 daltons

Nucleosome
6 nm
11 nm

Histone H1 may lie in the region joining adjacent nucleosomes. This would be consistent with old results that H1 is removed the most readily from chromatin, and that H1-depleted chromatin is more readily "solubilized"; and also with more recent results that show it is easier to obtain a stretched-out fiber of beads on a string when the H1 has been removed (see later).

THE NUCLEOSOME IS THE BASIC SUBUNIT OF ALL CHROMATIN

A simple diagnostic test for the presence of nucleosomes derives from the observation that originally indicated that DNA is organized into a regularly repeating structure. When isolated nuclei or chromatin are digested with the enzyme micrococcal nuclease (or in some cases simply isolated to allow the action of an endogenous enzyme), the DNA is cleaved into integral multiples of a unit length. Upon fractionation by gel electrophoresis, this forms the "ladder" presented in **Figure 29.4.** Such ladders extend for 10 or

Figure 29.4
Micrococcal nuclease digests chromatin in nuclei into a multimeric series of DNA bands that can be separated by gel electrophoresis.

Photograph kindly provided by Markus Noll.

605 →

405 →

205 →

more steps, and the unit length, determined by the increments between successive steps, is ~200 bp.

This observation can be correlated with the properties of individual nucleosomes, as shown in **Figure 29.5.** The analysis of isolated particles is given in the upper part of the figure. The preparation of nucleosomes was fractionated by sedimentation on a sucrose gradient to give monomers, dimers, trimers, etc. Then the DNA was purified from each of these fractions and analyzed by gel electrophoresis.

The electrophoresis of the DNA is shown in the lower part of the figure. Each fraction yields a band of DNA of a particular size. Each of these bands corresponds with the appropriate step on the ladder produced by digestion of chromatin. (The ladder is given for comparison in the rightmost panel.) The monomeric nucleosome contains DNA of the unit length, the nucleosome dimer contains DNA of twice the unit length, and so on.

So each step on the ladder represents the DNA derived from a discrete number of nucleosomes. Thus we can take the existence of the 200 bp ladder in any chromatin to indicate that the DNA is organized into nucleosomes. The micrococcal ladder is generated when only about 2% of the DNA in the nucleus is rendered acid-soluble (degraded to small fragments) by the enzyme. Thus a small proportion of the DNA is specifically attacked; this must represent especially susceptible regions.

When chromatin is spilled out of nuclei, or when nucleosomes are reconstituted *in vitro* (see later), we often see a series of nucleosomes connected by a thread of free DNA (the beads on a string). Is this the natural situation, with nucleosomes separated by free DNA, or is it an artifact of conditions *in vitro?* The need for tight packaging of DNA *in vivo* suggests *a priori* that probably there is usually little (if any) free DNA. This is confirmed by the fact that more than 90% of the DNA of chromatin can be recovered in the form of the 200 bp ladder. DNA must therefore be organized in, rather than between, nucleosomes. In their natural state, nucleosomes are likely to be closely packed, with DNA passing directly from one to the next. Free DNA is probably generated by the loss of some histone octamers during isolation.

The length of DNA present in the nucleosome may vary somewhat from the value of 200 bp characterized

Figure 29.5
Each multimer of nucleosomes contains the appropriate number of unit lengths of DNA.

Photograph kindly provided by John Finch.

in the first experiments. When entire genomes are characterized from particular cells, each has a fairly well-defined average value (± 5 bp). The average most often is between 180 and 200, but there are extremes as low as 154 bp (in a fungus) or as high as 260 bp (in a sea urchin sperm). The value may change during embryonic development. In the case of the sea urchin, it is reduced to a more typical level during the early embryonic cell divisions. The average value may be different in the individual tissues of the adult organism. And there can be differences between different parts of the genome in a single cell type; for example, known cases of variation from the genome average include satellite DNAs and 5S RNA genes.

THE CORE PARTICLE IS HIGHLY CONSERVED

The histone octamer forms the protein component of the nucleosome (probably) in all eucaryotes. This constancy of function and structure explains the striking conservation of histone sequences. The extreme conservation of H3 and H4 implies that their role may be central and unchanging, while H2A and H2B provide species-specific variation.

All nucleosomes consist of a histone octamer associated with a particular length of DNA. What is responsible for the variation in the length of the DNA in nucleosomes from different sources? It is easiest to answer by defining what is *not* responsible. The variation does not represent a change in the association of DNA with the histone octamer. This always forms a **core particle** containing 146 bp of DNA, irrespective of the total length of DNA in the nucleosome. So the variation in length of DNA present on the nucleosome is superimposed on this basic core structure.

The core particle was identified by the effects of micrococcal nuclease on the nucleosome monomer. The initial reaction of the enzyme is to cut between nucleosomes, but if it is allowed to continue after monomers have been generated, then it proceeds to digest some of the DNA of the individual nucleosome. This occurs by a reaction in which DNA is "trimmed" from the ends of the nucleosome.

The length of the DNA is reduced in discrete steps, as shown in **Figure 29.6.** With rat liver nuclei, for ex-

Figure 29.6
Micrococcal nuclease reduces the length of DNA of nucleosome monomers in discrete steps.

Photograph kindly provided by Roger Kornberg.

ample, first of all the nucleosome monomers have 205 bp of DNA. Then some monomers are found in which the length of DNA has been reduced to 160–170 bp. Then this is reduced to the length of the DNA of the core particle, 146 bp. (This is reasonably stable, but if digestion is continued further, cuts can be made within the core DNA to generate a **limit digest,** in which the longest fragments are the 146 bp DNA of the core, while the shortest are as small as 20 bp.)

This analysis suggests that the nucleosomal DNA can be divided into two regions. The **core DNA** has an invariant length of 146 bp, and is relatively resistant to digestion by nucleases. (Other nucleases as well as the micrococcal enzyme also stop at this length.) The **linker DNA** comprises the rest of the repeating

unit. It can vary in length *in vivo* from as little as 8 bp to as much as 114 bp per nucleosome.

The well-defined nature of the band of DNA generated by the initial cleavage with micrococcal nuclease suggests that the region immediately available to the enzyme is restricted. It represents only part of each linker. (If the entire linker DNA were susceptible, the band would range from 146 bp to more than the repeating length.) But once a cut has been made in the linker DNA, the rest of this region becomes susceptible, and can be removed relatively rapidly by further enzyme action.

The core particles have properties similar to those of the nucleosomes themselves, although they are smaller. Their shape and size are similar to nucleosomes, which suggests that the essential geometry of the particle is established by the interactions between DNA and the protein octamer in the core particle. Because core particles are more readily obtained as a homogeneous population, they have been used for many of the structural studies in preference to nucleosome preparations. The nucleosome monomers tend to vary more because it is difficult to obtain a preparation in which there has been no end-trimming of the DNA.

What is the physical nature of the core and the linker regions? These terms are *operational* definitions that describe the regions relatively more and relatively less susceptible to nuclease treatment. This does not make any implication about their actual structure; in particular, it does not imply that the linker DNA has a more extended conformation. On the one hand, the path of DNA in the nucleosome could be continuous, with no distinction evident between these regions of the monomer. Indeed, this is a convenient working assumption often made in attempts to extrapolate from core particle structure to nucleosome structure. On the other hand, it is possible that the path of the linker DNA does differ from the path of the core particle DNA, especially given the pronounced variations that occur in its length.

The existence of linker DNA depends on factors other than the four core histones. Reconstitution experiments *in vitro* show that they have an intrinsic ability to organize DNA into core particles, but do not form nucleosomes with the unit length of DNA characteristic of the *in vivo* state. Histone H1 and/or nonhistone proteins must be involved in increasing the amount of DNA that is associated with the histone octamer in a natural series of nucleosomes. The primary responsibility for this may rest with "assembly proteins" that are involved *in vivo* in constructing nucleosomes from histones and DNA, but that are not part of the nucleosome structure (see later).

We have mentioned that H1 is lost during the degradation of monomeric nucleosomes. It can be retained on monomers that still have 160–170 bp of DNA; but it is always lost with the final reduction to the 146 bp core particle. This suggests that H1 could be located in the region of the linker DNA immediately adjacent to the core DNA.

DNA IS COILED AROUND THE HISTONE OCTAMER

Two types of data suggest independently that DNA lies on the surface of the nucleosome, following a path around the outside of the histone octamer. Biophysical studies suggest that the diameter of the protein component of the nucleosome is smaller than that of the DNA component. Biochemical studies show that the DNA is accessible to nucleases at rather regular intervals (see later).

The shape of the nucleosome is generally ellipsoidal. Its axial ratio of 0.5 corresponds to dimensions of about 11 × 11 × 6 nm. This is usually represented as a flat cylinder, diameter 11 nm and height 6 nm. Even just from the dimensions, it seems likely that the DNA lies on the exterior. The circumference of the particle is about 34 nm, compared with the length of DNA of 67 nm (= 200 bp). It is difficult to see how the DNA could be squashed into the interior of such a particle.

The properties of the individual components can be measured by neutron scattering, a technique that allows the scatter due to DNA to be distinguished from that due to protein. This shows that the protein component (the histone octamer) has a radius of gyration of about 3.2 nm. But the radius of gyration of the DNA component is about 5.2 nm. The difference corresponds to the 2 nm diameter of the DNA double helix.

Figure 29.7
The nucleosome may be a cylinder with up to two turns of DNA around the surface.

This suggests that the protein forms a compact body, around which the DNA is wrapped.

There is a variety of models for the path that DNA follows on the nucleosome. The common features are that the structure should be symmetrical and that the DNA is likely to pass around the octamer twice. **Figure 29.7** is a diagrammatic illustration of a path in which the DNA lies in a helical coil that makes two turns. One consequence of this particular path is that the DNA "enters" and "leaves" the nucleosome at points that are close to each other. A question that has not received much attention is how this path might be modified to account for the variation in length of DNA on the nucleosome.

Considering this model in terms of a cross section through the nucleosome, as depicted in **Figure 29.8,** we see that the two circumferences made by the DNA lie close to each other. This has a possible functional consequence. One turn around the nucleosome is about 80 bp of DNA, so that points this distance apart in the free double helix may actually be rather close to each other on the nucleosome. Thus if a DNA-binding pro-

Figure 29.8
The two turns of DNA on the nucleosome must lie quite close together.

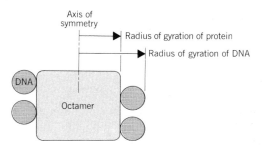

DNA binding sites
are close together
on nucleosome

〰〰 Rest of DNA

〰 Binding sites

⦿ DNA-binding
protein

Binding sites would be far apart on extended DNA duplex

Figure 29.9
A protein could contact sequences on the DNA that lie on different turns around the nucleo-some.

tein simultaneously contacted both turns of DNA as illustrated in **Figure 29.9,** the sequences it recognizes could lie farther apart in the duplex DNA than the apparent span of the polypeptide. One possibility that is often considered is whether the startpoint for transcription might be close in space to the −80 position, so that both are contacted simultaneously by RNA polymerase. (This would provide a particular example of the less detailed model illustrated in Figure 11.8.)

The packing ratio of the individual nucleosome is

~6 (since 67 nm of DNA is packaged into a particle of length about 11 nm). It is possible to pack a series of nucleosomes tightly enough to maintain this ratio. Much work on the structure of sets of nucleosomes has been carried out with the virus SV40.

The DNA of SV40 is a circular molecule of 5200 bp, that is, of contour length about 1500 nm. Both in the virion and in the infected nucleus, it is packaged into a series of nucleosomes. In this form, it is called the **minichromosome.** As usually isolated, the contour

Figure 29.10
The supercoils of the minichromosome can be relaxed to generate a circular structure, whose loss of histones then generates supercoils in the free DNA.

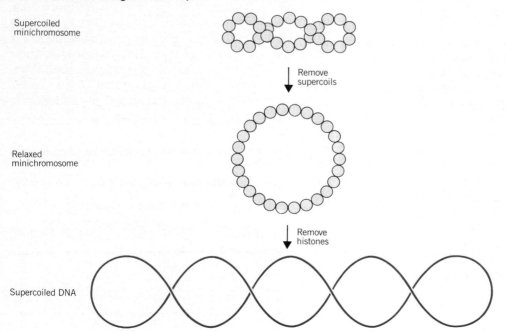

Supercoiled
minichromosome

Remove
supercoils

Relaxed
minichromosome

Remove
histones

Supercoiled DNA

length of the minichromosome is about 210 nm, corresponding to a packing ratio of ~7. Changes in the salt concentration can convert it to a flexible string of beads with a much lower overall packing ratio. This emphasizes the point that nucleosome strings can take more than one form *in vitro,* depending on the conditions.

An important parameter in describing the structure of chromatin is the degree of supercoiling, which can be created at several levels. First, there may be supercoiling as a result of the path that DNA follows on the nucleosome. Second, there may be supercoiling as a result of the path that the nucleosomes follow in higher-level structures. Much of this supercoiling may be restrained by proteins (according to the principle described in Chapter 28). Direct measurements of the supercoiling density therefore will show what average torsional tension applies to DNA as the result of free supercoils, but will not reveal any restrained supercoiling in the DNA path.

The degree of supercoiling on the nucleosome itself can be measured directly for the SV40 minichromosome. The minichromosome may have free supercoils in the string of nucleosomes, as well as restrained supercoiling on the nucleosome. The procedure to measure supercoiling due just to nucleosome structure is illustrated in **Figure 29.10.** First, the free supercoils of the minichromosome itself are relaxed, so that the nucleosomes form a circular string with a superhelical density of 0. Then the histone octamers are extracted. This releases the DNA to follow a free path. Thus every supercoil that was present but restrained in the minichromosome will appear in the deproteinized DNA as −1 turn. So now the total number of supercoils in the SV40 DNA is measured.

The value actually observed is close to the number of nucleosomes. The reverse result is seen when nucleosomes are assembled *in vitro* on to a supercoiled SV40 DNA: the formation of each nucleosome removes ~1 negative supercoil. Thus the DNA follows a path on the nucleosomal surface that generates ~1 negative supercoiled turn when the restraining protein is removed. We note that there is a potential discrepancy between the measured value of −1 turn per nucleosome and the model that shows DNA in a path equivalent to −2 superhelical turns (see later).

DNA IS SYMMETRICALLY EXPOSED TO NUCLEASES

The enzymes DNAase I and DNAase II make single-strand nicks in DNA. This means that they cleave a bond in one strand, but the other strand remains intact at this point. Thus no effect is visible in the double-stranded DNA. But upon denaturation, shorter fragments are released instead of full-length single strands. This is illustrated in **Figure 29.11.**

When DNA is free in solution, it is nicked at random. The DNA on nucleosomes also can be nicked by the enzymes, but only at certain points. Cutting occurs at regular intervals, so that when DNA is denatured and electrophoresed, a ladder is obtained. In this type of experiment, fragments of denatured DNA can represent the distance from the end of the nucleosome to the site of cutting, or can be generated by two internal

Figure 29.11
Nicks in double-stranded DNA are revealed by fragments when the DNA is denatured to give single strands.

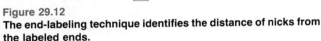

Figure 29.12
The end-labeling technique identifies the distance of nicks from the labeled ends.

Figure 29.13
Sites for nicking lie at regular intervals along core DNA, as seen in a DNAase I digest of nuclei.

Photograph kindly provided by Leonard Lutter.

nicks. The actual points of cutting can be determined by using radioactively end-labeled DNA, and identifying the fragments by autoradiography. *Only those fragments with a labeled end are detected*, as indicated in **Figure 29.12.** (This is exactly analogous to the sequencing technique shown in Figure 3.4 or to the footprinting technique of Figure 11.4.)

A ladder obtained by DNAase I treatment of nuclei is displayed in **Figure 29.13.** Similar results are obtained with chromatin. (Actually, core particles usually are used for such experiments *in vitro* because of the difficulty in obtaining nucleosome preparations with a homogeneous size distribution of DNA.) The interval between successive steps on the ladder is approximately 10 bases. The ladder extends for the full distance of core DNA, and the cleavage sites are numbered as S1 through S13 (where S1 is ~10 bases from the labeled 5′ end, S2 is ~20 bases from it, and so on). This result makes two important points. First, DNA is exposed on the nucleosome so that it is accessible to DNAase I; it is not "covered" by proteins. Second, there is a periodicity in its exposure.

Not all sites are cut with equal frequency: some are cut rather effectively, others are cut scarcely at all. The enzymes DNAase I and DNAase II generate the same ladder, although with some differences in the intensities of the bands. This shows that the pattern of cutting represents a unique series of targets in DNA,

Figure 29.14
Each nuclease cuts DNA with a characteristic separation between the sites on the two strands.

MN = micrococcal nuclease
I = DNAase I
II = DNAase II
• = Point of symmetry

determined by its organization, with only some slight preference for particular sites imposed by the individual enzyme.

Since there are two strands of DNA in the core particle, in an end-labeling experiment both 5' (or 3') ends are labeled, one on each strand. Thus the cutting pattern includes fragments derived from both strands. This is implied in Figure 29.12, where each of the fragments is derived from a different strand. The corollary is that, in an experiment, each labeled band in fact represents two fragments, generated by cutting the *same* distance from *either* of the labeled ends. How then should we interpret discrete preferences at particular sites? The usual view is that the two strands are treated symmetrically by the enzyme. Thus if (for example) no 80-base fragment is generated by DNAase I, this must mean that the position at 80 bases from the 5' end of *either* strand is not susceptible to the enzyme.

What is the nature of the target sites? The first point to resolve is the relationship between the sites of cutting on the two strands of DNA. This has been determined by using conditions in which double-strand cuts are produced by the enzyme. The ends of the released duplex material reveal the relationship between the cuts made on the two strands. This is summarized in **Figure 29.14.**

The enzyme DNAase I cuts the two strands at a distance of 2 bp apart. The enzyme DNAase II cuts them at a separation of 4 bp. Relative to the point of symmetry indicated in the figure, DNAase I cleaves one base away in either 3' direction, DNAase II cleaves two bases away in either 3' direction. Micrococcal nuclease makes a single cut across both strands, one base away in either 5' direction. Most likely the enzymes all recognize the same point, but actually cleave the DNA with an offset determined by their individual geometries.

The critical point established by these results is that a cutting site represents a short stretch of bonds on both strands, exposed to nuclease action over 3–4 base pairs. A similar point is made for the individual strand by a detailed examination on high-resolution gels of the sites cut by DNAase I or II in end-labeled core particles. An example is displayed in **Figure 29.15.** At each site, there are actually 3–4 positions at which cutting may occur; that is, the cutting site is defined ± 1 bp. The relative intensities indicate that some positions are preferred to others.

From this pattern, we can calculate the "average" point that is cut. This shows that pairs of sites from S1 to S4 lie apart a distance of 10.0 bases each, the separation from sites S4 to S10 is 10.7 bases, and for sites S10 to S13 the distance returns to 10.0 bases. (Because this analysis deals with *average* positions, sites need not lie an integral number of bases apart.)

Figure 29.15

High-resolution analysis shows that each site for DNAase I consists of several adjacent susceptible phosphodiester bonds as seen in this example of sites S4 and S5 analyzed in end-labeled core particles.

Photograph kindly provided by Leonard Lutter.

THE UNRESOLVED QUESTION OF THE PERIODICITY OF DNA

Experiments in which DNA on nucleosomes is analyzed for its sensitivity to nucleases are analogous to footprinting experiments. Thus we can assign the lack of reaction at particular target sites to the structure of the nucleosome, in which certain positions on DNA are rendered inaccessible. But what is the reason for the 10.0–10.7 periodicity of cutting?

This periodicity is close to the helical periodicity of B-form DNA. Does this mean that the recurrence of the exposed site simply reflects the immobilization of the DNA on the histone octamer? This could happen as shown in **Figure 29.16,** where a peak of exposed

sites occurs on each strand with a periodicity determined by the number of base pairs per turn of the double helix.

Some light on this question is cast by comparable studies with DNA immobilized on a flat surface (the situation envisaged in Figure 29.16). In this case, sites are cut evenly at a separation of 10.6 bases. This probably means that free DNA (that is, as obtained in solution) has a double-helical B-type structure in which the repeat length is 10.6 bp. Other measurements by independent techniques of the number of base pairs per turn have produced values close to this, notably 10.4.

We assume that when the DNA is immobilized on a flat surface, the **cutting periodicity** (the spacing between cleavage points) coincides with, indeed, is a reflection of, the **structural periodicity** (the number of base pairs per turn of the double helix). Because of the variation in cutting periodicity along the core DNA (10.0 at the ends, 10.7 in the middle), we are immediately faced with a problem in applying this interpretation.

One possibility is that the structural periodicity of core DNA actually varies, being close to its solution value in the middle, but being less tightly screwed at the ends. Another possibility is that the DNA has a constant structural periodicity, but that the geometry of the nucleosome causes a change in the average separation of cuts. If the approach of the enzyme is restricted at some positions, there could be a shift in its selection between the 3–4 available bonds. Then the cutting periodicity is no longer identical with the structural periodicity.

If this is true, what is the structural periodicity? In the absence of data, it is necessary to resort to model building to see which assumptions can be reconciled with the needs of the enzyme for access to DNA. Changing the structural periodicity, or fixing a periodicity of 10.6 bp, both fit less well than supposing there is a constant structural periodicity on the nucleosome of 10.0 bp per turn.

Some interesting implications follow from the possibility that the structural periodicity of DNA could be different on the nucleosome (10.0) from its value in solution (10.6). When DNA is released from the nucleosome, its structure would become more tightly wound, since there are more base pairs per turn. This

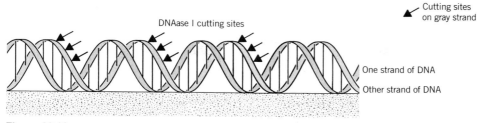

Figure 29.16
The most exposed positions on DNA recur with a periodicity that reflects the structure of the double helix (for clarity, sites are shown for only one strand).

change would reduce the apparent degree of supercoiling. Suppose that on the nucleosome the DNA follows a path equivalent to 2 superhelical turns. Then the histone octamer is removed. Some of the torsional strain goes into increasing the winding of DNA; only the rest is left to be measured as a supercoil. This is one possible way to reconcile models for -2 superhelical turns per nucleosome with data that identify only -1 superhelical turn, since the difference in periodicity (0.6 bp per turn of the helix) multiplied by the number of helical turns per nucleosome (>15) is roughly 10 bp, corresponding to 1 turn of the double helix, and thus potentially absorbing 1 negative superhelical turn.

One extremely important point remains to be made about the path of DNA and its possible degree of supercoiling. All of these data have been obtained with core particles. The models for reconciling discrepancies in the number of superhelical turns assume that linker DNA follows the same path as core DNA; but it could be different. An indication that the same pattern actually may extend beyond the core is given by some DNAase I ladders that extend for more than 200 bp. This implies that the same periodic exposure is maintained not just on the individual nucleosomes, but even between nucleosomes. In other words, DNA passes from one nucleosome to the next without any disruption in its path, at least as seen by the cutting periodicity.

THE ARRANGEMENT OF HISTONES AND DNA

So far we have considered the construction of the nucleosome from the perspective of how the DNA becomes organized into its repeating unit length on the nucleosome surface. From the perspective of protein, we need to know how the histones interact with each other and with DNA. Do histones react properly only in the presence of DNA, or do they possess an independent ability to form octamers?

We do not know very much about the structures of individual histones in the nucleosome, but we are beginning to deduce their relative locations. Most of the evidence about histone-histone interactions is provided by their abilities to form aggregates, and by cross-linking experiments with the nucleosome.

The abilities of the histones to aggregate with one another are quite well characterized. The arginine-rich histones can be obtained as a well-defined tetramer ($H3_2H4_2$). The slightly lysine-rich histones form a product that has been less clearly characterized, but that seems to be a dimer (H2A·H2B), with a tendency to aggregate further. One of the forms of aggregation could be the tetramer ($H2A_2H2B_2$). The two tetramers have been described as **homotypic** (the name reflects the content in each only of histones belonging to one of the original classes).

Intact histone octamers can be obtained either by extraction from chromatin or (with more difficulty) by association of the histones *in vitro* under conditions of high-salt and high-protein concentration. The octamer can dissociate to generate a hexamer of histones that has lost an H2A·H2B dimer. Then the other H2A·H2B dimer is lost separately, leaving the $H3_2H4_2$ tetramer. These forms all can be extracted from chromatin. This argues for a form of organization in which the nucleosome may have a central "kernel" consisting of the $H3_2H4_2$ tetramer, associated with two independent H2A·H2B dimers.

Cross-linking studies extend these relationships to show which pairs of histones lie near each other in the nucleosome. (A difficulty with such data is that usually only a small proportion of the histones become cross-linked, so it is necessary to be cautious in deciding whether the results typify the major interactions.) From these data, a model has been constructed for the organization of the nucleosome. This is shown in rather diagrammatic form in **Figure 29.17.**

Structural studies show that the overall shape of the isolated histone octamer is similar to that of the core particle. This suggests that it is the histone-histone interactions that establish the general structure. The positions of the individual histones have been assigned to regions of the octameric structure on the basis of their aggregation behavior and response to cross-linking.

The $H3_2H4_2$ tetramer accounts for the diameter of the octamer. This explains the ability of the tetramer to organize DNA in vitro into particles that display some of the properties of the core particle (see later). The proteins are arranged in the tetramer to account for the formation of the observed H3-H4 and H3-H4 cross-links. The H2A·H2B pairs fit in as two dimers. DNA is wound twice around the octamer. Histone H1 could "seal" the DNA in the nucleosome by binding at the point where it enters and leaves. This would explain the ability of H1 to be cross-linked to H3. The model displays twofold symmetry.

Although the overall shape of the octamer is now quite well defined (although not evident from the figure), the individual histones are shown as amorphous blobs, because we lack information on their shapes. The core histones all share a common pattern for the distribution of amino acids, with an N-terminal region that is highly charged; the rest of the molecule contains the hydrophobic amino acids that are likely to generate a globular structure and to be involved in protein-protein interactions. For this reason, histones sometimes have been visualized as globular proteins with charged N-terminal "tails." One would think that the tails are likely to possess the predominant DNA-binding activity, while the globular regions constitute the interior of the particle. But arguing against this model is the observation that the N-terminal regions can be cleaved from the histones of core particles (by treatment with trypsin) without greatly disrupting the

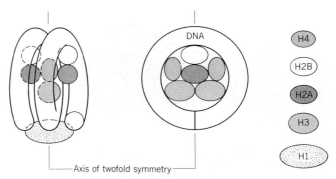

Figure 29.17
A model for the nucleosome as seen from the "front" or "top" is symmetrical, with the $H3_2H4_2$ tetramer providing a kernel for the shape.

structure. And the N-tailless histones can participate in nucleosome assembly in vitro. At present, we are left without any assignment of individual functions to particular regions of the histone molecules.

NUCLEOSOME ASSEMBLY VERSUS CHROMATIN REPRODUCTION

Core histones possess an inherent ability to wrap DNA around the protein octamer. But there are several routes by which DNA might associate with the histones.

The reconstitution reaction at first could be performed in vitro only under strikingly nonphysiological conditions, by mixing the components in high salt and urea, which were then dialyzed away. But now the reaction can be partially controlled by adjusting the order of addition and the proportions of histones and DNA. This allows self-assembly to occur under conditions that approach the physiological. It is a slow process, still limited by the tendency of the assembling particles to precipitate. Two possibilities for de novo nucleosome assembly pathways are summarized in **Figure 29.18.**

One pathway is revealed by the ability of the $H3_2H4_2$ tetramer to organize DNA into particles that partially resemble cores (as judged by their susceptibility to micrococcal nuclease). These bodies can be converted into core particles by the addition of H2A·H2B. This was the origin of the idea that there is a kernel of arginine-rich histones in the nucleosome structure. The

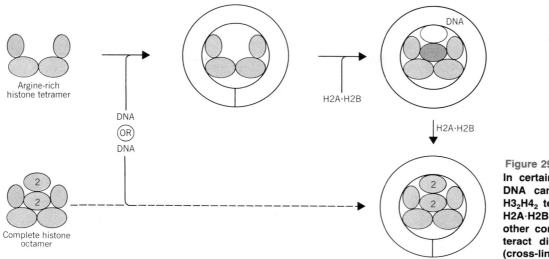

Figure 29.18

In certain conditions *in vitro*, DNA can assemble with the $H3_2H4_2$ tetramer, to which two H2A·H2B dimers are added. In other conditions, DNA can interact directly with an intact (cross-linked) histone octamer.

possible use of this pathway in assembly *in vivo* is consistent with observations that H3 and H4 are incorporated into replicating chromatin before H2A and H2B enter the product.

Another result makes use of cross-linked histone octamers that cannot dissociate into individual proteins. Still it is possible to bind DNA to form a particle. This shows that in principle it is possible for DNA to be wrapped around a previously intact octamer.

A further idea that has been considered is that, because of their symmetrical structure, nucleosomes might be able to separate into "half-nucleosomes." A possibly relevant observation is the ability of SV40 minichromosomes to change their structure at low salt concentration, generating twice the number of beads, each now of reduced size compared with the nucleosome. Could these be half-nucleosomes, each consisting of a **heterotypic tetramer** (H2A·H2B·H3·H4)? Could such an intermediate be employed when chromatin replicates, assembling with another set of four histones to reform the full nucleosome, as indicated speculatively in **Figure 29.19?**

Attempts to produce nucleosomes *in vitro* essentially consider the process of assembly as one involving free DNA and histones. But what actually happens *in vivo* is that *chromatin* is reproduced. A stretch of DNA *already associated with nucleosomes* is replicated, giving rise to two daughter duplexes. What

Figure 29.19

Can nucleosomes separate into half-nucleosomes (heterotypic tetramers) that can be converted into full nucleosomes by addition of a second heterotypic tetramer of histones?

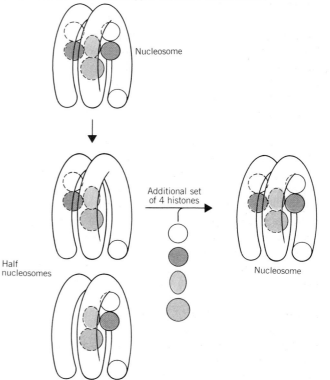

Nucleosome

Half nucleosomes

Additional set of 4 histones

Nucleosome

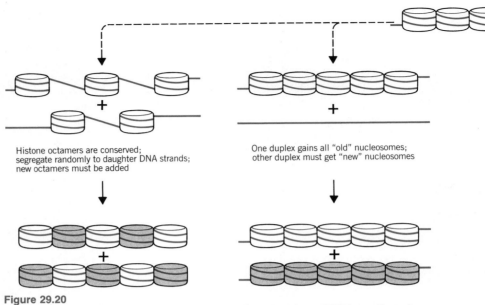

Figure 29.20
Many possibilities are open for the fate of nucleosomes when DNA is replicated.

happens to the preexisting nucleosomes at this point? Are the histone octamers dissociated into free histones for reuse, or do they remain assembled? Some possibilities are illustrated in **Figure 29.20.**

Cross-linking experiments suggest the possibility that the histone octamer might be **conserved,** surviving as such through cycles of replication. In this case, do the "old" octamers associate in any particular pattern with the duplicate DNAs? For example, all the old octamers might stay with one daughter duplex, while the "new" octamers assemble on the other. Or are octamers distributed at random to the daughter duplexes? Experiments on this question are not decisive, because of the difficulties of distinguishing the newly replicated material from the bulk of preexisting chromatin.

It is clear that the process of reproducing chromatin does not involve any protracted period during which the DNA is free of nucleosomes. Once DNA has been replicated, nucleosomes are quickly generated on both the duplicates, whatever the exact mechanism involved. This point is illustrated by the electron micrograph of **Figure 29.21,** which shows a recently replicated stretch of DNA, already covered with nucleosomes on both daughter duplex segments.

The structure of the point at which DNA is actually

being replicated does appear to differ from other regions. It is more resistant to micrococcal nuclease and is digested into bands that differ in size from nucleosomal DNA. This suggests that quite a large protein complex is engaged in replicating the DNA, but the

Figure 29.21
Replicated regions of chromatin contain nucleosomes on both daughter DNA duplexes.
Photograph kindly provided by Steven L. McKnight.

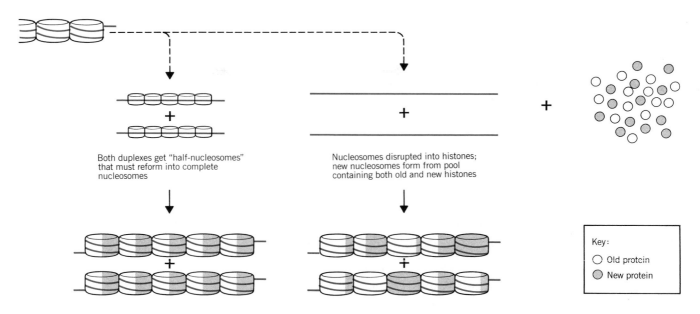

Both duplexes get "half-nucleosomes" that must reform into complete nucleosomes

Nucleosomes disrupted into histones; new nucleosomes form from pool containing both old and new histones

Key:
○ Old protein
◐ New protein

nucleosomes reform more or less immediately behind as it moves along.

Chromatin contains nonhistone proteins as well as the nucleosomal histones, and these additional components also must be reproduced. Since the nonhistone complement is likely to vary with the phenotype of the cell, its reproduction involves the maintenance of cell-specific features. Thus the possible existence of segregation patterns for proteins during DNA replication has a significance extending beyond nucleosome assembly. One of the principal questions that we should like to answer is how alternate states of chromatin structure are perpetuated through cell division.

Consider a gene that is activated (or repressed) by the binding to DNA of some specific regulator protein and/or by some change in chromatin structure. How is this particular state to be inherited by the duplicate chromosomes of the daughter cells produced by a division? If all the proteins dissociate from DNA during replication, the specific state virtually must be re-established de novo in every cell cycle. However, a definite pattern of segregation could be used to propagate information about the state of gene expression. One possibility is that a specific structure might be

perpetuated by segregation and duplication during DNA replication—for example, following a pattern formally equivalent to the half-nucleosomal segregation shown in Figure 29.20 (irrespective of whether the histones themselves follow this pathway). Thus a complex of nonhistone proteins might be established on the DNA, split into half-complexes at replication, and rebuild complete complexes on each daughter duplex.

The other side of this issue is how a *change* is made in the pattern of gene expression. One possibility often discussed is that changes might be connected with DNA replication, taking advantage of the transient disruption of chromatin structure. Once made, a change could be inherited via a segregation mechanism, unless and until another change is made (see also Chapter 30).

NUCLEOSOME ASSEMBLY REQUIRES NONHISTONE PROTEINS

Procedures for reconstituting histones with DNA generate a structure more akin to the core particle than to the complete nucleosome. At low concentrations, when individual reconstituted particles are connected by a thread of DNA, more than 146 bp are associated with the protein octamer. But this is not a firm characteristic, since as soon as the concentration is increased so that the particles become more tightly packed, each octamer is found to organize only the core length of 146 bp of DNA. It is as though, in the absence of an adjacent core particle, a histone octamer will opportunistically mobilize more DNA, but gives up all but the core length as soon as an immediate neighbor is formed.

What is responsible for modulating this reaction *in vivo* so that nucleosomes are formed with the spacing that is characteristic of their genome or even of their particular location within it?

The only system in which an authentic spacing of reconstituted nucleosomes has been obtained is provided by the *Xenopus* oocyte. When SV40 DNA is injected into the oocyte, the circular molecules can be assembled into minichromosomes. When a sufficient excess of DNA is used, the pool of endogenous histones becomes exhausted, and assembly depends on the injection of additional histones. The features of the system are retained in a cell-free extract, which assembles nucleosomes onto free DNA with a spacing of 195 bp. This is important information because it shows that proper assembly of nucleosomes can occur de novo with free DNA. Assembly is not necessarily associated with the act of replication; and it is independent of the sequence of the added DNA.

These features have not yet been reproduced entirely in a fractionated system, but some progress has been made toward identifying the components involved. An assembly protein has been purified from the *Xenopus* oocyte. It is a pentamer containing identical subunits of 29,000 daltons. This protein is the most abundant in the oocyte, and is located in the nucleoplasm. An antibody prepared against it reacts with proteins in the nucleoplasms of a wide range of eucaryotes, which suggests that it could represent a universal function that has been conserved in evolution. It is called **nucleoplasmin.**

In the presence of nucleoplasmin, the histones can associate with DNA to generate particles under physiological (low-salt) conditions. When digested with micrococcal nuclease, these particles generate bands of DNA at 146 bp and at 165 bp. This suggests that core particles are formed, some with an additional length of DNA, but not yet extending to the construction of the full nucleosome. Nucleoplasmin is therefore involved in controlling the reaction of histones with DNA in such a way that it becomes productive, and forms nucleosomes instead of random aggregates. However, it is not by itself sufficient to establish the spacing that is characteristic of the nucleosomes.

What is the function of the nucleoplasmin? It is an acidic protein that does not bind to free DNA or to intact nucleosomes, but it does bind to all of the individual histones. The reaction saturates at a level equivalent to one pentamer of nucleoplasmin per octamer of histones. Nucleoplasmin may be a "molecular chaperone," binding to the histones and releasing them to DNA in a more controlled manner than is possible when the DNA has no competitor for their attention. This idea is supported by the observation that the acidic polyglutamic acid can act similarly as an assembly factor, as also can RNA.

The feature common to all of these factors is their ability to bind to histones to reduce the net positive charge. The use of high salt concentration for histone octamer assembly *in vitro* may mimic this situation. In this context, we should also mention an early idea that modification of charged groups on histones might be involved in regulating the affinity of the protein for DNA (see Chapter 30). The consequence of such interactions may be to allow histones to form the thermodynamically most stable aggregate without becoming trapped in a kinetic intermediate.

This implies that there must be at least two features of nucleosome assembly that are extrinsic to the components of the nucleosome. The first is concerned with controlling the affinity of the histones for DNA; this is the role of nucleoplasmin. The second is involved in establishing the length of DNA that is contained in the nucleosome; this remains to be identified.

THE PATH OF NUCLEOSOMES IN THE CHROMATIN FIBER

When chromatin is examined in the electron microscope, two types of fiber are seen: the 10-nm fiber

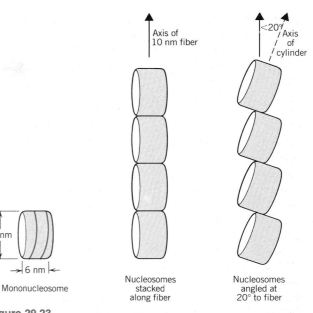

Figure 29.23
The 10-nm fiber consists of a series of nucleosomes organized edge to edge, stacked along the fiber or tilted less than 20° to the axis.

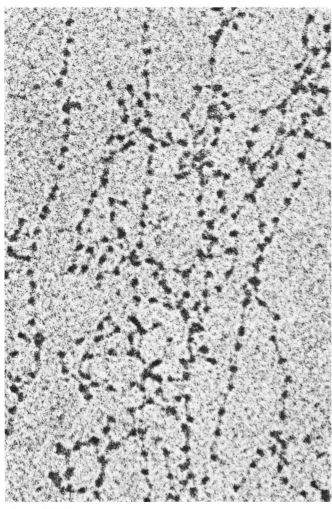

Figure 29.22
The 10-nm fiber in partially unwound state can be seen to consist of a string of nucleosomes.

Photograph kindly provided by Barbara Hamkalo.

and 30-nm fiber. They are described by the approximate diameter of the thread (that of the 30-nm fiber actually varies from about 25 nm to 30 nm).

The **10-nm fiber** is essentially a continuous string of nucleosomes. Sometimes, indeed, it runs continuously into a more stretched-out region in which nucleosomes are seen as a string of beads, as indicated in the example of **Figure 29.22.** The 10-nm fibril structure is obtained under conditions of low ionic strength and in the absence of histone H1. This means that it is a function strictly of the nucleosomes themselves.

How are the nucleosomes arranged in the 10-nm fiber? Viewing the particle itself as a somewhat flat cylinder, adjacent cylinders might be arranged either edge to edge or with their faces touching. These arrangements can be distinguished by biophysical techniques, such as neutron scattering or electric dichroism, that depend on the orientation of the individual subunit relative to the axis of the fiber. The results suggest the type of model illustrated in **Figure 29.23,** in which the cylinders are oriented edge to edge, with their faces parallel (or at least, not much inclined) to the axis of the fiber. The data imply that the maximum angle between the faces and the axis can be about 20°.

When chromatin is visualized in conditions of greater ionic strength, and H1 remains, the **30-nm fiber** is obtained. An example is given in **Figure 29.24.** The fiber can be seen to have an underlying coiled struc-

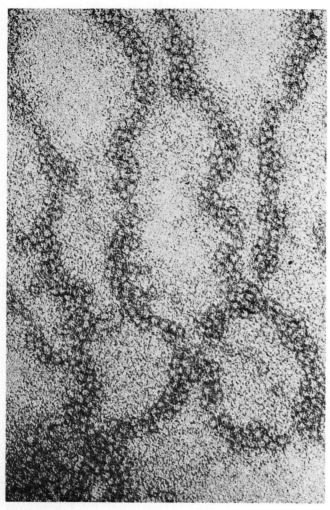

Figure 29.24
The 30-nm fiber has a coiled structure.

This fiber is shown at the same magnification as the 10-nm fiber of Figure 29.22. Photograph kindly provided by Barbara Hamkalo.

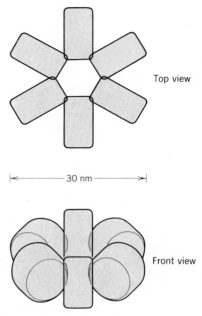

Top view

|← 30 nm →|

Front view

Figure 29.25
The 30-nm fiber may have a helical coil of 6 nucleosomes per turn, organized radially.

ture. It has about 6 nucleosomes for every turn, which corresponds to a packing ratio of 40 (that is, each μm along the axis of the fiber contains 40 μm of DNA). This fiber is the basic constituent of interphase chromatin and mitotic chromosomes (see, for example, Figure 28.7).

The 30-nm and 10-nm fibers can be reversibly converted by changing the ionic strength. This suggests that the linear array of nucleosomes in the 10-nm fiber is coiled into the 30-nm structure at higher ionic strength and in the presence of H1.

Again there are several possibilities for the packing of nucleosomes into the fiber. The most likely at the present is a radial model, illustrated in **Figure 29.25,** in which the nucleosomes turn in a helical array, with an angle of about 60° between the faces of adjacent nucleosomes.

It seems likely that the parameters of the 30-nm fiber are not rigidly fixed, but can vary. This would accommodate the variation in the length of DNA per nucleosome, as well as allowing for other changes in the density of packing. It is not certain whether the fiber has the identical structure in both interphase chromatin and mitotic chromosomes.

Although the presence of H1 is necessary for the formation of the 30-nm fiber, information about its location is conflicting. Its relative ease of extraction from chromatin seems to argue that it may be present on the outside of the superhelical fiber axis; but other data on its accessibility suggest it is harder to find in 30-nm fibers than in 10-nm fibers that retain it, which would argue for an inside location.

How do we get from the 30-nm fiber to the specific structures displayed in mitotic chromosomes? And is there any further specificity in the arrangement of interphase chromatin; do particular regions of 30-nm fibers bear a fixed relationship to one another or is their arrangement random? To such questions we have no answers at present.

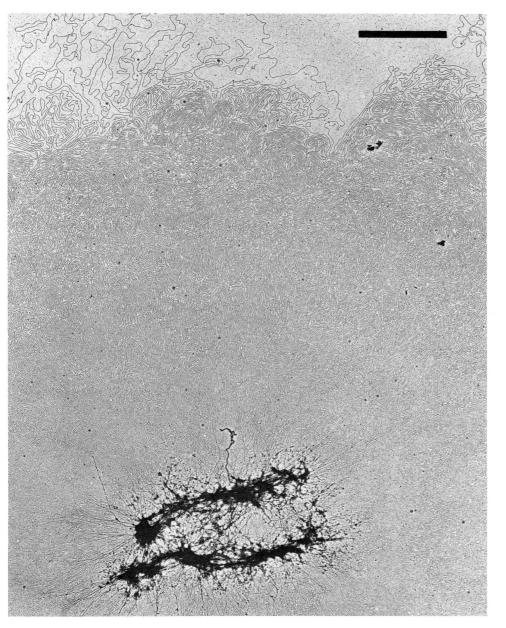

Figure 29.26

Histone-depleted chromosomes consist of a protein scaffold to which loops of DNA are anchored.

Photograph kindly provided by Ulrich K. Laemmli.

LOOPS, DOMAINS, AND SCAFFOLDS

Interphase chromatin appears to be a tangled mass occupying a large part of the nuclear volume, in contrast with the highly organized and reproducible ultrastructure of mitotic chromosomes. What controls the distribution of interphase chromatin within the nucleus?

Some indirect evidence on its nature is provided by the isolation of the genome as a single, compact body. Using the same technique described in Chapter 28 for isolating the bacterial nucleoid, nuclei can be lysed on top of a sucrose gradient. This releases the genome in a form that can be collected by centrifugation. As isolated from *D. melanogaster,* it can be visualized as a compactly folded 10-nm fiber consisting of DNA and the four core histones.

The supercoiling of the compact body can be measured by its response to ethidium bromide (see Chapter 28). The level corresponds to about one negative supercoil for every 200 bp. These supercoils can be removed by nicking with DNAase, although the nucleosomes remain present. This suggests that the supercoiling is caused by the arrangement of nucleosomes. It must represent torsion across the nucleosome junctions. Full relaxation of the supercoils requires one nick for every 85 kb. Thus the average length of "closed" DNA is ~85 kb. It could be a loop or domain similar in nature to those identified in the bacterial genome. We should like to know whether these loops correspond to specific sequences and whether they have functional significance.

Loops can be seen directly when the majority of proteins are extracted from mitotic chromosomes. The histones are removed by competition with the polyanions dextran sulfate and heparin. This also removes a large part of the nonhistone proteins. The resulting complex consists of the DNA associated with about 8% of the original protein content. As seen in **Figure 29.26,** the histone-depleted chromosomes take the form of a central **scaffold** surrounded by a halo of DNA.

The scaffold consists of a dense network of fibers. Threads of DNA emanate from the scaffold, apparently as loops of average length 10–30 μm (30–90 kb). If the loops were compacted 40-fold to fit into a 30-nm fiber, their average lengths would be in the range of 0.25–1.0 μm, not much greater than the diameter of the chromosome.

The loops can be visualized in another way. When divalent cations are removed from the chromosomes, cross sections show the loops in the form of radial arrays of the 10-nm fiber, average length 3–4 μm. This again is consistent with an organization in which loops of DNA of about 60 kb are anchored in a central proteinaceous scaffold.

The nonhistone proteins that constitute the scaffold may be metalloproteins. Chelation of metal ions disrupts the scaffold. It can be restored by the addition specifically of Cu^{2+} ions. Calcium ions also promote scaffold formation, although apparently less specifically.

The DNA can be digested without affecting the integrity of the scaffold. The scaffold looks like a mitotic pair of sister chromatids. The sister scaffolds usually are tightly connected, but sometimes are separate, joined only by a few fibers. Could this be the structure responsible for maintaining the shape of the mitotic chromosomes? Could it be generated by bringing together the protein components that usually secure the bases of the loops in interphase chromatin?

FURTHER READING

The development of the nucleosome can be traced from the somewhat flimsy evidence on which the model was originally propounded by **Kornberg** (*Science* **184,** 868–871, 1974) to the massive weight of evidence now assembled in reviews such as those of **Kornberg** (*Ann. Rev. Biochem.* **46,** 931–954, 1977), **McGhee** & **Felsenfeld** (*Ann. Rev. Biochem.* **49,** 1115–1156, 1980), and **Lewin** in *Gene Expression, 2, Eucaryotic Chromosomes* (Wiley, New York, 1980, pp. 332–393). The relationship between DNAase I digestion and nucleosomal DNA periodicity has been explored in the greatest bearable detail by **Klug** & **Lutter** (*Nuc. Acids Res.* **9,** 4267–4283, 1981). The path of DNA on the nucleosome has been succinctly analyzed by **Wang** (*Cell* **29,** 724–726, 1982). A model for the histone octamer was developed by **Klug et al.** (*Nature* **287,** 509–516, 1980), who also cite earlier structural analyses. The properties of nucleoplasmin were reported by **Earnshaw et al.** (*Cell* **21,** 373–383, 1980), who also reviewed nucleosome assembly (**Laskey** & **Earnshaw,** *Nature* **286,** 763–767, 1980).

CHAPTER 30
NUCLEOSOMES IN
ACTIVE CHROMATIN

The description of chromatin as a thread of duplex DNA coiled around a series of nucleosomes is the crucial first step toward visualizing the state of the genetic material in the nucleus. This somewhat static view accounts for the structure of the individual subunit and (to some degree) for its relationship with the next subunit. We have yet to consider whether the nucleosomes of this thread are deposited randomly or specifically with respect to the DNA sequence.

The organization of nucleosomes in reality must be *flexible* enough to satisfy the various structural and functional demands that are made on chromatin. For example, nucleosomes are present in both euchromatin and heterochromatin. Can we identify sets of nucleosomes whose different properties explain the structure or function of particular regions? And are nucleosomes the sole type of protein-DNA structure in the duplex thread, or are other structures present to delineate particular sites?

So far we have considered structural change in chromatin largely from the perspective of the need for euchromatin to become even more tightly packaged in mitotic chromosomes. This cyclical change affects all euchromatin more or less simultaneously, and most likely is controlled by changes in proteins that are widely distributed through chromatin. In the opposite direction of change, two types of local event seem likely to require some dispersion of structure because of their topological demands. These are replication and transcription.

During the interphase of a growing cell, all the chromatin must be reproduced. Replication occurs as a series of individual events in local regions (replicons), generating duplicate double-stranded DNA regions each associated with a set of histone octamers. The events involved in reproducing the nucleosome particle have yet to be defined (see Chapter 29), but it seems inevitable that the separation of parental DNA strands must disrupt the structure at least of the 30-nm fiber and probably also of the 10-nm fiber.

We should like to know the extent of this disruption. Is it confined to the immediate vicinity of the point where DNA is being synthesized, or does it extend farther? Are there discernible structural differences between regions that have replicated and those that have yet to do so? The transience of the replication event is a major difficulty in analyzing the structure of a particular region while it is being replicated.

Transcription also involves the unwinding of DNA, and presumably therefore requires unfolding of the fiber in restricted regions of chromatin. A simple-minded view suggests that some "elbow-room" must be needed for the process. The features of polytene and lampbrush chromosomes described in Chapter 28 offer hints that a more expansive structural organization may be associated with gene expression. We should like to know what structural changes occur when a gene is being transcribed. Does it remain in the form of nucleosomes, and if so, what happens to the nucleosome when RNA polymerase transcribes its DNA? Are there general changes in the structures of active transcription units extending beyond the sequence actually associated with RNA polymerase? What ensures that the promoter is initially accessible to the enzyme?

ARE NUCLEOSOMES ARRANGED IN PHASE?

We know that nucleosomes can be reconstituted *in vitro* without regard to DNA sequence, but this does not exclude the possibility that their formation *in vivo* is controlled in a sequence-dependent manner. Does a particular DNA sequence always lie in a certain position *in vivo* with regard to the topography of the nucleosome? Or are nucleosomes arranged randomly on DNA, so that a particular sequence may occur at any location, for example, in the core region in one copy of the genome and in the linker region in another?

To investigate this question, it is necessary to use a defined sequence of DNA; or, more precisely, we need to determine the position relative to the nucleosome of a defined point in the DNA. **Figure 30.1** illustrates the principle of a procedure that is used to achieve this.

Suppose that the DNA sequence is organized into nucleosomes in only one particular configuration, so that each site on the DNA always is located at a particular position on the nucleosome. This is called **nucleosome phasing.** In a series of phased nucleosomes, the linker regions of DNA comprise unique sites.

Consider the consequences for just a single nucleosome. Cleavage with micrococcal nuclease generates a monomeric fragment that constitutes a *specific sequence*. If the DNA is isolated and cleaved with a restriction enzyme that has only one target site in this fragment, it should be cut at a unique point. This produces two fragments, each of unique size.

The products of the micrococcal/restriction double

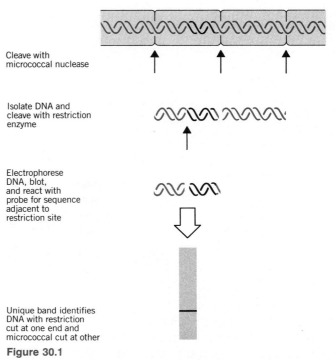

Cleave with micrococcal nuclease

Isolate DNA and cleave with restriction enzyme

Electrophorese DNA, blot, and react with probe for sequence adjacent to restriction site

Unique band identifies DNA with restriction cut at one end and micrococcal cut at other

Figure 30.1

Nucleosome phasing places restriction sites at unique positions relative to the linker sites cleaved by micrococcal nuclease.

digest are separated by gel electrophoresis. A probe representing the sequence immediately adjacent to (one side of) the restriction site is used to identify the corresponding fragment in the double digest. This technique is called **indirect end labeling** (not altogether an appropriate name).

Reversing the argument we have just made, the identification of a single sharp band demonstrates that the position of the restriction site is uniquely defined with respect to the end of the nucleosomal DNA (as defined by the micrococcal cut). Thus the nucleosome has a unique sequence of DNA.

What happens if the nucleosomes do *not* lie at a single position? The situation is illustrated in **Figure 30.2,** which shows just five possible nucleosomal locations with respect to DNA sequence; but actually there are as many as the number of base pairs in the repeating unit. Now the linkers consist of *different* DNA sequences in each copy of the genome. The restriction site lies at a different position each time; in fact,

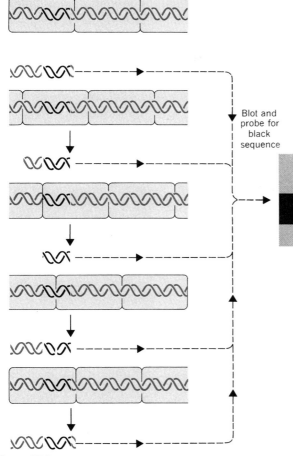

Figure 30.2

In the absence of nucleosome phasing, a restriction site lies at all possible locations in different copies of the genome.

The figure shows the result of making cleavages at the right end of the black sequence and at the junctions between nucleosomes.

it lies at all possible locations relative to the ends of the monomeric nucleosomal DNA.

When a monomeric fragment is generated by micrococcal cleavage, it contains a variety of sequences, and the restriction enzyme cleaves it into a continuous series of smaller fragments. If we take 20 bases to be the shortest length that can be identified in the hybridization protocol, the same probe as before now will reveal a broad smear ranging from 20 bases up to the length of the monomeric DNA.

With an actual sequence, it is a little more complex to analyze the results. With large genomes, it may be

impractical to identify a single unique fragment with sufficient precision. And we want to know whether nucleosomes are phased over some substantial region, not just over one particular short sequence. So attempts to investigate nucleosome phasing often have made use of tandemly repeated sequences. This approach effectively amplifies the fragment that is investigated: a single probe will detect whether the corresponding sequence is phased in every one of its genomic repeating units.

Thus probes have been used to investigate phasing on satellite DNA, tDNA, 5S DNA, and histone genes. In such cases, nucleosome phasing can occur only if there is a simple relationship between the nucleosomal repeat length and the repeating unit of the tandem array. This means that there must be an integral number of nucleosomes per tandem repeat.

In most cases, experimental data have not generated the clear type of result illustrated in Figure 30.1 for a unique disposition of nucleosomes; but sometimes they fall short of the wide bands or smears expected of random location. What is actually seen is a confinement of the micrococcal cutting sites to a small number of positions, relative to the defined restriction site. This suggests that nucleosomes may be limited so that they lie in only a few (2–4) alternative phases.

An intriguing situation has been noted in some tRNA genes of *X. laevis*. In erythrocyte chromatin, in which the genes are not expressed, they are organized into nucleosomes with a DNA repeat length of 198 bp. The nucleosomes lie mostly in a single phase over a distance of at least 3 kb (within which there is a stretch of 250 bp that may be nonnucleosomal). A different situation occurs in the chromatin of liver and kidney cells, in which the tRNA genes are expressed. The repeating length of the nucleosomes is altered to 185 bp. And now there is no phasing. This suggests that both the repeat length and the location of nucleosomes can be influenced by circumstance.

Nucleosome phasing might be accomplished in either of two ways. One is for *every* nucleosome to be deposited specifically at a particular DNA sequence. This is somewhat at odds with the view of the nucleosome as a subunit able to form between any sequence of DNA and a histone octamer. An alternative is for some particular sequence preferentially to assemble the first nucleosome in a region; then the adjacent nucleosomes may be assembled sequentially, with a defined repeat length. (If there is some variability in the construction of the nucleosomes—for example, if the length of the linker can vary by, say, 10 bp—the specificity of location would decline continuously, proceeding away from the first, defined nucleosome.) A starting point for nucleosome formation could be determined by binding of nonhistone proteins at a specific site of DNA.

Even accepting the existence of nucleosome phasing in some cases, its significance is not clear. One possible cause is that the apparent phasing results from the presence of a region from which nucleosomes happen to be excluded. Such regions might be created by complexes concerned with controlling gene expression or generating chromatin higher-order structure. The excluded region may provide a *boundary* that restricts the positions available to the adjacent nucleosomes. The statistical distribution of nucleosomes then may give the same results as phasing, although the individual positions are not necessarily fixed in any particular copy of the genome.

THE SPECIFICITY OF MICROCOCCAL NUCLEASE

In discussing these experiments, we have treated micrococcal nuclease as an enzyme that cleaves DNA at the exposed linker regions without any sort of sequence specificity. This indeed originally was thought to be the case. However, it has since turned out that the enzyme actually does have some sequence specificity (biased toward selection of certain A-T-rich sequences). So we cannot assume that the existence of a specific band in the indirect end-labeling technique represents the distance from a restriction cut to the linker region. It could instead represent the distance from the restriction cut to a preferred micrococcal nuclease site!

This possibility is dealt with by performing a control experiment in which the naked DNA is treated in exactly the same way as the chromatin. If there are preferred sites for micrococcal nuclease in the particular region, specific bands are found. Then this pattern of bands can be compared with the pattern generated from chromatin.

If there is a random arrangement of nucleosomes, all the preferred sites for micrococcal nuclease will at one time or another be available in a linker region. The pattern of cutting in chromatin and in DNA will be identical. But if the nucleosomes lie in a particular phase, some of the preferred micrococcal cleavage sites will be unavailable, because they lie in core regions. Some bands that are present in the DNA pattern will disappear from the chromatin pattern. Other bands may appear in the chromatin pattern if the distribution of nucleosomes makes new sites preferentially accessible. So a *difference* between the control DNA band pattern and the chromatin pattern provides evidence for nucleosome phasing. This is how experiments on nucleosome phasing actually have been performed.

These results also cast some light on the nature of the site cut by micrococcal nuclease between nucleosomes. Probably the enzyme does not cut exactly at the most exposed point in the linker DNA, but cuts the preferred site nearest to that position (if there is one). Since the relationship will vary, this increases the breadth of the bands. So the ability of preferred sites to seduce the enzyme may have increased the range of positions in linker DNA that appear to be most exposed to micrococcal nuclease.

ARE TRANSCRIBED GENES ORGANIZED IN NUCLEOSOMES?

Transcription involves the movement along DNA of a region of local unwinding that is generated by RNA polymerase (as described in Chapter 10). Usually both strands of DNA are envisaged as being intimately associated with, or even surrounded by, the enzyme, as imagined in Figure 10.3. About 50 bp or so of DNA are actually bound to the enzyme tightly enough to be protected against nucleases. The need to unwind the DNA makes it seem unlikely that the segment engaged by RNA polymerase could remain on the surface of the nucleosome. On the other hand, little is known about the stereochemistry of RNA polymerase action (especially in eucaryotes), so it is possible that the enzyme might be able to use one strand of DNA as template while the other strand remains bound to histones.

The mechanism by which the DNA unwinds to gen-

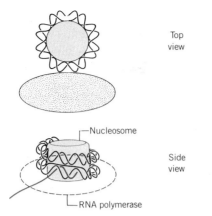

Figure 30.3

RNA polymerase is comparable in size to the nucleosome and might encounter difficulties in following the DNA around the nucleosome.

Figure 30.4

The extended axis of an rDNA transcription unit alternates with the only slightly less extended nontranscribed spacer.

Photograph kindly provided by Charles Laird.

erate a free template strand for RNA synthesis is not clear. Usually it is assumed to involve some sort of swivel that allows one strand of DNA to rotate relative to the other. The problem of unwinding would be exacerbated if DNA were immobilized on the nucleosome during transcription.

In considering these issues, we must bear in mind the relative sizes of RNA polymerase and the nucleosome. The eucaryotic enzymes are large proteins, typically >500,000 daltons. Compare this with the 264,000 daltons of the nucleosome. **Figure 30.3** illustrates the approach of RNA polymerase to nucleosomal DNA. Even without detailed knowledge of the interaction, it is evident that what is involved would be the approach of two comparable bodies. It seems inevitable that this must involve a structural change.

Nucleosomal DNA is relatively inaccessible to micrococcal nuclease; and even on isolated core particles, only some of the bonds in DNA are accessible to nicking by DNAase I (see Chapter 29). Remember that the nucleosome is not an isolated object but is adjacent to others, and consider the two turns that DNA makes around it. All of this highlights the question of whether RNA polymerase would have sufficient access to DNA if the nucleic acid were to follow its customary path on the nucleosome. It is hard to imagine that during transcription the polymerase could follow the DNA around the nucleosome.

So the first question to ask about the structure of active genes is whether DNA that is being transcribed remains organized in nucleosomes.

The genes coding for rRNA can be examined by electron microscopy. The structure of the DNA is hard to see, because of the intense packing of RNA polymerases. This is evident in Figure 23.4 and is shown in another example in **Figure 30.4.** The packing ratio of the DNA can be calculated by dividing the known length of the transcription unit by the measured length of the axis of the transcription matrix. This is about 1.2. Thus the DNA is almost completely extended; it cannot be organized in nucleosomes.

The nontranscribed spacers between the transcription matrices also appear to be quite extended, with a packing ratio of about 1.4. Under some conditions of preparation, the spacers contain beads that could be nucleosomes; but if this is what they are, there must be free spacer DNA connecting them. Under other conditions, the spacer appears to consist of free DNA. All of this argues that the state of DNA in the very intensively transcribed rRNA tandem genes is a far cry from the compact organization that would be seen even of a simple string of adjacent nucleosomes in a 10-nm fiber (which would have a packing ratio of ~6).

On the other hand, transcription complexes of SV40 minichromosomes can be extracted from infected cells. They contain the usual complement of histones and display a beaded structure. Chains of RNA can be seen to extend from the minichromosome, as in the example of **Figure 30.5.** This argues that transcription can proceed while the SV40 DNA is organized into nucleosomes. The SV40 minichromosome is transcribed less intensely than the rRNA genes. Less frequently transcribed cellular genes also can be visualized as partially extended structures containing fewer nucleosomes than nontranscribed regions.

Another approach is to digest chromatin with micrococcal nuclease, and then to use a probe to some specific gene or genes to determine whether the corresponding fragments are present in the usual 200 bp ladder. In such experiments, rRNA genes can be found in the nucleosomal fraction. However, it is always pos-

sible that these particular fragments are derived from some nontranscribed members of the tandem set.

A related problem is encountered with unique genes. With genes that are not under incessant transcription, there is the problem that only some copies of the gene in the cellular population are being transcribed at any time. To deal with this difficulty, the concentration of sequences that hybridize with the probe is compared in the monomeric fragments and in whole DNA. With probes that represent individual genes or populations of mRNA (or, for that matter, satellite DNA), the same result is obtained with either nucleosomal fragments or whole DNA.

The conclusions that we can draw from these experiments are limited but important. Genes that are being transcribed contain nucleosomes at the same frequency as nontranscribed sequences. Thus genes do not necessarily enter an alternative form of organization in order to be transcribed. But since the proportion of the gene associated with RNA polymerases may be rather small, this does not reveal what is happening at the sites actually engaged by the enzyme. It is possible that they retain their nucleosomes; it is possible that the nucleosomes are temporarily displaced as RNA polymerase passes by, but reform immediately afterward.

These experiments also show that active genes are more rapidly digested into monomeric fragments than are inactive sequences. This may mean that, although organized in nucleosomes, the structure of the active genes is in some way more exposed.

An indication that changes occur in the structure of the chromatin fiber during transcription is offered by the example of some heat-shock genes of *D. melanogaster.* These genes are transcribed rather infrequently prior to the heat shock, and in this condition they display the usual ladder when digested by micrococcal nuclease. Upon heat shock they are activated; and then they are intensely transcribed. **Figure 30.6** shows that this results in a considerable smearing of the ladder, implying that the nucleosomal organization has been changed.

The nature of the change is not known. Treatments such as shearing of chromatin can abolish the characteristic response to micrococcal nuclease, even though nucleosomes remain present. Thus we cannot tell whether the ladder disappears because it is ob-

Figure 30.5
An SV40 minichromosome can be transcribed.

Photograph kindly provided by Pierre Chambon.

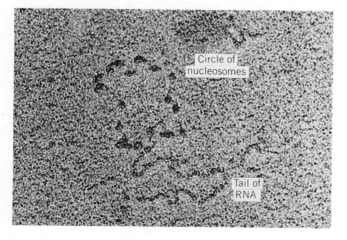

Figure 30.6

The micrococcal ladder is evident before activation of heat shock genes, but becomes smeared soon after intense transcription begins.

Photograph kindly provided by Sarah Elgin.

scured by the presence of RNA polymerase (or other proteins), because there is a disruption of higher-order structure, or because the nucleosomes have been modified or even displaced altogether.

It is clear that an important structural change occurs when a gene is intensely transcribed. In the case of the rRNA genes, it looks as though the nucleosomes are displaced. But this could be an exceptional case. It might be reconciled with the presence of nucleosomes in less heavily transcribed genes by supposing that RNA polymerase displaces the nucleosome at the point of transcription, but that the histone octamer immediately recaptures its position unless another RNA polymerase is present to prevent it from doing so.

THE DNAase-SENSITIVE DOMAINS OF TRANSCRIBABLE CHROMATIN

It is important to distinguish between two possible causes of change in the structure of transcribed regions of chromatin.

Since only a characteristic minority of the DNA sequences are transcribed in any particular cell, some feature must distinguish at least the promoters (and possibly the entire lengths) of these transcription units. The first change is therefore some event that indicates to the transcription apparatus that a locus is available for expression.

There must be some perturbation of structure in a gene that is being transcribed, if only as a result of the movement of RNA polymerase along the DNA. Thus structural changes could be a consequence of the act of transcription, rather than a cause of it. So in assessing the properties of transcribed loci, we need to determine which of their particular features occur prior to transcription as a prerequisite for it, and which are induced subsequently as a result of the events involved in the synthesis of RNA.

When chromatin is digested with DNAase I, it is eventually degraded into acid-soluble material (very

1. Digest chromatin with DNAase I

2. Extract DNA and cleave with restriction enzyme

3. Electrophorese fragments

4. Denature DNA and make Southern transfer

5. Hybridize with labeled probes for expressed and nonexpressed genes

Probe 1

Probe 2

6. Compare intensity of bands in preparations in which chromatin was digested with different concentrations of DNAase

Concentration of DNAase

7. Restriction fragment corresponding to probe 1 was preferentially digested, fragment responding to probe 2 was not preferentially digested

Figure 30.7

Sensitivity to DNAase I can be measured by determining the rate of disappearance of the material hybridizing with a particular probe.

small fragments of DNA). The progress of the overall reaction can be followed in terms of the proportion of DNA that is rendered acid soluble. The fate of individual genes can be followed by quantitating the amount of DNA that survives to react with a specific probe (using the usual techniques of restriction cleavage, blotting and hybridization). The type of protocol is outlined in **Figure 30.7.** The principle is that the loss of a particular band indicates that this region of DNA has been degraded by the enzyme.

When *only 10% of the total DNA* has become acid soluble, more than *50% of the DNA of an active gene* has been lost. This suggests that active genes are preferentially degraded. Their loss can be confirmed by comparing directly the fate of two genes, one active and one inactive. **Figure 30.8** shows what happens to β-globin genes and an ovalbumin gene in chromatin extracted from chicken red blood cells (in which globin genes are expressed and the ovalbumin gene is inactive).

The restriction fragments representing the β-globin genes are rapidly lost, while those representing the ovalbumin gene show little degradation. (The ovalbumin gene in fact is digested at the same rate as the bulk of DNA, although this is not evident from the blotting analysis at the early stage of the reaction shown in the figure.) The situation is reversed in oviduct tissue, in which the ovalbumin gene is preferentially degraded, while the globin genes are resistant like bulk chromatin. This correlates with the activity of ovalbumin and inactivity of globin.

So the bulk of chromatin is relatively resistant to DNAase I and contains nonexpressed genes (as well as other sequences). The ovalbumin and globin genes become relatively susceptible to the enzyme specifically in the tissues in which they are expressed.

Is preferential susceptibility a characteristic only of rather actively expressed genes, such as globin or ovalbumin, or of all active genes? Experiments using cDNA probes representing the entire cellular mRNA population suggest that all active genes, whether coding for abundant or for rare mRNAs, are preferentially susceptible to DNAase I. (However, it is possible that there may be differences in the degree of susceptibility.) Since the rarely expressed genes are likely to have very few RNA polymerase molecules actually engaged in transcription at any moment, this implies that the sensitivity to DNAase I does not result from the act of transcription, but is a feature of genes that are able to be transcribed.

How extensive is the region that is preferentially

Figure 30.8
In chromatin of 14-day erythroid cells, adult β-globin gene is highly sensitive to DNAase I, embryonic β-globin gene is sensitive, ovalbumin is not sensitive. In 5-day cells, embryonic gene is highly sensitive, adult gene is sensitive, ovalbumin is not sensitive.

Changes in the sensitivity of embryonic and adult β-globin genes are discussed later in the text. Data kindly provided by Harold Weintraub.

sensitive to DNAase I? This can be determined by using a series of probes representing the flanking regions as well as a transcription unit itself. The sensitive region always extends for a considerable distance on either side of the transcribed region. This has been taken to suggest that DNAase sensitivity defines a chromosomal **domain,** a region that includes at least one active transcription unit, but whose structural

change extends well beyond it. (Note that the use of the term "domain" does not imply any necessary connection with the structural domains identified by the loops of chromatin or chromosomes.)

There are some differences between the results of attempts to analyze whether the entire domain is uniformly sensitive to DNAase. In the β-globin gene cluster of the chicken, the sensitive domain extends for 6–7 kb on the 5′ side of the cluster and for at least 8 kb on the 3′ side. Within the gene cluster itself, regions that are being transcribed appear to be more sensitive to DNAase I than the flanking regions that are not actually expressed (and which are said to have "intermediate" sensitivity). Similarly, the ovalbumin gene and the related X and Y genes may lie in a DNAase-sensitive domain that extends for about 100 kb. But here there do not seem to be any differences in sensitivity between coding and noncoding DNA within a domain.

The critical concept implicit in the description of the domain is that a region of high sensitivity to DNAase I extends at least over the length of the transcription unit. Often we think of regulation as residing in events that occur at a discrete sequence of DNA—for example, in the ability to initiate transcription at the promoter. Even if this is true, such regulation must determine, or must be accompanied by, a more wide-ranging change in structure. This may be a difference between eucaryotes and procaryotes.

NONHISTONE PROTEINS CONFER
DNAase SENSITIVITY

What is responsible for the DNAase-sensitive state? Two nonhistone proteins, HMG14 and HMG17, are preferentially released from chromatin on digestion with DNAase I. (HMG is an abbreviation for high-mobility group, a set of small nonhistone proteins. These proteins are extracted from chromatin with 0.35 M NaCl. This means that they are relatively loosely bound.)

After the extraction of these proteins, a globin gene in red blood cell chromatin does not show preferential sensitivity to DNAase I. But sensitivity is restored when the proteins are added back to the salt-extracted chromatin. The proteins do *not* confer preferential susceptibility on globin genes when they are added to brain chromatin treated in the same way. This means that the HMG proteins are not solely responsible for the effect. Some other factor, resident in the NaCl-extracted chromatin, must be different in the globin genes of the red blood cell and brain. This difference allows only the red blood cell chromatin to bind the HMG14 and HMG17 proteins into a structure that makes the globin gene sensitive to DNAase.

Either HMG14 or HMG17 is effective by itself. About 1 mole of added HMG is sufficient to confer full sensitivity for every 10–20 nucleosomes extracted from bulk chromatin. This means that the HMG can pick out the appropriate target in a preparation of chromatin. The ratio is the same for genes that are transcribed actively, such as globin, or for genes that are transcribed only rarely.

Is DNAase sensitivity conferred at the level of the individual nucleosome? Monomers isolated by treatment with micrococcal nuclease show exactly the same sensitivity to DNAase I as whole nuclei or chromatin. The effect is lost when HMG14 and HMG17 are removed; it is regained when they are restored. The same result is obtained with *core particles*. This means that the core particles of active genes must have some particular feature that allows them to bind HMG14 and HMG17. Attempts to identify a distinctive property have not yet succeeded. Note the clear implication in this work that the globin gene remains in the form of nucleosomes when it is transcribed.

The intermediate level of DNAase sensitivity displayed by the regions flanking the transcribed chicken globin genes is lost upon treatments that tend to destroy higher-order structure in chromatin, and is not sustained in individual nucleosomes. It cannot be restored by addition of HMG14 and HMG17. When HMG14 and HMG17 are immobilized on columns, they can be used to isolate the salt-extracted nucleosomes capable of reaction. In such experiments, only the highly sensitive transcribed regions are recovered.

This suggests that the basis for the effect of DNAase sensitivity may lie in the creation of a highly sensitive region very closely surrounding the transcribable sequence. Apparently this extends less than 1 kb on either side. The adjacent regions could show intermediate sensitivity because of the spreading of a structural effect (although the distance of spreading is quite large relative to the length of the transcribed region itself).

What is the timing of the acquisition of the DNAase-sensitive state in relation to the occurrence of transcription? Although data are available with the chick globin system, their interpretation is not clear. Erythroid cells of 5-day-old embryos produce only embryonic globin, while erythroid cells of 14-day embryos produce the adult globins. But in the 5-day cells, the genes for embryonic and adult globins *both* are sensitive to DNAase I (though the embryonic may be more sensitive, as is evident in Figure 30.8). Similarly, in 14-day cells the adult gene is highly sensitive (more sensitive than it was in the 5-day cells), and the embryonic gene is of intermediate sensitivity.

At first it was thought that the DNAase-sensitive state might represent an activation of gene structure occurring before the onset of transcription. Sensitivity thus would be a hallmark of potentially transcribable genes in the cell lineage, as well as of genes actually being transcribed. On the other hand, if sensitivity spreads from the maximally sensitive region of the active gene to extend the domain into flanking material, the sensitivity of adult genes in embryonic cells (or vice versa) could be a feature of the clustering of these genes rather than having functional significance. To determine whether DNAase sensitivity appears before transcription starts, we need to investigate a gene that occurs in isolation, where there is no reason to expect a spreading effect from adjacent loci.

It is clear from two further situations that the sensitive state may persist after transcription has ceased. Chick erythrocytes are mature red blood cells that continue to translate globin mRNA, but do not transcribe the genes. However, the globin genes remain in the sensitive state. And withdrawal of estrogen from chickens brings transcription of the ovalbumin gene to a halt, but the coding region remains in the active state. We do not yet have data from a situation that reveals at what point a previously active gene returns to an inactive (DNAase-resistant) structure, for example, in the course of embryonic development.

The change in DNAase sensitivity evidently is necessary for transcription, but is neither a cause nor a consequence of it. The continuation of the sensitive state after transcription has ceased shows that the gene structure can be active without necessarily possessing active RNA polymerase molecules. This result, together with the sensitivity of rarely transcribed genes, suggests that the change in structure is independent of the act of transcription. DNAase sensitivity is therefore necessary, but not sufficient, for transcription. We should like to know more about the nature of the sensitive domain, in particular how the borders are defined. And we must ask why it is necessary to change the structure of every nucleosome in the transcription unit. A possible answer is that the change is needed to allow RNA polymerase to proceed along the template—for example, perhaps the histones can be displaced from the DNA only if the nucleosomes are modified.

HISTONES SUFFER TRANSIENT MODIFICATIONS

Each histone has an invariant sequence in the somatic cells of an adult organism. However, all of the histones are subject to modifications in which additional moieties are linked covalently to the free groups of certain amino acids. These modifications reduce the overall positive charge (that is, the basicity) of the protein molecule. This allows the unmodified and the various modified forms to be separated, for example, by gel electrophoresis.

Acetylation and methylation occur on the free (ϵ) amino group of lysine. As seen in **Figure 30.9,** this removes the positive charge that resides on the NH_3^+ form of the group. Methylation also occurs on arginine and on histidine.

Phosphorylation of histones occurs on the hydroxyl group of serine and also on histidine. This introduces a negative charge in the form of the phosphate group (see Figure 30.9).

All of these modifications affect internal residues and are transient. A contrast is provided by the stable acetylation of the N-termini of some histones that occurs at the time of synthesis.

The transient modifications occur at one point in the cell cycle and (usually) are reversed at another point. Because they reduce the positive charge of the protein molecule, they have been viewed as potentially able to change the functional properties of the histones. At present there is no evidence that these changes are related to chromatin functions, although there are some quite provocative correlations.

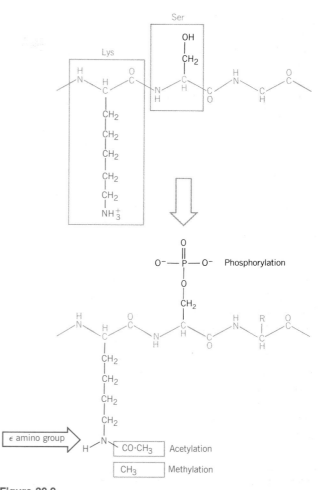

Figure 30.9
Acetylation of lysine or phosphorylation of serine reduces the overall positive charge of a protein.

A cycle of histone acetylation and deacetylation is well characterized in the trout testis (where the histones are displaced by other basic proteins called protamines). Here H3 and H4 are the principal targets, with up to four acetyl groups added to each of them; H2A and H2B are affected also, but to a lesser degree. The positions that can be acetylated (or methylated) have been identified in this and in some other situations. They are not identical in different organisms, although they do constitute only a small number of overlapping sites.

In synchronized cells in culture, both the preexisting and the newly synthesized core histones appear to be acetylated and methylated during S phase. (This is the part of the cell cycle during which DNA is replicated and when the histones also are synthesized.) During the cell cycle, the modifying groups are later removed. These events are repeated in each cell cycle, and their relative timing is summarized in **Figure 30.10.**

The coincidence of the modification with the reproduction of chromatin suggests that acetylation (and methylation) could be connected with nucleosome assembly. One speculation has been that the reduction of positive charges on histones might be useful in reducing their affinity for DNA. This could allow the reaction to be better controlled. The idea has lost some ground in view of the observation that nucleosomes can be reconstituted, at least *in vitro,* with unmodified histones (but see Chapter 29).

The transience of the acetylation event has been an obstacle to its study. This can be overcome by adding butyric acid to cells growing in culture. The treatment inhibits the enzyme histone deacetylase, so that acetylated nucleosomes accumulate. All the core histones are acetylated. Associated with the acetylation, there are changes in chromatin similar to those found on gene activation. The chromatin is more sensitive to DNAase I and (possibly) to micrococcal nuclease.

However, it has not been possible to demonstrate any decisive relationship; and there is no evidence for preferential acetylation of active genes. This result therefore tells us that acetylation can indeed affect the structure of chromatin, but the significance of the change remains to be seen.

A cycle of phosphorylation and dephosphorylation occurs with H1, but its timing is different from the modification cycle of the other histones. With cultured mammalian cells, one or two phosphate groups may be introduced at S phase. But the major phosphorylation event is the later addition of more groups, to bring the total number up to as many as six. This occurs at mitosis, as indicated in Figure 30.10. All the phosphate groups are removed at the end of the process of division. The introduction of some of the phosphate groups is catalyzed by a phosphokinase enzyme whose activity increases sharply at the very start of mitosis. Not much is known about the phosphatase that removes the groups later.

The timing of the major H1 phosphorylation has prompted speculation that it may be involved in mitotic condensation. Certainly this is consistent with the need for H1 to generate the 30-nm chromatin fiber (see Chapter 29). Some temperature-sensitive mutants that are defective in H1 phosphorylation are unable to complete DNA replication and cannot pass through

Figure 30.10
Core histone acetylation occurs during DNA synthesis and then is reversed; the major phosphorylation of HI occurs at the start of mitosis and is reversed at the end.

cell division. But this correlation does not reveal whether the phosphorylation is cause or effect of the mitotic condensation.

H2A IS CONJUGATED WITH UBIQUITIN ON A SUBSET OF NUCLEOSOMES

An extraordinary modification of a histone has been revealed by the characterization of a protein originally found in the nonhistone fraction of rat liver chromatin. Known at first as **A24,** this protein has the same C-terminal amino acid sequence as H2A. But it has two N-terminal sequences, as drawn in **Figure 30.11.** One is that of H2A. The other is that of the protein **ubiquitin,** whose name reflects its ubiquitous presence in cells from bacteria to mammals. The H2A-ubiquitin conjugate now is known as **UH2A.**

The ubiquitin protein itself is 76 residues long (compared with the ~130 of H2A). An isopeptide link is made between the C-terminus of a glycine in ubiquitin and the free ϵ–NH_2 of the lysine at position 119 of H2A. (It is called an *isopeptide* bond to indicate that the NH_2 group is not the usual amino group involved in peptide bond formation.) Ubiquitin is an acidic protein, whose content of glutamic and aspartic acids reduces the basic/acidic ratio of the conjugated protein.

Some 5–15% of the H2A may be in the form of UH2A. Usually only one of the two H2A molecules in a histone octamer carries ubiquitin, so that about 10–30% of the

nucleosomes may be ubiquitinated. The ubiquitin probably lies at the surface of the nucleosome. It does not have any discernible effect on nucleosome structure. A rather small proportion of the H2B also can be conjugated with ubiquitin.

Nucleosomes in which H2A is ubiquitinated can be separated from unmodified nucleosomes by two-dimensional electrophoresis. First the nucleosomes are electrophoresed on a single polyacrylamide strip; then they are electrophoresed again at right angles to this, through a slab gel. The relevant fractions are marked in **Figure 30.12.** There are nucleosomes containing about 170 bp of DNA and H1 (the MN2 fraction), core particles (the MN1 fraction), and core particles that contain either one or two UH2A conjugates.

Does the presence of UH2A identify a special class of nucleosomes? About 20% of the nucleosomes of chromatin of *D. melanogaster* have UH2A; and their DNA sequences can be examined by blotting the nucleosome fractions for hybridization with specific probes. With probes representing transcribed genes, about 50% of the material is in the ubiquitinated fraction. With a satellite DNA probe, less than 4% of the material is ubiquitinated. The UH2A therefore seems to be excluded from heterochromatin. (A nonhistone protein is found *only* in the nucleosomes of the *D. melanogaster* heterochromatin.) There is a tendency for the UH2A to be concentrated in the nucleosomes of transcribed sequences.

Very little is known about the function of the ubiquitin. It appears to be released from chromatin at mitosis. This is disconcerting, because it raises the question of how the ubiquitinated state of particular regions might be perpetuated through cell division.

A possible function has been reported for ubiquitin

Figure 30.11
UH2A consists of ubiquitin linked to H2A.

Figure 30.12
Nucleosomes, core particles, and ubiquitinated particles can be separated by two-dimensional gel electrophoresis.
Photograph kindly provided by Alex Varshavsky.

in the cytoplasm, where it is a participant in a system for protein degradation. One or more ubiquitin moieties are covalently linked to the "target" protein in a reaction that uses ATP. Then the target protein is degraded. So the ubiquitin provides a marker to identify the substrate for the degradation system. We do not know whether this is relevant to nuclear events.

GENE EXPRESSION IS ASSOCIATED WITH DEMETHYLATION

Between 2% and 7% of the C residues of animal cell DNA are methylated (the value varies with the species). Often satellite DNA is extensively methylated. The remaining methyl groups are present throughout the genome. Most of the methyl groups are found in CG "doublets," and, in fact, the majority of the CG sequences may be methylated. This short sequence is palindromic and usually the C residues on both strands are methylated, giving the structure

$$5' \quad {}^{m}CpG \quad 3'$$
$$3' \quad GpC^{m} \quad 5'$$

A doublet that instead is methylated on only one of the two strands is said to be **hemimethylated.**

The distribution of methyl groups can be examined by taking advantage of restriction enzymes that cleave target sites containing the CG doublet. Two types of restriction activity are compared in **Figure 30.13.** These **isoschizomers** are enzymes that cleave the same target sequence in DNA, but have a different response to its state of methylation.

In this pair, the enzyme HpaII cleaves the sequence CCGG (writing the sequence of only one strand of DNA). But if the second C is methylated, the enzyme can no longer recognize the site. However, the enzyme MspI cleaves the same target site *irrespective* of the state of methylation at this C. So MspI can be used to identify all the CCGG sequences; and HpaII can be used to determine whether or not they are methylated. Thus with a substrate of nonmethylated DNA, the two enzymes would generate the same restriction bands. But in DNA that is methylated at some sites, these positions are not cleaved by HpaII; and for every such position, one larger HpaII fragment replaces two MspI fragments. An example is given in **Figure 30.14.**

Is the pattern of methylation invariant or is it subject to alteration by circumstance? Individual sites have been analyzed in several cases, including cellular protein-coding genes, the tandem rDNA cluster, and sequences of several integrated or free viral genomes. Those sites identified by the use of restriction enzymes comprise only some of the methylated sequences, but we assume that their behavior is typical of the entire set.

A similar result is obtained in all these cases. Some sites are methylated in all tissues examined; some sites

Figure 30.13
The restriction enzyme MspI cleaves all CCGG sequences whether or not they are methylated, but HpaII cleaves only nonmethylated CCGG tetramers.

are unmethylated in all tissues. A *minority of sites* are methylated in tissues in which the gene is not expressed, but are not methylated in tissues in which the gene is active. Thus an active gene may be described as "undermethylated."

In the example of the human β-globin gene cluster, the correlation tends to be best in the embryonic genes. The majority (~15) of identified sites are unmethylated in expressed erythroid tissue, are partially methylated in adult erythroid tissue, and elsewhere are methylated. With the adult genes, there is only a partial reduction in methylation at some sites in the expressed tissue. ("Partial" methylation implies variability, with a given site methylated in some cells, but not methylated in others apparently of the same phenotype.)

In the rRNA genes of *X. laevis,* most of the CG sites remain methylated in all circumstances. Two sites (located near the promoter) become unmethylated at the time in embryonic development when transcription of rRNA starts.

As well as examining the state of methylation of resident genes, we can compare the results of introducing methylated or nonmethylated DNA into new host cells. First performed with viral genes, and since extended to several purified cellular genes, such experiments show a clear correlation: the methylated gene is inactive, but the nonmethylated gene is active.

The extent of the region undermethylated *in vivo* has been determined in the chicken α-globin gene cluster. In adult erythroid cells, the undermethylation is confined to sites that extend from 500 bp upstream of the

first of the two adult α genes to about 500 bp downstream of the second. Sites of undermethylation are present in the entire region, including the spacer between the genes. The region of undermethylation coincides rather well with the region of maximum sensitivity to DNAase I. This argues that undermethylation may be a feature of a domain that contains a transcribed gene or genes.

Our problem in interpreting the association between undermethylation and gene activation is that only a minority (sometimes a small minority) of the methylated sites are involved. One possibility is that the state of methylation is critical at specific sites; for example, demethylation at the promoter might be involved in making this region available for the initiation of transcription. Another possibility is that a reduction in the level of methylation is part of some structural change needed to permit transcription to proceed.

SOME MODELS FOR THE CONTROL OF METHYLATION

Much attention has been paid to how the state of methylation might be perpetuated or changed. Gamete DNA, as represented by sperm, displays the inactive state of each gene: it is methylated both at the constant sites (modified in all tissues) and at the variable-sites (those specifically unmethylated in expressed tissue). Thus the lack of certain methyl groups in the active state represents a *loss* of modifications that were previously present. We do not know whether cellular genes regain methyl groups if and when they cease to be expressed. A critical question we should like to answer is how sequences are selected as the targets for tissue-specific changes in the state of methylation.

A simple model for the perpetuation of methylated sites is to suppose that the DNA methylase acts on hemimethylated DNA. As can be seen from **Figure 30.15,** the replication of a fully methylated CG doublet produces two hemimethylated daughter duplexes. Recognizing each of the hemimethylated sites, the enzyme could convert it to the normal fully methylated state. Such a model accords with the observation that, when methylated DNA is introduced into a cell, it continues to be methylated through an indefinite number

Figure 30.14

The results of MspI and HpaII cleavage are compared by gel electrophoresis of the fragments.

The numbers of the bands refer to the fragments identified in Figure 30.13.

Figure 30.15

The state of methylated sites could be perpetuated by an enzyme that recognizes only hemi-methylated sites as substrates.

of replication cycles, with a fidelity approaching 95% per site. If nonmethylated DNA is introduced, it is not methylated de novo. This implies that the enzyme recognizes *only* the hemimethylated sites. Its action allows the condition of a $\frac{CG}{GC}$ doublet—methylated or nonmethylated—to be perpetuated.

If this model is correct, an entirely different enzyme activity must be involved in any creation of new sites of methylation (for example, if the methylated condition of a gene is restored if transcription should cease).

There are two ways that sites might lose the methylated condition. Methyl groups might be actively removed by a demethylase enzyme. Or methylation might simply fail to occur at a hemimethylated site generated by replication. One of the DNA duplexes produced by the next replication then would lack the methylated site.

Experiments with the drug 5-azacytidine produce indirect evidence that demethylation can result in gene expression. The drug is incorporated into DNA in place of cytidine, and cannot be methylated because the 5′ position is blocked. This leads to the appearance of demethylated sites in DNA. The phenotypic effects of 5-azacytidine include the induction of changes in the state of cellular differentiation; for example, muscle cells are induced to develop from nonmuscle cell precursors. The drug also activates genes on a silent X chromosome, which raises the possibility that inactivation of the chromosome could be due to its state of methylation.

Although at present we do not fully understand the effects of methylation, it is clear that the absence of methyl groups is associated with the gene expression. However, there are some difficulties in supposing that the state of methylation provides a general means for controlling gene expression. In the case of *D. melanogaster* (and other Dipteran insects), there is no methylation of DNA. The other differences between inactive and active chromatin appear to be the same as in species that display methylation. Thus in *Drosophila*, methylation either is superfluous or is replaced by some other mechanism.

DNAase HYPERSENSITIVE SITES LIE UPSTREAM FROM ACTIVE PROMOTERS

When chromatin is digested with a very low concentration of DNAase I, the first effect is the introduction

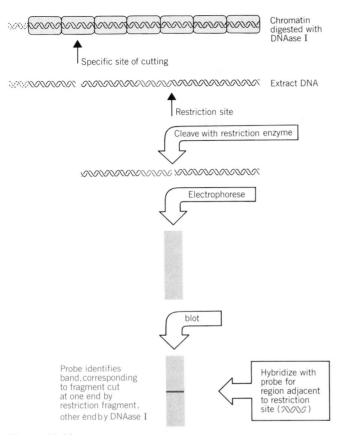

Specific site of cutting

Restriction site

Cleave with restriction enzyme

Electrophorese

blot

Probe identifies
band, corresponding
to fragment cut
at one end by
restriction fragment,
other end by DNAase I

Hybridize with
probe for
region adjacent
to restriction
site ( )

Chromatin
digested with
DNAase I

Extract DNA

Figure 30.16
Indirect end labeling identifies the distance of a DNAase hypersensitive site from a restriction cleavage site.

of breaks in the duplex at specific sites. For an entire genome, as in the example of *D. melanogaster,* the result is to generate a series of specific bands, ranging in size from 2 kb to 20 kb. These fragments represent the separation between the **hypersensitive sites.** The bands are not generated when DNA itself is digested; they result from the (tissue-specific) structure of chromatin.

Where are these hypersensitive sites located? Their positions can be determined for individual loci by the technique of indirect end labeling that we introduced earlier in the context of nucleosome phasing. The protocol is recapitulated in **Figure 30.16.** Chromatin is digested with DNAase I, the DNA is isolated, and then it is cleaved with a restriction enzyme. The material

produced by the double digest is electrophoresed and hybridized with a probe that represents a region adjacent to the restriction site. The existence of a particular cutting site for DNAase I generates a discrete fragment, identified as a band whose size indicates the distance of the DNAase I hypersensitive site from the restriction site.

Many of the hypersensitive sites are related to gene expression. Every active gene has a site of cutting, or sometimes more than one site, in the region immediately upstream from the promoter. Some examples of the locations of hypersensitive sites are summarized in **Figure 30.17.**

A 5′ cutting site is found *only* in chromatin of cells in which the gene is being expressed; it does not occur when the gene is inactive. In the globin genes, for example, hypersensitive sites are found upstream from embryonic genes in embryonic cells but not in adult cells, and vice versa. The sites may vary somewhat in sensitivity (as judged from the intensity of the corresponding band). When multiple sites lie close together, often some are appreciably more sensitive than others. In addition to these sites, there may be others located elsewhere, within a transcribed gene or downstream from it. The significance of the sites in these other regions is unclear.

The correlation between transcription and upstream hypersensitive sites is greatly strengthened by data on the *Sgs4* gene of *D. melanogaster.* This gene codes for the glue protein of salivary glands. In this tissue, the locus has two prominent hypersensitive sites, lying about 330 and 405 bp upstream from the 5′ end of the gene. A mutant strain of the fly has a deletion of about 100 bp that removes both the hypersensitive sites. In spite of the retention of the remaining region, which includes more than 250 bp of the sequence upstream from the startpoint, this mutant does not produce glue protein. This suggests that removal of the hypersensitive site prevents activation of the gene.

Although necessary, the presence of a hypersensitive site is not sufficient to ensure transcription. The timing of the appearance of the hypersensitive site relative to the onset of transcription is hard to establish in authentic situations, where it is difficult to obtain cells all at the same stage of development. However, this can be accomplished with certain chicken erythroid cells that have been immortalized in culture by

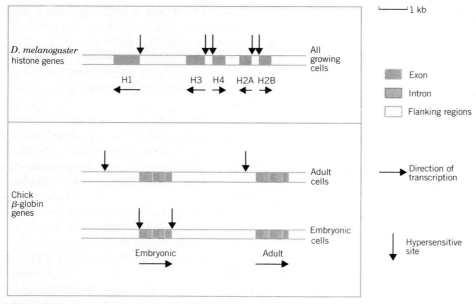

Figure 30.17

Hypersensitive sites lie near the 5' ends of active genes and (sometimes) also in other places.

transformation with an RNA tumor virus. Different cell lines represent different stages of erythroid development. One such line has the DNAase hypersensitive sites, but has not yet commenced transcription of the globin genes; it may have been captured at a stage between activation of the hypersensitive site and the start of gene expression.

The stability of hypersensitive sites is revealed by the properties of chick fibroblasts transformed with temperature-sensitive tumor viruses. These experiments take advantage of an unusual property: although fibroblasts do not belong to the erythroid lineage, transformation of the cells at the normal temperature leads to activation of the globin genes. The activated genes have hypersensitive sites. If transformation is performed at the higher (nonpermissive) temperature, the globin genes are not activated; and hypersensitive sites do not appear. When the globin genes have been activated by transformation at low temperature, they can be inactivated by raising the temperature. But the hypersensitive sites are retained through at least the next 20 cell doublings.

This result again demonstrates that acquisition of a

hypersensitive site is only one of the features necessary to initiate transcription; and it implies that the events involved in establishing a hypersensitive site are distinct from those concerned with perpetuating it. Once the site has been established, it is perpetuated through replication in the absence of the circumstances needed for induction. (Presumably a specific intervention therefore would be needed to abolish a hypersensitive site.)

How are hypersensitive (or other) sites propagated through DNA replication? A variety of mechanisms can be imagined. An inheritable change might be made in the DNA itself—for example, demethylation. The structure of the DNA might be different from the regular double helix (through the mediation of proteins allowing reformation after each replication). Highly cooperative proteins may form a complex (with duplex DNA) that is able to segregate at replication and then rebuild (see Figure 29.20).

We have described three changes that occur in active genes. A hypersensitive site(s) is established near the promoter. The nucleosomes of a domain including the transcribed region become more sensitive to

DNAase I. The DNA of the same region is under-methylated. All of these changes may be necessary for transcription.

NUCLEASE-SENSITIVE AND PROTECTED REGIONS

What is the nature of the hypersensitive sites? Each seems to represent a region of 100–200 bp to which access is not restricted in the manner typical of nucleosomes. Some of these sites have been investigated in detail.

The site at the 5' end of the chicken adult β-globin gene is preferentially digested by several enzymes, including DNAase I, DNAase II, and micrococcal nuclease. The map of **Figure 30.18** shows that these enzymes have preferred cleavage sites that lie at slightly different points in the same general region. The restriction enzyme MspI cleaves two sites in the region; a proportion of the fragments released by this cleavage take the form of free DNA. Thus a region extending from about -60 to -260 is preferentially accessible to nucleases when the gene is transcribable.

Similar experiments show that the 5' hypersensitive site of the thymidine kinase gene of herpes virus extends through the region from -4 to -182. Target sites in this region are preferentially cleaved by restriction enzymes.

The DNA of the hypersensitive sites of active chick globin genes is cleaved by S1 nuclease, the enzyme whose substrate is single-stranded DNA. This susceptibility suggests that the condition of the DNA duplex itself may be modified in a hypersensitive site.

A hypersensitive site therefore represents a structural change in chromatin. It is plausible to suppose that the site consists of sequences that do not form the usual nucleosomal structure, perhaps because other proteins are complexed with the DNA. If nucleosomes do intrude into the hypersensitive region, their structure must be different from usual.

A significant feature of the hypersensitive sites is that they appear able to exercise their function at varying distances from the startpoint for transcription. In the Sgs gene of D. melanogaster the site is more than 250 bp upstream, in the chicken β-globin gene it is just upstream, and in the thymidine kinase gene it includes all of the sequences necessary for promoter function. The hypersensitive sites could be involved directly in promoter function, most likely in providing upstream components needed for recognition by RNA polymerase (see Chapter 11). Alternatively, the sites may play a more general role in gene activation—for example, triggering a change in local structure that is necessary for RNA polymerase binding or even for synthetic activity.

A particularly well-characterized nuclease-sensitive region lies on the SV40 minichromosome. A short segment near the origin of replication, just upstream from the promoter for the late transcription unit, is cleaved preferentially by DNAase I, micrococcal nuclease, and other nucleases (including restriction enzymes). The stretch over which the preferential digestion occurs is about 400 bp long, and the segment can be released as a fragment of free DNA.

The state of the SV40 minichromosome can be visualized by electron microscopy. In up to 20% of the samples, a "gap" is visible in the nucleosomal organization. This is evident in **Figure 30.19.** The gap is a region of about 120 nm in length (about 350 bp), surrounded on either side by the nucleosomes that occupy the rest of the genome. The location of the gap can be determined by cleaving the circular minichromosome with a restriction enzyme that has a single

Figure 30.18
The hypersensitive site of a chicken β-globin gene is in an extended region that is susceptible to several nucleases.

Figure 30.19

The SV40 minichromosome may have a nucleosome-free gap.

Photograph kindly provided by Moshe Yaniv.

Figure 30.20

The SV40 gap includes hypersensitive sites, sensitive regions, and a protected region of DNA.

known target. The visible gap corresponds with the nuclease-sensitive region. This shows directly that increased sensitivity to nucleases is associated with the exclusion of nucleosomes. It does not imply that the DNA is free *in vivo,* because some other proteinaceous structure may have been lost in isolating the minichromosome.

A similar nuclease-sensitive gap is found in the related polyoma virus, and here it has been shown that the entire region is not uniformly sensitive to nucleases. Within a sensitive region of about 260 bp, there are two hypersensitive DNAase I sites and a "protected" region. The map is given in **Figure 30.20.** The protected region presumably reflects the association of protein with the DNA. Indeed, such an association might well be the cause of the exclusion of the nucleosomes from this region.

In a mutant of polyoma in which the sequences present in this region are substantially rearranged, *the same result is obtained.* This implies that the presence of the gap is not determined by the entire sequence of the sensitive region itself, but must be "measured" in some way from a fixed point that is either part of the region or is adjacent to it.

The region of SV40 or polyoma surrounding the nuclease-sensitive gap has several potential functions, and so the structure cannot simply be equated with activation of a promoter. The presence of the gap may

be associated with the sequences of the "enhancer" elements, which lie in this region and are necessary for promoter function (see Chapter 11).

Comparable structural alterations in chromatin may exist also in the region of the yeast centromere. In lieu of a visible kinetochore in yeast, it seems likely that the chromatin thread itself might be involved in centromeric function. The structure of the centromeric region has been investigated by using a short fragment of the DNA as a probe in the indirect end-labeling technique. The probe identifies two DNAase I hypersensitive sites, located on either side of regions I and III, two of the short highly conserved sequences in the centromeric DNA, whose sequences are summarized in Figure 28.11. The region of about 220–250 bp between the two hypersensitive sites is protected against nuclease digestion. Thus the role of the conserved region could be to bind specific centromeric proteins that form a nonnucleosomal structure.

Another effect is seen when the same experiment is performed using micrococcal nuclease to digest the chromatin. The same protected region is evident; and, in addition, a phased series of nucleosomes extends on either side. The phasing lasts for more than 12 nucleosomes, each containing 160 bp of DNA. This result contrasts with an apparent lack of nucleosome phasing elsewhere in the yeast genome.

Does the nucleosome phasing propagate from the centromeric sequence itself, or is it a property of the phased sequences themselves? In a plasmid that contains these sequences but lacks the centromeric region, the phasing is retained. This result suggests that the authentic centromeric region has two features. A conserved sequence establishes some nonnucleosomal structure. And the flanking sequences appear to possess an intrinsic ability to phase their nucleosomes.

The existence of hypersensitive sites may reflect a wider phenomenon. It may be possible for particular (unknown) proteins to modify the properties of a short region of DNA so that nucleosomes are excluded. The structures formed in each situation need not necessarily be similar (except that each, by definition, creates a site hypersensitive to DNAase I). We have seen that sites detectable by hypersensitivity are present upstream from transcribable genes, within the nuclease-sensitive gap in SV40 and polyoma, and at the yeast centromere; they may also be associated with transposition of yeast mating type genes (see Chapter 37). These various structures may serve a variety of purposes, and it will not be surprising if there are other such situations yet to be discovered.

FURTHER READING

The topics of this chapter were reviewed in **Lewin's** *Gene Expression, 2, Eucaryotic Chromosomes* (Wiley, New York, 1980, pp. 394–427). Functional aspects of chromatin were brought up to date by **Igo-Kemenes, Horz,** & **Zachau** (*Ann. Rev. Biochem.* **51,** 89–121, 1982). The existence of hypersensitive sites was summarized briefly by **Elgin** (*Cell* **27,** 413–415, 1981). Their propagation has been followed by **Groudine** & **Weintraub** (*Cell* **30,** 131–139, 1982). A series of experiments to distinguish the changes in transcribable genes culminated in the reports of **Weintraub, Larsen** & **Groudine** (*Cell* **24,** 333–344, 1981) and **Weisbrod, Groudine,** & **Weintraub** (*Cell* **19,** 289–301, 1980). Some speculations on methylation are the subject of a review from **Razin** & **Riggs** (*Science* **210,** 604–610, 1980). Results on the viral gap were brought together in the report by **Herbomel et al.** (*Cell* **25,** 651–658, 1981). The structure of the centromere was revealed by **Bloom** & **Carbon** (*Cell* **29,** 305–317, 1982).

PART 9
PERPETUATION OF DNA

If it be true that the essence of life is the accumulation of experience through the generations, then one may perhaps suspect that the key problem of biology, from the physicist's point of view, is how living matter manages to record and perpetuate its experiences. Look at a single bacterium in a large volume of fluid of suitable chemical composition. It assimilates substance, grows in length, divides in two. The two daughters do the same, like the broomstick of the Sorcerer's apprentice. Occasionally the replica will be slightly faulty and an individual arises with somewhat different properties, and it perpetuates itself in this modified form. It is quite easy to believe that the game of evolution is on once the trick of reproduction, covariant on mutation, has been discovered, and that the variety of types will be multiplied indefinitely.

Max Delbrück, 1949

CHAPTER 31
THE REPLICON:
UNIT OF REPLICATION

Whether a cell has only one chromosome (as in procaryotes) or has many chromosomes (as in eucaryotes), the entire genome must be replicated precisely once for every cell division. How is the act of replication linked to the cell cycle?

There are two points at which the state of replication and the condition of the cell must be compared. The *initiation* of replication commits the cell (procaryotic or eucaryotic) to a further division. From this standpoint, the number of descendants that a cell generates is determined by a series of decisions on whether or not to initiate DNA replication. If replication does occur, the consequent division cannot be permitted to occur until the replication event has been *completed*. Indeed, the completion of replication could provide a trigger for cell division. Then the duplicate genomes are segregated one to each daughter cell, via mitosis in eucaryotes, and by some other mechanism in procaryotes. The unit of segregation is the chromosome.

The unit in which the cell controls individual acts of replication is called the **replicon.** Each replicon "fires" only once in each cell cycle. The replicon is defined by its possession of the control elements needed for replication. It has an **origin** at which replication is initiated. It may have a **terminus** at which replication stops. *Any* sequence attached to an origin (or, more precisely, not separated from an origin by a terminus)

is replicated as part of that replicon. In genetic terms, the origin is a *cis*-acting site, able to affect only that molecule of DNA of which it is physically part.

The replication event is controlled at the stage of initiation. Once replication has started, it continues until the entire replicon has been duplicated. The frequency of initiation is controlled by the interaction of regulator protein(s) with the origin. Thus the original formulation of the replicon (in procaryotes) viewed it as the possessor of both the origin *and* the gene coding for the regulator protein. Now, however, "replicon" is often applied to eucaryotic chromosomes to describe a unit of replication that contains an origin; any necessary regulator protein(s) can be provided by another unit.

The bacterial chromosome constitutes a single replicon: so the units of replication and segregation coincide. Initiation at a single origin sponsors replication of the entire genome. In addition to the chromosome, bacteria may contain plasmids; each plasmid comprises an autonomous circular DNA genome that constitutes a separate replicon. Some plasmids share the stringency of replication of the bacterial chromosome: these **single-copy** plasmids are present as one copy for every copy of the bacterial chromosome. Other plasmids are **multicopy** and exist in a characteristic number > 1 per bacterial chromosome. Each phage or virus DNA also constitutes a replicon, able to initiate many times during an infectious cycle. Perhaps a better way to view the procaryotic replicon is to reverse the definition: any DNA molecule that contains an origin can be replicated autonomously in the cell. The number of replication events will depend on the interaction of the origin with the appropriate regulator proteins. (A caveat is that some independent replicons also can be physically integrated into the bacterial chromosome; in this case, the use of their own origin is suppressed, and the unit is replicated as part of the bacterial chromosome.)

A eucaryotic chromosome contains a large number of replicons. So the unit of segregation includes many units of replication. This adds another dimension to the problem of control. All the replicons on a chromosome must be fired during one cell cycle; however, they are not active simultaneously, but are fired over a fairly protracted period. Yet each of these replicons must be activated *only* once in the cell cycle. Some signal must exist to distinguish replicated from non-

replicated replicons and to indicate when the entire process has been completed.

We should like to know the set of sequences that function as origins and how they are recognized by the appropriate proteins of the apparatus for replication. From the perspective of the DNA, we want to know whether a duplex sequence of base pairs is recognized, or whether some alternative secondary structure is formed. From the perspective of the regulator proteins, we want to know how they manage to initiate just one cycle of replication for every cycle of cell division.

DNA SYNTHESIS IS SEQUENTIAL AND SEMICONSERVATIVE

The two strands of duplex DNA separate to act as templates for the synthesis of daughter strands (as we have seen already in considering the structure of DNA in Chapter 2). The products of a round of replication are two daughter duplexes, each of which consists of one parental DNA strand and one (newly synthesized) daughter strand. Thus the unit conserved from one generation to the next is one of the two strands comprising the parental duplex. This behavior is described as **semiconservative replication,** and is recapitulated in **Figure 31.1.**

The figure illustrates a prediction of this model. If the parental strands are distinguished from the newly synthesized strands by a density label—the old strands are "heavy," the new strands are "light"—the first replicated generation should consist of "hybrid" duplexes. The second generation should produce equal proportions of hybrid and light duplexes. The individual single strands of these duplexes remain heavy or

Figure 31.1
Replication of DNA is semiconservative.

Parental

Generation 1

Generation 2

——— Parental DNA
——— DNA synthesized in generation 1 or 2

Figure 31.2
The replication fork is the region of DNA between the unwound parental duplex and the newly replicated daughter duplexes.

Replication fork

Replicated (daughter) DNAs

Nonreplicated (parental) DNA

light. This pattern was confirmed experimentally in the Meselson-Stahl experiment of 1958, which followed the semiconservative replication of *E. coli* DNA through many generations.

Consider a molecule of DNA engaged in replication. Its nonreplicated region consists of the parental duplex, opening into the replicated region where the two daughter duplexes have formed. The point at which replication is occurring is called the **replication fork** (sometimes also known as the **growing point**). Its structure is illustrated in **Figure 31.2.** The replication fork moves *sequentially* along the DNA, from its starting point at the origin.

Replication may occur in either a **unidirectional** or **bidirectional** manner. **Figure 31.3** shows that the type of event is determined by whether one or two replication forks set out from the origin. In unidirectional replication, one replication fork leaves the origin and proceeds along the DNA. In bidirectional replication, two replication forks are formed, and they proceed away from the origin in opposite directions.

When replicating DNA is viewed by electron microscopy, the region that has been replicated appears as an **eye** within the nonreplicated DNA. However, its appearance does not distinguish between unidirec-

Figure 31.4
A replication eye can represent either a unidirectional or bidirectional replicon.

tional and bidirectional replication. As depicted in **Figure 31.4,** the eye can represent either of two structures. If generated by unidirectional replication, the eye represents a fixed origin and a moving replication fork. If generated by bidirectional replication, the eye represents a pair of replication forks. In either case, the progress of replication expands the eye until ultimately it encompasses the whole replicon.

When the replicon is circular, the presence of an eye forms the θ-structure drawn in **Figure 31.5.** The successive stages of replication of the circular DNA of polyoma virus are visualized by electron micros-

Figure 31.3
Replicons may be unidirectional or bidirectional, depending on whether one or two replication forks are formed at the origin.

Figure 31.5
A replication eye forms a theta structure in circular DNA.

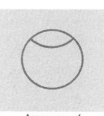

Replicating θ structure

Appearance of θ-structure by electron microscopy

Figure 31.6
The replication eye becomes larger as the replication forks proceed along the replicon.

Photograph kindly provided by Bernard Hirt.

copy in **Figure 31.6.** Note that the "eye" becomes larger than the nonreplicated segment. (In convoluted structures, the two sides of the eye can be defined because they are both the same length.)

Whether a replicating eye has one or two replication forks can be determined in two ways. The choice of method depends on whether the DNA is a defined molecule or an unidentified region of a cellular genome.

With a defined linear molecule, we can use electron microscopy to measure the distance of each end of the eye from the end of the DNA. Then the positions of the ends of the eyes can be compared in molecules that have eyes of different sizes. If replication is unidirectional, only one of the ends will move; the other is the fixed origin. If replication is bidirectional, both will move; the origin is the point midway between them. With a defined circular molecule, the DNA is converted before electron microscopy into a linear form (by treatment with a restriction enzyme that cleaves at a unique site).

With undefined regions of large genomes, two suc-

Unidirectional replication

Bidirectional replication

——— Not labeled (invisible on autoradiograph)

——— Line from light density label

━━━ Line from heavy density label

Figure 31.7

Different densities of radioactive labeling can be used to distinguish unidirectional and bidirectional replication.

cessive pulses of radioactivity can be used to label the movement of the replication forks. One pulse has a more intense label than the other, so the pulses can be distinguished by the relative intensities of labeling, which are visualized by autoradiography. **Figure 31.7** shows that unidirectional replication produces the appearance of one type of label followed by the other at *one* end of the eye. Bidirectional replication produces a (symmetrical) pattern at *both* ends of the eye.

THE BACTERIAL GENOME IS A SINGLE REPLICON

The genome of *E. coli* is replicated bidirectionally from a single origin, identified as the genetic locus *oriC*. The origin was originally identified by indirect techniques, but now has been isolated as a DNA fragment.

In a population of bacteria, the number of copies of genes near the origin is increased, because this region replicates sooner in the cell cycle. Thus the location of the origin and the mode of replication can be recognized by the increased frequency of genes in the vicinity. When replication is bidirectional, the frequency declines as the map position proceeds away from the origin in either direction.

The origin can be identified more precisely as the

site where a brief radioactive label is incorporated when cells are synchronized to start replication at the same time. This technique may be useful for identifying the actual startpoints for replication.

The DNA of the origin can be isolated by virtue of its ability to support replication of any DNA sequence to which it is joined. The principle of this approach is to clone DNA from the region of the origin into a DNA molecule (carrying suitable genetic markers) that lacks an origin. This reconstruction will create a plasmid capable of autonomous replication only if the DNA from the origin region contains all the sequences needed to identify itself as an authentic origin for replication. (A comparable approach has been used to identify centromeric or telomeric DNA in yeast, by virtue of their characteristic effects on plasmid survival; see Chapter 28.)

Some recombinant plasmids with about 1 kb of DNA from the region of *oriC* are maintained in *E. coli* at a stringent level of 1–2 copies per bacterial chromosome. Thus they behave just like the bacterial chromosome itself. (However, the presence of additional origins could cause some problems for the cell; if it is not possible to distinguish among the various origins, the plasmid and the authentic chromosome will divide the number of initiation events, with the result that the number of chromosomes may be reduced. The obvious corollary is that genuine plasmid origins may be expected to differ from the chromosomal origin.)

By reducing the size of the cloned fragment of *oriC*, the functional region has been equated with a fragment of only 422 bp. This is adequate to ensure survival of a plasmid that carries it. However, such plasmids exist in the cell at a copy number of about 20 per bacterial chromosome. Thus they have lost some feature that restricts the frequency of initiation events. The determination of copy number and the act of initiating replication could therefore depend on different sequences.

The origin of the bacterium *Salmonella typhimurium* has been located in a 296 bp fragment of DNA. Comparison with *E. coli* shows that 86% of the bases are conserved between the origins. Considering the sequence as a single strand, a secondary structure resembling a cloverleaf can be formed, as drawn in **Figure 31.8**. Its potential significance is indicated by the fact that almost all of the 42 bases that differ between

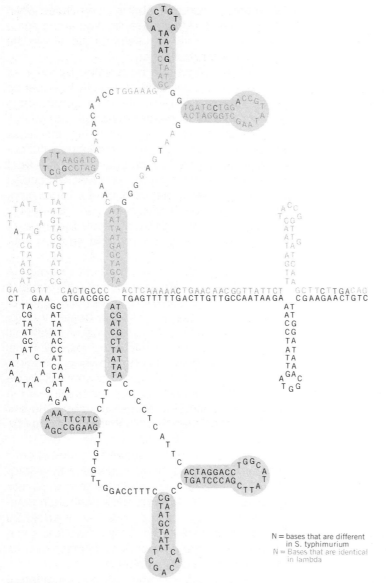

Figure 31.8
The origin for replication in *E. coli* has palindromic sequences that could form a complex secondary structure.

E. coli and *S. typhimurium* fall into the unpaired regions. This raises the possibility that, instead of lying in the regular duplex structure, the DNA of the origin could assume a distinctive secondary structure, with each strand in a cloverleaf.

The sequences of the origins of some lambdoid phages (of which lambda is one example) also have been determined. These origins are related to one another, but differ in sequence from the bacterial origin. However, they could (in theory) form a similar cloverleaf structure (with the arms a little closer together). A conserved sequence, as indicated in Figure 31.8, lies to the right of the potential cloverleaf. The sequence is present in *E. coli*, lambda, and another phage, G4. If we assume that the conserved features identify the functions of the origin, there may be two important aspects: a secondary structure that is recognized as such rather than by its sequence; and a short conserved sequence that probably remains in the duplex form.

The disruption of DNA structure to form the potential cloverleaf is unlikely to occur spontaneously, since this form of organization is intrinsically less stable than the regular duplex. However, it could be assisted by extraneous factors, such as pressure from supercoiling and stabilization by proteins. If this entire stretch of DNA is indeed necessary for origin function, it would certainly be the largest *cis*-acting site yet found. (Compare its length with the much shorter duplex stretches of promoters and operators.)

CONNECTIONS BETWEEN DNA REPLICATION AND CELL DIVISION

Bacteria have two links between replication and cell growth. The frequency with which cycles of replication are initiated is adjusted to fit the rate at which the cell is growing. And the completion of a replication cycle is connected with the division of the cell into two, in the course of which the daughter chromosomes are segregated.

Cells of *E. coli* can grow at a variety of rates, ranging from doubling times as quick as 18 minutes to slower than 180 minutes. Because the bacterial chromosome is a single replicon, the frequency of replication cycles is controlled by the number of initiation events at the single origin. The rate of DNA synthesis is more or less invariant at a constant temperature; it proceeds at the same speed unless and until the supply of precursors becomes limiting.

The replication cycle of *E. coli* can be defined in terms of two constants. **D** is the fixed time of about 20 minutes that elapses between the completion of a round of replication and the cell division with which it is connected. This period may represent the time required to assemble the components needed for division. **C**

is the fixed time of about 40 minutes required to replicate the entire bacterial chromosome. Its duration corresponds to a rate of movement by the individual replication fork of about 50,000 bp per minute. (The constants C and D can be viewed as representing the maximum speed of which the bacterium is capable; they apply for all growth rates between doubling times of 18 and 60 minutes, but both constant phases become longer when the cell cycle occupies more than 60 minutes.)

A cycle of chromosome replication must be initiated a fixed time, C + D = 60 minutes, before a cell division. For bacteria dividing more frequently than at 60-minute intervals, a cycle of replication must therefore be initiated before the end of the *preceding* division cycle. Consider the example of cells dividing every 35 minutes. The cycle of replication connected with a division must have been initiated 25 minutes before the preceding division. This situation is illustrated in **Figure 31.9,** which shows the chromosomal complement of a bacterial cell at 5-minute intervals throughout the cycle.

At division (35/0 minutes), the cell receives a par-

Figure 31.9

The fixed interval of 60 minutes between initiation of replication and cell division produces multiforked chromosomes in rapidly growing cells.

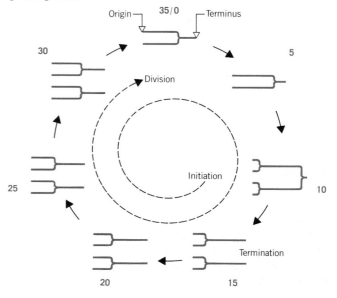

tially replicated chromosome. The replication fork continues to advance. At 10 minutes, when this "old" replication fork has not yet reached the terminus, initiation occurs at both origins on the partially replicated chromosome. The start of these "new" replication forks creates a **multiforked chromosome.** At 15 minutes— that is, at 20 minutes before the next division—the old replication fork reaches the terminus. Its arrival allows the two daughter chromosomes to separate; each of them has already been partially replicated by the new replication forks (which now are the *only* replication forks). These forks continue to advance. At the point of division, the two partially replicated chromosomes segregate. This recreates the point at which we started. The single replication fork becomes "old," it terminates at 15 minutes, and 20 minutes later there is a division. We see that the initiation event occurs $1^{25}/_{35}$ cell cycles before the division event with which it is associated.

The general principle of the link between initiation and the cell cycle is that as cells grow more rapidly (the cycle is shorter), the initiation event occurs an increasing number of cycles before the related division. There are correspondingly more chromosomes in the individual bacterium. This relationship can be viewed as the cell's response to its inability to accomplish the periods of C and D more rapidly to keep pace with the shorter cycle.

How does the cell know when to initiate the replication cycle? The initiation event occurs at a constant ratio of cell mass to the number of chromosome origins. Cells growing more rapidly are larger, have greater mass, and therefore possess a greater number of origins. In terms of Figure 31.9, it is at the point 10 minutes after division that the cell mass has increased sufficiently to support an initiation event at both available origins.

Two types of model have been proposed for the titration of cell mass. An initiator protein could be synthesized continuously throughout the cell cycle; accumulation of a critical amount would trigger initiation. Or an inhibitor protein might be synthesized at a fixed point, and diluted below an effective level by the increase in cell volume. There is evidence to suggest that a titration model actually does regulate initiation, but the data do not distinguish between accumulation of an initiator and dilution of an inhibitor. Current thinking favors an initiator, which is consistent with evidence that protein synthesis is needed for the initiation event.

Mutations in several *E. coli* genes affect the initiation of replication cycles; and there are many genes whose products are necessary for cell division. Analysis of these mutants, however, has not yet identified any proteins responsible for controlling the frequency of initiation or for linking replication to the division cycle.

Whichever type of titration model applies, the growth of the bacterium can be described in terms of the **unit cell,** an entity 1.7 μm long. A bacterium contains one origin per unit cell; a rapidly growing cell with two origins will be > 3.4 μm long. A topological link between the initiation event and the structure of the cell could take the form of a growth site, a physical entity in the cell that provides the only place at which initiation can occur. There should be one growth site per unit cell.

There have been suspicions for years that a physical link may exist between bacterial DNA and the membrane. Bacterial DNA can be found in membrane fractions, which tend to be enriched in genetic markers near the origin, the replication fork, and the terminus. The growth site could be a structure on the membrane to which the origin must be attached for initiation. Mammalian forks may be bound to the nuclear matrix.

A link between DNA and the membrane also could account for segregation. If daughter chromosomes are attached to the membrane, they could be physically separated as the membrane grows between them. **Figure 31.10** shows that then a septum might form to segregate each chromosome into a different daughter cell. If we recall the segregation event depicted in Figure 31.9, we would expect the terminus of the chromosome to be involved.

So far we have dealt with the bacterial chromosome as though it were linear. Because it is really circular, the two replication forks each move around the genome to a meeting point. We do not know what happens at this encounter. Do the forks crash into each other, or is there a specific terminus at which they stop? How do the enzymes involved in replication disengage from the chromosome, remembering that the

DNA must be replicated right across the region where the forks meet? A mutation of *E. coli* called *dnaT* has properties that suggest it lies in a gene that codes for a protein involved in terminating the round of replication, but the product has not been characterized.

A region of the *E. coli* chromosome has been identified that may provide a terminus at which the replication forks cease to move. It is not located exactly halfway around the chromosome; the disparity implies that the two forks must move different distances. In the plasmid R6K, a terminus has been identified as a region that stops replication from proceeding; this region remains effective when it is inserted into another DNA molecule. The potential terminus sequence does not seem to have the potential to form a secondary structure, so its possible model of action is unknown.

Figure 31.10
Attachment of bacterial DNA to the membrane could provide a mechanism for segregation.

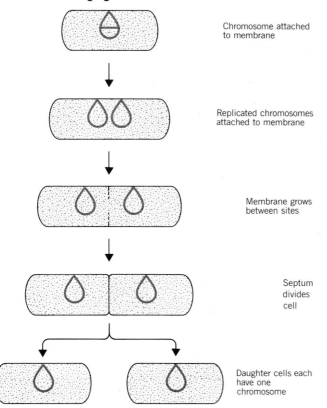

Chromosome attached to membrane

Replicated chromosomes attached to membrane

Membrane grows between sites

Septum divides cell

Daughter cells each have one chromosome

EACH EUCARYOTIC CHROMOSOME CONTAINS MANY REPLICONS

In eucaryotic cells, the replication of DNA is confined to part of the cell cycle. The period between mitoses (interphase) can be divided into the phases indicated in **Figure 31.11.** The typical cell starts its cycle in the diploid condition. It remains in this state through the **G1** period, which often comprises the major part of the cycle (and is the most variable in length between cells of different phenotype). The synthesis of DNA defines the period of **S phase,** which often lasts a few hours in a higher eucaryotic cell. Replication of DNA converts the cell into a tetraploid condition, in which it remains for the **G2** period. Then mitosis (M) reduces the chromosome complement to the diploid number in each daughter cell. (The periods preceding and succeeding S phase are called G1 and G2 to indicate that they represent "gaps" in DNA synthesis.)

The events responsible for the initiation of S phase occur during G1. Protein synthesis is needed, but the exact nature of the events remains unclear; in particular, there has been much discussion as to whether cumulative or sudden actions are involved. At the molecular level, little is known about how the cell takes

Figure 31.11
The eucaryotic cell cycle is divided into four phases, G1, S, G2, and M. The durations indicated for the phases represent a typical mammalian cell line growing in culture.

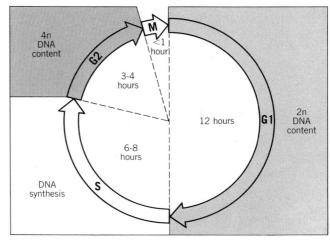

the decision to release the genome for replication. It is assumed that this step constitutes a regulatory event different in nature from the events involved in the subsequent synthesis of DNA.

Replication of the large amount of DNA contained in a eucaryotic chromosome is accomplished by dividing it into many individual replicons. These replicons are not all active simultaneously; at any point during S phase, only some are engaged in replication. Presumably each replicon is activated at a specific time during S phase, although the evidence on this issue is not decisive.

The important point is that the start of S phase is signaled by the activation of the first replicons. Then, over the next few hours, many further initiation events occur at other replicons. The control of S phase therefore involves two processes: release of the cell from G1; and initiation of replication at individual replicons in an ordered manner.

Much of our knowledge about the properties of the individual replicons is derived from autoradiographic studies, generally using the type of protocol illustrated in Figure 31.7. Chromosomal replicons usually display bidirectional replication, as seen by the existence of matched pairs of tracks.

How large is the average replicon, and how many are there in the genome? A difficulty in characterizing the individual unit is that adjacent replicons may fuse to give large replicated eyes, as illustrated in **Figure 31.12.** The approach usually used to distinguish individual replicons from fused eyes is to rely on measurements of stretches of DNA in which several replicons can be seen to be active, presumably captured at a stage when all have initiated around the same time, but before the forks of adjacent units have met. There is some evidence that "regional" controls might produce this sort of activation pattern, in which groups of replicons are initiated more or less coordinately, as opposed to a mechanism in which individual replicons are activated one by one in dispersed areas of the genome.

In groups of active replicons, the average size of the unit is measured by the distance between the origins (that is, between the midpoints of adjacent replicons). For many of the higher eucaryotes, this lies in the range of 100–200 kb. Thus in a haploid mammalian genome,

Figure 31.12

Measuring the size of the replicon requires a stretch of DNA in which adjacent replicons are active.

there should be 20,000–30,000 replicons. In *D. melanogaster* or in *S. cerevisiae,* the replicons are smaller, averaging about 40 kb. This corresponds to about 3500 replicons in the haploid set of the fruit fly, and to about 500 replicons in the yeast. The range of individual lengths is high, more than tenfold, so the average gives only an approximate value for the number of replicons.

The rate at which the replication fork moves can be estimated from the maximum distance that the autoradiographic tracks travel during a given time. This is in the range of 1000–3000 bp/minute for mammals; it is of the order of 1000 bp/minute for plants, possibly because of their lower temperatures. These values are very much slower than the rate of bacterial replication fork movement (~50,000 bp/minute).

We should like to know what constitutes the origin of each replicon. Is it a specific DNA sequence, possibly a member of a repetitive sequence family? Or could it be a feature of the higher order structure of the chromatin fiber, not necessarily associated with a single type of sequence? Then we must ask how origins are selected for initiation at different times during S phase. Does the structure of the origin carry some information that establishes its time of use, or is this a consequence of some ordered rearrangement of structure? An especially intriguing question is how the replication apparatus distinguishes origins that have already been replicated from those that have yet to be replicated. Is the utilization of each origin once and only once during S phase a property of the DNA (for example, as seen by its state of methylation) or of proteins associated with it?

We know of one case in which origins are not fixed, but depend on the cell type. The replicons of early embryonic divisions in *D. melanogaster* are much smaller than those of somatic cells. Their simultaneous activity reduces the time required to complete chromosome replication, a need presumably resulting from the rapidity of cell divisions at this time. A possible explanation is that the genome contains a set of tissue-specific origins, activated only in embryonic divisions, and not recognized as such in somatic cell divisions. We do not have enough information to know whether tissue-specific replication patterns are common.

It is not clear whether chromosomal replicons have discrete termini at which the replication forks cease movement and (presumably) dissociate from the DNA. The alternative is that a replication fork continues from its origin until it meets a fork proceeding toward it from the adjacent replicon. We have already mentioned the potential topological problem of joining the newly synthesized DNA at the junction of the replication forks.

There is a difference in the timing of the replication of euchromatin and heterochromatin. Both constitutive and facultative heterochromatin are **late replicating,** as seen by their later start and finish to replication. This may be a consequence of their more condensed structure. Usually (although not always) the last material in the cell to synthesize DNA during S phase is heterochromatic.

ISOLATING THE ORIGINS OF YEAST REPLICONS

Any segment of DNA that has an origin should be able to replicate. Thus although plasmids are rare in eucaryotes, it may be possible to construct such molecules by suitable manipulation *in vitro*. Potential plasmids can be introduced into suitable recipient cells. (We shall deal in Chapter 38 with the fate of DNA introduced into eucaryotic cells.)

One system in which extrachromosomal DNA molecules can replicate is the *Xenopus* egg. An injected DNA is replicated once in the cell cycle, which suggests that it is under proper control. However, *any* injected (circular) DNA is able to replicate, which argues against the need for a specific origin of replication in this system.

Another system offers the opportunity to isolate discrete origins. Cells of the yeast *S. cerevisiae* that are mutant in some function can be "transformed" by addition of DNA that carries a wild-type copy of the gene. The protocol is illustrated in **Figure 31.13.** The mutation of the host cell must lie in some gene whose product can be subjected to selection. The DNA of wild-type cells is extracted, cleaved into fragments, and cloned in *E. coli* plasmids. The chimeric plasmids are incubated with the mutant yeast cells, which are placed in conditions in which they can survive only if they express the function of the wild-type gene. Two types of transformation can be distinguished by the frequency with which transformed cells arise.

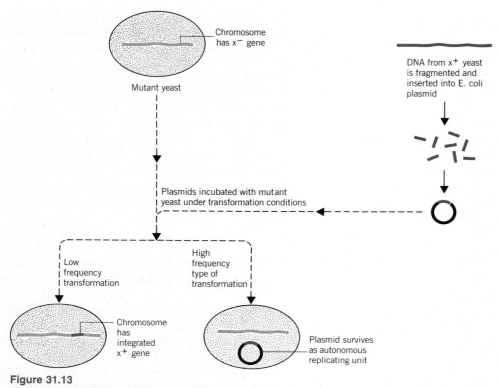

Figure 31.13
Transformation frequency in yeast is influenced by whether a fragment integrates into chromosomal DNA or survives as a plasmid.

Any gene can be transformed at a *low frequency.* The surviving cells possess a copy of the wild-type donor gene integrated into the chromosome. Probably it has replaced the original mutant gene by a homologous recombination. The absence of any other progeny argues that the donor gene cannot survive in the recipient cell *unless* it is integrated. Thus the reason for the low frequency of transformation appears to be the need for integration.

Some yeast DNA fragments are able to transform defective cells at much *higher* (~1000-fold) *frequencies.* These fragments can survive in the cell in the unintegrated (autonomous) state, that is, in the form of the originally added circular DNA molecule. The lack of any need for an (inefficient) integration event allows transformation to occur at a high frequency.

A high-frequency transforming fragment possesses a sequence that confers on a chimeric plasmid the ability to replicate efficiently in yeast. This segment is called an *ars* (for autonomously replicating sequence). We should like to know whether the *ars* sequences are authentic origins of replication. The minimum sequence necessary to provide *ars* function can be identified by progressively recloning shorter fragments from the active material. Two *ars* sequences that have been treated like this are *ars1* (available in the form of an 850 bp fragment) and *ars2* (available as a 100 bp fragment).

The only homology between the two *ars* sequences lies in two short sequences (one of 8 bp, one of 10 bp). Neither *ars* segment appears to be related to other sequences of the yeast genome; if the *ars* sequences represent authentic functions, yeast origins of replication cannot simply be members of a repetitive sequence family. Another known origin of replication is that of the yeast 2μ plasmid, and this also lacks extensive homology with the *ars* sequences. It is therefore possible that many different sequences may be able to provide origin function in yeast.

REPLICATION CAN PROCEED THROUGH EYES, ROLLING CIRCLES, OR D LOOPS

The structures generated by replication depend on the relationship between the template and the replication fork. The critical features are whether the template is circular or linear, and whether the replication fork is engaged in synthesizing both strands of DNA or only one.

The conventional picture we have assumed so far treats the replication fork as a site at which both new strands of DNA are synthesized. In a linear molecule, whether replicated unidirectionally or bidirectionally, the movement of the fork(s) generates an eye, as seen in Figure 31.4. If the template is a circular DNA, the replicating molecule takes the form of a θ structure, as drawn in Figure 31.5.

Another outcome is possible with a circular molecule. Suppose that replication of *one* strand of a duplex molecule starts at the origin. A nick opens one strand, and then the free 3′-OH end generated by the nick is extended by the DNA polymerase. The newly synthesized strand displaces the original parental strand. The ensuing events are depicted in **Figure 31.14.**

This type of structure is called a **rolling circle,** because the growing point can be envisaged as rolling around the circular template strand. It could in principle continue to do so indefinitely. As it moves, the replication fork extends the outer strand and displaces the previous partner. Because the newly synthesized material is covalently linked to the original material, the displaced strand has the original unit genome at its 5′ end. The original unit is followed by any number of unit genomes, synthesized by continuing revolutions of the template. Each revolution displaces the material synthesized in the previous cycle. The rolling circle is put to several uses *in vivo.*

In certain phages the genome consists of a single-stranded circle of DNA. This circle is first converted to a duplex form, which is then replicated by a rolling circle mechanism. The displaced tail generates a series of unit genomes; these can be cleaved and inserted into phage particles, or they can be used for further replication cycles. (These replication systems are discussed in Chapter 33.)

The displaced single strand of a rolling circle can be converted to duplex DNA by synthesis of a complementary strand. This action can be used to generate the concatemeric duplex molecules needed for maturation of certain phage DNAs (see Chapter 28). This provides the principal pathway through which phage lambda DNA matures. (It is not the only way that concatemeric molecules can be generated; recombination is another pathway, which appears to predominate with phage T4.)

Rolling circles also are used to generate amplified rDNA in the *Xenopus* oocyte. This explains why the amplified material consists of a large number of identical repeating units (see Figure 23.6). In this case, a single genomic repeating unit is converted into a rolling circle. The displaced tail is converted into duplex DNA; and at some point this is cleaved from the circle so that the two ends can be joined together to generate the amplified circle of rDNA.

Another type of action for a replication fork has been identified in some mitochondria. Replication starts at a specific origin in the circular duplex DNA. But initially only one of the two parental strands (the H strand in mammalian mitochondrial DNA) is used as a template for synthesis of a new strand. Synthesis proceeds for only a short distance, displacing the original partner (L) strand, which remains single-stranded, as illustrated in **Figure 31.15.** The condition of this region gives rise to its name as the **displacement** or **D loop.**

A single D loop is found as a stable structure of 500–600 bases in mammalian mitochondria. The short strand has a unique 5′ end and a variable 3′ end, probably as the result of termination of synthesis at any one of 3–4 discrete sites. This short strand is unstable and turns over; it is frequently degraded and resynthesized to maintain the opening of the duplex at this site. Some mitochondrial DNAs, such as *X. laevis,* possess a single but longer D loop. Others may possess several D loops; for example, there may be as many as six in the linear mitochondrial DNA of *Tetrahymena.* The same mechanism is employed in chloroplast DNA, where (in higher plants) there are two D loops.

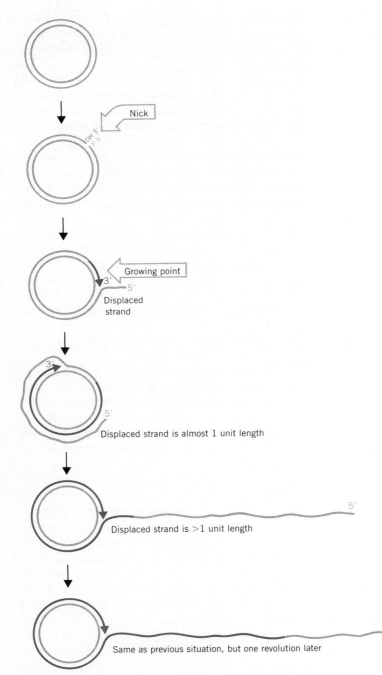

Figure 31.14
The rolling circle generates a multimeric single-stranded tail (which can be converted to a duplex form by synthesis of a complement; not shown).

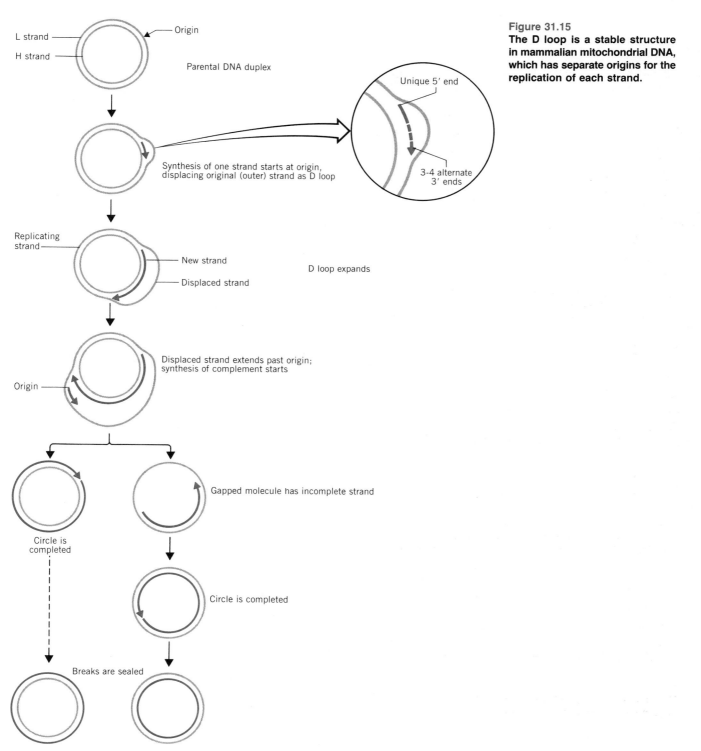

L strand

H strand

Origin

Parental DNA duplex

Synthesis of one strand starts at origin, displacing original (outer) strand as D loop

Unique 5′ end

3-4 alternate 3′ ends

Replicating strand

New strand

Displaced strand

D loop expands

Displaced strand extends past origin; synthesis of complement starts

Origin

Gapped molecule has incomplete strand

Circle is completed

Circle is completed

Breaks are sealed

Figure 31.15
The D loop is a stable structure in mammalian mitochondrial DNA, which has separate origins for the replication of each strand.

To replicate mammalian mitochondrial DNA, the short strand in the D loop is extended. The displaced region of the original L strand becomes longer, expanding the D loop. This expansion continues until it reaches a point about 67% of the way around the circle. Replication of this region exposes an origin in the displaced L strand. Synthesis of an H strand initiates at this site, proceeding around the displaced single-stranded L template in the opposite direction from L-strand synthesis. Because of the lag in its start, H-strand synthesis has proceeded only 30–40% of the way around the circle when L-strand synthesis finishes. This releases one completed duplex circle and one gapped circle, which remains partially single-stranded until synthesis of the H strand is completed. Finally, the new strands are sealed to become covalently intact.

The existence of rolling circles and D loops exposes a general principle. An origin can be a sequence of DNA that serves to initiate DNA synthesis using one strand as template. The opening of the duplex does not necessarily lead to the initiation of replication on the other strand. This might need its own origin for replication. A D loop results when the origins of the two strands are separated by some distance. The corollary is that an origin that serves to initiate replication of *both* strands may do so specifically because of its features; that is, because the origins of both strands are located at the same point. These features also will determine whether replication forks set out unidirectionally or bidirectionally.

PLASMID INCOMPATIBILITY IS CONNECTED WITH COPY NUMBER

Plasmids are selfish. Having obtained possession of a bacterium, the resident plasmid will seek to prevent any other plasmid of the same type from establishing residence. Two independent devices are used to establish these territorial rights: **surface exclusion** and **plasmid incompatibility.**

Surface exclusion describes the inability of a plasmid to enter cells already carrying another plasmid of the same type. The effect is mediated at the surface of the bacterium and is a feature of the system used to transfer the plasmid between bacteria. In the case

of the sex (F) factor of *E. coli,* surface exclusion is achieved by inhibiting the DNA from leaving its present bacterial host in order to enter another. Thus it is an emigration rather than immigration control.

Incompatibility is related to the regulation of plasmid copy number. A **compatibility group** is defined as a set of plasmids whose members are unable to coexist in the same bacterial cell. The presence of a member of one compatibility group does not directly affect the survival of a plasmid belonging to a different group. Thus only one replicon of a given compatibility group can be maintained in the bacterium, but it does not interact with replicons of other compatibility groups (although in limiting conditions they may compete for lebensraum).

The negative control model for plasmid incompatibility follows the idea that copy number control is achieved by synthesizing a repressor that measures the number of origins. (Formally this is the same as the titration model for regulating bacterial chromosomal replication.) The introduction of a new origin in the form of a second plasmid of the same compatibility group mimics the result of replication of the resident plasmid; two origins now are present. Thus any further replication is prevented until after the two plasmids have been segregated to different cells to create the correct prereplication copy number.

Mutations may either increase or decrease the copy number. Such mutants are important because they define the genetic locus responsible for controlling copy number; and their analysis opens the way to defining the mechanism involved. These mutations also influence incompatibility, demonstrating its connection with copy number.

The best characterized copy number and incompatibility system is that of the plasmid ColE1 of *E. coli.* Before considering it, we must digress briefly into the mechanism for initiating replication at the ColE1 origin. The relevant events are illustrated in **Figure 31.16.** Replication starts with the transcription of an RNA that initiates 555 bp upstream from the origin. Transcription continues through the origin. The enzyme RNAase H (whose name reflects its specificity for a substrate of RNA hybridized with DNA) cleaves the transcript at the origin. This generates a 3′-OH end that is used as the "primer" for synthesis of DNA.

Another RNA species is synthesized from ColE1 DNA

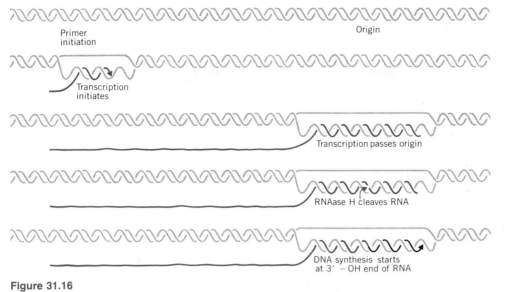

Figure 31.16
Replication of ColE1 DNA is initiated by cleaving the primer RNA to generate a 3′-OH end.

in the same region. This is RNA I, a molecule of about 108 bases, coded by the opposite strand from that specifying primer RNA. The relationship between the primer RNA and RNA I is illustrated in **Figure 31.17.** The RNA I molecule terminates very close to the site where the primer RNA initiates. Thus the RNA I species is complementary to the 5′-terminal region of the primer RNA.

Mutants that reduce or eliminate incompatibility can be obtained by selecting plasmids of the same group for their ability to coexist. Mutants in the ColE1 group map in the region of overlap between RNA I and primer RNA, four of them representing changes of G-C to A-T, one the reverse. Because this region is represented in two different RNAs, either or both might be involved in the effect.

When RNA I is added to a system for replicating

Figure 31.17
The sequence of RNA I is complementary to the 5′ region of primer RNA.

RNA I (108 bases)
←
Origin
→
Primer RNA (555 bases)

ColE1 DNA *in vitro,* it inhibits the formation of primer RNA. The presence of RNA I does not inhibit the initiation or elongation of primer RNA synthesis. This suggests that RNA I prevents RNAase H from generating the 3′ end of the primer RNA. The most likely basis for this effect lies in base pairing between RNA I and primer RNA.

Both RNA molecules have the same potential secondary structure in this region, with two duplex hairpins terminating in single-stranded loops. The incompatibility mutations are located in these loops. This suggests that an initial step in base pairing between RNA I and primer RNA may be contact between the unpaired loops. Their pairing could be impeded when the RNA I and primer RNA have different sequences because of a mutation in one of the parent plasmids. Each RNA I would continue to pair with the primer RNA coded by the same plasmid, but might be unable to pair with the primer RNA coded by the other plasmid.

How does pairing with RNA I prevent cleavage to form primer RNA? A model is illustrated in **Figure 31.18.** In the absence of RNA I, the primer RNA has a secondary structure in which its 5′ terminal region is paired with some other part of the molecule. The intervention of RNA I sequesters the 5′ end of the potential primer

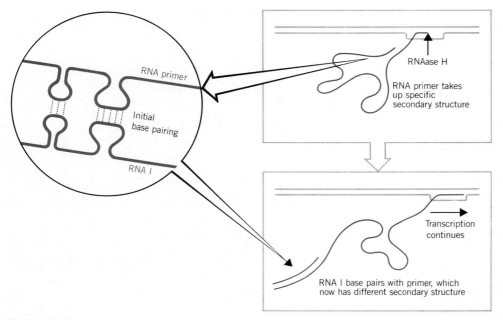

Figure 31.18
Base pairing with RNA I may change the secondary structure of the primer RNA sequence and thus prevent cleavage from generating a 3′-OH end.

RNA, altering its secondary structure. This new secondary structure in some way prevents the cleavage reaction, although we do not understand the basis for this effect. (The action of RNAase H has been thought to depend on recognition of RNA-DNA hybrid regions, not on RNA structure.)

The model is somewhat reminiscent of the mechanism involved in attenuation of transcription, in which the alternative pairings of an RNA sequence permit or prevent formation of the secondary structure needed for termination by RNA polymerase (see Chapter 15). Formally, the model is equivalent to postulating the presence in the cell of a repressor that prevents newly introduced DNA from functioning, analogous to the role of the lambda lysogenic repressor (see Chapter 16). Instead of a repressor protein that binds the new DNA, an RNA binds the newly synthesized precursor to the RNA primer. The ability of RNA I to inhibit initiation of replication could be part of a negative control circuit by which incompatibility is related to control of copy number. However, we do not yet know how these events are related to the maintenance of ColE1 DNA at its characteristic copy number (about 20 per cell).

FURTHER READING

General reviews of bacterial and eucaryotic replicons are to be found in **Lewin's** *Gene Expression,* respectively pp. 550–576 of **1**, *Bacterial Genomes,* and pp. 570–582 of **2**, *Eucaryotic Chromosomes* (Wiley, New York, 1974 & 1980). Bacterial and phage DNA origins were isolated by **Yasuda** & **Hirota** (*Proc. Nat. Acad. Sci. USA* **74**, 5458–5462, 1977), **Zyskind** & **Smith** (*Proc. Nat. Acad. Sci. USA* **77**, 2460–2464, 1980), and **Grosschedl** & **Hobom** (*Nature* **282**, 39–43, 1979). **Laskey** & **Harland** (*Cell* **24**, 283–284, 1981) have speculated about eucaryotic replication origins. Mitochondrial DNA replication has been reviewed by **Clayton** (*Cell* **28**, 693–705, 1982). The ColE1 system has been elucidated by **Tomizawa** & **Itoh** (*Proc. Nat. Acad. Sci. USA* **78**, 6096–6100, 1981).

CHAPTER 32
THE TOPOLOGY OF DNA REPLICATION

The topological manipulation of DNA is a crucial aspect of its functional activities (replication and transcription) as well as of the organization of higher order structure. The original description of the double-helical structure suggested that DNA synthesis involves separation of the parental strands, a proposal confirmed by the experimental demonstration that replication is semiconservative (see Chapter 31). The strands can be separated by input of sufficient energy to disrupt the noncovalent forces that stabilize the double helix. However, the strands do not simply lie side by side; they are intertwined. Their separation therefore requires the strands to rotate about each other in space. Some possibilities for the unwinding reaction are illustrated in **Figure 32.1.**

We might envisage the structure of DNA in terms of a free end that would allow the strands to rotate about the axis of the double helix for unwinding (part *A*). Given the length of the double helix, however, this would involve the separating strands in a considerable amount of flailing about, which seems unlikely in the confines of the cell. A similar result is achieved by placing an apparatus to control the rotation at the free end (part *B*). However, the effect must be transmitted over a considerable distance, again involving the rotation of an unreasonable length of material. Remember also that replication converts each of the rotating single strands into the more cumbersome structure of a daughter duplex.

The demonstration that DNA behaves as a closed structure lacking free ends (see Chapter 28) excludes these models as a matter of principle and brings home the severity of the topological problem. Consider the effects of separating the two strands in a molecule whose ends are not free to rotate (part *C* of the figure). When two intertwined strands are pulled apart from one end, the result is to increase their winding about each other farther along the molecule. Thus movement of a replication fork per se would generate increasing positive supercoiling ahead of it, rapidly generating insuperable resistance to further movement.

The problem can be overcome by introducing a transient nick in one strand. Part *D* of the figure shows that the internal free end allows the nicked strand to rotate about the intact strand, after which the nick can be sealed. The nicking and sealing reaction can be repeated as the replication fork advances. Such reactions occur with substrates of closed DNA; they are mediated by enzymes called topoisomerases, whose actions are crucial in replication (and also involved in transcription; see Chapter 11).

Enzymes that are able to synthesize new DNA strands on a template strand are called DNA polymerases. Both procaryotic and eucaryotic cells contain more

Figure 32.1

Separation of the strands of a DNA double helix could be achieved by one of

A rotation about a free end.
B an apparatus that holds the strands while it rotates.
C compensating positive supercoiling elsewhere.
D nicking, rotation, and ligation.

than one DNA polymerase activity. But only one particular enzyme provides the replicase function in either the bacterium or eucaryotic nucleus. The others are involved in subsidiary roles in replication and/or participate in "repair" synthesis of DNA to replace damaged sequences (see Chapter 34).

It is convenient to think of DNA-synthesizing activities in terms of the DNA polymerase enzymes; but it

is a moot point whether a DNA replicase enzyme exists as a discrete entity. (We have considered a similar issue for RNA polymerase in Chapters 10 and 13.) In bacteria, a protein complex with DNA replicase activity can be identified, but it is recovered in the form of several (overlapping) aggregates of various subunits. It is not clear which of these should properly be considered a component of the replicase per se and which constitute accessory factors needed for proper replicase function. In the normal process of replication, the DNA-synthesizing activity is only one of several associated functions; and perhaps we should not be unduly influenced to think of it as a separate function because of our ability to assay its particular step in the overall process.

No DNA polymerase activity yet has been discovered that can initiate a deoxyribonucleotide chain. All the enzymes extend previously started chains. So a necessary feature of the replication apparatus is the provision for "priming" activities that start off the DNA chain. This usually involves the synthesis of a short ribonucleotide sequence that is later removed.

DESCRIBING THE TOPOLOGY OF DNA

So far we have dealt with the organization of duplex DNA in space in terms of the superhelical density, the number of superhelical turns per turn of the double helix. A superhelical turn occurs when the DNA duplex makes one revolution about its own axis in space, either in the same direction as the winding of the double helix itself (positive supercoiling) or in the opposite direction (negative supercoiling). Positive supercoiling "overwinds" the double helix, negative supercoiling "underwinds" it (see Chapter 2).

Since the original definition of supercoiling, a new nomenclature has been adopted to describe this situation. The concept is based on the **linking number,** which specifies the number of times that the two strands of the double helix of a closed molecule cross each other in toto. Thus the linking number is the number of revolutions that one strand makes around the other when the DNA is considered (hypothetically) to lie flat on a plane surface. The convention is to count the linking number in terms of the number of right-hand crossings over left-hand crossings, so it is positive for the right-handed double helix. It is necessarily an integer.

The linking number has two components, the twist and the writhe, defined by the equation

$$L = W + T$$

The **twisting number,** T, is a property of the DNA double helix per se, representing the rotation of one strand about the other. It is determined by the number of base pairs per turn. For a relaxed closed circular DNA lying flat in a plane, the twist is the total number of base pairs divided by the number of base pairs per turn. Thus T is the number of turns of the duplex.

The **writhing number** represents the twisting of the *axis* of the duplex in space. It corresponds to the intuitive concept of supercoiling, but does not have exactly the same quantitative definition or measurement. For a relaxed molecule, W = 0. In this case, the linking number equals the twist.

The parameter measured in certain experiments is the *change* in linking number, ΔL. This is given by

$$\Delta L = \Delta W + \Delta T$$

The equation states that any change in the total number of revolutions of one DNA strand about the other can be expressed as the sum of the changes of the coiling of the duplex axis in space (ΔW) and changes of the screwing of the double helix itself (ΔT). In the terms of our previous discussions, ΔW is analogous to change in supercoiling, and ΔT represents underwinding or overwinding of DNA. A decrease in L, that is, a negative ΔL, corresponds to the introduction of some combination of negative supercoiling and/or underwinding, or to a reduction in positive supercoiling/overwinding. A positive ΔL increases positive supercoiling/overwinding or decreases negative supercoiling/underwinding.

The critical feature about the use of the linking number is that *this parameter is an invariant property of any individual closed DNA molecule* (that is, whose strands lack free ends). The linking number cannot be changed by any deformation except one that involves the breaking and rejoining of strands. So a circular molecule with a particular linking number can express this in terms of different combinations of T and W, but

cannot change their sum so long as the strands are unbroken. (In fact, the partition of L between T and W prevents the assignment of fixed values to the latter parameters for a DNA molecule in solution.)

We now see the utility of the linking number. *It is related to the actual enzymatic events by which changes are made in the topology of DNA.* An enzyme breaks a strand or strands, reorganizes their relationship in space, and rejoins them. In doing so, it must change the linking number by an integer; this value can be determined as a characteristic of the reaction. Then we can consider the effects of this change in terms of ΔW and ΔT.

TOPOLOGICAL MANIPULATION OF DNA

Molecules of DNA that are the same except for their linking numbers are called **topological isomers.** Enzymes that can interconvert topological isomers are called **DNA topoisomerases,** a name that has replaced a variety of earlier descriptions, such as *DNA relaxing enzyme, swivelase, untwisting enzyme, nicking-closing enzyme.* Some topoisomerases can relax (remove) only negative supercoils from DNA; others can relax both negative and positive supercoils. Some can introduce negative supercoils. The topoisomerases are divided into two classes, according to the nature of the mechanisms they employ. **Type I** enzymes act by making a transient break in one strand of DNA. **Type II** enzymes act by introducing a transient double-strand break.

The best characterized type I topoisomerase is the enzyme of *E. coli* (originally called the ω protein). This enzyme relaxes highly negatively supercoiled DNA with high efficiency; the rate of relaxation becomes slower as the linking number approaches that of completely relaxed DNA. The enzyme does not act on positively supercoiled DNA. It is coded by the gene *top;* mutations cause an increase in the level of supercoiling in the nucleoid (and may affect transcription, as described in Chapter 11).

In addition to the relaxation of negative supercoils in duplex DNA, the enzyme displays several interactions with single-stranded DNA. This suggests that the reason for its specificity toward negative supercoils is that they tend to stabilize single-stranded regions, which could provide the substrate bound by the enzyme. The enzyme does not recognize a particular target sequence, but does show nonrandom recognition of sites on DNA. The nature of its preference is not known.

When *E. coli* topoisomerase I binds to DNA, it forms a stable complex in which one strand of the DNA has been nicked and its 5'-phosphate end covalently linked to a tyrosine residue in the enzyme. This suggests a mechanism for the action of the enzyme; it transfers a phosphodiester bond in DNA to the protein, manipulates the structure of the two DNA strands, and then rejoins the original strand. The transfer of bonds from nucleic acid to protein explains how the enzyme can function without requiring any input of energy. There has been no irreversible hydrolysis of bonds; their energy has been conserved through the transfer reactions.

A model for the action of topoisomerase I is illustrated in **Figure 32.2.** The enzyme binds to a region in which duplex DNA becomes separated into its single strands; then it breaks one strand, pulls the other strand through the gap, and finally seals the gap. This changes the linking number in steps of 1. Each time one strand is passed through the break in the other, there is a ΔL of $+1$. The figure illustrates the enzyme activity in terms of moving the individual strands; but in a highly supercoiled molecule, the free interchangeability of W and T should let the change in linking number be taken up by a change of $\Delta W = +1$, that is, by one less turn of negative supercoiling. The reaction is equivalent to the rotation illustrated in part *D* of Figure 32.1, with the restriction that the enzyme limits the reaction to a single strand passage per event. (By contrast, the introduction of a nick in a supercoiled molecule allows free strand rotation to relieve all the tension by multiple rotations per event.)

The type I topoisomerase also can pass one segment of a single-stranded DNA through another. Two consequences of this **single-strand passage** reaction are illustrated in **Figure 32.3.**

Two single-stranded complementary circles may be permitted to renature into a duplex circle. Formation of a duplex region connecting the two circles must generate an equivalent underwinding, that is, negative T or W, elsewhere in each circle. In the absence of the enzyme, this prevents continuation of the pairing.

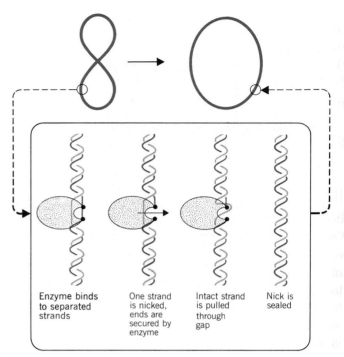

Figure 32.2
Bacterial type I topoisomerases recognize partially unwound segments of DNA and pass one strand through a break that is made in the other.

Enzyme binds to separated strands

One strand is nicked, ends are secured by enzyme

Intact strand is pulled through gap

Nick is sealed

Figure 32.3
The single-strand passage reaction of type I topoisomerases can assist renaturation of complementary circles of single-stranded DNA, can knot single-stranded DNA, and can catenate circles of (nicked) duplex DNA.

Renaturation

Knotting

Catenation

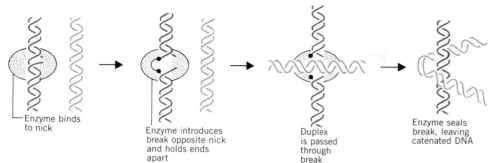

Enzyme binds to nick

Enzyme introduces break opposite nick and holds ends apart

Duplex is passed through break

Enzyme seals break, leaving catenated DNA

Figure 32.4
Type I topoisomerases can pass a duplex DNA through the intact single strand opposite a nick in another duplex DNA.

Topoisomerase I relieves the tension by passing the two strands about each other in the opposite direction from the rotation involved in pairing. This allows the reaction to continue.

If the single-strand passage reaction is performed with two parts of the *same* circular molecule, a **knot** is produced.

These actions of topoisomerase I are extended by another activity discovered more recently. The enzyme can act on two circular *duplex* molecules of DNA, provided that one of them has a nick in one strand. When one circle is passed through another, they become interlocked. The structure is called a **catenane;** the activity is described as **catenating** the DNA. (Formally, catenation and knotting are topologically similar, since the first involves interlocking of different circles, while the second involves interlocking within a single circle.)

The catenation of double-stranded circles suggests the model illustrated in **Figure 32.4,** in which the enzyme binds opposite the nick, introduces a break in the other strand, passes a duplex of DNA through the break, and reseals the break. The only difference from

the procedure involved in knotting single-stranded DNAs is that the passage reaction involves transfer of a duplex DNA. Since double-stranded and single-stranded DNA molecules have different structures, it seems surprising that the enzyme is able to handle passage of either. This could be a matter of appearance, however, if the enzyme actually passes one strand of the duplex through the break at a time (the predicted intermediates would have a complex topology).

The formal properties of eucaryotic type I topoisomerases are similar, but they can relax positive as well as negative supercoils.

Type II topoisomerases have been characterized from both procaryotes and eucaryotes. Most relax both negative and positive supercoils; the reaction is ATP-dependent. Probably one unit of ATP is hydrolyzed for each catalytic event. The reaction is mediated by making a double-stranded break in one DNA duplex, and passing another duplex region through it. This is illustrated in **Figure 32.5.** A formal consequence of two-strand transfer is that the linking number is always changed in multiples of two. The topoisomerase II ac-

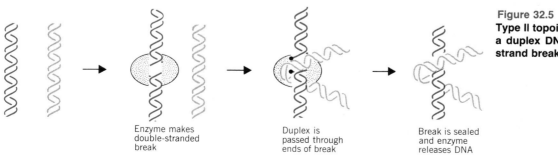

Enzyme makes double-stranded break

Duplex is passed through ends of break

Break is sealed and enzyme releases DNA

Figure 32.5
Type II topoisomerases can pass a duplex DNA through a double strand break in another duplex.

tivity can be used also to introduce or resolve catenated duplex circles and knotted molecules.

The reaction probably represents a nonspecific recognition of duplex DNA in which the enzyme binds any two double-stranded segments that cross each other. The hydrolysis of ATP may be used to drive the enzyme through conformational changes that provide the force needed to push one DNA duplex through the break made in the other. Because of the topology of supercoiled DNA, the relationship of the crossing segments will be such that the supercoils will be removed from either positively or negatively supercoiled circles. The best characterized example of this type of enzyme may be the T4 topoisomerase II, which is the product of three genes, and whose activity is essential for replication of phage DNA.

GYRASE INTRODUCES NEGATIVE SUPERCOILS IN DNA

Bacterial DNA gyrase (best characterized in *E. coli*) is a topoisomerase of type II that is able to *introduce* negative supercoils into a relaxed closed circular molecule. DNA gyrase binds to a circular DNA duplex and supercoils it processively and catalytically: it continues to introduce supercoils into the same DNA molecule. One molecule of DNA gyrase can introduce about 100 supercoils per minute.

The supercoiled form of DNA has a higher free energy than the relaxed form, and the energy needed to accomplish the conversion is supplied by the hydroly-

sis of ATP. In the absence of ATP, the gyrase can *relax* negative but not positive supercoils, although the rate is more than 10 times slower than the rate of introducing supercoils.

The *E. coli* DNA gyrase consists of two types of subunit. It is inhibited by two types of antibiotic, each of which acts on one of the subunits. Mutations that confer resistance to the antibiotics identify the loci that code for the subunits. The *gyrA* locus (formerly called *nalA*) is the site of resistance to nalidixic acid and oxilinic acid. The *gyrB* locus (formerly called *cou*) is the site of resistance to coumermycin A1 and novobiocin. These drugs inhibit replication, which suggests that DNA gyrase is necessary for DNA synthesis to proceed.

The active DNA gyrase probably has the structure A_2B_2, with an overall size of about 400,000 daltons. The GyrA subunit is ~105,000 daltons; the GyrB subunit is ~95,000 daltons. (*E. coli* cells possess about 10 times more of subunit A than subunit B. The reason is that subunit A is also associated with a 50,000-dalton protein called ν, which may be a fragment of subunit B. The Aν aggregate is called *E. coli* topoisomerase II'; it appears to be a natural component of the cell.)

The gyrase binds its DNA substrate around the outside of the protein tetramer. Gyrase protects about 140 bp of DNA from digestion by micrococcal nuclease (very similar to the protection afforded by the much smaller histone octamer). The **sign inversion** model for the action of DNA gyrase is illustrated in **Figure 32.6.** The enzyme binds the DNA in a cross-

Figure 32.6
DNA gyrase may introduce negative supercoils in duplex DNA by inverting a positive supercoil.

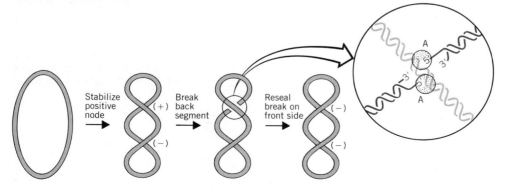

over configuration that is equivalent to a positive supercoil. This induces a compensating negative supercoil in the unbound DNA. Then the enzyme breaks the double strand at the crossover of the positive supercoil, passes the other duplex through, and reseals the break. This directly inverts the sign of the supercoil: it has been converted from a $+1$ turn to a -1 turn. Thus the linking number has changed by $\Delta L = -2$, conforming with the demand that all events involving double-strand passage must change the linking number by a multiple of two. The gyrase then releases one of the crossing segments of the now negative bound supercoil, the negative turns redistribute along DNA (as change in either T or W or both), and the cycle begins again. The same type of topological manipulation is responsible for catenation and knotting, although the roles of these rearrangements *in vivo* are unknown.

When the gyrase releases the inverted supercoil, its conformation changes. For the enzyme to undertake another cycle of supercoiling, its original conformation must be restored. This process is called **enzyme turnover.** It is thought to be driven by the hydrolysis of ATP, since the replacement of ATP by an analog that cannot be hydrolyzed allows gyrase to introduce only one inversion (-2 supercoils) per substrate. Thus it does not need ATP for the supercoiling reaction, but does need it to undertake a second cycle. Novobiocin interferes with the ATP-dependent reactions of gyrase, by preventing ATP from binding to the B subunit.

The (ATP-independent) relaxation reaction is inhibited by nalidixic acid. This implicates the A subunit in the breakage and reunion reaction. Treating gyrase with nalidixic acid allows DNA to be recovered in the form of fragments generated by a staggered cleavage across the duplex. The termini all possess a free 3'-OH group and a 4-base 5' single strand extension covalently linked to the A subunit (see Figure 32.6). The covalent linkage retains the energy of the phosphate bond; this can be used to drive the resealing reaction, explaining why gyrase can undertake relaxation without ATP. The sites of cleavage are fairly specific, occurring about once every 100 bp.

EUCARYOTIC DNA POLYMERASES

Eucaryotic DNA polymerases have been investigated the most thoroughly in mammals, where the first such enzyme was discovered as long ago as 1960 (in calf thymus). Most eucaryotic enzymes possess only the DNA-synthesizing activity, a contrast with the wide capacities of bacterial enzymes, most of which also possess exonucleolytic activity (see later).

Three classes of eucaryotic DNA polymerase have been identified. Their properties are summarized in **Table 32.1.** Can one be identified as the authentic replicase?

DNA polymerase α is the only one whose levels increase during S phase. It is resistant to 2'3'-dideoxythymidine triphosphate; the analog does not inhibit *in vitro* replication systems, but does inhibit polymerases β and γ. A fungal antibiotic, aphidicolin, inhibits eucaryotic DNA replication *in vivo;* the only enzyme that it inhibits *in vitro* is DNA polymerase α. Mutant cells resistant to aphidicolin contain a DNA polymer-

Table 32.1
There are Three Mammalian DNA Polymerases

	DNA Polymerase α	DNA Polymerase β	DNA Polymerase γ
Location	Nuclear (but leaks into cytoplasm during isolation)	Nuclear	Nuclear and mitochondrial
Proportion of activity	~80%	10–15%	2–15%
Size (daltons)	110–220,000	45,000	60,000
Subunits	Several	1	?

ase α that is resistant to the antibiotic *in vitro*. These results identify DNA polymerase α as the nuclear DNA replicase.

Other activities than replication in which DNA polymerases are involved include "repair" functions, in which segments of damaged DNA are replaced. Probably the DNA polymerase β has a repair function.

The role of DNA polymerase γ presumably is to undertake the replication of mitochondrial DNA. Its role in the nucleus (if any) is unknown.

DNA replication requires many activities in addition to the DNA polymerase. A difficulty in analyzing the role of the eucaryotic DNA polymerases is caused by the lack of systems containing all the components needed for replication. However, there is progress toward obtaining suitable *in vitro* systems that can replicate defined templates. Some nuclear systems have been obtained that contain all of the components necessary to continue replication fork movement. Initiation of synthesis at the origins of new replicons has not yet been obtained *in vitro*.

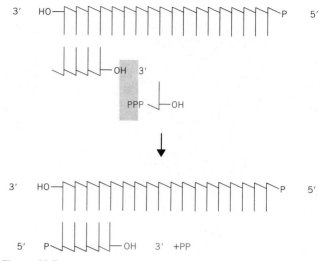

Figure 32.7
DNA synthesis occurs by addition of nucleotides to the 3′-OH end of the growing chain.

PROCARYOTIC DNA POLYMERASES HAVE SEVERAL ENZYMATIC ACTIVITIES

Three DNA polymerase enzymes have been characterized in *E. coli*. The first to be discovered, DNA polymerase I, is involved in the repair of damaged DNA and, in a subsidiary role, in semiconservative replication. It is coded by the locus *polA*. The second enzyme, DNA polymerase II, is coded by *polB,* but nothing is known of its function *in vivo*. The third enzyme, DNA polymerase III, is a multisubunit protein,

including one subunit coded by the locus *polC* (also known as *dnaE*). This enzyme is the replicase responsible for de novo synthesis of new strands of DNA. The enzymatic activities of the three enzymes are summarized in **Table 32.2.**

All procaryotic and eucaryotic DNA polymerases share the same fundamental type of synthetic activity. Each can extend a DNA chain by adding nucleotides one at a time to a 3′-OH end. Thus the new chain is synthesized in the usual direction, 5′-3′. The reaction is illustrated diagrammatically in **Figure 32.7.** The precursor for DNA synthesis is a nucleoside triphosphate,

Table 32.2
There are Three DNA Polymerases in *E. coli*

	DNA Polymerase I	DNA Polymerase II	DNA Polymerase III
Size (daltons)	109,000	120,000	>250,000
Composition	Single chain	—	Heteromultimeric complex
Number per cell	~400	—	10–20
Enzymatic activities	5′-3′ elongation from 3′-OH primer	5′-3′ elongation from 3′-OH primer	5′-3′ elongation from 3′-OH primer
	3′-5′ exonuclease	3′-5′ exonuclease	3′-5′ exonuclease
	5′-3′ exonuclease		5′-3′ exonuclease
Mutants	*polA* is defective in repair activities, has reduced viability only when 5′-3′ exonuclease affected	*polB* has no effect	*polC, dnaN, dnaX, dnaZ*, are conditional lethals in DNA replication

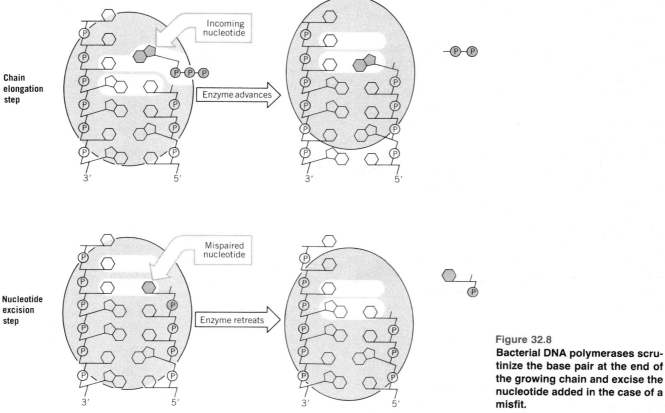

Figure 32.8
Bacterial DNA polymerases scrutinize the base pair at the end of the growing chain and excise the nucleotide added in the case of a misfit.

which loses the terminal two phosphate groups in the reaction. The choice of which nucleotide is added to the chain is dictated by base pairing with the template strand.

The fidelity of replication poses the same sort of problem we have already encountered in considering (for example) the fidelity of translation. It relies on the specificity of base pairing. Yet when we consider the interactions involved in base pairing, we would expect errors to occur with a frequency of 10^{-4} to 10^{-5} per base pair replicated. The actual rate in *E. coli* cells or in phage T4 replication seems to be in the range of 10^{-8} to 10^{-10} errors per base pair replicated. (This corresponds to approximately 1 error per 1000 bacterial replication cycles, or, in other terms, to a rate of $\sim 10^{-6}$ per gene per generation.)

The enzyme might improve the specificity of complementary base selection at either (or both) of two stages. It could scrutinize the incoming base for the proper complementarity with the template base; for example, by specifically recognizing matching chemical features. This would be a *presynthetic* error control. Or it could scrutinize the base pair *after* the new base has been added to the chain, and, in those cases in which a mistake has been made, remove the most recently added base. All of the bacterial enzymes possess a 3'-5' exonucleolytic activity that appears to exercise just such a **proofreading** function. The absence of this activity from the eucaryotic enzymes implies that they must have some other mechanism for controlling the error rate.

The 3'-5' exonucleolytic activity takes place in the reverse direction from DNA synthesis. The proofreading action is illustrated diagrammatically in **Figure 32.8.** In the chain elongation step, a precursor nucleotide enters the position at the end of the growing chain. A bond is formed. The enzyme moves one base pair farther, ready for the next precursor nucleotide to enter. If a mistake has been made, the enzyme moves backward, excising the last base that was added, and thus creating a site at which an alternative precursor nucleotide can enter to replace it. This activity has been the most fully characterized with DNA polymerase I, but is thought to function in an analogous manner with other procaryotic DNA polymerases.

DNA polymerase I is a talented enzyme with several activities. It is a single polypeptide of 109,000 daltons. The chain can be cleaved into two regions by proteolytic treatment. The large fragment (76,000 daltons) contains the 5'-3' polymerase and the 3'-5' exonuclease activities. The small fragment (36,000 daltons) possesses another activity of the enzyme, a 5'-3' exonucleolytic activity. This activity is therefore independent of the synthetic/proofreading activities. The 5'-3' exonuclease action excises small groups of nucleotides, up to ~10.

The DNA polymerase I enzyme has a unique ability to start replication *in vitro* at a nick in DNA. At a point where a phosphodiester bond has been broken in a double-stranded DNA, the enzyme extends the 3'-OH end. As the new segment of DNA is synthesized, it *displaces the existing homologous strand in the duplex*. This process of **nick translation** is illustrated in **Figure 32.9.** The displaced strand is degraded by the 5'-3' exonucleolytic activity of the enzyme. The result is to leave a DNA whose properties are unaltered, except that a segment of DNA has been replaced with newly synthesized material, and the position of the nick has been moved along the duplex. This is of great practical use, because it allows radioactively labeled nucleotides to be inserted into DNA *in vitro*. Nick translation is a major technique for producing radioactively labeled DNA.

Although there is a large amount of DNA polymerase I in the bacterium, it is not responsible for replication, which continues in *polA* mutants. The 5'-3' synthetic/3'-5' exonuclease action is probably used *in vivo* mostly for filling in short single-stranded regions in double-stranded DNA. These regions arise during replication (see later and Chapter 33), and when bases that have been damaged are removed from DNA (see Chapter 34).

Figure 32.9
Nick translation replaces part of a preexisting strand of duplex DNA with newly synthesized material.

The enzyme DNA polymerase III was originally discovered by the conditional lethality of a mutation in the *polC* locus, which codes for the 140,000 α subunit that is a component of all the enzyme preparations. Purified DNA polymerase III activities have several subunits. We use the plural "activities" because the enzyme has been obtained in several forms, and it is unclear which (if any) represents the unit involved in replication *in vivo*. These forms are summarized in **Table 32.3.**

All the subunits listed in the table are needed for the full activity of *in vitro* replication systems (see Chapter 33). There have been attempts to divide the subunits into components of the DNA replicase and "accessory proteins" needed to assist the replicase. However, the *in vitro* systems contain many additional activities, and it is not at all clear whether there is in fact such an entity as a "DNA replicase." The function of DNA synthesis might be undertaken only as part of a more complex replication apparatus in which the only discernible replicase activity would be that of the polypeptide actually involved in catalyzing elongation of the DNA chain.

The bacterium *Bacillus subtilis* also contains three DNA polymerase enzymes, which are numbered to show the resemblance with their counterparts in *E. coli*. There are some differences in their exonucleolytic activities relative to the *E. coli* enzymes. The subunit identified with the DNA synthetic activity of *B. subtilis* DNA polymerase III is stimulated by factors I and II of *E. coli* (subunits βδγ), which suggests that there may be a related division of labors in the replication apparatus of the two bacteria. The level of *B. subtilis*

Table 32.3
Different Combinations of Subunits Make Up Various Forms of DNA Polymerase III

Subunit	Size (daltons)	Other Names and/or Gene	Subunits Present In:			
			Core Pol III	Pol III′	Pol III*	Holoenzyme
α	140,000	*polC* (*dnaE*)	+	+	+	+
β	40,000	*dnaN* = factor I = coPol III*				+
δ	32,000	*dnaX* = factor III ⎱ factor II			+	+
γ	52,000	*dnaZ* ⎰			+	+
ε	25,000		+	+	+	+
θ	10,000		+	+	+	+
τ	83,000			+	+	+

The stoichiometry of the various complexes is not certain, although in most cases there is only one copy of each subunit.

DNA polymerase II is related to physiological events; it is absent from dormant spores, but appears about 30 minutes after the start of germination.

Some phages also code for DNA polymerases. These include T4, T5, T7, and SPO1. The enzymes all possess 5′-3′ synthetic activities and 3′-5′ exonuclease proofreading activities. In each case, a mutation in the gene that codes for a single-phage polypeptide prevents phage development. Each phage polymerase polypeptide associates with other proteins, of either phage or host origin. The examples of phages T4 and T7 are discussed in Chapter 33.

DNA SYNTHESIS IS SEMIDISCONTINUOUS

The antiparallel structure of the two strands of duplex DNA poses a problem for replication. As the replication fork advances, daughter strands must be synthesized on both of the exposed parental single strands. The fork is moving in the direction from 5′ to 3′ on one strand, and in the direction from 3′ to 5′ on the other strand. Yet nucleic acids are synthesized only from a 5′ end toward a 3′ end. The problem is solved by synthesizing the strand that grows overall from 3′ to 5′ in a series of short fragments, each actually synthesized with the customary 5′ to 3′ polarity.

Consider the region immediately behind the replication fork, as illustrated in **Figure 32.10**. On one strand, DNA synthesis can proceed continuously in the 5′ to 3′ direction as the parental duplex is unwound. This is called the **leading strand.** On the other strand, however, a stretch of single-stranded parental DNA must be exposed, and then a segment is synthesized in the reverse direction (relative to fork movement). This is called the **lagging strand.** A series of these fragments are synthesized, each 5′ to 3′, and then they are joined together to create an intact lagging strand. This behavior is called **discontinuous replication.**

Discontinuous replication is studied in terms of the fate of a very brief label of radioactivity. The label enters newly synthesized DNA in the form of short fragments, sedimenting in the range of 7–11S, corresponding (approximately) to sequences of 1000–2000 bases in length. These are called **Okazaki fragments.** They are found in all replicating DNA, both procaryotic and eucaryotic. After longer periods of incubation, the label enters larger segments of DNA. This transition represents the covalent linkage of Okazaki fragments.

The lagging strand *must* be synthesized in the form of Okazaki fragments. There has been a protracted controversy as to whether the leading strand is synthesized in the same way or is synthesized continuously.

In *E. coli*, all of the newly synthesized DNA is found in the form of short fragments. With phage T4, it has been shown that both strands are represented in the Okazaki fragments. Superficially, this suggests that both strands are synthesized discontinuously. However, it turns out that not all of the fragment population rep-

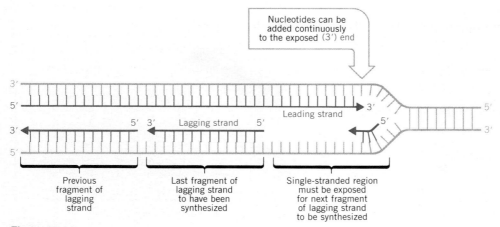

Figure 32.10
The leading strand is synthesized continuously while the lagging strand is synthesized discontinuously.

resents bona fide Okazaki fragments; some are pseudofragments, generated by breakage in a DNA strand that actually was synthesized as a continuous chain.

The source of this breakage is the incorporation of some uracil into DNA in place of thymine. At one time, it was thought that such misincorporation is a rare event. But now it seems that *E. coli* DNA polymerase III does not discriminate well between dUTP and dTTP *in vitro* and (presumably) *in vivo*.

Figure 32.11 shows the pathway for production of dUTP and dTTP. Some 75% of the dUTP is produced by deamination of dCTP, the rest by phosphorylation of dUDP. The dUTP is converted by the enzyme **dUTPase** into dUMP (which is also produced by deamination of dCMP). The dUMP is the intermediate used to synthesize dTMP, leading to the production of dTTP, the precursor for DNA synthesis.

Under normal conditions, the ratio of dUTP to dTTP in *E. coli* seems to be about 1:300. If there is no discrimination between dUTP and dTTP, this would correspond to the incorporation on average of a U residue once every 1200 bases of newly synthesized DNA. What happens to this uracil residue? It is removed by an enzyme, **uracil-DNA-glycosidase,** which is important as a defense against mutation as well as against replication errors. (The mutation pathway produces uracil by deaminating cytosine, as described in Chap-

ter 2.) The uracil-DNA-glycosidase activity has been identified in both procaryotes and eucaryotes. It is likely to be a ubiquitous system that protects DNA against the presence of uracil.

The removal of the uracil is accomplished by cleaving the base from the sugar. This reaction is followed by the introduction of a break in the phosphodiester backbone. The break may be widened into a gap by the removal of further nucleotides. These gaps are filled by synthesizing DNA to replace the lost material. This series of events is illustrated in **Figure 32.12.** Dur-

Figure 32.11
The synthesis of dUTP and dTTP is connected.

About 1
U residue
present per
1200 nucleotides

Gaps are
created when
the U bases
are removed

The gaps created
by the U excisions,
and between the
Okazaki fragments
are filled

Figure 32.12
Excision of uracil from newly synthesized DNA creates gaps in both the continuously synthe-
sized leading strand and the Okazaki fragments of the lagging strand. The gaps must be filled
and the Okazaki fragments must be linked to generate continuous DNA strands.

ing the period between the creation and filling of the gap, the breaks in the newly synthesized DNA generate fragments that are comparable in size to the Okazaki fragments. (The same events also reduce the size of the Okazaki fragments of the lagging strand.)

The importance of two enzymes, the uracil-DNA-glycosidase and the dUTPase, is shown by the effects of mutations in their genes.

The excision of uracil is prevented in *ung⁻* mutants of *E. coli,* which lack the glycosidase enzyme. In these mutants, about half of a radioactive label appears in newly synthesized DNA in the form of (genuine) Okazaki fragments; the other half directly enters much larger fragments. In this situation, we can see that the leading strand is synthesized continuously.

The cellular balance between dUTP and dTTP is greatly influenced by the dUTPase activity, since this simultaneously serves the purpose of removing dUTP and increasing passage along the pathway leading to synthesis of dTTP. The dUTPase is inactive in *dut⁻* (also called *sof⁻*) mutants of *E. coli.* Its absence greatly increases the amount of dUTP at the expense of dTTP, with the result that the mutant cells display large amounts of small fragments of newly synthesized DNA.

Thus the lagging strand is synthesized discontinuously and the leading strand is synthesized continuously in *E. coli.* The same is probably true in other systems. This mode of synthesis is called **semidiscontinuous replication.**

OKAZAKI FRAGMENTS
ARE PRIMED BY RNA

We have already mentioned that all of the known DNA polymerases of both procaryotes and eucaryotes can extend a deoxyribonucleotide chain from a free 3′-OH end; but none can initiate synthesis. Yet there must be continual initiation events, since each Okazaki fragment of the lagging strand requires its own start de novo. (This demand is distinct from the question of how the replication fork initiates synthesis at the origin, which we consider in Chapter 33.) Each Okazaki fragment starts with a **primer,** a (short) sequence that provides the 3′-OH end for extension by DNA polymerase. The primer is RNA, about 10 bases long.

We can now expand our consideration of the actions involved in joining Okazaki fragments, as illus-

RNA primer

Okazaki fragment

RNA removed

Gap filled
with DNA

Nick
sealed

Figure 32.13
**Linkage of Okazaki fragments requires removal
of RNA, gap filling, and nick ligation.**

trated in **Figure 32.13.** The order of events is uncertain, but must involve the removal of the RNA primer, its replacement by a stretch of DNA, and the covalent linking of adjacent Okazaki fragments. The figure suggests that synthesis of an Okazaki fragment terminates just before the start of the RNA primer of the preceding fragment. The enzyme DNA polymerase I is likely to be involved in filling the gap left by removal of the primer, since *polA* mutants may fail to join their Okazaki fragments properly. Indeed, this enzyme might use its 5'-3' exonuclease activity to remove the RNA primer while simultaneously replacing it with a DNA

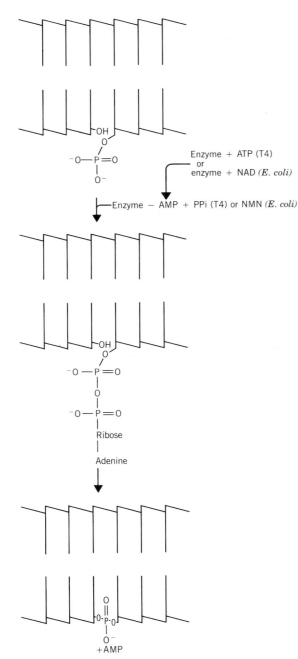

Figure 32.14

DNA ligase seals nicks between adjacent nucleotides by employing an enzyme-AMP intermediate.

sequence extended from the 3'-OH end of the next Okazaki fragment. This is equivalent to the activity of nick translation, except that the new DNA replaces a stretch of RNA rather than a segment of DNA.

Once the RNA has been removed and replaced, it remains necessary just to link the adjacent Okazaki fragments. With all the gaps filled, the 3'-OH end of one fragment will be adjacent to the 5'-phosphate end of the previous fragment. The responsibility for sealing this nick lies with the enzyme **DNA ligase** (which we have mentioned previously in connection with its use in the *in vitro* reconstruction of DNA). Ligases are present in both procaryotes and eucaryotes; the two most intensively studied are those of *E. coli* and phage T4. The enzymes are implicated in joining Okazaki fragments *in vivo* by the persistence of unconnected fragments in *lig⁻* mutants.

The *E. coli* and T4 enzymes share the property of sealing nicks that have 3'-OH and 5'-phosphate termini, as illustrated in **Figure 32.14,** although they use different cofactors. The *E. coli* enzyme uses NAD (nicotinamide adenine dinucleotide) as a cofactor, while the T4 enzyme uses ATP. Both enzymes undertake a two-step reaction, in which the first step is the formation of an enzyme-AMP complex. This is achieved in *E. coli* by splitting NMN from NAD, and in T4 by cleaving pyrophosphate from ATP. The AMP of the enzyme complex becomes attached to the 5'-phosphate of the nick; and then a phosphodiester bond is formed with the 3'-OH terminus of the nick, releasing the enzyme and the AMP.

The discovery of these events raises questions about the specificity of synthesis of the Okazaki fragments. What enzyme is responsible for synthesizing the RNA primer? Does it start and stop at specific sequences, or is its action determined by the organization of the replication fork? Does the synthesis of an Okazaki fragment continue until it reaches the adjacent fragment, or are there specific termination points?

FURTHER READING

Topoisomerases are much in vogue at present. The distinction between types has been considered by **Liu, Liu,** & **Alberts** (*Cell* **19,** 697–707, 1980) and by **Cozzarelli** (*Cell* **22,** 327–328, 1980). Gyrase actions have been reviewed by

Cozzarelli (*Science,* **207,** 953–960, 1981). A general review of all topoisomerases has been accomplished by **Gellert** (*Ann. Rev. Biochem.* **50,** 879–910, 1981). Eucaryotic DNA polymerases have been summarized by **Huberman** (*Cell* **23,** 647–648, 1981). References to the process of replication and the enzymes involved will be found in Chapter 33.

CHAPTER 33
THE ENZYMATIC APPARATUS
FOR DNA REPLICATION

Replication of duplex DNA is a complex endeavor that requires a conglomerate of enzyme activities. The parental strands must be separated and (temporarily) stabilized in the single-stranded state. The leading strand must be initiated and then elongated. Okazaki fragments of the lagging strand must be continually initiated, elongated, and linked. At the end of the replicon, joining and/or termination reactions are necessary. Although these various activities can be equated with individual proteins or groups of proteins, they do not function independently; all may be contained in a discrete multiprotein structure, sometimes called the **replisome.** This is not evident as a free cellular body (for example, analogous to the ribosome), but may be assembled from its components at the onset of replication. Thus the replisome may exist *only* as a protein complex associated with the particular structure that DNA takes at the replication fork.

With eucaryotes, we have at present made little more than a start toward defining the assembly of proteins involved in replication. With procaryotes, two approaches converge in defining the components of either the bacterial or a phage replication apparatus. First, mutants defective in replication identify genes whose products either are components of the apparatus for replication or are involved in peripheral functions. Second, the genomes of certain small phages can be used to provide templates that are replicated *in vitro* by a system of defined host (or phage-coded) proteins. There has therefore been appreciable progress toward defining a replication apparatus in which each component is available for study *in vitro* as a biochemically pure product, and is implicated *in vivo* by the effect of mutation in its gene.

THE COMPLEXITY OF THE BACTERIAL REPLICATION APPARATUS

The inability to replicate DNA is fatal for any cell. Mutants in replication must therefore be obtained as conditional lethals, able to accomplish replication under permissive conditions (provided by the normal temperature of incubation), and displaying their defect only under nonpermissive conditions (provided by the higher temperature of 42°C). These temperature-sensitive mutants identify a series of loci described as *dna* genes. The *dna* mutants are divided into two general classes on the basis of their behavior when the temperature is raised.

The **slow-stop mutants** complete the current round of replication, but cannot start another. They are therefore defective in the events involved in initiating a cycle of replication at the origin. Not a great deal is known about them (see Table 33.3 later). The main point to be made at present is that several different proteins are needed to initiate a cycle of replication.

The **quick-stop mutants** cease replication *immediately* on a temperature rise. Those concerned directly with replication are defective in the components of the replisome (see Table 33.1 later). Mutants assigned to the quick-stop class could also be defective in initiating cycles of replication, although this would be obscured by the quick-stop phenotype.

Some quick-stop mutants are defective in the production of the precursors needed for DNA synthesis. For example, mutations in *nrdA* were originally called *dnaF;* this gene codes for a subunit of ribonucleotide reductase, which is essential for providing precursors. Similarly, the *dut⁻* mutations that eliminate dUTPase activity were originally called *dnaS.* These mutants illustrate the importance of the precursor supply chain. Thus the isolation of a mutation with the Dna phenotype does not imply that its gene product is directly involved in replication.

Although efforts to isolate *dna* mutants have been intensive, the mutations do not (yet) identify all the functions needed for replication. Several additional proteins, coded by unknown genes, are needed for the function of *in vitro* replication systems (see Table 33.2 later).

When extracts of *E. coli* are assayed simply for their ability to synthesize DNA, the predominant enzyme activity is that of DNA polymerase I. In fact, its activity is so great as to make it impossible to detect the activities of the enzymes actually responsible for DNA replication. To develop *in vitro* systems in which replication can be followed, extracts have therefore been prepared from *polA* mutant cells.

A halfway house is presented by systems that allow the features of replication to be delineated more clearly than *in vivo,* although they cannot be dissected into individual components. One system treats bacteria with lipid solvents, to render the membrane permeable to the precursors for DNA synthesis (and other small molecules). The treated cells are no longer viable, but can replicate their DNA at rates barely lower than the usual *in vivo* rate. Another system lyses bacteria on a cellophane disk held on an agar surface. The replication apparatus is held together on top of the disk, to which further components can be added.

The characterization of *in vitro* replication systems in terms of their individual components requires a template that allows the nature of the DNA synthesis to be assayed. We must be sure that synthesis can be equated with an authentic *in vivo* replication process, not an alternative activity, such as the repair of damaged DNA. Different templates can be used to establish *in vitro* systems whose characteristics reflect the synthesis of either lagging strands or leading strands.

With a suitable *in vitro* system, the roles of individual components can be distinguished by two approaches. First, the complete system can be fractionated into components that are purified and reconstituted. Second, the system can be prepared from a *dna* mutant and operated under conditions in which the mutant gene product is inactive. Then extracts from wild-type cells can be tested for their ability to restore activity. This **in vitro complementation assay** can be used to purify the protein coded by the *dna* locus.

INITIATING SYNTHESIS OF A SINGLE DNA STRAND

The production of an Okazaki fragment on the lagging strand involves the synthesis of a DNA complement to the single-stranded region exposed by the movement of the replication fork. Three single-stranded DNA phages have been used as templates to provide a paradigm for the events involved in initiating and then extending the synthesis of such a DNA strand. Each of these phage genomes consists of a circular single strand of DNA, called the **viral** or (+) strand. **Figure 33.1** shows that synthesis of a **complementary** or (−) strand converts the genome into a duplex circular **replicative form** (**RF**) DNA. This may exist as either a relaxed circle (RFII) or supercoiled circle (RFI). Synthesis of the complementary strand is regarded as an analog for synthesis of the lagging strand of duplex DNA.

Although each of these phage genomes shows little intrastrand base pairing and therefore remains largely in the single-stranded state, the DNA itself does not

Viral single-
stranded (+) DNA

Duplex DNA of viral (+) strand
base paired with
complementary (−) strand as RF II

Supercoiled circular duplex, RF I

Figure 33.1
**Single-stranded circular phage DNA is converted into a duplex
by synthesis of the complementary strand.**

provide a suitable template for the replication reaction. The nucleic acid first must be coated by the host **single-strand binding protein (SSB)**. This is a tetramer of 74,000 daltons that binds cooperatively to single-stranded DNA, maintaining it in the extended state. The significance of the cooperative mode of binding is that the binding of one protein molecule makes it much easier for another to bind. So once the binding reaction has started on a particular DNA molecule, it is rapidly extended until all of the single-stranded DNA is covered with the SSB protein. Note that this protein is *not* a DNA-unwinding protein; its function is to stabilize DNA that is already in the single-stranded condition.

The simplest of the replication systems is presented by phage M13. The relevant stages are illustrated in **Figure 33.2.** An exception to the generally single-

stranded structure of the DNA exists in a sequence of 59 bases that forms a hairpin. This is the *only* part of the phage genome that is not bound by the SSB. The hairpin provides a signal for initiation.

The host RNA polymerase initiates the synthesis of RNA six bases before the start of the duplex hairpin. The RNA chain is extended for 20–30 residues, disrupting the hairpin. The SSB binds to the nontemplate half of the hairpin released by the formation of the RNA-DNA hybrid. This action is probably responsible for terminating transcription at the top of the hairpin. Because the reaction depends on RNA polymerase, it is inhibited by rifampicin.

We do not yet know how the RNA polymerase recognizes the hairpin as a site for initiating transcription; it is possible that other proteins may be involved in accessory capacities. Deletion of the entire sequence coding for the RNA primer does not render the phage inviable, although it does grow more slowly. Thus there must be other sequences that can be used as origins, albeit less efficiently in *E. coli*. (Perhaps they provide more effective origins in other bacterial hosts.)

The RNA primer is extended by DNA polymerase III, which synthesizes a DNA strand that extends all the way around the circle until it reaches the start of the RNA primer. The RNA is removed by an exonucleolytic action; it is not certain which enzyme is responsible. It may be DNA polymerase I, which also could extend the new DNA strand to fill the gap. This leaves only the nick between the start and the end of the complementary strand to be sealed by the ligase.

When the single-stranded circular DNA of the phage G4 provides the template, there is one significant difference in the reaction. Instead of the host RNA polymerase, the protein coded by the *dnaG* gene is needed. This enzyme is a single polypeptide of 60,000 daltons (much smaller than RNA polymerase). It is a **primase.** One molecule of the DnaG protein binds directly to the SSB-resistant duplex hairpin in the DNA. The primase synthesizes a 29-base RNA that is complementary to one strand of the hairpin. When DNA synthesis is directly coupled with the priming action (occurring simultaneously instead of following later as a separate step), the primer is shorter, sometimes as short as six bases. The priming reaction is not inhibited by rifamp-

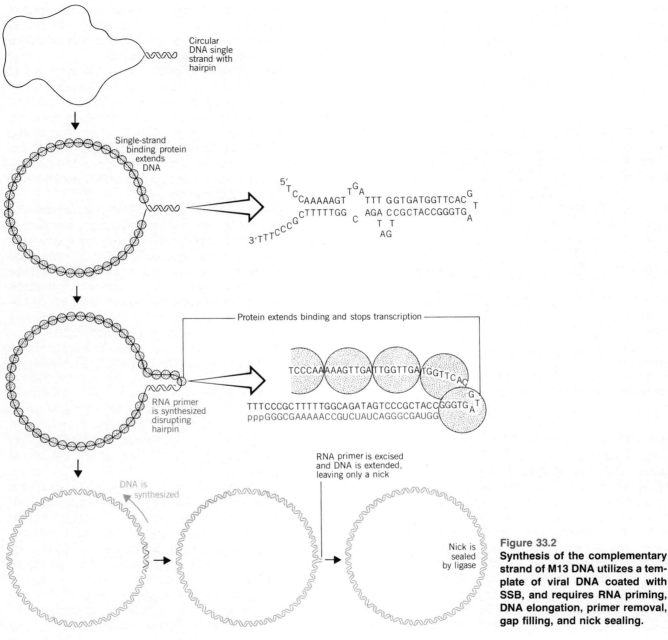

Figure 33.2
Synthesis of the complementary strand of M13 DNA utilizes a template of viral DNA coated with SSB, and requires RNA priming, DNA elongation, primer removal, gap filling, and nick sealing.

icin; the primase also is distinguished from RNA polymerase by its ability to use deoxyribonucleotides as well as ribonucleotide precursors *in vitro*.

The situation of M13 and G4 shows that an RNA primer may be synthesized by RNA polymerase or by primase, in either case at a unique site presumably indicated by the secondary structure of the DNA. The reaction may be terminated by the topology of the initiation region and/or by the action of DNA polymerase. Extension of a DNA strand from the 3'-OH end of the primer involves the same synthetic events for either phage.

MOVEMENT OF THE PRIMOSOME

The situation of phage φX174 may be a better paradigm for the initiation of Okazaki fragments, since DNA synthesis is initiated at several sites on the phage genome. The φX174 system is compared with the M13 and G4 systems in **Figure 33.3.** A much greater number of proteins are needed to replicate φX174 DNA. Some of these have been identified as the products of *dna* (or other) genes, as summarized in **Table 33.1;** others are available only as proteins needed for the *in vitro* reaction, as described in **Table 33.2.**

The first step with φX174 is the same as with M13 or G4. The DNA is coated with SSB. But then it is converted to a **prepriming** intermediate by six proteins: n, n', n", i, DnaB, and DnaC. The prepriming complex assembles at a specific site, a hairpin formed from a 55-base sequence (at a location analogous to the unique start point in G4 DNA). After the prepriming complex has been formed, the DnaG primase can recognize the template; it synthesizes a variety of short molecules, 15 to 50 bases long, starting at many different sites on φX174 DNA. A revealing feature of this reaction is the discrepancy between the formation of the prepriming complex at a unique location and the consequent ability of the primase to initiate RNA synthesis at many different locations.

Protein n' may be the component responsible for selecting the site at which the prepriming complex is assembled. The protein displays an ATPase activity that depends specifically on the φX174 55-base hairpin. (Formation of a hairpin per se is not adequate to constitute the n' recognition site, because the hairpins of neither M13 nor G4 DNA can substitute for the φX174

Table 33.1
Some Components of the Replication Apparatus Are Identified By Mutations

Gene	Function	Product (size in daltons)	Copies per Cell
dnaB	Prepriming	300,000 hexamer	~20
dnaC	Acts with dnaB	Unknown	Unknown
dnaG	Primase initiates Okazaki fragments	60,000 monomer	~75
polC (dnaE)	DNA polymerase III	140,000 subunit	~20
dnaZ	DNA polymerase III	52,000 subunit	~20
lig	Seals nicks between Okazaki fragments	75,000 monomer	~300
rep	ATP-dependent helicase	66,000 monomer	~50
ssb	Single-strand DNA binding protein	74,000 tetramer	~300

Table 33.2
Other Proteins Also Are Components of the Replication Apparatus

Protein	Function	Product (size in daltons)	Copies per Cell
i	Prepriming (1–2/fork)	66,000 trimer	~50
n' (= Y)	Site recognition, ATPase (1/fork)	55,000 monomer	~80
n"	Prepriming	Unknown	Unknown
n			

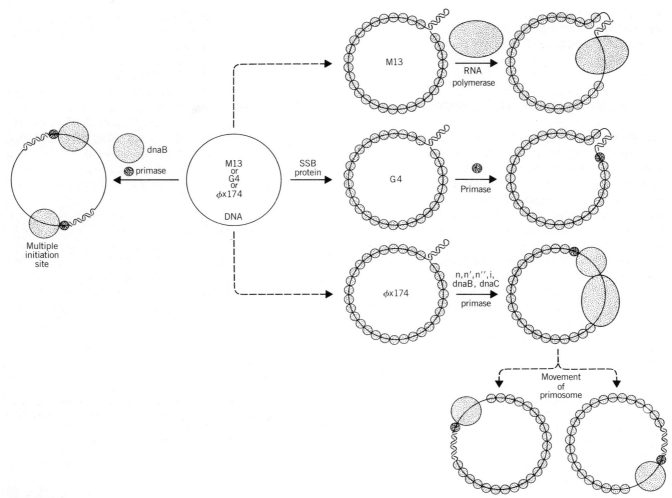

Figure 33.3
**Free DNA can be primed at multiple sites by DnaB•primase. SSB-coated M13 or G4 DNA is
primed at only one site. SSB-coated øX174 DNA requires several additional proteins to form a
prepriming complex at a unique site; then priming occurs at several sites.**

hairpin.) The amount of ATP hydrolyzed, and the use made of the released energy, is not known.

When øX174 DNA is *not* coated with SSB, the requirement for the prepriming proteins is lost. With pure single-stranded øX174 DNA, the primase needs only the presence of DnaB protein in order to synthesize an RNA primer. The DnaB•primase combination also is active with noncoated M13 or G4 DNA, or, for that matter, with a template of poly(dT).

Multiple primers are synthesized when DnaB and primase are added to øX174 DNA. The RNAs range in size from 10 to 60 bases, and probably represent selection of a number of specific sites. Thus DnaB•primase has the ability to recognize a series of initiation sites in free DNA; this ability is abolished by the presence of SSB. Restoration of priming activity requires the other proteins of the prepriming complex: n, n', n'', i, and DnaC. We know that the latter situation reflects the *in vivo* reality, because SSB is needed for replication in bacteria, as indicated by the quick-stop phenotype produced by mutation in its gene.

The complex of proteins involved in the priming reaction has been called the **primosome.** One interpretation of the discrepancy between the unique prepriming site and the multiple priming sites is to suppose that the primosome initially can assemble only at one location; but then some or all of the proteins can move along the single-stranded DNA to the sites at which priming actually occurs. The moving complex must include at least the DnaB and primase proteins; the n' protein is also likely to be a component, because it has the ability to displace SSB from DNA, probably a necessary function for movement.

The behavior of DnaB also may be important in movement. There is probably one hexamer of DnaB per replication fork. In the presence of ATP, the protein binds to single-stranded DNA with high affinity. The primase further stabilizes the complex of DnaB•ATP with single-stranded DNA. Replacement of the ATP by ADP effects an allosteric change that reduces the affinity of DnaB for the DNA. Thus the hydrolysis of ATP after the priming reaction could be necessary to allow DnaB to dissociate from DNA to move on to another priming site (where the ADP is replaced by ATP).

The use of linear fragments of φX174 DNA identifies the relationship between the unique prepriming site and the multiple priming sites. The primosome seems to move primarily in the antielongation direction—along the parental strand from 5' toward 3'. **Figure 33.4** shows that this would be suitable for initiating the Okazaki fragments of the lagging strand of a duplex DNA. As the replication fork advances, it creates a single-stranded region ahead of the primosome (on its 3' side). After each initiation event, the primosome moves along the single-stranded stretch to the site for starting the next Okazaki fragment. Thus the primosome moves in the same direction as the replication fork, but in the opposite direction from DNA synthesis of the lagging strand.

To bring the reaction characterized with φX174 DNA further into the context of the replication of duplex DNA, we must consider how the single-stranded template is prepared for synthesis of the lagging strand. The SSB presumably binds to DNA as the replication fork

Parental strand
Daughter DNA strand
RNA primer

Figure 33.4
The primosome could follow the replication fork, moving 5'-3' along the template for lagging strand synthesis, periodically initiating synthesis of Okazaki fragments.

advances, keeping the two parental strands separate so that they (or at least the lagging strand) are in the appropriate condition to act as template. The SSB is needed in stoichiometric amounts at the replication fork.

What is responsible for initially unwinding the two strands of the parental duplex? The probable function is identified by *rep* mutations, which slow the movement of the replication fork. The Rep protein is a single strand-dependent ATPase. In the presence of ATP, the combined action of Rep protein and SSB can separate a nicked duplex of φX174 RF DNA into its constituent single strands. Probably the Rep protein unwinds the strands, which are then trapped in the single-stranded state by the SSB.

Figure 33.5
Rep protein may move along one strand of DNA, displacing the partner strand.

A model for Rep action is illustrated in **Figure 33.5.** The protein binds to a single-stranded region within a duplex DNA, migrating along in a fixed direction. When it reaches the duplex segment, it displaces the 5′ end. Every base pair unwound requires the hydrolysis of two ATP molecules.

A problem in assigning Rep the role of unwinding at the replication fork is that *rep* mutants continue to replicate bacterial DNA, even though they cannot separate the strands of φX174 DNA. Survival of M13 DNA also depends on Rep; in *rep*⁻ mutant hosts, progeny (single-stranded) phage DNA cannot be generated from the circular RF DNA. However, if an *E. coli* origin is inserted into M13 DNA, replication can occur. Thus two phage DNAs require Rep for replication, but bac-

terial DNA can manage without it. The *E. coli* origin probably can initiate and then sustain strand separation by some other (unknown) mechanism.

INITIATING REPLICATION AT DUPLEX ORIGINS

Starting a cycle of replication of duplex DNA requires two types of activity in addition to the functions needed to synthesize Okazaki fragments. First, the two strands of DNA must suffer their initial separation. (There have been speculations that acquisition of the unusual secondary structure shown in Figure 31.8 could be involved in this step.) Second, synthesis of the leading strand must be initiated. This may or may not occur by the same or a similar mechanism as that involved in initiating Okazaki fragment synthesis.

The slow-stop mutations of *E. coli* identify several loci whose products may participate in the events at the origin, but very little is known about their molecular activities. Of the loci summarized in **Table 33.3,** information is available about the functions of only three. Mutations of either the slow-stop or quick-stop phenotype may occur in *dnaB* and *dnaC,* and we have some information about the function of the proteins in the continuation of replication (see earlier). We assume that their role at the origin is related. Some mutations in *dnaA* can be suppressed by mutations in RNA polymerase, which suggests that DnaA could be involved in some initiation step that requires RNA syn-

Table 33.3
Slow-Stop Mutations Are Not Well Characterized

Gene	Function
dnaA	Concerned with initiation of replication at origin
dnaB	One mutation has slow-stop phenotype; for others see Table 33.1
dnaC	Product acts in concert with dnaB
dnaI	Unknown
dnaJ	Unknown
dnaK	Unknown
dnaP	Resistance to phenethyl alcohol; concerned with membrane?

thesis. Attempts to identify the role of DnaA have been hindered by difficulties in purifying the protein.

An involvement of RNA polymerase in initiation at the origin also is implied by the observation that rifampicin can inhibit the start of bacterial replication at a time when inhibitors of protein synthesis are ineffective. This means that the synthesis of RNA per se is important (rather than its translation into protein). A suitable system for studying this function is provided by phage lambda, whose origin sponsors bidirectional replication, and could take a cloverleaf structure similar to that postulated for *E. coli*. Initiation of replication at the lambda origin requires "activation" by transcription.

Replication of phage lambda cannot initiate when repressor is bound to O_R, because the binding prevents transcription from starting at P_R in order to proceed through the origin. A map of the region is shown in **Figure 33.6.** The need for RNA synthesis does not necessarily imply that the RNA provides a primer for the leading strand. An alternative is that RNA synthesis could be involved in promoting some structural change in the region.

The replication of phage lambda requires the products of phage genes *O* and *P,* as well as several host functions, including the products of slow-stop genes. The phage O protein binds to the lambda origin; the phage P protein interacts with the O protein and with the bacterial proteins. The origin lies within gene *O,* so the protein acts close to its site of synthesis.

Variants of the phage called λ*dv* consist of shorter genomes that carry all the information needed to replicate, but lack infective functions; λ*dv* DNA survives in the bacterium in the form of plasmids. The λ*dv* DNA can be replicated *in vitro* by a system consisting of the phage-coded proteins O and P as well as bacterial replication functions. With this system, it should become possible to analyze the role of transcriptional activation.

The availability of plasmids carrying the *E. coli oriC* sequence makes it possible to develop a cell-free system for replication from this origin. Initiation of replication of the plasmid is inhibited by rifampicin and also by drugs that prevent DNA gyrase action. Antibodies against DnaB or SSB proteins inhibit the reaction. An extract prepared from *dnaA* mutant cells is inactive.

Two possibilities have been considered for the initiation of leading strand synthesis. One relies on a primer RNA, just like the initiation of Okazaki fragments on the lagging strand, except that the priming event is followed by continuous elongation. An alternative is that a nick is made in the parental strand, whose 3'-OH end is used as a primer, as illustrated in **Figure 33.7.** Eventually the covalent junction between the daughter strand and parental strand must be broken.

Nicking followed by extension of a parental strand is a well-established mechanism in the replication of some DNA phages, which proceed through the rolling circle illustrated previously in Figure 31.14. This has been investigated with the RF DNA of φX174. Replication via the rolling circle requires fewer enzymatic functions than are involved in producing RF DNA from the original viral single strand. Supercoiled circles of φX174 RF DNA can be replicated *in vitro* by a system containing the product of phage gene *A,* the host Rep protein, SSB, and DNA polymerase III.

The A protein nicks the viral (+) strand of the duplex DNA at a specific site that defines the origin for RF replication. The DNA can be nicked *only* when it is supercoiled. The A protein is able to bind to a single-stranded decamer fragment of DNA that surrounds the site of the nick. This suggests that the supercoiling may be needed to assist the formation of a single-stranded region that provides the A protein with its binding site. The nick generates a 3'-OH end and a

Figure 33.6
Transcription initiating at P_R is required to activate the origin of lambda DNA.

Figure 33.7
Leading strand synthesis could be initiated by nicking to generate a 3'-OH end that primes synthesis of a new DNA strand to displace the original strand.

5'-phosphate end, both of which have roles to play in φX174 replication.

In the presence of Rep protein, SSB, and ATP, the nicked DNA unwinds in the manner indicated in **Figure 33.8.** The Rep protein provides the helicase function that separates the strands; the SSB traps them in single-stranded form. The A protein remains covalently attached to the 5'-phosphate end; it is also as-

Figure 33.8
øX174 RF DNA can be separated into single strands by the combined effects of three functions: nicking with A protein, unwinding by Rep, and single strand stabilization by SSB.

sociated with the Rep protein, moving around the circle with it. Thus the displaced single viral strand is looped out from the site of unwinding. The completion of movement around the circle releases a circular complementary (−) strand and a linear viral (+) strand.

If DNA polymerase III is included in the system, the 3′-OH end of the nick is extended into a new chain. The chain is elongated around the circular (−) strand template, until it reaches the starting point and displaces the origin, as indicated in **Figure 33.9.** Since the A protein has remained associated with Rep·protein during the traverse of the circle, it is now in the vicinity as the growing point returns past the origin. Thus the same A protein is available again to recognize the origin and nick it, now attaching to the end generated by the new nick. The cycle can be repeated indefinitely.

Following this nicking event, the displaced single (+) strand is freed as a linear molecule that can recircularize. The A protein is involved in the circularization. In fact, the joining of the 3′ and 5′ ends of the (+) strand product may be accomplished by the A protein as part of the reaction by which it is released from the strand and attaches to the new nicked end.

The A protein has an unusual property that may be connected with these activities. It is *cis*-acting *in vivo*. (This behavior is not reproduced *in vitro,* as can be seen from its utilization by any DNA template when it is added to a cell-free system.) The implication is that *in vivo* the A protein synthesized by a particular genome can attach only to the DNA of that genome. We do not know how this is accomplished. However, its

activity *in vitro* shows how it may remain associated with the same parental (−) strand template.

For phage morphogenesis, a modification is made in the cycle shown in Figure 33.9. Instead of the displaced (+) strand being covered with SSB (and reentering the replication cycle), it enters the phage virion.

THE REPLICATION APPARATUS OF PHAGE T4

When phage T4 takes over an *E. coli* cell, it provides a large number of functions of its own that either replace or augment the host functions. Because the host DNA is degraded, the phage places little reliance on expression of host functions, except for utilizing those enzymes that normally are present in large amounts in the bacterium. The degradation of host DNA is important in releasing nucleotides that are reused in the synthesis of phage DNA. (The phage DNA differs in base composition from cellular DNA in using hydroxymethylcytosine instead of the customary cytosine.)

The phage-coded functions concerned with DNA synthesis in the infected cell can be identified by mutations that prevent or impede the production of mature phages. Essential phage functions are identified by conditional lethal mutations, which fall into three phenotypic classes. Those in which there is *no DNA synthesis* at all identify genes whose products either are components of the replication apparatus or are involved in the provision of precursors (especially the hydroxymethylcytosine). Those in which the onset of

Figure 33.9
øX174 RF DNA can be used as a template for synthesizing single-stranded viral circles. The A protein remains attached to the same genome through indefinite revolutions, each time nicking the origin on the viral (+) strand and transferring to the new 5′ end. At the same time, the released viral strand is circularized.

DNA synthesis is *delayed* are concerned with the initiation of replication. Those in which DNA synthesis starts but then is *arrested* include regulatory functions, the DNA ligase, and some of the enzymes concerned with host DNA degradation. There are also nonessential genes concerned with replication; for example, including those involved in glucosylating the hydroxymethylcytosine in the DNA.

Synthesis of T4 DNA is catalyzed by a multienzyme aggregate assembled from the products of seven essential genes. Each of the components has been purified by an *in vitro* complementation assay. The system is active with several DNA templates *in vitro,* it proceeds at a velocity close to the natural rate, and its frequency of errors is about 10^{-5} to 10^{-6} (compared with an *in vivo* error rate of $\sim 2 \times 10^{-8}$). *In vitro,* CTP is used as efficiently as hydroxymethyl-CTP.

The proteins of the T4 replication complex are described in **Table 33.4.**

The gene *32* protein is a highly cooperative single-strand binding protein, needed in stoichiometric amounts. In fact, it was the first example of its type to be characterized. The geometry of the T4 replication fork may specifically require the phage-coded protein, since the *E. coli* SSB cannot substitute. The gene *32* protein forms a complex with the T4 DNA polymerase; this interaction could be important in constructing the replication fork.

The T4 system depends on RNA priming. The nature of the primer, however, is different from that provided in *E. coli*. With single-stranded T4 DNA as template, the gene *41* and *61* products act together to synthesize short primers. These are all pentaribonucleotides with the general sequence pppApCpNpNpNp. Thus the dinucleotide starting sequence is specific, and the length is limited to five bases, but the last three positions can be occupied by any of the bases.

When the complete replication apparatus is present, these primers are extended into DNA chains. The primers may be formed at specific sites, although these sites are likely to occur more frequently than the average 2000-base spacing between the starts of Okazaki fragments. Possibly the replication fork passes a series of potential primer sites, only some of which are used, as dictated by the distance moved since the last initiation event.

The gene *41* protein has a GTPase activity stimulated by single-stranded DNA. When the protein binds to single-stranded DNA, it may be able to move from its initial binding site, migrating at a rate of about 400 nucleotides per second. Its role could be analogous to the host DnaB protein, finding periodic sites at which to halt and initiate primer synthesis. The hydrolysis of GTP is presumed to provide the energy for movement.

The gene *61* protein is needed in much smaller amounts than most of the T4 replication proteins. This has impeded its characterization. (It is required in such small amounts that originally it was missed as a necessary component, because enough was present as a contaminant of the gene *32* protein preparation.) It is possible that the gene *61* protein is the primase,

Table 33.4
The T4 Replication Apparatus Consists of Defined Components

Gene	Function	Product (size in daltons)	Relative Number
43	DNA polymerase	110,000	1
32	Single-strand binding protein	35,000	150
44	{ATPase, increases rate 3–4 fold}	$4 \times 34,000$ } 180,000	5
62		$2 \times 20,000$ } complex	
41	Single-strand-stimulated GTPase required for RNA priming	58,000	20
61	Required for RNA priming		1
45	Unknown, but assists DNA polymerase	$2 \times 27,000$	10

analogous to DnaG of *E. coli*. There may be as few as 10 copies of *61* protein per cell.

The gene *43* DNA polymerase has the usual 5'-3' synthetic activity associated with a 3'-5' exonuclease proofreading activity (see Chapter 32). The remaining three proteins are referred to as "polymerase accessory proteins." The gene *45* product is a dimer. The products of genes *44* and *62* form a tight complex, which has ATPase activity and increases the rate of movement of the DNA polymerase by 3–4-fold, from 250 nucleotides/second to about 800 nucleotides/second, close to the *in vivo* rate of about 1000 nucleotides/second. A similar increase is seen whether the template is single-stranded or double-stranded.

When T4 DNA polymerase uses a single-stranded DNA as template, its rate of progress is uneven. The enzyme moves rapidly through single-stranded regions, but proceeds much more slowly through regions that have a base-paired intrastrand secondary structure. The accessory proteins assist the DNA polymerase in passing these roadblocks. Thus their function in increasing the rate of replication may be to keep the DNA polymerase up to the same speed in "difficult" regions. The presence of the proteins increases the affinity of the DNA polymerase for the DNA, and also its processivity—its ability to stay on the same template molecule without dissociating. Possibly the proteins act as a "clamp," holding the DNA polymerase subunit more tightly on the template. The combined action of all three proteins is needed for this effect. As we have mentioned before, this type of intimate relationship makes it a moot point what is a component of DNA polymerase and what is an accessory factor.

We have dealt with DNA replication so far solely in terms of the progression of the replication fork. The need for other functions is shown by the DNA-delay and DNA-arrest mutants. The four genes of the DNA-delay mutants are *39, 52, 58,* and *60*. Using the *in vitro* complementation assay, the gene *39* and *52* products have been shown to be associated with another protein (of unknown origin) in a complex of three subunits (of 51,000, 64,000, and 110,000 daltons). This constitutes the T4 topoisomerase II activity. The essential role of this enzyme suggests that T4 DNA does not remain in a linear form, but lacks free ends. The

topoisomerase could be needed to unwind the DNA ahead of the replication fork.

THE REPLICATION APPARATUS OF PHAGE T7

As further replication systems are characterized, we see that the same types of activity are present in each, although they may be provided in different ways. Phage T7 has a replication apparatus whose activities are analogous to those of *E. coli* or phage T4, although apparently less complex.

The phage codes for one subunit of the DNA polymerase. A subunit of the host protein thioredoxin is sequestered to provide the second subunit of the T7 DNA polymerase. The host protein is a 12,000-dalton polypeptide that is used as a redox coenzyme in ribonucleotide reduction. Its function as a partner for T7 DNA polymerase is a mystery, but presumably represents a different activity from its role in the host bacterium.

The T7 DNA polymerase employs an auxiliary protein, the product of phage gene *4*. This combines the roles of unwinding the duplex template and synthesizing the RNA primer. The DNA polymerase and gene *4* protein together can accomplish replication *in vitro*, although *in vivo* the exonuclease coded by phage gene *6* may also be involved, in removing the RNA primer.

The replication fork probably contains ten molecules of gene *4* protein with one DNA polymerase per growing strand. The gene *4* product hydrolyzes NTPs over and above the extent of their use as substrates for RNA synthesis. Up to four additional hydrolytic events may occur for every base incorporated into the DNA. Probably this hydrolysis provides the energy for strand separation.

The gene *4* primase activity synthesizes tetraribonucleotide primers with the almost fixed sequence pppApCpCp^A_Cp. The sequence recognized for the priming reaction includes a preceding G residue. The primer sequence should occur in T7 DNA about once every 500 bases, compared with the Okazaki fragment length of 1000 to 6000 bases. So probably only some primer sites are used. The point of switching from the primer to DNA synthesis is not absolutely specified, because some primers extend into penta-ribonucleotides.

THE PROBLEM OF LINEAR REPLICONS

The ability of all known nucleic acid polymerases, DNA or RNA, to proceed only in the 5'-3' direction poses a problem for synthesizing DNA at the end of a linear replicon. Consider the two parental strands depicted in **Figure 33.10.** The lower strand presents no problem: it can act as template to synthesize a daughter strand that runs right up to the end, where presumably the polymerase falls off. But to synthesize a complement at the end of the upper strand, a priming event must occur right at the very last base. (This applies whether it is a lagging strand or a leading strand.)

We do not know whether terminal priming is feasible. We usually think of RNA polymerase as binding at a site *surrounding* the position at which a base is to be incorporated. Two types of solution have been proposed to accommodate the need to copy a terminus. The DNA may form an unusual structure—for example, by creating a hairpin at the terminus, so that there is in fact no free end. Formation of a cross-link is involved in replication of the linear mitochondrial DNA of *Paramecium.* (The problem is circumvented in some cases by converting a linear replicon into a circular or concatemeric molecule.) Or a protein may intervene to make a priming reaction possible.

Several linear viral nucleic acids have proteins that are *covalently linked to the 5' terminal base.* The best characterized examples are adenovirus DNA, phage φ29 DNA, and poliovirus RNA. Adenovirus DNA is a large linear duplex molecule; both 5' ends are covalently attached to a 55,000-dalton protein. The linkage involves a phosphodiester bond to serine, as indicated in **Figure 33.11.** The same type of arrangement is found in φ29, where a 27,000-dalton protein is affixed to each 5' end. In the single-stranded RNA poliovirus, the VPg protein of only 22 amino acids is linked via the hydroxyl group of tyrosine to the 5' terminal base. In each case, the attached protein is coded by the virus and is implicated in replication.

The DNA of adenovirus or φ29 replicates from its ends by the mechanism of **strand displacement** illustrated in **Figure 33.12.** The same events occur independently—that is, not simultaneously—at either end. Synthesis of a new strand starts, displacing the homologous strand that was previously paired in the duplex. Eventually, either the replication fork reaches the other end of the molecule, or two replication forks proceeding from opposite ends meet somewhere in the middle. Either event releases the parental strands from each other.

The protein present on replicating strands of adenovirus DNA is 80,000 daltons, larger than the pro-

Figure 33.11
The 5' terminal phosphate at each end of adenovirus DNA is covalently linked to serine in the 55,000-dalton Ad-binding protein.

Figure 33.10
Replication could run off the 3' end of a newly synthesized linear strand, but could it initiate a 5' end?

DNA synthesis
initiated at
5' end

Replication fork
proceeds,
displacing one
parental strand

DNA synthesis
initiated
at 5' end

Replication forks
pass, and
parental
strands separate

Replication
is
completed

Figure 33.12
Adenovirus DNA replication is initiated independently at the two ends of the molecule and proceeds by strand displacement.

tein found on mature DNA isolated from the virion. The larger protein is related to the smaller species. This suggests that the 80,000-dalton protein is involved in initiating replication, but, at some point during the maturation of the virus, is cleaved to the smaller size.

How could the attachment of a protein overcome the priming problem? The role of the 80,000-dalton adenovirus protein is suggested by its ability to become covalently linked to dCTP in extracts of adenovirus-infected cells. This suggests the model illustrated in **Figure 33.13.** The large protein, bearing a cytidine that has a free 3'-OH group, approaches the end of the adenovirus DNA. It displaces the current 5' end, which may be recognized by its possession of the 55,000-dalton version of the protein. Then the

free 3'-OH end is proffered to an incoming nucleotide, which pairs with the template strand under direction from DNA polymerase. The 80,000-dalton protein may be able to form a complex with the DNA polymerase. The entire series of reactions may be coordinated, because the 80,000-dalton protein can become linked to dCTP only in the presence of adenovirus DNA. A similar model probably applies to φ29 replication.

The RNA polymerase responsible for replicating poliovirus is strictly primer-dependent, like DNA polymerases. Here again, a larger precursor to the covalently linked VPg protein has been found. It is 12,000 daltons in size, bound to the membrane; the precursor could undertake the priming reaction and then be cleaved to the vestigial VPg. The VPg protein is re-

Figure 33.13

The 80,000-dalton adenovirus protein displaces the 5′ end of DNA (which is bound to the 55,000 dalton), and provides CTP to prime synthesis of a new DNA strand.

moved from poliovirus RNA that is destined to be used as mRNA and is found only on those genomes intended for replication.

FURTHER READING

An extensive account of the replication apparatus is given by **Kornberg's** *DNA Replication* (Freeman & Co., San Francisco, 1980). Various replication systems have been reviewed by **Tomizawa** & **Selzer** (*Ann. Rev. Biochem.* **48,** 999–1034, 1979). Among the research papers reporting developments in the systems, the *E. coli* system is dealt with by **Arai et al.** (*J. Biol. Chem.* **256,** 5239–5246, 1981), the primosome by **Arai** & **Kornberg** (*Proc. Nat. Acad. Sci. USA* **78,** 69–73, 1981), Rep protein by **Yarranton** & **Gefter** (*Proc. Nat. Acad. Sci. USA* **76,** 1658–1662, 1979), the action of ϕX174 A protein by **Scott et al.** (*Proc. Nat. Acad. Sci. USA* **74,** 193–197, 1977), and the replication of *oriC* by **Fuller, Kakunagi,** & **Kornberg** (*Proc. Nat. Acad. Sci. USA* **78,** 7370–7374, 1981). Reviews closely based on research deal with the T4 system from **Liu et al.** (*Cold Spring Harbor Symp. Quant. Biol.* **43,** 469–487, 1979), the T7 system from **Richardson et al.** (*Cold Spring Harbor Symp. Quant. Biol.* **43,** 427–440, 1978), and viral linked proteins from **Wimmer** (*Cell* **28,** 199–201, 1982).

CHAPTER 34
SYSTEMS THAT SAFEGUARD DNA

DNA fulfills its hereditary functions via replication and transcription. As we have seen in Chapters 10 and 33, each process involves the interaction with DNA of several enzymatic or regulatory activities in addition to the appropriate polymerase. As well as these components, a variety of enzymes interact with DNA to modify its structure or to repair damage that has occurred to it. These activities are important in understanding how DNA is perpetuated through indefinite numbers of generations; the act of replication itself is insufficient to safeguard its role in evolution.

Procaryotes and eucaryotes both contain enzymes that methylate DNA. In procaryotes, the particular enzyme(s) present in a bacterial strain determine the pattern of this **modification** of DNA; viewed in reverse, the pattern identifies the origin of the DNA. Thus the bacterium can distinguish between its own DNA and any invading "foreign" DNA, which lacks the characteristic modification pattern of the host. This difference renders the foreign DNA susceptible to attack by **restriction enzymes** that recognize the absence of methyl groups at the appropriate sites. Such **restriction and modification** systems are widespread in bacteria; in the context of safeguarding DNA, their object is somewhat xenophobic—to protect the resident DNA against contamination by sequences of foreign origin (although their presence is not obligatory;

some bacterial strains lack any restriction system). Methylation in eucaryotes has another purpose: distinguishing genes in different functional conditions as described in Chapter 30; this is not associated with restriction of unmodified sites.

A bacterium contains several systems that protect its DNA against the consequences of damage by external agents or faults in replication. Any event that causes the structure of DNA to deviate irreversibly from its regular double helix is recognized as inadmissible. The change can be a point mutation that converts one base into another, thus creating a pair of bases not related by the Watson-Crick rules. Or it can be a structural change that adds a bulky adduct to DNA or links two bases together.

The sites of mutation are recognized by special nucleases that excise the damaged region from DNA; then further enzymes synthesize a replacement sequence. Together, these activities form a **repair system.** As well as the direct repair of damage by excision and replacement, there are systems to cope with the adverse consequences of replicating damaged DNA. These **retrieval systems** are related to those involved in genetic recombination. Their importance is indicated by the effects of their elimination from *E. coli:* the bacteria become extremely susceptible to certain types of damage. Although the repair and retrieval systems have been characterized the most fully in *E. coli,* counterparts are likely to be equally important in eucaryotes. Certain human diseases seem to result from defects in particular repair systems.

At the molecular level, these various systems include enzymes with some remarkable abilities to recognize sequences or structures in DNA. Restriction enzymes bind to DNA at specific target sequences, offering another perspective on the nature of protein-nucleic acid interactions. Some of these enzymes bind to DNA at one site, but then cleave at another site far away, offering insights into the nature of the ability of proteins to move along duplex DNA. Some repair enzymes recognize damaged sites in DNA apparently by the distortion in the regular structure of the double helix. An enzyme involved in recombination can bind two molecules of DNA to promote base pairing between them.

From these various activities, we gain two types of information. We learn about the interactions of proteins with DNA in a sequence- or structure-specific manner, an issue at the crux of the biology of the gene. And we begin to obtain a general view of the multifarious nature of the systems needed to preserve and protect DNA against insults from the environment or the errors of other cellular systems.

THE OPERATION OF RESTRICTION AND MODIFICATION

Bacterial restriction endonucleases all recognize specific, rather short sequences of DNA as binding sites. Binding is followed by cleavage, at the recognition site itself or elsewhere, depending on the enzyme. We have already discussed the cleavage reaction in some detail from the perspective of its use in mapping and reconstructing DNA *in vitro* (see Chapters 3 and 19). Now we must ask how the natural function of these enzymes is exercised.

In addition to its restriction activity (or activities), a bacterial strain possesses a DNA methylase activity with the *same sequence specificity* as the restriction activity. The methylase adds methyl groups (to adenine or cytosine residues) in the same target sequence that constitutes the restriction enzyme binding site. The methylation renders the target site resistant to restriction. Thus methylation protects the DNA against cleavage.

In the bacterium, the resident DNA is methylated at the appropriate target sites, and so is immune from attack by the restriction activity, as indicated in **Figure 34.1.** This is clearly necessary to prevent the restriction enzyme from degrading the DNA of the cell in which it resides. But this protection does not extend to any foreign DNA that gains entry to the cell. Such DNA has unmodified target sites, and therefore is attacked by the restriction enzyme. Thus the combination of modification and restriction allows the cell to distinguish foreign DNA from its own, protected sequences.

In this context, "foreign" DNA means any DNA derived from a bacterial strain *lacking the same modification and restriction activity*. There is no distinction among *types* of DNA, that is, between bacterial, plasmid, or phage sequences. The same modification pattern is possessed by *all* DNA that is resident in, or has passed through, a particular strain of bacteria. Thus the restriction and modification system offers no hindrance to the passage of DNA between bacteria of the same strain.

Indeed, these systems were originally discovered by their effects on infecting phage DNAs. The phage DNA released from a bacterium of one strain can successfully prosecute an infection in another bacterium of the same strain, because it has the same modification pattern as the host DNA. But phage DNA that proceeds from one bacterial strain to another is attacked by the restriction activity. Thus the phage is "restricted" to one strain of bacteria, hence the origin of the term. The restriction is not absolute. Some infecting phages escape restriction, because of mutations in the target sites or negligence by the host system. In the latter case, they acquire the modification type of the new host.

Plasmids may contribute to the strain specificity of bacteria. Some plasmids (and phages) possess genes for restriction and modification systems, so a bacterium carrying one of these elements is defined as belonging to a different strain from a bacterium that lacks the element. At all events, the existence of these systems serves to counter the mobility of bacterial sequences conferred by the passage of plasmids and phages between bacteria.

Resident DNA is methylated (∗) at all target sites

Replication gives hemimethylated DNA

Hemimethylated DNA becomes fully methylated

Foreign DNA lacks methyl groups at target sites

Foreign DNA is degraded

Figure 34.1
Methylated sites are perpetuated indefinitely and are safe from restriction; unmethylated sites are cleaved.

The restriction enzymes fall into two general classes, which can be further subdivided. Their properties are summarized in **Table 34.1.**

The first class consists of the **type II** activities, which include the restriction enzymes we have discussed previously. A type II system is found in about one in three bacterial strains. Each of these enzymes is responsible only for the act of restriction; a separate enzyme is responsible for methylating the same target sequence. The protein structures tend to be uncomplicated. The best characterized examples are the components of the EcoRI system, where the restriction enzyme is a dimer of identical subunits and the methylase is a monomer. The recognition of a common target could have arisen either by duplication of the relevant part of the protein or by convergent evolution.

The target sites are often palindromes of 4–6 bp.

The symmetry means that the bases to be methylated occur on both strands of DNA. Thus a target site may be fully methylated (both strands are modified), hemimethylated (only one strand is methylated), or nonmethylated. A nonmethylated target site is a substrate for either restriction or modification *in vitro*. In the cell, unmodified DNA is more likely to be restricted.

The methylase adds only one methyl group at a time. Thus with a nonmethylated substrate, a single methylation is followed by dissociation from DNA. A separate binding and methylation event occurs to introduce the second group. Since replication of fully methylated DNA produces hemimethylated DNA (see Figure 34.1 and also Figure 31.15 earlier), the recognition of hemimethylated sites is likely to be the usual mode of action of the methylase *in vivo*. The restriction enzyme does not bind to methylated target sites.

Most of the type II restriction enzymes cleave the

Table 34.1
Restriction and Methylation Activities May Be Associated or May Be Separate

Class of Enzyme	Type II	Type III	Type I
Protein structure	Separate endonuclease and methylase	Bifunctional enzyme of 2 subunits	Bifunctional enzyme of 3 subunits
Recognition site	Short sequence (4–6 bp), often palindromic	Asymmetrical sequence of 5–7 bp	Bipartite and asymmetrical (e.g., $TGAN_8TGCT$)
Cleavage site	Same as or close to recognition site	24–26 bp downstream from recognition site	Nonspecific at least 1000 bp from recognition site
Restriction and methylation	Separate reactions	Simultaneous	Mutually exclusive
Is ATP needed for restriction?	No	Yes	Yes

DNA *at* an unmethylated target site. One bond is cleaved in each strand of DNA; the cleavages may occur sequentially. Some enzymes introduce staggered cuts, others generate blunt ends (see Chapter 19). A subclass consists of enzymes that cleave the DNA a few bases to one side of the recognition site. Probably as more enzyme mechanisms are investigated, there will prove to be further subclasses.

The second class of restriction and modification activities consists of the type I and type III enzymes, both of which comprise multimers that undertake *both* the endonuclease and methylation functions. Their mechanisms of action are somewhat different from each other and from the type II activities. It is uncertain what proportion of bacteria have type I or type III systems, but they are less common than type II.

THE ALTERNATE ACTIVITIES OF TYPE I ENZYMES

The **type I** enzymes, represented by the EcoK and EcoB activities of *E. coli* strains K and B, were the first to be discovered. A type I restriction and modification enzyme consists of three types of subunit, as indicated in **Figure 34.2.** The R subunit is responsible for restriction, the M subunit is responsible for methylation, and the S subunit is responsible for recognizing the target site on DNA. When a target site has been recognized by the S subunit, the enzyme's binding to DNA can be followed by *either* restriction *or* modifi-

cation; the activities of the R and M subunits are mutually exclusive.

Each subunit is coded by a single gene, denoted *hsdR, hsdM,* and *hsdS.* Mutants in *hsdR* are phenotypically r^-m^+; they cannot restrict DNA, but can modify it. Mutants in *hsdS* are r^-m^-; they can neither modify nor restrict DNA. Mutants in *hsdM* turn out to prevent restriction as well as modification. Thus the M subunit is involved in R subunit function. This may be a fail-safe mechanism, ensuring that *hsdM* mutants are r^-m^-,

Figure 34.2
Multifunctional type I enzymes have different subunits for restriction, modification, and recognition.

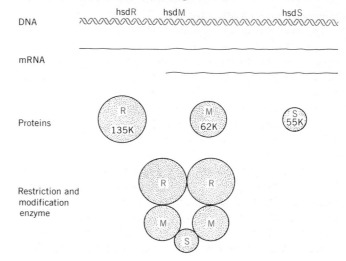

rather than the r^+m^- phenotype they would otherwise have, which presumably would be lethal (because the restriction activity would degrade the cell's own unmodified DNA).

The EcoB and EcoK enzymes are allelic. Their target sequences are different, but by constructing diploids from appropriate mutants, it is possible to show that the S subunit of one strain can direct the activities of the R and M subunits of the other bacterial strain. This confirms that the recognition step is independent of the succeeding cleavage or methylation events.

The EcoK enzyme has a molecular weight of about 400,000 daltons. This includes two copies of the 135,000-dalton R subunit, two copies of the 62,000-dalton M subunit, and one copy of the 55,000-dalton S subunit. The EcoB enzyme has subunits of similar size, but they may be present in different molar ratios. The M and S subunits of strain B can form a 1:1 complex that exercises the methylation function independently of the restriction function. This may be a natural occurrence, because the three genes lie in an operon, in the order: *P1-hsdR-P2-hsdM-hsdS,* where *P1* and *P2* are independent promoters (see Figure 34.2). Thus *hsdM* and *hsdS* can be expressed separately from *hsdR.*

The recognition sites for EcoB and EcoK are bipartite structures, each consisting of a specific sequence of 3 bp separated by a certain distance (8 bp for EcoB, 6 bp for EcoK) from a specific sequence of 4 bp. The distance between the two parts of the recognition site means that both lie on one face of the DNA; the sequence of the intervening region does not seem to be important. Note that neither of the recognition sites is symmetrical. However, each of them does possess one adenine for methylation on each strand, as indicated by the asterisks in the EcoB recognition sequence.

```
        *
T G A N N N N N N N N T G C T
A C T N N N N N N N N A C G A
                      *
```

Whether a given DNA is to be cleaved or modified is determined by the state of the target site, as indicated in **Figure 34.3.** If the target site is fully methylated, the enzyme may bind to it, but is released without any further action. If the target is hemimethylated, the enzyme methylates the unmethylated strand. This action perpetuates the state of methylation in host DNA, as noted previously. If the target site is unmethylated, its recognition triggers a cleavage reaction.

The groups for methylation are provided by the cofactor S-adenosyl-methionine (SAM), which is converted to S-adenosyl-homocysteine (SAH) in the reaction. Binding of SAM to the M subunit is needed for recognition of DNA by the S subunit. Mutations in *hsdM* can abolish the ability to bind SAM. In the initial stage of the reaction, the SAM acts as an allosteric effector that changes the conformation of the protein to allow it to bind to DNA. The conformational change involves the S subunit, as indicated by mutations in *hsdS* that prevent the transition to the active form.

After binding to DNA, the next step is a reaction with ATP. If the enzyme is bound at a completely methylated site, the arrival of ATP releases it from the DNA. At an unmethylated site, the ATP converts the enzyme into a state ready to sponsor cleavage. This depends on the R subunit, as indicated by mutations in *hsdR* that prevent the ATP-induced conformational change. Hydrolysis of ATP is needed to cleave DNA. The SAM is released from the enzyme before the restriction step occurs. The role of ATP when the substrate is hemimethylated is not clear; methylation of hemimethylated DNA can occur without ATP, but the reaction is more efficient in the presence of the nucleotide.

The type I enzymes display an extraordinary relationship between the sites of recognition and cleavage. The cleavage site is located more than 1000 bp away from the recognition site. Cleavage does not occur at a specific sequence, but the selection of a site for cutting does not seem to be entirely random, because there are regions of DNA that are preferentially cleaved. (The discrepancy between the sites of recognition and cleavage means that recognition sites cannot be defined by characterizing the broken ends of DNA; the target sites are identified by their modification in the methylation event, and by the locations of mutations that abolish recognition.)

The cleavage reaction itself involves two steps. First, one strand of DNA is cut; then the other strand is cut nearby. There may be some exonucleolytic degradation in the regions on either side of the site of cleavage. Extensive hydrolysis of ATP occurs; its function is not yet known.

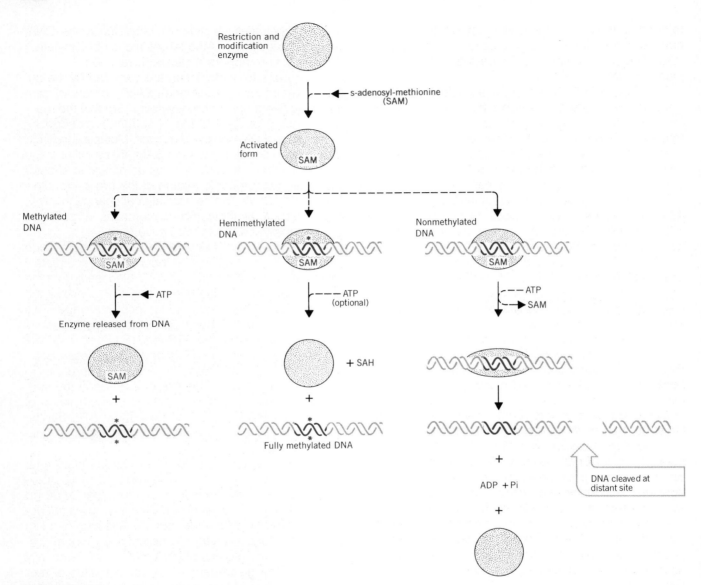

Figure 34.3

Type I enzymes bind to target sites, after which they are released from fully methylated sites, complete the methylation of hemimethylated sites, and move along DNA from nonmethylated sites to cleave the molecule elsewhere.

How does the enzyme recognize one site and cleave another so far away? An important feature is that the protein never lets go of the DNA molecule to which it initially binds. If the enzyme is incubated with a mixture of modified and unmodified DNA, it preferentially cleaves the unmodified DNA. This implies that, having recognized a binding site, it does not dissociate from the unmethylated DNA in order to find its cleavage site.

There are two types of model for the relationship

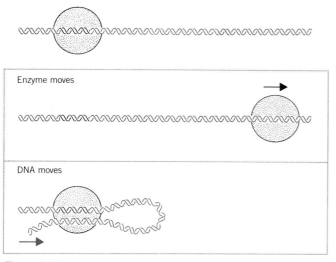

Figure 34.4

Does a type I enzyme move along DNA or does it remain at its target site, simultaneously pulling the DNA through the protein?

Figure 34.5

Type III enzymes have two subunits: recognition and methylation by the MS subunit occur at the target site; the restriction event may occur at a nearby site contacted by the R subunit.

between the recognition and cleavage sites: the enzyme moves, or the DNA moves. These are illustrated in **Figure 34.4.** If the enzyme moves, it could translocate along the DNA until (for some unknown reason) it exercises the cleavage option. If the DNA moves, the enzyme could remain attached at the recognition site, while it also pulls the DNA through a second binding site on the enzyme, winding along until it reaches a cleavage region (again undefined in nature). There is electron microscopic evidence that the enzyme generates loops in DNA, and it seems to remain attached to its recognition site after cleavage; this supports the second model.

THE DUAL ACTIVITIES OF TYPE III ENZYMES

Three type III restriction and modification enzymes have been investigated; EcoP1 and EcoP15 are coded by plasmids P1 and P15 in *E. coli,* and Hinf is found in *H. influenzae* of serotype R$_f$. Each enzyme consists of two types of subunit. The R subunit is responsible for restriction; *hsdR* mutants have the r$^-$m$^+$ phenotype.

The MS subunit is responsible for *both* modification and recognition. Mutations in the *hsdMS* gene generate the r$^-$m$^-$ phenotype if they are located in the recognition part of the sequence; they are reported to generate the r$^+$m$^-$ phenotype if located in the modification part (presumably this is lethal). The recognition part of the gene probably lies in the middle of its sequence, as judged by comparisons of *hsdMS* genes from the P1 and P15 systems. The R subunit is 106,000–110,000 daltons; the MS polypeptide is 73,000–80,000 daltons.

In the type III enzymes, the restriction and modification activities are expressed *simultaneously*. The enzyme first binds to its site on DNA, an action that requires ATP. (The subsequent dependence on ATP varies among the enzymes.) Then the methylation and restriction activities *compete* for reaction with the DNA. The methylation event takes place at the binding site, consistent with the combination of methylation and recognition in the MS subunit. The restriction cleavage occurs 24–26 bases on one side, possibly because the enzyme is large enough for the restriction subunit to contact DNA at this point, as depicted rather schematically in **Figure 34.5.** Restriction involves staggered cuts, 2–4 bases apart.

The enzymes methylate adenine residues, but the target sites of P1 and P15 have an intriguing feature: they can be methylated *only on one strand*. The sequences of the sites are

sP1	AGACC
	TCTGG
sP15	CAGCAG
	GTCGTC

This raises the question of how the state of methylation is perpetuated. **Figure 34.6** demonstrates that two types

Figure 34.6
Replication of a methylated sP1 sequence generates one methylated and one unmethylated replica.

of site are generated by replication. One replica has the original methylated strand; it is in fact indistinguishable from the parental site. The other replica is entirely unmethylated. It is therefore in principle a target for either restriction or modification. What determines that it is methylated rather than restricted? We do not know the answer, but one idea is that the modification reaction is linked to the act of replication.

DEALING WITH INJURIES IN DNA

Injury to DNA is minimized by damage-containment systems that recognize the occurrence of a change and then rectify it. Thus the measured mutation rate—spontaneous or induced—reflects a balance between the number of damaging events occurring in DNA and the number that have been corrected (or miscorrected). These repair systems are likely to be as complex as the replication apparatus itself, an indication of their importance for the survival of the cell.

"Damage" to DNA is defined as any change that introduces a deviation from the usual double-helical structure. Events in this category include:

Introduction of single strand breaks.

Removal of a base to leave its former partner unpaired.

Covalent addition of new groups to a base to modify its properties.

Conversion of one base into another that is not properly paired with the partner base.

Introduction of covalent links between bases on one strand of DNA or between bases on opposite strands.

Some repair systems are not specific for particular types of damage, but recognize a range of distortions in DNA as signals indicating the need for action. We can divide the changes listed above into two general classes.

Point mutations may be recognized because they disrupt proper pairing of the partner bases. This does not affect transcription or replication, when the strands of the DNA duplex are separated. Thus these changes exert their damaging effects on future generations through the consequences of the change in DNA sequence (see Chapter 2).

Structural distortions may provide a physical impediment to replication or transcription. A break or the absence of a base may prevent a strand from serving as a proper template for synthesis of RNA or DNA. The presence of intrastrand cross-linked structures may inhibit replication, causing the polymerase to stop at the site of damage. The enzyme resumes synthesis at a subsequent point, leaving a gap in the newly synthesized strand. Interstrand cross-links prevent the strands of DNA from separating. Transcription terminates prematurely at either type of cross-link. The most studied example is the effect of ultraviolet irradiation, which introduces covalent bonds between two adjacent thymine bases, giving the intrastrand **thymine dimer** drawn in **Figure 34.7.**

Three types of event may cope with the damage. The links in thymine dimers may be directly reversed by a light-dependent enzyme. This process is called **photoreactivation;** in *E. coli* it depends on the product of a single gene (*phr*). Alternatively, the damaged

Figure 34.7
A thymine dimer is generated by covalent links between adjacent bases.

Pathway for base mismatch

Excision—repair

Pathway for any damage

Base
is
damaged

Base
is
removed

Incision
Endonuclease cuts
on 5′ side of
damaged base

AP endonuclease
makes cut

Excision
Exonuclease removes
stretch of DNA

Polymerase
synthesizes
new DNA

Ligase
seals
nick

Figure 34.8
Excision-repair removes and replaces a stretch of DNA that includes the damaged base(s) but extends farther.

material may be removed and replaced by a **repair system.** Finally, the damage may remain at the original site, but the daughter strands generated by replication may be repaired by a **retrieval system.** A given cell may have several systems engaging in repair and retrieval.

Excision-repair systems remove the mispaired or damaged bases from DNA and then synthesize a new stretch of DNA to replace them. The main type of pathway for excision-repair is illustrated in **Figure 34.8.** In the **incision** step, the damaged structure is recognized by an endonuclease that cleaves the DNA strand on the 5′ side of the damage. In the **excision** step, a 5′-3′ exonuclease removes a stretch of this strand, extending past the site of damage. In the **repair** step, the resulting single-stranded region serves as a template for a DNA polymerase to synthesize a replacement for the excised sequence. Finally, DNA ligase covalently links the 3′ end of the new material to the old material.

If the damage in DNA is a structural distortion—such as a thymine dimer—the damaged bases are removed, leading to the restoration of wild-type sequence. However, if the feature recognized in DNA is a mispaired partnership of bases created when one of them was mutated, the repair system has no intrinsic means of knowing which is the wild-type base and which is the mutant. All it sees are two improperly paired bases, either of which can provide the target for excision-repair. If the mutated base is excised, the wild-type sequence is restored. But if it happens to be the original (wild-type) base that is excised, the new (mutant) sequence becomes fixed. In many cases, the direction of excision-repair is not random, but is influenced by the local sequence. This can be important in genetic recombination (see Chapter 35).

In the case of mismatching errors that escape the attention of DNA polymerase during replication, it may be possible to distinguish the original strand of DNA. Immediately after replication of methylated DNA, only the original parental strand carries the methyl groups (see Figure 34.1). In the period while the newly synthesized strand awaits the introduction of methyl groups, the nature of the two strands can be distinguished. This may provide the basis for a system to correct replication errors; *dam*⁻ mutants of *E. coli* are deficient in the methylation of adenine and show an increased rate of spontaneous mutation. Their behavior implies the existence of a repair system that specifically corrects the strand of DNA containing nonmethylated adenines, to ensure it is complementary with the methylated strand.

Another type of specific repair system that avoids the problem of distinguishing wild-type and mutant sequences is the uracil-glycosylase that specifically removes uracil bases from DNA (see Chapters 2 and 32). This functions by the side pathway for base mismatch illustrated in Figure 34.8. The base is removed

from the phosphodiester backbone, creating an apurinic site. This site is recognized specifically by an **AP endonuclease** (AP stands for apurinic or apyrimidinic, depending on the particular enzyme), which introduces a break on the 5′ side of the site. Then the subsequent steps in the pathway for general excision-repair can be followed.

Retrieval systems are provided by **recombination-repair** enzymes that function by retrieving material from one duplex of DNA to repair another. The name reflects an overlap with the activities involved in genetic recombination. Because such systems require the target DNA to have a replica, this type of activity is sometimes also called **postreplication-repair.** Such systems are effective in dealing with the defects produced in daughter duplexes by replication of a template that contains damaged bases. An example is illustrated in **Figure 34.9.**

Consider a structural distortion such as a thymine dimer on one strand of a double helix. When the DNA is replicated, the dimer prevents the damaged site from acting as a template to direct the insertion of bases in the daughter strand. Replication is forced to skip past the damaged site, leaving a gap. The probable series of events is for the enzyme to proceed up to or close to the point at which the thymine dimer provides the template. Then the polymerase ceases synthesis of the corresponding daughter strand. Replication restarts some distance farther along—for example, at the next initiation site for an Okazaki fragment in the lagging strand, or at some facsimile of an initiation site in the leading strand. The gap left in the newly synthesized strand may therefore be substantial.

The resulting daughter duplexes are different in nature. One of them has the parental strand containing the damaged adduct, facing a newly synthesized strand with a lengthy gap. The other duplicate has the undamaged parental strand, which has been copied into a normal complementary strand. The gap opposite the damaged site in the first duplex is filled by taking the homologous single strand of DNA from the other duplex. Following this **single strand exchange,** the recipient duplex has a parental (damaged) strand facing a wild-type strand. The donor duplex has a normal parental strand facing a gap; the gap can be filled by

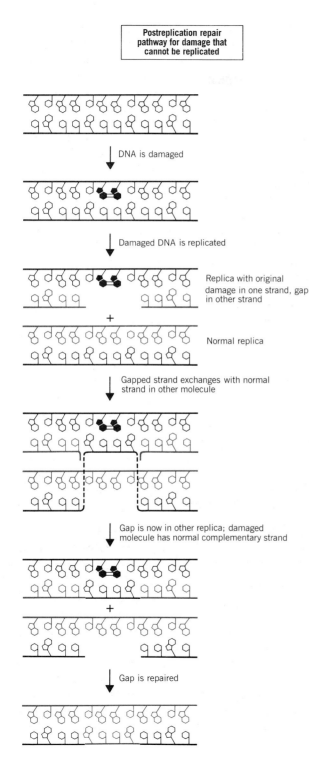

Postreplication repair pathway for damage that cannot be replicated

↓ DNA is damaged

↓ Damaged DNA is replicated

Replica with original damage in one strand, gap in other strand

+

Normal replica

↓ Gapped strand exchanges with normal strand in other molecule

↓ Gap is now in other replica; damaged molecule has normal complementary strand

+

↓ Gap is repaired

repair synthesis in the usual way, generating a normal duplex. Thus the damage is confined to the original distortion (although the same recombination-repair events must be repeated after every replication cycle unless and until the damage is removed by an excision-repair system).

EXCISION-REPAIR SYSTEMS IN *E. COLI*

Mutations in a large number of loci in *E. coli* affect the response to agents that damage DNA. Many of these genes are likely to code for enzymes that participate in DNA repair systems, although in the majority of cases the gene product and its function remain to be identified. When the repair systems are eliminated, cells become *exceedingly* sensitive to ultraviolet irradiation. The introduction of UV-induced damage has been a major test for repair systems, and so in assessing their activities and relative efficiencies, we should remember that the emphasis might be different if another damaged adduct were studied.

Table 34.2 summarizes the properties of some of the mutations that affect the ability of *E. coli* cells to engage in DNA repair. We know of at least three repair pathways. Two different excision-repair modes are identified by the heterogeneity of the lengths of the segments of repaired DNA. These pathways are described as **short-patch repair** and **long-patch repair.** A recombination-repair pathway (or pathways) overlaps extensively with the enzymes involved in recombination itself (see Chapter 35).

The uvr system is involved in both modes of excision-repair and includes three genes, *uvrA,B,C,* that code for the components of a repair endonuclease. Initial characterization of the enzyme activity *in vitro* suggests that the endonuclease recognizes the thymine dimer (or other distortion) and makes the initial incision on its 5′ side; another incision is made on the 3′ side about 12 nucleotides away.

Several *E. coli* enzymes can excise thymine dimers

Figure 34.9
Recombination-repair retrieves a normal strand of DNA to replace the gap left in a newly synthesized strand opposite a site of unrepaired damage.

Table 34.2
Many Gene Products Are Involved in Repairing DNA Damage in *E. coli*

Gene	Effect of Mutation *in vivo*	Gene Product	Function of Protein
uvrA		ATPase component of endonuclease	Initiates removal of thymine dimers from DNA
uvrB	Sensitivity to UV	Component of endonuclease	
uvrC		Component of endonuclease	
uvrE	Sensitivity to UV	Unknown	Unknown
recA	1. Deficiency in recombination 2. Does not induce repair pathways after damage	RecA protein of 40,000 daltons	1. DNA strand-exchange activity needed for recombination and recombination-repair 2. Protease activity initiates SOS induction of repair pathways
recB	Deficiency in recombination	Subunit of exonuclease V	Needed for recombination and recombination-repair
recC		Subunit of exonuclease V	
sbcB	Suppresses *recBC* mutation	Exonuclease I	Unknown
recF	Deficiency in recombination-repair; and deficiency in recombination in *recBCsbcB* mutants	Unknown	Unknown
recJ		Unknown	Unknown
recK		Unknown	Unknown
recE	Deficiency in recombination	Exonuclease VIII	Unknown
umuC	Abolishes mutagenic effects of error-prone repair	Unknown	Unknown
lon	UV-sensitive, septum-deficient	ATP-dependent protease that binds to DNA	One activity may be control of genes for capsular polysaccharide
mut	At least 6 genes whose mutation causes increased rate of mutation	Unknown	Unknown
dam	UV-sensitive, increased mutation rate	Could be methylase	Concerned with methylation of adenine
lexA	Abolishes SOS response	Repressor (22,000 daltons)	Acts on many genes

in vitro from DNA in which such incisions have been made. The enzymes include the 5'-3' exonuclease activities of DNA polymerase I and DNA polymerase II (see Table 32.2) and the single-strand specific exonuclease VII. However, mutation of any one of these putative excision nucleases does not result in a measurable diminution in the ability of *E. coli* cells to remove thymine dimers. Based on the behavior of various double mutants, current opinion is that DNA polymerase I is likely to perform most of the excision *in vivo*.

The average length of excised DNA is about 20 nucleotides, which gives rise to the description of this mode as the short-patch repair. The enzyme involved in the repair synthesis probably also is DNA polymerase I (although enzymes II and III are able to substitute for it).

Other potential components of the *uvr* system have been identified by mutations that impede short-patch repair, but none has yet been identified with a product or function in the pathway.

Short-patch repair accounts for 99% of the excision-repair events. The remaining 1% involve the replacement of stretches of DNA mostly of about 1500 nucleotides; a minority of the long patches extend for more than 9000 nucleotides. This mode also requires the *uvr* genes and involves DNA polymerase I. A difference between the two modes of repair is that short-patch repair is a constitutive function of the bacterial cell, but long patch repair must be induced by damage (see below). We have not yet characterized the differences between these modes in terms of the involvement of different gene products.

The existence of repair systems that engage in DNA synthesis raises the question of whether their quality control is comparable with that of DNA replication. So far as we know, most systems, including *uvr*-controlled excision repair, do not differ significantly from DNA replication in the frequency of mistakes. However, error-prone repair does occur in *E. coli* under certain circumstances.

The error-prone feature was first observed when it was found that the repair of damaged phage DNA is accompanied by the induction of mutations if the phage is introduced into cells that had previously been irradiated with UV. This suggests that the UV irradiation of the host has activated some protein(s) that do not function in the unirradiated cell, and whose activity in some repair pathway generates mutations. The mutagenic response also operates on the bacterial host DNA.

What is the actual error-prone activity? Current thinking focuses on the idea that the relevant component of the repair pathway may be responsible for permitting or compelling replication to proceed past the site of damage. When the DNA polymerase passes the thymine dimer, it does not insert the correct bases, thus causing mutations. There is evidence that the error-prone activity requires DNA polymerase III, the usual replicase, which is consistent with the idea that the relevant function acts in concert with the normal replication apparatus. Mutations in a gene called *umuC*

abolish UV-induced mutagenesis, but do not interfere with any known enzymatic functions. The *umuC* product is a potential component of the error-prone system.

RECOMBINATION-REPAIR SYSTEMS IN *E. COLI*

In cells deficient in excision-repair, mutation in the *recA* gene essentially abolishes all the remaining repair and recovery facilities. Thus attempts to replicate DNA in *uvr⁻ recA⁻* cells produce fragments of DNA whose size corresponds with the expected distance between thymine dimers. This result implies that the unremoved dimers provide a lethal obstacle to replication in the absence of RecA function. It explains why the double mutant cannot tolerate more than 1–2 dimers in its genome (compared with the ability of a wild-type bacterium to shrug off as many as 50).

The dramatic effects of mutation in *recA* bring us to a major feature of the response of *E. coli* to damage in its DNA. The first interpretation of this result was that the *recA* gene must be part of another pathway. Since *recA* mutants are almost completely deficient in genetic recombination as well as in the recovery response, this pathway seemed likely to be a recombination-repair system. This view was supported by the isolation of other *rec* mutants with similar (although lesser) deficiencies in both repair and recombination (see Table 34.2). The components involved in recombination-repair overlap with those involved in recombination, but are not entirely identical with them, since some mutations affect one but not the other activity.

The RecA protein has the function of exchanging strands between DNA molecules, a central activity in recombination (see Chapter 35), related to the proposed single strand exchange of recombination-repair. The properties of double mutants suggest that there may be two Rec pathways. To test whether two genes with related functions are part of the same or different pathways, the phenotypes of the single mutants are compared with the phenotype of the double mutant. If the genes are in the same pathway, the phenotype of the double mutant will be no different from that of the individual mutants. If the genes are in different pathways, the double mutant will lack both instead of one, and so has a more severe phenotype. By this criterion, one *rec* pathway involves the *recBC* genes; the other involves *recF*.

The *recBC* genes code for the two subunits of exonuclease V, whose action is limited by some other component of the pathway. (In *recA* mutants, the enzyme is unlimited, and degrades an excessive amount of DNA.) The function of *recF* is unknown. Other *rec* loci have been identified by the effects of mutations on recombination-repair or recombination. Several are identified by their ability to suppress other *rec* mutants. As with the additional *uvr* loci, they have not been equated with products or functions.

The designations of these genes are based on the phenotypes of the mutants; but sometimes a mutation isolated in one set of conditions and named as a *uvr* locus turns out to have been isolated in another set of conditions as a *rec* locus. This uncertainty makes an important point. We cannot yet define how many functions belong to each pathway or how the pathways interact. The *uvr* and *rec* pathways may not be entirely independent, because *uvr* mutants show reduced efficiency in recombination-repair. We must expect to find a network of nuclease, polymerase, and other activities, constituting repair systems that may be partially overlapping (or in which one enzyme usually used to provide some function can be substituted by another from a different pathway).

The direct involvement of RecA protein in recombination-repair is only one of its activities. This extraordinary protein also has another, quite distinct function. It is a protease activated by ultraviolet irradiation (or by other treatments that block replication). The protease activity is responsible for inducing the expression of many genes, whose products include repair functions. Among them, for example, is the long-path repair mode; so RecA is needed to induce this excision-repair system as well as to participate in recombination-repair. These dual activities of the RecA protein make it difficult to know whether a deficiency in repair in *recA* mutant cells is due to loss of the DNA strand-exchange function of RecA or to some other function whose induction depends on the protease activity.

AN SOS SYSTEM OF MANY GENES

Many treatments that damage DNA or inhibit replication in *E. coli* induce a complex series of phenotypic changes called the **SOS response.** This is set in train by the interaction of the RecA protein with the LexA repressor.

The damage can take the form of ultraviolet irradiation (the most studied case) or can be caused by cross-linking or alkylating agents. Inhibition of replication by several means, including deprivation of thymine, addition of drugs, or mutations in several of the *dna* genes, has the same effect.

The response takes the form of increased capacity to repair damaged DNA, achieved by inducing the synthesis of the components of both the long-patch excision-repair system and the Rec recombination-repair pathways. In addition, cell division is inhibited. Lysogenic prophages may be induced.

The initial event in the response is the activation of the RecA protease activity by the damaging treatment. We do not know very much about the relationship between the damaging event and the sudden change in RecA activity. Because a variety of damaging events can induce the SOS response, current work focuses on the idea that RecA may be stimulated by some common intermediate in DNA metabolism. The inducing signal could consist of a small molecule released from DNA; or it might be some structure formed in the DNA itself. *In vitro,* the protease activity of RecA requires the presence of single-stranded DNA and ATP. Thus the activating signal could be the presence of a single-stranded region at a site of damage.

Whatever form the signal takes, its interaction with RecA is rapid: the SOS response occurs within a few minutes of the damaging treatment. On activation, the RecA protease cleaves the 22,000-dalton repressor protein coded by the *lexA* gene. The LexA protein is relatively stable in untreated cells, where it functions as a repressor at many operons. Proteolytic cleavage of the repressor coordinately induces all these operons.

The target genes for LexA repression include many repair functions, only some of which are identified at present. A systematic screen for LexA-responsive genes has been constructed by fusing random operons to the *lacZ* gene, and then assaying for increased levels of the product β-galactosidase when cells are treated with mitomycin C. At least five responsive genes have been identified; they are called *din* (damage inducible). Genes previously known to be part of the SOS response by the activation of their products include *recA, lexA, uvrA, uvrB, umuC,* and *himA.*

Some of the SOS genes may be active only in treated cells; others are active in untreated cells, but the level of expression is increased by cleavage of LexA. In the case of *uvrB,* which is a component of the excision-repair system, the gene has two promoters; one functions independently of LexA, the other is subject to its control. Thus after cleavage of LexA, the gene can be expressed from the second promoter as well as from the first.

Purification of the LexA repressor allows its action to be characterized *in vitro.* It represses its target genes by binding to a stretch of DNA of about 20 bp, called an **SOS box;** this displays symmetry, and a copy is present at each target locus. The SOS boxes at different loci are not identical, but conform to a consensus sequence with 8 absolutely conserved positions. Like other operators, the SOS boxes overlap with the respective promoters. At the *lexA* locus, the subject of autogenous repression, there are two adjacent SOS boxes.

RecA and LexA are mutual targets in the SOS circuit: RecA cleaves LexA, which represses *recA.* The SOS response therefore causes amplification of both the RecA protein and the LexA repressor. The results are illustrated in **Figure 34.10;** they are not so contradictory as might at first appear.

The increase in expression of RecA protein is necessary (presumably) for its direct role in the recombination-repair pathways. It also means that there is sufficient protease to ensure that all the additional LexA protein is cleaved. This should prevent LexA from re-establishing repression of the target genes. However, the main importance of this circuit for the cell may lie

Figure 34.10

The LexA protein represses many genes, including repair functions, *recA* and *lexA*. Activation of the RecA protease cleaves LexA and induces all of these genes.

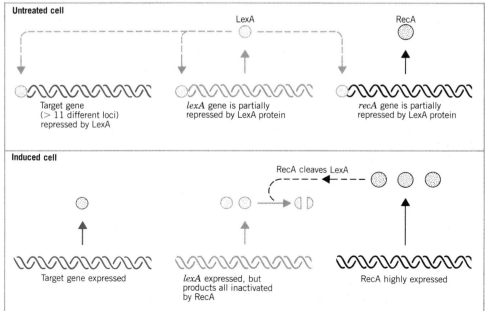

in the ability to return rapidly to normalcy. When the inducing signal ceases, the RecA protein loses its protease activity. At this moment, the *lexA* gene is being expressed at a high level; in the absence of RecA protease, the LexA proteins rapidly accumulate in the uncleaved form and turn off the SOS genes. This may explain why the SOS response is freely reversible.

The RecA protease also acts on some other repressor proteins, those of several prophages. Among these is the lambda repressor (with which the protease activity in fact was discovered). This explains why lambda is induced by ultraviolet irradiation; RecA cleaves the lysogenic repressor, releasing the phage to enter the lytic cycle. This reaction is not a cellular SOS response, but instead represents a recognition by the prophage that the cell is in trouble, so that survival is best assured by entering the lytic cycle to generate progeny phages. In this sense, the prophage induction is piggybacking onto the cellular system by responding to the same indicator (activation of RecA).

The RecA protease cleaves all its known target proteins at an Ala-Gly dipeptide sequence in the middle of the polypeptide chain. Mutants of the type *lexA*(Ind⁻) cannot be induced; one of them has the replacement sequence Ala-Asp. There is only limited amino acid homology on either side of the dipeptide, which suggests that the tertiary structure of the protein may be an important feature in target recognition.

The two activities of RecA may be relatively independent. The *recA441* mutation allows the SOS response to occur without inducing treatment, probably because the protease function is spontaneously active. Another mutation inactivates the protease activity. Neither mutation affects the ability of RecA to handle DNA. The reverse type of mutation, inactivating the recombination function but leaving intact the ability to induce the SOS response, would be useful in disentangling the direct and indirect effects of RecA in the repair pathways.

MAMMALIAN REPAIR SYSTEMS

Biochemical characterization of repair systems in eucaryotic cells is only primitive, for the most part confined to the isolation of the occasional enzyme preparation whose properties *in vitro* suggest that it could be part of a repair system. The existence of excision-repair pathways can be established in cultured cells by following the replacement of DNA segments in response to damaging treatments. One technique is to follow events in synchronized cells outside S phase, when the occurrence of **unscheduled DNA synthesis** can be entirely attributed to the operation of repair systems.

Mammalian cells show heterogeneity in the amount of DNA incorporated after damage. However, the longest patches seen in mammalian cells are comparable to those of the short-patch bacterial repair system. This pathway operates on damage caused by ultraviolet irradiation or treatments with related effects. Another pathway introduces only 3–4 repair bases at sites of damage generated by X-irradiation or alkylation.

An indication of the existence and importance of the mammalian repair systems is given by certain human hereditary disorders. The best investigated of these is Xeroderma pigmentosum (XP), a recessive disease resulting in hypersensitivity to sunlight, in particular to ultraviolet. The deficiency results in skin disorders (and sometimes more severe defects). The disease is explicable in terms of a failing in excision-repair; fibroblasts from XP patients are deficient in the excision of thymine dimers and other bulky adducts. Seven genetic complementation groups have been distinguished, many characterized by a deficiency at the incision step of repair.

Some indirect results suggest that mammalian cells may have recombination-repair systems. Again, these systems may be related to genetic recombination itself. An example is the recessive human disorder of Bloom's syndrome; an increased frequency of chromosomal aberrations, including sister chromatid exchanges, could be related to the operation of recombination systems.

FURTHER READING

A review including the separate (type II) restriction and modification activities is by **Modrich** (*Quart. Rev. Biophys.* **3,** 315–369, 1979). A valuable tour of multifunctional (type I and type III) enzymes, emphasizing the enzymatic activities, is provided by **Yuan** (*Ann. Rev. Biochem.* **50,** 285–315, 1981), who developed a model for type I enzyme translocation (*Cell* **20,** 237–244, 1980). The functions of AP endonucleases and associated systems have been reviewed by **Lindahl** (*Prog. Nucleic Acid Res.* **22,** 135–192, 1979). An integrated view of the systems and mechanisms for DNA repair in procary- otes and eucaryotes is presented by **Hanawalt et al.** (*Ann. Rev. Biochem.* **48,** 783–836, 1979). The SOS control network of *E. coli* has been encapsulated by **Gottesman** (*Cell* **23,** 1–2, 1981) and reviewed extensively by **Little** & **Mount** (*Cell* **29,** 11–22, 1982).

CHAPTER 35
RETRIEVAL AND RECOMBINATION OF DNA

Without genetic recombination, the content of each individual chromosome would be irretrievably fixed in its particular alleles, subject only to the changes caused by mutations. The length of the target for mutation damage would be increased from the gene to the chromosome. Deleterious mutations would accumulate, eliminating each chromosome (and therefore removing any favorable mutations that may have occurred). By shuffling the genes, recombination allows favorable and unfavorable mutations to be separated and tested as individual units in new assortments. It provides a means of escape and spreading for favorable alleles, and a means to eliminate an unfavorable allele without bringing down all the other genes with which this allele may have been associated in the past. From the long perspective of evolution, a chromosome is a bird of passage, a temporary association of particular alleles. Recombination is responsible for this flighty behavior.

This type of recombination, involving reaction between homologous sequences of DNA, is called **general recombination.** Its critical feature is that the enzymes responsible can use *any* pair of homologous sequences as substrates (although there is always the possibility that some types of sequence are favored over others). From the behavior of gene clusters outlined in Chapter 21, we know that the homology does not have to be complete; related nonallelic genes can undergo (unequal) recombination with a frequency that declines with the divergence between their sequences. In the genome as a whole, the frequency of recombination is not constant, but may be influenced by chromosome structure; for example, crossing-over may be suppressed in the vicinity of heterochromatin.

Another type of event sponsors recombination between *specific* pairs of sequences. **Site-specific recombination** is responsible for the integration of phage genomes into the bacterial chromosome. The recombination event involves a specific sequence of the phage DNA and a specific sequence of the bacterial DNA. Within these sequences, there is only a short stretch of homology, necessary for the recombination event, but not sufficient for it. The enzymes involved in this event cannot recombine other pairs of (homologous or nonhomologous) sequences, but act only on the particular pair of phage and bacterial sequences. This type of recombination is also called **conservative recombination,** because it relies entirely on the reaction of the two preexisting DNA molecules.

Another type of recombination event allows one DNA sequence to be inserted into another without reliance on sequence homology. **Replicative recombination** is responsible for the ability of certain elements to move from one location to another. The movable element of DNA is copied, and then the replica is transferred elsewhere. This is the subject of Chapter 36.

Figure 35.1

The synaptonemal complex of *Neotellia* shows that pairing of chromosomes does not bring the homologous DNAs into juxtaposition.

xt indicates chromatin; *le* indicates the lateral elements, each about 50 nm in diameter; *ce* indicates the central element, about 18 nm in diameter. The distance between the lateral elements is about 120 nm.

Photograph kindly provided by Drs. M. Westergaard and D. Von Wett-stein.

RECOMBINATION REQUIRES SYNAPSIS OF HOMOLOGOUS DUPLEX DNAs

There is old evidence that recombination between genes on a chromosome requires a physical exchange of parts (see Figure 1.10). The structure created by this exchange is visible at meiosis in the form of a chiasma (see Figure 1.9). The chiasma represents the results of a **breakage and reunion,** in which two nonsister chromatids (each containing a duplex of DNA) have been broken and then linked each with the other. This occurs between precisely corresponding sequences, so that not a single base pair is added to or lost from the recombinant chromosomes.

Yet this is the act of recombination itself. Its description begs the issue of the first, crucial step: two homologous duplex molecules of DNA must be brought into close contact so that the corresponding sequences can be exchanged. The nature of the problem may be somewhat different in eucaryotes and procaryotes.

In procaryotic systems, we are dealing largely with the DNA itself (albeit in the nucleoid). There is no vis-ible guide to the events responsible for recognition of homology, but probably they involve only those regions actually participating in the recombination crossover. The recognition may be part and parcel of the recombination mechanism, which may rely on the comparison of single strands of DNA to identify complementary sequences (see later).

In eucaryotic cells, the DNA is tightly packaged into the discrete structures of the chromosomes. How do the DNA molecules come into juxtaposition? Contact between a pair of parental *chromosomes* occurs early in meiosis. The process is called **synapsis** or **chromosome pairing.** Homologous chromosomes (each of which actually consists of the two sister chromatids produced by the prior replication) approach one another. They become laterally associated in the form of a **synaptonemal complex,** which has a characteristic structure in each species, although there is wide variation in the details between species.

An example of a synaptonemal complex is shown in **Figure 35.1.** Each chromosome at this stage appears as a mass of chromatin bounded by a **lateral element** (which in this case has a striated structure).

The two lateral elements are separated from each other by a **central element.** The triplet of parallel dense strands lies in a single plane that curves and twists along its axis. The distance between the homologous chromosomes is considerable in molecular terms, more than 200 nm (the diameter of DNA is 2nm).

The generation of the synaptonemal complex coincides with the presumed time of crossing-over, although there is no direct evidence that the recombination event occurs at the stage of synapsis. A major problem in understanding the role of the complex is that, although it brings homologous chromosomes into contact, it is far from bringing homologous DNA molecules into contact. The only link that can be seen between the two sides of the synaptonemal complex is provided by spherical or cylindrical structures observed in fungi and insects. They lie across the complex and are called **nodes** or **recombination nodules;** they occur with the same frequency and

distribution as the chiasmata. Their name reflects the hope that they may prove to be the sites of recombination.

At the next stage of meiosis, the chromosomes shed the synaptonemal complex; then the chiasmata become visible as points at which the chromosomes are connected. This is presumed to indicate the occurrence of a genetic exchange. Later in meiosis, the chiasmata may move toward the ends of the chromosomes. This flexibility suggests that they may represent some remnant of the recombination event, rather than providing the actual intermediate.

BREAKAGE AND REUNION OCCURS VIA HETERODUPLEX DNA

In contrast with the unknown basis for the ability of homologous duplex DNA sequences to recognize one

Paired DNA duplexes

Nicks made in homologous strands

Broken ends can move

Each strand crosses over to pair with complement in other duplex

Nicks are sealed

Crossover point moves by branch migration

Structure can generate planar molecule by rotation

Figure 35.2
Recombination between two paired duplex DNAs could be initiated by reciprocal single strand exchange, extended by branch migration, and resolved by nicking.

another, a mechanism of considerable precision—base pairing—exists for recognition between complementary single strands. This is used in the formation of recombination intermediates. One of the several models for the reaction is illustrated in **Figure 35.2.**

The process starts with breakage at the corresponding points of the homologous strands of two paired DNA duplexes. The breakage allows movement of the free ends created by the nicks. Each strand leaves its partner and crosses over to pair with its complement in the other duplex. This reciprocal exchange creates a connection between the two DNA duplexes. Initially, this is sustained only by hydrogen bonding; at some point, it is made covalent by sealing the nicks at the sites of exchange (the sealing could occur at a later stage than shown in the figure). The connected pair of duplexes is called a **joint molecule.**

Each duplex in the joint molecule has a region consisting of one strand from each of the originally separate parental DNA molecules. This region is called **hybrid DNA** or **heteroduplex DNA.** Model building shows that there is (surprisingly) little steric hindrance of the formation of the reciprocal heteroduplex regions

between the paired DNA molecules; thus virtually all of the bases can remain base paired (that is, in either the parental duplex or heteroduplex regions). It is not necessary to invoke the existence of extensive single-stranded regions.

Once the strand exchange has been initiated, it can move along the duplex. This ability is called **branch migration.** Branches in a DNA structure can involve either single or double strands; either type can migrate. **Figure 35.3** shows an example of single strand branch migration. The point of branching can migrate in either direction as one strand is displaced by the other. The reaction is important for both theoretical and practical reasons. As a matter of principle, it confers a dynamic property on recombinant structures. As a practical feature, its existence means that the point of branching cannot be established by examining a molecule *in vitro* (because the branch may have migrated since the molecule was isolated).

The same type of movement could allow the point of crossover in the recombination intermediate to move in either direction. The rate of branch migration appears to be fast enough to support the formation of extensive regions of heteroduplex DNA (> 1000 bp)

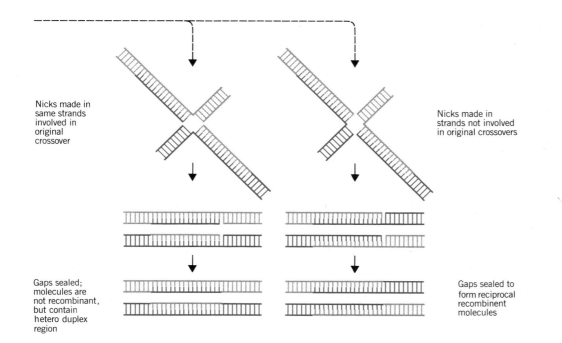

Nicks made in same strands involved in original crossover

Nicks made in strands not involved in original crossovers

Gaps sealed; molecules are not recombinant, but contain hetero duplex region

Gaps sealed to form reciprocal recombinent molecules

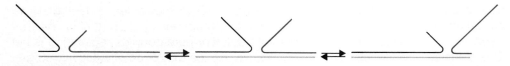

Figure 35.3
Branch migration can occur in either direction when an unpaired single strand displaces a paired strand.

This structure could be created by a renaturation event in which a single DNA strand anneals at one end with one complementary strand, and anneals at the other end with an independent complementary strand.

in natural conditions. When the reaction involves duplex DNA molecules, topological manipulation may be required; either the DNA duplex must be free to rotate, or equivalent relief from topological restraint must be provided, for example, by a topoisomerase.

The joint molecule formed by strand exchange must be **resolved** into two separate duplex molecules. This requires a further pair of nicks. The alternatives for this reaction are visualized in the lower part of Figure 35.2 in terms of the planar molecule generated by rotating one of the duplexes of the recombination intermediate. The consequences depend on which pair of strands is nicked.

If the nicks are made in the pair of strands that were *not* originally nicked (the pair that did not initiate the strand exchange), all four of the original strands have been nicked. This releases recombinant DNA molecules. The duplex of one DNA parent is covalently linked to the duplex of the other DNA parent, via a stretch of heteroduplex DNA. There has been a conventional recombination event between markers located on either side of the heteroduplex region.

If the *same* two strands involved in the original nicking are nicked again, the other two strands remain intact. The nicking releases the original parental duplexes, intact except that each has a residuum of the event in the form of a length of heteroduplex DNA.

These alternative resolutions of the joint molecule establish the principle that a strand exchange between duplex DNAs always leaves behind a region of heteroduplex DNA, but the exchange may or may not be accompanied by recombination of the flanking regions.

Several variations on the use of heteroduplex DNA can be envisaged. The strand exchange could be initiated by only one strand instead of by both; branch

migration could be responsible for extending the reaction to displace the other strand. The reaction could occur by unwinding and crosswise pairing of strands, without any nicking (although this imposes greater topological requirements). In the model of Figure 35.2, the heteroduplex DNA in each molecule is the same length. Genetic studies in fungi (see later) suggest that the regions of heteroduplex DNA in recombinant chromosomes can vary in length. This can be accommodated by a modification of the model in which a new stretch of DNA is synthesized to replace part of one of the exchanging strands. All of these factors affect the details of the mechanism and the order of events, but all models rely on the same general principle: the formation of an intermediate involving heteroduplex DNA that can be extended by branch migration.

ISOLATION OF RECOMBINATION INTERMEDIATES

A property of circular DNA offers an approach to isolating recombination intermediates. So far, we have considered the recombining DNA duplexes as linear molecules, but many genomes (especially in viruses and plasmids) are circular. **Figure 35.4** illustrates the consequences of a reciprocal recombination between homologous sites on two circular DNA molecules. The expanded part of the figure shows that the structure at the recombinant joint is the same as that shown in Figure 35.2; the only difference is that the ends of each duplex parental DNA are joined together. This has an important topological consequence. Resolution of the structure by the recombination route generates a dimeric circle, with the two original parental sequences joined head to tail.

Figure 35.4
Reciprocal recombination between homologous duplex circles proceeds through a figure-eight
intermediate to generate a dimeric circle.

What does this mean for the life-style of circular DNA elements? If they indulge in recombination, the original (monomeric) circular genome can be maintained only by *pairs* of recombination events. A single reciprocal recombination always generates the dimer shown in Figure 35.4. This dimer can return to the monomeric condition by a second recombination involving any pair of homologous sequences, as illustrated in **Figure 35.5.** In any population of circular genomes, we may therefore expect to find some multimeric circles if recombination is occurring.

A practical consequence of recombination between circular DNAs is that the recombination intermediate should take the form of a figure eight, as depicted in the expanded structure of Figure 35.4. Can we isolate such molecules? Dimeric DNA molecules with this appearance have been detected among the DNAs of several phages in infected cells and also in a yeast plasmid population. However, the outline of these structures does not unequivocally identify them as re-

Figure 35.5
A dimeric circle can generate two monomeric circles by a reciprocal recombination between homologous sequences.

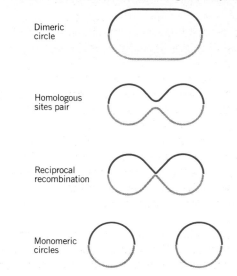

Dimeric circle

Homologous sites pair

Reciprocal recombination

Monomeric circles

combination intermediates; for example, a figure eight could represent two interlocked (catenated) monomeric circles.

The nature of the figure eight can be distinguished by cleaving the isolated molecule with a restriction enzyme that cuts each monomeric circle only once. The two parts of the figure eight then will fall apart *unless they are covalently connected.* If the figure eight is a recombination intermediate, cleavage will generate a structure in which the parental duplex molecules are held together by a region of heteroduplex DNA at the point of fusion. The cleaved molecule has four arms, each pair of the arms corresponding to each of the original monomeric circles.

In the cleaved form, this molecule is called a **chi structure,** because of its resemblance to the greek letter χ. An example is shown in **Figure 35.6** together with an interpretive drawing. The four arms of duplex DNA are connected by a region in which the heteroduplex segments have been pulled apart into their constituent single strands (probably by the conditions of preparation). This structure corresponds precisely with the predicted intermediate of Figures 35.2 and 35.4; its isolation directly demonstrates the existence of heteroduplex DNA *in vivo.*

Chi structures are found only in wild-type cells, not in *recA⁻* mutants. Thus their presence is correlated with the ability of the cell to undertake genetic recombination. This strengthens the case that they are authentic intermediates in the process and not some irrelevant oddity.

These observations offer an assay for purifying the

Figure 35.6

Cleavage of a figure-eight at homologous sites on each circle generates a chi structure in which the four duplex arms are held together at a site of strand exchange.

Photograph kindly provided by David Dressler.

recombination enzymes. Cell-free extracts can be assayed for their ability to generate dimeric circles from monomeric circles. Extracts from wild-type cells can accomplish this reaction, generating figure eight molecules that can be cleaved into chi structures. Extracts from *recA⁻* cells are inactive. In principle, it should be possible to fractionate this system into its components and to use *in vitro* complementation assays, along lines similar to the approach developed previously for DNA replication (see Chapter 33).

THE STRAND-EXCHANGE FACILITY OF RecA

Another approach is to investigate the properties of enzymes implicated in genetic recombination by the occurrence of *rec⁻* mutations in their genes. The paucity of known products means that this approach so far has concentrated on characterizing the ability of RecA to exchange DNA strands. We have already mentioned that RecA requires single-stranded DNA and ATP for its protease activity (see Chapter 34). The same substrates are required for its ability to manipulate DNA molecules.

RecA promotes base pairing between a single strand of DNA and its complement in a supercoiled duplex circle. The single strand invades the circle to displace the original partner to its complement, forming a D loop, as illustrated in the upper row of **Figure 35.7.** This reaction is called **single strand uptake** or **single strand assimilation.** Single strand assimilation is potentially related to the model for heteroduplex formation in Figure 35.2. The difference is that the recombination model invokes two invasions, with each single strand reciprocally displacing its homologue in the partner duplex.

The RecA protein can bind both single-stranded DNA (an action stimulated by the presence of ATP) and double-stranded DNA (an action requiring the presence of ATP). The assimilation reaction requires hydrolysis of the ATP. Assimilation works at an optimum rate with 1 monomer of ATP for every 5–10 nucleotides of single-stranded DNA. Thus the reaction is driven by the binding of RecA to single-stranded DNA. We do not yet understand the mechanism of the prelimi-

Figure 35.7
RecA promotes the assimilation of invading single strands into duplex DNA so long as one of the reacting strands has a free end.

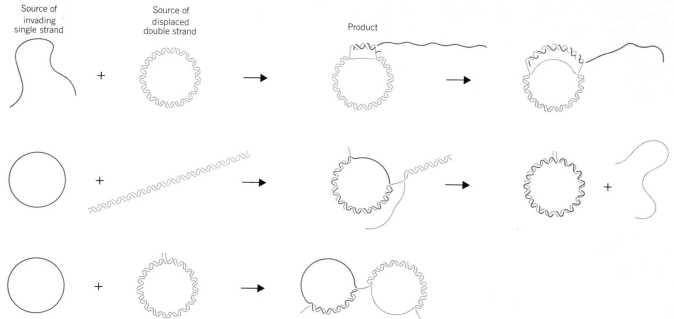

Source of
invading
single strand

Source of
displaced
double strand

Product

nary stages that precede D-loop formation, but probably the RecA protein first binds to single-stranded DNA, and then binds duplex DNA to search for a complement to the single-stranded region. The main query about this procedure is how the duplex DNA is melted to allow examination of its potential complementarity.

In the D-loop reaction, the displacement of the homologous circular strand is only partial; the structure of the covalently closed duplex circle may impose topological constraints on the progress of the reaction. A more extensive reaction can occur if we reverse the structures of the double- and single-stranded molecules. A circular single-stranded molecule can react with a homologous duplex molecule to displace one strand in its entirety. The reaction is shown in the center row of Figure 35.7. It starts at one end of the duplex and continues to the other end, forming a heteroduplex up to several thousand base pairs long.

This pair of substrates has been used to ask some questions about the mechanisms of the reaction. Does it proceed in a specific direction? **Figure 35.8** shows that this can be tested by using a linear duplex DNA that is complementary with the single strand only at one end. This substrate is generated by adding a foreign DNA fragment to one end of the linear molecule. When the fragment is located at the 5' end of the complementary strand—the strand that must pair with the invading single strand—the reaction proceeds normally (upper row of the figure). But when the 3' end of the complementary strand consists of foreign DNA, there is no reaction (center row of figure). This means that single strand assimilation can proceed only with a fixed polarity, from the 3' end of the complement toward the 5' end, or, in terms of the invading strand, from 5' to 3'.

What happens if two sequences are partially complementary, but do not match perfectly? What degree of divergence can RecA accommodate? When foreign DNA is inserted in the center of the linear duplex (lower row of figure), the strand-assimilation reaction halts at the boundary. Thus its continuation, as well as initiation, depends on complementary base pairing. However, strand assimilation can proceed through a modest degree of mismatching. For example, the DNAs of phages fd and M13 are about 97% homologous and can undergo strand assimilation together. But φX174

Figure 35.8
RecA can assimilate a circular single strand into a duplex only if there is a complementary 3' end with which the single strand can base pair. If foreign DNA is present within the duplex, the assimilation reaction stops at the end of the complementary sequence.

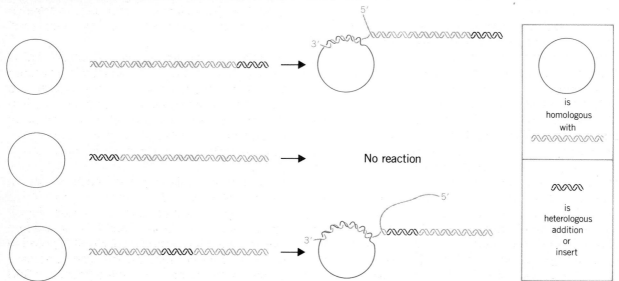

DNA and G4 DNA, which are 70% homologous, cannot do so. The maximum tolerable divergence is not yet known.

RecA AND THE CONDITIONS OF RECOMBINATION

We know that RecA is required to *initiate* the single strand assimilation reaction, but the existence of spontaneous branch migration leaves open the question of whether RecA also is needed subsequently. In conditions more like those prevailing in the cell, when the SSB (single-strand binding protein) is present, RecA is required to extend as well as to initiate heteroduplex formation. The hydrolysis of ATP is needed for the extension reaction.

A mechanism for the activity of RecA in stimulating branch migration is suggested by its ability to aggregate into filaments. The protein can form lengthy filaments with single-stranded DNA. If provided with a nucleotide cofactor, RecA can polymerize by itself or can incorporate duplex DNA into filaments. These filaments are longer than the original duplex, possibly reflecting some unwinding of the DNA. The RecA filaments do not form in the presence of ATP, but can be formed in the presence of nonhydrolyzable analogs.

One possibility is that molecules of RecA extend strand assimilation by binding cooperatively to DNA to form a structure related to the filament. This structure might be disrupted by hydrolysis of ATP. Such a model explains the requirement of the reaction for stoichiometry between RecA and single-stranded DNA (which would not be found if individual molecules of RecA moved along the nucleic acid). The presence of SSB stimulates both single strand assimilation and the formation of filaments, possibly by cooperating in some manner with the RecA protein to construct an appropriate framework for heteroduplex formation.

Single strand assimilation requires a free end of DNA. As **Figure 35.7** shows, the free end can be on the invading strand (upper row) or on the duplex molecule (center row). If both molecules are circular, the duplex circle must have a break in the strand that is to be displaced (lower row). The need for a free end could reflect an inherent property of the RecA protein, or it could result from topological constraints. A single strand

with a free end can wrap around its complement to form a duplex, but if there is no free end, the molecules are rather quickly constrained from further pairing.

The importance of the topology of the molecules is shown by the ability of a single-stranded circle to invade a covalently closed supercoiled duplex circle only when topoisomerase I is present. The topoisomerase presumably solves the unwinding/rewinding problem by making transient breaks that obviate the need for a free end. An implication of this result is that recombination could occur by unpairing and cross-pairing without the initial strand breakages shown in Figure 35.2.

All of the reactions we have discussed so far represent only a part of the potential recombination event: the invasion of one duplex by a single strand. Two duplex molecules can interact with each other under the sponsorship of RecA, provided that one of them has a single-stranded region of at least 50 bases. The single-stranded region can take the form of a tail on a linear molecule or of a gap in a circular molecule.

The reaction between a partially duplex molecule and an entirely duplex molecule leads to the exchange of strands. An example is illustrated in **Figure 35.9**. Assimilation starts at one end of the linear molecule, where the invading single strand displaces its homologue in the duplex in the customary way. But when the reaction reaches the region that is duplex in both molecules, the invading strand unpairs from its partner, which then pairs with the other displaced strand.

At this stage, the molecule has a structure that is indistinguishable from the recombinant joint in Figure 35.2. When the reacting molecules are circular, the product can be visualized as a chi structure. The isolation of such products of the reaction sponsored *in vitro* by RecA demonstrates the ability of the protein to mediate reciprocal strand transfer in these conditions.

RecA requires a single-stranded region in one of the reacting DNA molecules. How is this provided *in vivo*? One possibility is that there is a limited degradation of one strand of DNA. Another is that some recombination activity unwinds the DNA. It is possible that the RecBC product may be involved. Originally identified as a potent nuclease, in the presence of SSB

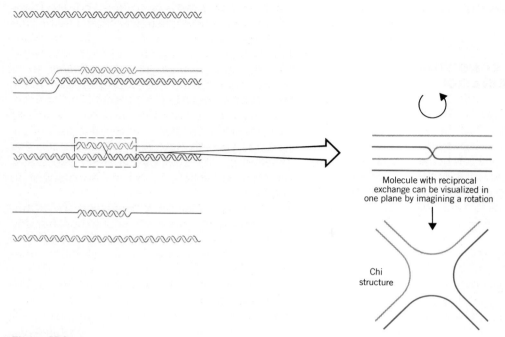

Figure 35.9
RecA-mediated strand exchange between partially duplex and entirely duplex DNA generates a joint molecule with the same structure as a recombination intermediate.

the enzyme also is able to unwind DNA to generate single-stranded regions.

We have dealt with the actions of RecA in a context independent of DNA sequence. However, there are hotspots that stimulate the RecA recombination system. They were discovered in phage lambda through a somewhat roundabout route.

The maturation of lambda DNA requires concatemeric molecules. Usually these are formed by a switch in the replication mode, from the initial pattern of the θ structure to the rolling circle. The tail of the rolling circle provides the concatemeric DNA. The switch needs the products of lambda genes involved in general recombination of the phage DNA, *red* and *gam*. One function of the Gam product is to inhibit the action of exonuclease V (the RecBC product), which otherwise prevents the conversion of θ molecules into the rolling circles. In *red⁻ gam⁻* infections, lambda not only is restricted to the θ form, but also fails to provide any recombination capacity of its own.

To provide an alternative substrate for phage maturation, dimeric circles must be generated by recombination between monomeric circles (see Figure 35.4). The absence of phage-coded recombination functions makes this event entirely dependent on the RecA function of the host cell. The occurrence of the recombination (originally as shown by an increase in the yield of phages) is greatly increased by certain mutations of lambda, called *chi*. The *chi* mutants turn out to have single base-pair changes creating sites that stimulate recombination.

These sites share the same nonsymmetrical sequence of 8 bp. From the presence of related sequences that lack the stimulatory effect, we may infer that any base change in this sequence abolishes its ability to act as a hotspot. The *chi* sequence occurs naturally in *E. coli* DNA about once every 10 kb. Its absence from wild-type lambda DNA, and also from other genetic elements, shows that it is not essential for RecA-mediated recombination. The *chi* sequence is able to stimulate recombination in its general vicinity, say within a distance of up to 10 kb from the site. We do not know what role it plays in recombination, but it is possible that the *chi* sequence is needed to initiate or to resolve recombination intermediates.

GENE CONVERSION ACCOUNTS FOR INTERALLELIC RECOMBINATION

The involvement of heteroduplex DNA in recombination explains the characteristics of recombination between alleles; indeed, these data provided the impetus for the development of the heteroduplex model. When recombination between alleles was discovered,

the natural assumption was that it takes place by the same mechanism of reciprocal recombination that applies to more distant loci. That is to say that an individual breakage and reunion event occurs between the loci to generate a reciprocal pair of recombinant chromosomes. However, in the close quarters of a single gene, the formation of heteroduplex DNA itself is usually responsible for the recombination event.

Individual recombination events can be studied in the *Ascomycetes* fungi, because the products of a single meiosis are held together in a large cell, the ascus. Even better, the four haploid nuclei produced by meiosis are arranged in a linear order. Actually, a mitosis occurs after the production of these four nuclei, giving a linear series of eight haploid nuclei. **Figure 35.10** shows that each of these nuclei effectively

Figure 35.10

Spore formation in the *Ascomycetes* allows determination of the genetic constitution of each of the DNA strands involved in meiosis.

A recombination event forms hybrid DNA in the region indicated by the shaded box. Follow the genetic fate of a locus in this region, indicated by the vertical line. If one of the hybrid DNA molecules is corrected, the spores form a 3:5 segregation. If both are corrected the same way, the ratio is 2:6. Each spore represents one DNA strand, because a further replication occurs between meiosis and spore formation.

represents the genetic character of one of the eight strands of the four chromosomes produced by the meiosis.

Meiosis in a heterozygote should generate four copies of each allele. This is seen in the majority of spores. But there are some spores with abnormal ratios. These are explained by the formation and correction of heteroduplex DNA in the region in which the alleles differ.

Suppose that two alleles differ by a single point mutation. When a strand exchange occurs to generate heteroduplex DNA, the two strands of the heteroduplex will be mispaired at the site of mutation. In effect, each strand of DNA carries different genetic information. If no change is made in the sequence, the strands separate at the ensuing replication, each giving rise to a duplex that perpetuates its information. This leads to the abnormal 4:4 ratio, in which the *order* of the spores is altered, because of the single strand exchanges. This ratio is called **postmeiotic segregation,** since its occurrence depends on the separation of the DNA strands at the replication that succeeds meiosis.

Some asci display the ratios of 3:5 or 2:6, in which one or two spores, respectively, that should have been of one allelic type actually are of the other type. This may be caused by the correction of mismatches in heteroduplex DNA. When a repair system recognizes the mispaired apposition of two bases in heteroduplex DNA, it may excise and replace one of the strands to restore complementarity. Such an event changes the strand of DNA representing one allele into the sequence of the other allele. If this happens at only one of the reciprocal heteroduplexes, a 3:5 or 5:3 ratio results (depending on the direction of conversion). If both heteroduplexes are corrected in the same way, a 2:6 or 6:2 ratio results. The correction process is called **gene conversion;** it may be recognized in the form of any one of the aberrant ratios. (Either direction of conversion may be equally likely, or there may be allele-specific effects causing a preference for one direction.)

Gene conversion does not depend on crossing-over, but is correlated with it. A large proportion of the aberrant asci show genetic recombination between two markers on either side of a site of interallelic gene conversion. This is exactly what would be predicted if the aberrant ratios result from the formation of heteroduplex DNA, with an approximately equal probability of resolving the structure with or without recombination (as indicated in Figure 35.2). The implication is that fungal chromosomes initiate crossing-over about twice as often as would be expected from the frequency of recombination between distant genes.

Within a gene, a recombination between two sites occurs when heteroduplex DNA is formed or is corrected at one site but not at the other. Thus the interallelic recombination event shows the same behavior as gene conversion: it need not necessarily be associated with reciprocal recombination between flanking markers.

Why does this process yield a linear genetic map for recombination within a gene? It seems that the closer two sites lie together, the greater the probability that they will suffer **coconversion** in the same stretch of heteroduplex DNA. In this case, both sites are corrected in the same direction, upsetting the ratios of genotypes emerging from the cross, but not resulting in a recombination between the markers.

Heteroduplex DNA may extend for appreciable distances. Some information about the extent of gene conversion is provided by the sequences of members of gene clusters. Usually, the products of a recombination event will separate and become unavailable for analysis at the level of DNA sequence. However, if an unequal exchange takes place between two members of a gene cluster, as illustrated previously in Figure 21.3, a heteroduplex may be formed between the two nonallelic genes. This heteroduplex may suffer gene conversion to the sequence of one of its strands. In effect, this converts one of the nonallelic genes to the sequence of the other. A similar heteroduplex could occur by a direct reaction between the two copies of the gene on the same chromosome.

The presence of more than one gene copy on the same chromosome provides a footprint to trace these events. For example, if heteroduplex formation and gene conversion occurred over part of one gene, this part may have a sequence identical with or very closely related to the other gene, while the remaining part shows more divergence. Available sequences suggest that gene conversion events may extend for up to a few thousand bases.

SPECIALIZED RECOMBINATION RECOGNIZES SPECIFIC SITES

The conversion of lambda DNA between its different life forms involves two types of event. First, the pattern of gene expression is regulated as described in Chapter 16. Repression of the operators is established and maintained for lysogeny; while repression is abolished for passage through the lytic cycle. Second, the physical condition of the DNA is different in the lysogenic and lytic states; interconversion of these states involves a site-specific recombination.

In the lytic life-style, lambda DNA exists as an independent molecule; it is circular in the infected bacterium. In the lysogenic state, the phage DNA is an integral part of the bacterial chromosome (called prophage). To enter the lysogenic condition , free lambda DNA must be **integrated** into the host DNA. To be released from lysogeny into the lytic cycle, the prophage DNA must be **excised** from the chromosome.

Integration and excision occur by recombination at specific loci of the bacterial and phage DNAs called **attachment (att) sites.** The attachment site on the bacterial chromosome is called att^λ in bacterial genetics; the locus is defined by its possession of prophage λ in lysogenic strains and by mutations that prevent integration of lambda. For considering the recombination reaction illustrated in **Figure 35.11,** the bacterial attachment site is called *attB*, consisting of the sequence components *BOB'*. The attachment site on the phage, *attP*, consists of the components *POP'*. The sequence *O* is common to *attB* and *attP* and is the site of a reciprocal recombination between them. Because the phage DNA is circular, the recombination event results in its insertion into the bacterial chromosome as a linear sequence. The prophage is bounded by two new *att* sites, the products of the recombination. As the map is usually written, at the left is *attL* consisting of *BOP'*, and at the right is *attR* consisting of *POB'*.

In the earlier formulations of the recombination model, the two sites *attB* and *attP* were represented as half sites, *B.B'* and *P.P'*, where the dot (.) identified the point of crossing-over. The reaction was envisaged to proceed between sites entirely lacking in homology. Later it was realized on genetic grounds that there

must be some homology between the *att* sites, so the *O* was introduced to represent a common **core** sequence, within which the recombination crossover occurs. The flanking regions *B, B'* and *P, P'* are referred to as the **arms,** and each is distinct in sequence.

An important consequence of the constitution of the *att* sites is that the integration and excision reactions do not involve the same pair of reacting sequences. Integration requires recognition between *attP* and *attB;* while excision requires recognition between *attL* and *attR*. Thus although the recombination event is reversible, different conditions prevail for each direction of the reaction. This may be important in the life of the phage, since it offers a means to ensure that an integration event is not immediately reversed by an excision, and vice versa.

Figure 35.11
Circular phage DNA is converted to an integrated prophage by a reciprocal recombination between *attP* and *attB;* the prophage is excised by reciprocal recombination between *attL* and *attR*.

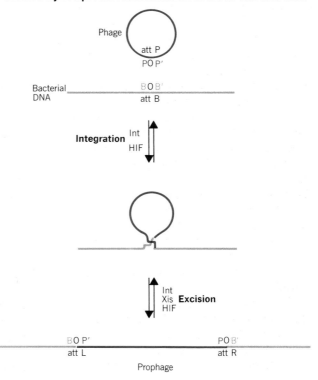

The difference in the pairs of sites reacting at integration and excision is reflected by a difference in the proteins that mediate the two reactions. Integration requires the product of the phage gene *int* and a bacterial protein called integration host factor (IHF). The IHF consists of two subunits, one of which is coded by the SOS-regulated locus *himA*. In addition to Int and IHF, the excision reaction requires the product of phage gene *xis*. (Under certain conditions, excision can occur rather inefficiently in the absence of Xis, probably reflecting a residual ability of Int and IHF to reverse the integration reaction directly.)

All four type of *att* site are available for study. They are provided in a convenient form by "illegitimate recombination," in which the lambda genome is released by reciprocal recombination between a sequence in the prophage and a sequence in the flanking bacterial DNA. The reaction occurs by the generalized (RecA-dependent) pathway, probably reflecting a spurious homology between the recombining sites. The consequences are illustrated in **Figure 35.12.**

Illegitimate recombination generates a defective phage in which some of the authentic phage DNA has been replaced by some of the flanking bacterial DNA. The excised genome carries *attL* if the recombination involves a bacterial site on the left; it carries *attR* if the event occurs on the other side. The presence of some bacterial DNA in the phage is a very useful feature. The transport of bacterial genes via such a phage is called **transduction;** before the advent of recombinant DNA technology, this provided an important technique for obtaining the bacterial genes residing on either side of the *att*^λ site.

Using phages of this sort, it is possible to obtain similar DNA molecules carrying each type of *att* site. These substrates can be used to test the abilities of all pairwise combinations of *att* sites in site-specific recombination. The Int protein is needed for *all* combinations; it is aided by Xis most dramatically in the *attL* × *attR* reaction, but Xis has a stimulatory effect on some other combinations. We do not yet understand the differences in the various reactions that control the need for Xis.

STAGGERED BREAKAGE AND REUNION IN THE CORE

The nature of the core sequence became evident as soon as the *att* sites were sequenced. All the sites are A-T-rich regions that possess a common sequence of 15 bp. The occurrence of recombination within this sequence is shown by the exchange of the arms in *attL* and *attR* compared with *attP* and *attB*.

Does recombination occur at any position within the core sequence or at a specific point? Tracing the fate of a radioactive label in the phosphate groups of the *attB* site has shown that material is transferred directly in the recombination—there is a precise breakage and reunion in the absence of any synthesis of DNA. The pattern of transfer of the label shows that specific points exist for recombination. The points of exchange are different on each strand of DNA, suggesting that breakage and reunion is staggered.

The model illustrated in **Figure 35.13** shows that if *attP* and *attB* sites each suffer the same staggered cleavage, complementary single-stranded ends could be available for crosswise hybridization. This is reminiscent of the reaction between the sticky ends generated by some restriction enzymes (see Chapter 19).

Figure 35.12
Illegitimate recombination between a site in prophage and a site in bacterial DNA excises a transducing phage in which bacterial genes have replaced some phage genes.

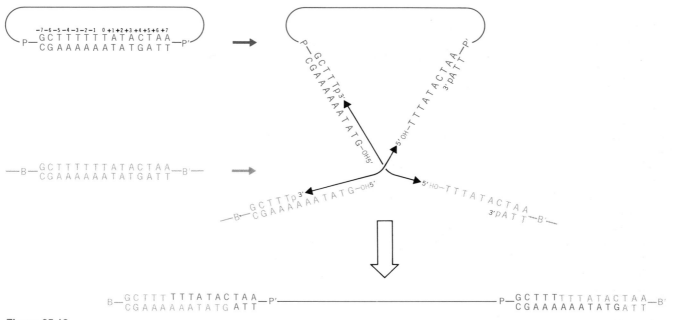

Figure 35.13

Staggered cleavages in the common core sequence of *attP* and *attB* allow crosswise reunion to generate reciprocal recombinant junctions.

The sequence is numbered relative to the center of the core.

Because the *phosphate* groups were labeled, the results do not unequivocally identify the distance between the staggered cuts unless we know which side of the phosphodiester bond is nicked. The intermediates with single-stranded ends have not been isolated. The data show that the stagger must be 5 or 7 bp, and other information is consistent with the latter. A 7 bp stagger implies that the reaction proceeds by generating 3'-phosphate and 5'-OH ends.

The nature of the recombination event means that a mutation occurring on either side of the 7 bp overlap is destined to remain in the arm in which it is located. Thus it can never be transferred from *attB* to *attP* (or from *attL* to *attR*). A mutation within the 7 bp overlap, however, can be transferred by gene conversion between *attP* and *attB* (and thus can move between all *att* sites). The discovery of such transfer was the basis of the original conception that the *att* sites must have a common core, although the common sequence (15 bp) turns out to be longer than the region involved in the stagger.

When the *att*$^\lambda$ site is deleted from the *E. coli* chro-

mosome, an infecting lambda phage can establish lysogeny by integrating elsewhere, although the efficiency of the reaction is less than 0.1% of the frequency of integration at *att*$^\lambda$. This inefficient integration occurs at **secondary attachment sites.** These sites have sequences resembling those within and adjacent to the 15 bp common core. The recombinants generated by insertion at these sites have sequences consistent with a 7 bp separation between the sites of crossover on the two strands. In these secondary reactions, an important feature in the core region is the homology between the particular pair of recombining sites in the 7bp overlap; another feature is homology with the sequences that Int recognizes around the sites of cutting (see later).

The site-specific recombination system is amenable to analysis of enzymatic mechanism since the reaction can be performed *in vitro* by Int and IHF. The roles of the arms can be investigated by making deletions on either side. The function of *attP* is abolished by deletions reducing its size beyond the limits of -152 (in *P*) to + 82 in (*P'*). To determine the role of individual regions within this stretch of 240 bp, it will be neces-sary to introduce changes within each segment. The *attB* site is smaller; its function can be exercised by the 23 bp fragment extending from -11 to $+11$, in which there are only 4 bp on either side of the core. A fragment of *attP* of similar size behaves like an *attB* site in this assay. The disparity in their sizes suggests that *attP* and *attB* play different roles in the recombi-nation, with *attP* providing additional information nec-essary to distinguish it from *attB*.

When the reaction is performed *in vitro* between two supercoiled DNA molecules, almost all of the super-coiling is retained by the products. Thus there cannot be any free intermediates in which strand rotation could occur. Either the free ends generated by breakage are held in a fixed configuration by the recombination en-zyme(s), or one end is transferred directly to the other end via some four-stranded intermediate structure. The breakage and reunion could be accomplished by an activity resembling topoisomerase I, except that the nicked strand is sealed together with a nicked strand of a different instead of the same duplex. Int protein may be involved in such a function, since it has a (rather ineffectual) topoisomerase I ability to relax negatively supercoiled DNA.

Both the Int and IHF proteins bind to specific sites in *att* DNA *in vitro*. For recombination *in vitro,* about 20–40 molecules of the 40,000-dalton Int protein are needed per recombinant DNA. Even more of the IHF protein is needed, about 70 molecules of the 20,000-dalton protein per recombinant. (IHF consists of sub-units of 11,000 and 9500 daltons.) The high stoichi-ometry suggests that the proteins do not function cat-alytically, but form some structure that supports only a single recombination event.

Both the Int and IHF proteins bind to specific sites in the *att* region. The binding sites in *attP* are sum-marized in **Figure 35.14.** Int binds to a segment of 30 bp encompassing the core region, to a sequence of similar length in *P'*, and to two separate shorter 15 bp sequences in *P*. The differences in the lengths of the individual sites probably reflect the number of Int mon-omers bound. (Int binding to *attB* resembles the bind-ing in the core region of *attP*.) IHF binds to sequences of about 20 bp in the arms of *attP;* the IHF binding

Figure 35.14
Int and IHF bind to different sites in *attP*. The Int recognition sequences in the core region include the sites of cutting.

sites are approximately adjacent to the Int sites. To-gether, Int and IHF cover most of *attP*.

The basis for Int binding in the arms and the core is different. First, in the arms the binding sites are related to a consensus sequence of 8 bp that is ab-sent from the core region. Second, Int can bind to DNA fragments containing the sequence CAACTTNTT. Sequences related to this are found in inverted orien-tation at the core-arm junctions of *attP* and *attB*. These latter sites encompass the sites of cutting in the core, as indicated in Figure 35.14, placing Int in the right location to accomplish recombination; we do not yet understand the role of Int and IHF binding at the sites in the arms.

FURTHER READING

Recombination has been a focus for many reviews over the past few years. A splendid view of mechanisms, delving into the role of RecA, is given by **Dressler** & **Potter** (*Ann. Rev. Biochem.* **51,** 727–761, 1982). A review relating mecha-nisms to earlier observations in fungi, and also considering the role of hotspots, is from **Stahl** (*Ann. Rev. Genet.* **13,** 7–24, 1979). The mechanism of lambda site-specific recom-bination has been discussed by **Nash** (*Ann. Rev. Genet.* **15,** 143–167, 1981).

PART 10
THE DYNAMIC GENOME: DNA IN FLUX

Elements carried in the maize chromosomes . . . serve to control gene action and to induce, at the site of the gene, heritable modifications affecting this action. These elements were initially discovered because they do not remain at one position in the chromosome complement. They can appear at new locations and disappear from previously determined locations. The presence of one such element at or near the locus of a known gene may affect the action of this gene. In so doing, it need not alter the action potential of the genic substances at the locus. Therefore, these elements were called controlling elements. . . . It might be considered that a controlling element represents some kind of extrachromosomal substance that can attach itself or impress its influence in some manner at various positions in the chromosome complement and so affect the action of the genic substances at these positions. The modes of operation of controlling elements do not suggest this, however. Rather they suggest that controlling elements are integral components of the chromosomes themselves, and that they have specific activities and modes of accomplishing them, much as the genes are presumed to have. . . . Transpositions of controlling elements either arise from some yet-unknown mechanism or occur during the chromosome reduplication process itself and are a consequence of it.

Barbara McClintock, 1956

CHAPTER 36
TRANSPOSABLE ELEMENTS
IN BACTERIA

Until recently, the genomes of both procaryotes and eucaryotes were regarded as generally static sequences, subject to change only on the leisurely time scale of evolution. We are accustomed to the idea that the construction of a genetic map identifies the loci at which known genes reside; by implication, other (unidentified) sequences also may be expected to remain at constant positions in the population of genomes. The stability of genetic organization is indicated by the retention of linkage relationships even after speciation—for example, between man and the apes. The difference of generation times between procaryotes and eucaryotes suggests that their evolutionary scales might be different in terms of real time; but even in the procaryotes, the overall organization of the genome changes only relatively slowly. For example, a very similar genetic map describes the different bacterial species *E. coli* and *S. typhimurium.*

The "conventional" evolution of the genome relies on mechanisms that have been familiar for many years. In procaryotes, extrachromosomal elements are responsible for mediating genetic exchanges. Plasmids sponsor conjugation of bacteria, while phages are released by infection. Both may occasionally transfer host genes along with the autonomous replicon. In eucaryotes, reciprocal recombination occurs each generation between corresponding sites on homologous chromosomes; very infrequently, it is accompanied by duplication or rearrangement of loci. Such reorganization is essentially a side effect of the usual mechanisms involved in genetic recombination and/or DNA synthesis (such as unequal crossing-over or gene conversion). The mechanisms responsible for translocations between nonhomologous chromosomes remain unknown.

Within the framework of the genome, however, are mechanisms that allow the movement of certain sequences. Only in the past few years have we obtained evidence for this type of rearrangement, involving **transposable elements** or **transposons.** Many such elements have been found in the DNA of procaryotes. Bacterial transposition involves duplication of the element; one copy remains at the original "donor" site, while the other copy appears at a new "recipient" or "target" site. Transposition does not rely on any relationship between the sequences at the donor and recipient sites, an important difference from all other mechanisms involved in reconstruction of DNA. Each transposon carries the genes required for its own transposition, although it may also require functions of the genome in which it resides (such as DNA polymerase and DNA gyrase). The occurrence of comparable events in eucaryotes may be inferred from the general structural similarity and variety in location of analogous elements (see Chapters 37 and 38).

What consequences follow from the presence in a genome of transposable sequences able to reproduce and move to new locations? A possible advantage for the organism lies in the ability of transposition to sponsor rearrangements, directly or indirectly. The transposition event itself may cause deletions or inversions or may lead to the transport of a host sequence to a new location. Transposons could provoke cellular systems by functioning as "portable regions of homology"; two copies of a transposon at different locations (even on different chromosomes) may provide sites for reciprocal recombination. Such exchanges could result in deletions, insertions, inversions, or translocations.

Transposable elements have indeed played an active role in the evolution of some bacterial plasmids. The components of plasmids may be brought together by reactions involving recognition of transposon sequences. The integration of the F (sex) plasmid of *E. coli* into the bacterial chromosome often occurs by a host (RecA)-mediated reciprocal recombination between a transposon in the plasmid DNA and a homologous transposon in the bacterial DNA.

On some occasions, a cellular or phage genome may have imprisoned a transposable function for its own purposes. The element becomes static and plays a role in regulation, engaging in specific transposition-like events. Examples in *Salmonella* and in phage Mu show that regulation may involve inversion of a specific segment of DNA, apparently by a recombination mechanism related to transposition functions. These processes also display some resemblance to phage lambda integration. Thus transposition mechanisms may run the gamut from the ability to link nonhomologous DNA sequences to the sponsorship of specific recombination events.

Considering the transposon itself, the advantages conferred by its intermittent activities provide somewhat nebulous targets for natural selection. This view has prompted suggestions that (at least some) transposable elements may confer neither advantage nor disadvantage on the phenotype, but could constitute "selfish DNA," concerned only with their own propagation. According to this concept, the relationship of the transposon to the genome resembles that of a parasite with its host. Presumably the spreading of the elements by duplication and transposition is balanced by the harm done if a transposition event inactivates a necessary gene, or if the number of transposons becomes a burden on cellular systems. Yet we must remember that any transposition event conferring a selective advantage—for example, a genetic rearrangement—will lead to preferential survival of the genome carrying the active transposon.

THE DISCOVERY OF TRANSPOSITION IN BACTERIA

Transposable elements were discovered by insertions of new material within bacterial operons. Located within a gene, such an insertion prevents transcription and/or translation of the gene, and may also have a polar effect in reducing the expression of subsequent genes in the operon. Initially detected as a negative mutant, the insertion is identified by properties unlike those of a point mutation; the only form of reversion is the deletion of the inserted material. By comparing the sequences present in various insertion mutants, it is possible to classify several discrete elements. Each type of element may be inserted in either orientation at a given site.

Many transposons have been followed through more than one cycle of transposition. **Figure 36.1** summarizes a protocol that allows both the donor and the recipient to be examined. A plasmid carrying the transposon is introduced into a cell that has another plasmid lacking the transposon. Recipients are selected for transposition (for example, by the inactivation of a drug resistance marker on the target plasmid). Then the recipient and donor plasmid DNA molecules are isolated. Their characterization confirms that the recipient indeed has gained an insertion of the transposon; and it also shows that the donor has not lost its copy. Thus transposition is associated with the generation of an *additional copy* of the transposon at the recipient site.

Transposition is often studied by isolating many independent insertions of an element in a selected target region. Fine-structure analysis then reveals whether there is any pattern in the selection of sites for inser-

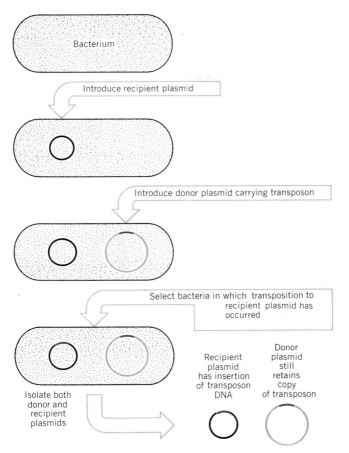

Figure 36.1
Transposition generates a new copy of the transposon at the target site while the original copy remains at the donor site.

ent elements, but typically is in a range comparable to the spontaneous mutation rate, usually within 10^{-5} to 10^{-7} per generation. Reversion (by deletion) is infrequent, with a range of rates of 10^{-6} to 10^{-10} per generation, up to some 10^3 times less frequent than insertion.

Although transposons vary in type, certain features are common to all transposable elements. Each element is an autonomous unit that codes for the proteins needed to sponsor transposition. The reaction involves recognition of the ends of the transposable element. The smallest transposons code only for the genes involved in transposition. (They fit the definition of selfish DNA.) Larger transposons carry additional genetic markers.

Why do transposons have polar effects on genes distal to a site of insertion in an operon? The inserted sequence becomes part of the polycistronic mRNA of the operon. Thus it may interfere with transcription or translation. In some cases, polarity depends on rho-mediated termination of transcription; it can therefore be relieved in *rho*⁻ strains (see Chapter 13). In other cases, polarity results from termination of translation at a nonsense codon in the transposon. Since transposons have regulatory signals pertinent to their own genes, these signals sometimes may influence events in an operon; some transposons have promoters near their boundaries that sponsor the transcription of flanking material, thus activating genes adjacent to the element.

INSERTION SEQUENCES ARE BASIC TRANSPOSONS

The first set of transposons to be isolated were called **insertion sequences** by virtue of the way in which they were detected. Each type is given the prefix **IS,** followed by a number that identifies the type. (The original classes were numbered IS1-4; later classes have numbers reflecting the history of their isolation, but not corresponding to the total number of elements so far isolated!) The IS elements are normal constituents of bacterial chromosomes and plasmids. For example, one standard strain of *E. coli* has eight copies of IS1 and five copies of IS2. To describe an insertion into a particular site, a double colon is used; thus

tion. Most transposons have multiple sites of insertions; some show (varying degrees of) preference for hotspots. Transposons whose sites of insertion are random within a small region may show a preference for one general region compared with another. For example, insertions may be more common over some particular stretch of 3 kb or so, but sites may be chosen at random within that stretch. The basis for this selection is unknown; it is described in Tables 36.1 and 36.3 as a "regional" pattern. The most important point is that, at the molecular level, selection of a target is independent of nucleotide sequence (with one prominent exception).

The frequency of transposition varies among differ-

Table 36.1
Some Transposons Consist of or Contain Individual Modules

	Length (bp)	Inverted Terminal Repeat (bp)	Direct Repeat at Target (bp)	Possible Number of Proteins	Target Selection
IS elements					
IS1	768	23	9	2	Regional
IS2	1327	41	5	2	Hotspots
IS4	1428	18	11 or 12	2	Hotspots
IS5	1195	16	4	4	Hotspots
IS-like elements					
IS10R	1329	22	9	3	Hotspot
IS50R	1531	9	9	3	Hotspots
IS903	1057	18	9	2	Unknown

lambda::IS1 describes an IS1 element inserted into phage lambda.

The IS elements identify the simplest class of transposons, since their genetic functions are concerned only with the ability to transpose. Although each IS element is different in sequence, all share a common form of organization. The parameters of some common IS elements are summarized in the upper part of **Table 36.1.**

Each element possesses short **inverted terminal repeats.** Most commonly these are between 15 and 25 bp long. Usually the two copies of the repeat are closely related rather than identical. **Figure 36.2** shows that the presence of the inverted terminal repeats means that the same sequence is encountered proceeding toward the element from the flanking DNA on either side of it.

When an IS element transposes, a sequence of host DNA at the site of insertion is duplicated. The nature of the duplication is revealed by comparing the sequence of the target site before and after an insertion has occurred. The IS DNA is always flanked by very

Figure 36.2
Transposons have inverted terminal repeats and generate direct repeats of flanking DNA at the target site.

In this example, the target is a 5 bp sequence. The ends of the transposon consist of inverted repeats of 9 bp, where the numbers 1 through 9 indicate a sequence of base pairs.

```
                                    Target
                                     site
N N N N N N N N N N N N N N N N N N A T G C A M M M M M M M M M M M M M M M M M M
N N N N N N N N N N N N N N N N N N T A C G T M M M M M M M M M M M M M M M M M M
```

```
N N N N N N N N N N N N N N N N N N A T G C A 1 2 3 4 5 6 7 8 9 ————————— 9 8 7 6 5 4 3 2 1 A T G C A M M M M M M M M M M M M M
N N N N N N N N N N N N N N N N N N T A C G T 1 2 3 4 5 6 7 8 9 ——Transposon—— 9 8 7 6 5 4 3 2 1 T A C G T M M M M M M M M M M M M M
                                    Inverted                                      Inverted
                                     repeat                                        repeat
```

short **direct repeats.** (In this context, "direct" indicates that two copies of a sequence are repeated in the same orientation, not that the repeats are adjacent.) The target site has the sequence of only *one* of these repeats. In the hypothetical example of Figure 36.2, the target site consists of the sequence $^{ATGCA}_{TACGT}$. After transposition, one copy of this sequence is present on either side of the transposon.

For most transposons, the *sequence* of the direct repeat is different in each individual transposition event, but the *length* is constant (a reflection of the mechanism of transposition). The most common lengths for the direct repeats are 5 bp and 9 bp.

As can be seen from Figure 36.2, a transposon therefore displays a characteristic structure in which its ends are identified by the inverted terminal repeats, while the adjacent ends of the flanking host DNA are identified by the short direct repeats. When observed in a sequence of DNA, this type of organization is taken to be diagnostic of a transposon, and makes a prima facie case that the sequence originated in a transposition event.

Transposition always involves the insertion of a non-permutated linear sequence. Thus all copies of a given transposon have the same inverted terminal repeats at the junction with host DNA. The importance of the ends of the element is confirmed by the terminal location of mutations that prevent an element from transposing. The *cis*-acting nature of such mutations shows that the ends are recognized by the protein(s) responsible for transposition.

The shortest transposon, IS1, contains two open reading frames in the same direction; they are translated into the proteins InsA and InsB (whose functions have yet to be delineated). All the other IS elements contain a single long coding region, starting just inside the inverted repeat at one end, and terminating just before or within the inverted repeat at the other end. This long reading frame is virtually certain to be translated in each case; the product has been isolated in several cases. Usually there are also some shorter potential coding regions on the other strand of DNA— that is, reading the same sequence in the opposite direction. Whether these overlapping reading frames give rise to proteins is less certain, but they are included in the count of possible products in Table 36.1.

COMPOSITE TRANSPOSONS HAVE *IS* MODULES

Some transposons carry drug resistance (or other) markers as well as the functions concerned with transposition. These transposons are named **Tn** followed by a number. One class of larger transposons comprises **composite elements** that consist of a central region carrying the drug marker(s) flanked on either side by **arms** whose sequences are related. (These arms sometimes have been called **lengthy terminal repeats,** but this description risks confusion with the short inverted terminal repeats of the IS elements.)

The arms may be in either the same or (more commonly) an inverted orientation. Thus a composite transposon with arms that are direct repeats has the structure

If the arms are inverted repeats, the structure is

Arm L	Central region	Arm R

The arrows indicate the orientation of the arms, which are identified as L and R according to an (arbitrary) orientation of the genetic map of the transposon from left to right. The structure of a composite transposon is illustrated in more detail in **Figure 36.3.**

The arms may consist of identical IS elements. In

Figure 36.3
A composite transposon has a central region carrying markers unconnected with transposition (such as drug resistance) flanked by IS modules. The modules have short inverted terminal repeats. If the modules themselves are in inverted orientation (as drawn), the short inverted terminal repeats at the ends of the transposon are identical.

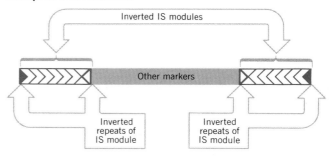

the case of Tn9, a direct repeat of an IS1 element is present at each end of the transposon.

In other cases, the arms resemble IS elements. Each consists of a single continuous open reading frame, flanked by the usual short inverted terminal repeats. However, because these elements are not found as individual components in bacterial DNA, they are described as **IS-like** elements. They are named in the same way as IS elements, and usually are given a number related to the transposon in which they are found. The properties of IS-like elements are compared with those of the IS elements in the lower part of Table 36.1.

The properties of some composite transposons are summarized in **Table 36.2.** Since the arms on the one hand are part of the composite transposon, but on the other resemble IS elements or even are IS elements, they are sometimes described as IS or IS-like **modules.** In some cases, the modules of a composite transposon are identical, such as Tn9 (direct repeats of IS1) or Tn903 (inverted repeats of IS903). In other cases, the modules are closely related, but not identical. Thus we can distinguish the modules such as IS10L and IS10R. (In this case, the L and R terminology identifies *different modules;* when the modules are identical, the same terminology merely indicates their location relative to the central region of the transposon.)

Since each module has the usual structure ending in inverted repeats, the composite transposon also ends in the same short inverted repeats. (When the arms are present in inverted orientation, the short inverted repeats at the ends of the transposon are likely to be identical rather than just related.)

Although the IS-like modules are not usually found as independent elements in bacterial DNA (hence their name), a functional module may be able to transpose itself, as shown by the examples of IS10R of Tn10 or IS50R of Tn5. When the modules of a composite transposon are identical, presumably either module can sponsor movement of the transposon. When the modules are different, they may differ in functional ability, so transposition will depend entirely or principally on one of the modules.

We assume that composite transposons evolved by an association of two originally independent modules with the central region. Such a situation could arise when an IS element transposes to a recipient site close to the donor site. Originally identical, therefore, the modules may have remained identical or may have diverged. The ability of a single module to transpose the entire composite element may explain the lack of selective pressure for both modules to remain active.

We should like to know how two modules cooperate to transpose a composite transposon instead of just the individual module. This question is especially pressing in cases where both the modules *are* functional. In the example of Tn9, where the modules are IS1 elements, presumably each is active in its own right as well as on behalf of the composite transposon. Of course, since transposition involves the generation of a new copy at the target site, the existence of a composite transposon is not imperiled if one of its modules sponsors an independent transposition of itself (although this does not further the aim of spreading the entire transposon).

Two IS elements in fact are able to transpose any sequence residing between them, as well as them-

Table 36.2
Some Transposons Are Composites Bracketed by Individual Modules

Element	Length (bp)	Genetic Markers	Terminal Modules	Module Orientation	Module Relationship	Module Function
Tn10	9300	tet^R	IS10R IS10L	Inverted	2.5% divergence	Fully functional Reduced function
Tn5	5700	kan^R	IS50R IS50L	Inverted	1 bp change	Functional Nonfunctional
Tn903	3100	kan^R	IS903	Inverted	Identical	Both functional
Tn9	2500	cam^R	IS1	Direct	Presumed identical	Presumed functional

Figure 36.4
The IS10 modules can cooperate to transpose any region of DNA that lies between them.

When Tn10 is part of a small circular molecule, the IS10 repeats can transpose either side of the circle. Transposition of *tet*ʳ corresponds to the movement of Tn10. Transposition of the markers on the other side creates a new "inside-out" transposon.

selves. An experiment that demonstrates this ability is summarized in **Figure 36.4.** If Tn10 resides on a circular replicon, its two modules can be considered to flank *either* the *tet*ʳ gene of the original Tn10 *or* the sequence in the other part of the circle. Thus a transposition event can involve either the original Tn10 transposon or the "inside-out" transposon with the alternate central region. Note that both transposons have inverted modules, but these modules evidently can function in either orientation relative to the central region. In events of this nature, the frequency of transposition declines with the distance between the modules. The length dependence probably reflects problems that occur during the process of transposition. It may have helped determine the sizes of the common composite transposons.

To summarize the features that are necessary for transposition of IS or IS-like elements, all except IS1 have a single major gene product, corresponding to a reading frame that runs the length of the element. This single product is probably responsible for creating a target site and for recognizing the ends of the transposon. *Only* the ends are needed for a transposon to serve as a substrate for transposition. Assistance from host functions may be needed to replicate the transposon. We know little about the detailed processes and order of events involved in these reac-

tions, but we surmise that they may be similar for all the IS elements and composite transposons (perhaps excepting IS1).

ONLY ONE MODULE OF Tn10 IS FUNCTIONAL

The relationship between modules that are no longer identical may offer insights into the evolution of the composite transposon. The examples of Tn10 and Tn5 are the best characterized.

Only a few bases at each end of Tn10 are needed to recognize the element as a substrate for transposition. The inverted terminal repeats are 22 bp long; probably only the last 13 are needed for the transposition reaction. Mutations in this region prevent transposition and are *cis*-acting.

Tn10 is an exceptional element that has a specific target sequence. The 9 bp direct repeats of flanking DNA generated by transposition display a consensus of a 6 bp sequence symmetrically disposed within the target. Thus the repeats on either side of Tn10 often take the form $^{NGCTNAGCN}_{NCGANTCGN}$, where N identifies any base pair. The stronger the hotspot, the more closely it conforms to the consensus. Probably the same single Tn10-

coded function recognizes the target sequence and the ends of the transposon.

The element of IS10R provides the active module of Tn10. The IS10L module is functionally defective and provides only 1–10% of the transposition activity of IS10R. The accumulation of mutations in IS10L relative to IS10R corresponds to about 2.5% divergence between the modules.

The organization of IS10R is summarized in **Figure 36.5.** It has a continuous reading frame that represents an essential protein of 47,000 daltons; the significance of another, overlapping reading frame is unknown. The level of the essential protein is enhanced by "high-hopper" mutations that increase expression of the IS10R gene. Thus the amount of this protein may limit the rate of transposition.

Mutants in this gene can be complemented in *trans* by another, wild-type element only with some difficulty. This reflects a strong preference of the protein for *cis*-action. In some unknown way, the protein functions efficiently only with the DNA template from which it was transcribed and translated. This is a common feature of proteins involved in IS-mediated transposition. (We have described the *cis*-preference of another function, the A protein involved in φX174 replication, in Chapter 33.)

Two promoters are found close to the outside boundary of IS10R. The promoter P_{IN} is responsible for transcription of IS10R. The promoter P_{OUT} causes transcription to proceed into the adjacent flanking DNA,

an effect that could be responsible for activating bacterial genes adjacent to Tn10.

The phenomenon of "multicopy inhibition" reveals that expression of the IS10R gene is regulated. Transposition of a Tn10 element on the bacterial chromosome is reduced when additional copies of IS10R are introduced via a multicopy plasmid. The inhibition requires the sequence of the P_{OUT} promoter; and it is effective only for an IS10R gene expressed from the P_{IN} promoter (the effect is lost if the gene is placed under control of a different promoter). However, multicopy inhibition is exercised at the level of translation, not transcription. The basis for the effect may lie with the 40 bp overlap in the 5′ terminal regions of the transcripts from P_{IN} and P_{OUT}. Because P_{OUT} is a much stronger promoter, a large amount of the OUT RNA may be present; it could base pair with the IN RNA to prevent translation of the protein needed for transposition.

THE MODULES OF Tn5 ARE ALMOST IDENTICAL BUT VERY DIFFERENT

The inverted modules of Tn5 provide a striking example of the ability of small mutational changes to produce major functional effects. The module IS50R is functional; module IS50L is nonfunctional with regard to transposition. The sole difference between the two modules lies in the substitution of a single base

Figure 36.5
Two promoters in opposite orientation lie near the outside boundary of IS10R. The strong promoter P_{OUT} sponsors transcription into the flanking host DNA. The weaker promoter P_{IN} starts transcription of an RNA with a coding region that extends the length of IS10R. The "OUT" and "IN" transcripts have a 40-base overlap.

Figure 36.6

A single base-pair change in IS50L prematurely terminates synthesis of proteins 1 and 2 and creates a promoter for transcription of the gene for neomycin (and kanamycin) resistance.

pair. The effects of this substitution are illustrated in **Figure 36.6.**

Two proteins are produced from the same reading frame in IS50R. Their only difference is that protein **1** has an additional N-terminal ~40 amino acids that are absent from protein **2**. The precise events involved in transcription and translation of these proteins have yet to be defined; all we know for certain is that more is produced of protein **2** than of protein **1**.

The single base-pair change in IS50L simultaneously affects translation of these proteins and controls transcription of the central region. The substitution creates an ochre codon that prematurely terminates translation of both proteins **1** and **2**. The truncated proteins (sometimes called **3** and **4**) lack transposition activity. The same substitution also creates a promoter for transcription of the gene of the central region that codes for neomycin phosphotransferase II, the enzyme responsible for resistance to antibiotics such as neomycin and kanamycin. The change is therefore necessary for this function of the transposon.

The functions of proteins **1** and **2** are related but different. Protein **1** is essential for transposition of either the IS50 module or the intact Tn5 transposon. The wild-type protein complements a defective module in

trans only exceedingly poorly. Thus protein **1** may be a typical *cis*-acting transposition function.

Protein **2** is an inhibitor of transposition. Its action is exercised at some currently unknown stage of the transposition process itself (that is, rather than by regulating gene expression). Protein **2** is *trans*-acting; one possibility for its action is that it preempts the transposition function of protein **1** by binding to the same sites that protein **1** must recognize in order to sponsor transposition. An alternative is that proteins **1** and **2** form some oligomeric complex in which the activity of protein **1** is inhibited.

The behavior of protein **2** may explain a property of Tn5. On entering a new host bacterium, Tn5 transposes at a high frequency. Once the element has become established, the frequency decreases. An established element also can inhibit transposition by an incoming element. This *trans*-acting function may reflect the ability of protein **2** to prevent transposition; when a new element enters the cell, sufficient amounts of protein **2** already are present to inhibit the *cis*-acting function of the protein **1** of the new as well as the old Tn5 element. The truncated proteins of IS50L also may be involved in this effect.

The relationship between protein **2** and protein **1** of

Tn5 is naturally different from the relationship between the P_{OUT} and P_{IN} transcripts of Tn10. Yet these interactions share a common feature: they limit the frequency of transposition. Mechanisms to limit transposons from multiplying ad infinitum may be important to prevent them from interfering too much with cellular functions.

TRANSPOSITION INVOLVES REPLICATIVE RECOMBINATION

"Transposition" is almost a misnomer, in the sense that a transposon does not leave its former site in the course of transposition. Since the process involves duplication of the transposon, one copy remains at the original site, while the other is found at the new site. Thus transposition is accompanied by an increase in the number of copies of the transposon.

The basic nature of the transposition event appears to be similar for all transposons, as indicated by the diagnostic repetition of a short sequence of target DNA. Transposition involves two types of reaction, which in principle could occur in either order. One involves replication of the transposon without any replication of adjacent chromosomal sequences. The other requires breakage of the target DNA sequence to generate a site into which the transposon is inserted. At no stage in this process is a free molecule of transposon DNA generated; all the events of transposition proceed by manipulation of the DNA within which the transposon resides.

The general nature of the insertion reaction is illustrated in **Figure 36.7.** It consists of making staggered breaks in the target DNA, joining the transposon to the protruding single-stranded ends, and filling in the gaps. The generation and filling of the staggered ends explain the occurrence of the direct repeats of target DNA at the site of insertion. The stagger between the cuts on the two strands determines the length of the direct repeats; thus the target repeat characteristic of each transposon reflects the geometry of the enzyme involved in cutting target DNA.

The common outcome of transposition disguises the existence of more than one mechanism for transposon duplication and insertion. The existence of at least two general classes of transposition is indicated both by

differences between the products of the reaction and by virtue of variety in other reactions, as well as the transposition event itself.

In addition to the "simple" intermolecular transposition that results in insertion of an additional copy at a new site, transposons promote other types of DNA rearrangements. Some of these events may represent alternative outcomes of the transposition mechanism, and they leave clues about the nature of the underlying events. Others are consequences of the relationship between the multiple copies of the transposon.

Rearrangements of host DNA may result when a transposon inserts a copy at a second site near its original location. Host systems may undertake reciprocal recombination between the two copies of the transposon; the consequences are determined by whether the repeats are the same or in inverted orientation.

Figure 36.8 illustrates the general rule that recombination between any pair of direct repeats will result in a deletion of the material between them. The intervening region is excised as a circle of DNA (which is

Figure 36.7

The direct repeats of target DNA flanking a transposon are generated by the introduction of staggered cuts whose protruding ends are linked to the transposon.

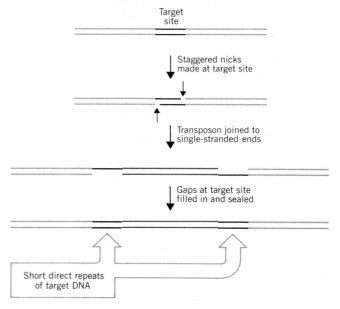

Target site

Staggered nicks made at target site

Transposon joined to single-stranded ends

Gaps at target site filled in and sealed

Short direct repeats of target DNA

lost from the cell); the chromosome retains a single copy of the direct repeat. Note that a recombination between the directly repeated IS1 modules of the composite transposon Tn9 would therefore replace the transposon with a single IS1 module. **Deletion** of sequences adjacent to a transposon could result from a two-stage process; transposition generates a direct repeat of a transposon, and recombination occurs between the repeats. However, the majority of deletions that arise in the vicinity of transposons probably result from a variation in the pathway followed in the transposition event itself.

Figure 36.9 depicts the consequences of a reciprocal recombination between a pair of inverted repeats. The region between the repeats becomes inverted; the repeats themselves remain available to sponsor further inversions. **Duplicative inversions** are associated with transposition, and take the form of transposons in inverted orientation on either side of a central region that has been inverted from its usual orientation. Note that a composite transposon whose modules are inverted is a stable component of the genome, although the direction of the central region with regard to the modules could be inverted by recombination.

A significant omission from the list of events sponsored by transposons is **excision.** The loss of a transposon can be detected by restoration of function at the site of insertion. This must involve **precise excision;** it occurs at a frequency of $\sim 10^{-6}$ for Tn5. Loss can also be detected by relief of polarity in adjacent genes; this can be accomplished by **imprecise excision** (which leaves a remnant of the transposon, sufficient to prevent reactivation of the target gene, but insufficient to cause polar effects). Imprecise excision occurs at a frequency of $\sim 10^{-5}$ for Tn5. Neither type of excision is associated with transposition to new sites. The process does not rely on transposon-coded functions; it is RecA-independent and could occur by some cellular mechanism that generates spontaneous deletions between closely spaced repeated sequences.

The most revealing of the transposon-mediated reactions is **replicon fusion** to form a **cointegrate** structure. A replicon containing a transposon may become fused with a replicon lacking the element, as illustrated in **Figure 36.10.** The resulting cointegrate has two copies of the transposon, one at each junction between the original replicons, oriented as direct repeats. The cointegrates generated by composite

Figure 36.8
Reciprocal recombination between direct repeats excises the material between them; each product of recombination has one copy of the direct repeat.

Figure 36.9
Reciprocal recombination between inverted repeats inverts the region between them.

Figure 36.10
Transposition may fuse a donor and recipient replicon into a cointegrate.

transposons may have a duplication either of the whole element (Tn cointegrates) or of the IS module (IS cointegrates).

The relative frequencies with which cointegrates are generated by different transposons indicate that there may be different pathways for transposition. For example, Tn903 and IS1 generate "simple" transpositions and cointegrate structures at about the same frequency. But Tn5 and Tn10 generate transpositions 50–200 times more frequently than cointegrates. The cointegrates generated by Tn5 are stable, which suggests that cointegrate formation is an alternative (albeit rare) outcome of Tn5 transposition, not an intermediate in the reaction.

A model for a transposition pathway in which a cointegrate is an obligatory intermediate is illustrated in **Figure 36.11.** Four single-strand cleavages initiate the process. The donor molecule is cleaved at either end of the transposon by a site-specific enzyme that recognizes the termini. The target molecule is cleaved at

sites staggered (most commonly) by 5 or 9 bases. The order and time at which the cleavages are made during the transposition reaction could be changed.

The donor and target strands are ligated at the nicks. Each end of the transposon sequence is joined to one of the protruding single strands generated at the target site. The linkage generates a chi-shaped structure held together at the duplex transposon. The formation of this structure is responsible for the movement of the transposon; the enzyme activity responsible for it is called the **transposase.**

The chi structure contains a single-stranded region at each of the staggered ends. These regions are pseudoreplication forks that provide a template for DNA synthesis. (Use of the ends as primers for replication implies that the strand breakage must occur with a polarity that generates a 3′-OH terminus at this point.) If replication continues from both the pseudoreplication forks, it will proceed through the transposon, separating its strands, and terminating at its ends. Replication is probably accomplished by host-coded functions. At this juncture, the structure has become a cointegrate, possessing direct repeats of the transposon at the junctions between the replicons (as can be seen by tracing the path around the cointegrate).

A site-specific recombination between the two copies of the transposon can regenerate the original donor replicon, releasing a target replicon that has gained a transposon flanked by short direct repeats of the host target sequence. This reaction is called **resolution;** the enzyme activity responsible is called the **resolvase.**

TRANSPOSITION OF Tn3 PROCEEDS BY COINTEGRATE RESOLUTION

Some large (~5 kb) transposons are not composites relying on IS-type transposition modules, but comprise independent units carrying genes for transposition as well as for features such as drug resistance. The TnA family is the best characterized. Originally thought to be a single transposon, it is now known to include many related transposons, some of which are described in **Table 36.3.**

Transposition of these elements proceeds in two

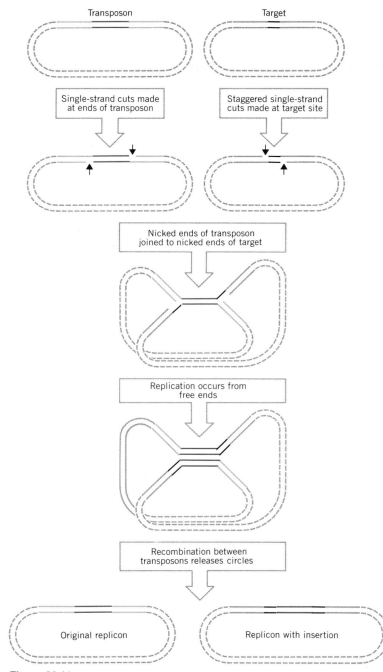

Figure 36.11
One pathway for transposition proceeds through the formation of a cointegrate that is resolved by recombination.

Table 36.3
The TnA Family of Transposons Constitutes Large but Individual Units

Element	Length (bp)	Genetic Markers	Inverted Terminal Repeats (bp)	Flanking Repeat at Target (bp)	Target Selection
Tn3	4957	*amp*^R	38	5	Regional
Tn1	5000	*amp*^R	?	5	Regional
γδ	5800	?	37	5	Regional
Tn501	8200	Hg^R	38	5	Unknown

stages: first a cointegrate is formed; then it is resolved. Each stage is accomplished by a different product of the transposon (in contrast with the putative single-transposon protein of IS elements). The transposition stage involves the ends of the element, as it does in IS-type elements. Resolution requires a specific internal site, a feature unique to the TnA family.

The TnA transposons have the usual terminal feature of closely related inverted repeats, generally around 38 bp in length. *Cis*-acting deletions in either repeat prevent transposition of an element. A 5 bp direct repeat is generated at the target site, which is selected randomly, although with regional preference.

The two genes whose products are needed for transposition are identified by recessive mutations. Since the mutant genes can be complemented in *trans* by wild-type genes of another copy of the transposon, their actions do not show the *cis*-acting preference characteristic of the IS-type transposition process.

Mutants in *tnpA* gene are unable to transpose. The gene product is a transposase of 1021 amino acids (~120,000 daltons). The purified protein binds to single-stranded DNA. Little else is known about its function, although it is needed presumably both for recognizing the ends of the element and for making the staggered 5 bp breaks to which they are connected in target DNA. The transposases of Tn3 and the related element γδ work efficiently only on their own type of transposon, although the ends of these two elements are the same in 27 of the 38 bp inverted repeats. Some of the positions that are different therefore must be critical points for recognition of the ends.

The *tnpR* gene codes for a protein of 185 amino acids (23,000 daltons) whose dual functions are revealed by the diverse effects of mutations. The protein both acts as a repressor of gene expression and provides the resolvase function.

Mutations in *tnpR* increase the transposition frequency. The reason is that TnpR represses the transcription of both *tnpA* and its own gene. Thus inactivation of TnpR protein allows increased synthesis of TnpA, which results in an increased frequency of transposition. This implies that the amount of the TnpA transposase must be a limiting factor in transposition.

The *tnpA* and *tnpR* genes are expressed divergently from a 163 bp A-T-rich intercistronic control region, as indicated in the map of Tn3 given in **Figure 36.12.** (The organization of other TnA-type transposons is similar.) Both the effects of the TnpR protein are mediated by its binding in this region.

In its capacity as the resolvase, TnpR is involved in recombination between the direct repeats of Tn3 in a cointegrate structure. The cointegrate is an obligatory intermediate in transposition; resolution is blocked in *tnpR*⁻ mutants, which therefore give rise only to cointegrates.

A cointegrate can in principle be resolved by a homologous recombination between any corresponding pair of points in the two copies of the transposon (see Figure 36.11). In fact, however, the resolution reaction occurs only at a specific site. This site was originally identified by *cis*-acting deletions in an internal region that block completion of transposition, causing the accumulation of cointegrates. The site was originally called *IRS* (internal resolution sequence); now it is known as *res*. In the absence of *res*, the resolution reaction can

Figure 36.12
Transposons like Tn3 have inverted terminal repeats, an internal *res* site, and three known genes.

Arrows indicate the direction of expression of the coding regions. Numbers in the upper diagram indicate the length in base pairs of each region. Numbers in the lower diagram give the distance in base pairs from the crossover point (0).

be substituted by RecA-mediated general recombination, but this is much less efficient.

The sites bound by the TnpR resolvase have been determined by footprinting the DNA-protein complex. Their locations are summarized in the lower part of Figure 36.12. Binding occurs independently at each of three sites, each 30–40 bp long. Site I includes the region genetically defined as the *res* site; in its absence, the resolution reaction does not proceed at all. However, resolution also involves binding at sites II and III, since the reaction proceeds only poorly if either of these sites is deleted. Site I appears to overlap with the startpoint for *tnpA* transcription. Site II overlaps with the startpoint for *tnpR* transcription; an operator mutation maps just at the left end of the site.

The three binding sites in Tn3 share a sequence homology that defines a consensus sequence; and there is dyad symmetry in the sites. What interactions occur between the sites? One possibility is that binding at all three is required to hold the DNA in an appropriate topology; the resolution reaction could be triggered when the protein-DNA complex of one transposon interacts with that of another transposon. Binding at a single set of sites may repress *tnpA* and *tnpR* transcription without introducing any change in the DNA.

The region in which the site-specific recombination occurs was first identified by comparing the reactions of transposons Tn3 and γδ, whose *tnpR* genes are highly homologous and can substitute for each other in complementation assays. The sequences of recombinants show that the resolution reaction occurs within a 19 bp part of site I that is identical in both plasmids.

More information about the reaction has been gained from an *in vitro* resolution assay. This uses a cointegrate-like DNA molecule as substrate. (The substrate was constructed by cloning *res* to form direct repeats at two locations in the plasmid pBR322.) The substrate must be supercoiled; its resolution produces two catenated circles, each containing one *res* site. The reaction requires large amounts of the TnpR resolvase (produced in quantity by standard cloning techniques); resolution works best at enzyme to DNA ratios of ~20:1. No host factors are needed (contrast with lambda site-specific recombination).

Resolution is a conservative reaction; bonds are broken and rejoined without demand for input of energy. Magnesium ions are needed for completion of the reaction; if they are omitted, a partial reaction occurs with low efficiency. The products identify an intermediate stage in cointegrate resolution; they consist of resolvase covalently attached to both 5′ ends of double-stranded cuts made at the *res* site. The cleavage occurs symmetrically at a short palindromic region to generate two base extensions. Expanding the view of the crossover region located in site I, we can describe the cutting reaction as

$$
\begin{array}{llll}
5' & \text{T T A T A A} & 3' \\
3' & \text{A A T A T T} & 5' \\
 & \qquad \downarrow \\
5' & \text{T T A T} \\
3' & \text{A A} - \text{protein} \\
 & \qquad + \\
 & \text{protein} - \text{A A} & 3' \\
 & \qquad \text{T A T T} & 5'
\end{array}
$$

A similarity with the lambda site-specific recombination system is implied by the relationship of the *res* site with the lambda *att* core described in Chapter 35. Immediately surrounding the point of crossover, there is homology between the two sites at 10 of the 15 bp. The sequences are

The arrows identify the sites cleaved in *res* by TnpR and in *att* by Int. The reactions themselves are analogous, although resolution occurs only between intramolecular sites, whereas the recombination between *att* sites is intermolecular and directional (as seen by the differences in *att* sites). The resemblance to the cutting and rejoining reactions of topoisomerases is evident in both cases.

Finally, we should mention a remarkable effect of Tn3 called **transposition immunity.** A plasmid carrying Tn3 is immune to further insertions from any Tn3 element carried on a different DNA molecule (although the Tn3 on the plasmid appears able itself to transpose within the plasmid). There is a lag between the insertion of Tn3 into a new plasmid and the establishment of immunity. The effect is specific; it applies only to Tn3 and does not prevent other transposons from inserting into the plasmid. The basis for this *cis*-acting long-range effect is mysterious.

SOME UNUSUAL FEATURES OF TRANSPOSING PHAGE MU

Mu is a phage with the alternate life-styles of lytic cycle and lysogeny. It was discovered by its ability to cause mutations in *E. coli*. The mutations result from insertion of the 37 kb phage genome into random sites (with regional specificity) in the bacterial DNA. The insertions do not revert at a detectable frequency.

The map of phage Mu is shown in **Figure 36.13.** The integrated Mu genome has the same gene order as the free phage DNA, which is a linear molecule. (Note the contrast with phage lambda, whose linear free DNA circularizes on infection; thus integration generates a linear prophage whose gene order is a permutation of that of the free λ DNA.) The linear Mu genomes have neither sticky ends nor repeats, which raises the question of how the ends act in concert

during integration. Possibly the reaction involves homologies that occur up to ~100 bp from each end. The existence of a mechanism to recognize the specific ends of an integrated Mu sequence was revealed by the discovery of Mu excision, a reaction that occurs only when Mu prophage itself has gained an insertion of IS1. The mechanism is unknown, but involves Mu-coded functions as well as IS1-coded functions.

The ends of free Mu DNA possess additional material that is not part of the phage genome. The actual termini consist of sequences obtained from the host genome, about 100 bp long at the left (*c*) end and about 1500 bp long at the right (*S*) end. These sequences are different in individual molecules of Mu; they are acquired as a consequence of the packaging of Mu DNA from precursors in which the phage genome is flanked by host sequences. The additional regions are shed when a phage genome integrates into the bacterial chromosome.

An infecting parental Mu DNA is *not* efficiently integrated into the host genome. Most of the parental DNA retains its free ends during the lytic cycle. Less than 10% of the infecting phage genomes integrate. However, Mu DNA sequences are efficiently integrated during the reproductive cycle. This paradox suggests that the *copies* of Mu produced by replication have the ability to integrate. In a related reaction, induction of a Mu prophage to enter the lytic cycle is not accompanied by excision of phage DNA from the host chromosome. It is (again) the replication product that is important.

A connection between integration and DNA replication is revealed by the nature of mutants of Mu that are defective in integration. These mutants are unable to replicate; they lie in either of the two genes *A* and *B*, needed for phage replication. Thus integration seems to depend on the replication functions rather than on separate recombination enzymes (as seen with lambda).

There may be two pathways for Mu insertion. First, lysogens contain prophages generated by simple insertions of Mu. The similarity of this event to transposition is indicated by the presence of direct repeats of 5 bp of target DNA on either side of the prophage. Second, during lytic growth, almost all of the products of Mu-mediated transposition are cointegrate structures. We do not know whether the simple insertions

Figure 36.13

The order of Mu genes is the same in the prophage and free DNA, but the ends are different. Prophage DNA is flanked by 5 bp direct repeats of a target host sequence. Free phage DNA has different additional sequences of ~100 bp at the left end and ~1500 bp at the right end.

and cointegrates represent alternative interactions of Mu DNA with its target or are alternative fates of a common intermediate. The basic mechanisms of Mu transposition have yet to be worked out.

Another unusual feature of phage Mu is found near the right end of the genome. The **G** or **invertible segment** comprises a 3 kb segment found in different orientations in different molecules of the Mu DNA. In phage grown by lytic infection of *E. coli* strain K12, the G segment is always in the orientation called G(+). However, the DNA of phages generated by induction may contain the G segment in either this or the opposite G(−) orientation. A Mu gene called *gin*, located just beyond the G segment, is required for the inversion reaction.

The G segment contains the genes coding for proteins involved in phage adsorption. The orientation of the segment controls the expression of these genes, with the result that G(+) and G(−) phages have different specificities for bacterial strains. The dependence of strain specificity on the G segment is illustrated diagrammatically in **Figure 36.14.**

The G segment carries alternate sets of genes. In the G(+) orientation, genes *S* and *U* are expressed. Their products allow the phage to adsorb to *E. coli* K12, but not to *E. coli* C. In the G(−) orientation, genes *S'* and *U'* are expressed. Their products allow adsorption to *E. coli* C but not K.

The proteins have been tentatively identified: S is 56,000 daltons, U is 21,000 daltons, S' is 48,000 daltons, and U' is 26,000 daltons. A significant feature is that their combined lengths would require a coding sequence of about 4 kb, substantially longer than the G·segment itself. This suggests that the genes must overlap.

Pending determination of the sequence of the in-

Figure 36.14

Inversion of the G segment determines whether the S and U proteins or the S' and U' proteins are synthesized. One model supposes that their expression is initiated to the left of the invertible segment, in which case S and U must be coded by one strand of DNA while S' and U' are coded by the complementary strand.

vertible region, we do not know precisely how the two gene sets are organized. The model shown in Figure 36.14 supposes that the alternate sets are carried on the complementary strands of the DNA. Which set is expressed is determined by the orientation of the segment, because the promoter lies outside and sponsors transcription of only one strand within the segment.

SALMONELLA PHASE VARIATION OCCURS BY INVERSION

The movement of bacteria is accomplished by the waving of their flagella. Many *Salmonella* species are **diphasic** because they possess two nonallelic genes for flagellin (the protein subunit of the flagellum). A given clone of bacteria may express either the **H1** type (it is said to be in **phase 1**) or the **H2** type (the bacteria are in **phase 2**). Transition from one phase to the other

occurs about once in every 1000 bacterial divisions and is called **phase variation.**

The genes for the two types of flagellin reside at different chromosomal locations. The circuit for control of flagellin synthesis is illustrated in **Figure 36.15.** The H2 gene is closely linked to another gene (*rh1*) that codes for a repressor of H1 synthesis. These two genes are coordinately expressed. Thus when H2 is expressed, the repressor also is expressed and prevents any synthesis of H1. When H2 is not expressed, neither is the repressor, so synthesis of H1 occurs. The phase of the bacterium is determined by whether the *H2-rh1* transcription unit is active.

Expression of this transcription unit is controlled by the orientation of a segment of DNA adjacent to it. The segment is 995 bp long and is bounded by 14 bp repeats (*IRL* and *IRR*). The initiation codon for H2 lies 16 bp to the right of the adjacent inverted repeat.

Figure 36.15
***Salmonella* phase is determined by the activity of the H2 transcription unit.**
In phase 1, the H2 unit is inactive, so the H1 flagellin is synthesized. In phase 2, the H2 unit is active, and synthesizes H2 flagellin and a repressor that prevents expression of the H1 flagellin gene.

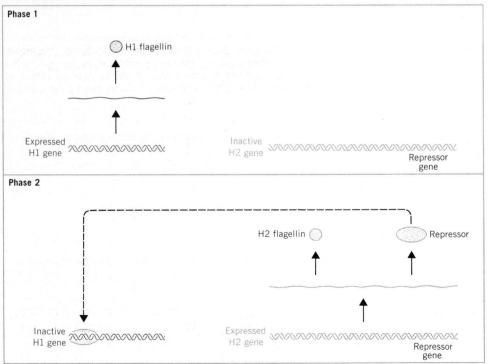

The segment of DNA between *IRL* and *IRR* contains the *hin* gene, whose product is a *trans*-acting protein that mediates the inversion of the entire segment by a reciprocal recombination between the inverted repeats (see Figure 36.9 previously). Mutations in the *hin* gene reduce the frequency of inversion by 10^4 times.

The consequences of the inversion reaction are illustrated in **Figure 36.16.** The promoter for the *H2-rh1* transcription unit lies within the invertible segment. In one orientation, transcription initiates at the promoter and continues through *H2-rh1,* resulting in phase 2 expression. In the other orientation, the promoter faces in the other direction, and the transcription unit is not expressed (although transcription probably occurs in the other direction, with unknown consequences). The absence of promotion of *H2-rh1* results in phase 1 expression.

Phase variation and Mu strain specificity therefore both use inversion to regulate gene expression. In phase variation, an on/off switch is provided by moving the promoter relative to the transcription unit. In Mu strain specificity, alternate genes are expressed by inverting the coding segment relative to the promoter.

Some interesting parallels emerge when these reactions are compared with one another and with Tn3 transposition. The *hin* function of *Salmonella* and the *gin* function of Mu can substitute for each other in complementation assays. (Another function interchangeable with these is the *cin* gene responsible for inversion of the *C* segment of phage P1.) Although the

Figure 36.16

Expression of H2 flagellin and the phase 1 repressor is controlled by the orientation of a 995 bp invertible region.

Phase 2 occurs when the promoter at position 950 sponsors transcription through *IRR* into the H2 unit. Phase 1 occurs when the promoter faces in the opposite direction. In both phases, *hin* is expressed from its own promoter (whose location is not determined). Numbers indicate base pairs within the invertible segment.

Figure 36.17
The H2 flagellin gene is homologous with the H1 gene, and the upstream regulatory region *hin* is homologous with the *gin* function of Mu.

The diversion of the mechanism for recombinational inversion can be seen by comparing the sequence around H2 with the other flagellin gene (H1) and with the Mu G segment. The situation is summarized in **Figure 36.17.** The H2 and H1 flagellin genes are quite well related and their homology extends upstream from each coding region. Farther upstream, the region including the *hin* gene shows strong homology with the *gin* function of Mu. Thus an ancestral fusion occurred between the H2 flagellin gene and a sequence related to the current Mu inversion function. This evolved into the present system for control of phase variation.

Tn3 *tnpR* gene cannot replace either of these sequences, the amino acid sequence of the resolvase shows considerable homology with that of the Hin protein.

The region of crossover for resolvase action shows sequence homology with *IRL* and *IRR*. A 34 bp sequence including *IRL* has only seven mismatches with inverted repeats that lie at either end of the G segment. This suggests that inversion of the *hin* segment of *Salmonella* and the G segment of Mu occurs by a site-specific recombination between the inverted repeats, related in mechanism to the reaction involved in resolution of Tn3 cointegrates. Thus these static inversion events could have evolved by interchanges between transposons and bacterial loci, possibly even by some ancestral immobilization of a transposable element, whose functions became limited.

FURTHER READING

The development of views on transposons can be followed through the reviews of **Kleckner** (*Cell* **11,** 11–23, 1977), **Calos** & **Miller** (*Cell* **20,** 579–595, 1980), and **Kleckner** (*Ann. Rev. Genet.* **15,** 341–404, 1981). Their evolutionary significance is considered by **Campbell** (*Ann. Rev. Microbiol.* **35,** 55–83, 1981). The original model for transposition via cointegrates was presented by **Shapiro** (*Proc. Nat. Acad. Sci. USA* **76,** 1933–1937, 1979); the sequence and function of Tn3 were analyzed by **Heffron et al.** (*Cell* **18,** 1153–1163). Two chapters in *Mobile Genetic Elements* (ed. Shapiro, Academic Press, New York, 1983) deal with individual transposons: **Heffron** on Tn3 and **Kleckner** on Tn10; a chapter by **Silverman** & **Simon** relates phase variation and Mu inversion, which also were previously explored by **Simon et al.** (*Science* **209,** 1370–1374, 1980). The sequence involved in phase inversion was reported by **Zieg** & **Simon** (*Proc. Nat. Acad. Sci. USA* **77,** 4196–4200, 1980). The invertible G segment was reviewed briefly by **Howe** (*Cell* **21,** 605–606, 1980).

CHAPTER 37
MOBILE ELEMENTS IN EUCARYOTES

The ubiquity of transposable elements could be taken either to indicate their importance for eucaryotes as well as procaryotes or to imply that they are equally successful in propagating in the context of either type of genome. In conjunction with the ability of plasmids to transfer between bacteria, procaryotic transposons may confer mobility on some host genes (possibly a useful substitute for the lack of sexual interaction). In a few instances, transposition-like mechanisms are involved in gene regulation. In eucaryotes, on the other hand, several situations are already known in which movement of sequences from or to specific sites plays a role in gene regulation. And in the genetic background are elements comparable to bacterial transposons, with the ability to move to randomly selected new locations in the genome.

Elements most directly comparable to bacterial transposons have been found in *S. cerevisiae* and *D. melanogaster*. They are small, have terminal repeats, and are found at different locations in different genomes. We assume that they carry genes needed to code for transposition; we do not know whether they carry accessory functions unconnected with the transposition event itself.

The existence of transposition was in fact first deduced many years ago in McClintock's genetic studies of maize. Although her results remained in an in-tellectual void until the more recent discovery of transposition in bacteria, now these events can be studied at the molecular level in eucaryotes also. In maize, considerable information exists about the genetic relationships between individual members of a transposon family as well as on the effects of insertion at particular loci and the various outcomes of the transposition event.

Analogous transposable elements have not yet been detected as components of animal cell genomes, but footprints of ancient transposition events are found in the form of direct target repeats flanking dispersed repetitive sequences. A current use of transposition-like mechanisms is presented by the ability of retroviruses to insert DNA copies of the RNA viral genome into the chromosomes of a host cell. These cases are discussed together with other examples of variability of eucaryotic DNA in Chapter 38.

The molecular mechanisms of eucaryotic transposition are not defined in the same detail as those of bacteria; we have yet to identify the relevant gene products of a eucaryotic transposon. However, the presence of the diagnostic feature of the short direct repeats of target DNA suggests an underlying similarity of mechanism in which this "generalized" transposition selects random sites for insertion. We should like to know what sort of relationship exists between the transposable elements of different species and whether transposition functions can be retained in passage of an element from one species to another.

Another type of event is represented by the "directed" transposition of sequences whose movement from one specific site to another has regulatory functions. In bacteria, there is a clear relationship between the mechanisms of site-specific inversion and transposition. In eucaryotes, directed movement of DNA may occur by other mechanisms. Yeast displays unidirectional transfer of copies of genes for mating type from "silent" to "active" loci. The process resembles bacterial transposition in leaving the donor locus intact while changing the recipient; however, there is only a single recipient locus, which initiates the transfer instead of being an innocent target. Another situation in which the content of the genome changes is provided by the immunoglobulin genes, in which recombination is used to construct active genes in the appropriate somatic cells (see Chapter 39). The gamut of changes in the genome thus runs from random transpositions to tissue-specific reconstructions.

YEAST TY ELEMENTS
RESEMBLE BACTERIAL TRANSPOSONS

The Ty elements of yeast comprise a family of dispersed repetitive DNA sequences. Each element is 6.3 kb long; the last 330 bp at each end constitute direct repeats, called δ. The number of Ty elements varies in different strains of *S. cerevisiae,* but in com-mon laboratory strains is in the range of 30 to 35. In addition to the intact Ty elements, there are about 100 independent delta elements, called solo δ's. Ty is an abbreviation for transposon yeast; the existence of a transposition function is suggested by the observation that Ty elements are found at different sites in different strains. We cannot yet estimate the frequency of Ty transposition, but it seems to be less than that of bacterial transposons. Movement of Ty elements has been followed largely by selecting mutations created by insertions into regulatory loci, so most of our information about the results of insertion concerns the effects of Ty on genes adjacent to it.

There is considerable divergence between individual Ty elements. Most elements fall into one of two major classes, called Ty1 and Ty917. Their relationship is illustrated in **Figure 37.1.** The elements share the presence of the delta repeats, a long region at the left end, another region in the center, and a short region adjacent to the right delta. The two regions of major substitution show no homology as judged by restriction mapping or heteroduplex formation. Individual Ty elements of each type have many changes from the prototype of their class, including base-pair substitutions, insertions, and deletions.

The delta sequences also show considerable heterogeneity of sequence, although the two repeats of one Ty element are likely to be identical or at least very closely related. (The most extreme known case of divergence is limited to 6 differences out of the 330 bp.) We do not know how the identity of the terminal deltas is generated; it could involve gene conversion

Figure 37.1
Ty elements terminate in short direct repeats and are transcribed into two overlapping RNAs.

during or after transposition, or even the generation of both repeats of a "new" Ty element from one of the repeats of the donor. The delta sequences associated with Ty elements show greater conservation of sequence than the solo delta elements, which suggests that recognition of the repeats may be involved in transposition.

A transposition event leaves the same footprint as a bacterial transposition: 5 bp of target DNA are repeated on either side of the inserted Ty element. We do not know the source of whatever enzyme(s) are needed for Ty transposition; some doubt about its origin has been expressed because of the wide variety in Ty sequences.

The presence of Ty elements provides regions of portable homology that are targets for recombination events mediated by host systems. The usual tendency of such an event is to mess up the chromosome, either by causing a deletion or inversion (by recombination between two Ty elements on a chromosome), or by causing more dramatic changes when recombination occurs between two Ty elements on different chromosomes. Recombination between Ty elements seems to occur in bursts; when one event is detected, there is an increased probability of finding others. Among the interactions of Ty elements, gene conversion occurs between those at different locations.

Ty elements can excise by homologous recombination between the directly repeated delta sequences. The large number of solo delta elements may be footprints of such events. An excision of this nature may be associated with reversion of a mutation caused by the insertion of Ty; the level of reversion may depend on the exact delta sequences left behind.

The Ty element is transcribed into two poly(A)$^+$ RNA species. As indicated in Figure 37.1, both initiate ~95 bp from the right end of the element. One terminates after 5 kb, within the common region at the left end; the other terminates after 5.7 kb, within 40 bp of the left boundary. This longer RNA therefore both starts and stops within the delta sequence and as a result has repeats at its ends. A paradox is that both delta elements have the same sequence, yet a promoter is active in the right delta and a terminator is active in the left delta. (A similar feature is found in other transposable elements, including the retroviruses described in Chapter 38.) We do not know whether the Ty RNA is translated into protein.

SEVERAL TYPES OF TRANSPOSABLE ELEMENTS RESIDE IN *D. MELANOGASTER*

The existence of transposable elements in *D. melanogaster* was first inferred from observations analogous to those that identified the first insertion sequences in *E. coli*. Unstable mutations are found that revert to wild type or generate deletions of the flanking material with an endpoint at the original site of mutation. Direct evidence for the occurrence of several types of transposable sequence was provided by studies of the location of certain components of repetitive DNA.

The best-characterized family is called *copia*. Its name reflects the presence of a large number of closely related sequences that code for abundant mRNAs. The *copia* family is taken as a paradigm for several other types of elements whose sequences are unrelated, but whose structure and general behavior appear to be similar. These are described as "*copia*-like" elements; they are listed in Table 37.1. Together the *copia*-like families account for more than 1% of the DNA of *D. melanogaster*.

The number of copies of the *copia* element depends on the strain of fly; usually it is between 20 and 60. The members of the family are widely dispersed. When the genomes of different strains of flies are compared by restriction analysis, the fragments hybridizing with a *copia*-specific probe show a different (although overlapping) spectrum of sizes in each strain. Hybridization *in situ* with polytene chromosomes shows that *copia* sequences are present at different (again overlapping) series of sites. The variability of the *copia* pattern is indicated by the fact that no one site is occupied by a *copia* sequence in every single strain examined.

These differences have developed over evolutionary periods. Comparisons of strains that have diverged recently (over the past 40 years or so) as the result of their propagation in the laboratory reveal few changes. At present, we have no estimate of the rate of change. However, the nature of the underlying events is indicated by the result of growing cells in culture.

The number of *copia* elements per genome then increases substantially, up to 2–3 times. The additional elements represent insertions of *copia* sequences at new sites. Adaptation to culture in some unknown way transiently increases the rate of transposition to a range of 10^{-3} to 10^{-4} events per generation.

The structure of the *copia* element is illustrated in **Figure 37.2** and detailed in **Table 37.1.** The element is 5000 bp long, with identical direct terminal repeats of 276 bp. Each of the direct repeats itself ends in related inverted repeats. A direct repeat of 5 bp of target DNA is generated at the site of insertion. No common sequence has been discerned among known sites of insertion, although there could be regional preferences equivalent to the behavior of some bacterial transposons. The divergence between individual members of the *copia* family is slight, less than 5%; variants often contain small deletions. All of these features are common to the other *copia*-like families, although their individual members may display greater divergence, as seen in the maximum of ~20% for element *297*.

The identity of the two direct repeats of each *copia* element implies either that they interact to permit correction events, or that both are generated from one of the direct repeats of a progenitor element during transposition. (We have already described the same feature for yeast Ty elements; it is also characteristic of retroviruses; see Chapter 38.) The *copia* elements in the genome are always intact; individual copies of the terminal repeats have not been detected (although we would expect them to be generated if recombination deleted the intervening material). On the other

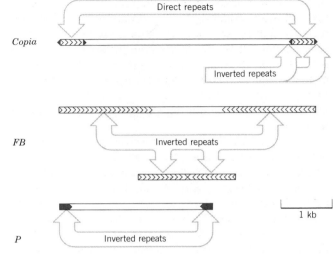

Figure 37.2
Three types of transposable element in *D. melanogaster* have different structures.

hand, *copia* elements sometimes are found in the form of free circular DNA, a contrast with the absence of free forms of bacterial transposons. The two major types of molecule in the circular *copia* population are 5000 bp and 4700 bp long. The latter has a deletion of one of the terminal repeats (which would be expected if it had originated by excision from the chromosome via reciprocal recombination).

Transcripts of *copia* are found in the form of abundant poly(A)$^+$ mRNAs, representing both full-length and part-length transcripts. The mRNAs have a common 5′ terminus, resulting from initiation in the middle

Table 37.1
Several Different Types of Transposable Elements Are Found in *D. melanogaster*

Element	Copies per Genome	Length (bp)	Terminal Direct Repeat	Terminal Inverted Repeat	Target Direct Repeat
copia	20–60	5000	276	17	5
FB	~30	500–5000	None	250–1250	9
P	~50 or 0	500–2900	None	31	Not known

Elements that resemble *copia* in organization are *412*, *297*, *mgd1*, *mgd3*, *B104*, and *gypsy*. Their copy numbers vary from 10 to 100 per genome, their lengths from 5500 to 8500 bp, and the direct repeats from 269 to 571 bp. The target repeat lengths are usually 4 or 5 bp.

of one of the terminal repeats. The part-length 2 kb species has messenger activity for a 51,000-dalton protein of unknown function. Since the direct repeats are identical, it is possible that both function as promoters, one reading into the element to produce the *copia* RNAs, the other reading into the adjacent DNA to initiate transcription of genes downstream from the site of insertion. (Again, a similar feature is found with retroviruses.)

The members of another family of transposable elements, called *FB* (an abbreviation for foldback), have inverted terminal repeats of variable length. Some FB elements consist solely of juxtaposed inverted repeats; in others the inverted repeats are separated by a region of nonrepetitive DNA. In spite of the variation in length, the inverted repeats of all members of the FB family are homologous. This feature is explained by their structure, which consists of tandem copies of a simple-sequence DNA, separated by longer stretches of more diverse sequences. Proceeding from the end into the element, the length of the simple-sequence unit increases; initially it is 10 bp, then expands to 20 bp, and finally expands again to 31 bp. The two copies of the inverted repeat in a single element are not identical. The inverted repeats superficially resemble satellite DNA, and it will be interesting to establish the relationship between the repeats of different members of the family.

The structure of the ends poses a puzzle about FB elements; we have no knowledge of what confers their ability to transpose. Sometimes two (nonidentical) FB elements apparently cooperate to transpose a large intervening segment of DNA, possibly in a manner reminiscent of composite bacterial transposons (although the length of the DNA between FB elements can be much greater). Two examples of such units have been found; one is 200 kb long and is called TE (for transposable element). It is possible that any sequence of DNA flanked on either end by FB elements could behave as such a unit.

Another type of transposable element is represented by the much smaller P elements, whose maximum length is 2.9 kb. They terminate in short inverted repeats and transpose only in certain conditions (see later).

A potential relic of a transposable element is identified by the organization of the rRNA genes. About two thirds of the genes for 28S rRNA are interrupted in *D. melanogaster*. The interruptions fall into two classes, located at positions some 50–80 bp apart in the gene. Neither type of interruption can be classed as an intron, because the interrupted genes are not transcribed. The connection between the presence of the interruption and the failure of these genes to initiate at the promoter (which apparently is identical with that of the noninterrupted genes) is provocative but obscure.

Type I insertions are found in rRNA genes on the X chromosome. The insertions are homologous, but vary in length from 0.5 kb to 5.0 kb; the shorter insertions may have been produced by internal deletions of an original 5 kb insert. Their relative structures are illustrated in **Figure 37.3.** The 0.5 or 1.0 kb insertions display a duplication of the flanking coding sequence of 14 or 11 bp, respectively. The same 14 bp target site duplication is present in the type I interrupted genes of the related fly *D. virilis*. Did type I insertions originate in a transposition event? The footprint of the target site duplication seems to have been lost from the 5 kb insertion, which instead of a duplication has a deletion of 9 bp of the coding sequence on the left side of the insertion.

Sequences homologous to the type I insertion occur in other locations besides the rDNA, which is consist-

Figure 37.3

Type I rDNA insertions (red segments) are homologous sequences of various lengths, probably derived by internal deletions. At the point of interruption in the coding sequence, the 5 kb insertion has a 9 bp deletion of rDNA, the 1 kb insertion has a 14 bp duplication of rDNA, and the 0.5 kb insertion has an 11 bp duplication of rDNA.

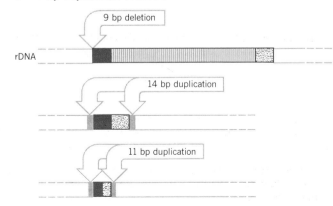

ent with the idea that they represent an ancient transposon. They often occur in the form of tandem repeats interspersed with nonhomologous DNA.

Type II insertions are found in the rRNA genes of both X and Y chromosomes. They vary in length from 1.5 to 4.0 kb. Although also complex in structure, their sequences are not related to the type I insertion. They present no evidence for any ancestral transposition event.

ABOUT COMPLEX LOCI AND CHROMOSOME WALKING

We now digress to consider some features of the *Drosophila* genome whose analysis has been made possible by taking advantage of transposable elements. Several loci in *D. melanogaster* behave in a manner that is difficult to reconcile with the notion that each contains a single gene coding for a unique product. They are called **complex loci.** A conventional gene—even an interrupted one—is identified at the level of the genetic map by a tightly linked cluster of noncomplementing mutations. A hallmark of complex loci is the presence of rather well-spaced groups of mutations, occupying a relatively large map distance, and often displaying a complex pattern of complementation—for example, the mutations may fall into several overlapping complementation groups. The individual mutations may have different and complex morphological effects on the phenotype.

One example of a complex locus is *rudimentary;* mutations produce far-reaching changes in the morphology of the wing. *Rudimentary* has a map of about 0.07 units (which would correspond to ~35 kb if the average frequency of recombination applied). Another example is *decapentaplegic,* which affects development of the imaginal disks (larval tissues that metamorphose into adult body parts). The classic locus is *bithorax,* defined by several groups of mutations that affect sequential stages of development of the thorax, causing major morphological changes in the abdomen. A current model supposes that *bithorax* includes several genes—at one time, these were called position pseudoalleles—coding for products that regulate thoracic development.

The *w (white)* locus was identified by the very first mutation isolated in the fruit fly, in the form of a male whose eye appeared bleached or translucent instead of displaying the usual wild-type red color. The *white* locus turns out to be interesting itself, as well as informative about the events involved in transposition of various elements.

Mutations of the *w* locus fall over a region of about 0.04 map units and identify only a single complementation group. On the basis of this genetic behavior, *w* is a doubtful example of a complex locus. However, a question about the function of the locus is raised by the absence from the *w* eye of all color. The wild-type eye possesses both red and brown pigments, unrelated in structure, and synthesized by different pathways. Several conventional genes are identified by their ability to interfere with one or the other pathway (*vermilion* mutants lack brown pigment, *brown* mutants lack red pigment, and so on). The absence of the products of *both* pathways from *w* mutant eyes suggests that w^+ does not itself code for an enzyme involved in pigment production, but regulates, either directly or indirectly, the production or deposition of the pigments.

The *w* mutants are classified phenotypically by the magnitude of their effect on pigment formation. Null or bleached alleles—such as the original isolate w^1—entirely lack both red and brown pigments. Other alleles have intermediate phenotypes in which reduced amounts of one or both pigments are produced. The alleles usually are identified by superscripts whose names reflect the phenotype. For example, w^a is the apricot allele, w^i is the ivory allele, w^e is the eosin allele, and so on. Most of the alleles affect the quality of pigment, but the w^{sp} group also affect its distribution within the eye. This is consistent with the view that *w* governs the deposition of pigment.

The DNA of the *w* locus has been cloned by the procedure illustrated in **Figure 37.4,** which takes advantage of the presence of a *copia* element in the w^a allele (see later). A genome library prepared from a w^a fly was screened to isolate vectors carrying *copia* elements. These chimeric phages represent all loci at which *copia* is present. Individual vectors are then hybridized *in situ* with the polytene chromosomes of a w^+ fly. Because this fly lacks *copia* at the *w* locus,

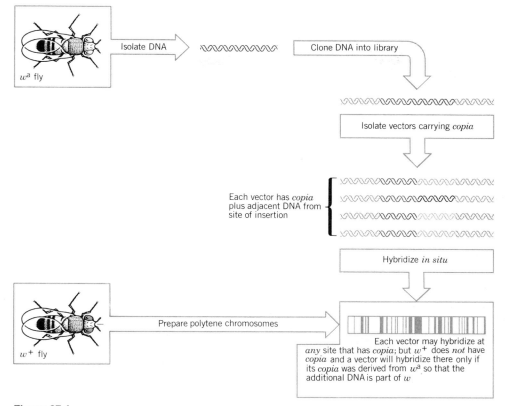

Figure 37.4
Transposable elements can be used to isolate regions of the genome flanking a site of insertion.

hybridization at *w* occurs only with the particular vector derived from the *w*ᵃ allele, because it carries *w* sequences flanking the site of insertion of the *copia* element. Roughly 1 in 50 vectors should fall into this category.

Whatever segment of the *w* locus has been obtained in this vector can be used to isolate the entire locus by the technique of **chromosome walking.** This represents an extension of the usual techniques of restriction mapping to take advantage of the presence in a genome library of overlapping fragments, derived from different breakages of the genome in the same general region. The principle is illustrated in **Figure 37.5.**

We start with a clone representing a site in or near the region of interest. Then we identify other vectors that carry sequences overlapping with the first clone. These new vectors will extend on one side or the other of the fragment carried in the first clone. We can determine the direction of extension by making a restriction map of each fragment. Because these fragments may be quite large, the process is made more controllable by using a subfragment of nonrepetitive DNA from the end of one clone to isolate the next clone. The process can be repeated indefinitely. In principle, it is possible to walk for hundreds of kb, although a problem is encountered in passing through any extensive region of repetitive DNA, because then many hybridizing clones may be obtained, each representing an entirely different region of the chromosome.

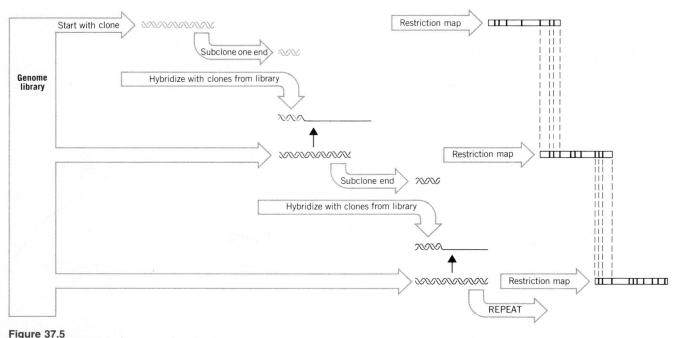

Figure 37.5
Chromosome walking is accomplished by successive hybridizations between overlapping genomic clones.

INSERTIONS AT THE *w* LOCUS DEFINE A COMPLEX TARGET

Figure 37.6 shows a map of a region of about 15 kb that includes all known sites of *w* mutations. Mutations are classified according to whether they cause a change in the restriction map. The type of mutation and the location of the different types identify two areas with potentially different functions.

Many mutations prove to result from insertions of material. In several cases, the insertions can be identified with known transposable elements. In others, the inserted DNA is a member of a repetitive family and may well turn out to represent an unidentified transposable element. The only detectable deletions are the alleles w^{sp3} and w^{sp4}, which occur at or close to the site of the w^{sp1} insertion.

Potential point mutations are identified by *w* loci in which no change can be detected in the restriction map. Genetic mapping locates all of these mutations to the left of the w^a insertion, in the region of the map shown in white. This is denoted the left segment of the locus. We will need to obtain the sequence of the region to confirm that these mutations are base substitutions. Many of them entirely abolish the function of the locus (the eye is bleached); others retain some residual pigment. The left segment identifies a potential coding region. This assignment is supported by the observation that the known insertions in this segment cause the bleached null-phenotype that would be expected if a protein-coding region were disrupted.

The mutants mapping to the right of (and including) w^a almost all are leaky, displaying some residual pigment. No potential point mutations are found in this region; all mutants have changes in the restriction map. The great majority are insertions of known or presumed transposable elements. (Ironically, the atypical allele in this region is w^1, the original isolate with the bleached phenotype.) These mutations define the right segment, which stretches from w^a to w^{DZL}.

The leakiness of insertions in the right segment suggests that they do not interrupt a coding region, but interfere with some other function—for example, an

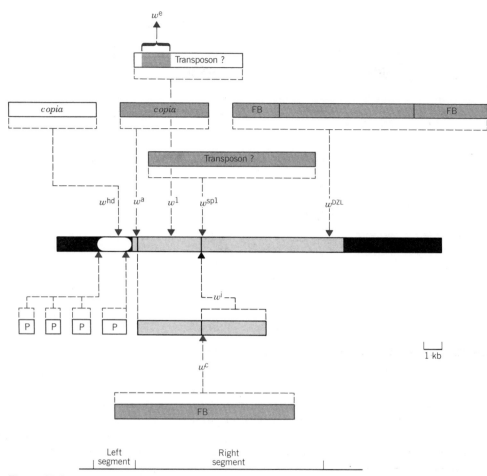

Figure 37.6

The *white* locus may be mutated by insertions, deletions, or (potential) point mutations.

The region in which potential point mutations occur is white. Some of these mutations have a null phenotype. Insertions that cause null phenotype are indicated by red outlines. Insertions that allow residual pigment formation are shown by shaded boxes. Names are given for insertions that are members of known transposable families; others are indicated as transposon followed by a query.

adjacent regulatory region. The insertions map at a few discrete sites. One may be w^a, which is on the edge of the potential coding region. Next is the w^1 site, at which there is a series of insertions and their derivatives. At the w^{sp} site are several insertions, deletions, and combinations of them. The w^{DZL} insertion is the farthest on the right, possibly identifying the limits of the locus. It is possible that these sites of insertion identify individual regulatory regions; insertions at other sites in the right segment might not produce detect-

able mutant phenotypes. All mutations that affect potential regulatory functions, such as timing or pattern of deposition of pigment, or dosage compensation, map in the right segment.

The idea that insertions in the right segment have an indirect effect on the activity of the gene product is supported by the nature of revertants. Often a change in the function of the locus is associated with a change in the inserted segment. For example, the deletion of a small part of the w^1 insertion gives the w^e allele, which has recovered some pigment expression. The many examples of this type of effect imply that the loss of function is not due merely to the fact of insertion, but can be influenced by either the nature or the length of the inserted segment.

Reversion does not necessarily occur by loss of any part of the inserted material, as highlighted by the derivatives of the w^i mutation. Unlike other changes that increase the size of the locus, w^i is not due to insertion of extraneous material, but represents a tandem duplication of a region of 3 kb from within w. The mutation reverts by precisely losing the duplication at the relatively high rate of ~10^{-5}, probably by intrachromosomal recombination.

An exceptional revertant has been found in the form of the w^c mutant, which has a crimson eye; it results from the insertion of an 11 kb FB element into the w^i duplication. The w^c allele is unstable, reverting at a frequency of ~10^{-3} to wild type, back to w^i, or to other forms, including stable derivatives with deletions ending in w; it seems likely that these various rearrangements are sponsored by the FB element. It seems extraordinary that an insertion as large as this FB element can restore the phenotype that had been abolished by the internal duplication. Perhaps the FB element substitutes a new promoter for a damaged site, although one would think that the increased distance to other sites in the right segment would be deleterious.

An intriguing regulatory effect in complex loci is identified by the phenomenon of **transvection**. This term describes the ability of chromosomal rearrangements that prevent synapsis between homologous regions (as seen in polytene chromosomes) to change phenotype, although the overall genotype is unaltered. The basis for the effect lies in the ability of a mutation in one allele to influence the activity of the *other* allele only so long as the two alleles are synapsed. Thus transvection may depend on some structural effect mediated by physical contact. Its existence raises the possibility that chromosome pairing (or some related effect) occurs in normal diploid cells.

An opportunity for insight into the basis for transvection is presented by the w^{DZL} mutation (DZL is an abbreviation for dominant zeste-like). So far as the w region to its left is concerned, the DZL insertion exerts a *cis*-acting inhibition of function. The unusual feature

of this effect is illustrated in **Figure 37.7;** the w^{DZL} allele also prevents expression of a w^+ allele synapsed with it. This is not a conventional dominant *trans*-acting effect, because the same w^+ allele is expressed normally if it is translocated to another location where it no longer comes under the influence of a synapsed w^{DZL}.

The w^{DZL} allele has an insertion of a 13 kb sequence consisting of two nonidentical FB elements flanking a nonhomologous 6 kb segment of DNA (a nonrepetitive sequence from chromosome II). Like the other FB insertion (in w^c), it is unstable, reverting at a frequency of 10^{-2} to 10^{-3}. Revertants to wild type have an excision of part of the DZL insertion, always including the central 6 kb region. This partial excision probably occurs by recombination between the FB elements at the ends of the insertion; the revertants remain unstable because of the presence of the FB ends. However, since the presence of a mere FB element at this site does not cause the DZL phenotype, the central 6 kb sequence rather than the insertion per se may be responsible for the transvection effect. Could this region code for a *cis*-acting product that also recognizes contact between DNAs?

Transvection can be exercised over long distances. The original example is the effect of the z mutation,

Figure 37.7
The w^{DZL} mutation prevents expression of a w^+ allele synapsed with it, but cannot act on a w^+ allele that has been translocated to a different location.

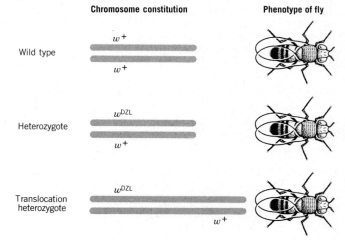

Chromosome constitution	Phenotype of fly

Wild type — w^+ / w^+

Heterozygote — w^{DZL} / w^+

Translocation heterozygote — w^{DZL} / w^+

0.75 map units to the left of *w*, a distance probably corresponding to 300–400 kb. The *z* mutation is recessive. It has no effect in males (which have only one X chromosome); but in females that are homozygous for *z*, it represses a pair of synapsed w^+ alleles, giving a pale-colored (zeste) eye. The effect is abolished if one of the w^+ alleles is translocated elsewhere, or even if it is deleted. It may be altered by certain alleles of *w* created by mutations in the right segment of the locus.

The nature of the interaction between *z* and *w* suggests not only that it extends over synapsed alleles, but further that synapsis is necessary for the effect to occur at all. The propagation of this effect over such a long distance implies also that it may be a function of different loci (*z* and *w*); the corollary is that the DZL mutation need not necessarily be considered to comprise part of the *w* locus itself, but may be adjacent to it.

THE ROLE OF TRANSPOSABLE ELEMENTS IN HYBRID DYSGENESIS

Certain strains of *D. melanogaster* encounter difficulties in interbreeding. When flies from two of these strains are crossed, the progeny may display "dysgenic traits," a series of defects inimical to propagation of the organism, including mutations, chromosomal aberrations, distorted segregation at meiosis, and sterility. The appearance of these correlated defects is called **hybrid dysgenesis.**

Two systems responsible for hybrid dysgenesis have been identified in *D. melanogaster*. In the first, flies are divided into the types I (inducer) and R (reactive). Reduced fertility is seen in crosses of I males with R females, but not in the reverse direction. In the second system, flies are divided into the two types P (paternal contributing) and M (maternal contributing). **Figure 37.8** illustrates the asymmetry of the system; a cross between a P male and an M female causes dysgenesis, but the reverse cross does not.

Dysgenesis is principally a phenomenon of the germ cells. In crosses involving the P-M system, the F1 hybrid flies have normal somatic tissues. However, their gonads do not develop. The morphological defect in gamete development dates from the stage at which rapid cell divisions commence in the germ line.

Any one of the chromosomes of a P male can induce dysgenesis in a cross with an M female. The construction of recombinant chromosomes shows that several regions within each P chromosome are able to cause dysgenesis. This suggests that a P male has a large number of **P factors,** sequences occupying many different chromosomal locations. The locations differ between individual P strains. The P factors are absent from chromosomes of M flies.

The M flies make another type of contribution. Hybrid dysgenesis is triggered only when the P factors become exposed to cytoplasm inherited from an M mother. The maternal contribution is called the M cytotype; its basis is not understood, but does not seem to be a simple maternal effect, because it is inherited

Figure 37.8

Hybrid dysgenesis is asymmetrical; it is induced by P male × M female crosses, but not by M male × P female crosses.

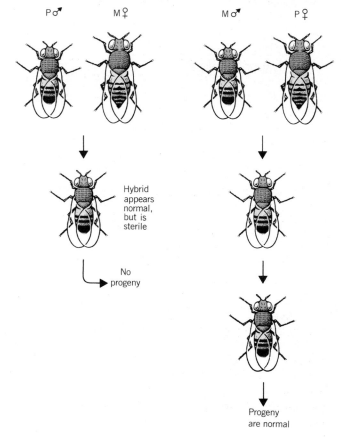

for more than one generation when there is an M female in the line of descent. In rare lines descended from P male × M female crosses, the P factors convert the line to P cytotype—that is, to the state in which P factors are not active.

The events responsible for the induction of mutations in dysgenesis have been examined by mapping the DNA of bleached (null) *w* mutants found among the dysgenic hybrids. All the mutations result from the insertion of DNA into the *w* locus. Three of the four characterized P insertions occur at the same site, as noted in Figure 37.6. The P insertions remain stable when the chromosome is perpetuated in P cytotype, but become unstable, as detected by their excision, on exposure to M cytotype. In addition to the P insertions, insertions of *copia* elements also are found in dysgenic flies; one is shown as w^{hd} in the figure.

The P insertions vary in length but are homologous in sequence. The length variation may arise, apparently rather frequently, by internal deletions of a full-length P factor. The longest P elements are ~2.9 kb long and have direct terminal repeats, as indicated in Figure 37.2 and detailed in Table 37.1. A current model for the dysgenic interaction is to suppose that the intact P factors code for a transposase that is inactive in P cytotype, but becomes active in M cytotype. At least some of the shorter P elements may have lost the capacity to produce the transposase, but may be activated in *trans* by the enzyme coded by a complete P element. The ends of the P factor may indicate that it is a suitable substrate for transposition. The P-M dysgenic interaction may also activate the movement of *copia* (and other transposable elements); these insertions, however, remain stable irrespective of the subsequent cytotypes to which the chromosome is exposed.

In situ hybridization identifies 30–50 copies of the P factor in each P strain; the factors are absent from M strains. Chromosomes from P-M hybrid dysgenic flies have P factors inserted at many new sites. The chromosome breaks typical of hybrid dysgenesis occur at hotspots that are the sites of residence of P factors. The average rate of transposition of P elements to M chromosomes is different in male and female dysgenic hybrids, but is of the order of one event per generation. A chromosome that has acquired P elements becomes dysgenic when it is crossed into

M cytotype; the capacity of a chromosome to induce sterility is related to the number of P elements it has gained. Independence of the abilities to transpose and to induce sterility is suggested by the existence of the Q strain, which behaves like P strains in all ways except that it does not induce sterility in the hybrids.

Strains of *D. melanogaster* descended from flies caught in the wild more than 30 years ago are always M. Strains descended from flies caught in the past 10 years are almost always P. Does this mean that the P element family has invaded wild populations of *D. melanogaster* in recent years? P elements are indeed highly invasive when introduced into a new population; the source of the invading element would have to be another species. On the other hand, the P element may have been lost from populations maintained in the laboratory over protracted periods.

Because hybrid dysgenesis reduces interbreeding, it is a step on the path to speciation. Suppose that a dysgenic system is created by a transposable element in some geographic location. Another element may create a different system in some other location. Flies in the two areas will be dysgenic for two (or possibly more) systems. If this renders them intersterile and the populations become genetically isolated, further separation may occur. Multiple dysgenic systems may therefore lead to inability to mate—and to speciation.

CONTROLLING ELEMENTS IN MAIZE ARE TRANSPOSABLE

Genetic studies of maize initiated in the 1940s identified changes in the genome that occur during somatic cell division. These changes are brought about by **controlling elements,** which are recognized by their ability to move from one site to another. The insertion of a controlling element at a given site may control the activity of adjacent genes. The frequency and timing of events associated with the controlling elements are subject to developmental regulation; and their actions may be tissue-specific.

We can now recognize the controlling elements as transposable elements with several functions analogous to those described for other systems. Insertion may generate an unstable allele at a locus. Originally described as "mutable alleles," the terminology still is reflected by the use of *m* to describe such alleles. Loss of the controlling element itself or of its ability to transpose converts a mutable allele to a stable allele. Deletions, duplications, inversions, and translocations all may occur at the sites where controlling elements are present. Chromosome breakage is a common consequence of the presence of some elements.

The maize genome contains several families of controlling elements. There are extensive differences in the numbers, types, and sites of presence of the elements in different maize strains. The members of each family may be divided into two classes. **Autonomous** elements have the ability to excise and transpose; their insertion gives rise to unstable (*m*) alleles. **Nonautonomous** elements become unstable *only* when an autonomous member of the same family is present elsewhere in the genome. When complemented in *trans* by an autonomous element, a nonautonomous element may display the usual range of activities associated with autonomous elements. It seems likely that nonautonomous elements are derived from autono-

mous elements by loss of *trans*-acting functions needed for transposition (and sometimes other functions also).

Autonomous elements are subject to "changes of state," heritable but relatively unstable alterations in their properties. These changes can influence the timing and frequency of transposition. Some alterations appear to cause autonomous elements to lose their autonomy. *Cis*-acting defects may occur in nonautonomous elements that render them impervious to autonomous elements; thus a nonautonomous element may become permanently stable because it can no longer be activated to transpose.

In lieu of sequence information, the families of controlling elements have been defined by the interactions between autonomous and nonautonomous elements. A nonautonomous element can be activated in *trans* only by certain autonomous elements; these are considered to be members of the same family. The three best-characterized families of controlling elements, *Ac-Ds, Spm,* and *Dt,* are summarized in **Figure 37.9.**

The abilities of controlling elements can be distin-

Figure 37.9

Each controlling element family has both autonomous and nonautonomous members.

Autonomous elements are capable of transposition and also have other activities (for example, influencing gene expression). Nonautonomous elements are deficient at least in transposition. The names originally given to pairs of autonomous and nonautonomous elements are classified by families. There may be more than one element in each classification; for example, *Spm-s* and *Spm-w* are different versions of *Spm*.

Family		
Ds—Ac	*Ac* (activator) *Mp* (modulator)	*Ds* (dissociation) *Ds* (dissociation)
Spm	*Spm* (Suppressor—mutator) *En* (enhancer)	Unnamed element *I* (inhibitor)
Dt	*Dt* (dotted)	Unnamed element

guished by mutations. For example, the dual properties of members of the *Spm* family are indicated by its name: suppressor-mutator. The effects are shown by the result of introducing *Spm* into a genome to activate a nonautonomous element at a given locus. "Suppressor" describes the ability to inhibit expression of a structural gene at the locus. "Mutator" describes the ability to revert (transpose) so that the gene regains full activity. Both properties are held by the element

Figure 37.10

Clonal analysis identifies a group of cells descended from a single ancestor in which a transposition-mediated event altered the phenotype. Timing of the event during development is indicated by the number of cells; tissue specificity of the event may be indicated by the location of the cells.

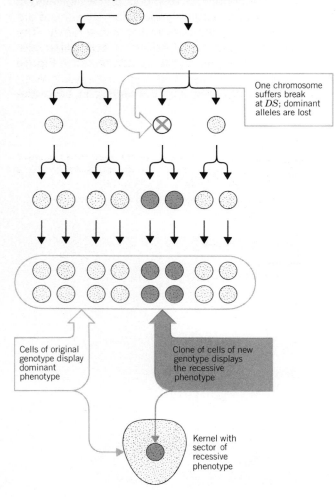

since they transpose together; but they can be distinguished by $sp^+ \, m^-$ mutants that have lost one but not the other property. Since the ability of a nonautonomous element to suppress adjacent functions is controlled in *trans* by an autonomous element, the effect cannot result merely from insertion of the nonautonomous element, but must have a specific regulatory basis.

Two features of maize have helped to follow the results of transposition events. Controlling elements often insert near genes that have visible but nonlethal effects on the phenotype. And because maize displays clonal development, the occurrence and timing of a transposition event can be visualized as depicted diagrammatically in **Figure 37.10.** The nature of the event does not matter: it might be an insertion, excision, or (as assumed in the figure) a chromosome break. What is important is that it occurs in a heterozygote to alter the expression of one allele. (The next section shows an example in which a break leads to loss of dominant alleles, allowing the recessive phenotype to be expressed.) Thus the descendants of a cell that has suffered an element-dependent event display a new phenotype, while the descendants of cells not affected by the event continue to display the original phenotype.

Mitotic descendants of a given cell remain in the same location and thus give rise to a **sector** of tissue. A change in phenotype during somatic development is called **variegation;** it is revealed by a sector of the new phenotype residing within the tissue of the original phenotype. The size of the sector depends on the number of divisions in the lineage giving rise to it; so the size of the area of the new phenotype is determined by the timing of the change in genotype. The earlier its occurrence in the cell lineage, the greater the number of descendants and thus the size of patch in the mature tissue.

Ds MAY TRANSPOSE OR CAUSE CHROMOSOME BREAKAGE

The presence of the nonautonomous element *Ds* has several possible consequences for an adjacent locus. Upon activation by an autonomous *Ac* element, *Ds* may either transpose to a new site or cause chromosome breakage. Transposition is accompanied by excision from the donor site, so it converts an unstable allele to a stable allele (not necessarily of wild type). An inverse relationship between reversion to stability and the occurrence of chromosome breakage suggests that transposition and breakage are alternative outcomes of the activation of *Ds*.

While present at a locus, *Ds* can alter the level of gene expression. A feature of transposable elements at present seen only in maize is that the effect may include alteration of the developmental timing of gene expression. A striking feature of the interaction of *Ds* with a structural gene is its ability to alter the structure of the gene product itself.

Our best present approach to structural analysis of *Ds* is provided by its insertions near the *Sh* locus. The *Sh* alleles code for sucrose synthetase, whose absence is reflected in the shrunken kernel phenotype that gave rise to the name of the locus. The sucrose synthetase messenger has been isolated as a 3 kb mRNA. Its availability has made it possible to isolate the corresponding genomic DNA. *Sh-m* loci in which *Ds* is present have a different restriction map in the 5′ region of the gene, corresponding to insertion or perhaps rearrangement of the DNA. The exact location of the change in sequence is different for individual *sh-m* alleles.

The presence of *Ds* alters *sh-m* expression. In some mutants, the *sh-m* gene is barely expressed. In other mutants, the locus is transcribed into abundant aberrant RNAs that are shorter than *Sh* mRNA and consist in part of *Sh* sequences and in part of new sequences. They are translated into a protein immunologically related to sucrose synthetase, but having an altered electrophoretic mobility. As the structures of the various *Sh* and *sh-m* loci become elucidated, we shall be able to see which sites within or adjacent to the structural gene are influenced by the presence of *Ds* and how this affects gene expression.

The *Ds* element was originally identified by its ability to provide a site for chromosome breakage upon activation by *Ac*. The consequences are illustrated in **Figure 37.11.** Consider a heterozygote in which *Ds* lies on one homologue between the centromere and a series of dominant markers *(C-I, Bz, Wx)* whose effects can be visualized by the color of the cells or by appropriate staining. *(C-I* is a regulator that causes

colorless aleurone; *Bz* glycosylates the anthocyanid pigments; *Wx* is starchy and stains blue with I_2-KI.) The other homologue lacks *Ds* and has recessive markers *(C, bz, wx)*. Breakage at *Ds* generates an **acentric fragment** carrying the dominant markers. Because of its lack of a centromere, this fragment is lost at mitosis. Thus the descendant cells have only the recessive markers carried by the intact chromosome. This gives the type of situation whose results are depicted in Figure 37.10.

Breakage at *Ds* leads to the formation of two unusual chromosomes. These are generated by joining the broken ends of the products of replication. One is a U-shaped acentric fragment consisting of the joined sister chromatids for the region distal to *Ds* (on the left as drawn in Figure 37.11). The other is a U-shaped **dicentric chromosome** comprising the sister chromatids proximal to *Ds* (on its right in the figure). The latter structure leads to the classic **breakage-fusion-bridge** cycle illustrated in **Figure 37.12.**

Consider the fate of the dicentric chromosome when it attempts to segregate on the mitotic spindle. Be-

Figure 37.11

A break at a controlling element causes loss of an acentric fragment; if the fragment carries the dominant markers of a heterozygote, its loss changes the phenotype.

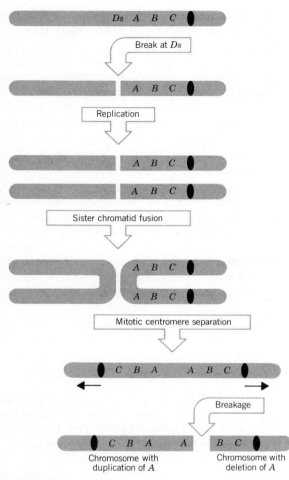

Figure 37.12
***Ds* provides a site to initiate the chromatid fusion-bridge-break-age cycle (on activation by *Ac*). The products can be followed by clonal analysis.**

cause it has two centromeres, it is pulled by each centromere toward an opposite pole. The tension breaks the chromosome at a random site between the centromeres. In the example of the figure, breakage occurs between loci *A* and *B,* with the result that one daughter chromosome has a duplication of *A,* while the other has a deletion. If *A* is a dominant marker, the cells with the duplication will retain **A** phenotype, but cells with the deletion will display the recessive **a** phenotype.

The breakage-fusion-bridge cycle continues through further cell generations, allowing genetic changes to continue in the descendants. For example, consider the deletion chromosome that has lost *A*. In the next cycle, a break may occur between *B* and *C,* so that the descendants are divided into those with a duplication of *B* and those with a deletion. Successive losses of dominant markers may be revealed by subsectors within sectors.

Ds TRANSPOSITION IS CONNECTED WITH REPLICATION

We have mentioned that transposition of *Ds* usually is accompanied by its disappearance from the donor location, a marked contrast with the mode of transposition in bacteria. The results of clonal analysis suggest that transposition of *Ds* almost always occurs after the donor element has been replicated. The most revealing information is provided by transpositions from a donor to a recipient site on the same chromosome. Studies of this sort have been performed with *Mp,* an element of the *Ac* type.

Two possible outcomes of the transposition are distinguished by whether the recipient site has replicated. They are illustrated in the top two rows of **Figure 37.13.** If the recipient site has not replicated before transposition, the element will be replicated in its new site. Thus it is present at the recipient site on both daughter chromosomes. If the recipient site has previously replicated, only the chromatid physically gaining the element will display a change at the recipient locus. In either case, the element is lost from the donor site of one chromatid and remains at the donor site of the other chromatid.

In the two first situations illustrated in the figure, cell division produces daughter cells whose genotypes are different from each other as well as from the parent cell. Clonal analysis reveals their descendants in the form of **twin sectors:** the appearance of a sector of one new genotype is accompanied by the appearance of another sector representing the other new genotype. Note that twin sectors are the inevitable result of transposition to an unreplicated recipient site.

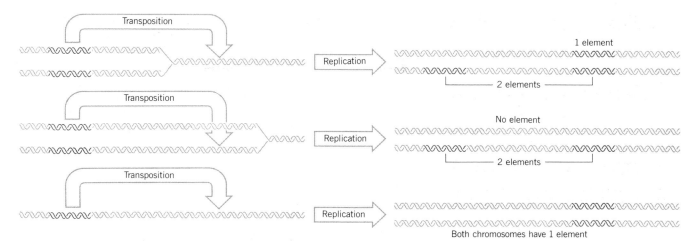

Figure 37.13

Maize transposition involves movement of an element from the donor site after it has been replicated; the nature of the products depends on whether the recipient site has been replicated prior to the transposition.

Top: If the recipient site has not been replicated, the element is duplicated after transposition, so each daughter chromosome gains an element. Only one of the daughter chromosomes has lost the element. Thus one daughter chromosome has an element only at the recipient site; the other has elements at both the donor and recipient sites.

Center: If the recipient site has been replicated, only one daughter chromosome gains an element at the recipient position. Usually an element moves from a donor site on one chromatid to a recipient site on the *other* chromatid, so that both chromatids have a change in genotype as shown. (If the element moved from a donor to a recipient site on the *same* chromatid, the genetic result would be indistinguishable from that shown in the bottom drawing.)

Bottom: If transposition were to occur before replication of the donor site, both the progeny chromosomes would have the element at the recipient site.

When the recipient site has been replicated, they result only if the target lies on the opposite chromatid from the donor.

The predominance of twin sectors (80% of all maize transpositions) demonstrates that these are the customary events. The minority of transpositions generating single sectors could represent insertions at recipient sites on the same chromatid as the donors, could result from transposition not connected with replication (shown in the lowest line of the figure), or could be artifacts resulting from the loss of one of the products of transposition (for example, because the cell entered another lineage). Although it is therefore clear that transposition usually is associated with replication of the donor site, we do not know whether this connection is obligatory.

YEAST HAS SILENT AND ACTIVE LOCI FOR MATING TYPE

The yeast *S. cerevisiae* can be happily propagated in either the haploid or diploid condition. Interconversion between these states takes place by mating (fusion of haploid spores to give a diploid) and by sporulation (meiosis of diploids to give haploid spores). The ability to engage in these activities is determined by the **mating type** of the strain. The properties of the two mating

Table 37.2
Mating Type Controls Several Activities

Property	MATa or MATa/MATa	MATα or MATα/MATα	MATa/MATα
Mating	Yes	Yes	No
Sporulation	No	No	Yes
Secretion	**a** factor	**α** factor	None

types are summarized in **Table 37.2.** We may view them as resting on the teleological proposition that there is no point in mating unless the haploids are of different genetic types; and there is no point sporulating unless the diploid is heterozygous and thus can generate recombinants.

The two mating types are determined by the alleles MATa and MATα of the mating type locus. Haploid cells *must* be either MATa or Matα. Cells of opposite type can mate; cells of the same type cannot. Recognition of cells of opposite mating type is accomplished by the secretion of pheromones. MATα cells secrete the small polypeptide α factor; MATa cells secrete **a** factor. Each factor acts on cells of the opposite mating type to coordinate cell and nuclear fusion. Although usually we deal with the mating of haploid cells, diploids that are homozygous (MATa/MATa or MATα/MATα) behave like their haploid counterparts in the ability to mate with cells of the opposite type.

Homozygous diploids cannot sporulate. On the other hand, diploids of the type MATa/Matα, produced by mating of haploids, can sporulate. **Figure 37.14** demonstrates how this design maintains the normal haploid/diploid life cycle. Several other differences also distinguish heterozygotes from diploid homozygotes, including the pattern of budding (mitotic release of new daughter cells), sensitivity to irradiation, and some cell cycle functions. Since the heterozygote is phenotypically distinct from the homozygote, MATa and MATα formally are codominant alleles: each must specify a different product; neither is a defective version of the other.

A remarkable feature that led to present studies of the genetic basis of yeast mating type is the ability of some yeast strains to **switch** their mating types. These **homothallic** strains carry a dominant allele HO and *change* their mating type frequently, as often as once

every generation. **Heterothallic** strains with the recessive allele ho have a stable mating type, subject to change with a frequency ~10^{-6}.

Expression of the HO gene is itself under mating type control, since it does not function in MATa/MATα diploids—teleologically, we may think that there is no need for switching when *both* MAT alleles are expressed anyway. When HO is present in a yeast population, irrespective of the initial mating type, in very few generations there are large numbers of cells of

Figure 37.14
The yeast life cycle proceeds through mating of MATa and MATα haploids to give heterozygous diploids that sporulate to generate haploid spores.

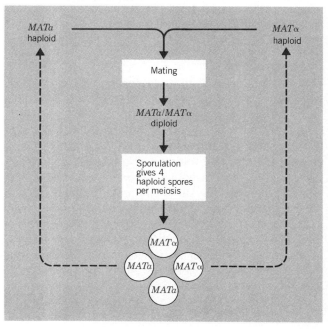

both mating types, leading to the formation of *MATa/MATα* diploids that take over the population. The production of stable diploids from a haploid population can be viewed as the raison d'être for switching.

The existence of switching suggests that all cells contain the *potential* information needed to be either *MATa* or *MATα*, but express only one type. This conclusion is confirmed by analysis of mutants in either type of *MAT* allele. Switching in *HO* strains allows the mutant allele to be replaced by either a wild-type *MATa* or *MATα* gene.

Where does the wild-type information come from? Two additional loci are needed for switching. *HMLα* is needed for switching to give a *MATα* type; *HMRa* is needed for switching to give a *MATa* type. These loci lie on the same chromosome that carries *MAT*; *HML* is far to the left, *HMR* is far to the right.

The **cassette model** for mating type proposes that *MAT* has an **active cassette** of either type α or type a. *HML* and *HMR* have **silent cassettes** that are physically substituted for the cassette at MAT by the switching process. Usually *HML* carries an α cassette, while *HMR* carries an *a* cassette. The cassettes constitute transposable coding information.

The process of switching is illustrated in **Figure 37.15.** Switching is nonreciprocal, because the copy at *HML* or *HMR replaces* the allele at *MAT*. We know this because a mutation at *MAT* is lost permanently when it is replaced by switching—it does not exhange with the copy that replaces it.

The copies present at *HML* or *HMR* can also be mutated. In this case, switching introduces a mutant allele into the *MAT* locus. The mutant copy at *HML* or *HMR* remains there through an indefinite number of switches. This immediately identifies a resemblance with replicative transposition, in which the donor element generates a new copy at the recipient site, while itself remaining inviolate. Mating type switching corresponds to a directed transposition, since there is only one recipient *(MAT)* and two donors *(HML* and *HMR)*.

As illustrated in Figure 37.15, the usual event observed in switching is the replacement of *MATa* by the copy at *HMLα* or replacement of *MATα* by the copy at *HMRa*. However, studies with mutant loci show that *HMLα* occasionally replaces *MATα*, and *HMRa* occasionally replaces *MATa*. Such switches pass unob-

Figure 37.15
Changes of mating type occur when silent cassettes replace active cassettes of opposite genotype; when transpositions occur between cassettes of the same type, the mating type remains unaltered.

served unless a mutation is present at one of the loci to mark it genetically. In 80–90% of switches, the *MAT* allele is replaced by one of opposite type, an effect apparently determined by the phenotype of the cell. Cells of **a** phenotype preferentially choose *HML* as donor; cells of **α** phenotype preferentially choose *HMR*. It is possible to obtain yeast strains in which the usual orientation of the silent cassettes is reversed. When their genotypes are *HMLa* and *HMRα*, 92% of the switches are homologous, because the choice of donors remains unaltered.

An interesting feature of the switching event is its dependence on cell lineage. Switching is detected only in the products of a division; *both daughter cells have the same mating type,* switched from that of the parent. This suggests a connection with replication in the sense that switching must occur *before* the MAT locus is replicated. (Compare this with the connection between transposition and replication in maize shown in Figure 37.13.)

SILENT AND ACTIVE CASSETTES HAVE THE SAME SEQUENCES

Both the silent and active mating type loci have been cloned onto plasmids that replicate in yeast. The genes have the structures shown in **Figure 37.16.** Each type of allele has a distinct region, denoted Ya or Yα. This region is flanked on either side by the X and Z1 regions that are virtually identical in all cassettes. (Only four differences, all point mutations, have been identified in individual sequences.) The further flanking regions W and Z2 are common to *HML* and *MAT,* but are not present at *HMR.* (The organization is the same irrespective of the orientation of *a* and α alleles at *HML, HMR,* and *MAT.*)

Mutations in *MAT*α fall into two complementation groups (α1 and α2), while mutations in *MAT*a fall into a single group *(a1).* The ability of the plasmid-carried cassettes to complement mutant loci at *MAT* allows individual regions of each locus to be equated with the complementation groups (by introducing damage at specific sites in the cloned copies). The relationship between the complementation groups and the transcripts corresponding to *MAT* alleles is indicated in Figure 37.16. Note that the proteins are deduced from the coding sequence and have not yet been characterized.

The *a1* complementation group is equated with the a1 transcript, which is coded entirely within the region of Ya and therefore is unique to the *a*-type locus. The *a1* sequence has two potential products. The shorter terminates at a UGA codon; however, we know that readthrough must occur to give the longer protein, because mutations beyond the termination codon prevent *a1* function. We do not know whether the shorter protein plays a different role from the longer. No function has been detected for the a2 transcript, which is transcribed from a promoter at least part of which lies in Ya. The transcript lies within the common region X and could be translated into either of two short polypeptides.

Each of the transcripts generated at *MAT*α corresponds to one of the known complementation groups. The transcripts initiate divergently within Yα and each extend into the adjacent common region. The α1 transcript codes for a protein largely coded by Yα. The α2 transcript can be translated from an initiation codon in Yα, but most of the sequence corresponds to X.

The functions of the products, as deduced from the deficiencies of mutants in each complementation group, are illustrated in **Figure 37.17.** In haploids, **a** mating functions are expressed constitutively. The function of *a1* is needed, in cooperation with α2, to induce sporulation in diploids. The α2 group also has responsibility for repressing **a** mating functions in haploids. The α1 product induces **α** mating functions in haploids.

The transcription map in Figure 37.16 reveals an intriguing feature. Transcription of both *MAT*a and *MAT*α initiates within the Y region. Only the MAT locus is expressed; yet the same Y region is present in the corresponding nontranscribed cassette, *HML* or *HMR.* This implies that regulation of expression is not accomplished by direct recognition of some site overlapping with the promoter. A site *outside* the cassettes must distinguish *HML* and *HMR* from *MAT.*

Can we find the basis for the control of cassette activity by identifying other loci that act on the cassettes? Some are revealed by mutations that prevent

Figure 37.16
Silent cassettes have the same sequences as the corresponding active cassettes. Only the Y region changes between *a* and α types. *HMR* lacks the W and Z2 sequences common to *HML* and *MAT*.

expression of the allele occupying *MAT*. Other mutations allow the usually silent cassettes at *HML* and *HMR* to be expressed. (When this results in expression of both **a** and **α** functions, the cells behave like *MATa*/*MATα* diploids.) Four complementation groups have been identified in which mutations lead to expression of *HML* and *HMR*. They are called *SIR* (silent information regulator). The four wild-type *SIR* loci are needed to maintain *HML* and *HMR* in the repressed state; mutation in any one of these loci to give a *sir⁻* allele has two effects. Both *HML* and *HMR* can be transcribed. And both the silent cassettes become targets for replacement by switching. Thus the same regulatory event is involved in repressing a silent cassette and in preventing it from being a recipient for transposition.

The different activities of the silent and active cassettes may result from the structure of chromatin. The use of probes representing the regions adjacent to each of the cassettes allow chromatin structure to be visualized by the end-labeling technique described in Chapter 30. All cassettes possess sites hypersensitive

to DNAase I, but there are significant differences between the active and silent cassettes. The afflicted sites are summarized in **Figure 37.18.**

The general pattern of hypersensitivity is similar in *MATa* and *MATα*. In each case, there is a site at the 3′ end of the leftward transcript (a2 or α2). There is either a single (*a*) or double (*α*) site in the region between the two transcripts. And there is a doublet of hypersensitive sites around the Y-Z boundary.

All of these sites are absent from *HML* and *HMR* except for the doublet between α1 and α2. However, the introduction of a *sir⁻* mutation restores all the missing hypersensitive sites. Thus activation of the cassettes is associated with the presence of hypersensitive sites. The various sites may have different functions. The sites within Y may be necessary (although not sufficient) to initiate transcription. The function of the sites within X is unknown. The doublet at the X-Y boundary may be important in initiating the transposition event (see next section).

Some sites are found in the flanking DNA near *HML* and *HMR*. Their presence does not depend on *SIR* function; in fact, these sites could identify the silent cassettes as targets for SIR action. Deletion mapping shows that the flanking site on the left of *HML* is essential for the response to *SIR*; the flanking site on the right also is important. Their recognition by SIR must in some unknown way be transmitted within the cassettes to prevent formation of the hypersensitive sites. In lieu of this action, the various hypersensitive sites

Figure 37.17
The α1 and α2 functions cooperate to induce sporulation in diploids. In haploids, the α2 function represses *a* mating functions; α1 induces α mating functions.

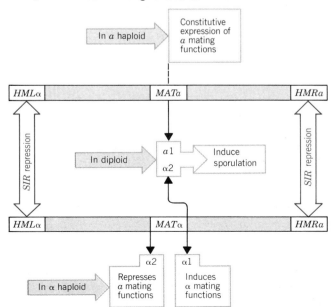

Figure 37.18
Most of the DNAase I hypersensitive sites found in *MAT* are not present in *HML* or *HMR*; but they are activated in *sir⁻* mutants. Some permanent hypersensitive sites are found in the regions flanking *HML* and *HMR*.

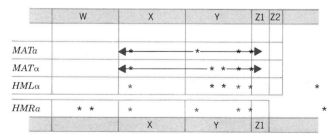

	W	X	Y	Z1	Z2
MATa					
MATα					
HMLα					
HMRa					
		X	Y	Z1	

* Hypersensitive sites are always present

* Hypersensitive sites only present in *sir⁻* strains

allow the locus to be both expressed and subject to switching.

UNIDIRECTIONAL TRANSPOSITION IS INITIATED BY THE RECIPIENT (*MAT*) LOCUS

In populations undergoing switching, recombination between *MAT* and either *HML* or *HMR* occurs at a low frequency. The occurrence of this (aberrant) event suggests that switching involves pairing between *MAT* and one of the silent cassettes. The usual outcome of their interaction is a switch in mating type, but occasionally reciprocal recombination occurs instead. The recombination event essentially fuses the recombining loci and excises the material between; this indicates that all three cassettes lie in the same orientation on the chromosome.

Sites needed for transposition may be identified by mutations that prevent switching. Deletions at the right end of Y do not have any effect until they cross the boundary into Z1; then they abolish switching. *Cis*-acting point mutations called *inc* (for inconvertible) have been found at both *MATa* and *MATα*. They result from base-pair substitutions in Z1, close to the boundary with Y. In the rare case when switching does occur, the mutation is lost from *MAT,* so it may lie in the region that is replaced. Another class of mutation is called *stk* (for stuck); these are retained after a switching event, so they appear to lie outside the region actually replaced. Thus replacement of *MAT* involves a site in the vicinity of the Y-Z1 boundary, and sequences both within and outside the replaced region may be needed for the event. An indication of the unidirectional nature of the process is the lack of mutations in *HML* or *HMR* that prevent switching; and if *inc* mutations are transferred by guile into the silent cassettes, they do not prevent switching. In fact, they are transposed efficiently to the *MAT* locus, whereupon the *inc* mutation reappears.

The Y-Z boundary is the site of a change in DNA that may be connected with transposition. In populations of cells undergoing switching, 1–3% of the DNA of the *MAT* locus has a double-stranded cut at this site. The cut lies within 10 bp of the boundary and coincides with the DNAase hypersensitive site.

We can now suggest a possible series of events for switching. The hypersensitive site at the Y-Z1 boundary of *MAT* may be accessible because it lacks a nucleosome. It is recognized by an endonuclease controlled by the *HO* locus. In fact, this is equivalent to suggesting that hypersensitivity to DNAase I *in vitro* reflects a natural sensitivity to the *HO* endonuclease *in vivo*!

The ensuing reaction is illustrated schematically in terms of single strands in **Figure 37.19.** (Of course, it

Figure 37.19

Cassette substitution is initiated by the recipient locus and may involve pairing on either side of the Y region with the donor locus. The sequence that is removed and replaced may extend into the X region. The figure shows the reaction in terms of a single recipient and single donor strand; actually it involves duplex regions of DNA.

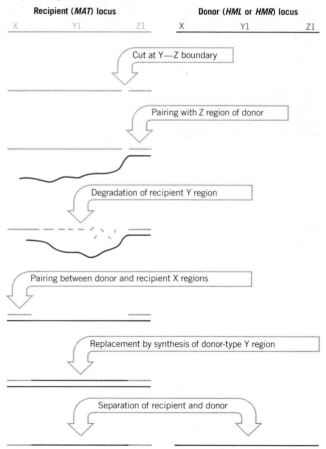

must be more complex in terms of double strands, and several pathways are possible.) Suppose that the free Z end of *MAT* invades either the *HML* or *HMR* locus and pairs with the Z region. The Y region of *MAT* is degraded until a region with homology to X is exposed. At this point, *MAT* is paired with *HML* or *HMR* at both the left side (X) and the right side (Z1.) The Y region of *HML* or *HMR* is copied to replace the region lost from *MAT* (which might extend beyond the limits of Y itself). The paired loci separate. (The order of events could be different.) The stages following the initial cut require the enzymes involved in general recombination.

Two features are notable. Unlike other transpositions that we have described, the switching event involves pairing at both ends of the transposed segment, with degradation and replacement of the material between. And the process is initiated by *MAT, the locus that is to be replaced.* In this sense, the description of *HML* and *HMR* as donor loci refers to their ultimate role, but not to the mechanism of the process, which is the reverse of the usual initiation of transposition by the donor. The sequences of the loci do not identify any short direct repeats in the relevant regions, which argues that the mechanism of joining may be different from the use of staggered breaks found in other transposition events.

FURTHER READING

Eucaryotic transposable elements are extensively reviewed in *Mobile Genetic Elements* (ed. Shapiro, Academic Press, New York, 1983). Chapters of particular interest are: **Fink** discusses Ty elements of yeast; **Rubin** reviews the tour de force of characterizing *D. melanogaster* transposons and mapping the *white* locus; **Federoff** brings the topic of maize-controlling elements into the molecular era; and **Haber** provides a comprehensive summary of yeast mating type. The determination of mating type has been reviewed from a biological perspective by Herskowitz (*Science,* in press, 1983). The rDNA of *D. melanogaster* is discussed briefly by **Glover** (*Cell,* **26,** 297–298, 1981).

CHAPTER 38
ELEMENTS THAT MOVE IN
AND OUT OF THE GENOME

Transposons apparently have no life outside the genome. Whether valuable components playing a role in cellular survival or selfish parasites concerned only with their own survival, they do not generate free molecules of DNA. In this chapter, we shall be concerned with elements that can exist both within and outside the genome; in some cases these are authentic elements, in others they owe their existence to laboratory recombinations.

In several situations, an RNA sequence has been implicated as the progenitor of a genomic (DNA) sequence. We think that the RNA must have been converted into a duplex DNA copy that was inserted into the genome by a transposition-like event. Direct evidence for the existence of this pathway derives largely from studies of the retroviruses, in which it constitutes part of the regular infective cycle. The structures of some pseudogenes and other sequences (including members of the Alu family) imply that an analogous pathway may (rather rarely) be followed by a cellular sequence. In both cases, an extrachromosomal element gives rise to a genomic sequence.

Alterations in the relative proportions of components of the genome sometimes occur during somatic development, most notably as characterized in insect larvae to increase the number of copies of certain genes. The occasional amplification of genes in cultured mammalian cells is indicated by our ability to select variant cells that have an increased copy number of some gene. Initiated within the genome, the amplification event can create additional copies of the gene that survive in either intrachromosomal or extrachromosomal form.

Pressing on the discovery of the existence of transposition and other rearrangements of DNA is the knowledge that DNA sequences are surprisingly adjustable. Extraneous DNA can be introduced into eucaryotic cells, where it survives and may be expressed. In some cases it remains extrachromosomal; in others it is integrated into the genome. The relationship between the extrachromosomal and genomic forms is irregular, depending on chance and to some degree unpredictable events, rather than resembling the interchange between free and integrated forms of bacterial plasmids. Yet, however accomplished, the process may lead to stable change in the genome; following its injection into animal eggs, DNA may even be incorporated into the genome and inherited thereafter as a normal component, sometimes continuing to function. Considerable manipulation of DNA sequences therefore is achieved both in authentic situations and by experimental fiat.

632

THE RETROVIRUS LIFE CYCLE INVOLVES TRANSPOSITION-LIKE EVENTS

Retroviruses have genomes of single-stranded RNA that are replicated through a double-stranded DNA intermediate. The class of retroviruses about which we have the most information are those that give rise to the **C-type** particles in birds and mammals. Two copies of the RNA genome are packaged into each virion, making the individual virus particle effectively diploid. When a cell is simultaneously infected by two different C-type viruses, it is possible to generate heterozygous virus particles carrying one genome of each type. The diploidy may be important in allowing the virus to acquire cellular sequences.

A typical retroviral sequence contains three "genes," the term here identifying coding regions each of which actually gives rise to multiple proteins by processing reactions. The order of the genes is *gag-pol-env,* as indicated in the upper part of Figure 38.1. The viral RNA is translated as a normal messenger—it is capped at the 5′ end and polyadenylated at the 3′ end—into the *gag* polyprotein. One spliced version of the RNA (lacking the *gag* termination codon) is translated into a *gag-pol* protein. Splicing also generates a shorter **subgenomic** messenger that is translated into the *env* polyprotein, a precursor for two proteins.

The *gag* gene gives rise to the protein components of the nucleoprotein core of the virion. The *env* gene codes for components of the envelope of the particle, which also sequesters components from the cellular cytoplasmic membrane. The *pol* gene specifies the enzyme reverse transcriptase, which consists of two subunits, one comprising a processed fragment of the other. As its name implies, the enzyme is responsible for converting the RNA genome into a complementary DNA strand. It also catalyzes subsequent stages in the production of duplex DNA; it has a DNA polymerase activity, behaves like RNAase H (can degrade the RNA part of an RNA-DNA hybrid), and is an endonuclease.

Reverse transcriptase is carried with the genome in the viral particle. The enzyme converts the RNA into a linear duplex in the cytoplasm of the infected cell. The linear DNA makes its way (by unknown means) to the nucleus, where it is converted into circular forms. One or more DNA copies become integrated into the host genome. The integrated **proviral DNA** is transcribed by the host machinery to produce viral RNAs, which serve both as mRNAs and as genomes for packaging into virions. Integration is a normal part of the life cycle and is necessary for transcription.

The structures of the DNA forms of the virus are compared with the RNA in **Figure 38.1.** The viral RNA

Figure 38.1
Retrovirus RNA ends in direct repeats (R), the free linear DNA ends in LTRs, and the provirus ends in LTRs shortened by two bases each.

has direct repeats at its ends. These **R** segments vary in different virus strains from 10–80 nucleotides. Following the R segment at the 5′ end of the virus is the **U5** region of 80–100 nucleotides, whose name indicates that it is unique to the 5′ end. Preceding the R segment at the 3′ terminus is the **U3** segment of 170–1250 nucleotides, which is unique to the 3′ end.

The ends of the linear DNA have additional sequences. A U3 segment is added to the 5′ end; a U5 segment is added to the 3′ end. As a result, each end of the DNA has the sequence U5-R-U3; this is called the **long terminal repeat (LTR).** Its generation involves the reaction illustrated in **Figure 38.2** in which the reverse transcriptase *switches templates,* carrying the nascent DNA with it to the new template. The figure shows the generation of one LTR; a similar event is required to generate another at the other end. One feature of the retroviral genome is that the LTRs at each end always are identical, and this feature must be accommodated by models for their generation. The 3′ end of U5 consists of a short inverted repeat relative to the 5′ end of U3, so the LTR itself ends in short inverted repeats. Thus the organization of the retroviral DNA resembles that of transposable elements like *copia* (see Figure 37.2).

The retroviral life cycle is illustrated diagrammatically in **Figure 38.3.** The stages between the production of linear DNA and its integration to generate provirus are not well understood. After its entry into the nucleus, the linear DNA gives rise to two forms of circular DNA. (The number of circular molecules ranges from 0.1 to 5 copies per cell in different systems.) Two

Figure 38.2
The LTR is generated by switching templates during reverse transcription. The new template can be the other end of the same molecule (as drawn) or can be a different molecule.

Reverse transcriptase initiates synthesis of DNA strand

Enzyme reaches end of template

5′ terminal region of RNA is degraded

Single-stranded DNA region pairs with RNA at 3′ terminus

— RNA template strand
═ DNA product strand

LTR

Reverse transcriptase continues

Figure 38.3
Retrovirus integration is accompanied by the formation of two types of circular DNA molecule.

RNA

Linear DNA   LTR   LTR

Circular DNA   LTR   LTR   LTR

Integrated provirus DNA   Host   Host

major forms of monomeric closed circular DNA are found. The more abundant form contains a single copy of the LTR, as though it had originated by a homologous recombination between the LTRs. (We do not know whether this actually is the mechanism for its production, but solitary LTRs that could be relics of the excision are present in some cellular genomes.) The less abundant form contains two copies of the LTR in tandem; sometimes they are joined directly, as though by blunt-end ligation; sometimes nucleotides are missing from one or both LTRs at the sites of junction. The role(s) of these circular molecules in integration has not yet been worked out. We note again the similarities with the circular forms of *copia* found in *D. malanogaster* (see Chapter 37).

The viral DNA integrates into the host genome at randomly selected sites. A successfully infected cell gains between one and ten copies of the provirus. At each site of insertion, a short direct repeat of target DNA is generated. The length of the target repeat depends on the particular virus; it may be 4, 5, or 6 bp. The occurrence of the direct repeats is taken to imply that the mechanism of integration may involve the generation of staggered cuts in host DNA, generally analogous to the mechanism of bacterial transposition. The site of integration is specific with regard to the virus, and falls short of each end of unintegrated linear DNA by 2 bp. Thus the integrated viral DNA has lost 2 bp from the left end of the 5′ terminal U3 and has lost 2 bp from the right end of the 3′ terminal U5.

The U3 region of each LTR carries a promoter. The promoter in the left LTR is responsible for initiating transcription of the provirus. Sometimes (probably rather rarely), the promoter in the right LTR sponsors transcription of whatever host sequences happen to be adjacent to the site of integration. The integration of a retroviral genome may be responsible for some classes of cell transformation—the conversion of the host cell into a tumorigenic state—by activating certain types of cellular genes. The ability to switch on flanking host sequences is analogous to the behavior of some bacterial transposons, although at least in some cases it may be due to provision of enhancer rather than promoter sequences. However, the activity of a provirus genome may be influenced by the site at which the virus is integrated in the host genome.

We have dealt so far with retroviruses in terms of

the infective cycle, in which integration is necessary for the production of further copies of the RNA. However, when a virus integrates in a germ-line cell and ceases to be efficiently expressed, it may become an inherited "endogenous virus" of the organism. This situation has been best investigated in the mouse and chicken, whose genomes both carry several inactive endogenous viruses. Sometimes the endogenous viruses are activated by external events, such as infection with another virus.

RETROVIRUSES MAY TRANSDUCE CELLULAR SEQUENCES

An interesting light on the viral life cycle is cast by the occurrence of transducing viruses, genomes that have acquired cellular sequences in the form illustrated in **Figure 38.4.** Part of the viral sequence—in this case, the *env* gene—has been replaced by the *v-onc* gene. The resulting virus is **replication defective;** it cannot sustain an infective cycle by itself. However, it can be perpetuated in the company of a **helper virus** that provides the missing viral functions.

Onc is an abbreviation for oncogenesis, the ability to transform cultured cells so that the usual regulation of growth is released to allow unrestricted division. Both viral and cellular *onc* genes may be responsible for creating tumorigenic cells. The *onc* genes carried by retroviruses are called *v-onc,* and they confer upon the virus the ability to transform a certain type of host cell. Loci with homologous sequences found in the host genome are called *c-onc* genes. In most cases, we have no direct information about the function of the cellular genes. However, in some instances, active *c-onc* genes have been identified by transfection protocols, in which their addition to a recipient cell generates the transformed state. They may also be activated by integration nearby of a retroviral genome, as

Figure 38.4
Replication-defective transforming viruses have a cellular sequence substituted for part of the viral sequence. The length of the cellular sequence and the viral region that is replaced are characteristic for each virus.

described in the previous section. The *onc* genes consititute a large class; individual members are given names derived from the name of the virus in which they were first found (*src* from Rous sarcoma virus causes sarcoma, *fes* is from feline sarcoma virus, and so on).

How are the *onc* genes acquired by the retroviruses? A revealing feature is the discrepancy in the structures of *c-onc* and *v-onc* genes. The *c-onc* genes usually are interrupted by introns and are expressed at low levels in the cell. The *v-onc* genes are uninterrupted and are expressed at a high level as part of the viral transcription unit. Their relative structures suggest that the *v-onc* genes originate from spliced RNA copies of the *c-onc* genes.

First thoughts about the origin of the transforming viruses were that a cellular mRNA might accidentally

have been packaged into a virion together with a viral genome; then a nonhomologous recombination between the RNAs would be required to generate the transforming viral genome. This model has been reconsidered because of two difficulties. First, the *c-onc* genes are represented in scarce mRNAs, which lack packaging signals and would be unlikely to be sequestered. Second, there is no simple model for the recombination event at the level of RNA.

A current model for the formation of transforming viruses is illustrated in **Figure 38.5.** A retrovirus has integrated near a *c-onc* gene. A deletion occurs to fuse the provirus to the *c-onc* gene; then transcription generates a joint RNA, containing viral sequences at one end and cellular *onc* sequences at the other end. Splicing occurs to remove the introns in both the viral and cellular parts of the RNA. The RNA has the ap-

Figure 38.5
Replication-defective viruses may be generated through integration and deletion of a viral genome to generate a fused viral-cellular transcript that is packaged with a normal RNA genome. Nonhomologous recombination is necessary to generate the replication-defective transforming genome.

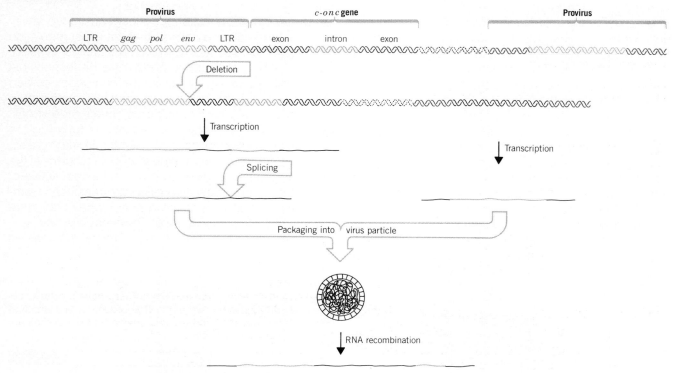

propriate signals for packaging into the virion; virions will be generated if the cell also contains another, intact copy of the provirus. Then some of the diploid virus particles may contain one fused RNA and one viral RNA.

A recombination between these sequences could generate the transforming genome, in which the viral repeats are present at *both* ends. (Recombination occurs at a high frequency during the retroviral infective cycle; presumably it employs DNA intermediates or additional strand-transfer reactions. We do not know anything about its demands for homology in the substrates, but we assume that the nonhomologous reaction between a viral genome and the cellular part of the fused RNA proceeds by the same mechanism responsible for viral recombination.)

Did retroviruses evolve from cellular elements? The common features of the entire retroviral class suggest that there may have been a single ancestor. Its resemblance to transposable elements suggests that primordial IS-like elements could have surrounded a host gene for a nucleic acid polymerase; the resulting unit would have the form *LTR-pol-LTR*. It might evolve

into a retrovirus by acquiring further nucleic acid polymerizing activities, enabling it to handle both DNA and RNA substrates, followed by the incorporation of genes whose products allowed packaging of the RNA. Other functions, such as transforming genes, might be incorporated later. (There is no reason to suppose that the mechanism involved in acquisition of cellular functions is unique for transforming genes; but these viruses may have a selective advantage because of their effects on the growth of the cells they infect.)

RNA-DEPENDENT TRANSPOSITIONS MAY HAVE OCCURRED IN THE CELL

Some pseudogenes display external and internal features suggesting that they may have originated in RNA sequences, although in these cases we can only speculate on how a DNA copy may have been generated. The characteristic features of such a pseudogene are compared in **Figure 38.6** with the features of the original gene and the mRNA. The figure shows *all* the relevant diagnostic features, only some of which may be found in any individual example.

Figure 38.6
Pseudogenes could arise by reverse transcription of RNA to give duplex DNAs that become integrated into the genome.

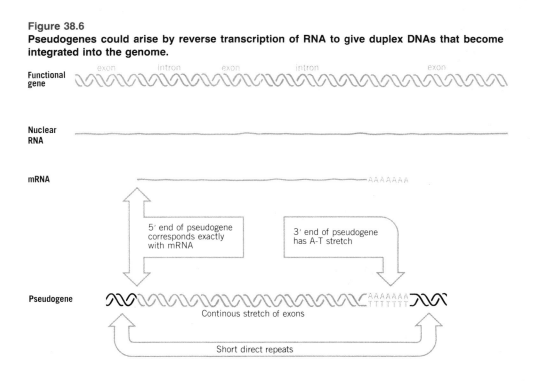

The pseudogene may start at the point equivalent to the 5′ terminus of the RNA, which would be expected only if the DNA had originated from the RNA. Several pseudogenes consist of precisely joined exon sequences; we know of no mechanism to recognize introns in DNA, so this feature argues for an RNA-mediated stage. The pseudogene may end in a short stretch of A-T base pairs, presumably derived from the poly(A) tail of the RNA. On either side of the pseudogene is a short direct repeat, presumed to have been generated by a transposition-like event. Supporting the idea that these pseudogenes originated by a mechanism different from those responsible for generating pseudogenes found near the active members of gene clusters, the "processed" pseudogenes reside at locations unrelated to their presumed sites of origin.

As Figure 38.6 implies, if a pseudogene is derived from an mRNA sequence, its homology at the 5′ end cannot extend upstream from the initiation site. However, a mouse α globin processed pseudogene does have some upstream homology with the active gene. The processing model could be retained by supposing that initiation occurred farther upstream to generate a longer RNA (either because of an aberrant initiation event or because the promoter was then located differently).

Other sequences that show features resembling transposons include members of the Alu family, which are flanked by short direct repeats. However, these display the curious feature that the lengths of the repeats are different for individual members of the family. An insertion into an Alu sequence that was responsible for creating a small nuclear RNA also appears to have been the result of a transposition event. Evidence for ancient transposition events also is found in the genes for several small nuclear RNAs.

All of these examples have an important difference from the situation of the retroviruses: the presumptively transposed sequences do not themselves carry any information that might be used to sponsor a transposition event (or carry out the preceding reverse transcription of the RNA). Could the process have been mediated by a retrovirus? Was it accomplished by an aberrant cellular system? Perhaps the ends of the transposed sequence fortuitously resembled sequences at the ends of a transposon. Are transposition events currently occurring in these genomes or are we seeing only the footprints of ancient systems?

TISSUE-SPECIFIC VARIATIONS OCCUR IN THE DROSOPHILA GENOME

The content of the genome remains inviolate in most somatic cells. However, we now know of some situations in which changes are made in the relative *proportions* of certain sequences; the best-characterized examples occur during larval development in insects. The adjustment takes the form of underreplicating or overreplicating specific sequences.

Underreplication occurs during development of the polytene tissues of the fly. In these tissues, the giant chromosomes described in Chapter 28 are formed by multiple successive duplications of the original (synapsed) diploid set.

About a quarter of the genome of D. melanogaster is contained in the heterochromatic regions of the chromosomes, which aggregate to form a chromocenter (a large part of which consists of satellite DNA sequences). The relative amount of heterochromatin is much less in polytene cells compared with diploid cells. The explanation is that this part of the genome has failed to duplicate during the polytenization of the euchromatic DNA sequences. Measurements of satellite DNA content suggest that these sequences duplicate no or very few times, compared with the 9 duplications of euchromatic DNA in the salivary gland.

The rRNA genes also are underreplicated in polytene tissues. The region of rDNA passes through only 6–7 duplications (actually reaching the same final level whether only one or both of the nucleolar organizers are present. This type of dosage control may be specific for rDNA.)

The genome of D. melanogaster can therefore be divided into three types of region according to the control of replication in polytene tissues. A natural assumption is that the origins for replication in each type of region control the number of initiation events. Origins in satellite DNA may fail to be recognized; origins in rDNA stop functioning in response to some feedback from the number of rRNA genes; and general euchromatic origins continue initiating to the bitter end.

Differential amplification of particular protein-coding

sequences also occurs in insects. In the salivary gland of *Rhyncosciara* and other Sciarids, **DNA puffs** are generated. Their appearance is superficially similar to the puffs that represent active bands of Dipteran polytene chromosomes, but the Sciarid puffs contain locally amplified DNA as well as RNA. The amplified sequences may have undergone up to 4 additional initiations of replication, increasing the copy number 16-fold.

Insight into possible mechanisms for differential amplification is provided by the state of the chorion genes in the development of *D. melanogaster.* The proteins that make up the chorion (eggshell) are synthesized and secreted by the polyploid ovarian follicle cells. Insect chorion genes tend to be clustered, and two groups have been identified in *D. melanogaster.* Prior to their expression in the follicle cell, chorion genes on the X chromosome are amplified by up to 16-fold (4 additional doublings), while the genes on chromo-some III are amplified up to 60 times (6 additional doublings).

In each case, amplification extends for a distance of about 45–50 kb on either side of the chorion genes. The level of maximum amplification represents a plateau of about 20 kb surrounding the chorion genes. The extent of amplification shows a gradient of decline on either side of the plateau.

What is responsible for the amplification? The gradient of decline suggests that the endpoints of the amplified regions in individual molecules are heterogeneous. This could be explained by the model illustrated in **Figure 38.7.** Multiple initiations of bidirectional replication occur at an origin in the center of the region. The replication forks progress for distances varying from 10 to 50 kb. This model views the entire amplified region as the replicon, present in multiple but only partially replicated copies. (The relationship between underreplicated and more-replicated regions

Figure 38.7

Amplification of a local region could be accomplished by multiple initiations within a single replicon.

Each replicated region is joined to the flanking unreplicated regions by static replication forks. If the forks cease to move at imprecisely determined positions, they generate a gradual gradient, as indicated on the left. If several forks terminate coordinately at a fixed site, they generate sudden decreases in the level of amplification, as indicated on the right.

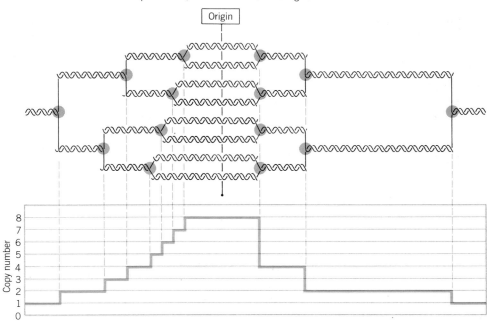

in salivary gland chromosomes could be explained by a similar model, in which the junctions between regions are represented by static replication forks.)

Does the amplified region contain a *cis*-acting origin responsible for replication? In this case, amplification should be abolished for any chromosome whose origin is mutated. In a heterozygote, the failure should apply only to the mutated chromosome. And if the origin is translocated elsewhere, it should sponsor the amplification of whatever replicon it finds itself in.

The properties of the *ocelliless* mutant conform with these predictions. This mutation causes complex changes in the phenotype; it is an inversion of about three bands, containing the region of the chorion genes on the X chromosome. The chorion genes lie within 3 kb of the left end of the inversion.

The inversion causes a major change in the pattern of amplification. **Figure 38.8** compares the regions amplified in wild-type and *ocelliless* files. Of the original amplified region, the 40 kb on the left side of the inversion breakpoint fail to be amplified at all. The 50 kb on the other side are amplified, although at a reduced level, reaching only about half the usual maximum. (The reason for the reduction in level is unknown.) Also, the amplification now spreads into new regions, as seen by the increase in copy number of sequences on the right that are not amplified in the wild type.

These results suggest that an origin able to sponsor bidirectional amplification lies in the vicinity of the chorion genes. The extent of the amplified region remains similar whether the origin lies in its usual position in

Figure 38.8
The *ocelliless* inversion changes the region of DNA that is amplified.

The left end of the inversion lies within 3 kb of the left boundary of the chorion genes; the right end is not known, but for the purposes of illustration the inversion is assumed to be about 85 kb long.

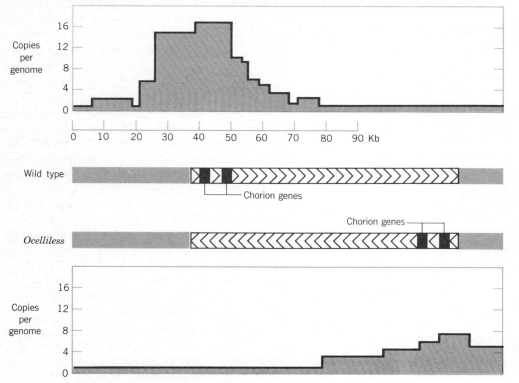

the wild type or is transferred to a new position by the *ocelliless* inversion.

The classic description of the type of effect produced by the inversion is encompassed by the phenomenon of the **position effect.** This term describes the ability of location to influence the expression of a gene, as seen by the change in activity resulting from a translocation. A well-known example is the inactivity that results when a gene is translocated from its wild-type position in euchromatin to a new location close to heterochromatin. Although the gene itself is unchanged, its activity is influenced by its surroundings. In the case of *ocelliless,* the inversion causes a failure in the amplification of some genes, while others are amplified instead. The dosage effects produce complex changes in phenotype, without any alteration of the genes themselves.

SELECTION OF AMPLIFIED GENOMIC SEQUENCES

The eucaryotic genome has the capacity to accommodate additional sequences of either exogenous or endogenous origin. Whether added to cells or originating in the chromosomes, DNA sequences may give rise to multiple copies that survive as a tandem array in either extrachromosomal or chromosomal location. In the extrachromosomal form, the additional material is inherited in an irregular way (it does not segregate evenly at division in the manner of an authentic plasmid). In the chromosomal form, however, the material has been integrated into the resident genome and becomes a component of the genotype.

The procedure for introducing exogenous donor DNA into recipient cells is called **transfection** (see next section). The **amplification** of endogenous sequences is provoked by selecting cells for resistance to certain agents. The same general pattern of flexibility is seen in both transfection and amplification, which very likely involve overlapping mechanisms. We know nothing about the enzymatic activities involved in this manipulation, but we suppose that they are engaged in related activities in the normal course of perpetuating DNA.

The best-characterized example of amplification re-

sults from the addition of **methotrexate** (mtx) to any one of several cultured cell lines. The reagent blocks folate metabolism. Resistance to it may be conferred by mutations that change the activity of the enzyme dihydrofolate reductase (DHFR). As an alternative to change in the enzyme itself, the amount of enzyme may be increased. The cause of this increase is an amplification of the number of *dhfr* structural genes.

Although amplification represents a relatively rare mechanism for acquiring resistance to a toxic reagent, it is not unique to DHFR. Resistance to an inhibitor of the enzyme aspartyltranscarbamylase also is conferred by an amplification of the corresponding structural gene. Cells resistant to the heavy metal cadmium have an increased copy number of the gene coding for the detoxifying protein metallothionein. A common feature in all three systems is that highly resistant cells are not obtained in a single step, but instead appear when the cells are adapted to gradually increasing doses of the toxic reagent. Thus gene amplification may require several stages.

The number of *dhfr* genes in a cell line resistant to methotrexate varies from 40 to 400, depending on the stringency of the selection and the individual cell line. The mtx^r lines fall into two classes, distinguished by their response when cells are grown in the absence of methotrexate, removing the former pressure for high levels of DHFR activity. In **stable** lines, the amplified genes are retained. In **unstable** lines, the amplified genes are at least partially lost when the selective pressure is released. The cause of this difference is the condition of the amplified *dhfr* genes in the two lines. The situation responsible for this state of affairs is summarized in **Figure 38.9.**

In a stable mtx^r line, the amplified *dhfr* genes are chromosomal, occupying the usual site of *dhfr* on one of the chromosomes. Usually the other chromosome retains its normal single copy of *dhfr*. Thus amplification generally occurs at only one of the two *dhfr* alleles; and increased resistance to methotrexate is accomplished by further increases in the degree of amplification at this locus. (However, sometimes the chromosome with the amplified *dhfr* locus itself is duplicated, presumably through nondisjunction at mitosis.)

Gene amplification has a visible effect on the chro-

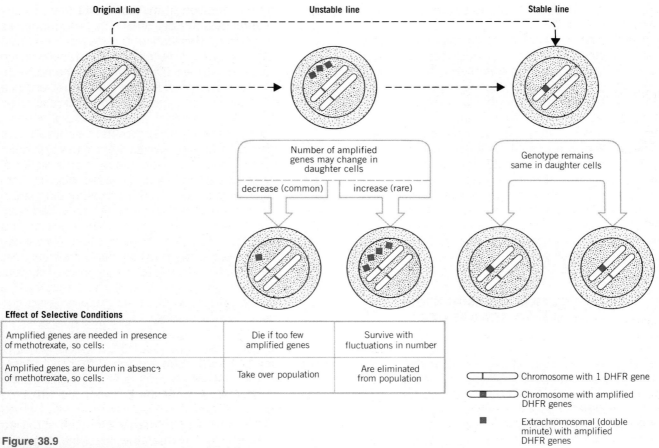

Effect of Selective Conditions		
Amplified genes are needed in presence of methotrexate, so cells:	Die if too few amplified genes	Survive with fluctuations in number
Amplified genes are burden in absence of methotrexate, so cells:	Take over population	Are eliminated from population

⊂▭▭⊃ Chromosome with 1 DHFR gene

⊂▭■▭⊃ Chromosome with amplified DHFR genes

■ Extrachromosomal (double minute) with amplified DHFR genes

Figure 38.9
The *dhfr* gene can be amplified to give additional copies that are extrachromosomal (unstable) or chromosomal (stable).

mosome; the locus can be visualized in the form of a **homogeneously staining region (HSR).** An example is shown in **Figure 38.10.** The HSR takes its name from the presence of an additional region that lacks any chromosome bands after treatments such as C-banding (see, for example, Figure 28.9). This suggests that some region of the chromosome between bands has undergone an expansion.

In unstable cell lines, no change is seen in the chromosomes carrying *dhfr*. However, large numbers of elements called **double-minute chromosomes** are visible. An example is shown in **Figure 38.11.** In a typical cell line, each double minute carries 2–4 *dhfr* genes. The double minutes appear to be self-repli-

cating; but they lack centromeres. As a result, they do not attach to the mitotic spindle and therefore segregate erratically, frequently being lost from the daughter cells. Notwithstanding the original name, the actual status of the double minutes is regarded as extrachromosomal.

The irregular inheritance of the double minutes explains the instability of methotrexate resistance in these lines. Double minutes will be lost continuously during cell divisions; and in the presence of methotrexate, the cells with reduced numbers of *dhfr* genes will die. Only those cells that have retained a sufficient number of double minutes will appear in the surviving population. The presence of the double minutes reduces

Figure 38.10
Amplified copies of the *dhfr* gene produce a homogeneously staining region (HSR) in the chromosome.
Photograph kindly provided by Robert Schimke.

the rate at which the cells proliferate. Thus when the selective pressure is removed, cells lacking the amplified genes have an advantage; they generate progeny more rapidly and soon take over the population. This explains why the amplified state is retained in the

Figure 38.11
Amplified extrachromosome *dhfr* genes take the form of double-minute chromosomes, as seen in the form of the small white dots.

Photograph kindly provided by Robert Schimke.

cell line only so long as cells are grown in the presence of methotrexate.

Both stable and unstable lines are found after long periods of selection for methotrexate resistance. What is the initial step in gene amplification? After short periods of selection, most or all of the resistant cells are unstable. The formation of extrachromosomal copies clearly is a more frequent event than amplification within the chromosome. We do not know whether intrachromosomal amplification simply proceeds less often as a de novo step or requires extrachromosomal amplification to occur as an intermediate step. Because of the erratic segregation of the double minutes, increases in the copy number can occur relatively quickly as cells are selected at each division for progeny that have gained more than their fair share of the *dhfr* genes. Cells with greater numbers of copies are found in response to increased levels of methotrexate. The behavior of the double minutes explains the stepwise evolution of the mtx^r condition and the incessant fluctuation in the level of *dhfr* genes in unstable lines.

How do the extrachromosomal copies arise? We know that their generation occurs without loss of the

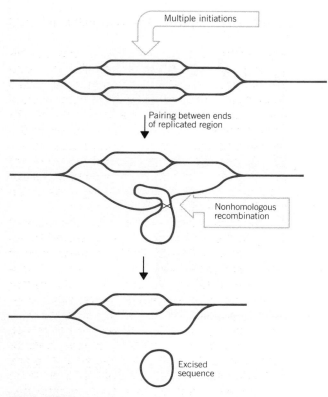

Multiple initiations

Pairing between ends
of replicated region

Nonhomologous
recombination

Excised
sequence

Figure 38.12
**Extrachromosomal copies of DNA could be generated by rein-
itiation of replication followed by nonhomologous recombination between the replicas.**

original chromosomal copy. One possibility is illustrated in **Figure 38.12.** Additional cycles of replication
are initiated in the vicinity of the *dhfr* gene. The extra
copies could be released from the chromosome, possibly by some recombination-like event. Depending on
the nature of this event, it could generate an extrachromosomal DNA molecule containing one or several
copies. If the double minutes contain circular DNA,
recombination between them in any case is likely to
generate multimeric molecules.

The origin of amplified chromosomal copies may be
more difficult to explain. The HSR consists of a large
number of tandemly repeated units. One problem in
supposing that they arise by integration of double-
minute sequences is that the number of chromosomal
repeats is many times greater than the number of re-

peats in the individual double minutes. The large number of integrated copies is also an impediment to constructing models for a strictly intrachromosomal
amplification event.

The amplified region is much longer than the *dhfr*
gene itself. The gene has a length of ~31 kb, but the
average length of the repeated unit is 500–1000 kb in
the chromosomal HSR. The amount of DNA contained
in a double minute seems to lie in a range from 100
to 1000 kb. The chromosomal region that gives rise to
the amplified genes might correspond with the replicon in which they are located.

The same amplification events presumably occur in
the absence of methotrexate, but the products are lost
from the cell population in the absence of any selective pressure. Why is this phenomenon seen with so
few genes? Perhaps it occurs only with genes that lie
in an appropriate location relative to a replicon or when
appropriate repetitive sequences occur at locations
that permit recombination to excise the region.

Some information about the events involved in perpetuating the double minutes is given by an unstable
cell line whose amplified genes code for a mutant DHFR
enzyme. The mutant enzyme is not present in the original (diploid) cell line; thus the mutation must have
arisen at some point during the amplification process.
Despite variations in the number of amplified genes,
these cells display *only* the mutant enzyme. Thus the
wild-type chromosomal genes cannot be continuously
generating large numbers of double minutes anew,
because these amplified copies would produce normal enzyme.

Once amplified extrachromosomal genes have
arisen, therefore, changes in the state of the cell are
mediated through these genes and not through the
original chromosomal copies. When methotrexate is
removed, the cell line loses its double minutes in the
usual way. On reexposure to the reagent, *normal* genes
are amplified to give a new population of double minutes. This shows that none of the extrachromosomal
copies of the mutant gene had integrated into the
chromosome. Another striking implication of these results is that the double minutes of the mutant line carried *only* mutant genes—so if there is more than one
dhfr gene per double minute, all must be of the mutant
type. This suggests that multicopy double minutes can
be generated from individual extrachromosomal genes.

EXOGENOUS SEQUENCES CAN BE INTRODUCED BY TRANSFECTION

Transfection experiments began with the addition of preparations of metaphase chromosomes to cell suspensions. The chromosomes are taken up rather inefficiently by the cells and give rise to unstable variants at a low frequency. Intact chromosomes rarely survive the procedure; the recipient cell usually gains a fragment of a donor chromosome (which is unstable because it lacks a centromere). Rare cases of stable lines may have resulted from integration of donor material into a resident chromosome.

Similar results are obtained when purified DNA is added to a recipient cell preparation. However, with purified DNA it is possible to add particular sequences rather than rely on random fragmentation of chromosomes. Transfection with DNA yields stable as well as unstable lines, with the former relatively predominant. (These experiments are directly analogous to those performed in bacterial transformation, as indicated in Figure 2.3, but are described as transfection because of the historical use of transformation to describe changes that allow unrestrained growth of eucaryotic cells.)

The low frequencies of transfection make it necessary to use donor markers whose presence in the recipient cells can be selected for. Note that the transfected sequence is expressed. Most transfection experiments have used markers representing readily assayed enzymatic functions, but in principle, any marker that can be selected can be assayed. This allows the isolation of genes responsible for morphological phenomena. For example, as we have already mentioned, transfected cells can be selected for acquisition of the transformed (tumorigenic) phenotype. Then we can identify the DNA responsible for conferring the phenotype. This type of protocol has led to the isolation of several cellular *onc* genes.

Cotransfection with more than one marker has proved informative about the events involved in transfection and has extended the range of questions that we can ask with this technique. When tk^- cells are transfected with a DNA preparation containing both a purified tk^+ gene and the ϕX174 genome, the tk^+ transformants all have both donor sequences. This is a useful observation, because it allows unselected markers to be introduced routinely by cotransfection with a selected marker.

The arrangement of *tk* and ϕX174 sequences is different in each transfected line, but remains the same during propagation of that line. Often multiple copies of the donor sequences are present, the number varying with the individual line. Revertants lose the ϕX174 sequences together with *tk* sequences. Amplification of transfected sequences under selective pressure results in the increase of copy number of all donor sequences pari passu. Thus the two types of donor sequence become physically linked during transfection and suffer the same fate thereafter.

To perform a transfection experiment, the mass of DNA added to the recipient cells is increased by including an excess of "carrier DNA," a preparation of some other DNA (often from salmon sperm). Transfected cells prove to have sequences of the carrier DNA flanking the selected sequences on either side. Transfection therefore appears to be mediated by a large unit, consisting of a linked array of all sequences present in the donor preparation. Since revertants for the selected marker lose all of this material, it seems likely that the transfected cell gains only a *single* large unit. The unit may be formed by a concatemeric linkage of donor sequences in a reaction that is rapid relative to the other events involved in transfection. This transfecting package may be of the order of 1000 kb in length.

Because of the size of the donor unit, we cannot tell from blotting experiments whether it is physically linked to recipient chromosomal DNA (the relevant end fragments are present in too small a relative proportion). From the characteristics of the transfected lines, it seems possible that the first stage is the establishment of an unstable extrachromosomal unit, followed by the acquisition of stability via integration. In some cell lines, *in situ* hybridization has been used to show that transfected cells have donor material integrated into the resident chromosomes. Any given cell line has only a single site of integration; but the site is different in each line. Probably the selection of a site for integration is a random event; sometimes it is associated with a gross chromosomal rearrangement.

Some interesting results obtained by cotransfection with two selectable markers are summarized in **Figure 38.13.** An APRT$^-$ TK$^-$ recipient cell line was trans-

Figure 38.13
When transfected genes are amplified, the entire cluster responds coordinately to further selective pressure.

fected with DNA carrying the *aprt* and *tk* genes. The *aprt* gene was the active allele *aprt*⁺, which codes for the enzyme adenosine phosphoribosyltransferase. The *tk* gene was an inactive allele, retaining the coding region for the enzyme thymidine kinase, but lacking the promoter. The transfected cells were selected for APRT⁺ function, but both donor sequences were acquired.

Now these APRT⁺ TK⁻ cells were selected for their ability to provide *tk*⁺ function. Its acquisition depends on two events. The *tk* must be transcribed from some site substituting for its promoter; and because this achieves only a low level of expression, the number of copies must be amplified. The selected cells be-

come APRT⁺ TK⁺ and they have about 40 copies of the *aprt-tk* transfected unit, integrated at a single chromosomal site.

Finally these cells were selected for the APRT⁻ TK⁺ phenotype. Thus they are required to eliminate the expression of the multiple copies of the *aprt*⁺ gene without affecting the expression of the *tk* gene. The interesting feature is that *all copies of the* aprt *gene in these cells acquired the same negative mutation*. This result implies the existence of some mechanism that allows the entire gene cluster to be regenerated from a single member or corrected to reflect the sequence of one member; apparently this happens within a single cell generation.

TRANSFECTED DNA CAN ENTER THE GERM LINE

An exciting development of transfection techniques is their application to introduce genes into animals. An approach that has proved successful with the mouse is summarized in **Figure 38.14.** Plasmids carrying the gene of interest are injected into the germinal vesicle (nucleus) of the oocyte or into the pronucleus of the fertilized egg. The egg is implanted into a pseudopregnant mouse. Such experiments have been performed with globin genes and (more revealingly) with a "fusion gene" that has the promoter for metallothionein linked to the thymidine kinase coding sequence. The MT promoter is derived from a natural mouse gene and can be tested for its ability to respond to the customary induction by heavy metals or glucocorticoids; the response can be assayed by measuring thymidine kinase activity.

In a series of 69 injected eggs, 10 of the resulting mice carried the injected sequence as shown by the analysis of kidney DNA. Usually, multiple copies of the plasmid appear to have been integrated in tandem into a single chromosomal site. The number of copies varies from 1 to 150. They are inherited by the progeny of the injected mouse as expected of a Mendelian locus.

Seven of the mice expressed the thymidine kinase gene in liver and kidney (tissues in which MT usually is active). Expression did not occur in brain, a tissue in which the gene usually does not function. Expres-

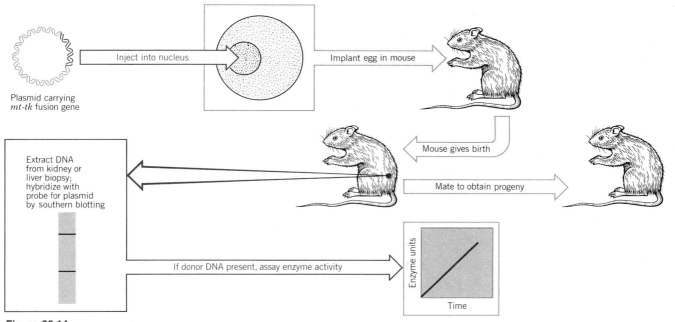

Figure 38.14
Transfection can introduce DNA into the germ lines of animals.

sion is induced by the addition of heavy metals, although not by glucocorticoids.

In the progeny of the injected mice, expression of the donor gene is extremely variable; it may be extinguished entirely, reduced somewhat, or even increased. Even in the original parents, the level of gene expression does not correlate with the number of tandemly integrated genes. Probably only a few of the genes are active. (Also note that if all the additional genes were active, it would be surprising if the normal regulatory response were retained, since the large number of promoters might well dilute out the regulatory molecules.)

What could be responsible for the variation in gene expression? One possibility that has often been discussed for transfected genes, and that applies also to integrated retroviral genomes, is that the site of integration may be important. Perhaps a gene is expressed if it integrates within an active domain, but not if it integrates in another area of chromatin. Another possibility is the occurrence of epigenetic modification. Some of the transfected genes in the mouse are methylated. (Since the donor DNA lacked methyl

groups, this observation implies the existence of an enzyme able to engage in de novo methylation.) Perhaps changes in the pattern of methylation might be responsible for changes in activity. Alternatively, the genes that happened to be active in the parents may have been deleted or amplified in the progeny.

With these experiments, we see the possibility of extending from cultured cells to animals the option of examining the regulatory features. The ability to introduce DNA into the genotype allows us to make changes in it, to add new genes that may have had particular modifications introduced *in vitro*. Thus it may become possible to delineate the features responsible for tissue-specific gene expression. Ultimately it may be possible to replace defective genes in the genotype.

FURTHER READING

Retroviruses have been powerfully analyzed as both viruses and transposable elements by **Varmus** in *Mobile Genetic Elements* (ed. Shapiro, Academic Press, New York, 1983). The genome of *D. melanogaster* was reviewed by **Spradling** & **Rubin** (*Ann. Rev. Genet.* **15,** 219–264, 1981); the effects of *ocelliless* were discovered by **Spradling, Spradling, & Mahowald** (*Cell* **27,** 193–202 and 203–210, 1981). The events involved in amplification of *dhfr* were brought up to date by **Schimke et al.** (*Cold Spring Harbor Symp. Quant. Biol.* **45,** 785–797, 1981). The uses and insights offered by transfection have been reviewed by **Pellicer et al.** (*Science* **209,** 1414–1422, 1980). The injection of genes into mice was developed by **Palmiter et al.** (*Cell* **29,** 701–710, 1982).

CHAPTER 39
GENERATION OF
ANTIBODY DIVERSITY

The **immune response** of the mammal is exercised by an apparently limitless repertoire of **antibodies,** each synthesized in response to provocation by a different **antigen.** Usually the antigen is a protein (or protein-attached moiety) of foreign origin that has entered the bloodstream of the animal—for example, it may be the coat protein of an infecting virus. Exposure to an antigen initiates production of an antibody that *specifically recognizes the antigen.* Antibody production occurs in the B lymphocytes and enables them (in conjunction with T lymphocytes and macrophages, the other white blood cells) to remove the offending antigen. Once an organism has synthesized a particular antibody, the manufacturing capacity remains available to combat any future reexposure. One of the mysterious features of this process has been the animal's ability to produce an appropriate antibody whenever it is exposed to a new antigen. How can the organism be prepared to produce antibody proteins each designed specifically to recognize an antigen whose structure cannot be anticipated?

For practical purposes, we usually reckon that the organism has the ability to produce $> 10^6$ different antibodies. Each antibody is a tetrameric **immunoglobulin** consisting of two identical **light (L) chains** and two identical **heavy (H) chains.** The theory of combinatorial association supposes that any type of

light chain can associate with any type of heavy chain. To produce 1,000,000 potential antibodies therefore requires > 1000 types of light chain and > 1000 types of heavy chain.

The structure of the immunoglobulin tetramer is illustrated in **Figure 39.1.** Each protein chain consists of two principal regions: the N-terminal **variable (V) region;** and the C-terminal **constant (C) region.** They are defined by comparing the amino acid sequences of different immunoglobulin chains. As the names suggest, the variable regions show considerable changes in sequence from one protein to the next, while the constant regions show substantial homology. Combination of the corresponding regions of the light and heavy chains generates distinct domains in the immunoglobulin protein.

The variable domain is generated by association between the variable region of the light chain and the variable region of the heavy chain. The V domain is responsible for recognizing the antigen. Thus the protein displays the maximum versatility in the region whose target differs in each individual antibody-antigen reaction. Production of V domains of appropriate specificity creates the ability to respond to diverse antigens.

Within the variable region, some parts are more variable than others. The most variable stretches are called **hypervariable (HV) regions.** The relatively less vari-

Figure 39.1
Heavy and light chains combine to generate an immunoglobulin with several discrete domains.

able stretches are called **framework (FR) regions.** A variable region can be written as an alternating series of these types of subregion: FR1-HV1-FR2-HV2-FR3-HV3-FR4. The hypervariable segments are shorter than the framework segments. The total number of variable regions for either light- or heavy-chain proteins is measured in hundreds.

The number of different constant regions is rather small. Defined by their constant regions, light chains fall into two types: kappa and lambda. The kappa type is characterized by a single constant region; the lambda type has a few related types of constant region.

The heavy chains have larger constant regions. Different classes of antibodies are defined by the type of heavy-chain constant region, of which there are about eight. The association of constant regions generates several individual domains in the molecule. The first domain results from association of the constant region of the light chain with the CH1 part of the heavy-chain constant region. The structure of the rest of the heavy

chain is (slightly) different in the various classes. In the example shown in Figure 38.1, a short hinge region connects the first half of the molecule to the other two constant domains, each generated by association of the corresponding regions (CH2 and CH3) of the heavy chains. The constant domains mediate the effector functions responsible for executing the immune response. Thus the regions of the molecule with constant functions are conserved. Actually, different classes of immunoglobulins have related but not necessarily identical effector functions; the type of heavy-chain constant region determines the type of effector function.

Comparing the characteristics of the variable and constant regions, we see a central dilemma in immunoglobulin gene structure. How does the genome produce proteins in which the variable regions have over 1000 possible sequences, while the constant regions of the *same molecule* have less than 10 possible sequences? It turns out that the number of coding sequences for each type of region reflects its variability. There are many genes coding for V regions; only a few genes coding for C regions.

In this context, "gene" means a sequence of DNA coding for a discrete *part* of the final immunoglobulin polypeptide (heavy or light chain). Thus **V genes** code for variable regions and **C genes** code for constant regions, although neither type of gene is expressed as an independent unit. To construct a unit that can be expressed in the form of an authentic light or heavy chain, a V gene must be joined physically to a C gene. In this system, two "genes" code for one polypeptide.

Within each of the three immunoglobulin chain families—kappa, lambda, and heavy—any one of many V genes usually may be joined to any one of the few C genes. The process occurs by **somatic recombination** of the appropriate DNA sequences *in the B lymphocyte in which the antibody is expressed.* In most cases, the number of V genes is very high, probably a few hundred, organized in a huge gene cluster. (We note that the mouse lambda type is an exception, consisting of only a few genes.) The variety in choice of V genes is responsible for some part of the diversity of immunoglobulins. However, not all diversity is coded in the genome; some is generated by changes that occur during the process of constructing a functional gene.

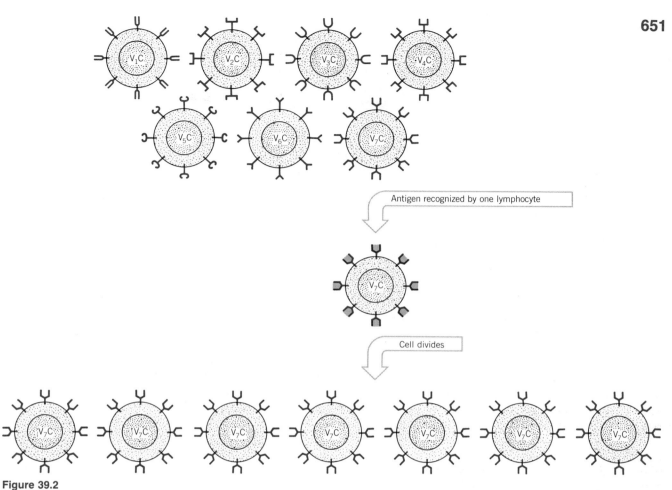

Figure 39.2
Reaction with an antigen leads to clonal expansion of the lymphocyte with the corresponding antibody.

The immunoglobulin V gene clusters are not the only example of huge regions (by molecular standards) devoted to synthesis of a single type of gene. The major histocompatibility complex also is concerned with generating proteins involved in the immune response; it consists of a very long region within which are many genes of related structure and function. (These genes are expressed directly, without rearrangement of sequences.) The importance of the immune response for the mammal is indicated by the large number of genes devoted to the immunoglobulin and histocompatibility functions; and this offers us the opportunity to characterize these massive gene clusters.

The construction of a functional immunoglobulin gene

might seem superficially to be a Lamarckian process, representing a change in the genome associated with a particular feature of the genotype (the antigen). At birth, the organism does not possess the functional gene for producing a particular antibody. The subsequent construction of the gene allows the antibody to be synthesized so that it is available to react with the antigen.

The feedback from antigen to the production of antibody is a provocative phenomenon. However, it may be explained without resort to Lamarckian heresy by the **clonal selection theory** summarized in **Figure 39.2.** Recombination of V and C genes to give functional loci occurs in a population of immature B lymphocytes. *Each cell produces only a single antibody;* it undertakes only one productive rearrangement of light-chain genes and one of heavy-chain genes. The antibody is different in each cell, because different V genes and C genes have been joined in each reconstruction. On exposure to antigen, a cell whose antibody is able to bind the antigen will be stimulated to divide, probably by virtue of some feedback from the surface of the cell, where the antibody-antigen reaction occurs. Now large numbers of this lymphocyte are produced, its antibody is secreted in large quantities, and it may even come to dominate the antibody population. We note that the entire process occurs in somatic cells and does not affect the germ line; so the response to an antigen is not inherited.

IMMUNOGLOBULIN GENES ARE ASSEMBLED FROM THEIR PARTS

Probes for immunoglobulin genes have been obtained from myelomas or plasmacytomas—tumor cell lines derived from immunoglobulin-expressing ancestors. Each tumor cell line usually synthesizes a single antibody. The protein is a major product of the cell, and

its mRNA may occupy up to 10% of the poly(A)$^+$ mRNA. The mRNA can be isolated in the usual way; and cloned probes representing the V and C regions can be used to identify the corresponding genomic sequences. The arrangement of these sequences is *different in the cells producing the immunoglobulins* (usually as represented by the myeloma or plasmacytoma, sometimes as characterized in the lymphocyte itself) from all other somatic cells or germ cells. Most of our information is about the immunoglobulin genes of mouse and man.

In the cell expressing the antibody, each immunoglobulin chain is coded by a single interrupted gene. The exons of the gene correspond precisely with the functional domains of the protein. In the variable region, the first exon codes for the signal sequence (involved in membrane attachment), and the second exon codes for the major part of the variable region itself (< 100 codons long). The structure of the constant region depends on the type of chain. For light chains, the constant region is coded by a single exon (the third of the intact gene). For heavy chains, the constant region is coded by several exons; corresponding with the protein chain shown in Figure 38.1, there would be separate exons for the regions CH1, hinge, CH2,

Figure 39.3

The germ line carries the V and C genes of the lambda family on the same chromosome; V-J recombination generates a functional lambda light-chain gene in the lymphocyte.

and CH3. Each CH exon is about 100 codons long; the hinge is shorter. The introns usually are relatively small (~300 bp). Expression of the immunoglobulin genes proceeds in the usual way by transcription of a nuclear RNA that is spliced to give an uninterrupted mRNA (which, in addition to the coding regions, contains nontranslated leader and trailer regions, coded as part of the terminal exons.)

In nonexpressing cells, the V and C probes identify sequences whose flanking regions are different from those found in the functional gene. These sequences correspond to separate V and C genes. We know that the V and C genes belonging to any one immunoglobulin family—lambda, kappa, or heavy—are carried on a single chromosome; we do not know the distance between the separate V and C genes. The genes for each of the three Ig chain families are carried on a different chromosome. The organization of the separate V and C genes is described as the **germline** pattern (it is also found in all somatic cells of lineages other than the immune system).

A lambda light chain is assembled from two parts, as illustrated in **Figure 39.3.** The V gene consists of the leader exon (L) separated by a single intron from the variable (V) segment. The C gene consists of the

J segment separated by a single intron from the constant (C) exon. The name of the **J segment** is an abbreviation for joining, since it identifies the region to which the V segment becomes connected. The J segment is short and actually codes for the last few (13) amino acids of the variable region, as defined by amino acid sequences. In the intact gene, the V-J segment therefore constitutes a single exon coding for the entire variable region. It is interesting that the C gene brings in the J segment to provide the final sequence of the V region. A V_λ gene may be joined to either of two C_λ genes, each of which has the same J-C dipartite structure. In view of this multipartite structure, we shall try to avoid confusion about the nature of the DNA sequences by referring to V, J, or C **segments** of the genome or **regions** of the polypeptide.

A kappa light chain also is assembled from two parts, but there is a difference in the structure of the C region. A group of five J segments is spread over a region of 500–700 bp, separated by an intron of 2–3 kb from the single C_κ exon. In the mouse, the central J segment is nonfunctional (ψJ3). A V_κ segment may be joined to any one of the J segments. The consequences of the kappa joining reaction are illustrated in **Figure 39.4.**

Figure 39.4
The kappa C gene is preceded by multiple J segments in the germ line; V-J joining in the lymphocyte may recognize any one of the J segments, which is then spliced to the C region during RNA processing.

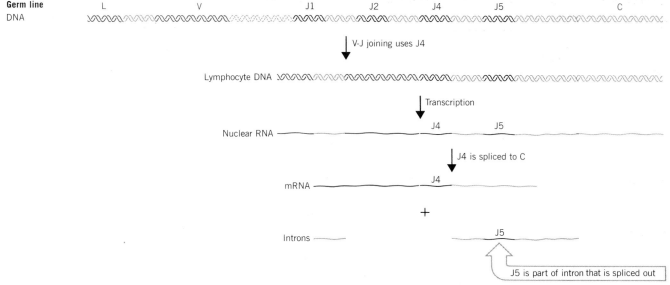

Whichever J segment is used becomes the terminal part of the intact variable exon. Any J segments on the left of the recombining J segment are lost (J1 and J2 have been lost in the figure). Any J segment on the right of the recombining J segment is treated as part of the intron between the variable and constant exons (J5 is included in the region that is spliced out in the figure). Thus all functional J segments possess a signal at the left boundary that makes it possible to recombine with the V segment; and they possess a signal at the right boundary that can be used for splicing to the C exon. Whichever J segment is recognized in DNA recombination uses its splicing signal in RNA processing.

An additional segment is involved in the construction of a heavy-chain gene, as illustrated in **Figure 39.5.** The **D** (for diversity) segment was discovered by the presence in the protein of a few (2-13) amino acids between the sequences coded by the V segment and the J segment. An array of D segments lies on the chromosome between the V_H segments and the J_H segments. A V_H segment is recombined with one of the D segments; and the D segment is recombined with one of four J_H segments (which vary in length between 4 and 6 codons). The reconstruction leads to expression of the adjacent C_H segment (which consists of several exons).

The organization of the D segments is not entirely clear, but many seem to lie in a tandem array. The mouse heavy-chain locus contains about ten D segments of variable length. Some unknown mechanism must ensure that the *same* D segment is involved in the V-D joining and D-J joining reactions.

THE DIVERSITY OF GERM-LINE INFORMATION

Now we must examine the different types of V and C genes to see how much diversity can be accounted for by the variety of the coding regions carried in the germ line. Immunoglobulin genes are most completely characterized in the mouse, and we have almost as much information about the human loci. The structures of the two light Ig gene families of the mouse genome are summarized in **Figure 39.6.**

In the mouse, almost all the light chains are of the kappa type; only about 5% are of the lambda type. In man, about 60% of the light chains are kappa and about 40% are lambda. A given lymphocyte generates either a kappa or a lambda light chain to associate with the heavy chain.

The structures of the kappa loci in mouse and man are similar. We have seen that several 13-codon long J segments precede the single C_κ gene. Many different variable sequences can be joined to the J-C locus.

Figure 39.5
Heavy genes are assembled by joining a V gene to a D segment and joining the D segment to one of the J segments preceding the C gene.

Figure 39.6
The kappa and lambda light-chain families have different complexities in the mouse.

What is the origin of the V_κ diversity? Historically, several ideas have been considered. All the V_κ sequences may be contained in the germ line. New V_κ genes could be generated by somatic recombination between the germ-line V_κ genes. These ideas turn out to be incorrect. Instead, somatic recombination involving (randomly selected) pairs of individual V and J segments creates a variety of intact coding regions; and somatic mutation may generate new V_κ segments by introducing changes into the germ-line segments.

Sequencing of the protein chains identifies groups of variable regions that are closely related, derived from one another by small changes in amino acid sequence. Each such group is called an **isotype.** Known V_κ protein sequences can be classified into about 26 isotypes; we think that ultimately there will prove to be about 50 isotypes. The divergence between isotypes is too great to be explained by somatic changes, which suggests that each isotype is coded by at least one gene in the germ line.

Probes for genes representing some of the V_κ isotypes have been used to identify the number of hybridizing fragments in the mouse germ line. The number varies from one to ten. Thus some isotypes may be coded by a single gene; others may be coded by several genes. In this sense, the isotype is a somewhat arbitrary classification, based on amino acid sequences. Its relationship with the underlying genetic structure depends on the variation between genes.

When a V_κ gene has no close relative, it identifies a single isotype. When there are several well-related V_κ genes, their proteins may fall into a single isotype.

Because of the varying degrees of divergence between germ-line genes, it is difficult to make a direct estimate of the number of V_κ genes in the germ line. An individual probe may react with several genes; when one of these genes is used as a probe, it may react with several others; and so on. This is the classic situation of a repetitive gene family, some members of which are closely related and some of which are distantly related. If we assume that there are 50 isotypes each coded by an average of 6 genes, there will be about 300 V_κ germ-line genes.

By combining any one of 300 V_κ genes with any one of 4 J segments, the mouse genome has the potential to produce some 1200 kappa chains. However, in several instances there appear to be more protein members of an isotype than can be accounted for by the number of corresponding V_κ genes. The new members are created by somatic changes in individual germ-line genes (not by recombination between V_κ genes). We shall consider the relevant mechanisms in later sections.

We do not have much information about the relative organization of the germ-line genes. Usually we assume that the V_κ genes occupy a large cluster on the chromosome, an unknown distance upstream from the constant region. It will not be surprising if the more closely related genes form subclusters, generated by duplication and divergence of individual ancestral members.

The lambda genes show an interesting difference between mouse and man. The diversity of the murine V_λ regions is enormously reduced compared with that of the human V_λ regions (which are comparable to kappa). The reason is that the mouse genome contains only two V_λ genes. We assume that at some time in the past, the mouse suffered a catastrophic deletion of most of its germ-line V_λ genes. This may explain why most of its light chains are of the kappa type.

The mouse lambda genes fall into two groups, presumably generated by an ancestral duplication. Each of the V_λ genes is linked to two constant regions, each of which comprises a J-C structure. The $V_{\lambda 1}$ gene can be linked to either J-$C_{\lambda 3}$ or J-$C_{\lambda 1}$. The V gene therefore has a choice of J segments for joining. (The difference from the kappa gene is that each J segment is associated with a different C gene instead of with the same one.) The $V_{\lambda 2}$ gene usually joins only with J-$C_{\lambda 2}$. The $C_{\lambda 4}$ region is not used, because its J segment has an aberrant structure. Occasionally $V_{\lambda 2}$ joins with J-$C_{\lambda 1}$ or J-$C_{\lambda 3}$. We do not know the physical relationship between the two lambda groups; presumably they lie near one another. In man, not only are there more V_λ genes, but at least six C_λ genes have been identified in a 25 kb cluster. (Additional diversity is generated by somatic mutation; for example, several variants of $V_{\lambda 1}$ have been found; as described later.)

The single locus for heavy-chain production consists of several discrete sections, whose structure in the mouse genome is summarized in **Figure 39.7.** There are probably about 300 V_H genes. At an unknown distance from them lies the cluster of D segments. Probably not much farther along comes the cluster of J segments. Over the next 170 kb lie all the C_H genes. By combining any one of 300 V_H genes, 10 D segments, and 4 J segments, the mouse genome potentially can produce 12,000 variable regions to accompany any C_H gene.

The eight C_H genes of the mouse all are functional; usually only a single one is expressed in any given cell. The cluster has diverged by the usual processes of duplication, divergence, and reassortment of exons by means of unequal crossing-over or gene conversion. The four gamma genes appear to be the most closely related on the basis of protein structure.

There are more C_H genes in man. The mu and delta genes remain single, the four active gamma genes are numbered 1 through 4, and there is also a gamma pseudogene; there are three epsilon genes of which at least one is a pseudogene, and there are two alpha genes.

JOINING REACTIONS GENERATE ADDITIONAL DIVERSITY

As well as the diversity made possible by different combinations of the component segments of light or heavy chains, the joining reaction itself is a source of somatic change. In the case of light chains, the junction between the V and J segments corresponds to the third hypervariable region. The first codon differs among the J segments, but generally they are well conserved, with homology at 10 of the 13 positions (allowing for silent third-base changes). The end of the J segment corresponds to the fourth framework of the variable region.

Kappa V-J joining takes place in the sequences coding for amino acids 95 and 96. If the joining reac-

Figure 39.7
A single gene cluster in the mouse contains all the information for heavy-chain gene assembly.

V_κ41 sequence   C C T C C C

Recombination Frames   1 2 3 4 5

J5 segment   T G G T G G

1   C C G T G G
     Pro   Trp

2   C C T T G G
     Pro   Trp

3   C C T C G G
     Pro   Arg

4   C C T C C G
     Pro   Pro

5   C C T C C C
     Pro   Pro

Amino acid   95   96

Figure 39.8

Use of alternative sites for recombination between aligned V and J segments creates new coding sequences in amino acids 95 and 96 of the kappa light chain. Similar events occur at the V-J and V-D-J junctions of other families.

tion is not fixed in position, but can occur between different base pairs, several different amino acids can be generated at the site of each potential V-J recombination. An example is illustrated in **Figure 39.8.** The use of five potential frames for recombination generates three different amino acids at position 96, including one (arginine) *not coded in the germ line*. Since other $V_κ$ and $J_κ$ segments have different codons at these positions, great diversity becomes possible at the point of junction. It is interesting that the amino acid at position 96 forms part of the antigen-binding site and also is involved in making contacts between the light and heavy chains. A similar conclusion applies to the junction with the J segments of the lambda light chain or the heavy chain.

Even greater diversity becomes possible if recombination is not restricted to positions in the *corresponding* reading frames, but can occur between any pair of points in the relevant regions of V and J. There is evidence for the occurrence of such events. Some delete a codon from the somatic gene. Others join the

V-J region so that the J segment is out of phase and is translated in the wrong reading frame. The resulting gene is aberrant, since its expression is terminated by a nonsense codon in the incorrect frame. We do not know the proportion of joining events that create aberrant genes; this will depend on whether there is preference for particular joining reactions. However, we may think of the formation of aberrant genes as comprising the price the cell must pay for the increased diversity that it gains by being able to adjust the site of the recombination event.

Similar although even greater diversity is generated in the joining reactions that involve the D segment of the heavy chain. The same problem remains of generating nonproductive genes by recombinations that place J and C out of phase.

A puzzle about D segments is that some heavy chains have amino acids in the corresponding region that do not correspond to any of the known D segments. Two ideas have been proposed to explain this observation. One possibility is that new D segments are generated in the somatic cell by D-D recombination. This recombination would involve a different mechanism from that involved in V-D-J joining. Another possibility is that additional nucleotides are incorporated into the V-D or D-J or both junctions during the recombination process. A speculation is that their insertion could occur via the activity of the enzyme deoxynucleoside transferase (known to be an active component of lymphocytes) at a free 3′ end generated during the joining process. At all events, the existence of new sequences identifies another mechanism for generating diversity.

RECOMBINATION OF V AND C GENES GENERATES DELETIONS AND REARRANGEMENTS

Assembly of the parts of both light- and heavy-chain genes seems to occur by the same mechanism, although the number of parts is different in each case. The employment of the same mechanism is indicated by the presence of the same consensus sequences at the boundaries of all germ-line segments that participate in joining reactions. Each consensus sequence consists of a palindromic heptamer separated by either 12 or 23 bp from a nanomer.

Figure 39.9 illustrates the relationship between the consensus sequences at the mouse Ig loci. At the kappa locus, each V_κ gene is followed by a consensus sequence with a 12 bp spacing. Each J_κ segment is preceded by a consensus sequence with a 23 bp spacing. The V and J consensus sequences are inverted in orientation. The reverse arrangement is found at the lambda locus; each V_λ gene is followed by a consensus sequence with 23 bp spacing, while each J_λ gene is preceded by a consensus of the 12 bp spacer type.

Comparing these two situations suggests a rule for the joining reaction: a consensus sequence with one type of spacing can be joined only to a consensus sequence with the other type of spacing. Since the consensus sequences at V and J segments can lie in either order, the different spacings do not impart any directional information, but serve to prevent one V gene from recombining with another or one J segment from recombining with another.

This concept is borne out by the structure of the components of the heavy genes. Each V_H gene is followed by a consensus sequence of the 23 bp spacer type. The D segments are flanked on either side by consensus sequences of the 12 bp spacer type. The J_H segments are preceded by consensus sequences of the 23 bp spacer type. Thus the V gene must be

joined to a D segment; and the D segment must be joined to a J segment. A V gene cannot be joined directly to a J segment, because both possess the same type of consensus sequence.

The spacing between the components of the consensus sequences corresponds almost to one or two turns of the double helix. This feature may reflect the use of some geometric relationship in the recombination reaction. For example, the recombination protein(s) may approach the DNA from one side in a manner reminiscent of the possible interactions with other recognition elements, such as promoters and operators (see Chapters 11 and 14).

In using the term "recombination" to describe the joining of the components of immunoglobulin genes, we do not imply that the reaction involves a reciprocal recombination between homologous sequences. It does involve a physical rearrangement of sequences, involving breakage and reunion (rather than transposition-like events), but we do not yet know the mechanism of the molecular events. Three possibilities are illustrated schematically in Figures 39.10 to 39.12.

The deletion model is illustrated in **Figure 39.10.** It supposes that one of the V genes is directly fused with one of the J segments. All the material between the two reacting sequences is excised and presumably lost. The reacting sequences could be brought

Figure 39.9
Consensus sequences are present in inverted orientation at each pair of recombining sites. One member of each pair has a spacing of 12 bp between its components; the other has 23 bp spacing.

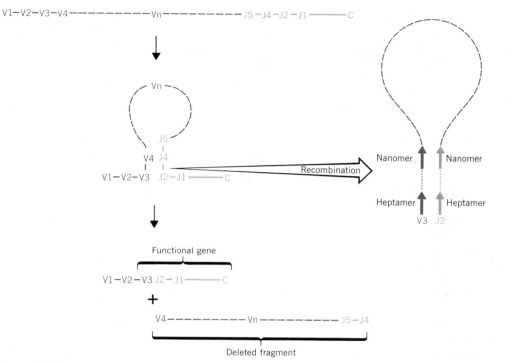

Figure 39.10
The consensus sequences could be used to bring the V and J regions into juxtaposition so that a breakage and reunion excises the material between them.

together by pairing between the inverted consensus sequences (in terms of the separated strands of duplex DNA, this would generate a cruciform). Note that a reciprocal recombination within the consensus sequences would invert the region between V and J instead of deleting it; so if pairing of the consensus sequences leads to a deletion, we must be dealing with a different type of breakage and reunion. One possibility is that the loop created by the pairing could be removed by cleavage at the base, with the ends of the V and J regions subsequently ligated together.

Several inversion models are possible, and one form is illustrated in **Figure 39.11.** We have supposed so far that all the V genes lie in the same orientation as the J segments, although we do not yet have information linking the V-coding regions to the J segments. If a V gene were present in reverse orientation, a recombination reaction could generate a functional gene by inverting the intervening material to fuse V to J. If

some V genes are in the same orientation as J but others are inverted, some V-J joinings might be accompanied by deletions while others are accompanied by inversions.

Figure 39.11
A joining reaction involving a V gene in inverse orientation to the J segment could generate a functional V-J-C gene by inversion of the region between V and J.

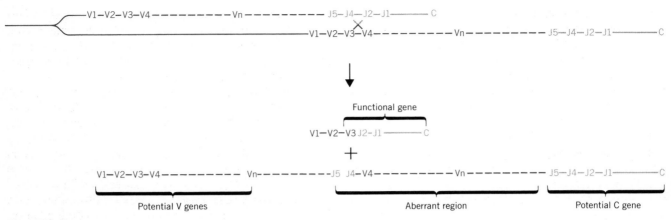

Figure 39.12
Unequal crossing-over between sister chromatids could generate one recombinant with a functional gene and a deletion of material between V and J, while the other recombinant has a duplication of the region between V and J.

The unequal crossing-over model is illustrated in **Figure 39.12.** It involves an unequal recombination between sister chromatids (replicated copies) of the gene. (The model begs the question of what role the consensus sequences may play.) The recombination event produces one chromatid that has a functional gene, accompanied by deletion of the material between the recombining V and J segments. The other chromatid has a duplication of the region between V and J; on either side of the aberrant duplicated region are V and C genes that may remain available for further recombinations. The two chromatids segregate to daughter cells at mitosis.

The structure of the active gene is the same in both the deletion and unequal crossing-over model. The models can be distinguished by the fate of the progeny of the cell in which the event occurred. In the deletion model, all descendants have the deletion. In the unequal crossing-over model, half of the descendants have the deletion, but the other half have the potentially reactive duplication. Pedigree analysis of cells in which the recombination reaction is occurring should be able to resolve the actual situation. The inversion model is readily distinguished from both of these models by the retention in the inverted segment of the material between the V and C genes.

SOME POSSIBLE CAUSES OF ALLELIC EXCLUSION

Each V-J or V-D-J recombination is a unique event. A single productive rearrangement of each type occurs in a given lymphocyte, to produce one light and one heavy chain gene. Because each event involves the genes of only *one* of the homologous chromosomes, the alleles on the other chromosome are not expressed in the same cell. This phenomenon is called *allelic exclusion.*

The occurrence of allelic exclusion introduces a complication in analyzing the process of somatic recombination. A probe reacting with a region that has rearranged on one homologue will also detect the allelic sequences on the other homologue. We may therefore be compelled to analyze the different fates of the two chromosomes together. However, the usual pattern displayed by a rearranged active gene can be interpreted in terms of a deletion of the material between the recombining V and C loci.

Sometimes an active cell displays *both* the rearranged and germ-line patterns of organization for a given gene. We assume then that joining has occurred on one chromosome, while the other chromosome has remained unaltered. The rearrangement could have occurred by either deletion or unequal crossing-over.

In some lines, both chromosomes have suffered independent rearrangements. In some of these instances, material between the recombining V and C genes is entirely absent from the cell line. This is most easily explained by the occurrence of independent deletions on each chromosome, but could be accommodated by the unequal crossing-over model.

When two chromosomes both lack the germ-line pattern, only one of them has passed through a **productive rearrangement** to generate a functional gene. The other has suffered a **nonproductive rearrangement;** this may take several forms, but in each case the gene sequence is aberrant and cannot be expressed as an immunoglobulin chain. If an out-of-frame rearrangement has occurred, the structure of the aberrant gene may be superficially normal. Two types of gross misarrangement are seen. Incomplete rearrangements represent a partial joining reaction; for example, D has joined to J, but V has not been connected. Abnormal rearrangements result from joining reactions in which one of the recombining sites is not an authentic consensus sequence; for example, in one case a V_κ gene is joined to a point in the intron between the J_κ group and the C_κ gene, creating a pseudogene sequence that lacks a J segment.

Nonproductive rearrangements are common in myelomas and plasmacytomas. The assumption that their presence is not an aberration of the tumor cells, but reflects the usual pattern of rearrangement, suggests a stochastic model for rearrangements. Each cell starts with two loci of the unrearranged germ-line configuration Ig^0. Either of these loci may be rearranged to generate a productive gene Ig^+ or a nonproductive gene Ig^-. If the rearrangement is productive, the synthesis of an active chain provides a trigger to prevent rearrangement of the other allele. The active cell has the configuration Ig^0/Ig^+.

If the rearrangement is nonproductive, it creates a cell with the configuration Ig^0/Ig^-. There is no impediment to rearrangement of the remaining germ-line allele. If this rearrangement is productive, the expressing cell has the configuration Ig^+/Ig^-. Again, the presence of an active chain suppresses the possibility of further rearrangements.

Two successive nonproductive rearrangements produce the cell Ig^-/Ig^-. (Indeed, one prediction of the stochastic model is that we should be able to find such cells.) However, it is possible that in some cases an Ig^-/Ig^- cell can try yet again. In some cases, the observed patterns of DNA can only have been generated by successive rearrangements.

If one of the Ig^- alleles has a gene aberrantly joined to an early J segment, it might be possible for one of the remaining upstream V genes to "hop over" by joining one of the remaining downstream J segments. If joining occurs by unequal crossing-over, an Ig^- locus generated by the aberrant duplication may still be able to support joining of V and C genes that lie on either side of the duplication. Some unusual structures found in nonproductively rearranged loci could be explained by this model. They could also be explained by an alternating series of intrachromosomal deletions and inversions.

The crux of the model is that the cell keeps trying to recombine V and C genes until a productive rearrangement is achieved. Allelic exclusion is caused by the suppression of further rearrangement as soon as an active chain is produced. This feedback is independent for the heavy- and light-chain loci (heavy-chain genes usually rearrange first), but in the case of the light chains must apply equally to both families (cells may have either active kappa or lambda light chains). It is likely that the cell rearranges its kappa genes first, and only tries to rearrange lambda if both kappa attempts are unsuccessful.

What is the connection between joining of V and C genes and their activation? Unrearranged V_κ genes are not represented in RNA. Whenever a V_κ gene is joined productively to a C_κ gene, the resulting unit is transcribed. However, since the sequence upstream from a V gene is not altered by the joining reaction, the promoter must be the same in unrearranged, nonproductively rearranged, and productively rearranged genes.

Does this mean that activation of a V promoter depends on some effect mediated by the sequences downstream? If the nonexpressed C_κ allele remains in the germ-line configuration, it may be transcribed into an unprocessed 8.4 kb RNA that contains upstream sequences. A possible interpretation is that the V_κ promoter is activated by its relocation to the C_κ region. Most gross nonproductive rearrangements are not represented in RNA; we do not understand the reason—possibly some necessary site is lost during the rearrangement. In a few cases, an aberrant RNA is produced.

FURTHER DNA RECOMBINATION CAUSES CLASS SWITCHING

The class of immunoglobulin is defined by the type of C_H region. IgM has the C_μ sequence, IgD has the C_δ sequence, IgG has one of the C_γ sequences, IgE has the C_ϵ sequence, and IgA has the C_α sequence. A lymphocyte generally produces only a single class of immunoglobulin at any one time, but the class may change during the cell lineage. This is accomplished by a substitution in the type of C_H region that is expressed; the change in expression is called **class switching.** It involves *only* the C_H gene; the same V_H gene continues to be expressed. Thus a given V_H gene may be expressed successively in combination with more than one C_H gene. The same light chain continues to be expressed throughout the lineage of the cell. Class switching therefore allows the type of effector response (mediated by the C_H region) to change, while maintaining a constant antigen-recognition facility (mediated by the V regions).

All lymphocytes start productive life as immature cells engaged in synthesis of IgM. Cells expressing IgM have the germ line arrangement of C_H genes shown in Figure 39.7. The V-D-J joining reaction is sufficient to trigger expression of the C_H gene.

Changes in the expression of C_H genes are made in two ways. Some occur at the level of RNA processing (see next section). The majority occur via further DNA recombinations, involving a system different from that concerned with V-D-J joining (and able to operate only at a subsequent point in time).

Cells expressing later C_H genes usually have deletions of C_μ and the other genes preceding the expressed C_H gene. Thus class switching may be accomplished by a recombination to bring a new C_H gene into juxtaposition with the expressed V-D-J unit. The sequence of the switched V-D-J-C_H units shows that the sites of switching lie upstream from the C_H genes themselves. The switching sites are called **S regions. Figure 39.13** shows a (hypothetical) example of two successive switches.

In the first switch, expression of C_μ is succeeded by expression of $C_{\gamma 1}$. The $C_{\gamma 1}$ gene is brought into the expressed position by recombination between the sites S_μ and $S_{\gamma 1}$, deleting the material between. The S_μ site lies between V-D-J and the C_μ gene. The $S_{\gamma 1}$ site lies upstream from the $C_{\gamma 1}$ gene. The region between V-

Figure 39.13
Class switching of heavy genes may occur by recombination between switch regions (S), deleting the material between the recombining S sites. Successive switches may occur.

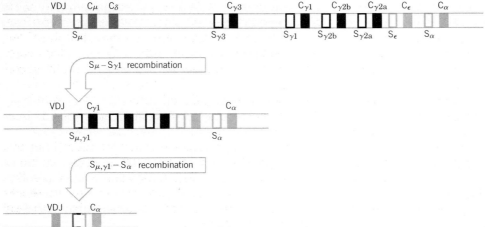

D-J and the $C_{\gamma 1}$ gene is removed as an intron during processing of the RNA.

The linear deletion model imposes a restriction on the heavy-gene locus: once a class switch has been made, it becomes impossible to express any C_H gene that used to reside between C_μ and the new C_H gene. In the example of Figure 39.13, cells expressing $C_{\gamma 1}$ should be unable to give rise to cells expressing $C_{\gamma 3}$, which has been deleted.

However, it should in principle be possible to undertake another switch to any C_H gene downstream from the expressed gene. The figure shows a second switch to C_α expression, accomplished by recombination between S_α and the switch region $S_{\mu,\gamma 1}$ generated by the original switch.

We assume that all of the C_H genes have S regions upstream from the coding sequences. We do not know whether there are any restrictions on the use of S regions. Sequential switches do occur, but we do not know whether they are optional or an obligatory means to proceed to later C_H genes. We should like to know whether IgM can switch directly to *any* other class.

In addition to the deletion of material upstream from an expressed C_H gene, a cell that has undergone a switch may also have rearrangements downstream. (These could affect the ability to support subsequent switches.) One model is to suppose that the first switch between S_μ and another S region activates the recombination system, which then may cause rearrangements as well as further switches.

We know that switch sites are not uniquely defined, because different cells expressing the same C_H gene prove to have recombined at different points. When enough of these S sites have been sequenced, we may be able to define the limits of the regions within which switching can occur. In the meantime, we can scrutinize the sequences in the area for common features that might be involved.

The S regions lie about 2 kb upstream from the C_H genes. Three distinct S_μ sites have been characterized; two of them show homology, but the other does not. There are homologies with some of the S regions of other genes, but not all. Tandem repeats of short sequences are present in several S regions, but we have yet to discern their significance.

As with the joining of V and C genes, we know that deletions occur during C_H switching; but this does not prove that the underlying mechanism is an intramolecular deletion; the reaction could occur by sister chromatid exchanges. A sister chromatid exchange appears to have occurred in one case, in which a switch occurred from C_α expression to $C_{\gamma 1}$ expression. This is not permitted by the linear deletion model. On the other hand, this could have been an illicit retrieval of information from the sister chromatid. Our problem in interpreting data obtained with myelomas is some uncertainty whether the observed switching reflects the regular pathway (most tumor cells do not switch at all). Thus we will need to characterize the switching process in authentic lymphocytes.

EARLY HEAVY-CHAIN EXPRESSION CAN BE CHANGED BY RNA PROCESSING

The period of IgM synthesis that begins lymphocyte development falls into two parts, during which different versions of the mu constant region are synthesized. As a stem cell differentiates to a pre-B lymphocyte, an accompanying light chain is synthesized, and the IgM molecule ($\mu_2 L_2$) appears at the surface of the cell. This form of IgM contains the μ_m version of the constant region (*m* indicates that IgM is located in the membrane). When the B lymphocyte differentiates further into a plasma cell, the μ_s version of the constant region is expressed. (The IgM actually is secreted as a pentamer IgM$_5$J, in which J is a joining polypeptide that forms disulfide linkages with mu chains.)

The μ_m and μ_s versions of the mu heavy chain are identical for most of their length. However, the μ_m chain is longer than the μ_s chain; the difference results from a change in the amino acid sequence at the C-terminal end. The μ_m chain ends in a 41-residue hydrophobic sequence that may be responsible for securing its place in the membrane. This sequence is replaced by a 20-residue hydrophilic sequence in μ_s; the substitution may allow the mu heavy chain to pass through the membrane.

The μ_m and μ_s chains are coded by different mRNAs. The two mRNAs share an identical sequence up to the end of the last constant domain. Then they differ; μ_m has 41 more codons followed by a nontranslated trailer; and μ_s has 20 codons followed by a different nontranslated trailer.

The genome sequence shows that the terminal regions of μ_m and μ_s are coded in different exons. The relationship between the structure of the gene and the mRNAs is illustrated in **Figure 39.14.**

At the membrane-bound stage, the constant region is produced by splicing together six exons. The first four code for the four domains of the constant region. The last two, M1 and M2, code for the 41-residue hydrophobic C-terminal region and its nontranslated trailer.

At the secreted stage, the constant region is generated by joining only the first four exons. The last of these exons extends farther than it did at the previous (membrane) stage; it brings in the last 20 codons and its nontranslated trailer.

The difference between the two mRNAs therefore hinges on whether a splicing junction *within* the exon for the last constant domain is spliced to M1 and M2 (shortening the exon and generating the coding sequence for μ_m) or is ignored (lengthening the exon and generating the coding sequence for μ_s). How is the use of this splicing site controlled?

A plausible model is illustrated in Figure 39.14. At the membrane-bound stage, the nuclear RNA has a polyadenylated end after M2 (probably by virtue of cleavage at this site, possibly by termination). Because the nuclear RNA contains the splicing site at the beginning of M1, it uses the splicing site within the last constant exon. At the secreted stage, the nuclear RNA ends at an earlier site, after the last constant

Figure 39.14

The site of termination or cleavage and polyadenylation may control the use of splicing junctions so that alternate forms of the heavy gene are expressed.

Exons

Intron in μn; exon in μ_s

Intron

exon. Because no subsequent exons are present in the nuclear RNA, the splicing site within the exon cannot be utilized. By placing the onus for regulation on the choice of a site for cleavage and polyadenylation (or for termination), we avoid the dilemma of needing to choose between splicing alternatives for a single RNA. To confirm this model, we will need to characterize the nuclear RNA.

A similar transition from membrane to secreted forms is found with other constant regions. The conservation of exon structures suggests that the mechanism is the same. The M1 and M2 exons of three gamma chains are closely related, coding for a sequence of 66 amino acids, the first 41 of which show homology with the mu membrane sequence. The conserved sequence may provide a signal recognized by a receptor protein that anchors the m form of the immunoglobulin in the membrane. Perhaps the anchorage is related to the need to initiate cell proliferation in response to the initial recognition of an antigen.

An exception to the rule that only one immunoglobulin type is synthesized by any one cell is presented by the simultaneous production of IgM and IgD in mature B lymphocytes. The two immunoglobulins are identical except for the substitution between mu and delta constant regions in the heavy chain. It seems likely that this is the outcome of alternative pathways for RNA processing. The delta constant region is close to mu; if transcription sometimes continues through the region, the VDJ exon could be spliced to the series of delta exons. This is reminiscent of the situation involved in late adenovirus expression, and we may invoke the same type of model illustrated previously in Figure 20.20; again, the generation of different polyadenylated ends for the nuclear RNA controls the selection of coding regions for splicing to the initial exon. (One case in which choice of polydenylation sites may not be adequate is the switch from δ_m to δ_s, where direct choice of different splicing sites may be involved.)

An even more striking extension of the use of splicing to express constant regions is provided by a murine leukemic cell line that appears to switch from μ to γ2b synthesis while retaining the mu gene. A possibility is that transcription continues all the way through the γ2b region, followed by splicing of the very long (~50 kb) transcript. (However, the γ2b gene could

have been brought closer to mu, even if not right up to VDJ.) We do not know whether this type of event represents a respectable alternative to DNA recombination or is an aberration of the particular cell in which it was detected.

SOMATIC MUTATION GENERATES ADDITIONAL DIVERSITY

Comparisons between the sequences of expressed immunoglobulin genes and the corresponding V genes of the germ line show that new sequences appear in the expressed population. We have seen that some of this additional diversity results from shifts in the site of recombination during assembly of the V-J light-chain exons or the V-D-J heavy-chain exons. However, some changes occur at preceding locations within the variable domain; these represent **somatic mutations** induced specifically in the active lymphocyte.

With kappa light chains or heavy chains, many potential germ-line V genes could be used to construct the variable domain. A probe representing an expressed V gene can be used to identify all the corresponding fragments in the germ line. Their sequences should identify the complete repertoire available to the organism. Any expressed gene whose sequence is different must have been generated by somatic changes. The main experimental problem in this analysis lies in ensuring that every potential contributor in the germ-line V genes actually has been identified. This problem is overcome by the fortuitous simplicity of the mouse lambda chain system. A survey of several myelomas producing λ_1 chains showed that many have the sequence of the single germ-line gene. But others have new sequences that must have been generated by mutation of the germ-line gene.

To determine the frequency of somatic mutation in other cases, we need to examine a large number of cells in which the same V gene is expressed. A practical procedure for identifying such a group is to characterize the immunoglobulins of a series of cells, all of which express an immune response to a particular antigen. (Antigens used for this purpose are small molecules whose discrete structure is likely to provoke a consistent response, unlike a large protein, parts of which may provoke potentially different antibodies. The small molecules are called haptens and are conjugated with a nonreactive protein to form the antigen. The cells are obtained by immunizing mice with the antigen, obtaining the reactive lymphocytes, and sometimes fusing these lymphocytes with a myeloma cell to generate a **hybridoma** that continues to express the desired antibody indefinitely.)

A survey of 19 different cell lines producing antibody directed against the hapten phosphorylcholine showed that ten have the same V_H sequence (expressed in conjunction with one of the mu, gamma, or alpha constant regions). The V_H sequence can be identified in the germ line as T15, one of four V_H genes hybridizing with a probe for the expressed V_H sequence. The other nine expressed genes differ from each other and from all four germ-line members of the family. They are more closely related to the T15 germ-line sequence than to any of the others, and their flanking sequences are the same as those around T15. This suggests that they have arisen from the T15 member by somatic mutation.

The sequences of these expressed genes vary from the germ-line sequence in all regions of the variable domain, including frameworks as well as hypervariable regions. Sequence changes are found in the downstream flanking regions as well as in the coding regions; they do not seem to extend far upstream. All the variation is due to substitutions of individual nucleotide pairs. The variation is different in each case. One gene shows 3.8% divergence from the germ line; another shows 1.4% divergence. This translates to 8 and 3 amino acid substitutions, respectively. Only a minority of the mutations affect the amino acid sequence, since many lie in third-base coding positions as well as in nontranslated regions.

The large proportion of ineffectual mutations suggests that somatic mutation occurs more or less at random in a region including the V gene and extending beyond it. In the heavy-chain genes, class switching may provide the trigger for mutation. The V_H genes associated with mu constant chains are rarely mutated; the (same) V_H genes associated with gamma or alpha constant chains may be mutated.

Random mutations have unpredictable effects on protein function; some may inactivate the protein, others may confer high specificity for a particular antigen. Thus a critical feature of the process may be selection among the lymphocyte population for those cells bearing antibodies in which chance mutation has created a suitable V domain to bind whatever antigen is at hand.

This does not entirely solve our dilemma about the generation of antibody diversity. Consider the large number of V genes, not to mention the additional diversity generated by alternative joinings to D and J segments. We see in the case of phosphorylcholine that in many separate antigenic provocations, the same V_H gene is used to generate the antibody. For this particular V_H gene always to be available, in either original or mutated sequence, an immense range of lymphocytes must have suffered rearrangements while waiting for an antigen. Does this mean that so many rearrangements occur that every potential V gene actually is available in expressed form?

COMPLEXITY OF MAJOR HISTOCOMPATIBILITY LOCI

The major histocompatibility locus occupies a small segment of a single chromosome in the mouse (where it is called the H2 locus) and in man (called the HLA locus). Within this segment are many genes coding for functions concerned with the immune response. At those individual gene loci whose products have been identified, many alleles have been found in the population; the locus is described as highly **polymorphic,** meaning that individual genomes are likely to be different from one another.

The several types of function identified in the H2 and the adjacent region are summarized on the map of **Figure 39.15.** The classic H2 region occupies 0.3 map units and includes three classes of genes. The adjacent region extends for another map unit; within it are genes related to those present in the H2 region. In molecular terms, this "small segment" of the chromosome is sizable; the 1.3 map units together potentially represent > 2500 kb of DNA.

Class I genes code for **transplantation antigens,** proteins present on every cell of the mammal. As their name suggests, these proteins are responsible for the rejection of foreign tissue, which is recognized as such by virtue of its particular array of transplantation antigens. In the immune system, their presence on certain

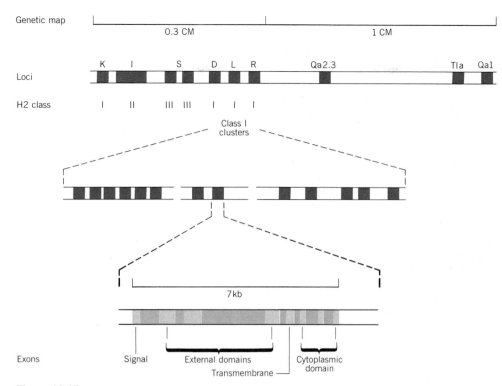

Figure 39.15
The histocompatibility locus contains several genetic regions; class I genes are organized in large structures; the structure of the individual gene corresponds with the structure of the protein.

(cytotoxic) T lymphocytes plays a role in killing target cells. The K, D, L, and R regions all code for proteins of this type; these functions are defined serologically (by their antigenic properties). Each mouse strain has one of several possible alleles for each of these functions.

The class II genes code for proteins found on the surfaces of both B and T lymphocytes as well as macrophages. These proteins are involved in the communication between cells that is necessary to execute the immune response. The I region of the class II genes is divided into subregions A, B, J, E, and C (not shown on the map).

The class III genes code for **complement proteins.** This part of the locus is also known as the S region; S stands for serum, indicating that the proteins are components of the serum. Their role is to interact with antibody-antigen complexes to cause the lysis of cells.

To the right of the H2 region lie the Qa and Tla loci,

which code for proteins found on hematopoietic cells. The proteins are described as differentiation antigens, because each is found only on a particular subset of the blood cells, presumably related to their function. Since these proteins are structurally related to the class I H2 proteins, we may regard these genes as extending the region of the chromosome devoted to functions concerned with the development of lymphocytes and macrophages.

The transplantation antigens are transmembrane proteins consisting of two chains. One chain is the 12,000-dalton β2 microglobulin (coded by a single gene located on another chromosome). This component is needed for the dimeric protein to reside at the cell surface. The transmembrane component is a 45,000-dalton chain coded at the histocompatibility locus. The protein has three **external domains** (each about 90 amino acids long, one of which interacts with β2 microglobulin), a **transmembrane region** of about 40 residues, and a short **cytoplasmic domain** of about 30 residues that resides within the cell.

The organization of class I genes coincides with the protein structure. The first exon codes for a signal sequence (cleaved from the protein during membrane passage). The next three exons code for each of the external domains. The fifth exon codes for the transmembrane domain. And the last three rather small exons together code for the cytoplasmic domain. The only difference from this structure in a gene for a human transplantation antigen is that the cytoplasmic domain is coded by only two exons.

The class I mouse genes reside in clusters. A total of 36 genes lying over some 850 kb of DNA have been identified so far (in the form of individual clusters that have yet to be connected). The largest individual group has 7 genes spread over 190 kb. All of the genes in each cluster are oriented in the same direction; adjacent genes tend to be more closely related, which suggests their origin by ancestral tandem duplications.

Differences in the restriction patterns of the DNA of different strains of mice can be used to relate the locations of the individual clusters with the genetic loci. (Essentially this involves using the restriction differences as markers in genetic mapping.) By this means, the cluster of 7 genes has been located in the *Qa2,3* region; other genes have been identified in the *Tla* region; and an H2-*L* allele has been identified. The relationship between these genes suggests that *Qa* and *Tla* can be considered to comprise part of the class I group together with the H2 loci. This implies that class I genes are present at both ends of the vast cluster; the class II and class III genes either evolved within the cluster or were translocated to their present positions. Pseudogenes are present as well as functional genes; at present we have some way to go before estimating the total number of active genes in the region.

The exon coding for the third external domain of the class I genes is highly conserved relative to the other exons. The conserved domain probably represents the region that interacts with β2 microglobulin, which explains the need for constancy of structure. This domain also exhibits homologies with the constant region domains of immunoglobulins.

The gene for β2 microglobulin itself has been identified. It has four exons, the first coding for a signal sequence, the second for the bulk of the protein (from amino acids 3 to 95), the third for the last four amino acids and some of the nontranslated trailer, and the last for the rest of the trailer. The length of β2 microglobulin is similar to that of an immunoglobulin gene; there are certain similarities in amino acid constitution; and there are some (limited) homologies of nucleotide sequence between β2 microglobulin and Ig constant domains or type I gene third external domains. It is possible that all three groups of genes are descended from a common ancestor that coded for a primitive domain.

The structures of the immunoglobulin and histocompatibility loci highlight the distance we have come from the concept of the gene as an isolated and fixed locus coding for a single protein. Genes may reside in huge clusters containing related sequences; they may be expressed in terms of alternate versions of a protein; they may be reconstructed by physical rearrangement of sequences during somatic development. The shibboleths of genetics have been tumbled during the past decade.

FURTHER READING

The initial characterizations of Ig gene structure are reviewed in **Lewin's** *Gene Expression, 2, Eucaryotic Chromosomes* (Wiley, New York,1980, pp. 847–861). For further work it is necessary to resort to research papers, ameliorated by some short reviews in *Cell*. A deletion model for V-J joining was presented by **Max, Seidman,** & **Leder** (*Proc. Nat. Acad. Sci. USA* **76**, 3450–3454, 1979); a model based on sister chromatid exchange was proposed by **Van Ness et al.** (*Proc. Nat. Acad. Sci. USA* **79**, 262–266, 1982). The structure of the heavy chain genes was elucidated by **Sakano et al.** (*Nature* **286**, 676–683, 1980); the role of consensus sequences in V-(D)-J joining was analyzed by **Sakano et al.** (*Nature* **286**, 676–683, 1980) and **Early et al.** (*Cell* **19**, 981–992, 1980). The nature of rearrangements and allelic exclusion has been reviewed by **Early** & **Hood** (*Cell* **24**, 1–3, 1981); aberrant joining was reported by **Max et al.** (*Cell* **21**, 793–799, 1980); the connection with transcription was explored by **Perry et al.** (*Proc. Nat. Acad. Sci. USA* **77**, 1937–1941, 1980). Class switching was reviewed by **Davis, Kim,** & **Hood** (*Cell* **22**, 1–2, 1980); the linear deletion model was originally proposed by **Honjo** & **Kataoka** (*Proc. Nat. Acad. Sci. USA* **75,** 2140–2144, 1978). Alternate processing of RNA was reported by **Early et al.** (*Cell* **20,** 313–319, 1980), **Maki et al.** (*Cell* **24,** 353–365, 1981), and **Kataoka** et al. (*Proc. Nat. Acad. Sci. USA* **79,** 2623–2627, 1982). Somatic mutation has been reviewed by **Baltimore** (*Cell* **26,** 295–296, 1981). The organization of histocompatibility genes was summarized by **Hood et al.** (*Cell* **28,** 685–687, 1982).

GLOSSARY

Abundance of an mRNA is the average number of molecules per cell.

Abundant mRNAs consist of a small number of individual species, each present in a large number of copies per cell.

Acceptor splicing site is the junction between the right end of an intron and the left end of an exon.

Acentric fragment of a chromosome (generated by breakage) lacks a centromere and is lost at cell division.

Allele is one of several alternate forms of a gene occupying a given locus on the chromosome.

Allelic exclusion describes the expression in any particular lymphocyte of only one of the alleles coding for the expressed immunoglobulin.

Allosteric control refers to the ability of an interaction at one site of a protein to influence the activity of another site.

Alu family is a set of dispersed and related sequences, each ~300 bp long, in the human genome.

Alu-equivalent family is a set of sequences in any mammalian genome, related to the human Alu family.

α-Amanitin is a bicyclic octapeptide derived from the mushroom *Amanita phalloides* that inhibits transcription by certain eucaryotic RNA polymerases, especially RNA polymerase II.

Amber codon is the nucleotide triplet UAG, one of three *nonsense* codons that cause termination of protein synthesis.

Amber mutation describes any change in DNA that creates an amber codon at a site previously occupied by a codon representing an amino acid in some protein.

Amber suppressors are mutant genes that code for tRNAs whose anticodons have been altered so that they can respond to UAG codons as well as or instead of to their previous codons.

Aminoacyl-tRNA synthetases are enzymes responsible for covalently linking amino acids to the 2'- or 3'-OH position of tRNA.

Amplification refers to the production of additional copies of a chromosomal sequence, found as either intrachromosomal or extrachromosomal DNA.

Anticoding strand of duplex DNA is used as template to direct the synthesis of RNA and is complementary to it.

Anticodon is a triplet of nucleotides in a constant position in the structure of tRNA that is complementary to the codon(s) in mRNA to which the tRNA responds.

Antigen is any molecule whose entry into an organism provokes synthesis of an antibody (immunoglobulin).

Antitermination proteins allow RNA polymerase to transcribe through certain terminator sites.

AP endonucleases make incisions in DNA on the 5' side of either apurinic or apyrimidinic sites.

Apoinducer is protein that binds to DNA to switch on transcription by RNA polymerase.

671

att sites are the loci on a phage and the bacterial chromosome whose recombination integrates the phage into, or excises it from, the bacterial chromosome.

Attenuation is the regulation of termination of transcription involved in controlling expression of some bacterial operons.

Attenuator is the terminator sequence at which attenuation occurs.

Autogenous control describes the action of a gene product that either inhibits (negative autogenous control) or activates (positive autogenous control) expression of the gene coding for it.

Autoradiography detects radioactively labeled molecules by their effect in creating an image on photographic film.

Autosomes are all chromosomes except the sex chromosomes; a diploid cell has two copies of each autosome.

Backcross involves crossing an unknown genotype to a recessive homozygote so that the phenotypes of the progeny correspond directly to the chromosomes carried by the parent of unknown genotype.

Back mutation reverses the effect of a mutation that had inactivated a gene; thus it restores wild phenotype.

Bacteriophages are viruses that infect bacteria; often abbreviated as *phages*.

Balbiani ring is an extremely large puff at a band of a polytene chromosome.

Bands of *polytene chromosomes* are visible as dense regions that contain the majority of DNA; *bands of normal chromosomes* are relatively much larger and are produced as regions that retain a stain on certain chemical treatments.

Base pair (bp) is a partnership of A with T or of C with G in a DNA double helix; other pairs are possible in RNA under some circumstances.

Bidirectional replication is accomplished when two replication forks move away from the same origin in opposite directions.

Bivalent is the structure containing all four chromatids (two representing each homologue) at the start of meiosis.

Blunt-end ligation is a reaction that joins two DNA duplex molecules directly at their ends.

Branch migration describes the ability of a DNA strand partially paired with its complement in a duplex to extend its pairing by displacing the resident strand with which it is homologous.

Breakage and reunion is a mode of genetic recombination in which two DNA duplex molecules are broken at corresponding points and then rejoined crosswise.

C banding is a technique for generating stained regions around centromeres.

C genes code for the constant regions of immunoglobulin protein chains.

C value is the total amount of DNA in a haploid genome.

CAAT box is part of a conserved sequence at about 75 bp upstream from the startpoint of eucaryotic transcription units; it may be involved in binding RNA polymerase II.

Cap is the structure at the 5′ end of eucaryotic mRNA introduced after transcription by linking the 5′ end of a guanine nucleotide to the terminal base of the mRNA and methylating at least the additional G; the structure is $^{7Me}G5′ppp5′Np.\ldots$

CAP (CRP) is a positive regulator protein activated by cyclic AMP, needed for RNA polymerase to initiate transcription of certain (catabolite-sensitive) operons of *E. coli*.

Capsid is the external protein coat of a virus particle.

Catabolite repression describes the decreased expression of many bacterial operons that results from addition of glucose; caused by decrease in intracellular level of cyclic AMP that inactivates CAP regulator.

cDNA is a single-stranded DNA complementary to an RNA, synthesized from it by *in vitro* reverse transcription.

cDNA clone is a duplex DNA sequence representing an RNA, carried in a cloning vector.

Cell cycle is the period from one division to the next.

Centromere is a constricted region of a chromosome that includes the site of attachment to the mitotic or meiotic spindle.

Chemical complexity is the amount of a DNA component measured by chemical assay.

Chi sequence is an octameric sequence that provides a hotspot for RecA-mediated genetic recombination in *E. coli*.

Chi structure is a joint between two duplex molecules of DNA revealed by cleaving an intermediate of two joined circles to generate linear ends in each circle.

Chiasma (*pl.* chiasmata) is a site at which two homologous chromosomes appear to have exchanged material during meiosis.

Chromatids are the copies of a chromosome produced by replication.

Chromatin is the complex of DNA and protein in the nucleus of the interphase cell, originally recognized by its reaction with stains specific for DNA.

Chromocenter is an aggregate of heterochromatin from different chromosomes.

Chromomere is a densely staining granule visible as an apparent component of the chromosome under certain conditions, especially early in meiosis.

Chromosome is a discrete unit of the genome carrying many genes, consisting of proteins and a very long molecule of DNA, visible as a morphological entity only during the act of cell division.

Chromosome walking describes the sequential isolation of clones carrying overlapping sequences of DNA to span large regions of the chromosome (often in order to reach a particular locus).

***Cis*-acting** locus affects the activity only of DNA sequences on its own molecule of DNA; usually implies that the locus does not code for protein.

***Cis*-acting** protein has the exceptional property of acting only on the molecule of DNA from which it was expressed.

***Cis*-configuration** describes two sites on the same molecule of DNA.

***Cis/trans* test** assays the effect of relative configuration on expression of two mutations. In a double heterozygote, two mutations in the same gene show mutant phenotype in *trans* configuration, wild-type in *cis* configuration.

Cistron is the genetic unit defined by the *cis/trans* test; equivalent to *gene* in comprising a unit of DNA representing a protein.

Class switching is a change in the expression of the C region of an immunoglobulin chain during lymphocyte differentiation.

Clone describes a large number of cells or molecules identical with a single ancestral cell or molecule.

Closed reading frame contains termination codons that prevent its translation into protein.

Coconversion is the simultaneous correction of two sites during gene conversion.

Coding strand of DNA has the same sequence as mRNA.

Codominant alleles both contribute to the phenotype; neither is dominant over the other.

Codon is a triplet of nucleotides that represents an amino acid or a termination signal.

Coevolution —*see* concerted evolution.

Cognate tRNAs are those recognized by a particular aminoacyl-tRNA synthetase.

Coincidental evolution —*see* concerted evolution.

Cointegrate structure is produced by fusion of two replicons, one originally possessing a transposon, the other lacking it; the cointegrate has copies of the transposon present at both junctions of the replicons, oriented as direct repeats.

Cold-sensitive mutant is defective at low temperature but functional at normal temperature.

Colony hybridization is a technique for using *in situ* hybridization to identify bacteria carrying chimeric vectors whose inserted DNA is homologous with some particular sequence.

Compatibility group of plasmids contains members unable to coexist in same bacterial cell.

Complementation refers to the ability of independent (non-allelic) genes to provide diffusible products that produce wild phenotype when two mutants are tested in *trans* configuration in a heterozygote.

In vitro **complementation assay** consists of identifying a component of a wild-type cell that can confer activity on an extract prepared from some mutant cell; identifies the component rendered inactive by the mutation.

Complementation group is a series of mutations unable to complement when tested in pairwise combinations in *trans*; defines a genetic unit (the cistron) that might better be called a *noncomplementation group*.

Complex locus (of *D. melanogaster*) has genetic properties inconsistent with the function of a gene representing a single protein.

Complexity is the total length of different sequences of DNA present in a given preparation.

Composite transposons have a central region flanked on each side by insertion sequences, either or both of which may enable the entire element to transpose.

Concatemer of DNA consists of a series of unit genomes repeated in tandem.

Concatenated circles of DNA are interlocked like rings on a chain.

Concerted evolution describes the ability of two related genes to evolve together as though constituting a single locus.

Conditional lethal mutations kill a cell or virus under certain (nonpermissive) conditions, but allow it to survive under other (permissive) conditions.

Consensus sequence is an idealized sequence in which each position represents the base most often found when many actual sequences are compared.

Conservative recombination involves breakage and re-union of preexisting strands of DNA without any synthesis of new stretches of DNA.

Constitutive genes are expressed as a function of the interaction of RNA polymerase with the promoter, without additional regulation; sometimes also called household genes in the context of describing functions expressed in all cells at a low level.

Constitutive heterochromatin describes the inert state of permanently nonexpressed sequences, usually satellite DNA.

Constitutive mutations cause genes that usually are regulated to be expressed without regulation.

Controlling elements of maize are transposable units originally identified solely by their genetic properties.

Coordinate regulation refers to the common control of a group of genes.

Cordycepin is 3' deoxyadenosine, an inhibitor of polyadenylation of RNA.

Core DNA is the 146 bp of DNA contained on a core particle.

Core particle is a digestion product of the nucleosome that retains the histone octamer and has 146 bp of DNA; its structure appears similar to that of the nucleosome itself.

Corepressor is a small molecule that triggers repression of transcription by binding to a regulator protein.

Cosmids are plasmids into which phage lambda *cos* sites have been inserted; as a result, the plasmid DNA can be packaged *in vitro* in the phage coat.

Cot is the product of DNA concentration and time of incubation in a reassociation reaction.

Cot$_{1/2}$ is the Cot required to proceed to half completion of the reaction; it is directly proportional to the unique length of reassociating DNA.

Crossing-over describes the reciprocal exchange of material between chromosomes that occurs during meiosis and is responsible for genetic recombination.

Crossover fixation refers to a possible consequence of unequal crossing-over that allows a mutation in one member of a tandem cluster to spread through the whole cluster (or to be eliminated).

Cruciform is the structure produced at inverted repeats of DNA if the repeated sequence pairs with its complement on the same strand (instead of with its regular partner in the other strand of the duplex).

Cryptic satellite is a satellite DNA sequence not identified as such by a separate peak on a density gradient; that is, it remains present in main-band DNA.

Cyclic AMP is a molecule of AMP in which the phosphate group is joined to both the 3' and 5' positions of the ribose; its binding activates the CAP, a positive regulator of procaryotic transcription.

Cytological hybridization —*see in situ* hybridization.

Cytoplasmic inheritance is a property of genes located in mitochondria or chloroplasts (or possibly other extranuclear organelles).

Cytoplasmic protein synthesis is the translation of mRNAs representing nuclear genes; it occurs via ribosomes attached to the cytoskeleton.

Cytoskeleton is a network of fibers in the cytoplasm of the eucaryotic cell.

D loop is a region within mitochondrial DNA in which a short stretch of RNA is paired with one strand of DNA, displacing the original partner DNA strand in this region. The same term is used also to describe the displacement of a region of one strand of duplex DNA by a single-stranded invader in the reaction catalyzed by RecA protein.

Degeneracy in the genetic code refers to the lack of an effect of many changes in the third base of the codon on the amino acid that is represented.

Deletions constitute the removal of a sequence of DNA, the regions on either side being joined together.

Denaturation of DNA or RNA describes its conversion from the double-stranded to the single-stranded state; separation of the strands is most often accomplished by heating.

Denaturation of protein describes its conversion from the physiological conformation to some other (inactive) conformation.

Derepressed state is the same as *induced* when describing the normal state of a gene; it has the same meaning as *constitutive* in describing the effect of mutation.

Dicentric chromosome is the product of fusing two chromosome fragments, each of which has a centromere. It is unstable and may be broken when the two centromeres are pulled to opposite poles in mitosis.

Diploid set of chromosomes contains two copies of each autosome and two sex chromosomes.

Direct repeats are identical (or related) sequences present in two or more copies in the same orientation in the same molecule of DNA; they are not necessarily adjacent.

Discontinuous replication refers to the synthesis of DNA in short (Okazaki) fragments that are later joined into a continuous strand.

Disjunction describes the movement of members of a chromosome pair to opposite poles during cell division. At mitosis and the second meiotic division, disjunction applies to sister chromatids; at first meiotic division it applies to sister chromatid pairs.

Divergence is the percent difference in nucleotide sequence between two related DNA sequences or in amino acid sequences between two proteins.

Divergent transcription refers to the initiation of transcription at two promoters facing in the opposite direction, so that transcription proceeds away in both directions from a central region.

dna mutants of bacteria cannot synthesize DNA at 42°C but can do so at 37°C.

DNA-driven hybridization involves the reaction of an excess of DNA with RNA.

Domain of a chromosome may refer *either* to a discrete structural entity defined as a region within which supercoiling is independent of other domains; *or* to an extensive region including an expressed gene that has heightened sensitivity to degradation by the enzyme DNAase I.

Domain of a protein is a discrete continuous part of the amino acid sequence that can be equated with a particular function.

Dominant allele determines the phenotype displayed in a heterozygote with another (recessive) allele.

Donor splicing site —*see* left splicing junction.

Down promoter mutations decrease the frequency of initiation of transcription.

Downstream identifies sequences proceeding farther in the direction of expression for example, the coding region is downstream from the initiation codon.

Early development refers to the period of a phage infection before the start of DNA replication.

Elongation factors are proteins that associate with ribosomes cyclically, for addition of each amino acid to the polypeptide chain.

End labeling describes the addition of a radioactively labeled group to one end (5′ or 3′) of a DNA strand.

End-product inhibition describes the ability of a product of a metabolic pathway to inhibit the activity of an enzyme that catalyzes an early step in the pathway.

Endonucleases cleave bonds within a nucleic acid chain; they may be specific for RNA or for single-stranded or double-stranded DNA.

Enhancer element is a viral sequence of 72 bp repeats that increases the utilization of (some) eucaryotic promoters in *cis* configuration, but that can function in any location, upstream or downstream, relative to the promoter.

Epigenetic changes influence the phenotype without altering the genotype.

Episome is a plasmid able to integrate into bacterial DNA.

Euchromatin comprises all of the genome in the interphase nucleus except for the heterochromatin.

Evolutionary clock is defined by the rate at which mutations accumulate in a given gene.

Excision of phage or episome or other sequence describes its release from the host chromosome as an autonomous DNA molecule.

Excision-repair systems remove a single-stranded sequence of DNA containing damaged or mispaired bases and replace it in the duplex by synthesizing a sequence complementary to the remaining strand.

Exon is any segment of an interrupted gene that is represented in the mature RNA product.

Exonucleases cleave nucleotides one at a time from the end of a polynucleotide chain; they may be specific for either the 5′ or 3′ end of DNA or RNA.

Extranuclear genes reside in organelles such as mitochondria and chloroplasts outside the nucleus.

Facultative heterochromatin describes the inert state of sequences that also exist in active copies—for example, one mammalian X chromosome in females.

Fast component of a reassociation reaction is the first to renature and contains highly repetitive DNA.

Figure eight describes two circles of DNA linked together by a recombination event that has not yet been completed.

Filter hybridization is performed by incubating a denatured DNA preparation immobilized on a nitrocellulose filter with a solution of radioactively labeled RNA or DNA.

Foldback DNA consists of inverted repeats that have renatured by intrastrand reassociation of denatured DNA.

Footprinting is a technique for identifying the site on DNA bound by some protein by virtue of the protection of bonds in this region against attack by nucleases.

Forward mutations inactivate a wild-type gene.

Frameshift mutations arise by deletions or insertions that are not a multiple of 3 bp, and that therefore change the frame in which triplets are translated into protein.

G banding is a technique that generates a striated pattern in metaphase chromosomes that distinguishes the members of a haploid set.

G1 is the period of the eucaryotic cell cycle between the last mitosis and the start of DNA replication.

G2 is the period of the eucaryotic cell cycle between the end of DNA replication and the start of the next mitosis.

Gamete is either type of reproductive (germ) cell—sperm or egg—with haploid chromosome content.

Gap in DNA is the absence of one or more nucleotides in one strand of the duplex.

Gene (cistron) is the segment of DNA that is involved in producing a polypeptide chain; it includes regions preceding and following the coding region (leader and trailer) as well as intervening sequences (introns) between individual coding segments (exons).

Gene conversion is the alteration of one strand of a heteroduplex DNA to make it complementary with the other strand at any position(s) where there were mispaired bases.

Genomic (chromosomal) **DNA clones** are sequences of the genome carried by a cloning vector.

Genotype is the genetic constitution of an organism.

Gratuitous inducers resemble authentic inducers of transcription but are not substrates for the induced enzymes.

GT-AG rule describes the presence of these constant dinucleotides at the first two and last two positions of introns of nuclear genes.

Gyrase is a type II topoisomerase of *E. coli* with the ability to introduce negative supercoils into DNA.

Hairpin describes a double-helical region formed by base pairing between adjacent (inverted) complementary sequences in a single strand of RNA or DNA.

Haploid set of chromosomes contains one copy of each autosome and one sex chromosome; the haploid number *n* is characteristic of gametes.

Hapten is a small molecule that acts as an antigen when conjugated to a protein.

Helper virus provides functions absent from a defective virus, enabling the latter to complete the infective cycle during a mixed infection.

Heterochromatin describes regions of the genome that are permanently in a highly condensed condition and are not genetically expressed.

Heteroduplex (hybrid) DNA is generated by base pairing between complementary single strands derived from the different parental duplex molecules; it occurs during genetic recombination.

Heterogametic sex has the chromosome constitution 2A + XY.

Heterogeneous nuclear (hn) RNA comprises transcripts of nuclear genes made by RNA polymerase II; it has a wide size distribution and low stability.

Heteromultimeric proteins consist of nonidentical subunits (coded by different genes).

Heterozygote is an individual with different alleles at some particular locus.

Highly repetitive DNA is the first component to reassociate and is equated with satellite DNA.

Histones are conserved DNA binding proteins of eucaryotes that form the nucleosome, the basic subunit of chromatin.

Hogness (TATA) box is a conserved A-T-rich septamer about 25 bp upstream from the startpoint of nuclear genes; may be involved in positioning RNA polymerase II for initiation.

Homogametic sex has the chromosome constitution 2A + XX.

Homologues are chromosomes carrying the same genetic loci; a diploid cell has two copies of each homologue, one derived from each parent.

Homomultimeric protein consists of identical subunits.

Homozygote is an individual with the same allele at corresponding loci on the homologous chromosomes.

Hotspot is a site at which the frequency of mutation (or recombination) is very much increased.

Housekeeping (constitutive) genes are those (theoretically) expressed in all cells because they provide basic functions needed for sustenance of all cell types.

Hybrid-arrested translation is a technique that identifies the cDNA corresponding to an mRNA by relying on the ability to base pair with the RNA *in vitro* to inhibit translation.

Hybrid dysgenesis describes the inability of certain strains of *D. melanogaster* to interbreed, because the hybrids are sterile (although otherwise they may be phenotypically normal).

Hybridization is the pairing of complementary RNA and DNA strands to give an RNA-DNA hybrid.

Hybridoma is the cell line produced by fusing a myeloma with a lymphocyte; it continues indefinitely to express the immunoglobulins of both parents.

Hyperchromicity is the increase in optical density that occurs when DNA is denatured.

DNAase I **hypersensitive site** is a short region of chromatin detected by its extreme sensitivity to cleavage by DNAase I and other nucleases; probably comprises an area from which nucleosomes are excluded.

Idling reaction is the production of pppGpp and ppGpp by ribosomes when an uncharged tRNA is present in the A site; triggers the stringent response.

Immunity refers to the ability of a prophage to prevent another phage of the same type from infecting a cell or to the ability of a plasmid to prevent another of the same type from becoming established in a cell. Different mechanisms are involved in the two types of immunity. Also, transposition immunity refers to the ability of certain transposons to prevent others of the same type from transposing to the same DNA molecule.

***In situ* hybridization** is performed by denaturing the DNA of cells squashed on a microscope slide so that reaction is possible with an added single-stranded RNA or DNA; the added preparation is radioactively labeled and its hybridization is followed by autoradiography.

Incompatibility is the inability of certain bacterial plasmids to coexist in the same cell.

Incomplete dominance —*see* codominance.

Indirect end labeling is a technique for examining the organization of DNA by making a cut at a specific site and isolating all fragments containing the sequence adjacent to one side of the cut; it reveals the distance from the cut to the next break(s) in DNA.

Induced mutations result from the addition of a mutagen.

Inducer is a small molecule that triggers gene transcription by binding to a regulator protein.

Induction refers to the ability of bacteria (or yeast) to synthesize certain enzymes only when their substrates are present; applied to gene expression, refers to switching on transcription as a result of interaction of the inducer with the regulator protein.

Induction of prophage describes its excision from the host genome and entry into the lytic (infective) cycle as a result of destruction of the lysogenic repressor.

Initiation factors (IF) are proteins that associate with the small subunit of the ribosome specifically at the stage of initiation of protein synthesis.

Insertions are identified by the presence of an additional stretch of base pairs in DNA.

Integration of viral or another DNA sequence is its insertion into a host genome as a region covalently linked on either side to the host sequences.

Interallelic complementation describes the change in the properties of a heteromultimeric protein brought about by the interaction of subunits coded by two different mutant alleles; the mixed protein may be more or less active than the protein consisting of subunits only of one or the other type.

Interbands are the relatively dispersed regions of polytene chromosomes that lie between the bands.

Intercistronic region is the distance between the termination codon of one gene and the initiation codon of the next gene in a polycistronic transcription unit.

Intermediate component(s) of a reassociation reaction are those reacting between the fast (satellite DNA) and slow (nonrepetitive DNA) components; contain moderately repetitive DNA.

Interphase is the period between mitotic cell divisions; divided into G1, S, and G2.

Intervening sequence is an intron.

Intron is a segment of DNA that is transcribed, but is removed from within the transcript by splicing together the sequences (exons) on either side of it.

Inverted repeats comprise two copies of the same sequence of DNA repeated in opposite orientation on the same molecule. Adjacent inverted repeats constitute a palindrome.

Inverted terminal repeats are the short related or identical sequences present in reverse orientation at the ends of some transposons.

IS is an abbreviation for **insertion sequence,** a small bacterial transposon carrying only the genetic functions involved in transposition.

Isoaccepting tRNAs represent the same amino acid.

Isotype is a group of closely related immunoglobulin chains.

kb is an abbreviation for 1000 base pairs of DNA or 1000 bases of RNA.

Kinetic complexity is the complexity of a DNA component measured by the kinetics of DNA reassociation.

Kinetochore is the structural feature of the chromosome to which microtubules of the mitotic spindle attach.

Lagging strand of DNA must grow overall in the 3'-5' direction and is synthesized discontinuously in the form of short fragments (5'-3') that are later connected covalently.

Late period of phage development is the part of infection following the start of DNA replication.

Leader is the nontranslated sequence at the 5' end of mRNA that precedes the initiation codon.

Leading strand of DNA is synthesized continuously in the 5'-3' direction.

Leaky mutations allow some residual level of gene expression.

Left splicing junction is the boundary between the right end of an exon and the left end of an intron.

Lethal locus is any gene in which a lethal mutation can be obtained.

Library is a set of cloned fragments together representing the entire genome.

Ligation is the formation of a phosphodiester bond to link two adjacent bases separated by a nick in one strand of a double helix of DNA. (The term can also be applied to blunt-end ligation and to joining of RNA.)

Linkage describes the tendency of genes to be inherited together as a result of their location on the same chromosome; measured by percent recombination between loci.

Linkage group includes all loci that can be connected (directly or indirectly) by linkage relationships; equivalent to a chromosome.

Linker DNA is all DNA contained on a nucleosome in excess of the 146 bp core DNA.

Linker fragment is short synthetic duplex oligonucleotide containing the target site for some restriction enzyme; may be added to ends of a DNA fragment prepared by cleavage with some other enzyme during reconstructions of recombinant DNA.

Linking number is the number of times the two strands of a closed DNA duplex cross over each other.

Liquid (solution) hybridization is a reaction between complementary nucleic acid strands performed in solution.

Locus is the position of a chromosome at which the gene for a particular trait resides; locus may be occupied by any one of the alleles for the gene.

Long-period interspersion is a pattern in the genome in which long stretches of moderately repetitive and nonrepetitive DNA alternate.

Loop is a single-stranded region at the end of a hairpin in RNA (or single-stranded DNA); corresponds to the sequence between inverted repeats in duplex DNA.

LTR is an abbreviation for **long-terminal repeat,** a sequence directly repeated at both ends of a retroviral DNA.

Luxury genes are those coding for specialized functions synthesized (usually) in large amounts in particular cell types.

Lysis describes the death of bacteria at the end of a phage infective cycle when they burst open to release the progeny of an infecting phage.

Lysogen is a bacterium that possesses a repressed prophage as part of its genome.

Lysogenic immunity is the ability of a prophage to prevent another phage genome of the same type from becoming established in the bacterium.

Lysogenic repressor is the protein responsible for preventing a prophage from reentering the lytic cycle.

Lysogeny describes the ability of a phage to survive in a bacterium as a stable prophage component of the bacterial genome.

Lytic infection of bacteria by a phage ends in destruction of bacteria and release of progeny phage.

Main band of genomic DNA consists of a broad peak on a density gradient, excluding any visible satellite DNAs that form separate bands.

Major histocompatibility locus is a large chromosome region containing a giant cluster of genes that code for transplantation antigens and other proteins found on the surfaces of lymphocytes.

Map distance is measured as cM (centiMorgans) = percent recombination (sometimes subject to adjustments).

Marker (DNA) is a fragment of known size used to calibrate an electrophoretic gel.

Marker (genetic) is any allele of interest in an experiment.

Maternal inheritance describes the preferential survival in the progeny of a cross of genetic markers provided by one parent.

Meiosis occurs by two successive divisions (meiosis I and II) that reduce the starting number of 4n chromosomes to 1n in each of four product cells. Products may mature to germ cells (sperm or eggs).

Melting of DNA means its denaturation.

Melting temperature (T_m) is the midpoint of the temperature range over which DNA is denatured.

Micrococcal nuclease is an endonuclease that cleaves DNA; in chromatin, DNA is cleaved preferentially between nucleosomes.

Minichromosome of SV40 or polyoma is the nucleosomal form of the viral circular DNA.

Mitosis is the division of a eucaryotic somatic cell.

Modification of DNA or RNA includes all changes made to the nucleotides after their initial incorporation into the polynucleotide chain.

Modified bases are all those except the usual four from which DNA (T, C, A, G) or RNA (U, C, A, G) are synthesized; they result from postsynthetic changes in the nucleic acid.

Monocistronic mRNA codes for one protein.

Multicopy plasmids are present in bacteria at amounts greater than one per chromosome.

Multiforked chromosome (in bacterium) has more than one replication fork, because a second initiation has occurred before the first cycle of replication has been completed.

Multimeric proteins consist of more than one subunit.

Mutagens increase the rate of mutation by causing changes in DNA.

Mutation describes any change in the sequence of genomic DNA.

Myeloma is a tumor cell line derived from a lymphocyte; usually produces a single type of immunoglobulin.

Negative complementation occurs when interallelic complementation allows a mutant subunit to suppress the activity of a wild-type subunit in a multimeric protein.

Negative regulators function by switching off transcription or translation.

Negative supercoiling comprises the twisting of a duplex of DNA in space in the opposite sense to the turns of the strands in the double helix.

Neutral substitutions in a protein are those changes of amino acids that do not affect activity.

Nick in duplex DNA is the absence of a phosphodiester bond between two adjacent nucleotides on one strand.

Nick translation describes the ability of *E. coli* DNA polymerase I to use a nick as a starting point from which one strand of a duplex DNA can be degraded and replaced by resynthesis of new material; is used to introduce radioactively labeled nucleotides into DNA *in vitro*.

Nondisjunction describes failure of chromatids (duplicate chromosomes) to go to opposite poles during mitosis or meiosis.

Nonpermissive conditions do not allow conditional lethal mutants to survive.

Nonrepetitive DNA shows reassociation kinetics expected of unique sequences.

Nonsense codon is any one of three triplets (UAG, UAA, UGA) that cause termination of protein synthesis. (UAG is known as amber; UAA as ochre.)

Nonsense mutation is any change in DNA that causes a nonsense (termination) codon to replace a codon representing an amino acid.

Nonsense suppressor is a gene coding for a mutant tRNA able to respond to one or more of the nonsense codons.

Nontranscribed spacer is the region between transcription units in a tandem gene cluster.

Northern blotting is a technique for transferring RNA from an agarose gel to a nitrocellulose filter on which it can be hybridized to a complementary DNA.

Nuclear matrix is a network of fibers surrounding and penetrating the nucleus.

Nucleolar organizer is the region of a chromosome carrying genes coding for rRNA.

Nucleoid is the compact body that contains the genome in a bacterium.

Nucleolus is a discrete region of the nucleus created by the transcription of rRNA genes.

Nucleosome is the basic structural subunit of chromatin, consisting of ~200 bp of DNA and an octamer of histone proteins.

Ochre codon is the triplet UAA, one of three nonsense codons that cause termination of protein synthesis.

Ochre mutation is any change in DNA that creates a UAA codon at a site previously occupied by another codon.

Ochre suppressor is a gene coding for a mutant tRNA able to respond to the UAA codon to allow continuation of protein synthesis.

Okazaki fragments are the short stretches of 1000–2000 bases that are produced during discontinuous replication; they are later joined into a covalently intact strand.

Open reading frame contains a series of triplets coding for amino acids without any termination codons; sequence is (potentially) translatable into protein.

Operator is the site on DNA at which a repressor protein binds to prevent transcription from initiating at the adjacent promoter.

Operon is a complete unit of bacterial gene expression and regulation, including structural genes, regulator gene(s), and control elements in DNA recognized by regulator gene product(s).

Origin (*ori*) is a sequence of DNA at which replication is initiated.

Orphons are isolated individual genes that are related to members of a gene cluster.

Overwinding of DNA is caused by positive supercoiling (which applies further tension in the direction of winding of the two strands about each other in the duplex).

Packing ratio is the ratio of the length of DNA to the unit length of the fiber containing it.

Pairing of chromosomes—*see* synapsis.

Palindrome describes a sequence of DNA that is the same when one strand is read left to right or the other is read right to left; consists of adjacent inverted repeats.

Papovaviruses are a class of animal viruses with small genomes, including SV40 and polyoma.

pBR322 is one of the standard plasmid cloning vectors.

Periodicity of DNA is the number of base pairs per turn of the double helix.

Permissive conditions allow conditional lethal mutants to survive.

Petite strains of yeast lack mitochondrial function.

Phage (bacteriophage) is a bacterial virus.

Phase variation describes an alternation in the type of flagella produced by a bacterium.

Phenotype is the appearance or other characteristics of an organism, resulting from the interaction of its genetic constitution with the environment.

Plasmid is an autonomous self-replicating extrachromosomal circular DNA.

Playback experiment describes the retrieval of DNA that has hybridized with RNA to check that it is nonrepetitive by a further reassociation reaction.

Pleiotropic gene affects more than one (apparently unrelated) characteristic of the phenotype.

Point mutations are substitutions of single base pairs.

Polarity refers to the effect of a mutation in one gene in influencing the expression (at transcription or translation) of subsequent genes in the same transcription unit.

Polyadenylation is the addition of a sequence of polyadenylic acid to the 3' end of a eucaryotic RNA after its transcription.

Polycistronic mRNA includes coding regions representing more than one gene.

Polymorphism refers to the simultaneous occurrence in the population of genomes showing allelic variations (as seen either in alleles producing different phenotypes or—for example—in changes in DNA affecting the restriction pattern).

Polyploid cell has more than two sets of the haploid genome.

Polyprotein is a gene product that is cleaved into several independent proteins.

Polysome (polyribosome) is an mRNA associated with a series of ribosomes engaged in translation.

Polytene chromosomes are generated by successive replications of a chromosome set without separation of the replicas.

Position effect refers to a change in the expression of a gene brought about by its translocation to a new site in the genome; for example, a previously active gene may become inactive if placed near heterochromatin.

Positive regulator proteins are required for the activation of a transcription unit.

Positive supercoiling describes the coiling of the double helix in space in the same direction as the winding of the two strands of the double helix itself.

Postmeiotic segregation describes the segregation of two strands of a duplex DNA that bear different information (created by heteroduplex formation during meiosis) when a subsequent replication allows the strands to separate.

Pribnow box is the consensus sequence TATAATG centered about 10 bp before the startpoint of bacterial genes. It is a part of the promoter especially important in binding RNA polymerase.

Primary transcript is the original unmodified RNA product corresponding to a transcription unit.

Primer is a short sequence (often of RNA) that is paired with one strand of DNA and provides a free 3'-OH end at which a DNA polymerase starts synthesis of a deoxyribonucleotide chain.

Primosome describes the complex of proteins involved in the priming action that initiates synthesis of each Okazaki fragment during discontinuous DNA replication; the primosome may move along DNA to engage in successive priming events.

Procaryotic organisms (bacteria) lack nuclei.

Processive enzymes continue to act on a particular substrate, that is, do not dissociate between repetitions of the catalytic event.

Promoter is a region of DNA involved in binding of RNA polymerase to initiate transcription.

Proofreading refers to any mechanism for correcting errors in protein or nucleic acid synthesis that involves scrutiny of individual units *after* they have been added to the chain.

Prophage is a phage genome covalently integrated as a linear part of the bacterial chromosome.

Provirus is a duplex DNA sequence in the eucaryotic chromosome corresponding to the genome of an RNA retrovirus.

Pseudogenes are inactive but stable components of the genome derived by mutation of an ancestral active gene.

Puff is an expansion of a band of a polytene chromosome associated with the synthesis of RNA at some locus in the band.

Pulse-chase experiments are performed by incubating cells very briefly with a radioactively labeled precursor (of some pathway or macromolecule); then the fate of the label is followed during a subsequent incubation with a nonlabeled precursor.

Quick-stop *dna* mutants of *E. coli* cease replication immediately when the temperature is increased to 42°C.

R loop is the structure formed when an RNA strand hybridizes with its complementary strand in a DNA duplex, thereby displacing the original strand of DNA in the form of a loop extending over the region of hybridization.

Rapid lysis (r) mutants display a change in the pattern of lysis of *E. coli* at the end of an infection by a T-even phage.

Reading frame is one of three possible ways of reading a nucleotide sequence as a series of triplets.

Reassociation of DNA describes the pairing of complementary single strands to form a double helix.

RecA is the product of the *recA* locus of *E. coli*; a protein with dual activities, acting as a protease and also able to exchange single strands of DNA molecules. The protease activity controls the SOS response; the nucleic acid handling facility is involved in recombination-repair pathways.

Recessive allele is obscured in the phenotype of a heterozygote by the dominant allele, often due to inactivity or absence of the product of the recessive allele.

Recessive lethal is an allele that is lethal when the cell is homozygous for it.

Reciprocal recombination is the production of new genotypes with the reverse arrangements of alleles according to maternal and paternal origin.

Recombinant progeny have a different genotype from that of either parent.

Recombinant joint is the point at which two recombining molecules of duplex DNA are connected (the edge of the heteroduplex region).

Recombination nodules (nodes) are dense objects present on the synaptonemal complex; could be involved in crossing-over.

Recombination-repair is a mode of filling a gap in one strand of duplex DNA by retrieving a homologous single strand from another duplex.

Regulatory gene codes for an RNA or protein product whose function is to control the expression of other genes.

Relaxed mutants of *E. coli* do not display the stringent response to starvation for amino acids (or other nutritional deprivation).

Relaxed replication control refers to the ability of some plasmids to continue replicating after bacteria cease dividing.

Release (termination) factors respond to nonsense codons to cause release of the completed polypeptide chain and the ribosome from mRNA.

Renaturation is the reassociation of denatured complementary single strands of a DNA double helix.

Repeating unit in a tandem cluster is the length of the sequence that is repeated; appears circular on a restriction map.

Repetition frequency is the (integral) number of copies of a given sequence present in the haploid genome; equals 1 for nonrepetitive DNA, ≥2 for repetitive DNA.

Repetitive DNA behaves in a reassociation reaction as though many (related or identical) sequences are present in a component, allowing any pair of complementary sequences to reassociate.

Replacement sites in a gene are those at which mutations alter the amino acid that is coded.

Replication-defective virus has lost one or more genes essential for completing the infective cycle.

Replication eye is a region in which DNA has been replicated within a longer, unreplicated region.

Replication fork is the point at which strands of parental duplex DNA are separated so that replication can proceed.

Replicon is unit of genome in which DNA is replicated; contains an origin for initiation of replication.

Replisome is the multiprotein structure that assembles at the bacterial replicating fork to undertake synthesis of DNA. Contains DNA polymerase and other enzymes.

Repression is the ability of bacteria to prevent synthesis of certain enzymes when their products are present; more generally, refers to inhibition of transcription (or translation) by binding of repressor protein to specific site on DNA (or mRNA).

Repressor protein binds to operator on DNA or to RNA to prevent transcription or translation, respectively.

Resolvase is enzyme activity involved in site-specific recombination between two transposons present as direct repeats in a cointegrate structure.

Restriction enzymes recognize specific short sequences of (usually) unmethylated DNA and cleave the duplex (sometimes at target site, sometimes elsewhere, depending on type).

Restriction map is a linear array of sites on DNA cleaved by various restriction enzymes.

Retroregulation describes the ability of a sequence downstream to regulate translation of an mRNA.

Retrovirus is an RNA virus that propagates via conversion into duplex DNA.

Reverse transcription is synthesis of DNA on a template of RNA; accomplished by reverse transcriptase enzyme.

Reversion of mutation is a change in DNA that either reverses the original alteration (true reversion) or compensates for it (second site reversion in the same gene).

Revertants have been derived by reversion of a mutant cell or organism.

Rho factor is a protein involved in assisting *E. coli* RNA polymerase to terminate transcription at certain (rho-dependent) sites.

Rho-independent terminators are sequences of DNA that cause *E. coli* RNA polymerase to terminate in the absence of rho factor.

Rifamycins (including rifampicin) inhibit transcription in bacteria.

Right splicing junction is the boundary between the right end of an intron and the left end of the adjacent exon.

RNA-driven hybridization reactions use an excess of RNA to react with all complementary sequences in a single-stranded preparation of DNA.

Rolling circle is a mode of replication in which a replication fork proceeds around a circular template for an indefinite number of revolutions; the DNA strand newly synthesized in each revolution displaces the strand synthesized in the previous revolution, giving a tail containing a linear series of sequences complementary to the circular template strand.

Rot is the product of RNA concentration and time of incubation in an RNA-driven hybridization reaction.

S phase is the restricted part of the eucaryotic cell cycle during which synthesis of DNA occurs.

S1 nuclease is an enzyme that specifically degrades unpaired (single-stranded) sequences of DNA.

Saltatory replication is a sudden lateral amplification to produce a large number of copies of some sequence.

Satellite DNA consists of many tandem repeats (identical or related) of a short basic repeating unit.

Saturation hybridization experiment has a large excess of one component, causing all complementary sequences in the other component to enter a duplex form.

Scaffold of a chromosome is a proteinaceous structure in the shape of a sister chromatid pair, generated when chromosomes are depleted of histones.

Scarce (complex) mRNA consists of a large number of individual mRNA species, each present in very few copies per cell.

scRNA is any one of several small cytoplasmic RNAs, molecules present in the cytoplasm and (sometimes) nucleus. Of unknown function.

Selection describes the use of particular conditions to allow survival only of cells with a particular phenotype.

Semiconservative replication is accomplished by separation of the strands of parental duplex, each then acting as a template for synthesis of a complementary strand.

Semidiscontinuous replication is mode in which one new strand is synthesized continuously while the other is synthesized discontinuously.

Sex chromosomes are those whose content are different in the two sexes; usually labeled X and Y (or W and Z), one sex has XX (or WW), the other sex has XY (or WZ).

Sex linkage is pattern of inheritance shown by genes carried on a sex chromosome (usually the X).

Shine-Dalgarno sequence is part or all of the polypurine sequence AGGAGG located on mRNA just prior to an AUG initiation codon; is complementary to the sequence at the 3′ end of 16S rRNA; involved in binding of ribosome to mRNA.

Short-period interspersion is a pattern in a genome in which moderately repetitive DNA sequences of ~300 bp alternate with nonrepetitive sequences of ~1000 bp.

Shotgun experiment is cloning of an entire genome in the form of randomly generated fragments.

Sigma factor is the subunit of bacterial RNA polymerase needed for initiation; is the major influence on selection of binding sites (promoters).

Signal hypothesis describes the role of the N-terminal sequence of a secreted protein in attaching nascent polypeptide to membrane; that is, mRNA and ribosome are attached to membrane via the N-terminal end of the protein under synthesis.

Silent mutations do not change the product of a gene.

Silent sites in a gene describe those positions at which mutations do not alter the product.

Simple-sequence DNA equals satellite DNA.

Single-copy plasmids are maintained in bacteria at a ratio of one plasmid for every host chromosome.

Single-strand assimilation describes the ability of RecA protein to cause a single strand of DNA to displace its homologous strand in a duplex; that is, the single strand is assimilated into the duplex.

Single-strand exchange is a reaction in which one of the strands of a duplex of DNA leaves its former partner and instead pairs with the complementary strand in another molecule, displacing its homologue in the second duplex.

Sister chromatids are the copies of a chromosome produced by its replication.

Site-specific recombination occurs between two specific (not necessarily homologous) sequences, as in phage integration/excision or resolution of cointegrate structures during transposition.

Slow component of a reassociation reaction is the last to reassociate; usually consists of nonrepetitive DNA.

Slow-stop *dna* mutants of *E. coli* complete the current round of bacterial replication but cannot initiate another at 42°C.

snRNA (small nuclear RNA) is one of many small RNA species confined to the nucleus; roles are unknown, but U1 snRNA might be involved in splicing.

Solution hybridization is the same as liquid hybridization.

Somatic cells are all the cells of an organism except those of the germ line.

SOS box is the DNA sequence (operator) of ~20 bp recognized by LexA repressor protein.

SOS response in *E. coli* describes the coordinate induction of many enzymes, including repair activities, in response to irradiation or other damage to DNA; results from activation of RecA protease to cleave LexA repressor.

Southern blotting describes the procedure for transferring denatured DNA from an agarose gel to a nitrocellulose filter where it can be hybridized with a complementary nucleic acid.

Spheroplast is a bacterial or yeast cell whose wall has been largely or entirely removed.

Spindle describes the reorganized structure of a eucaryotic cell passing through division; the nucleus has been dissolved and chromosomes are attached to the spindle by microtubules.

Splicing describes the removal of introns and joining of exons in RNA; thus introns are spliced out, while exons are spliced together.

Splicing junctions are the sequences immediately surrounding the exon-intron boundaries.

Spontaneous mutations are those that occur in the absence of any added reagent to increase the mutation rate.

Sporulation is the generation of a spore by a bacterium (by morphological conversion) or by a yeast (as the product of meiosis).

SSB is the single-strand protein of *E. coli*, a protein that binds to single-stranded DNA.

Staggered cuts in duplex DNA are made when two strands are cleaved at different points near each other.

Startpoint (startsite) refers to the position on DNA corresponding to the first base incorporated into RNA.

Stem is the base-paired segment of a hairpin.

Sticky ends are complementary single strands of DNA that protrude from opposite ends of a duplex or from ends of different duplex molecules; can be generated by staggered cuts in duplex DNA.

Strand displacement is a mode of replication of some viruses in which a new DNA strand grows by displacing the previous (homologous) strand of the duplex.

Streptolydigins inhibit the elongation of transcription by bacterial RNA polymerase.

Stringent replication describes the limitation of single-copy plasmids to replication pari passu with the bacterial chromosome.

Stringent response refers to the ability of a bacterium to shut down synthesis of tRNA and ribosomes in a poor-growth medium.

Structural gene codes for any RNA or protein product other than a regulator.

Supercoiling describes the coiling of a closed duplex DNA in space so that it crosses over its own axis.

Superrepressed means the same as uninducible.

Suppression describes the occurrence of changes that eliminate the effects of a mutation without reversing the original change in DNA.

Suppressor (extragenic) is usually a gene coding a mutant tRNA that reads the mutated codon either in the sense of the original codon or to give an acceptable substitute for the original meaning.

Suppressor (intragenic) is a compensating mutation that restores the original reading frame after a frameshift.

Synapsis describes the association of the two pairs of sister chromatids representing homologous chromosomes that occurs at the start of meiosis; resulting structure is called a bivalent.

Synaptonemal complex describes the morphological structure of synapsed chromosomes.

Syntenic genetic loci lie on the same chromosome.

T_m is the abbreviation for melting temperature.

Tandem repeats are multiple copies of the same sequence lying in series.

TATA (Hogness) box is a conserved A-T-rich septamer found about 25 bp before the startpoint of each eucaryotic RNA polymerase II transcription unit; may be involved in positioning the enzyme for correct initiation.

Telomere is the natural end of a chromosome.

Temperature-sensitive mutation creates a gene product that is functional at low temperature but inactive at higher temperature (the reverse relationship is usually called cold-sensitive).

Terminal redundancy describes the repetition of the same sequence at both ends of (for example) a phage genome.

Termination codon is one of three triplet sequences, UAG (amber), UAA (ochre), or UGA that cause termination of protein synthesis; they are also called nonsense codons.

Terminator is a sequence of DNA, represented at the end of the transcript, that causes RNA polymerase to terminate transcription.

Thalassemia is disease of red blood cells resulting from lack of either α or β globin.

Topoisomerase is an enzyme that can change the linking number of DNA (in steps of 1 by type I; in steps of 2 by type II).

Topological isomers are molecules of DNA that are identical except for a difference in linking number.

Tracer is a radioactively labeled nucleic acid component included in a reassociation reaction in amounts too small to influence the progress of reaction.

Trailer is a nontranslated sequence at the 3' end of an mRNA following the termination codon.

Trans configuration of two sites refers to their presence on two different molecules of DNA (chromosomes).

Transcribed spacer is the part of an rRNA transcription unit that is transcribed but discarded during maturation; that is, it does not give rise to part of rRNA.

Transcription is synthesis of RNA on a DNA template.

Transcription unit is the distance between sites of initiation and termination by RNA polymerase; may include more than one gene.

Transduction refers to the transfer of a bacterial gene from one bacterium to another by a phage; phage carrying host as well as its own genes is called transducing phage.

Transfection of eucaryotic cells is the acquisition of new genetic markers by incorporation of added DNA.

Transformation of bacteria describes the acquisition of new genetic markers by incorporation of added DNA.

Transformation of eucaryotic cells refers to their conversion to a state of unrestrained growth in culture, resembling or identical with the tumorigenic condition.

Transition is a mutation in which one pyrimidine is substituted by the other or in which one purine is substituted for the other.

Translation is synthesis of protein on the mRNA template.

Translocation of a chromosome describes a rearrangement in which part of a chromosome is detached by breakage and then becomes attached to some other chromosome.

Translocation of the ribosome is movement one codon along mRNA after the addition of each amino acid to the polypeptide chain.

Transplantation antigen is protein coded by a major histocompatibility locus, present on all mammalian cells, involved in interactions between lymphocytes.

Transposase is the enzyme activity involved in insertion of transposon at a new site.

Transposition immunity refers to the ability of certain transposons to prevent others of the same type from transposing to the same DNA molecule.

Transposon is a DNA sequence able to replicate and insert one copy at a new location in the genome.

Transvection describes the ability of a locus to influence activity of an allele on the other homologue only when two chromosomes are synapsed.

Transversion is a mutation in which a purine is replaced by a pyrimidine or vice versa.

True-breeding organisms are homozygous for the trait under consideration.

Twisting number of a DNA is the number of base pairs divided by the number of base pairs per turn of the double helix.

Underwinding of DNA is produced by negative supercoiling (because the double helix is itself coiled in the opposite sense from the intertwining of the strands).

Unequal crossing-over describes a recombination event in which the two recombining sites lie at nonidentical locations in the two parental DNA molecules.

Unidirectional replication refers to the movement of a single replication fork from a given origin.

Uninducible mutants cannot be induced.

Unscheduled DNA synthesis is any DNA synthesis occurring outside the S phase of the eucaryotic cell.

Up promoter mutations increase the frequency of initiation of transcription.

Upstream identifies sequences proceeding in the opposite direction from expression; for example, the bacterial promoter is upstream from the transcription unit, the initiation codon is upstream from the coding region.

V gene is sequence coding for the major part of the variable (N-terminal) region of an immunoglobulin chain.

Variegation of phenotype is produced by a change in genotype during somatic development.

Cloning **vector** is any plasmid or phage into which a foreign DNA may be inserted to be cloned.

Virion is the external protein coat of a virus particle.

Virulent phage mutants are unable to establish lysogeny.

Wobble hypothesis accounts for the ability of a tRNA to recognize more than one codon by unusual (non-G-C, A-T) pairing with the third base of a codon.

Writhing number is the number of times a duplex axis crosses over itself in space.

Zero time-binding DNA enters the duplex form at the start of a reassociation reaction; results from intramolecular reassociation of inverted repeats.

Zygote is produced by fusion of two gametes—that is, it is a fertilized egg.

INDEX

INDEX